Current Topics in Behavioral Neurosciences

Volume 28

About this Series

Current Topics in Behavioral Neurosciences provides critical and comprehensive discussions of the most significant areas of behavioral neuroscience research, written by leading international authorities. Each volume offers an informative and contemporary account of its subject, making it an unrivalled reference source. Titles in this series are available in both print and electronic formats.

With the development of new methodologies for brain imaging, genetic and genomic analyses, molecular engineering of mutant animals, novel routes for drug delivery, and sophisticated cross-species behavioral assessments, it is now possible to study behavior relevant to psychiatric and neurological diseases and disorders on the physiological level. The *Behavioral Neurosciences* series focuses on "translational medicine" and cutting-edge technologies. Preclinical and clinical trials for the development of new diagnostics and therapeutics as well as prevention efforts are covered whenever possible.

More information about this series at http://www.springer.com/series/7854

Trevor W. Robbins · Barbara J. Sahakian
Editors

Translational Neuropsychopharmacology

Springer

Editors
Trevor W. Robbins
University of Cambridge
Cambridge
UK

Barbara J. Sahakian
University of Cambridge
Cambridge
UK

ISSN 1866-3370 ISSN 1866-3389 (electronic)
Current Topics in Behavioral Neurosciences
ISBN 978-3-319-33911-5 ISBN 978-3-319-33913-9 (eBook)
DOI 10.1007/978-3-319-33913-9

Printed on acid-free paper

This Springer imprint is published by Springer Nature
The registered company is Springer International Publishing AG Switzerland

This volume is dedicated to Athina Markou's enormous contribution to translational neuropsychopharmacology

Preface

Mental health disorders currently exert an enormous socioeconomic burden, greater than those of other medical conditions arising from cardiovascular disease or cancer, and yet there have been very few therapeutic advances in recent years in the form of novel effective drug treatments in psychiatry. Indeed, the results of Phase 3 trials have been so disappointing and unsuccessful that many companies have withdrawn from neuroscience research related to psychiatry, as it has been thought to be somehow 'too difficult'. Various causes for that difficulty have been raised including regulatory stringency (as well as perhaps rigidity), the nosological heterogeneity of psychiatric disorders and the unavailability of predictive animal models. The first of these problems could perhaps eventually be addressed by the demonstration of a more successful drug discovery strategy. The heterogeneity of psychiatric disorders could perhaps be addressed by employing transdiagnostically more accurate and precise neurobehavioural measurements according to a 'Research Domain Criteria' type approach of the form recently advanced by the U. S. National Institute of Mental Health—but this development will not concern us directly here. The third problem, of animal models, has been considered to be replaced by superior predictive tests based on suitable 'biomarkers', but this strategy, although useful is unlikely by itself to replace the ultimate assays for psychiatric symptoms which are likely mainly to be behavioural or cognitive in nature

In the case of animal models, the defence has been offered (by Professor Mark Geyer, San Diego) that companies frequently are unable to predict the outcome of Phase 2 trials from (proof of concept and human dose-response) Phase 2 trials, let alone from the animal models alone. This insight raises the issue of whether there has been sufficiently effective 'translation' of the animal models even to human studies, and whether much more attention has to be paid to this particular 'translational gap', which could arise for example from a failure to ask similar behavioural or cognitive questions across the species—due to the use for example of clinical scales depending on subjective responses or impressions, rather than on objectively measured behavioural or cognitive signs. An alternative approach would validate animal models by 'back-translation', i.e. by feeding back the results of human studies with compounds to arbitrate amongst the various animal models and test paradigms in order to optimize them and encourage an iterative, 'bidirectional' translational process. This volume surveys some of the best developed examples of how investigators have tried to achieve this goal. It also addresses peripherally the second problem of translation, namely relating such cross-species bidirectional studies to clinical utilization.

Chapter "Translational Mouse Models of Autism: Advancing Toward Pharmacological Therapeutics" by Kazdoba et al. well exemplifies the cross-species approach to modelling a particular complex human disorder with behavioural, cognitive and social dimensions, autism, using rodent studies. In contrast, chapter "Translatable and Back-Translatable Measurement of Impulsivity and Compulsivity: Convergent and Divergent Processes" (Voon & Dalley) though also employing rodents, takes the dimensional approach to modelling psychiatric symptoms that may extend transdiagnostically, for example to attention deficit/hyperactivity disorder to addiction, and thence to eating disorders and obsessive-compulsive disorder. Chapter "Translational Models of Gambling-Related Decision Making" (Winstanley & Clark) continues this analysis specifically by examining these and additional dimensions based on explorations of the reward system and decision-making mechanisms that characterize risk-taking and compulsive gambling behaviour. Other forms of addiction are considered in chapter "Translational Research on Nicotine Dependence" (Falcone et al., nicotine dependence) and chapter "The Need for Treatment Responsive Translational Biomarkers in Alcoholism Research" (alcoholism) Heilig et al). The latter takes a biomarker approach echoed elsewhere in the volume (chapters "Animal Models of Deficient Sensorimotor Gating in Schizophrenia: Are They Still Relevant?" and "Relating Translational Neuroimaging and Amperometric Endpoints: Utility for Neuropsychiatric Drug Discovery") as a possible solution to frustrated attempts to "bridge the valley of death" of translational activity for the pharmacological treatment of alcoholism. Falcone et al. in contrast describe several optimistic approaches to treating the different facets of nicotine dependence, using a classical 'model' approach. Chapter "On the Road to Translation for PTSD Treatment: Theoretical and Practical Considerations of the Use of Human Models of Conditioned Fear for Drug Development" (Risbrough et al.) addresses post-traumatic stress disorder (PTSD) whereas chapter "Translational Approaches Targeting Reconsolidation" (Kroes et al.) introduces the general concept of memory reconsolidation as a route to remediation of conditions such as PTSD (and also addiction). Chapters "Translational Assessment of Reward and Motivational Deficits in Psychiatric Disorders" (Der-Avakian et al.) and "Affective Biases in Humans and Animals" (Robinson & Roiser) take complementary approaches to the special problems posed by modelling human affective disorders−whereas chapter "Translational Assessment of Reward and Motivational Deficits in Psychiatric Disorders" considers reward and effort-based approaches to measuring, e.g. anhedonia, chapter "Affective Biases in Humans and Animals" analyses affective biases, negative as well as positive, that predispose towards depression and its symptomatic heterogeneity. Chapters "Locomotor Profiling from Rodents to the Clinic and Back Again" and "Animal Models of Deficient Sensorimotor Gating in Schizophrenia: Are They Still Relevant?" deal with approaches to modelling the different forms of psychosis in bipolar and schizophrenia disorders. Chapter "Locomotor Profiling from Rodents to the Clinic and Back Again" (Young & Geyer) uses sophisticated quantitative measures of the pattern of locomotor activity in patients with bipolar disorder and rodents; quite striking parallels are found. Chapter "Animal Models of Deficient Sensorimotor Gating in Schizophrenia: Are They Still Relevant?" (Swerdlow & Light) re-evaluates the utility of the pre-pulse inhibition paradigm for schizophrenia, arriving at some new perspectives on the search for new therapeutic breakthroughs, with a memorable and perhaps radical conclusion, "For animal models to remain relevant in the search for schizophrenia therapeutics, they will need to focus less on what is valid, and focus more on what is useful". Chapter "Attention and the Cholinergic System: Relevance to Schizophrenia" (Lustig and Sarter) well illustrates how basic investigation of the functioning of an important chemical neurotransmitter system in experimental animals, namely that using acetylcholine in neurons originating in the basal forebrain, can lead to new insights into how this system may operate in healthy humans and how it may go wrong in disorders such as schizophrenia, with attendant therapeutic indications. Another approach to measuring attention is highlighted in the elegant translation in chapter "Attentional Set-Shifting Across Species" by Brown and Tait of the primate CANTAB intra-dimensional/extra-dimensional attentional set-shifting paradigm to rodent (rat and mouse) models. Their paradigm has been much used in industry as well as in academia to measure 'cognitive flexibility' and fronto-executive function and a substantial neuropsychopharmacological literature has resulted. Nevertheless, industry is now often taking an approach more akin to biomarkers for predicting future drug discovery that depends, for example, on electrophysiological and brain imaging measures. Chapter "Relating Translational Neuroimaging and Amperometric Endpoints: Utility for Neuropsychiatric Drug Discovery" by Li et al. from an industrial setting shows how it is now feasible to compare human psychopharmacological functional imaging paradigms with those in rodents by using the amperometry technique in rats, providing essentially another measure of the BOLD response in functional settings, including vigilant attention and reward-related behaviour—being very useful for Phase 2 type studies by pharma. Chapter "Cognitive Translation Using the Rodent Touchscreen Testing Approach" (Hvosfelt-Eide et al.) introduces an innovative new method of testing rodents using touch-sensitive screens to assess attention, learning and memory in a computerized tests—several exciting examples of direct animal–human translation are described, including in mice and humans with common genetic polymorphisms. This methodology sprang out of the original invention of touch-screen-sensitive cognitive tests in the CANTAB battery, which is the subject of chapter "The Paired Associates Learning (PAL) Test: 30 Years of CANTAB Translational Neuroscience from Laboratory to Bedside in Dementia Research". Using the same type of tests in humans and animals is surely the key to achieving translation across the animal–human boundary that is so important for integration of pre-clinical and clinical (i.e. experimental medicine) studies. Chapter “The Paired Associates Learning (PAL) Test: 30 Years of CANTAB Translational Neuroscience from Laboratory to Bedside in Dementia Research” (Barnett et al.) illustrates the bidirectional translational approach taken by the invention of the CANTAB battery—focusing on the evolution of a visuospatial Paired Associates Learning Test which is highly sensitive to detection of early Alzheimer’s disease in patients with Mild Cognitive Impairment. This chapter not only illustrates the prospects for ‘back-translation’ to animal models using such a battery, but also bridges a second translational ‘gap’, by having the tests adopted in an I-Pad format by GP clinics for screening memory dysfunction. Finally, chapter “Experimental Medicine in Psychiatry New Approaches in Schizophrenia, Depression and Cognition” (Dawson) shows how experimental medicine studies may provide an interface between Phase 1 and 2 trials to bridge the gap between animal and human studies.

We would like to thank all of the contributors to this volume, which we hope will have some impact in enabling scientists coming either from academia or industry, or alternatively, from pre-clinical or clinical backgrounds, perhaps to find a more common language, methodology and even motivation, for carrying out translational research. Additionally, we thank the Editors of the Current Topics in Behavioral Neuroscience series, as well as the Susan Dathé and the staff of Springer Verlag, for their nurturing patience in making this volume possible.

Cambridge
November 2015

Trevor W. Robbins
Barbara J. Sahakian

Contents

Translational Mouse Models of Autism: Advancing Toward Pharmacological Therapeutics 1
Tatiana M. Kazdoba, Prescott T. Leach, Mu Yang, Jill L. Silverman, Marjorie Solomon and Jacqueline N. Crawley

Translatable and Back-Translatable Measurement of Impulsivity and Compulsivity: Convergent and Divergent Processes 53
Valerie Voon and Jeffrey W. Dalley

Translational Models of Gambling-Related Decision Making. 93
Catharine A. Winstanley and Luke Clark

Translational Research on Nicotine Dependence. 121
Mary Falcone, Bridgin Lee, Caryn Lerman and Julie A. Blendy

The Need for Treatment Responsive Translational Biomarkers in Alcoholism Research 151
Markus Heilig, Wolfgang H. Sommer and Rainer Spanagel

On the Road to Translation for PTSD Treatment: Theoretical and Practical Considerations of the Use of Human Models of Conditioned Fear for Drug Development 173
Victoria B. Risbrough, Daniel E. Glenn and Dewleen G. Baker

Translational Approaches Targeting Reconsolidation 197
Marijn C.W. Kroes, Daniela Schiller, Joseph E. LeDoux and Elizabeth A. Phelps

Translational Assessment of Reward and Motivational Deficits in Psychiatric Disorders 231
Andre Der-Avakian, Samuel A. Barnes, Athina Markou and Diego A. Pizzagalli

Affective Biases in Humans and Animals 263
E.S.J. Robinson and J.P. Roiser

Locomotor Profiling from Rodents to the Clinic and Back Again 287
Jared W. Young, Arpi Minassian and Mark A. Geyer

Animal Models of Deficient Sensorimotor Gating in Schizophrenia: Are They Still Relevant? 305
Neal R. Swerdlow and Gregory A. Light

Attention and the Cholinergic System: Relevance to Schizophrenia 327
Cindy Lustig and Martin Sarter

Attentional Set-Shifting Across Species 363
Verity J. Brown and David S. Tait

Relating Translational Neuroimaging and Amperometric Endpoints: Utility for Neuropsychiatric Drug Discovery 397
Jennifer Li, Adam J. Schwarz and Gary Gilmour

Cognitive Translation Using the Rodent Touchscreen Testing Approach 423
M. Hvoslef-Eide, S.R.O. Nilsson, L.M. Saksida and T.J. Bussey

The Paired Associates Learning (PAL) Test: 30 Years of CANTAB Translational Neuroscience from Laboratory to Bedside in Dementia Research 449
Jennifer H. Barnett, Andrew D. Blackwell, Barbara J. Sahakian and Trevor W. Robbins

Experimental Medicine in Psychiatry New Approaches in Schizophrenia, Depression and Cognition 475
Gerard Dawson

Translational Mouse Models of Autism: Advancing Toward Pharmacological Therapeutics

Tatiana M. Kazdoba, Prescott T. Leach, Mu Yang, Jill L. Silverman, Marjorie Solomon and Jacqueline N. Crawley

Abstract Animal models provide preclinical tools to investigate the causal role of genetic mutations and environmental factors in the etiology of autism spectrum disorder (ASD). Knockout and humanized knock-in mice, and more recently knockout rats, have been generated for many of the de novo single gene mutations and copy number variants (CNVs) detected in ASD and comorbid neurodevelopmental disorders. Mouse models incorporating genetic and environmental manipulations have been employed for preclinical testing of hypothesis-driven pharmacological targets, to begin to develop treatments for the diagnostic and associated symptoms of autism. In this review, we summarize rodent behavioral assays relevant to the core features of autism, preclinical and clinical evaluations of pharmacological interventions, and strategies to improve the translational value of rodent models of autism.

Keywords Autism · Mice · Rats · Genes · Mutant models · Social behavior · Sociability · Repetitive behavior · Cognition · Ultrasonic vocalization · Pharmacological treatment · Mouse · Preclinical · Translational · Clinical trials · Face validity · Construct validity · Predictive validity

Contents

1 Introduction ... 2
2 Animal Models to Understand the Causes of Autism ... 3
3 Mouse Behavioral Assays Relevant to the Diagnostic and Associated Symptoms of Autism ... 5
3.1 Social Tests ... 5
3.2 Social Communication ... 8

T.M. Kazdoba · P.T. Leach · M. Yang · J.L. Silverman · M. Solomon · J.N. Crawley (✉)
MIND Institute, Department of Psychiatry and Behavioral Sciences, University of California Davis School of Medicine, Room 1001A Research 2 Building 96, 4625 2nd Avenue, Sacramento CA 95817, USA
e-mail: crawley@ucdavis.edu

Curr Topics Behav Neurosci (2016) 28: 1–52
DOI 10.1007/7854_2015_5003

3.3 Motor Stereotypies, Repetitive Behaviors, and Restricted Interests 10
3.4 Associated Symptoms 12
4 Evaluating Pharmacological Therapeutics in Animal Models with High Construct Validity and Strong Face Validity for ASD 15
5 Conclusions 30
References 32

1 Introduction

Autism spectrum disorder (ASD) includes common, impairing neurodevelopmental disorders that are present from early childhood and occur in approximately 1 % of the population (Kim et al. 2011; Elsabbagh et al. 2012). To receive an ASD diagnosis, one must exhibit symptoms from two core domains: (1) social interaction and social communication; and (2) restricted repetitive patterns of behaviors, interests, and activities. (American Psychiatric Association 2013). Associated symptoms, appearing in varying percentages of individuals, include intellectual disability, executive dysfunction, anxiety, seizures, attention deficits and hyperactivity, hyper- and hyporeactivity to sensory stimuli, and sleep disruption. The current standard of care for children is early intensive behavioral intervention (Rogers et al. 2012; Lord and Jones 2013). Early intensive behavioral intervention is highly effective in teaching young children to overcome their social challenges, although it does not work for all, and its benefits wane with the appearance of age-related challenges in middle childhood and adolescence. Further, these behavior therapies are expensive and time-intensive, and not uniformly widely available. There is an unmet need for medical therapeutics that can be given in combination with a behavioral intervention or alone. No approved medical treatments exist for reducing or preventing the diagnostic symptoms of autism. Efficacious medications that effectively treat ASD symptoms, and specifically target social deficits, are currently under investigation.

The decision to use the term ASD in DSM-5 reflects the current thinking about the heterogeneous causes and clinical presentations of autism. A large number of de novo single gene mutations and copy number variants (CNVs) are associated with autism, each in a small number of individuals (Parikshak et al. 2013; Coe et al. 2014; Pinto et al. 2014). Environmental risk factors have been implicated, including parental age (Kong et al. 2012) and atypical maternal autoantibodies (Braunschweig et al. 2013). Analogous to "cancers," there may be multiple "autisms," to be defined by clustered genetic mutations with common mechanisms and treated with different classes of therapeutics. No definitive biomarkers have yet been identified across all diagnosed cases. Intensive searches are underway to define abnormalities in neurophysiology, neuroanatomy, brain chemistry, immune markers, and other potential biological abnormalities that may stratify individuals with autism, and offer outcome measures for future clinical trials (Ecker et al. 2013).

Rodent models offer preclinical tools to understand the role of genetic mutations and environmental factors in producing the diagnostic and associated symptoms of autism. Knockout (KO) and humanized knock-in mice have been generated for many of the mutations and CNVs detected in ASD and comorbid neurodevelopmental disorders such as fragile X syndrome and tuberous sclerosis (Silverman et al. 2010b; Ey et al. 2011; Baudouin et al. 2012; Zoghbi and Bear 2012; Gross et al. 2015). Several of these genetic mouse models are in use for the preclinical testing of pharmacological targets to treat the core symptoms of autism (Spooren et al. 2012; Silverman and Crawley 2014; Vorstman et al. 2014; Gross et al. 2015).

One fundamental conundrum is defining mouse behavioral assays with high relevance to the diagnostic symptoms of autism, which is a uniquely human disorder (Crawley 2004). Modeling ASD in rodents is challenging in that the clinical phenotype is heterogeneous and encompasses a wide range of behaviors. Researchers focused on developing animal models based on ASD-related behaviors benefit greatly from participating in clinical observations to obtain a comprehensive understanding of the clinical phenotypes found in individuals with ASD. We have been fortunate to observe diagnostic interviews of children with autism at the University of California Davis MIND Institute. Knowledge gained through these sessions and from lectures and conversations with many generous colleagues working with children, adolescents, and adults with autism guided our thinking in the development of analogous behavioral assays to evaluate mouse models of autism. This chapter presents state-of-the-art assays for mouse social and repetitive behaviors and reviews the preclinical progress in evaluating hypothesis-driven pharmacological interventions, employing these behavioral assays in selected mouse models of autism.

2 Animal Models to Understand the Causes of Autism

The causes of autism are under intense investigation. Evidence supporting a large number of risk genes and CNVs at chromosomal loci is strong. Twin and family studies suggest that the genetic heritability of ASD is very high, ranging from 50 to 90 % (Ritvo et al. 1985; Smalley et al. 1988; Hallmayer et al. 2011; Miles 2011; Nordenbaek et al. 2014; Sandin et al. 2014). Genetic causes, primarily de novo mutations, have been identified in approximately 20–30 % of ASD cases, with no identified gene mutation in the majority of ASD cases (Miles 2011; Devlin and Scherer 2012; Murdoch and State 2013). Of the known genetic abnormalities associated with ASD, at least 5 % are caused by single gene mutations (Lim et al. 2013; De Rubeis et al. 2014; Iossifov et al. 2014), and at least 10 % are due to CNVs that cause structural variation, including duplications, deletions, inversions, and translocations (Marshall et al. 2008; Rosenfeld et al. 2010; Matsunami et al. 2013; Poultney et al. 2013). A remarkable preponderance of genetic mutations in ASD code for proteins mediating synaptic functions, such as those coding for the synaptic protein families SHANK (Durand et al. 2007), CNTNAP (Alarcon et al.

2008; Arking et al. 2008; Bakkaloglu et al. 2008), NLGN (Jamain et al. 2003; Laumonnier et al. 2004; Yan et al. 2005a; Talebizadeh et al. 2006; Lawson-Yuen et al. 2008), and NRXN (Kim et al. 2008). Examples of CNVs associated with ASD include chromosomal loci 15q11-q13 (Christian et al. 2008), 16p11.2 (Fernandez et al. 2010), and 22q11.21, and the *UBE3A*, *NRXN1,* and *CNTN4* genes (Fernandez et al. 2008; Kim et al. 2008; Glessner et al. 2009; Roohi et al. 2009). A subset of single gene mutations associated with ASD are responsible for other neurodevelopmental disorders, including *FMR1* in fragile X syndrome, *TSC* in tuberous sclerosis, and *MECP2* in Rett syndrome.

Genetic and environmental risk factors identified in ASD have led to the development of many useful model systems. The best animal models display all three types of validity: construct, face, and predictive (Crawley 2004). The initial development of a new animal model may determine the extent to which construct validity leads to face validity in these models, and offers predictive validity. Construct validity requires that the animal model is generated with the same underlying biological cause, e.g., a genetic mutation, neuroanatomical abnormality, or environmental factor implicated in ASD. Face validity requires that symptoms displayed in the animal model are analogous to the human symptoms, such as social deficits and repetitive behaviors that define ASD. Predictive validity requires that treatments that are efficacious for treating the human syndrome are similarly efficacious in reversing symptoms in the animal models, such as improving social deficits or reducing repetitive behaviors. As no drug treatment has been approved for the effective treatment of the diagnostic symptoms of autism, predictive validity cannot yet be determined in animal models of ASD.

Construct validity in mouse models of autism has most frequently addressed risk genes by generating targeted mutations in the syntenic genes in the mouse genome. The number of different genetic mutations identified in ASD, each in only a few individuals (De Rubeis et al. 2014; Iossifov et al. 2014; O'Roak et al. 2014), suggests that each of these mutations may be worthwhile to explore in mice with homologous mutations (Abrahams and Geschwind 2008; Silverman et al. 2010b; Ey et al. 2011; Spooren et al. 2012; Silverman and Crawley 2014; Wohr 2014). More recently, technological advances have enabled the development of genetically modified rats. Knockout rats (Engineer et al. 2014; Hamilton et al. 2014), as well as other species with sophisticated social behavioral repertoires, such as voles (Bales and Carter 2003a; Modi and Young 2012) and non-human primates (Bauman et al. 2014), provide additional research tools to determine how specific gene abnormalities, neurotransmission, neuroanatomical correlates, and environmental influences contribute to autism-relevant phenotypes across species.

In addition to the genetically modified rodent models of ASD, several inbred mouse strains incorporate face validity as ASD models, because they display robust and well-replicated social deficits and repetitive behaviors. These inbred strains are considered to be models of idiopathic autism, as their ASD-relevant behaviors are not caused by known genetic mutations. In assays of sociability, discussed below, the inbred strains A/J, BALB/cByJ (BALB), BTBR $T^{+}Itpr3^{tf}/J$ (BTBR), C58/J (C58), and 129S1/SvImJ mice exhibited lack of sociability, as compared to inbred mouse strains with high sociability, such as C57BL/6J (B6) and FVB/NJ mice (Brodkin 2007; Moy et al. 2007; Yang et al. 2007; McFarlane et al. 2008; Moy et al. 2008b). Additionally, several mouse strains, such as BTBR and C58, also display overt motoric stereotypies or repetitive behaviors, including jumping, digging, and high levels of self-grooming and marble burying (Bolivar et al. 2007; Moy et al. 2007; Panksepp et al. 2007; McFarlane et al. 2008; Moy et al. 2008b; Yang et al. 2009; Pobbe et al. 2010; Ryan et al. 2010; Silverman et al. 2010a; Wohr et al. 2011a; Yang et al. 2012a; Burket et al. 2013; Fairless et al. 2013; Silverman et al. 2013; Han et al. 2014). Of these, BTBR has been the most extensively characterized and well-replicated for ASD-related behaviors. In addition to abnormal sociability and repetitive behaviors, BTBR mice deposit fewer scent marks and emit fewer ultrasonic vocalizations (USVs) during social interactions, display an unusual repertoire of call categories during their USVs, exhibit a lower number of complex calls (Scattoni et al. 2008; Roullet et al. 2010; Scattoni et al. 2010), and are impaired on social transmission of food preference (McFarlane et al. 2008). These inbred strains add to the genetic mouse models, along with the rat, vole, and non-human primate models of ASD, which are available to evaluate therapeutics.

3 Mouse Behavioral Assays Relevant to the Diagnostic and Associated Symptoms of Autism

3.1 Social Tests

Several behavioral assays have been developed to assess various aspects of sociability in rodents. Like humans, both mice and rats are social species that display a wide repertoire of social behaviors, engaging in intraspecies reciprocal social interactions, parenting and mating behaviors, and scent marking and aggressive behaviors (Carter et al. 1992; Miczek et al. 2001; Terranova and Laviola 2005; Arakawa et al. 2008; Silverman et al. 2010b; Kaidanovich-Beilin et al. 2011). Behavioral phenotyping can utilize many of these species-specific behaviors to address whether preclinical animal models exhibit social deficits relevant to those seen in ASD.

Reciprocal social interactions. When placed together in a confined arena, juvenile and adult pairs of mice will engage in reciprocal social interactions, participating in various types of social sniffing and physical play (Terranova and Laviola 2005; McFarlane et al. 2008; Silverman et al. 2010b). Depending on the testing parameters, juvenile or adult mice of either the same sex or opposite sex can be evaluated in dyads. Additionally, genetically modified mice can be tested with partners of the same or different genotypes. Types of social partner investigation include nose-to-nose sniffing, nose-to-body sniffing, and nose-to-anogenital sniffing. Interactions include front approach, following, chasing, physical contact such as crawling over and under each other, wrestling, and pushing past each other. Because the complex interactions of these reciprocal social interactions cannot be fully captured by automated software, individual social behaviors are typically scored by investigators using event-recording software. Several ASD-relevant genetic mouse models have been evaluated using this paradigm and were found to exhibit reduced reciprocal social interactions, including *Engrailed2* (*En2*) null mutants (Cheh et al. 2006; Brielmaier et al. 2012), conditional *Pten* mutants (Kwon et al. 2006), *Shank3* heterozygotes (Bozdagi et al. 2010; Yang et al. 2012b), and *Tsc1* heterozygotes (Goorden et al. 2007; Tsai et al. 2012). Reduced reciprocal social interactions are also seen in two inbred strains, BTBR and BALB (Bolivar et al. 2007; Panksepp et al. 2007; Yang et al. 2007; McFarlane et al. 2008).

3-chambered social approach. A well-characterized automated test of sociability is our simplified three-chambered social approach task, which offers a high-throughput approach for assessing sociability (Nadler et al. 2004; McFarlane et al. 2008; Yang et al. 2011; Silverman et al. 2012, 2013). In this task, a subject mouse is assessed for its exploration of a novel mouse (e.g., a novel social stimulus) versus a novel object. The novel mouse is typically confined by an inverted wire pencil cup, which allows for visual, auditory, olfactory, and some tactile stimuli between the novel mouse and the subject mouse. An identical inverted wire pencil cup serves as the novel object, either alone or with an inanimate object inside. Mice that display species-typical sociability will spend more time in the side chamber with the novel mouse than in the side chamber with the novel object. Sociability is further defined more specifically by more time sniffing the novel mouse than sniffing the novel object. Chamber time is calculated automatically in a photocell-equipped apparatus, where beam breaks count chamber entries as a measure of locomotor activity. Videotracking systems can perform the same functions by defining zones around the cup or similar container (Ahern et al. 2009; Silverman et al. 2015). Many lines of mice with targeted mutations in risk genes for autism, as well as inbred strains, have been evaluated in the three-chambered social approach task (Moy et al. 2006; Moy and Nadler 2008; Moy et al. 2009; Silverman et al. 2010b; Patterson 2011; Qiu et al. 2012; Jiang and Ehlers 2013). Many genetic models of ASD were reported to exhibit low sociability in this assay including $GABA_A$ receptor *Gabrb3* KO mice (DeLorey et al. 2008), conditional *Pten* KO mice (Kwon et al. 2006), haploinsufficient *Pten* mutant mice (Page et al. 2009; Clipperton-Allen and Page 2014), *Ube3a* triplication mice (Smith et al. 2011), *Cntnap2* KO mice (Penagarikano et al. 2011), 15q11-13 duplication mice (Nakatani et al. 2009), and 17p11.2 duplication mice (Molina et al. 2008). In addition, BTBR, BALB, and C58 mice display low levels of sociability in the social approach assay (Brodkin et al. 2004; Brodkin 2007; Moy et al. 2007; Yang et al. 2007; McFarlane et al. 2008; Moy et al. 2008b; Yang et al. 2009; Ryan et al. 2010; Silverman et al. 2010a, 2012, 2013).

Partition test. The partition task is another straightforward assay for assessing sociability in mice, utilizing a perforated partition to separate a subject mouse from a target mouse. Similar to social approach, the subject mouse is exposed to visual, auditory, and olfactory stimuli from the target mouse, but the two mice do not physically interact. Social interest is represented by the time spent near the partition by the subject mouse. Paylor and coworkers often conduct the partition test first and then remove the partition to evaluate reciprocal social interactions in a habituated environment (Spencer et al. 2005).

Social recognition and social memory can be evaluated through the sequential use of different social partners in the partition task and in the three-chambered social approach apparatus (Moy et al. 2007; Arakawa et al. 2008). Given that mice are novelty-seeking, the subject mouse displays recognition of social novelty if it approaches and spends more time at the partition near the novel mouse as compared to the partition near the familiar mouse (Kudryavtseva 2003; Spencer et al. 2011). Similarly, in the three-chambered social approach task, social recognition is demonstrated if the subject mouse spends more time with a second novel mouse than with the previously novel but now familiar mouse. Adding delay periods of minutes or hours between presentations of the same and novel partners permits evaluation of social memory (Bielsky and Young 2004). Several genetically modified mice that exhibited reduced reciprocal social interactions or low sociability in three-chambered social approach also displayed a lack of preference for social novelty. Others were normal on social approach but failed on preference for social novelty (Moy et al. 2006; Moy and Nadler 2008; Moy et al. 2009; Silverman et al. 2010b; Patterson 2011; Qiu et al. 2012; Jiang and Ehlers 2013), including *Fgf17* KO mice (Scearce-Levie et al. 2008), *Gabrb3* KO mice (DeLorey et al. 2008), and *Nlgn4* KO mice (Jamain et al. 2008). Other genetic mouse models, such as *Nlgn3* KO mice (Radyushkin et al. 2009), demonstrated reduced social novelty, but did not have deficits in other aspects of sociability. Qualitatively divergent findings on social approach versus social recognition and social memory in several models reinforce the interpretation that sociability is distinct from social recognition memory, especially in the 3-chambered assay.

Visible burrow. Mice will typically form colonies that include shared nests composed of underground burrow and tunnel complexes (Lloyd 1975; Bouchard and Lynch 1989). Large visible burrow systems are enclosures that capitalize on the mouse social structure to investigate social interactions in a seminatural habitat using a series of tunnels, burrows, and a large open surface area (Blanchard et al. 1995, 2001). Compared to the social B6 strain, BTBR mice participate in fewer interactive behaviors, such as huddling and following, in the visible burrow system while spending more time alone and engaging in increased self-grooming (Pobbe et al. 2010).

Social transmission of food preference occurs when a subject mouse, after interacting with a cagemate that recently consumed a novel food, eats more of that novel food (Galef 2003; Wrenn et al. 2003; Wrenn 2004; Ryan et al. 2008). In addition to low sociability in several social tasks, BTBR mice also exhibit reduced social transmission of food preference (McFarlane et al. 2008).

Social dominance is measured in a tube task. Mice of two different genotypes with approximately similar body weights are placed in opposite ends of a long, narrow plastic tube. A socially dominant mouse is characterized as the mouse that advances past the halfway point of the tube or pushes the opposing mouse out of the tube. Tube test deficits in social dominance have been detected in mice with mutations in *Dvl1* (Lijam et al. 1997; Long et al. 2004), the serotonin transporter (Kerr et al. 2013), *Fmr1* (Spencer et al. 2005) and others, while 17p11.2 duplication mice exhibited increased dominant behavior in this assay (Molina et al. 2008).

Assessment of sociability in two or more cohorts of animals using multiple assays increases the strength of findings, by generating a more complete behavioral profile, assessing generalizability, and evaluating robustness and replicability. Robust, easily replicated social deficits in mutant lines of mice can then serve as primary preclinical models for the development of novel therapeutics.

3.2 Social Communication

Communication impairments are a hallmark of autism (Lord et al. 2000; Kim et al. 2014b). Depending on the intellectual ability of the individual, communication deficits can manifest as the absence of speech, language delay, the use of odd prosody and intonation, stereotyped speech, perseverative phrases, and difficulties with language pragmatics such as those involved in initiating and maintaining appropriate and meaningful conversations (Rapin and Dunn 2003).

Rodents communicate primarily through olfactory pheromones. However, mice and rats also emit vocalizations in the ultrasonic range during social interactions, and also in non-social contexts (Chabout et al. 2012). Extensive research has been done to identify components of rodent USVs that might be analogous to human language communication. The utility of USV emissions for modeling aspects of social communication deficits in autism is being extensively investigated by several laboratories. Determining whether mouse USV calls have a communication function during specific types of social interactions is a work in progress.

Mouse and rat pups emit USVs when separated from the mother and the nest (Ehret 2005). Pup USVs reliably elicit maternal retrieval (D'Amato et al. 2005; Fischer and Hammerschmidt 2011; Okabe et al. 2013) and are therefore thought to represent a communicatory signal emitted by pups at an age when they solely depend on the dam for thermoregulation and feeding. Separated pups emit even more USVs after a brief reunion period with the mother, followed by a second separation. This phenomenon, called "maternal potentiation", has been found in both mice and rats and has been used as a measure of attachment (Shair et al. 2014). Mouse pups with a null mutation in the μ-opioid receptor gene (*Orpm*$^{-/-}$) emitted fewer USVs when separated from the mother and did not exhibit maternal potentiation, reflecting deficits in attachment (Moles et al. 2004). In mice, pup call numbers peak between postnatal days (PND) 7 and 9 and diminish around the age of hearing onset (PND12) (Ehret 2005; Adise et al. 2014), suggesting that pup USVs are produced by innate mechanisms without a requirement for auditory feedback. It may be reasonable to suggest that pup USVs are a useful measure of physical development, reactivity to stress, anxiety, and attachment. However, since pup calls are likely more analogous to infant crying, quantitative and qualitative components of pup USVs are less likely to serve as a useful proxy for human language communication.

Juvenile and adult mice emit USVs during same-sex social interactions (Maggio and Whitney 1985; D'Amato and Moles 2001; Panksepp et al. 2007; Scattoni et al. 2011; Hammerschmidt et al. 2012). Pretest social isolation is usually a prerequisite for eliciting USVs in same-sex pairs. Currently, there is no practical method to differentiate calls from the two interacting animals. In juveniles, emission of USVs was positively correlated with social behaviors during juvenile social interaction (Panksepp et al. 2007), suggesting that USVs may be an affiliative component of the juvenile social repertoire. Adult mice emit large numbers of calls during same-sex interactions, following a short period of isolation. Female mice with null mutations in the *Shank2* gene emitted fewer calls as compared to wild-type females (Poultney et al. 2013). Adult male and female mice with null mutations of *Neuroligin4* emitted similar numbers of calls as compared to the wild-type controls (Ey et al. 2012). Calls emitted by the resident female during the resident–intruder paradigm have been used as a measure of social memory (D'Amato and Moles 2001).

Male–female social interactions have the advantages of not requiring pretest social isolation and a greater certitude that most calls are emitted by the male (Whitney et al. 1973; White et al. 1998; Wang et al. 2008; Sugimoto et al. 2011). The number of USVs emitted by a subject male in the presence of an estrus female has been widely used as an assay for social communication in mouse genetic models of autism (Ey et al. 2012; Yang et al. 2012b; Sowers et al. 2013).

Fresh female urine and other social odors are similarly effective in eliciting USVs from adult male mice (Nyby et al. 1977; Whitney and Nyby 1979; Byatt and Nyby 1986; Holy and Guo 2005; Hoffmann et al. 2009; Malkesman et al. 2010; Roullet et al. 2011; Wohr et al. 2011b). Playback studies indicate that female mice prefer male USVs over pup USVs, artificial control sounds, or silence (Hammerschmidt et al. 2009; Shepard and Liu 2011) and prefer vocalizing males over devocalized males (Pomerantz et al. 1983), suggesting that male USVs may have a role in facilitating courtship. Recent evidence indicates that male mice exhibit abrupt changes in call repertoires when the female stimulus mouse was removed (Hanson and Hurley 2012; Yang et al. 2013), suggesting that vocal flexibility may reflect the ability to detect sudden changes in salient social cues.

Distinct call categories have been cataloged within the highly complex structures of USVs (Holy and Guo 2005; Scattoni et al. 2011). The pioneering study by Holy and Guo (2005) catalyzed recent research on categorical analysis of mouse USVs. Most investigators classify calls by visually inspecting spectrograms of recorded USVs. Currently, there is no consensus on the number of categories or the definition of each category, with the number of categories ranging from three (Hammerschmidt et al. 2012) to fifteen (Mahrt et al. 2013). Recent electrophysiological recording studies have demonstrated that neurons in the mouse auditory midbrain respond differently to different call types (Mayko et al. 2012), highlighting the importance of categorizing calls in a manner that is biologically meaningful to mice.

Are USVs in adult mice relevant to human language? Recent studies indicate that call patterns are similar between deaf mice and hearing controls

(Hammerschmidt et al. 2012; Mahrt et al. 2013) and that cross-fostering failed to change strain-specific call patterns (Kikusui et al. 2011), suggesting that mouse USVs are not acquired through auditory feedback. It may be more reasonable to suggest that USVs are an important indication of responsivity to social stimuli during social interactions, but are not highly analogous to communicatory functions of complex human language.

3.3 Motor Stereotypies, Repetitive Behaviors, and Restricted Interests

The second ASD diagnostic symptom domain includes motor stereotypies, repetitive behaviors, insistence on sameness, and restricted interests (American Psychiatric Association 2013). Motor stereotypies in ASD include hand flapping and toe walking. **Stereotypies** in mice are species-typical behaviors such as circling and jumping, which occur with frequencies considerably higher than typical levels. Behavioral stereotypies can be assessed in the home cage or observed in an empty cage, by a trained investigator using an event recorder (Crawley 2012). Many genetic models of autism exhibit motor stereotypies. For instance, *Nlgn4* KO mice exhibited increased circling behavior (El-Kordi et al. 2013) and C58 mice exhibited high levels of jumping behavior (Moy et al. 2008b; Ryan et al. 2010; Silverman et al. 2012). *Gabrb3* KO mice showed high levels of circling behaviors (Homanics et al. 1997; DeLorey et al. 2008).

Repetitive self-grooming in mice has face validity to repetitive behaviors in ASD, such as assembling the same puzzle or playing one video game repeatedly. Normal patterns but unusually long bouts of self-grooming have been demonstrated in several mutant mouse models of autism, including *Shank3* (Peca et al. 2011), *Cntnap2* (Penagarikano et al. 2011), *Neurexin1α* (Etherton et al. 2009), and *Neuroligin1* (Blundell et al. 2010). High levels of self-grooming have been well-replicated in the BTBR mouse model of idiopathic autism (Yang et al. 2007; McFarlane et al. 2008; Yang et al. 2009; Pobbe et al. 2010; Silverman et al. 2010a; Amodeo et al. 2012, 2014b; Zhang et al. 2015), while the BALB inbred mouse line does not display repetitive self-grooming (Silverman et al. 2010b). Recent work in transgenic rats reported perseverative chewing behavior in *Fmr1* KO rats (Hamilton et al. 2014). Higher levels of **marble burying** are considered to reflect a repetitive behavior (Thomas et al. 2009). Marble burying relies on the species-typical burying of small objects placed into the cage. Higher marble burying was detected in BTBR (Amodeo et al. 2012; Silverman et al. 2012) and several mutant models (Silverman et al. 2010b), including *Tsc2* KO mice (Reith et al. 2013) and monoamine oxidase (MAO) A and A/B KO mice (Bortolato et al. 2013).

Versions of **open field holeboard exploration** are under development to model autism-relevant restricted interest/perseverative behaviors. Unusual hole board exploration was reported in BTBR and NMDA receptor (*Grin1*) mutant mice using olfactory cues (Moy et al. 2008a), and in MAO A and A/B knockout mice without olfactory cues (Bortolato et al. 2013).

Cognitive rigidity in autism has been modeled in several rodent models of autism. Morris water maze reversal learning assesses the ability of a mouse trained to locate a hidden platform in a pool of water to inhibit its previously learned navigation responses and learn a new platform location. Mice first learn the location of a hidden platform in a large pool of opaque water over the course of several days. After mice reach a criterion level of performance (i.e., latency under 15 s), the hidden platform is moved to the opposite side of the pool so that attempts to find the platform in the previous location must be suppressed and a new goal-directed behavior emerges for successful escape from the water. Two other versions of **maze reversal** are available: spontaneous alternation on a Y-maze, where reduced numbers of alternations between the two arms might represent perseverative behavior, and **rewarded T-maze reversal**, where the rewarded response shifts from the initial location of a food reinforcement located at one end of the T to the other end of the T. Other related tasks include extinction of fear conditioning, where a discrete cue previously paired with an aversive footshock is presented continuously without a footshock pairing, until the species-typical freezing response is attenuated. Deficits on some of these reversal tasks have been reported in BTBR (Moy et al. 2007; Yang et al. 2012a), 15q11-13 duplication (Nakatani et al. 2009), MAO A and A/B KO mice (Bortolato et al. 2013), and in eIF4E overexpressing mice (Santini et al. 2013). Similar to results of Morris water maze reversal tasks, MAO A and A/B KO mice also had decreased alternations in a forced-choice alteration T-maze (Bortolato et al. 2013) and BTBR showed deficits in water T-maze reversal (Guariglia and Chadman 2013).

Intellicages offer a home cage approach to test conditioned place preference learning and reversal, which showed a significant reversal-specific effect of valproic acid (VPA) in B6 mice, but not BALB mice (Puscian et al. 2014). Further, a **set-shifting** assay (Birrell and Brown 2000) showed a compound discrimination reversal deficit in Reeler heterozygous mice (Macri et al. 2010). An assay which employed alternation learning, followed by non-alternation learning, followed by reversal learning, used an H-shaped maze to demonstrate that tryptophan hydroxylase 2 mutants showed perseveration when the reinforcement contingencies changed (Del'Guidice et al. 2014).

The **five-choice serial reaction time task** (5-CSRTT) affords a robust measure of perseveration. The subject mouse pokes its nose into one of five holes at the front of an operant chamber, based on a stimulus presentation located in one of the five possible locations. Perseverative behavior is defined as choosing the previously rewarded stimulus location instead of choosing the currently active location. Mice with mutations in genes coding for the muscarinic acetylcholine receptor M1 and the NMDA receptor subunit Grin1 displayed perseverative deficits in 5-CSRTT (Bartko et al. 2011; Finlay et al. 2014). Despite the broad range of autism-relevant phenotypes displayed by BTBR mice, BTBR did not show perseverative behavior as assessed by the 5-CSRTT (McTighe et al. 2013).

3.4 Associated Symptoms

In addition to the core deficits associated with an autism diagnosis, there are several associated symptoms that commonly occur as comorbid conditions. A recent meta-analysis found that around 40 % of individuals with an ASD had elevated and clinically relevant symptoms of an **anxiety disorder** (van Steensel et al. 2011). Specific phobias were the most common anxiety disorder, occurring in approximately 30 % of autistic individuals, while obsessive–compulsive disorder and social anxiety disorder/agoraphobia occurred in 17 % of autistic individuals (van Steensel et al. 2011). Common rodent behavioral tasks for the assessment of anxiety-like behaviors are the elevated plus-maze and light ↔ dark exploration. These tasks rely on the conflict between the tendency of mice to explore a novel environment versus avoidance of brightly lit open areas. Mice generally enter and spend less time in the two open arms of an elevated plus-maze as compared to the two enclosed maze arms. Mice generally spend less time in the brightly lit compartment of the light ↔ dark apparatus and make fewer transitions between the brightly lit and dark compartments. Anxiolytic drugs selectively increase the number of open arm entries and time in the open arms in the elevated plus-maze, and increase time in the light compartment and number of transitions between compartments in the light ↔ dark apparatus, confirming predictive validity (Crawley 1985; Cryan and Sweeney 2011). Other less widely used tests that detect effects of anxiolytic drugs include the operant-based Geller-Seifter and Vogel conflict assays, vocalizations emitted by pups separated from their dams to model separation anxiety (Insel et al. 1986), and marble burying, which has been described as a model of obsessive–compulsive disorder (Thomas et al. 2009).

Seizure disorders are very common in autism. At least 20 % of individuals who meet the diagnostic criteria for autism experience seizures (Volkmar and Nelson 1990). Several genetic mouse models of autism recapitulate aspects of the increased seizure susceptibility, including mice with mutations in *Synapsin1* (Greco et al. 2013), *En2* (Tripathi et al. 2009), *Cntnap2* (Penagarikano et al. 2011), *Tsc1* (Meikle et al. 2007) and *Tsc2* (Zeng et al. 2011), *Gabrb3* (DeLorey et al. 2011; Homanics et al. 1997), and *Fmr1* (Chen and Toth 2001).

Intellectual disability is present in approximately 30–40 % of ASD subjects (Matson and Shoemaker 2009; Perou et al. 2013). Learning and memory deficits have been demonstrated in several mouse models of autism, often along with electrophysiological abnormalities detected in hippocampal slice assays. Water maze and fear conditioning deficits were reported in mice with mutations in *Pten*, *Tsc1*, *Shank3*, *Cntnap2*, *En2*, and in the BTBR inbred strain, among others (Upchurch and Wehner 1988; The Dutch-Belgian Fragile et al. 1994; D'Hooge et al. 1997; Paradee et al. 1999; Goorden et al. 2007; Moy et al. 2007; MacPherson et al. 2008; Baker et al. 2010; Penagarikano et al. 2011; Brielmaier et al. 2012; Sperow et al. 2012; Yang et al. 2012a, b; Scattoni et al. 2013).

Sleep disorders are common in children with ASD. As many as two-thirds of autistic individuals may have some kind of sleep disorder (Richdale 1999). Sleep patterns and circadian rhythms have not been extensively reported in mouse models of autism. Mutant mice lacking *Cadps2,* located in the 7q autism susceptibility locus, showed an aberration in intrinsic sleep-wake cycle maintenance (Sadakata et al. 2007). *Fmr1* KO mice demonstrated abnormal circadian activity patterns, which may suggest alterations in sleep–wake cycle stability (Baker et al. 2010). *Gbrb3* KO mice exhibited differences in activity-rest neural activity as assessed by EEG (DeLorey et al. 1998).

Attention deficits and hyperactivity are a commonly associated symptom of autism. Several mutant mouse models of autism display higher exploratory locomotion in the open field test, including *Fmr1* (Kramvis et al. 2013), *Cntnap2* (Penagarikano et al. 2011), *ProSAP1/Shank2* (Schmeisser et al. 2012), and a 16p11.2 deletion (Portmann et al. 2014).

Sensory symptoms, including under-and over-responsivity to sensory stimuli, are frequently found in those with ASD (Rogers and Ozonoff 2005). Idiosyncratic overreaction to a sudden loud noise can be tested in mice by assessing response to acoustic stimuli at various decibel levels. An increased response to sensory stimuli was observed in *Fmr1* mice (Chen and Toth 2001). Reduced acoustic startle was reported in several other mutant mouse models of autism including *Gabrb3* (DeLorey et al. 2011), *EphrinA* (Wurzman et al. 2014), and female *Mecp2* heterozygotes (Samaco et al. 2013). Idiosyncratic underreaction to painful stimuli can be assessed in mice with hot plate or tail flick thermal stimuli. Genetic models of autism have revealed increased sensitivity in these nociceptive tasks in *Gabrb3* KO mice (DeLorey et al. 2011).

Mouse behavioral assays described above have proven useful in phenotyping genetic mouse models of autism. Approaches to develop ideal models of ASD may utilize multiple species to ensure that the same outcomes are present across species, to best advance the potential for an integration of systems neuroscience with the human syndrome. Successful multiple species approaches will contribute to fast-forwarding our progress to develop effective mechanism-based therapeutics. Mouse models provide relatively low cost, high-throughput, valid phenotypes in various behavioral assays relevant to the diagnostic symptoms of ASD.

Comparative studies utilizing rodent vole models are another powerful approach for modeling social behavior relevant to ASD. Prairie and pine voles (*Microtus ochrogaster and Microtus pinetorum*, respectively) are a monogamous species living in highly social burrows (Carter and Getz 1993; Carter et al. 1995). In contrast, montane and meadow voles (*Microtus montanus and Microtus pennsylvanicus,* respectively) are non-monogamous and often live in social isolation. Differences in oxytocin peptide and receptor binding have been reported between these species of vole and are functionally related to their differences in social behavior (Winslow et al. 1993; Young et al. 2002). Carter, Bales, and colleagues have reported both facilitation and deleterious effects of oxytocin administration in voles in the partner preference pair bonding assay. These effects were both sexually dimorphic and developmentally specific (Bales and Carter 2003a, b; Carter et al. 2009; Bales et al. 2013). Intranasal oxytocin paradigms developed in the vole have recently been examined in mouse models, with reports of either adverse or minimally beneficial behavioral outcomes, dependent on length of exposure (Bales et al. 2014; Huang et al. 2014). Novel pharmacology using vole models recently illustrated that d-cycloserine, a partial agonist of the *N*-methyl-D-aspartate (NMDA) glutamate receptor that enhances receptor activation in the presence of glutamate, dose dependently enhanced partner preference in female prairie voles (Modi and Young 2011).

Rats have sophisticated behavioral repertoires which make this rodent species excellent for modeling the nuances of complex social behavior. Recent advances in genetic technologies allow for manipulation of rat gene expression. Two genetic models with relevance to ASD have been generated. One example is a rat knockout of the *Fmr1* gene, which is associated with fragile X syndrome. Behavioral phenotyping revealed that *Fmr1* KO rats have low levels of social play behavior and higher levels of a repetitive block chewing (Hamilton et al. 2014). Other genetic ASD-relevant rat KO models are the neuroligin-3 (*Nlgn3*) null and the neurexin-1α (*Nrxn1-α*) KO rats model. *Nlgn3* KO rats display reduced juvenile social play (Hamilton et al. 2014), while *Nrxn1-α* KO rats exhibit hyperactivity, exaggerated startle responses, and impairments in latent inhibition and spatial-dependent learning (Esclassan et al. 2015). Genetic rat models of autism offer a new set of tools for evaluating pharmacological interventions.

Several studies suggest a role for environmental factors, in combination with genetic susceptibility, in the etiology of ASD. An impressive population-based Danish study in 2013 outlined prenatal exposure to the anticonvulsant VPA, but not to other anti-seizure medications, nearly tripled the risk of ASD (Christensen et al. 2013). The larger study confirmed an earlier smaller report that exposure to VPA during gestation increased relative risk for ASD and maladaptive ASD-related behavioral dysfunction in children born to women who took VPA to treat their epilepsy (Bromley et al. 2008). Mouse models exposed to gestational VPA recapitulate selective behavioral and electrophysiological deficits analogous to those seen in the clinic (Wagner et al. 2006; Gandal et al. 2010; Mehta et al. 2011). Similarly, rats exposed to VPA in utero show increased frequency of motor stereotypies in adolescence, reduced social exploration, and low levels of juvenile rough and tumble play supporting the validity of this model (Schneider and Przewlocki 2005). Although the mechanisms underlying the link between VPA and autism are not fully understood, prenatal exposure to VPA alters GABA and monoamine systems, induces a loss of specific subsets of neurons, and acts through epigenetic mechanisms via histone deacetylase inhibition (Bambini-Junior et al. 2014).

Excitatory–inhibitory imbalance is a prominent hypothesis for the etiology of ASD. Pharmacological interventions that shift the balance closer to normal are under consideration. Acute exposure to the glutamate antagonist, MPEP, reduced marble burying phenotypes in offspring of dams treated with VPA, but did not alleviate anxiety-like behavior (Mehta et al. 2011). GABAergic neurons switch from excitatory to inhibitory during key developmental processes. This sequence was reported to be absent in hippocampal CA3 neurons of offspring of VPA-treated rat dams (Tyzio et al. 2014). Moreover, VPA-treated offspring emitted low numbers of isolation-induced pup USVs. Bumetanide pretreatment to dams rescued the

GABA developmental impairments and restored call emissions in VPA rodent offspring (Tyzio et al. 2014).

The first non-human primate model of ASD involved the bilateral removal of the medial temporal lobe of young rhesus macaque monkeys. Normal infant monkeys develop strong affiliative bonds. Lesioned subjects displayed atypical dyadic social interactions at 2 and 6 months and exhibited aberrant stereotypies (Bachevalier 1994; Bachevalier et al. 2001). Other lesion studies produced selective amygdala lesions in 2-week-old macaques. By 6–8 months of age, the lesioned animals demonstrated substantial fear behaviors during dyadic social interactions while maintaining much of the age-appropriate repertoire of social behavior (Prather et al. 2001).

Other reported non-human primate models of ASD have tested the hypothesis that exposure of the fetal brain to maternal autoantibodies during gestation increases ASD risk. Rhesus monkeys exposed to human immunoglobulin collected from mothers of multiple children diagnosed with ASD consistently demonstrated increased whole-body stereotypies and hyperactivity across multiple testing paradigms (Martin et al. 2008). In extended studies, these monkeys consistently deviated from species-typical social norms by more frequently approaching familiar peers in a social approach paradigm (Bauman et al. 2013).

Oxytocin administration in rhesus macaques was reported to significantly increase plasma oxytocin concentrations when administered using the aerosol or intranasal routes (Modi et al. 2014). Social perception in the dot-probe task in monkeys receiving intranasal oxytocin detected selectively reduced attention to negative facial expressions, but not neutral faces or nonsocial images (Parr et al. 2013). This first pharmacological report using non-human primates provides promising evidence for oxytocin-based compound efficacy in clinical populations.

4 Evaluating Pharmacological Therapeutics in Animal Models with High Construct Validity and Strong Face Validity for ASD

Clinical trials for ASD core symptoms are challenged by the heterogeneity of the disorder, which can limit study design parameters and statistical power for outcome measures. Currently, there are no pharmacotherapies approved by the US Food and Drug Administration specifically for social interaction, communication deficits, and repetitive behaviors. The only FDA-approved pharmacological treatments for autism are the antipsychotics risperidone and aripiprazole, which treat the associated irritability symptoms of aggression, self-injury, and temper tantrums. Greater than 50 % of children diagnosed with ASD in the USA are using at least one psychoactive drug (Spencer et al. 2013), as prescribed for irritability (Siegel and Beaulieu 2012), or given off-label. Risperidone, which modulates dopamine and serotonin systems, had a significant effect on stereotyped behavior in children with ASD (McCracken et al. 2002; McDougle et al. 2005; Chavez et al. 2006), although this was not seen in all studies (Ghaeli et al. 2014). Risperidone studies that included behavioral scales measuring aspects of sociability, such as social relationships and language, had large effect sizes, but failed to reach statistical significance (McDougle et al. 2005). Other studies that utilized additional behavioral scales, such as the Aberrant Behavior Checklist Social Withdrawal subscale and the Childhood Autism Rating Scale (CARS), found that risperidone treatment was effective compared to placebo (Scahill et al. 2013; Ghaeli et al. 2014). The lack of consistency for risperidone's effects on aspects of social behavior may be due to clinical heterogeneity within the studies' ASD subject population, differences in treatment duration, as well as differences in the tools used for sociability outcome measures. Treatment studies with antidepressants, such as selective serotonin reuptake inhibitors (SSRIs) and tricyclic antidepressants, have yielded mixed results on improvement of repetitive behaviors. For example, SSRI treatment with fluoxetine or citalopram did not produce a clinically significant improvement on repetitive behaviors in children (Hollander et al. 2005; King et al. 2009). However, additional studies with fluoxetine and fluvoxamine demonstrated improvement on repetitive thoughts, repetitive actions, and scores of an obsessive–compulsive scale in adults (McDougle et al. 1996; Hollander et al. 2012), suggesting that SSRI treatment approach may depend on the age of individuals with ASD.

As described in Table 1, additional classes of compounds have been evaluated for their efficacy in treating ASD core symptoms, although large-scale, randomized, double-blind, placebo-controlled trials are lacking. Administration of oxytocin, a neuropeptide involved in social pair bonding, social memory, and affiliative behaviors (Gimpl and Fahrenholz 2001), increased social awareness and emotional recognition in both neurotypical individuals and those with ASD in pilot studies (Hollander et al. 2007; Bartz and Hollander 2008; Rimmele et al. 2009; Bartz et al. 2010; Guastella et al. 2010). Interestingly, functional neuroimaging results from a randomized, double-blind cross-over study in children with ASD found that brain structures associated with sociability (e.g., striatum, posterior cingulate, and premotor cortex) showed greater recruitment after intranasal oxytocin administration, suggesting that this neuropeptide enhanced the saliency of social stimuli (Gordon et al. 2013).

STX209 (Arbaclofen), a selective $GABA_B$ agonist thought to stimulate inhibitory neurotransmission, was evaluated as a treatment for fragile X syndrome, a neurodevelopmental disorder with a high incidence of ASD comorbidity (Berry-Kravis et al. 2012). Although there were no statistically significant differences in the primary outcome (Aberrant Behavior Checklist-Irritability subscale), male subjects were noted as having positive improvements on several global measures including socialization scores. Additionally, in a study with individuals with ASD, Arbaclofen was well tolerated and improved scores on social responsiveness, social withdrawal, and clinical global impression scales (Erickson et al. 2014a).

D-cycloserine, a partial agonist of the ionotropic glutamatergic NMDA receptor, has been evaluated in one single-blind, placebo-controlled trial, where the majority of children with ASD treated with D-cycloserine improved their scores on the

Table 1 Examples of preclinical and clinical evaluations of drug treatments for autism

Mechanistic class	Preclinical model	Compound	Preclinical phenotype rescues	Reference
mGluR5 modulation	BTBR C58/J	MPEP MTEP GRN529	• Improved sociability • Reduced repetitive behavior • Improved cognition	Silverman et al. (2010a), Silverman et al. (2012) and Seese et al. (2014)
	Fmr1	AFQ056 CTEP MTEP MPEP Fenobam JNJ16259685	• Rescued abnormal dendritic spine morphology • Corrected excessive protein synthesis • Normalized altered long-term depression • Reduced seizure susceptibility • Decreased hyperactivity • Rescued cognitive deficits • Rescued sensorimotor gating • Reduced repetitive behavior	Yan et al. (2005b), De Vrij et al. (2008), Busquets-Garcia et al. (2013), Michalon et al. (2012), Gantois et al. (2013), Thomas et al. (2012), Gandhi et al. (2014) and Pop et al. (2014)
	Shank2	CDPPB	• Restored abnormal long-term potentiation and long-term depression • Improved sociability	Won et al. (2012)
	Valproic acid	MPEP	• Reduced repetitive behavior	Mehta et al. (2011)
	Clinical population	**Treatment**	**Phase**	**Reference***
	ASD—5 to 17 years old	Acamprosate	Phases 2 and 3; single-blind placebo lead-in trial	NCT01813318; Erickson et al. (2014b)

(continued)

Table 1 (continued)

Mechanistic class	Preclinical model	Compound	Preclinical phenotype rescues	Reference
	ASD—6 to 13	Acamprosate	Open label	Erickson et al. (2011a)
	Fragile X males with ASD—18 to 23 years old	Acamprosate	Open label	Erickson et al. (2010)
	Fragile X—5 to 17 years old	Acamprosate	Phase 3; open label	NCT01300923; Erickson et al. (2013)
	Fragile X—3 to 11 years old	AFQ056	Phase 1	NCT01482143
	Fragile X—12 to 17 years old	AFQ056	Phases 2 and 3	NCT01357239, NCT01433354
	Fragile X—18 to 45 years old	AFQ056	Phase 2	NCT01253629, NCT01348087, NCT00718341
	Fragile X—5 to 13 years old	RO4917523	Phase 2	NCT01750957
	Fragile X—14 to 50 years old	RO4917523	Phase 2	NCT01517698, NCT01015430

(continued)

Table 1 (continued)

Mechanistic class	Preclinical model	Compound	Preclinical phenotype rescues	Reference
$GABA_B$ modulation	BTBR C58/J	R-Baclofen	• Improved sociability • Reduced repetitive behavior	Silverman et al. (2015)
	Fmr1	STX209 (Arbaclofen)	• Normalized normal dendritic spine morphology • Corrected excessive protein synthesis • Reduced seizure susceptibility	Henderson et al. (2012)
	NMDA NR1 subunit knockout mice	Racemic baclofen	• Improved excitation/inhibition balance • Rescued gamma EEG band deficits • Reduced hyperactivity • Rescued sensorimotor gating deficits	Gandal et al. (2012)
	Clinical population	**Treatment**	**Phase**	**Reference***
	ASD—5 to 21 years old	STX209 (Arbaclofen)	Phases 2 and 3	NCT01706523, NCT01288716; Frye (2014)
	ASD—6 to 17 years old	STX209 (Arbaclofen)	Phase 2; open label	NCT00846547; Erickson et al. (2014a) and Frye (2014)
	Fragile X—6 to 40 years old	STX209 (Arbaclofen)	Phase 2	NCT00788073; Berry-Kravis et al. (2012)
	Fragile X—12 to 50 years old	STX209 (Arbaclofen)	Phase 3	NCT01282268

(continued)

Table 1 (continued)

Mechanistic class	Preclinical model	Compound	Preclinical phenotype rescues	Reference
$GABA_A$ modulation	BTBR	Diazepam Low dose of benzodiazepines L-838,417	• Increased GABAergic inhibitory neurotransmission • Improved social interactions • Ameliorated cognitive deficits	Pobbe et al. (2011) and Han et al. (2014)
	Clinical population	**Treatment**	**Phase**	**Reference***
	ASD—18 to 45 years old	Pregnanolone	Phase 2	NCT01881737
	High functioning ASD—18 to 35 years old	AZ7325	Phase 2	NCT01966679
mTOR inhibitors	BTBR	Rapamycin	• Improved sociability	Burket et al. (2014)
	Pten	Rapamycin RAD001 (Everolimus)	• Improved macrocephaly • Inhibits neuronal hypertrophy • Improved abnormal sociability • Reduced seizures	Zhou et al. (2009)
	Tsc1		• Improved survival rates and weight gain • Prevented seizures • Ameliorated abnormal EEG • Improved neuronal morphology • Prevented cell loss	Meikle et al. (2008), Zeng et al. (2008), Sato et al. (2012) and Tsai et al. (2012)

(continued)

Table 1 (continued)

Mechanistic class	Preclinical model	Compound	Preclinical phenotype rescues	Reference
			• Restored myelination abnormalities • Improved motor phenotypes • Improved sociability • Ameliorated cognitive deficits	
	Tsc2		• Improved cognition • Improved sociability	Ehninger et al. (2008) and Sato et al. (2012)
	Fmr1	Temsirolimus	• Rescued cognitive impairment • Reduced seizure susceptibility	Busquets-Garcia et al. (2013)
	Clinical population	**Treatment**	**Phase**	**Reference***
	Tuberous sclerosis complex—4 to 15 years old	Rapamycin (Sirolimus) RAD001 (Everolimus)	Phases 2 and 3	NCT01730209
	Tuberous sclerosis complex—2 to 61 years old		Phases 1, 2, and 3	NCT01929642, NCT00789828, NCT00790400, NCT00411619; Krueger et al. (2010), Tillema et al. (2012), Bissler et al. (2013) and Franz et al. (2013), Krueger et al. (2013)

(continued)

Table 1 (continued)

Mechanistic class	Preclinical model	Compound	Preclinical phenotype rescues	Reference
Neuropeptides	BTBR	Oxytocin	• No behavioral effects	Bales et al. (2014)
	C57BL6/J		• Reduced several social behaviors	Huang et al. (2014)
	C58/J		• Improved sociability	Teng et al. (2013)
	Oxytocin knockout mice		• Improved sociability	Ferguson et al. (2001)
	Oxytocin receptor mice		• Improved sociability • Restored cognitive inflexibility	Sala et al. (2011)
	Clinical population	**Treatment**	**Phase**	**Reference***
	ASD—3 to 17 years old	Oxytocin	Phases 2 and 3	NCT01944046, NCT01308749, NCT01624194; Tachibana et al. (2013)
	ASD—12 to 17 years old			NCT01417026, NCT02090829, NCT01931033, NCT02007447, ACTRN12609000513213; Guastella et al. (2014)
	ASD—18 to 60 years old			NCT00490802, NCT01337687, NCT01788072; Hollander et al. (2003) and Lin et al. (2014)
	ASD—21 to 38 years old		Randomized cross-over double-blind study	UMIN000002241, UMIN000004393; Aoki et al. (2014a, b) and Watanabe et al. (2014)
	ASD—19 to 56 years old		Randomized, placebo-controlled, double-blind study	Hollander et al. (2007)
	ASD—6 to 12 years old	Vasopressin	Phase 2	NCT01962870
	ASD—18 to 55 years old	RG-7314	Phase 2	NCT01793441

(continued)

Table 1 (continued)

Mechanistic class	Preclinical model	Compound	Preclinical phenotype rescues	Reference
Growth factors	*Fmr1*	BDNF	• Rescued long-term potentiation abnormalities	Lauterborn et al. (2007)
	Shank3	IGF1	• Improved long-term potentiation	Bozdagi et al. (2013)
	Mecp2-/-		• Reversed respiration deficits	Tropea et al. (2009)
	Clinical population	**Treatment**	**Phase**	**Reference***
	ASD—5 to 12 years old	IGF1	Phase 2	NCT01970345
	22q13 deletion (Phelan-McDermid syndrome)			NCT01525901
	Rett syndrome—4 to 11 years old	IGF1	Pilot study; case study	Pini et al. (2012, 2014)
NMDA glutamate receptor modulation	*Fmr1*	Memantine	• Corrected spine morphology	Wei et al. (2012)
	Shank2	D-cycloserine	• Improved sociability	Won et al. (2012)
	Clinical population	**Treatment**	**Phase**	**Reference**
	ASD—3 to 12 years old	D-cycloserine	Phase 3; Prospective, open label study	NCT00198120; Owley et al. (2006)
	ASD—14 to 25 years old		Pilot study; double-blind randomized trial	Posey et al. (2004) and Urbano et al. (2014)

(continued)

Table 1 (continued)

Mechanistic class	Preclinical model	Compound	Preclinical phenotype rescues	Reference
	ASD—2 to 26 years old	Memantine	Open label; retrospective study	Chez et al. (2007) and Erickson et al. (2007)
	ASD—6 to 12 years old		Phase 2	NCT01592786, NCT01592773
	ASD—13 to 17 years old		Phase 3	NCT01972074
	ASD—18 to 85 years old		Phase 4	NCT01078844, NCT01333865
AMPA glutamate receptor modulation	BTBR	AMPAkines CX546 CX1837 CX1739	• Improved facets of sociability	Silverman et al. (2013)
	Mecp2		• Reversed respiration deficits	Ogier et al. (2007)
	Clinical population	**Treatment**	**Phase**	**Reference***
	Fragile X or ASD—18 to 50 years old	CX516	Phase 2	NCT00054730; Berry-Kravis et al. (2006)
Atypical antipsychotics	BTBR	Risperidone M100907	• Improved reversal learning	Amodeo et al. (2014a)
		Risperidone	• Failed to improve sociability	Chadman (2011) and Gould et al. (2011)
	Cntnap2	Risperidone	• Reduced hyperactivity • Decreased repetitive behavior	Penagarikano et al. (2011)

(continued)

Table 1 (continued)

Mechanistic class	Preclinical model	Compound	Preclinical phenotype rescues	Reference
	Clinical population	**Treatment**	**Phase**	**Reference***
	ASD—30 months to 17 years old	Aripiprazole	Phase 2, 3 and 4	NCT00619190, NCT01617447, NCT00337571, NCT02069977, NCT00198107
	ASD—12 to 18 years old	Aripiprazole	Phase 2	NCT00208533
		Ziprasidone	Phase 2	NCT00208559
	ASD—5 to 18 years old	Aripiprazole	Phases 3 and 4; Pilot study; Open label, Chart review	NCT00332241, NCT00337571, NCT01227668, NCT00365859;, Marcus et al. (2009), Owen et al. (2009), Blankenship et al. (2010), Marcus et al. (2011a, b), Robb et al. (2011), Varni et al. (2012), Ishitobi et al. (2013), Mankoski et al. (2013), Findling et al. (2014), Maloney et al. (2014) and Adler et al. (2015)
		Risperidone	Phases 2, 3, and 4; Open label; Randomized, double-blind trial	NCT01171937, NCT00576732, NCT01624675, NCT00005014; McDougle et al. (2000), McCracken et al. (2002), McDougle et al. (2005), Rausch et al. (2005), Desousa (2010), Handen et al. (2013), Kent et al. (2013a, b), Scahill et al. (2013), Ghaeli et al. (2014) and Ghanizadeh and Ayoobzadehshirazi (2015)
		Lurasidone	Phase 3	NCT01911442
	Fragile X—6 to 25 years old	Aripiprazole	Open label	Erickson et al. (2011b)

(continued)

Table 1 (continued)

Mechanistic class	Preclinical model	Compound	Preclinical phenotype rescues	Reference
Serotonin reuptake inhibitors	BTBR	Busprione	• Enhanced social interactions	Gould et al. (2011)
		Fluoxetine	• Increased sociability	Chadman (2011) and Gould et al. (2011)
	Fmr1		• Reduced anxiety • Reduced locomotor activity	Uutela et al. (2014)
	Clinical population	**Treatment**	**Phase**	**Reference***
	ASD—2 to 6 years old	Buspirone	Phase 2	NCT00873509
	ASD—6 to 17 years old		Open label	NCT01850355, IRCT201307303930N28; Ghanizadeh and Ayoobzadehshirazi (2015)
	ASD—5 to 17 years old	Citalopram	Phase 2; Randomized controlled trial	NCT00086645, NCT00211770; King et al. (2009)
		Fluoxetine	Phase 3; Open label; Placebo-controlled crossover trial	NCT00515320, ACTRN12608000173392; Buchsbaum et al. (2001), Hollander et al. (2005), Desousa (2010), Chantiluke et al. (2014a, b) and Mouti et al. (2014)
	ASD—3 to 10 years old	Fluvoxamine Sertraline	Phase 3	NCT00655174
	ASD—3 to 12 years old	Olanzapine	Phase 2	NCT00057408
	ASD—6 to 16 years old		Open pilot study; double-blind, placebo-controlled trial	Potenza et al. (1999), Malone et al. (2001), Kemner et al. (2002) and Hollander et al. (2006)

(continued)

Table 1 (continued)

Mechanistic class	Preclinical model	Compound	Preclinical phenotype rescues	Reference
	ASD—18 to 39 years old	Sertraline	Open label	McDougle et al. (1998)
	ASD—18 to 53 years old	Fluvoxamine	Double-blind, placebo-controlled trial	McDougle et al. (1996)
	ASD—18 to 65 years old	Fluoxetine	Double-blind, placebo-controlled trial; open trial	NCT00027404; Fatemi et al. (1998) and Hollander et al. (2012)
	Fragile X—12 to 50 months old	Sertraline	Retrospective chart review	Indah Winarni et al. (2012)
Acetylcholine modulation	BTBR	Donepezil	• Improved cognitive flexibility • Enhanced sociability	Karvat and Kimchi (2014)
	Valproic acid		• Improved sociability • Reduced repetitive behavior • Reduced hyperactivity	Kim et al. (2014a)

(continued)

Table 1 (continued)

Mechanistic class	Preclinical model	Compound	Preclinical phenotype rescues	Reference
	Clinical population	**Treatment**	**Phase**	**Reference***
	ASD—22 to 44 months	Donepezil	Phase 2	NCT01887132
	ASD—2 to 7 years old		Open label	Buckley et al. (2011)
	ASD—7 to 19 years old		Double-blind, placebo-controlled trial	Hardan and Handen (2002) and Handen et al. (2011)
	ASD—10 to 18 years old		Phase 4	NCT01098383
	ASD—4 to 12 years old	Mecamylamine	Phase 2	NCT00773812; Arnold et al. (2012)
	Fragile X males—6 to 15 years old	Donepezil	Randomized, double-blind, placebo-controlled pilot study	CTRI-2008-000229; Sahu et al. (2013)

[a]Identifier numbers for clinical trials were indexed from clinicaltrials.gov (NCT), anzctr.org.au (ACTRN), umin.ac.jp/ctr (UMIN), irct.ir (IRCT), and ctri.nic.in/Clinicaltrials (CTRI) Web sites

Autistic Behavior Checklist Lethargy and Social Withdrawal subscale (Posey et al. 2004). Memantine, an NMDA receptor antagonist approved for Alzheimer's disease, has been assessed in several open label studies and retrospective reports. Some studies found that more than half of clinical responders had improvements in Clinical Global Impression scores or language and social behaviors (Chez et al. 2007; Erickson et al. 2007), although not all studies found similar effects (Owley et al. 2006; Niederhofer 2007). Open label studies with children with ASD using cholinesterase inhibitors suggest that there may be some improvement in expressive language, parent reports, and CARS scores (Niederhofer et al. 2002; Chez et al. 2004; Nicolson et al. 2006).

Many of these early clinical trials were based on hypotheses generated from mouse models. Of particular interest is Rubenstein's proposed excitatory inhibitory imbalance, which arose from electrophysiological assays in mutant mouse models (Rubenstein 2010). Both forward translation, to discover new pharmacological targets using mouse models, and back translation, to test compounds in mutant mouse models of ASD that are used off-label or have moved into clinical trials, are described below and in Table 1.

Drugs that increase GABAergic inhibition have been tested in several mouse models of autism. Using *Fmr1* mice, in which mGluR5 expression and AMPA receptors are elevated and dendritic spines are abnormal, r-baclofen corrected basal protein synthesis, reduced AMPA receptor internalization and increased spine density in *Fmr1* KO mice (Henderson et al. 2012). In the few studies that evaluated classical benzodiazepines, reduction in repetitive behaviors was reported in BTBR mice treated with clonazepam (Han et al. 2014), which also showed efficacy in social and cognitive deficits in *Scn1* heterozygous mice, a mouse model of Dravet's syndrome that exhibits ASD symptoms (Han et al. 2014). Further, acute intraperitoneal administration of r-baclofen reduced repetitive self-grooming and improved sociability in BTBR mice, and reduced stereotyped vertical jumping in C58 mice (Silverman et al. 2015).

Another strategy to reduce excitatory neurotransmission is to inhibit mGluR receptors with negative allosteric modulators. The mGluR antagonist MPEP was evaluated in the BTBR mouse model. Acute MPEP treatment reduced repetitive behaviors, including self-grooming and marble burying (Silverman et al. 2010a), and improved cognition in BTBR (Seese et al. 2014), and demonstrated anti-epileptic effects in *Fmr1* mice (Yan et al. 2005b). The mGluR5 receptor inverse agonist CTEP showed efficacy in ameliorating cognitive deficits, signaling abnormalities, and dendritic spine deficits in the *Fmr1* KO mouse (Michalon et al. 2012). The mGluR5 negative allosteric modulator GRN-529 rescued social deficits and repetitive self-grooming in BTBR mice and reduced stereotyped jumping in C58 mice (Silverman et al. 2012).

A large number of novel pharmacological targets are being tested in mouse models. Table 1 provides a partial summary of compounds tested in various mouse models. Some of these strategies have been evaluated in clinical investigations.

Others may be under consideration. Well-replicated results with a compound that reverses autism-relevant phenotypes, both behavioral and biological, in multiple animal models, may contribute to decisions about pursuing a clinical trial for ASD.

5 Conclusions

The summary above and in Table 1 provides descriptions of behavioral assays relevant to the symptoms of autism, representative results of behavioral phenotypes in many rodent models, and drug treatment outcomes in several mouse models of autism. Initial hypotheses for pharmacological targets derived from animal studies that documented (1) elevated excitatory neurotransmission or excess mGluR5 receptors, (2) reduced GABAergic inhibitory physiology, circuitry, or interneurons in genetic mouse models of autism, along with (3) oxytocin modulation of social behaviors and growth factors that mediate brain development. Preclinical findings of improvements by drug treatments in mouse models of fragile X and autism have led to a small number of clinical trials. Unfortunately, the Arbaclofen trial by Seaside Therapeutics did not detect significant improvement in its primary outcome measures, and the mGluR5 antagonist trial by Roche was terminated due to lack of initial efficacy. Central questions at present include (a) whether the animal results did not incorporate sufficient predictive validity and (b) whether the clinical trials were not optimally designed in terms of dose, age, treatment regimen, patient population, or outcome measures.

Many concerns have been raised about the predictive usefulness of results from animal models in the discovery of treatments for neuropsychiatric disorders (Markou et al. 2009; Belzung 2014). The autism field is similarly facing this dilemma. Our view is that animal studies must incorporate a high level of validity and reproducibility. Assays in rodents can be designed to maximize face validity, for maximal analogy to the behavioral and biological symptoms of autism. However, results from animal studies need to be interpreted cautiously, without exaggeration or hyperbole about relevance to the diagnostic symptoms. Requirements for replication of positive results in two cohorts of mice would greatly increase the strength of findings. Preclinical drug studies may be most predictive when dose–response relationships have been explicated, acute and chronic treatment regimens have been tested, and clinically relevant routes of administration have been used in two or more species. These expectations represent a great deal more effort than is often invested in early preclinical studies with animal models. More complete preclinical data may be needed to provide the confidence needed to move forward into a clinical trial.

In the autism field, where no pharmacological interventions have definitively improved the core diagnostic symptoms of social interaction and communication deficits and repetitive behaviors, early failures are to be expected. Without a gold standard therapeutic, mouse models cannot be tested a priori for predictive validity. The process will be iterative. Pharmacological target discovery is benefitting from mouse models with mutations in synaptic genes and signaling pathways identified in individuals with autism, especially in cases where the gene codes for a protein in a biological pathway which is susceptible to pharmacological intervention with available compounds. Back translation, to test compounds that are being used clinically for phenotypic reversal in animals, will help to establish whether a mutant rodent model is sufficiently predictive. The current trend for autism symposia and consortia to mix clinical and basic researchers working on pharmacological interventions is encouraging, to promote this iterative discovery process.

One major hurdle is the need for simple, real-life outcome measures of appropriate social interaction, social communication, and repetitive behaviors to use as discrete endpoints for human drug trials. Gold standard instruments for the diagnostic assessment of ASD are complex and expensive, limiting their usefulness for large-scale multi-site clinical trials of new medications. Simplified, shortened rating scales are in use and under development.

Another major hurdle is the behavioral heterogeneity which characterizes ASD, which presents a huge challenge for both clinical trials and preclinical animal models. One strategy would be to stratify the ASD population based on behavioral subgroups with specific associated symptoms, e.g. seizures, aggression, anxiety, repetitive behaviors, language skills, or IQ. Another strategy is to employ proposed biomarkers, e.g., EEG gamma activity (Bosl et al. 2011; Rojas and Wilson 2014), or delayed auditory responses (Edgar et al. 2014). Parsing ASD symptoms into more tractable endophenotypes would further allow the use of animal models to illuminate genetic underpinnings and relevant molecular pathways. Both behavioral and biomarker subcategorization of the ASD behavioral spectrum would be valuable for preclinical drug discovery, to provide sufficiently robust cross-species biological phenotypes to complement behavioral phenotypes and permit rigorous preclinical evaluation of pharmacological interventions.

In conclusion, we return to our initial discussion of construct, face, and predictive validity in animal models. Collaboration of clinical and basic science researchers will be required at each level. Progress in the refinement of construct validity will require both clinical observation and genetics research to hasten the identification of strong risk genes for ASD and endophenotypes with relevance to its symptoms. To improve the face validity of animal models, basic scientists need to discuss the meaning of species-typical behaviors in rodents and non-human primates with clinical scientists working with individuals with ASD. In this way, there will be better assurance that the behaviors selected and assessed in animal models are relevant to humans with ASD. At the level of predictive validity, the first step will be the discovery of hypothesis-driven compounds that improve endophenotypes in both rodents and humans, particularly through the use of simpler, more real-life single outcome measures of appropriate social interaction and repetitive behaviors. While effective medical treatments for autism are greatly needed, the knowledge base about the most relevant pharmacological targets is currently at an early stage. Appropriate choices of animal model constructs, assays with strong face validity, and rigorous analysis of drug effects will contribute to the maturation of therapeutic development.

Acknowledgments This work was supported by the MIND Institute Autism Research Training Program MH0073124 (TK), Autism Speaks Targeted Award #8703 (JC, JS) NINDS 1R01NS085709 (JC, PL), the MIND Institute Intellectual and Developmental Disabilities Research Center 1U54HD079125, and NIMH 1R01 MH103284 (MS).

References

Abrahams BS, Geschwind DH (2008) Advances in autism genetics: on the threshold of a new neurobiology. Nat Rev Genet 9:341–355

Adise S, Saliu A, Maldonado N, Khatri V, Cardoso L, Rodriguez-Contreras A (2014) Effect of maternal care on hearing onset induced by developmental changes in the auditory periphery. J Neurosci 34:4528–4533

Adler BA, Wink LK, Early M, Shaffer R, Minshawi N, McDougle CJ, Erickson CA (2015) Drug-refractory aggression, self-injurious behavior, and severe tantrums in autism spectrum disorders: a chart review study. Autism 19:102–106

Ahern TH, Modi ME, Burkett JP, Young LJ (2009) Evaluation of two automated metrics for analyzing partner preference tests. J Neurosci Methods 182:180–188

Alarcon M, Abrahams BS, Stone JL, Duvall JA, Perederiy JV, Bomar JM, Sebat J, Wigler M, Martin CL, Ledbetter DH, Nelson SF, Cantor RM, Geschwind DH (2008) Linkage, association, and gene-expression analyses identify CNTNAP2 as an autism-susceptibility gene. Am J Hum Genet 82:150–159

Amodeo DA, Jones JH, Sweeney JA, Ragozzino ME (2012) Differences in BTBR T+ tf/J and C57BL/6J mice on probabilistic reversal learning and stereotyped behaviors. Behav Brain Res 227:64–72

Amodeo DA, Jones JH, Sweeney JA, Ragozzino ME (2014a) Risperidone and the 5-HT2A receptor antagonist M100907 improve probabilistic reversal learning in BTBR T+ tf/J mice. Autism Res 7:555–567

Amodeo DA, Yi J, Sweeney JA, Ragozzino ME (2014b) Oxotremorine treatment reduces repetitive behaviors in BTBR T+ tf/J mice. Front Synaptic Neurosci 6:17

Aoki Y, Watanabe T, Abe O, Kuwabara H, Yahata N, Takano Y, Iwashiro N, Natsubori T, Takao H, Kawakubo Y, Kasai K, Yamasue H (2014a) Oxytocin's neurochemical effects in the medial prefrontal cortex underlie recovery of task-specific brain activity in autism: a randomized controlled trial. Mol Psychiatry 20(4):447–453

Aoki Y, Yahata N, Watanabe T, Takano Y, Kawakubo Y, Kuwabara H, Iwashiro N, Natsubori T, Inoue H, Suga M, Takao H, Sasaki H, Gonoi W, Kunimatsu A, Kasai K, Yamasue H (2014b) Oxytocin improves behavioural and neural deficits in inferring others' social emotions in autism. Brain 137:3073–3086

Arakawa H, Arakawa K, Blanchard DC, Blanchard RJ (2008) A new test paradigm for social recognition evidenced by urinary scent marking behavior in C57BL/6J mice. Behav Brain Res 190:97–104

Arking DE, Cutler DJ, Brune CW, Teslovich TM, West K, Ikeda M, Rea A, Guy M, Lin S, Cook EH, Chakravarti A (2008) A common genetic variant in the neurexin superfamily member CNTNAP2 increases familial risk of autism. Am J Hum Genet 82:160–164

Arnold LE, Aman MG, Hollway J, Hurt E, Bates B, Li X, Farmer C, Anand R, Thompson S, Ramadan Y, Williams C (2012) Placebo-controlled pilot trial of mecamylamine for treatment of autism spectrum disorders. J Child Adolesc Psychopharmacol 22:198–205

Association AP (2013) Diagnostic and statistical manual of mental disorders. American Psychiatric Publishing, Washington, DC

Bachevalier J (1994) Medial temporal lobe structures and autism: a review of clinical and experimental findings. Neuropsychologia 32:627–648

Bachevalier J, Malkova L, Mishkin M (2001) Effects of selective neonatal temporal lobe lesions on socioemotional behavior in infant rhesus monkeys (*Macaca mulatta*). Behav Neurosci 115:545–559

Baker KB, Wray SP, Ritter R, Mason S, Lanthorn TH, Savelieva KV (2010) Male and female Fmr1 knockout mice on C57 albino background exhibit spatial learning and memory impairments. Genes, Brain, Behav 9:562–574

Bakkaloglu B, O'Roak BJ, Louvi A, Gupta AR, Abelson JF, Morgan TM, Chawarska K, Klin A, Ercan-Sencicek AG, Stillman AA, Tanriover G, Abrahams BS, Duvall JA, Robbins EM, Geschwind DH, Biederer T, Gunel M, Lifton RP, State MW (2008) Molecular cytogenetic analysis and resequencing of contactin associated protein-like 2 in autism spectrum disorders. Am J Hum Genet 82:165–173

Bales KL, Carter CS (2003a) Developmental exposure to oxytocin facilitates partner preferences in male prairie voles (*Microtus ochrogaster*). Behav Neurosci 117:854–859

Bales KL, Carter CS (2003b) Sex differences and developmental effects of oxytocin on aggression and social behavior in prairie voles (*Microtus ochrogaster*). Horm Behav 44:178–184

Bales KL, Perkeybile AM, Conley OG, Lee MH, Guoynes CD, Downing GM, Yun CR, Solomon M, Jacob S, Mendoza SP (2013) Chronic intranasal oxytocin causes long-term impairments in partner preference formation in male prairie voles. Biol Psychiatry 74:180–188

Bales KL, Solomon M, Jacob S, Crawley JN, Silverman JL, Larke RH, Sahagun E, Puhger KR, Pride MC, Mendoza SP (2014) Long-term exposure to intranasal oxytocin in a mouse autism model. Transl Psychiatry 4:e480

Bambini-Junior V, Baronio D, MacKenzie J, Zanatta G, Riesgo R, Gottfried C (2014) Prenatal exposure to valproate in animals and autism. In: VB Patel et al. (eds) Comprehensive guide to autism, pp 1779–1793. Springer, New York

Bartko SJ, Romberg C, White B, Wess J, Bussey TJ, Saksida LM (2011) Intact attentional processing but abnormal responding in M1 muscarinic receptor-deficient mice using an automated touchscreen method. Neuropharmacology 61:1366–1378

Bartz JA, Hollander E (2008) Oxytocin and experimental therapeutics in autism spectrum disorders. Prog Brain Res 170:451–462

Bartz JA, Zaki J, Bolger N, Hollander E, Ludwig NN, Kolevzon A, Ochsner KN (2010) Oxytocin selectively improves empathic accuracy. Psychol Sci 21:1426–1428

Baudouin SJ, Gaudias J, Gerharz S, Hatstatt L, Zhou K, Punnakkal P, Tanaka KF, Spooren W, Hen R, De Zeeuw CI, Vogt K, Scheiffele P (2012) Shared synaptic pathophysiology in syndromic and nonsyndromic rodent models of autism. Science 338:128–132

Bauman MD, Iosif AM, Ashwood P, Braunschweig D, Lee A, Schumann CM, Van de Water J, Amaral DG (2013) Maternal antibodies from mothers of children with autism alter brain growth and social behavior development in the rhesus monkey. Transl Psychiatry 3:e278

Bauman MD, Iosif AM, Smith SE, Bregere C, Amaral DG, Patterson PH (2014) Activation of the maternal immune system during pregnancy alters behavioral development of rhesus monkey offspring. Biol Psychiatry 75:332–341

Belzung C (2014) Innovative drugs to treat depression: did animal models fail to be predictive or did clinical trials fail to detect effects? Neuropsychopharmacology 39:1041–1051

Berry-Kravis E, Krause SE, Block SS, Guter S, Wuu J, Leurgans S, Decle P, Potanos K, Cook E, Salt J, Maino D, Weinberg D, Lara R, Jardini T, Cogswell J, Johnson SA, Hagerman R (2006) Effect of CX516, an AMPA-modulating compound, on cognition and behavior in fragile X syndrome: a controlled trial. J Child Adolesc Psychopharmacol 16:525–540

Berry-Kravis EM, Hessl D, Rathmell B, Zarevics P, Cherubini M, Walton-Bowen K, Mu Y, Nguyen DV, Gonzalez-Heydrich J, Wang PP, Carpenter RL, Bear MF, Hagerman RJ (2012) Effects of STX209 (arbaclofen) on neurobehavioral function in children and adults with fragile X syndrome: a randomized, controlled, phase 2 trial. Sci Transl Med 4:152ra127

Bielsky IF, Young LJ (2004) Oxytocin, vasopressin, and social recognition in mammals. Peptides 25:1565–1574

Birrell JM, Brown VJ (2000) Medial frontal cortex mediates perceptual attentional set shifting in the rat. J Neurosci 20:4320–4324

Bissler JJ, Kingswood JC, Radzikowska E, Zonnenberg BA, Frost M, Belousova E, Sauter M, Nonomura N, Brakemeier S, de Vries PJ, Whittemore VH, Chen D, Sahmoud T, Shah G, Lincy J, Lebwohl D, Budde K (2013) Everolimus for angiomyolipoma associated with tuberous sclerosis complex or sporadic lymphangioleiomyomatosis (EXIST-2): a multicentre, randomised, double-blind, placebo-controlled trial. Lancet 381:817–824

Blanchard DC, Spencer RL, Weiss SM, Blanchard RJ, McEwen B, Sakai RR (1995) Visible burrow system as a model of chronic social stress: behavioral and neuroendocrine correlates. Psychoneuroendocrinology 20:117–134

Blanchard RJ, Dulloog L, Markham C, Nishimura O, Nikulina Compton J, Jun A, Han C, Blanchard DC (2001) Sexual and aggressive interactions in a visible burrow system with provisioned burrows. Physiol Behav 72:245–254

Blankenship K, Erickson CA, Stigler KA, Posey DJ, McDougle CJ (2010) Aripiprazole for irritability associated with autistic disorder in children and adolescents aged 6–17 years. Pediatric Health 4:375–381

Blundell J, Blaiss CA, Etherton MR, Espinosa F, Tabuchi K, Walz C, Bolliger MF, Sudhof TC, Powell CM (2010) Neuroligin-1 deletion results in impaired spatial memory and increased repetitive behavior. J Neurosci 30:2115–2129

Bolivar VJ, Walters SR, Phoenix JL (2007) Assessing autism-like behavior in mice: variations in social interactions among inbred strains. Behav Brain Res 176:21–26

Bortolato M, Godar SC, Alzghoul L, Zhang J, Darling RD, Simpson KL, Bini V, Chen K, Wellman CL, Lin RC, Shih JC (2013) Monoamine oxidase A and A/B knockout mice display autistic-like features. Int J Neuropsychopharmacol 16:869–888

Bosl W, Tierney A, Tager-Flusberg H, Nelson C (2011) EEG complexity as a biomarker for autism spectrum disorder risk. BMC Med 9:18

Bouchard PR, Lynch CB (1989) Burrowing behavior in wild house mice: variation within and between populations. Behav Genet 19:447–456

Bozdagi O, Sakurai T, Papapetrou D, Wang X, Dickstein DL, Takahashi N, Kajiwara Y, Yang M, Katz AM, Scattoni ML, Harris MJ, Saxena R, Silverman JL, Crawley JN, Zhou Q, Hof PR, Buxbaum JD (2010) Haploinsufficiency of the autism-associated Shank3 gene leads to deficits in synaptic function, social interaction, and social communication. Mol Autism 1:15

Bozdagi O, Tavassoli T, Buxbaum JD (2013) Insulin-like growth factor-1 rescues synaptic and motor deficits in a mouse model of autism and developmental delay. Mol Autism 4:9

Braunschweig D, Krakowiak P, Duncanson P, Boyce R, Hansen RL, Ashwood P, Hertz-Picciotto I, Pessah IN, Van de Water J (2013) Autism-specific maternal autoantibodies recognize critical proteins in developing brain. Transl Psychiatry 3:e277

Brielmaier J, Matteson PG, Silverman JL, Senerth JM, Kelly S, Genestine M, Millonig JH, DiCicco-Bloom E, Crawley JN (2012) Autism-relevant social abnormalities and cognitive deficits in engrailed-2 knockout mice. PLoS ONE 7:e40914

Brodkin ES (2007) BALB/c mice: low sociability and other phenotypes that may be relevant to autism. Behav Brain Res 176:53–65

Brodkin ES, Hagemann A, Nemetski SM, Silver LM (2004) Social approach-avoidance behavior of inbred mouse strains towards DBA/2 mice. Brain Res 1002:151–157

Bromley RL, Mawer G, Clayton-Smith J, Baker GA, Liverpool and Manchester Neurodevelopment Group (2008) Autism spectrum disorders following in utero exposure to antiepileptic drugs. Neurology 71:1923–1924

Buchsbaum MS, Hollander E, Haznedar MM, Tang C, Spiegel-Cohen J, Wei TC, Solimando A, Buchsbaum BR, Robins D, Bienstock C, Cartwright C, Mosovich S (2001) Effect of fluoxetine on regional cerebral metabolism in autistic spectrum disorders: a pilot study. Int J Neuropsychopharmacol 4:119–125

Buckley AW, Sassower K, Rodriguez AJ, Jennison K, Wingert K, Buckley J, Thurm A, Sato S, Swedo S (2011) An open label trial of donepezil for enhancement of rapid eye movement sleep in young children with autism spectrum disorders. J Child Adolesc Psychopharmacol 21:353–357

Burket JA, Benson AD, Tang AH, Deutsch SI (2013) D-Cycloserine improves sociability in the BTBR T+ Itpr3tf/J mouse model of autism spectrum disorders with altered Ras/Raf/ERK1/2 signaling. Brain Res Bull 96:62–70
Burket JA, Benson AD, Tang AH, Deutsch SI (2014) Rapamycin improves sociability in the BTBR T(+)Itpr3(tf)/J mouse model of autism spectrum disorders. Brain Res Bull 100:70–75
Busquets-Garcia A, Gomis-Gonzalez M, Guegan T, Agustin-Pavon C, Pastor A, Mato S, Perez-Samartin A, Matute C, de la Torre R, Dierssen M, Maldonado R, Ozaita A (2013) Targeting the endocannabinoid system in the treatment of fragile X syndrome. Nat Med 19:603–607
Byatt S, Nyby J (1986) Hormonal regulation of chemosignals of female mice that elicit ultrasonic vocalizations from males. Horm Behav 20:60–72
Carter CS, Boone EM, Pournajafi-Nazarloo H, Bales KL (2009) Consequences of early experiences and exposure to oxytocin and vasopressin are sexually dimorphic. Dev Neurosci 31:332–341
Carter CS, DeVries AC, Getz LL (1995) Physiological substrates of mammalian monogamy: the prairie vole model. Neurosci Biobehav Rev 19:303–314
Carter CS, Getz LL (1993) Monogamy and the prairie vole. Sci Am 268:100–106
Carter CS, Williams JR, Witt DM, Insel TR (1992) Oxytocin and social bonding. Ann N Y Acad Sci 652:204–211
Chabout J, Serreau P, Ey E, Bellier L, Aubin T, Bourgeron T, Granon S (2012) Adult male mice emit context-specific ultrasonic vocalizations that are modulated by prior isolation or group rearing environment. PLoS ONE 7:e29401
Chadman KK (2011) Fluoxetine but not risperidone increases sociability in the BTBR mouse model of autism. Pharmacol Biochem Behav 97:586–594
Chantiluke K, Barrett N, Giampietro V, Brammer M, Simmons A, Murphy DG, Rubia K (2014a) Inverse effect of fluoxetine on medial prefrontal cortex activation during reward reversal in ADHD and autism. Cereb Cortex
Chantiluke K, Barrett N, Giampietro V, Santosh P, Brammer M, Simmons A, Murphy DG, Rubia K (2014b) Inverse fluoxetine effects on inhibitory brain activation in non-comorbid boys with ADHD and with ASD. Psychopharmacology 232(12):2071–2082
Chavez B, Chavez-Brown M, Rey JA (2006) Role of risperidone in children with autism spectrum disorder. Ann Pharmacother 40:909–916
Cheh MA, Millonig JH, Roselli LM, Ming X, Jacobsen E, Kamdar S, Wagner GC (2006) En2 knockout mice display neurobehavioral and neurochemical alterations relevant to autism spectrum disorder. Brain Res 1116:166–176
Chen L, Toth M (2001) Fragile X mice develop sensory hyperreactivity to auditory stimuli. Neuroscience 103:1043–1050
Chez MG, Aimonovitch M, Buchanan T, Mrazek S, Tremb RJ (2004) Treating autistic spectrum disorders in children: utility of the cholinesterase inhibitor rivastigmine tartrate. J Child Neurol 19:165–169
Chez MG, Burton Q, Dowling T, Chang M, Khanna P, Kramer C (2007) Memantine as adjunctive therapy in children diagnosed with autistic spectrum disorders: an observation of initial clinical response and maintenance tolerability. J Child Neurol 22:574–579
Christensen J, Gronborg TK, Sorensen MJ, Schendel D, Parner ET, Pedersen LH, Vestergaard M (2013) Prenatal valproate exposure and risk of autism spectrum disorders and childhood autism. JAMA 309:1696–1703
Christian SL, Brune CW, Sudi J, Kumar RA, Liu S, Karamohamed S, Badner JA, Matsui S, Conroy J, McQuaid D, Gergel J, Hatchwell E, Gilliam TC, Gershon ES, Nowak NJ, Dobyns WB, Cook EH Jr (2008) Novel submicroscopic chromosomal abnormalities detected in autism spectrum disorder. Biol Psychiatry 63:1111–1117
Clipperton-Allen AE, Page DT (2014) Pten haploinsufficient mice show broad brain overgrowth but selective impairments in autism-relevant behavioral tests. Hum Mol Genet 23:3490–3505

Coe BP, Witherspoon K, Rosenfeld JA, van Bon BW, Vulto-van Silfhout AT, Bosco P, Friend KL, Baker C, Buono S, Vissers LE, Schuurs-Hoeijmakers JH, Hoischen A, Pfundt R, Krumm N, Carvill GL, Li D, Amaral D, Brown N, Lockhart PJ, Scheffer IE, Alberti A, Shaw M, Pettinato R, Tervo R, de Leeuw N, Reijnders MR, Torchia BS, Peeters H, O'Roak BJ, Fichera M, Hehir-Kwa JY, Shendure J, Mefford HC, Haan E, Gecz J, de Vries BB, Romano C, Eichler EE (2014) Refining analyses of copy number variation identifies specific genes associated with developmental delay. Nat Genet 46:1063–1071

Crawley JN (1985) Exploratory behavior models of anxiety in mice. Neurosci Biobehav Rev 9:37–44

Crawley JN (2004) Designing mouse behavioral tasks relevant to autistic-like behaviors. Ment Retard Dev Disabil Res Rev 10:248–258

Crawley JN (2012) Translational animal models of autism and neurodevelopmental disorders. Dialogues Clin Neurosci 14:293–305

Cryan JF, Sweeney FF (2011) The age of anxiety: role of animal models of anxiolytic action in drug discovery. Br J Pharmacol 164:1129–1161

D'Amato FR, Moles A (2001) Ultrasonic vocalizations as an index of social memory in female mice. Behav Neurosci 115:834–840

D'Amato FR, Scalera E, Sarli C, Moles A (2005) Pups call, mothers rush: does maternal responsiveness affect the amount of ultrasonic vocalizations in mouse pups? Behav Genet 35:103–112

D'Hooge R, Nagels G, Franck F, Bakker CE, Reyniers E, Storm K, Kooy RF, Oostra BA, Willems PJ, De Deyn PP (1997) Mildly impaired water maze performance in male Fmr1 knockout mice. Neuroscience 76:367–376

De Rubeis S, He X, Goldberg AP, Poultney CS, Samocha K, Cicek AE, Kou Y, Liu L, Fromer M, Walker S, Singh T, Klei L, Kosmicki J, Shih-Chen F, Aleksic B, Biscaldi M, Bolton PF, Brownfeld JM, Cai J, Campbell NG, Carracedo A, Chahrour MH, Chiocchetti AG, Coon H, Crawford EL, Curran SR, Dawson G, Duketis E, Fernandez BA, Gallagher L, Geller E, Guter SJ, Hill RS, Ionita-Laza J, Jimenz Gonzalez P, Kilpinen H, Klauck SM, Kolevzon A, Lee I, Lei I, Lei J, Lehtimaki T, Lin CF, Ma'ayan A, Marshall CR, McInnes AL, Neale B, Owen MJ, Ozaki N, Parellada M, Parr JR, Purcell S, Puura K, Rajagopalan D, Rehnstrom K, Reichenberg A, Sabo A, Sachse M, Sanders SJ, Schafer C, Schulte-Ruther M, Skuse D, Stevens C, Szatmari P, Tammimies K, Valladares O, Voran A, Li-San W, Weiss LA, Willsey AJ, Yu TW, Yuen RK, Cook EH, Freitag CM, Gill M, Hultman CM, Lehner T, Palotie A, Schellenberg GD, Sklar P, State MW, Sutcliffe JS, Walsh CA, Scherer SW, Zwick ME, Barett JC, Cutler DJ, Roeder K, Devlin B, Daly MJ, Buxbaum JD (2014) Synaptic, transcriptional and chromatin genes disrupted in autism. Nature 515:209–215

de Vrij FM, Levenga J, van der Linde HC, Koekkoek SK, De Zeeuw CI, Nelson DL, Oostra BA, Willemsen R (2008) Rescue of behavioral phenotype and neuronal protrusion morphology in Fmr1 KO mice. Neurobiol Dis 31:127–132

Del'Guidice T, Lemay F, Lemasson M, Levasseur-Moreau J, Manta S, Etievant A, Escoffier G, Dore FY, Roman FS, Beaulieu JM (2014) Stimulation of 5-HT2C receptors improves cognitive deficits induced by human tryptophan hydroxylase 2 loss of function mutation. Neuropsychopharmacology 39:1125–1134

DeLorey TM, Handforth A, Anagnostaras SG, Homanics GE, Minassian BA, Asatourian A, Fanselow MS, Delgado-Escueta A, Ellison GD, Olsen RW (1998) Mice lacking the beta3 subunit of the GABAA receptor have the epilepsy phenotype and many of the behavioral characteristics of Angelman syndrome. J Neurosci 18:8505–8514

DeLorey TM, Sahbaie P, Hashemi E, Homanics GE, Clark JD (2008) Gabrb3 gene deficient mice exhibit impaired social and exploratory behaviors, deficits in non-selective attention and hypoplasia of cerebellar vermal lobules: a potential model of autism spectrum disorder. Behav Brain Res 187:207–220

DeLorey TM, Sahbaie P, Hashemi E, Li WW, Salehi A, Clark DJ (2011) Somatosensory and sensorimotor consequences associated with the heterozygous disruption of the autism candidate gene, Gabrb3. Behav Brain Res 216:36–45

Desousa A (2010) An open-label trial of risperidone and fluoxetine in children with autistic disorder. Indian J Psychol Med 32:17–21

Devlin B, Scherer SW (2012) Genetic architecture in autism spectrum disorder. Curr Opin Genet Dev 22:229–237

Durand CM, Betancur C, Boeckers TM, Bockmann J, Chaste P, Fauchereau F, Nygren G, Rastam M, Gillberg IC, Anckarsater H, Sponheim E, Goubran-Botros H, Delorme R, Chabane N, Mouren-Simeoni MC, de Mas P, Bieth E, Roge B, Heron D, Burglen L, Gillberg C, Leboyer M, Bourgeron T (2007) Mutations in the gene encoding the synaptic scaffolding protein SHANK3 are associated with autism spectrum disorders. Nat Genet 39:25–27

Ecker C, Spooren W, Murphy D (2013) Developing new pharmacotherapies for autism. J Intern Med 274:308–320

Edgar JC, Lanza MR, Daina AB, Monroe JF, Khan SY, Blaskey L, Cannon KM, Jenkins J 3rd, Qasmieh S, Levy SE, Roberts TP (2014) Missing and delayed auditory responses in young and older children with autism spectrum disorders. Front Hum Neurosci 8:417

Ehninger D, Han S, Shilyansky C, Zhou Y, Li W, Kwiatkowski DJ, Ramesh V, Silva AJ (2008) Reversal of learning deficits in a Tsc2± mouse model of tuberous sclerosis. Nat Med 14:843–848

Ehret G (2005) Infant rodent ultrasounds—a gate to the understanding of sound communication. Behav Genet 35:19–29

El-Kordi A, Winkler D, Hammerschmidt K, Kastner A, Krueger D, Ronnenberg A, Ritter C, Jatho J, Radyushkin K, Bourgeron T, Fischer J, Brose N, Ehrenreich H (2013) Development of an autism severity score for mice using Nlgn4 null mutants as a construct-valid model of heritable monogenic autism. Behav Brain Res 251:41–49

Elsabbagh M, Divan G, Koh YJ, Kim YS, Kauchali S, Marcin C, Montiel-Nava C, Patel V, Paula CS, Wang C, Yasamy MT, Fombonne E (2012) Global prevalence of autism and other pervasive developmental disorders. Autism Res 5:160–179

Engineer CT, Centanni TM, Im KW, Rahebi KC, Buell EP, Kilgard MP (2014) Degraded speech sound processing in a rat model of fragile X syndrome. Brain Res 1564:72–84

Erickson CA, Early M, Stigler KA, Wink LK, Mullett JE, McDougle CJ (2011a) An open-label naturalistic pilot study of acamprosate in youth with autistic disorder. J Child Adolesc Psychopharmacol 21:565–569

Erickson CA, Mullett JE, McDougle CJ (2010) Brief report: acamprosate in fragile X syndrome. J Autism Dev Disord 40:1412–1416

Erickson CA, Posey DJ, Stigler KA, Mullett J, Katschke AR, McDougle CJ (2007) A retrospective study of memantine in children and adolescents with pervasive developmental disorders. Psychopharmacology 191:141–147

Erickson CA, Stigler KA, Wink LK, Mullett JE, Kohn A, Posey DJ, McDougle CJ (2011b) A prospective open-label study of aripiprazole in fragile X syndrome. Psychopharmacology 216:85–90

Erickson CA, Veenstra-Vanderweele JM, Melmed RD, McCracken JT, Ginsberg LD, Sikich L, Scahill L, Cherubini M, Zarevics P, Walton-Bowen K, Carpenter RL, Bear MF, Wang PP, King BH (2014a) STX209 (arbaclofen) for autism spectrum disorders: an 8-week open-label study. J Autism Dev Disord 44:958–964

Erickson CA, Wink LK, Early MC, Stiegelmeyer E, Mathieu-Frasier L, Patrick V, McDougle CJ (2014b) Brief report: pilot single-blind placebo lead-in study of acamprosate in youth with autistic disorder. J Autism Dev Disord 44:981–987

Erickson CA, Wink LK, Ray B, Early MC, Stiegelmeyer E, Mathieu-Frasier L, Patrick V, Lahiri DK, McDougle CJ (2013) Impact of acamprosate on behavior and brain-derived neurotrophic factor: an open-label study in youth with fragile X syndrome. Psychopharmacology 228:75–84

Esclassan F, Francois J, Phillips KG, Loomis S, Gilmour G (2015) Phenotypic characterization of nonsocial behavioral impairment in neurexin 1α knockout rats. Epub 2014 Nov 24. Behav Neurosci 129(1):74–85. doi: 10.1037/bne0000024.

Etherton MR, Blaiss CA, Powell CM, Sudhof TC (2009) Mouse neurexin-1alpha deletion causes correlated electrophysiological and behavioral changes consistent with cognitive impairments. Proc Natl Acad Sci 106:17998–18003

Ey E, Leblond CS, Bourgeron T (2011) Behavioral profiles of mouse models for autism spectrum disorders. Autism Res 4:5–16

Ey E, Yang M, Katz AM, Woldeyohannes L, Silverman JL, Leblond CS, Faure P, Torquet N, Le Sourd AM, Bourgeron T, Crawley JN (2012) Absence of deficits in social behaviors and ultrasonic vocalizations in later generations of mice lacking neuroligin4. Genes, Brain, Behav 11(8):928–841

Fairless AH, Katz JM, Vijayvargiya N, Dow HC, Kreibich AS, Berrettini WH, Abel T, Brodkin ES (2013) Development of home cage social behaviors in BALB/cJ vs. C57BL/6J mice. Behav Brain Res 237:338–347

Fatemi SH, Realmuto GM, Khan L, Thuras P (1998) Fluoxetine in treatment of adolescent patients with autism: a longitudinal open trial. J Autism Dev Disord 28:303–307

Ferguson JN, Aldag JM, Insel TR, Young LJ (2001) Oxytocin in the medial amygdala is essential for social recognition in the mouse. J Neurosci 21:8278–8285

Fernandez BA, Roberts W, Chung B, Weksberg R, Meyn S, Szatmari P, Joseph-George AM, Mackay S, Whitten K, Noble B, Vardy C, Crosbie V, Luscombe S, Tucker E, Turner L, Marshall CR, Scherer SW (2010) Phenotypic spectrum associated with de novo and inherited deletions and duplications at 16p11.2 in individuals ascertained for diagnosis of autism spectrum disorder. J Med Genet 47:195–203

Fernandez T, Morgan T, Davis N, Klin A, Morris A, Farhi A, Lifton RP, State MW (2008) Disruption of Contactin 4 (CNTN4) results in developmental delay and other features of 3p deletion syndrome. Am J Hum Genet 82:1385

Findling RL, Mankoski R, Timko K, Lears K, McCartney T, McQuade RD, Eudicone JM, Amatniek J, Marcus RN, Sheehan JJ (2014) A randomized controlled trial investigating the safety and efficacy of aripiprazole in the long-term maintenance treatment of pediatric patients with irritability associated with autistic disorder. J Clin Psychiatry 75:22–30

Finlay JM, Dunham GA, Isherwood AM, Newton CJ, Nguyen TV, Reppar PC, Snitkovski I, Paschall SA, Greene RW (2014) Effects of prefrontal cortex and hippocampal NMDA-NR1 subunit deletion on complex cognitive and social behaviors. Brain Res 1600:70–83

Fischer J, Hammerschmidt K (2011) Ultrasonic vocalizations in mouse models for speech and socio-cognitive disorders: insights into the evolution of vocal communication. Genes Brain Behav 10:17–27

Franz DN, Belousova E, Sparagana S, Bebin EM, Frost M, Kuperman R, Witt O, Kohrman MH, Flamini JR, Wu JY, Curatolo P, de Vries PJ, Whittemore VH, Thiele EA, Ford JP, Shah G, Cauwel H, Lebwohl D, Sahmoud T, Jozwiak S (2013) Efficacy and safety of everolimus for subependymal giant cell astrocytomas associated with tuberous sclerosis complex (EXIST-1): a multicentre, randomised, placebo-controlled phase 3 trial. Lancet 381:125–132

Frye RE (2014) Clinical potential, safety, and tolerability of arbaclofen in the treatment of autism spectrum disorder. Drug Healthc Patient Saf 6:69–76

Galef BG, Jr (2003) Social learning of food preferences in rodents: rapid appetitive learning. Current protocols in neuroscience Chapter 8:Unit 8 5D

Gandal MJ, Edgar JC, Ehrlichman RS, Mehta M, Roberts TP, Siegel SJ (2010) Validating gamma oscillations and delayed auditory responses as translational biomarkers of autism. Biol Psychiatry 68:1100–1106

Gandal MJ, Sisti J, Klook K, Ortinski PI, Leitman V, Liang Y, Thieu T, Anderson R, Pierce RC, Jonak G, Gur RE, Carlson G, Siegel SJ (2012) GABAB-mediated rescue of altered excitatory-inhibitory balance, gamma synchrony and behavioral deficits following constitutive NMDAR-hypofunction. Transl Psychiatry 2:e142

Gandhi RM, Kogan CS, Messier C (2014) 2-Methyl-6-(phenylethynyl) pyridine (MPEP) reverses maze learning and PSD-95 deficits in Fmr1 knock-out mice. Front Cell Neurosci 8:70

Gantois I, Pop AS, de Esch CE, Buijsen RA, Pooters T, Gomez-Mancilla B, Gasparini F, Oostra BA, D'Hooge R, Willemsen R (2013) Chronic administration of AFQ056/Mavoglurant restores social behaviour in Fmr1 knockout mice. Behav Brain Res 239:72–79

Ghaeli P, Nikvarz N, Alaghband-Rad J, Alimadadi A, Tehrani-Doost M (2014) Effects of risperidone on core symptoms of autistic disorder based on childhood autism rating scale: an open label study. Indian J Psychol Med 36:66–70

Ghanizadeh A, Ayoobzadehshirazi A (2015) A randomized double-blind placebo-controlled clinical trial of adjuvant buspirone for irritability in autism. Pediatr Neurol 52:77–81

Gimpl G, Fahrenholz F (2001) The oxytocin receptor system: structure, function, and regulation. Physiol Rev 81:629–683

Glessner JT, Wang K, Cai G, Korvatska O, Kim CE, Wood S, Zhang H, Estes A, Brune CW, Bradfield JP, Imielinski M, Frackelton EC, Reichert J, Crawford EL, Munson J, Sleiman PM, Chiavacci R, Annaiah K, Thomas K, Hou C, Glaberson W, Flory J, Otieno F, Garris M, Soorya L, Klei L, Piven J, Meyer KJ, Anagnostou E, Sakurai T, Game RM, Rudd DS, Zurawiecki D, McDougle CJ, Davis LK, Miller J, Posey DJ, Michaels S, Kolevzon A, Silverman JM, Bernier R, Levy SE, Schultz RT, Dawson G, Owley T, McMahon WM, Wassink TH, Sweeney JA, Nurnberger JI, Coon H, Sutcliffe JS, Minshew NJ, Grant SF, Bucan M, Cook EH, Buxbaum JD, Devlin B, Schellenberg GD, Hakonarson H (2009) Autism genome-wide copy number variation reveals ubiquitin and neuronal genes. Nature 459:569–573

Goorden SM, van Woerden GM, van der Weerd L, Cheadle JP, Elgersma Y (2007) Cognitive deficits in Tsc1± mice in the absence of cerebral lesions and seizures. Ann Neurology 62:648–655

Gordon I, Vander Wyk BC, Bennett RH, Cordeaux C, Lucas MV, Eilbott JA, Zagoory-Sharon O, Leckman JF, Feldman R, Pelphrey KA (2013) Oxytocin enhances brain function in children with autism. Proc Natl Acad Sci 110:20953–20958

Gould GG, Hensler JG, Burke TF, Benno RH, Onaivi ES, Daws LC (2011) Density and function of central serotonin (5-HT) transporters, 5-HT1A and 5-HT2A receptors, and effects of their targeting on BTBR T+ tf/J mouse social behavior. J Neurochem 116:291–303

Greco B, Manago F, Tucci V, Kao HT, Valtorta F, Benfenati F (2013) Autism-related behavioral abnormalities in synapsin knockout mice. Behav Brain Res 251:65–74

Gross C, Hoffmann A, Bassel GJ, Berry-Kravis EM (2015) Therapeutic strategies in Fragile X syndrome: From bench to bedside and back. Neurother 12:584–608

Guariglia SR, Chadman KK (2013) Water T-maze: a useful assay for determination of repetitive behaviors in mice. J Neurosci Methods 220:24–29

Guastella AJ, Einfeld SL, Gray KM, Rinehart NJ, Tonge BJ, Lambert TJ, Hickie IB (2010) Intranasal oxytocin improves emotion recognition for youth with autism spectrum disorders. Biol Psychiatry 67:692–694

Guastella AJ, Gray KM, Rinehart NJ, Alvares GA, Tonge BJ, Hickie IB, Keating CM, Cacciotti-Saija C, Einfeld SL (2014) The effects of a course of intranasal oxytocin on social behaviors in youth diagnosed with autism spectrum disorders: a randomized controlled trial. J Child Psychol Psychiatry Allied Disciplines 56(4):444–452

Hallmayer J, Cleveland S, Torres A, Phillips J, Cohen B, Torigoe T, Miller J, Fedele A, Collins J, Smith K, Lotspeich L, Croen LA, Ozonoff S, Lajonchere C, Grether JK, Risch N (2011) Genetic heritability and shared environmental factors among twin pairs with autism. Arch Gen Psychiatry 68:1095–1102

Hamilton SM, Green JR, Veeraragavan S, Yuva L, McCoy A, Wu Y, Warren J, Little L, Ji D, Cui X, Weinstein E, Paylor R (2014) Fmr1 and Nlgn3 knockout rats: novel tools for investigating autism spectrum disorders. Behav Neurosci 128:103–109

Hammerschmidt K, Radyushkin K, Ehrenreich H, Fischer J (2009) Female mice respond to male ultrasonic 'songs' with approach behaviour. Biol Lett 5:589–592

Hammerschmidt K, Reisinger E, Westekemper K, Ehrenreich L, Strenzke N, Fischer J (2012) Mice do not require auditory input for the normal development of their ultrasonic vocalizations. BMC Neurosci 13:40

Han S, Tai C, Jones CJ, Scheuer T, Catterall WA (2014) Enhancement of inhibitory neurotransmission by GABAA receptors having alpha2,3-subunits ameliorates behavioral deficits in a mouse model of autism. Neuron 81:1282–1289

Handen BL, Johnson CR, Butter EM, Lecavalier L, Scahill L, Aman MG, McDougle CJ, Arnold LE, Swiezy NB, Sukhodolsky DG, Mulick JA, White SW, Bearss K, Hollway JA, Stigler KA, Dziura J, Yu S, Sacco K, Vitiello B (2013) Use of a direct observational measure in a trial of risperidone and parent training in children with pervasive developmental disorders. J Dev Phys Disabil 25:355–371

Handen BL, Johnson CR, McAuliffe-Bellin S, Murray PJ, Hardan AY (2011) Safety and efficacy of donepezil in children and adolescents with autism: neuropsychological measures. J Child Adolesc Psychopharmacol 21:43–50

Hanson JL, Hurley LM (2012) Female presence and estrous state influence mouse ultrasonic courtship vocalizations. PLoS ONE 7:e40782

Hardan AY, Handen BL (2002) A retrospective open trial of adjunctive donepezil in children and adolescents with autistic disorder. J Child Adolesc Psychopharmacol 12:237–241

Henderson C, Wijetunge L, Kinoshita MN, Shumway M, Hammond RS, Postma FR, Brynczka C, Rush R, Thomas A, Paylor R, Warren ST, Vanderklish PW, Kind PC, Carpenter RL, Bear MF, Healy AM (2012) Reversal of disease-related pathologies in the fragile X mouse model by selective activation of GABAB receptors with arbaclofen. Sci Transl Med 4:152ra128

Hoffmann F, Musolf K, Penn DJ (2009) Freezing urine reduces its efficacy for eliciting ultrasonic vocalizations from male mice. Physiol Behav 96:602–605

Hollander E, Bartz J, Chaplin W, Phillips A, Sumner J, Soorya L, Anagnostou E, Wasserman S (2007) Oxytocin increases retention of social cognition in autism. Biol Psychiatry 61:498–503

Hollander E, Novotny S, Hanratty M, Yaffe R, DeCaria CM, Aronowitz BR, Mosovich S (2003) Oxytocin infusion reduces repetitive behaviors in adults with autistic and Asperger's disorders. Neuropsychopharmacology 28:193–198

Hollander E, Phillips A, Chaplin W, Zagursky K, Novotny S, Wasserman S, Iyengar R (2005) A placebo controlled crossover trial of liquid fluoxetine on repetitive behaviors in childhood and adolescent autism. Neuropsychopharmacology 30:582–589

Hollander E, Soorya L, Chaplin W, Anagnostou E, Taylor BP, Ferretti CJ, Wasserman S, Swanson E, Settipani C (2012) A double-blind placebo-controlled trial of fluoxetine for repetitive behaviors and global severity in adult autism spectrum disorders. Am J Psychiatry 169:292–299

Hollander E, Wasserman S, Swanson EN, Chaplin W, Schapiro ML, Zagursky K, Novotny S (2006) A double-blind placebo-controlled pilot study of olanzapine in childhood/adolescent pervasive developmental disorder. J Child Adolesc Psychopharmacol 16:541–548

Holy TE, Guo Z (2005) Ultrasonic songs of male mice. PLoS Biol 3:e386

Homanics GE, DeLorey TM, Firestone LL, Quinlan JJ, Handforth A, Harrison NL, Krasowski MD, Rick CE, Korpi ER, Makela R, Brilliant MH, Hagiwara N, Ferguson C, Snyder K, Olsen RW (1997) Mice devoid of gamma-aminobutyrate type A receptor beta3 subunit have epilepsy, cleft palate, and hypersensitive behavior. Proc Natl Acad Sci 94:4143–4148

Huang H, Michetti C, Busnelli M, Manago F, Sannino S, Scheggia D, Giancardo L, Sona D, Murino V, Chini B, Scattoni ML, Papaleo F (2014) Chronic and acute intranasal oxytocin produce divergent social effects in mice. Neuropsychopharmacology 39:1102–1114

Winarni TI, Chonchaiya W, Adams E, Au J, Mu Y, Rivera SM, Nguyen DV, Hagerman RJ (2012) Sertraline may improve language developmental trajectory in young children with fragile x syndrome: a retrospective chart review. Autism Res Treat 2012:104317

Insel TR, Hill JL, Mayor RB (1986) Rat pup ultrasonic isolation calls: possible mediation by the benzodiazepine receptor complex. Pharmacol Biochem Behav 24:1263–1267

Iossifov I, O'Roak BJ, Sanders SJ, Ronemus M, Krumm N, Levy D, Stessman HA, Witherspoon KT, Vives L, Patterson KE, Smith JD, Paeper B, Nickerson DA, Dea J, Dong S, Gonzalez LE, Mandell JD, Mane SM, Murtha MT, Sullivan CA, Walker MF, Waqar Z, Wei L, Willsey AJ, Yamrom B, Lee YH, Grabowska E, Dalkic E, Wang Z, Marks S,

Andrews P, Leotta A, Kendall J, Hakker I, Rosenbaum J, Ma B, Rodgers L, Troge J, Narzisi G, Yoon S, Schatz MC, Ye K, McCombie WR, Shendure J, Eichler EE, State MW, Wigler M (2014) The contribution of de novo coding mutations to autism spectrum disorder. Nature 515:216–221

Ishitobi M, Kosaka H, Takahashi T, Yatuga C, Asano M, Tanaka Y, Ueno K, Okazaki R, Omori M, Hiratani M, Tomoda A, Wada Y (2013) Effectiveness and tolerability of switching to aripiprazole from risperidone in subjects with autism spectrum disorders: a prospective open-label study. Clin Neuropharmacol 36:151–156

Jamain S, Quach H, Betancur C, Rastam M, Colineaux C, Gillberg IC, Soderstrom H, Giros B, Leboyer M, Gillberg C, Bourgeron T (2003) Mutations of the X-linked genes encoding neuroligins NLGN3 and NLGN4 are associated with autism. Nat Genet 34:27–29

Jamain S, Radyushkin K, Hammerschmidt K, Granon S, Boretius S, Varoqueaux F, Ramanantsoa N, Gallego J, Ronnenberg A, Winter D, Frahm J, Fischer J, Bourgeron T, Ehrenreich H, Brose N (2008) Reduced social interaction and ultrasonic communication in a mouse model of monogenic heritable autism. Proc Natl Acad Sci 105:1710–1715

Jiang YH, Ehlers MD (2013) Modeling autism by SHANK gene mutations in mice. Neuron 78:8–27

Kaidanovich-Beilin O, Lipina T, Vukobradovic I, Roder J, Woodgett JR (2011) Assessment of social interaction behaviors. J Vis Exp 48

Karvat G, Kimchi T (2014) Acetylcholine elevation relieves cognitive rigidity and social deficiency in a mouse model of autism. Neuropsychopharmacology 39:831–840

Kemner C, Willemsen-Swinkels SH, de Jonge M, Tuynman-Qua H, van Engeland H (2002) Open-label study of olanzapine in children with pervasive developmental disorder. J Clin Psychopharmacol 22:455–460

Kent JM, Hough D, Singh J, Karcher K, Pandina G (2013a) An open-label extension study of the safety and efficacy of risperidone in children and adolescents with autistic disorder. J Child Adolesc Psychopharmacol 23:676–686

Kent JM, Kushner S, Ning X, Karcher K, Ness S, Aman M, Singh J, Hough D (2013b) Risperidone dosing in children and adolescents with autistic disorder: a double-blind, placebo-controlled study. J Autism Dev Disord 43:1773–1783

Kerr TM, Muller CL, Miah M, Jetter CS, Pfeiffer R, Shah C, Baganz N, Anderson GM, Crawley JN, Sutcliffe JS, Blakely RD, Veenstra-Vanderweele J (2013) Genetic background modulates phenotypes of serotonin transporter Ala56 knock-in mice. Mol Autism 1; 4(1):35. doi: 10.1186/2040-2392-4-35

Kikusui T, Nakanishi K, Nakagawa R, Nagasawa M, Mogi K, Okanoya K (2011) Cross fostering experiments suggest that mice songs are innate. PLoS ONE 6:e17721

Kim HG, Kishikawa S, Higgins AW, Seong IS, Donovan DJ, Shen Y, Lally E, Weiss LA, Najm J, Kutsche K, Descartes M, Holt L, Braddock S, Troxell R, Kaplan L, Volkmar F, Klin A, Tsatsanis K, Harris DJ, Noens I, Pauls DL, Daly MJ, MacDonald ME, Morton CC, Quade BJ, Gusella JF (2008) Disruption of neurexin 1 associated with autism spectrum disorder. Am J Hum Genet 82:199–207

Kim JW, Seung H, Kwon KJ, Ko MJ, Lee EJ, Oh HA, Choi CS, Kim KC, Gonzales EL, You JS, Choi DH, Lee J, Han SH, Yang SM, Cheong JH, Shin CY, Bahn GH (2014a) Subchronic treatment of donepezil rescues impaired social, hyperactive, and stereotypic behavior in valproic acid-induced animal model of autism. PLoS ONE 9:e104927

Kim SH, Junker D, Lord C (2014b) Observation of spontaneous expressive language (OSEL): a new measure for spontaneous and expressive language of children with autism spectrum disorders and other communication disorders. J Autism Dev Disord 44:3230–3244

Kim YS, Leventhal BL, Koh YJ, Fombonne E, Laska E, Lim EC, Cheon KA, Kim SJ, Kim YK, Lee H, Song DH, Grinker RR (2011) Prevalence of autism spectrum disorders in a total population sample. Am J Psychiatry 168:904–912

King BH, Hollander E, Sikich L, McCracken JT, Scahill L, Bregman JD, Donnelly CL, Anagnostou E, Dukes K, Sullivan L, Hirtz D, Wagner A, Ritz L, Network SP (2009) Lack of efficacy of citalopram in children with autism spectrum disorders and high levels of repetitive behavior: citalopram ineffective in children with autism. Arch Gen Psychiatry 66:583–590

Kong A, Frigge ML, Masson G, Besenbacher S, Sulem P, Magnusson G, Gudjonsson SA, Sigurdsson A, Jonasdottir A, Wong WS, Sigurdsson G, Walters GB, Steinberg S, Helgason H, Thorleifsson G, Gudbjartsson DF, Helgason A, Magnusson OT, Thorsteinsdottir U, Stefansson K (2012) Rate of de novo mutations and the importance of father's age to disease risk. Nature 488:471–475

Kramvis I, Mansvelder HD, Loos M, Meredith R (2013) Hyperactivity, perseveration and increased responding during attentional rule acquisition in the Fragile X mouse model. Front Behav Neurosci 7:172

Krueger DA, Care MM, Holland K, Agricola K, Tudor C, Mangeshkar P, Wilson KA, Byars A, Sahmoud T, Franz DN (2010) Everolimus for subependymal giant-cell astrocytomas in tuberous sclerosis. N England J Med 363:1801–1811

Krueger DA, Wilfong AA, Holland-Bouley K, Anderson AE, Agricola K, Tudor C, Mays M, Lopez CM, Kim MO, Franz DN (2013) Everolimus treatment of refractory epilepsy in tuberous sclerosis complex. Ann Neurol 74:679–687

Kudryavtseva NN (2003) Use of the "partition" test in behavioral and pharmacological experiments. Neurosci Behav Physiol 33:461–471

Kwon CH, Luikart BW, Powell CM, Zhou J, Matheny SA, Zhang W, Li Y, Baker SJ, Parada LF (2006) Pten regulates neuronal arborization and social interaction in mice. Neuron 50:377–388

Laumonnier F, Bonnet-Brilhault F, Gomot M, Blanc R, David A, Moizard MP, Raynaud M, Ronce N, Lemonnier E, Calvas P, Laudier B, Chelly J, Fryns JP, Ropers HH, Hamel BC, Andres C, Barthelemy C, Moraine C, Briault S (2004) X-linked mental retardation and autism are associated with a mutation in the NLGN4 gene, a member of the neuroligin family. Am J Hum Genet 74:552–557

Lauterborn JC, Rex CS, Kramar E, Chen LY, Pandyarajan V, Lynch G, Gall CM (2007) Brain-derived neurotrophic factor rescues synaptic plasticity in a mouse model of fragile X syndrome. J Neurosci 27:10685–10694

Lawson-Yuen A, Saldivar JS, Sommer S, Picker J (2008) Familial deletion within NLGN4 associated with autism and Tourette syndrome. Eur J Hum Genet EJHG 16:614–618

Lijam N, Paylor R, McDonald MP, Crawley JN, Deng CX, Herrup K, Stevens KE, Maccaferri G, McBain CJ, Sussman DJ, Wynshaw-Boris A (1997) Social interaction and sensorimotor gating abnormalities in mice lacking Dvl1. Cell 90:895–905

Lim ET, Raychaudhuri S, Sanders SJ, Stevens C, Sabo A, MacArthur DG, Neale BM, Kirby A, Ruderfer DM, Fromer M, Lek M, Liu L, Flannick J, Ripke S, Nagaswamy U, Muzny D, Reid JG, Hawes A, Newsham I, Wu Y, Lewis L, Dinh H, Gross S, Wang LS, Lin CF, Valladares O, Gabriel SB, dePristo M, Altshuler DM, Purcell SM, State MW, Boerwinkle E, Buxbaum JD, Cook EH, Gibbs RA, Schellenberg GD, Sutcliffe JS, Devlin B, Roeder K, Daly MJ (2013) Rare complete knockouts in humans: population distribution and significant role in autism spectrum disorders. Neuron 77:235–242

Lin IF, Kashino M, Ohta H, Yamada T, Tani M, Watanabe H, Kanai C, Ohno T, Takayama Y, Iwanami A, Kato N (2014) The effect of intranasal oxytocin versus placebo treatment on the autonomic responses to human sounds in autism: a single-blind, randomized, placebo-controlled, crossover design study. Mol Autism 5:20

Lloyd JA (1975) Social structure and reproduction in two freely growing populations of house mice (*Mus musculus L.*). Anim Behav 23:413–424

Long JM, LaPorte P, Paylor R, Wynshaw-Boris A (2004) Expanded characterization of the social interaction abnormalities in mice lacking Dvl1. Genes Brain Behav 3:51–62

Lord C, Jones RM (2013) New strategies and findings for behavioral interventions in autism spectrum disorders. Ann NY Acad Sci 1304:70–76

Lord C, Risi S, Lambrecht L, Cook EH Jr, Leventhal BL, DiLavore PC, Pickles A, Rutter M (2000) The autism diagnostic observation schedule-generic: a standard measure of social and communication deficits associated with the spectrum of autism. J Autism Dev Disord 30:205–223

MacPherson P, McGaffigan R, Wahlsten D, Nguyen PV (2008) Impaired fear memory, altered object memory and modified hippocampal synaptic plasticity in split-brain mice. Brain Res 1210:179–188
Macri S, Biamonte F, Romano E, Marino R, Keller F, Laviola G (2010) Perseverative responding and neuroanatomical alterations in adult heterozygous reeler mice are mitigated by neonatal estrogen administration. Psychoneuroendocrinology 35:1374–1387
Maggio JC, Whitney G (1985) Ultrasonic vocalizing by adult female mice (*Mus musculus*). J Comp Psychol 99:420–436
Mahrt EJ, Perkel DJ, Tong L, Rubel EW, Portfors CV (2013) Engineered deafness reveals that mouse courtship vocalizations do not require auditory experience. J Neurosci 33:5573–5583
Malkesman O, Scattoni ML, Paredes D, Tragon T, Pearson B, Shaltiel G, Chen G, Crawley JN, Manji HK (2010) The female urine sniffing test: a novel approach for assessing reward-seeking behavior in rodents. Biol Psychiatry 67:864–871
Malone RP, Cater J, Sheikh RM, Choudhury MS, Delaney MA (2001) Olanzapine versus haloperidol in children with autistic disorder: an open pilot study. J Am Acad Child Adolesc Psychiatry 40:887–894
Maloney A, Mick EO, Frazier J (2014) Aripiprazole decreases irritability in 12 out of 14 youth with autism spectrum disorders. J Child Adolesc Psychopharmacol 24:357–359
Mankoski R, Stockton G, Manos G, Marler S, McQuade R, Forbes RA, Marcus R (2013) Aripiprazole treatment of irritability associated with autistic disorder and the relationship between prior antipsychotic exposure, adverse events, and weight change. J Child Adolesc Psychopharmacol 23:572–576
Marcus RN, Owen R, Kamen L, Manos G, McQuade RD, Carson WH, Aman MG (2009) A placebo-controlled, fixed-dose study of aripiprazole in children and adolescents with irritability associated with autistic disorder. J Am Acad Child Adolesc Psychiatry 48:1110–1119
Marcus RN, Owen R, Manos G, Mankoski R, Kamen L, McQuade RD, Carson WH, Corey-Lisle PK, Aman MG (2011a) Aripiprazole in the treatment of irritability in pediatric patients (aged 6–17 years) with autistic disorder: results from a 52-week, open-label study. J Child Adolesc Psychopharmacol 21:229–236
Marcus RN, Owen R, Manos G, Mankoski R, Kamen L, McQuade RD, Carson WH, Findling RL (2011b) Safety and tolerability of aripiprazole for irritability in pediatric patients with autistic disorder: a 52-week, open-label, multicenter study. J Clin Psychiatry 72:1270–1276
Markou A, Chiamulera C, Geyer MA, Tricklebank M, Steckler T (2009) Removing obstacles in neuroscience drug discovery: the future path for animal models. Neuropsychopharmacology 34:74–89
Marshall CR, Noor A, Vincent JB, Lionel AC, Feuk L, Skaug J, Shago M, Moessner R, Pinto D, Ren Y, Thiruvahindrapduram B, Fiebig A, Schreiber S, Friedman J, Ketelaars CE, Vos YJ, Ficicioglu C, Kirkpatrick S, Nicolson R, Sloman L, Summers A, Gibbons CA, Teebi A, Chitayat D, Weksberg R, Thompson A, Vardy C, Crosbie V, Luscombe S, Baatjes R, Zwaigenbaum L, Roberts W, Fernandez B, Szatmari P, Scherer SW (2008) Structural variation of chromosomes in autism spectrum disorder. Am J Hum Genet 82:477–488
Martin LA, Ashwood P, Braunschweig D, Cabanlit M, Van de Water J, Amaral DG (2008) Stereotypies and hyperactivity in rhesus monkeys exposed to IgG from mothers of children with autism. Brain Behav Immun 22:806–816
Matson JL, Shoemaker M (2009) Intellectual disability and its relationship to autism spectrum disorders. Res Dev Disabil 30:1107–1114
Matsunami N, Hadley D, Hensel CH, Christensen GB, Kim C, Frackelton E, Thomas K, da Silva RP, Stevens J, Baird L, Otterud B, Ho K, Varvil T, Leppert T, Lambert CG, Leppert M, Hakonarson H (2013) Identification of rare recurrent copy number variants in high-risk autism families and their prevalence in a large ASD population. PLoS ONE 8:e52239
Mayko ZM, Roberts PD, Portfors CV (2012) Inhibition shapes selectivity to vocalizations in the inferior colliculus of awake mice. Front Neural Circuits 6:73
McCracken JT, McGough J, Shah B, Cronin P, Hong D, Aman MG, Arnold LE, Lindsay R, Nash P, Hollway J, McDougle CJ, Posey D, Swiezy N, Kohn A, Scahill L, Martin A,

Koenig K, Volkmar F, Carroll D, Lancor A, Tierney E, Ghuman J, Gonzalez NM, Grados M, Vitiello B, Ritz L, Davies M, Robinson J, McMahon D (2002) Risperidone in children with autism and serious behavioral problems. N England J Med 347:314–321
McDougle CJ, Brodkin ES, Naylor ST, Carlson DC, Cohen DJ, Price LH (1998) Sertraline in adults with pervasive developmental disorders: a prospective open-label investigation. J Clin Psychopharmacol 18:62–66
McDougle CJ, Epperson CN, Pelton GH, Wasylink S, Price LH (2000) A double-blind, placebo-controlled study of risperidone addition in serotonin reuptake inhibitor-refractory obsessive-compulsive disorder. Arch Gen Psychiatry 57:794–801
McDougle CJ, Naylor ST, Cohen DJ, Volkmar FR, Heninger GR, Price LH (1996) A double-blind, placebo-controlled study of fluvoxamine in adults with autistic disorder. Arch Gen Psychiatry 53:1001–1008
McDougle CJ, Scahill L, Aman MG, McCracken JT, Tierney E, Davies M, Arnold LE, Posey DJ, Martin A, Ghuman JK, Shah B, Chuang SZ, Swiezy NB, Gonzalez NM, Hollway J, Koenig K, McGough JJ, Ritz L, Vitiello B (2005) Risperidone for the core symptom domains of autism: results from the study by the autism network of the research units on pediatric psychopharmacology. Am J Psychiatry 162:1142–1148
McFarlane HG, Kusek GK, Yang M, Phoenix JL, Bolivar VJ, Crawley JN (2008) Autism-like behavioral phenotypes in BTBR T+ tf/J mice. Genes Brain Behav 7:152–163
McTighe SM, Neal SJ, Lin Q, Hughes ZA, Smith DG (2013) The BTBR mouse model of autism spectrum disorders has learning and attentional impairments and alterations in acetylcholine and kynurenic acid in prefrontal cortex. PLoS ONE 8:e62189
Mehta MV, Gandal MJ, Siegel SJ (2011) mGluR5-antagonist mediated reversal of elevated stereotyped, repetitive behaviors in the VPA model of autism. PLoS ONE 6:e26077
Meikle L, Pollizzi K, Egnor A, Kramvis I, Lane H, Sahin M, Kwiatkowski DJ (2008) Response of a neuronal model of tuberous sclerosis to mammalian target of rapamycin (mTOR) inhibitors: effects on mTORC1 and Akt signaling lead to improved survival and function. J Neurosci 28:5422–5432
Meikle L, Talos DM, Onda H, Pollizzi K, Rotenberg A, Sahin M, Jensen FE, Kwiatkowski DJ (2007) A mouse model of tuberous sclerosis: neuronal loss of Tsc1 causes dysplastic and ectopic neurons, reduced myelination, seizure activity, and limited survival. J Neurosci 27:5546–5558
Michalon A, Sidorov M, Ballard TM, Ozmen L, Spooren W, Wettstein JG, Jaeschke G, Bear MF, Lindemann L (2012) Chronic pharmacological mGlu5 inhibition corrects fragile X in adult mice. Neuron 74:49–56
Miczek KA, Maxson SC, Fish EW, Faccidomo S (2001) Aggressive behavioral phenotypes in mice. Behav Brain Res 125:167–181
Miles JH (2011) Autism spectrum disorders—a genetics review. Genet Med 13:278–294
Modi ME, Connor-Stroud F, Landgraf R, Young LJ, Parr LA (2014) Aerosolized oxytocin increases cerebrospinal fluid oxytocin in rhesus macaques. Psychoneuroendocrinology 45:49–57
Modi ME, Young LJ (2011) D-cycloserine facilitates socially reinforced learning in an animal model relevant to autism spectrum disorders. Biol Psychiatry 70:298–304
Modi ME, Young LJ (2012) The oxytocin system in drug discovery for autism: animal models and novel therapeutic strategies. Horm Behav 61:340–350
Moles A, Kieffer BL, D'Amato FR (2004) Deficit in attachment behavior in mice lacking the mu-opioid receptor gene. Science 304:1983–1986
Molina J, Carmona-Mora P, Chrast J, Krall PM, Canales CP, Lupski JR, Reymond A, Walz K (2008) Abnormal social behaviors and altered gene expression rates in a mouse model for Potocki-Lupski syndrome. Hum Mol Genet 17:2486–2495
Mouti A, Reddihough D, Marraffa C, Hazell P, Wray J, Lee K, Kohn M (2014) Fluoxetine for autistic behaviors (FAB trial): study protocol for a randomized controlled trial in children and adolescents with autism. Trials 15:230

Moy SS, Nadler JJ (2008) Advances in behavioral genetics: mouse models of autism. Mol Psychiatry 13:4–26
Moy SS, Nadler JJ, Magnuson TR, Crawley JN (2006) Mouse models of autism spectrum disorders: the challenge for behavioral genetics. Am J Med Genet Part C, Semin Med Genet 142C:40–51
Moy SS, Nadler JJ, Poe MD, Nonneman RJ, Young NB, Koller BH, Crawley JN, Duncan GE, Bodfish JW (2008a) Development of a mouse test for repetitive, restricted behaviors: relevance to autism. Behav Brain Res 188:178–194
Moy SS, Nadler JJ, Young NB, Nonneman RJ, Grossman AW, Murphy DL, D'Ercole AJ, Crawley JN, Magnuson TR, Lauder JM (2009) Social approach in genetically engineered mouse lines relevant to autism. Genes Brain Behav 8:129–142
Moy SS, Nadler JJ, Young NB, Nonneman RJ, Segall SK, Andrade GM, Crawley JN, Magnuson TR (2008b) Social approach and repetitive behavior in eleven inbred mouse strains. Behav Brain Res 191:118–129
Moy SS, Nadler JJ, Young NB, Perez A, Holloway LP, Barbaro RP, Barbaro JR, Wilson LM, Threadgill DW, Lauder JM, Magnuson TR, Crawley JN (2007) Mouse behavioral tasks relevant to autism: phenotypes of 10 inbred strains. Behav Brain Res 176:4–20
Murdoch JD, State MW (2013) Recent developments in the genetics of autism spectrum disorders. Curr Opin Genet Dev 23:310–315
Nadler JJ, Moy SS, Dold G, Trang D, Simmons N, Perez A, Young NB, Barbaro RP, Piven J, Magnuson TR, Crawley JN (2004) Automated apparatus for quantitation of social approach behaviors in mice. Genes Brain Behav 3:303–314
Nakatani J, Tamada K, Hatanaka F, Ise S, Ohta H, Inoue K, Tomonaga S, Watanabe Y, Chung YJ, Banerjee R, Iwamoto K, Kato T, Okazawa M, Yamauchi K, Tanda K, Takao K, Miyakawa T, Bradley A, Takumi T (2009) Abnormal behavior in a chromosome-engineered mouse model for human 15q11-13 duplication seen in autism. Cell 137:1235–1246
Nicolson R, Craven-Thuss B, Smith J (2006) A prospective, open-label trial of galantamine in autistic disorder. J Child Adolesc Psychopharmacol 16:621–629
Niederhofer H (2007) Glutamate antagonists seem to be slightly effective in psychopharmacologic treatment of autism. J Clin Psychopharmacol 27:317–318
Niederhofer H, Staffen W, Mair A (2002) Galantamine may be effective in treating autistic disorder. BMJ 325:1422
Nordenbaek C, Jorgensen M, Kyvik KO, Bilenberg N (2014) A Danish population-based twin study on autism spectrum disorders. Eur Child Adolesc Psychiatry 23:35–43
Nyby J, Wysocki CJ, Whitney G, Dizinno G (1977) Pheromonal regulation of male mouse ultrasonic courtship (*Mus musculus*). Anim Behav 25:333–341
O'Roak BJ, Stessman HA, Boyle EA, Witherspoon KT, Martin B, Lee C, Vives L, Baker C, Hiatt JB, Nickerson DA, Bernier R, Shendure J, Eichler EE (2014) Recurrent de novo mutations implicate novel genes underlying simplex autism risk. Nat Commun 5:5595
Ogier M, Wang H, Hong E, Wang Q, Greenberg ME, Katz DM (2007) Brain-derived neurotrophic factor expression and respiratory function improve after ampakine treatment in a mouse model of Rett syndrome. J Neurosci 27:10912–10917
Okabe S, Nagasawa M, Kihara T, Kato M, Harada T, Koshida N, Mogi K, Kikusui T (2013) Pup odor and ultrasonic vocalizations synergistically stimulate maternal attention in mice. Behav Neurosci 127:432–438
Owen R, Sikich L, Marcus RN, Corey-Lisle P, Manos G, McQuade RD, Carson WH, Findling RL (2009) Aripiprazole in the treatment of irritability in children and adolescents with autistic disorder. Pediatrics 124:1533–1540
Owley T, Salt J, Guter S, Grieve A, Walton L, Ayuyao N, Leventhal BL, Cook EH Jr (2006) A prospective, open-label trial of memantine in the treatment of cognitive, behavioral, and memory dysfunction in pervasive developmental disorders. J Child Adolesc Psychopharmacol 16:517–524

Page DT, Kuti OJ, Prestia C, Sur M (2009) Haploinsufficiency for Pten and Serotonin transporter cooperatively influences brain size and social behavior. Proc Natl Acad Sci 106:1989–1994

Panksepp JB, Jochman KA, Kim JU, Koy JJ, Wilson ED, Chen Q, Wilson CR, Lahvis GP (2007) Affiliative behavior, ultrasonic communication and social reward are influenced by genetic variation in adolescent mice. PLoS ONE 2:e351

Paradee W, Melikian HE, Rasmussen DL, Kenneson A, Conn PJ, Warren ST (1999) Fragile X mouse: strain effects of knockout phenotype and evidence suggesting deficient amygdala function. Neuroscience 94:185–192

Parikshak NN, Luo R, Zhang A, Won H, Lowe JK, Chandran V, Horvath S, Geschwind DH (2013) Integrative functional genomic analyses implicate specific molecular pathways and circuits in autism. Cell 155:1008–1021

Parr LA, Modi M, Siebert E, Young LJ (2013) Intranasal oxytocin selectively attenuates rhesus monkeys' attention to negative facial expressions. Psychoneuroendocrinology 38:1748–1756

Patterson PH (2011) Modeling autistic features in animals. Pediatr Res 69:34R–40R

Peca J, Feliciano C, Ting JT, Wang W, Wells MF, Venkatraman TN, Lascola CD, Fu Z, Feng G (2011) Shank3 mutant mice display autistic-like behaviours and striatal dysfunction. Nature 472:437–442

Penagarikano O, Abrahams BS, Herman EI, Winden KD, Gdalyahu A, Dong H, Sonnenblick LI, Gruver R, Almajano J, Bragin A, Golshani P, Trachtenberg JT, Peles E, Geschwind DH (2011) Absence of CNTNAP2 leads to epilepsy, neuronal migration abnormalities, and core autism-related deficits. Cell 147:235–246

Perou R, Bitsko RH, Blumberg SJ, Pastor P, Ghandour RM, Gfroerer JC, Hedden SL, Crosby AE, Visser SN, Schieve LA, Parks SE, Hall JE, Brody D, Simile CM, Thompson WW, Baio J, Avenevoli S, Kogan MD, Huang LN (2013) Mental health surveillance among children–United States, 2005–2011. MMWR Surveill Summ 62(Suppl 2):1–35

Pini G, Scusa MF, Benincasa A, Bottiglioni I, Congiu L, Vadhatpour C, Romanelli AM, Gemo I, Puccetti C, McNamara R, O'Leary S, Corvin A, Gill M, Tropea D (2014) Repeated insulin-like growth factor 1 treatment in a patient with rett syndrome: a single case study. Front Pediatr 2:52

Pini G, Scusa MF, Congiu L, Benincasa A, Morescalchi P, Bottiglioni I, Di Marco P, Borelli P, Bonuccelli U, Della-Chiesa A, Prina-Mello A, Tropea D (2012) IGF1 as a potential treatment for rett syndrome: safety assessment in six rett patients. Autism Res Treat 2012:679801

Pinto D, Delaby E, Merico D, Barbosa M, Merikangas A, Klei L, Thiruvahindrapuram B, Xu X, Ziman R, Wang Z, Vorstman JA, Thompson A, Regan R, Pilorge M, Pellecchia G, Pagnamenta AT, Oliveira B, Marshall CR, Magalhaes TR, Lowe JK, Howe JL, Griswold AJ, Gilbert J, Duketis E, Dombroski BA, De Jonge MV, Cuccaro M, Crawford EL, Correia CT, Conroy J, Conceicao IC, Chiocchetti AG, Casey JP, Cai G, Cabrol C, Bolshakova N, Bacchelli E, Anney R, Gallinger S, Cotterchio M, Casey G, Zwaigenbaum L, Wittemeyer K, Wing K, Wallace S, van Engeland H, Tryfon A, Thomson S, Soorya L, Roge B, Roberts W, Poustka F, Mouga S, Minshew N, McInnes LA, McGrew SG, Lord C, Leboyer M, Le Couteur AS, Kolevzon A, Jimenez Gonzalez P, Jacob S, Holt R, Guter S, Green J, Green A, Gillberg C, Fernandez BA, Duque F, Delorme R, Dawson G, Chaste P, Cafe C, Brennan S, Bourgeron T, Bolton PF, Bolte S, Bernier R, Baird G, Bailey AJ, Anagnostou E, Almeida J, Wijsman EM, Vieland VJ, Vicente AM, Schellenberg GD, Pericak-Vance M, Paterson AD, Parr JR, Oliveira G, Nurnberger JI, Monaco AP, Maestrini E, Klauck SM, Hakonarson H, Haines JL, Geschwind DH, Freitag CM, Folstein SE, Ennis S, Coon H, Battaglia A, Szatmari P, Sutcliffe JS, Hallmayer J, Gill M, Cook EH, Buxbaum JD, Devlin B, Gallagher L, Betancur C, Scherer SW (2014) Convergence of genes and cellular pathways dysregulated in autism spectrum disorders. Am J Hum Genet 94:677–694

Pobbe RL, Defensor EB, Pearson BL, Bolivar VJ, Blanchard DC, Blanchard RJ (2011) General and social anxiety in the BTBR T+ tf/J mouse strain. Behav Brain Res 216:446–451

Pobbe RL, Pearson BL, Defensor EB, Bolivar VJ, Blanchard DC, Blanchard RJ (2010) Expression of social behaviors of C57BL/6J versus BTBR inbred mouse strains in the visible burrow system. Behav Brain Res 214:443–449

Pomerantz SM, Nunez AA, Bean NJ (1983) Female behavior is affected by male ultrasonic vocalizations in house mice. Physiol Behav 31:91–96
Pop AS, Levenga J, de Esch CE, Buijsen RA, Nieuwenhuizen IM, Li T, Isaacs A, Gasparini F, Oostra BA, Willemsen R (2014) Rescue of dendritic spine phenotype in Fmr1 KO mice with the mGluR5 antagonist AFQ056/Mavoglurant. Psychopharmacology 231:1227–1235
Portmann T, Yang M, Mao R, Panagiotakos G, Ellegood J, Dolen G, Bader PL, Grueter BA, Goold C, Fisher E, Clifford K, Rengarajan P, Kalikhman D, Loureiro D, Saw NL, Zhengqui Z, Miller MA, Lerch JP, Henkelman RM, Shamloo M, Malenka RC, Crawley JN, Dolmetsch RE (2014) Behavioral abnormalities and circuit defects in the basal ganglia of a mouse model of 16p11.2 deletion syndrome. Cell Reports 7:1077–1092
Posey DJ, Kem DL, Swiezy NB, Sweeten TL, Wiegand RE, McDougle CJ (2004) A pilot study of D-cycloserine in subjects with autistic disorder. Am J Psychiatry 161:2115–2117
Potenza MN, Holmes JP, Kanes SJ, McDougle CJ (1999) Olanzapine treatment of children, adolescents, and adults with pervasive developmental disorders: an open-label pilot study. J Clin Psychopharmacol 19:37–44
Poultney CS, Goldberg AP, Drapeau E, Kou Y, Harony-Nicolas H, Kajiwara Y, De Rubeis S, Durand S, Stevens C, Rehnstrom K, Palotie A, Daly MJ, Ma'ayan A, Fromer M, Buxbaum JD (2013) Identification of small exonic CNV from whole-exome sequence data and application to autism spectrum disorder. Am J Hum Genet 93:607–619
Prather MD, Lavenex P, Mauldin-Jourdain ML, Mason WA, Capitanio JP, Mendoza SP, Amaral DG (2001) Increased social fear and decreased fear of objects in monkeys with neonatal amygdala lesions. Neuroscience 106:653–658
Puscian A, Leski S, Gorkiewicz T, Meyza K, Lipp HP, Knapska E (2014) A novel automated behavioral test battery assessing cognitive rigidity in two genetic mouse models of autism. Front Behav Neurosci 8:140
Qiu S, Aldinger KA, Levitt P (2012) Modeling of autism genetic variations in mice: focusing on synaptic and microcircuit dysfunctions. Dev Neurosci 34:88–100
Radyushkin K, Hammerschmidt K, Boretius S, Varoqueaux F, El-Kordi A, Ronnenberg A, Winter D, Frahm J, Fischer J, Brose N, Ehrenreich H (2009) Neuroligin-3-deficient mice: model of a monogenic heritable form of autism with an olfactory deficit. Genes Brain Behav 8:416–425
Rapin I, Dunn M (2003) Update on the language disorders of individuals on the autistic spectrum. Brain Dev 25:166–172
Rausch JL, Sirota EL, Londino DL, Johnson ME, Carr BM, Bhatia R, Miller S (2005) Open-label risperidone for Asperger's disorder: negative symptom spectrum response. J Clin Psychiatry 66:1592–1597
Reith RM, McKenna J, Wu H, Hashmi SS, Cho SH, Dash PK, Gambello MJ (2013) Loss of Tsc2 in Purkinje cells is associated with autistic-like behavior in a mouse model of tuberous sclerosis complex. Neurobiol Dis 51:93–103
Richdale AL (1999) Sleep problems in autism: prevalence, cause, and intervention. Dev Med Child Neurol 41:60–66
Rimmele U, Hediger K, Heinrichs M, Klaver P (2009) Oxytocin makes a face in memory familiar. J Neurosci 29:38–42
Ritvo ER, Freeman BJ, Mason-Brothers A, Mo A, Ritvo AM (1985) Concordance for the syndrome of autism in 40 pairs of afflicted twins. Am J Psychiatry 142:74–77
Robb AS, Andersson C, Bellocchio EE, Manos G, Rojas-Fernandez C, Mathew S, Marcus R, Owen R, Mankoski R (2011) Safety and tolerability of aripiprazole in the treatment of irritability associated with autistic disorder in pediatric subjects (6–17 years old): results from a pooled analysis of 2 studies. Prim Care Companion CNS Disord 13:1
Rogers SJ, Estes A, Lord C, Vismara L, Winter J, Fitzpatrick A, Guo M, Dawson G (2012) Effects of a brief early start denver model (ESDM)-based parent intervention on toddlers at risk for autism spectrum disorders: a randomized controlled trial. J Am Acad Child Adolesc Psychiatry 51:1052–1065

Rogers SJ, Ozonoff S (2005) Annotation: what do we know about sensory dysfunction in autism? A critical review of the empirical evidence. J Child Psychol Psychiatry 46:1255–1268
Rojas DC, Wilson LB (2014) gamma-band abnormalities as markers of autism spectrum disorders. Biomark Med 8:353–368
Roohi J, Montagna C, Tegay DH, Palmer LE, DeVincent C, Pomeroy JC, Christian SL, Nowak N, Hatchwell E (2009) Disruption of contactin 4 in three subjects with autism spectrum disorder. J Med Genet 46:176–182
Rosenfeld JA, Ballif BC, Torchia BS, Sahoo T, Ravnan JB, Schultz R, Lamb A, Bejjani BA, Shaffer LG (2010) Copy number variations associated with autism spectrum disorders contribute to a spectrum of neurodevelopmental disorders. Genet Med 12:694–702
Roullet FI, Wohr M, Crawley JN (2010) Female urine-induced male mice ultrasonic vocalizations, but not scent-marking, is modulated by social experience. Behav Brain Res 216:19–28
Roullet FI, Wohr M, Crawley JN (2011) Female urine-induced male mice ultrasonic vocalizations, but not scent-marking, is modulated by social experience. Behav Brain Res 216:19–28
Ryan BC, Young NB, Crawley JN, Bodfish JW, Moy SS (2010) Social deficits, stereotypy and early emergence of repetitive behavior in the C58/J inbred mouse strain. Behav Brain Res 208:178–188
Ryan BC, Young NB, Moy SS, Crawley JN (2008) Olfactory cues are sufficient to elicit social approach behaviors but not social transmission of food preference in C57BL/6J mice. Behav Brain Res 193:235–242
Sadakata T, Washida M, Iwayama Y, Shoji S, Sato Y, Ohkura T, Katoh-Semba R, Nakajima M, Sekine Y, Tanaka M, Nakamura K, Iwata Y, Tsuchiya KJ, Mori N, Detera-Wadleigh SD, Ichikawa H, Itohara S, Yoshikawa T, Furuichi T (2007) Autistic-like phenotypes in Cadps2-knockout mice and aberrant CADPS2 splicing in autistic patients. J Clin Investig 117:931–943
Sahu JK, Gulati S, Sapra S, Arya R, Chauhan S, Chowdhury MR, Gupta N, Kabra M, Gupta YK, Dwivedi SN, Kalra V (2013) Effectiveness and safety of donepezil in boys with fragile x syndrome: a double-blind, randomized, controlled pilot study. J Child Neurol 28:570–575
Sala M, Braida D, Lentini D, Busnelli M, Bulgheroni E, Capurro V, Finardi A, Donzelli A, Pattini L, Rubino T, Parolaro D, Nishimori K, Parenti M, Chini B (2011) Pharmacologic rescue of impaired cognitive flexibility, social deficits, increased aggression, and seizure susceptibility in oxytocin receptor null mice: a neurobehavioral model of autism. Biol Psychiatry 69:875–882
Samaco RC, McGraw CM, Ward CS, Sun Y, Neul JL, Zoghbi HY (2013) Female Mecp2(±) mice display robust behavioral deficits on two different genetic backgrounds providing a framework for pre-clinical studies. Hum Mol Genet 22:96–109
Sandin S, Lichtenstein P, Kuja-Halkola R, Larsson H, Hultman CM, Reichenberg A (2014) The familial risk of autism. JAMA 311:1770–1777
Santini E, Huynh TN, MacAskill AF, Carter AG, Pierre P, Ruggero D, Kaphzan H, Klann E (2013) Exaggerated translation causes synaptic and behavioural aberrations associated with autism. Nature 493:411–415
Sato A, Kasai S, Kobayashi T, Takamatsu Y, Hino O, Ikeda K, Mizuguchi M (2012) Rapamycin reverses impaired social interaction in mouse models of tuberous sclerosis complex. Nat Commun 3:1292
Scahill L, Hallett V, Aman MG, McDougle CJ, Eugene Arnold L, McCracken JT, Tierney E, Deng Y, Dziura J, Vitiello B (2013) Brief report: social disability in autism spectrum disorder: results from research units on pediatric psychopharmacology (RUPP) autism network trials. J Autism Dev Disord 43:739–746
Scattoni ML, Gandhy SU, Ricceri L, Crawley JN (2008) Unusual repertoire of vocalizations in the BTBR T+ tf/J mouse model of autism. PLoS ONE 3:e3067
Scattoni ML, Martire A, Cartocci G, Ferrante A, Ricceri L (2013) Reduced social interaction, behavioural flexibility and BDNF signalling in the BTBR T+ tf/J strain, a mouse model of autism. Behav Brain Res 251:35–40
Scattoni ML, Ricceri L, Crawley JN (2010) Unusual repertoire of vocalizations in adult BTBR T+ tf/J mice during three types of social encounters. Genes Brain Behav 10:44–56

Scattoni ML, Ricceri L, Crawley JN (2011) Unusual repertoire of vocalizations in adult BTBR T+ tf/J mice during three types of social encounters. Genes Brain Behav 10:44–56

Scearce-Levie K, Roberson ED, Gerstein H, Cholfin JA, Mandiyan VS, Shah NM, Rubenstein JL, Mucke L (2008) Abnormal social behaviors in mice lacking Fgf17. Genes Brain Behav 7:344–354

Schmeisser MJ, Ey E, Wegener S, Bockmann J, Stempel AV, Kuebler A, Janssen AL, Udvardi PT, Shiban E, Spilker C, Balschun D, Skryabin BV, Dieck S, Smalla KH, Montag D, Leblond CS, Faure P, Torquet N, Le Sourd AM, Toro R, Grabrucker AM, Shoichet SA, Schmitz D, Kreutz MR, Bourgeron T, Gundelfinger ED, Boeckers TM (2012) Autistic-like behaviours and hyperactivity in mice lacking ProSAP1/Shank2. Nature 486:256–260

Schneider T, Przewlocki R (2005) Behavioral alterations in rats prenatally exposed to valproic acid: animal model of autism. Neuropsychopharmacology 30:80–89

Seese RR, Maske AR, Lynch G, Gall CM (2014) Long-term memory deficits are associated with elevated synaptic ERK1/2 activation and reversed by mGluR5 antagonism in an animal model of autism. Neuropsychopharmacology 39:1664–1673

Shair HN, Rupert DD, Rosko LM, Hofer MA, Myers MM, Welch MG (2014) Effects of maternal deprivation and the duration of reunion time on rat pup ultrasonic vocalization responses to isolation: possible implications for human infant studies. Dev Psychobiol 57(1):63–72

Shepard KN, Liu RC (2011) Experience restores innate female preference for male ultrasonic vocalizations. Genes Brain Behav 10:28–34

Siegel M, Beaulieu AA (2012) Psychotropic medications in children with autism spectrum disorders: a systematic review and synthesis for evidence-based practice. J Autism Dev Disord 42:1592–1605

Silverman JL, Crawley JN (2014) The promising trajectory of autism therapeutics discovery. Drug Discov Today 19:838–844

Silverman JL, Oliver CF, Karras MN, Gastrell PT, Crawley JN (2013) AMPAKINE enhancement of social interaction in the BTBR mouse model of autism. Neuropharmacology 64:268–282

Silverman JL, Smith DG, Rizzo SJ, Karras MN, Turner SM, Tolu SS, Bryce DK, Smith DL, Fonseca K, Ring RH, Crawley JN (2012) Negative allosteric modulation of the mGluR5 receptor reduces repetitive behaviors and rescues social deficits in mouse models of autism. Sci Transl Med 4:131ra151

Silverman JL, Tolu SS, Barkan CL, Crawley JN (2010a) Repetitive self-grooming behavior in the BTBR mouse model of autism is blocked by the mGluR5 antagonist MPEP. Neuropsychopharmacology 35:976–989

Silverman JL, Yang M, Lord C, Crawley JN (2010b) Behavioural phenotyping assays for mouse models of autism. Nat Rev 11:490–502

Silverman JL, Pride MC, Hayes JE, Puhger KR, Butler-Struben HM, Baker S, Crawley JN (2015) GABAB receptor agonist r-baclofen reverses social deficits and reduces repetitive behavior in two ,ouse models of autism. Neuropsychopharmacology 40:2228–2239

Smalley SL, Asarnow RF, Spence MA (1988) Autism and genetics. A decade of research. Arch Gen Psychiatry 45:953–961

Smith SE, Zhou YD, Zhang G, Jin Z, Stoppel DC, Anderson MP (2011) Increased gene dosage of Ube3a results in autism traits and decreased glutamate synaptic transmission in mice. Sci Transl Med 3:103ra197

Sowers LP, Loo L, Wu Y, Campbell E, Ulrich JD, Wu S, Paemka L, Wassink T, Meyer K, Bing X, El-Shanti H, Usachev YM, Ueno N, Manak JR, Shepherd AJ, Ferguson PJ, Darbro BW, Richerson GB, Mohapatra DP, Wemmie JA, Bassuk AG (2013) Disruption of the non-canonical Wnt gene PRICKLE2 leads to autism-like behaviors with evidence for hippocampal synaptic dysfunction. Mol Psychiatry 18:1077–1089

Spencer CM, Alekseyenko O, Hamilton SM, Thomas AM, Serysheva E, Yuva-Paylor LA, Paylor R (2011) Modifying behavioral phenotypes in Fmr1KO mice: genetic background differences reveal autistic-like responses. Autism Res 4:40–56

Spencer CM, Alekseyenko O, Serysheva E, Yuva-Paylor LA, Paylor R (2005) Altered anxiety-related and social behaviors in the Fmr1 knockout mouse model of fragile X syndrome. Genes Brain Behav 4:420–430

Spencer D, Marshall J, Post B, Kulakodlu M, Newschaffer C, Dennen T, Azocar F, Jain A (2013) Psychotropic medication use and polypharmacy in children with autism spectrum disorders. Pediatrics 132:833–840

Sperow M, Berry RB, Bayazitov IT, Zhu G, Baker SJ, Zakharenko SS (2012) Phosphatase and tensin homologue (PTEN) regulates synaptic plasticity independently of its effect on neuronal morphology and migration. J Physiol 590:777–792

Spooren W, Lindemann L, Ghosh A, Santarelli L (2012) Synapse dysfunction in autism: a molecular medicine approach to drug discovery in neurodevelopmental disorders. Trends Pharmacol Sci 33:669–684

Sugimoto H, Okabe S, Kato M, Koshida N, Shiroishi T, Mogi K, Kikusui T, Koide T (2011) A role for strain differences in waveforms of ultrasonic vocalizations during male-female interaction. PLoS ONE 6:e22093

Tachibana M, Kagitani-Shimono K, Mohri I, Yamamoto T, Sanefuji W, Nakamura A, Oishi M, Kimura T, Onaka T, Ozono K, Taniike M (2013) Long-term administration of intranasal oxytocin is a safe and promising therapy for early adolescent boys with autism spectrum disorders. J Child Adolesc Psychopharmacol 23:123–127

Talebizadeh Z, Lam DY, Theodoro MF, Bittel DC, Lushington GH, Butler MG (2006) Novel splice isoforms for NLGN3 and NLGN4 with possible implications in autism. J Med Genet 43: e21

Teng BL, Nonneman RJ, Agster KL, Nikolova VD, Davis TT, Riddick NV, Baker LK, Pedersen CA, Jarstfer MB, Moy SS (2013) Prosocial effects of oxytocin in two mouse models of autism spectrum disorders. Neuropharmacology 72:187–196

Terranova ML, Laviola G (2005) Scoring of social interactions and play in mice during adolescence. Curr Protoc Toxicol Chap 13(Unit13):10

The Dutch-Belgian Fragile XC, Bakker CE, Verheij C, Willemsen R, van der Helm R, Oerlemans F, Vermey M, Bygrave A, Hoogeveen A, Oostra BA, Reyniers E, De Boule K, D'Hooge R, Cras P, van Velzen D, Nagels G, Martin J-J, De Deyn PP, Darby JK, Willems PJ (1994) Fmr1 knockout mice: a model to study fragile X mental retardation. Cell 78:23–33

Thomas A, Burant A, Bui N, Graham D, Yuva-Paylor LA, Paylor R (2009) Marble burying reflects a repetitive and perseverative behavior more than novelty-induced anxiety. Psychopharmacology 204:361–373

Thomas AM, Bui N, Perkins JR, Yuva-Paylor LA, Paylor R (2012) Group I metabotropic glutamate receptor antagonists alter select behaviors in a mouse model for fragile X syndrome. Psychopharmacology 219:47–58

Tillema JM, Leach JL, Krueger DA, Franz DN (2012) Everolimus alters white matter diffusion in tuberous sclerosis complex. Neurology 78:526–531

Tripathi PP, Sgado P, Scali M, Viaggi C, Casarosa S, Simon HH, Vaglini F, Corsini GU, Bozzi Y (2009) Increased susceptibility to kainic acid-induced seizures in Engrailed-2 knockout mice. Neuroscience 159:842–849

Tropea D, Giacometti E, Wilson NR, Beard C, McCurry C, Fu DD, Flannery R, Jaenisch R, Sur M (2009) Partial reversal of Rett Syndrome-like symptoms in MeCP2 mutant mice. Proc Natl Acad Sci 106:2029–2034

Tsai PT, Hull C, Chu Y, Greene-Colozzi E, Sadowski AR, Leech JM, Steinberg J, Crawley JN, Regehr WG, Sahin M (2012) Autistic-like behaviour and cerebellar dysfunction in Purkinje cell Tsc1 mutant mice. Nature 488:647–651

Tyzio R, Nardou R, Ferrari DC, Tsintsadze T, Shahrokhi A, Eftekhari S, Khalilov I, Tsintsadze V, Brouchoud C, Chazal G, Lemonnier E, Lozovaya N, Burnashev N, Ben-Ari Y (2014) Oxytocin-mediated GABA inhibition during delivery attenuates autism pathogenesis in rodent offspring. Science 343:675–679

Upchurch M, Wehner JM (1988) Differences between inbred strains of mice in morris water maze performance. Behav Genet 18:55–68
Urbano M, Okwara L, Manser P, Hartmann K, Herndon A, Deutsch SI (2014) A trial of D-cycloserine to treat stereotypies in older adolescents and young adults with autism spectrum disorder. Clin Neuropharmacol 37:69–72
Uutela M, Lindholm J, Rantamaki T, Umemori J, Hunter K, Voikar V, Castren ML (2014) Distinctive behavioral and cellular responses to fluoxetine in the mouse model for Fragile X syndrome. Front Cell Neurosci 8:150
van Steensel FJ, Bogels SM, Perrin S (2011) Anxiety disorders in children and adolescents with autistic spectrum disorders: a meta-analysis. Clin Child Fam Psychol Rev 14:302–317
Varni JW, Handen BL, Corey-Lisle PK, Guo Z, Manos G, Ammerman DK, Marcus RN, Owen R, McQuade RD, Carson WH, Mathew S, Mankoski R (2012) Effect of aripiprazole 2 to 15 mg/d on health-related quality of life in the treatment of irritability associated with autistic disorder in children: a post hoc analysis of two controlled trials. Clin Ther 34:980–992
Volkmar FR, Nelson DS (1990) Seizure disorders in autism. J Am Acad Child Adolesc Psychiatry 29:127–129
Vorstman JA, Spooren W, Persico AM, Collier DA, Aigner S, Jagasia R, Glennon JC, Buitelaar JK (2014) Using genetic findings in autism for the development of new pharmaceutical compounds. Psychopharmacology 231:1063–1078
Wagner GC, Reuhl KR, Cheh M, McRae P, Halladay AK (2006) A new neurobehavioral model of autism in mice: pre- and postnatal exposure to sodium valproate. J Autism Dev Disord 36:779–793
Wang H, Liang S, Burgdorf J, Wess J, Yeomans J (2008) Ultrasonic vocalizations induced by sex and amphetamine in M2, M4, M5 muscarinic and D2 dopamine receptor knockout mice. PLoS ONE 3:e1893
Watanabe T, Abe O, Kuwabara H, Yahata N, Takano Y, Iwashiro N, Natsubori T, Aoki Y, Takao H, Kawakubo Y, Kamio Y, Kato N, Miyashita Y, Kasai K, Yamasue H (2014) Mitigation of sociocommunicational deficits of autism through oxytocin-induced recovery of medial prefrontal activity: a randomized trial. JAMA Psychiatry 71:166–175
Wei H, Dobkin C, Sheikh AM, Malik M, Brown WT, Li X (2012) The therapeutic effect of memantine through the stimulation of synapse formation and dendritic spine maturation in autism and fragile X syndrome. PLoS ONE 7:e36981
White NR, Prasad M, Barfield RJ, Nyby JG (1998) 40- and 70-kHz vocalizations of mice (*Mus musculus*) during copulation. Physiol Behav 63:467–473
Whitney G, Coble JR, Stockton MD, Tilson EF (1973) Ultrasonic emissions: do they facilitate courtship of mice. J Comp Physiol Psychol 84:445–452
Whitney G, Nyby JG (1979) Cues that elicit ultrasounds from adult male mice. Am Zool 19:457–464
Winslow JT, Shapiro L, Carter CS, Insel TR (1993) Oxytocin and complex social behavior: species comparisons. Psychopharmacol Bull 29:409–414
Wohr M (2014) Ultrasonic vocalizations in Shank mouse models for autism spectrum disorders: detailed spectrographic analyses and developmental profiles. Neurosci Biobehav Rev 43:199–212
Wohr M, Roullet FI, Crawley JN (2011a) Reduced scent marking and ultrasonic vocalizations in the BTBR T+ tf/J mouse model of autism. Genes Brain Behav 10:35–43
Wohr M, Roullet FI, Hung AY, Sheng M, Crawley JN (2011b) Communication impairments in mice lacking Shank1: reduced levels of ultrasonic vocalizations and scent marking behavior. PLoS ONE 6:e20631
Won H, Lee HR, Gee HY, Mah W, Kim JI, Lee J, Ha S, Chung C, Jung ES, Cho YS, Park SG, Lee JS, Lee K, Kim D, Bae YC, Kaang BK, Lee MG, Kim E (2012) Autistic-like social behaviour in Shank2-mutant mice improved by restoring NMDA receptor function. Nature 486:261–265

Wrenn CC (2004) Social transmission of food preference in mice. Curr Protoc Neurosci Chap 8, 8–5

Wrenn CC, Harris AP, Saavedra MC, Crawley JN (2003) Social transmission of food preference in mice: methodology and application to galanin-overexpressing transgenic mice. Behav Neurosci 117:21–31

Wurzman R, Forcelli PA, Griffey CJ, Kromer LF (2014) Repetitive grooming and sensorimotor abnormalities in an ephrin-A knockout model for autism spectrum disorders. Behav Brain Res 278c:115–128

Yan J, Oliveira G, Coutinho A, Yang C, Feng J, Katz C, Sram J, Bockholt A, Jones IR, Craddock N, Cook EH Jr, Vicente A, Sommer SS (2005a) Analysis of the neuroligin 3 and 4 genes in autism and other neuropsychiatric patients. Mol Psychiatry 10:329–332

Yan QJ, Rammal M, Tranfaglia M, Bauchwitz RP (2005b) Suppression of two major Fragile X Syndrome mouse model phenotypes by the mGluR5 antagonist MPEP. Neuropharmacology 49:1053–1066

Yang M, Abrams DN, Zhang JY, Weber MD, Katz AM, Clarke AM, Silverman JL, Crawley JN (2012a) Low sociability in BTBR T+ tf/J mice is independent of partner strain. Physiol Behav 107:649–662

Yang M, Bozdagi O, Scattoni ML, Wohr M, Roullet FI, Katz AM, Abrams DN, Kalikhman D, Simon H, Woldeyohannes L, Zhang JY, Harris MJ, Saxena R, Silverman JL, Buxbaum JD, Crawley JN (2012b) Reduced excitatory neurotransmission and mild autism-relevant phenotypes in adolescent Shank3 null mutant mice. J Neurosci 32:6525–6541

Yang M, Clarke AM, Crawley JN (2009) Postnatal lesion evidence against a primary role for the corpus callosum in mouse sociability. Eur J Neurosci 29:1663–1677

Yang M, Loureiro D, Kalikhman D, Crawley JN (2013) Male mice emit distinct ultrasonic vocalizations when the female leaves the social interaction arena. Front Behav Neurosci 7:159

Yang M, Scattoni ML, Zhodzishsky V, Chen T, Caldwell H, Young WS, McFarlane HG, Crawley JN (2007) Social approach behaviors are similar on conventional versus reverse lighting cycles, and in replications across cohorts, in BTBR T+ tf/J, C57BL/6J, and vasopressin receptor 1B mutant mice. Front Behav Neurosci 1:1

Yang M, Silverman JL, Crawley JN (2011) Automated three-chambered social approach task for mice. Curr Protoc Neurosci Chap 8, 8–26

Young LJ, Pitkow LJ, Ferguson JN (2002) Neuropeptides and social behavior: animal models relevant to autism. Mol Psychiatry 7(Suppl 2):S38–S39

Zeng LH, Rensing NR, Zhang B, Gutmann DH, Gambello MJ, Wong M (2011) Tsc2 gene inactivation causes a more severe epilepsy phenotype than Tsc1 inactivation in a mouse model of tuberous sclerosis complex. Hum Mol Genet 20:445–454

Zeng LH, Xu L, Gutmann DH, Wong M (2008) Rapamycin prevents epilepsy in a mouse model of tuberous sclerosis complex. Ann Neurol 63:444–453

Zhang WQ, Smolik CM, Barba-Escobedo PA, Gamez M, Sanchez JJ, Javors MA, Daws LC, Gould GG (2015) Acute dietary tryptophan manipulation differentially alters social behavior, brain serotonin and plasma corticosterone in three inbred mouse strains. Neuropharmacology 90:1–8

Zhou J, Blundell J, Ogawa S, Kwon CH, Zhang W, Sinton C, Powell CM, Parada LF (2009) Pharmacological inhibition of mTORC1 suppresses anatomical, cellular, and behavioral abnormalities in neural-specific Pten knock-out mice. J Neurosci 29:1773–1783

Zoghbi HY, Bear MF (2012) Synaptic dysfunction in neurodevelopmental disorders associated with autism and intellectual disabilities. Cold Spring Harb Perspect Biol 4(3):a009886

Translatable and Back-Translatable Measurement of Impulsivity and Compulsivity: Convergent and Divergent Processes

Valerie Voon and Jeffrey W. Dalley

Abstract Impulsivity and compulsivity have emerged as important dimensional constructs that challenge traditional psychiatric classification systems. Both are present in normal healthy populations where the need to act quickly and repeatedly without hesitation can be highly advantageous. However, when excessively expressed, impulsive and compulsive behavior can lead to adverse consequences and spectrum disorders exemplified by attention-deficit/hyperactivity disorder (ADHD), obsessive compulsive disorder (OCD), autism, and drug addiction. Impulsive individuals have difficulty in deferring gratification and are inclined to 'jump the gun' and respond prematurely before sufficient information is gathered. Compulsivity involves repetitive behavior often motivated by the need to reduce or prevent anxiety, thus leading to the maladaptive perseveration of behavior. Defined in this way, impulsivity and compulsivity could be viewed as separate entities or 'traits' but overwhelming evidence indicates that both may be present in the same disorder, either concurrently or even separately at different time points. Herein we discuss the neural and cognitive heterogeneity of impulsive and compulsive endophenotypes. These constructs map onto distinct fronto-striatal neural and neurochemical structures interacting both at nodal convergent points and as opponent processes highlighting both the heterogeneity and the commonalities of function. We focus on discoveries made using both translational research methodologies and studies exclusively in humans, and implications for treatment

V. Voon (✉) · J.W. Dalley
Behavioral and Clinical Neuroscience Institute, University of Cambridge, Cambridge CB2 3EB, UK
e-mail: voonval@gmail.com

V. Voon · J.W. Dalley
Department of Psychiatry, University of Cambridge, Addenbrooke's Hospital, Cambridge CB2 2QQ, UK

V. Voon
Cambridgeshire and Peterborough NHS Foundation Trust, Cambridge, UK

J.W. Dalley
Department of Psychology, University of Cambridge, Downing Street, Cambridge CB2 3EB, UK

Curr Topics Behav Neurosci (2016) 28: 53–91
DOI 10.1007/7854_2015_5013

intervention in disorders in which impulsive and compulsive symptoms prevail. We emphasize the relevance of these constructs for understanding dimensional psychiatry.

Keywords Impulse control disorders • DSM-5 • ADHD • OCD • Addiction • Endophenotypes • Impulsivity • Compulsivity

Contents

1 Introduction 54
2 Impulsivity 55
2.1 Waiting Impulsivity 56
2.2 Stop-Signal Reaction Time 60
2.3 Delay Discounting 63
2.4 Reflection Impulsivity 65
3 Compulsivity 67
3.1 Cognitive Inflexibility 67
3.2 Stimulus-Response (Habit) Learning 69
3.3 Motor Stereotypy 73
4 Synthesis and Future Perspectives 74
4.1 Neural Considerations: Fronto-striatal Nodes and Opponency Processes 75
4.2 Neurochemical Considerations: Dopamine and Serotonin 76
5 Conclusion 77
References 77

1 Introduction

How is behavior that is rash and disinhibited different from behavior that is inflexible and repetitive? Such 'failures' in behavioral control, typically considered impulsive and compulsive in character, are present in all people and thus represent dimensional constructs or endophenotypes (Gottesman and Gould 2003) where only extreme 'outliers' inform clinical diagnoses and intervention (Fineberg et al. 2010; Robbins et al. 2012; Berlin and Hollander 2014). Understanding the neurobiology and psychology of impulsivity and compulsivity has become increasingly important in recent years in light of the widely recognized and much discussed inadequacies of a categorical-based diagnostic system for psychiatric disorders (Robbins et al. 2012; Insel et al. 2010; Millan et al. 2015). Of particular note, impulsivity and compulsivity are widely documented 'neurocognitive endophenotypes' present in a remarkably broad range of psychiatric disorders, including drug addiction (Brewer and Potenza 2008; de Wit 2009; Ersche et al. 2010), personality and mood disorder (Perry and Korner 2011; Lombardo et al. 2012), schizophrenia (Kaladjian et al. 2011), problem gambling (Verdejo-Garcia et al. 2008), suicide (Dougherty et al. 2004), attention-deficit/hyperactivity disorder (ADHD) (Sonuga-Barke et al. 1992), and obsessive compulsive disorder (OCD) (Fineberg et al. 2010).

Impulsivity can be defined as the tendency toward rash decisions without adequate forethought and often results in mistimed and premature actions (Evenden 1999; Durana and Barnes 1993). Such behavior is clearly different from compulsivity (Latin '*compellere*': feeling forced or being cornered) where inflexible choices and actions are repeatedly carried out irrespective of changes in setting and the growing and obvious undesirable consequences of such behavior [e.g., in the context of OCD (Denys 2011)]. Historically, impulsivity and compulsivity have been regarded as widely contrasting constructs along a single continuum. Thus, whereas impulsivity may arise through abnormalities in reward seeking (positive reinforcement) compulsive behavior is thought to arise and persist by the need to avoid harm or unpleasant subjective feelings (negative reinforcement), a view championed by Koob and Le Moal in their opponent process theory of addiction (Koob and Le 2008). However, just as it is reasonable to question the singularity of the impulsivity construct and the inter-relatedness of different impulsivity subtypes (Winstanley et al. 2006; Dalley et al. 2011), compulsivity can be deconstructed in several ways from rigid stereotyped movements, maladaptive stimulus-response (SR) habits, attentional biases, perseveration, and a failure to extinguish responding when rewards are omitted (Robbins et al. 2012). As reviewed in this article, supporting evidence for the multifaceted nature of impulsivity and compulsivity constructs is compelling and represents tractable endophenotypes for translational research. The concept of endophenotypes (i.e., measurable heritable traits) in psychiatric research is not new (Gottesman and Gould 2003). However, with the increasing availability and sophistication of high-throughput translatable methodologies (e.g., behavioral/cognitive screening, brain imaging, genomics, proteomics, and metabolomics), the delivery of fine-grain mechanistic explanations of psychopathology is now a realistic prospect. By continuing to actively research their considerable phenotypic heterogeneity and underlying neurobiological mechanisms, further significant advances are anticipated in our understanding of complex polygenic mental disorders such as ADHD, OCD, and addiction (Fineberg et al. 2010). In this article, we review the latest research in this field and examine the implications of this work for the biological origins and treatment of impulsive and compulsive disorders.

2 Impulsivity

Impulsivity can be broadly divided into motor and decisional subtypes reviewed below. Motor impulsivity includes (i) waiting impulsivity or premature anticipatory responding prior to a cue predicting reward and (ii) response inhibition or stopping inhibition of a prepotent response. Decisional impulsivity includes (iii) delay and probabilistic discounting of reward and (iv) reflection impulsivity—the tendency to make rapid decisions without adequate accumulation and consideration of the available evidence.

2.1 Waiting Impulsivity

This form of impulsivity requires action restraint during the waiting period leading up to an expected reward. Traditionally, waiting impulsivity is assessed using so-called differential reinforcement of low rates of responding (DRL) schedules, first described by Ferster and Skinner (1957). DRL schedules set a minimum period of time between responses in order for reinforcement to occur and in rodents typically involve delays of 15–30 s (Evenden 1999). In humans, DRL schedules can be used to curb excessive behavior such as fast eating (Lennox et al. 1987) and even the number of times children seek assistance from classroom teachers (Austin and Bevan 2011).

Action restraint on DRL tasks is analogous to premature responding on serial-reaction time tasks where subjects must wait for several seconds for a reward-predictive cue to respond. In a popular variant of this task—the 5-choice serial-reaction time task (5CSRTT)—subjects (typically mice and rats) are trained to detect brief visual targets and to refrain from responding prior to their onset (Robbins 2002) (Fig. 1). High trait-like levels of impulsivity on this task predict the escalation of cocaine and nicotine self-administration (Dalley et al. 2007a; Diergaarde et al. 2008), an increased propensity to relapse following voluntary drug abstinence (Economidou et al. 2009), and the subsequent development of compulsive cocaine self-administration (Belin et al. 2008). Compulsivity in this context was assessed by the emergent tendency of high-impulsive (HI) rats to discount cocaine-associated adversity (i.e., the delivery of a mild electric shock rather than cocaine on roughly 50 % of trials). This inflexible form of behavior is outwardly similar to drug addicts who despite acknowledging the deleterious impact of chronic drug abuse (i.e., personal and wider harms) rarely ever achieve spontaneous voluntary abstinence. In addition being a trait marker for addiction, premature responding can be modulated in a state-dependent manner (i.e., a secondary consequence of drug exposure). For example, opiate and stimulant drugs, given acutely, strongly increase premature responding (Cole and Robbins 1987; Pattij et al. 2009), similar to methamphetamine after a forced protracted period of withdrawal (Dalley et al. 2007b).

Evidence for a role for premorbid impulsivity in predicting the development of alcohol use disorders is less clear with mixed results depending on the precise animal strain. Premature responding is associated with alcohol-preferring mouse strains compared with non-alcohol-preferring strains (Sanchez-Roige et al. 2014a) and is also associated with greater withdrawal severity from chronic alcohol among different mouse strains (Gubner et al. 2010). Acute alcohol exposure (Oliver et al. 2009) and early but not late abstinence following chronic alcohol exposure is also associated with increased premature responding in mice (Walker et al. 2011) suggesting that premature responding can be secondary to both acute and chronic alcohol exposure. However, alcohol-preferring rats do not differ in premature responding compared with non-alcohol-preferring rats either at baseline or following chronic exposure to alcohol (Pena-Oliver et al. 2015). Interestingly, high premorbid premature responding in rats

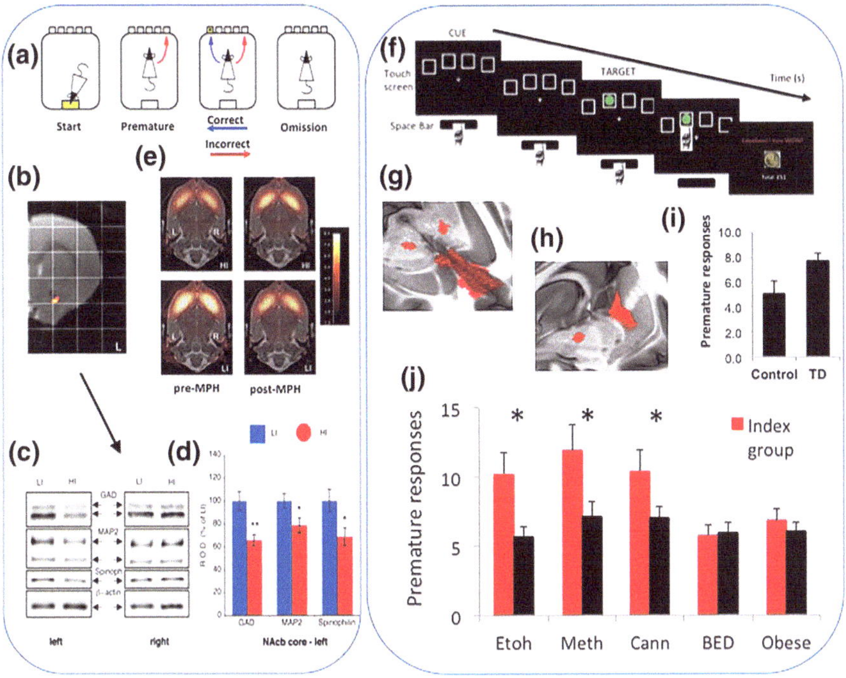

Fig. 1 Translational studies of waiting impulsivity in rodents and humans. **a** Rodent 5-choice serial-reaction time task (5-CSRTT). Rats are trained to wait for the occurrence of a brief visual stimulus presented in one of five open apertures in order to earn a food reward. Responding early ('premature' response), in a non-illuminated aperture ('incorrect' response), or not at all ('omission') is signaled by a loss of food reward and the houselight being briefly extinguished. **b–e** represent studies conducted using the 5-CSRTT in rats. **b–d** Reduced *gray* matter density in the NAcb core of high-impulsive (HI) rats compared with low-impulsive (LI) rats, measured using voxel-based morphometry, and associated with reduced levels of glutamic acid decarboxylase (GAD), microtubule-associated protein (MAP2), and spinophilin in the left NAcb core. E. [18F] fallypride-PET coregistered with horizontal MRI scans showing reduced D2/3 receptor availability in the ventral striatum of HI rats compared with LI rats prior to the oral administration of methylphenidate ('pre-MPH'). Following MPH administration ('post-MPH'), D2/3 receptor availability in the ventral striatum increased in HI rats but decreased in LI rats, according to the model of rate dependency. **f** Human 4-choice serial-reaction time task (4-CSRT). Subjects hold down the space bar with their index finger and release to press the box on the touch screen in which the green target appears. A premature response is measured as a release of the space bar before target onset. **g–i** represent studies conducted using the 4-CSRT. G and H. Resting state functional connectivity correlating with waiting impulsivity in humans: subgenual cingulate and subthalamic nucleus (**g**) and ventral striatum and subthalamic nucleus (**h**). **i** Enhanced waiting impulsivity with central 5-HT depletion in healthy humans (TD = tryptophan depletion). **j** Elevated waiting impulsivity in abstinent alcohol (Etoh) and methamphetamine (Meth) dependence and current cannabis users (Cann) with no differences observed in obese subjects with and without binge eating disorder (BED)

is also associated with greater escalation of sucrose-seeking behavior and reinstatement following extinction in rodents (Diergaarde et al. 2009).

In humans, enhanced waiting impulsivity can be assessed using the analogous 4-choice serial-reaction time task across multiple substance use disorders (e.g., abstinent methamphetamine and alcohol use disorders) along with current cannabis users and current smokers (Voon et al. 2014) (Fig. 1). In this task, a premature response is defined as early anticipatory release of the space bar prior to the onset of a target green stimulus within one of 4 boxes on a touch screen (Voon 2014). That waiting impulsivity is associated with current, but not ex-smokers, or never-smokers, suggest either that nicotine exerts state-dependent effects or that those with higher levels of waiting impulsivity find it difficult to quit smoking. College-age binge drinkers at elevated risk for the development of later alcohol use disorders also show enhanced waiting impulsivity as tested using either the 4-CSRT (Morris et al. 2015) or the Sussex-5-CSRT (Sanchez-Roige et al. 2014b). Differences exist between the two tasks with the former associated with a target cue predicting reward, whereas the latter does not (Voon 2014). In contrast, obese subjects with and without binge eating disorder (BED) did not show impaired waiting impulsivity (Voon et al. 2014). This may reflect a diminished sensitivity of obese people to monetary rewards but further studies (e.g., with food outcomes used instead) would be needed to test this possibility.

2.1.1 Neural Networks and Neurochemistry

The neural network underlying premature responding in the 5CSRTT has been extensively mapped in rodents using excitotoxins, intracerebral pharmacology, and selective immunotoxins [for review, see (Robbins 2002; Dalley et al. 2008)]. These studies indicate key roles of the infralimbic cortex—probably equivalent to the human subgenual anterior cingulate cortex (ACC)—nucleus accumbens (NAcb), and subthalamic nucleus (STN) (Chudasama et al. 2003; Baunez et al. 1995; Baunez and Robbins 1997; Aleksandrova et al. 2013) (Fig. 2) with dissociable contributions from dopaminergic, norepinephrinergic, serotonergic, GABA, and glutamatergic mechanisms (Winstanley et al. 2006; Dalley et al. 2011; Pattij et al. 2007; Hayes et al. 2014). Salient findings from research in rodents and humans are reviewed briefly below.

Acute amphetamine increases premature responding in rodents, an effect attenuated by dopamine (DA)-depleting 6-hydroxydopamine lesions of the NAcb and D1/2 receptor antagonists (Cole and Robbins 1989). More specifically, stimulants, such as amphetamine, nicotine, and cocaine, and DA reuptake inhibitors increased premature responding, which was blocked by intra-NAcb D1 and D2 receptor antagonists (Pattij et al. 2007). HI rats have lower ventral striatal D2/3 receptor availability (Dalley et al. 2007a) and lower left gray matter density, and markers of GABA and dendritic spine function, in the NAcb core (Caprioli et al. 2013) (see Fig. 1). In rodents, the influence of methylphenidate on premature responding appears to be mediated by beta-adrenergic and D4 receptors (Milstein et al. 2010)

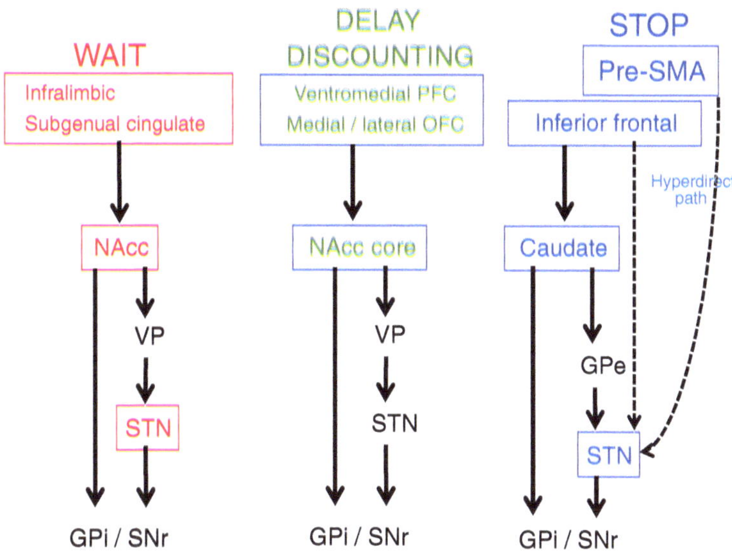

Fig. 2 Cortico-striatal circuitry of subtypes of impulsivity. A simplified box and arrow model of relevant fronto-striatal regions are shown to illustrate key anatomical commonalities and differences between subtypes of impulsivity. Abbreviations: *NAcc* nucleus accumbens; *STN* subthalamic nucleus; *pre-SMA* presupplementary motor area; *VP* ventral pallidum; *GPe* globus pallidus externa; *GPi* globus pallidus interna; *SNr* substantia nigra pars reticulate

and is influenced by dose (Navarra et al. 2008) and baseline impulsivity (Tomlinson et al. 2014). Atomoxetine, a selective norepinephrine (NE)-reuptake inhibitor, also dose-dependently decreases premature responding (Robinson et al. 2008a; Baarendse et al. 2013) in rodents and zebrafish (Parker et al. 2014) and decreases the likelihood of transition toward compulsivity as measured using a schedule-induced polydipsia procedure in HI rats (Ansquer et al. 2014). Crucially, convergent evidence indicates that the NAcb core and shell are functionally opposed with deep brain stimulation (DBS) of these regions producing opposite effects on premature responding (Sesia et al. 2008). DBS targeting the shell but not the core increases premature responding and DA levels in the NAcb, presumably via anti-dromic stimulation of ventral tegmental area projections (Sesia et al. 2010). Moreover, infusions of methylphenidate, a mixed DA/NE reuptake inhibitor, in the core but not the shell increased premature responding, whereas infusions of atomoxetine into the shell but not the core decreased premature responding (Economidou et al. 2012). In rodents, infusion of quinpirole, a D2/3 agonist, into the NAcb core increased premature responding in HI rats but within the shell increased locomotor activity instead (Moreno et al. 2013). Morphine, a mu-opioid receptor (MOR) agonist increases premature responding in the 5CSRTT, an effect blocked by the MOR antagonist naloxone (Pattij et al. 2009). MOR antagonism but not kappa-opioid receptor antagonism also selectively remediates amphetamine-induced impairments in premature responding (Wiskerke et al. 2011). This effect appears to be mediated in the NAcb shell and may interact with the mesolimbic dopaminergic system. Acute methylphenidate challenge is also associated with greater waiting impulsivity in healthy humans (Voon et al. 2015a) consistent with rodent studies of methylphenidate infusion in the NAcb core (Economidou et al. 2012), mediated most parsimoniously by increased DA neurotransmission in this region.

Central serotonin (5-HT) depletion and prefrontal and intra-NAcb 5-HT2C receptor antagonism is associated with greater premature responding in rodents, whereas prefrontal and intra-NAcb 5-HT2A receptor antagonism is associated with a decrease in premature responding (Winstanley et al. 2004a; Robinson et al. 2008b). Similarly, 5HT2A receptor antagonists have been shown to decrease cocaine-induced premature responding in rodents (Anastasio et al. 2011) and 5HT2C receptor agonists decrease nicotine-induced premature responding (Higgins et al. 2012). In humans, tryptophan depletion or acute central 5-HT depletion enhances waiting impulsivity (Worbe et al. 2014) thus corroborating findings in rodents (Winstanley et al. 2004a).

Greater premature responding in the human 4-CSRT task is associated with decreased resting state functional connectivity of the bilateral STN with bilateral subgenual cingulate and right ventral striatum (Morris et al. 2015) (Fig. 1). These findings provide translational evidence in humans for a similar network implicated in rodents. Furthermore, these findings indicate a dissociation with motor response inhibition or action cancelation as captured by the stop-signal task (see Sect. 2.2), which instead was associated with lower connectivity between hyper-direct projections of the right presupplementary motor area (SMA) and left STN together with dorsal caudate and STN connectivity. Furthermore, this network was shown to be dimensionally relevant across alcohol misuse with impairment as a function of alcohol severity in social drinkers and in binge drinkers and alcohol use disorders. Using machine learning techniques, STN connectivity successfully classified those who misuse alcohol from healthy volunteers. Connectivity of the STN and ventral striatum also improved with abstinence suggesting a state effect of alcohol on connectivity strength. Thus, human studies converge on rodent studies implicating a role in substance use disorders possibly mediated by dopaminergic, norepinephrinergic, and serotonergic mechanisms and similar underlying neural networks. Key translational findings of premature responding are shown in Fig. 1.

2.2 *Stop-Signal Reaction Time*

Response inhibition describes the capacity to inhibit a prepotent response. In human studies, differing subtypes of stopping behavior have been described including: fast reactive stopping in response to an external stop signal; proactive stopping in response to a cue predicting a stop signal (Aron 2011; Jaffard et al. 2008), and stopping in response to an internal signal (Schel et al. 2014). Fast reactive stopping is the most commonly tested form with extensive translational evidence as action restraint or cancelation and includes (i) stopping prior to movement initiation (action restraint) as measured using the Go/NoGo task in which subjects must inhibit responding to an infrequently presented stop signal while responding rapidly to a frequently presented stream of Go signals; and (ii) stopping after movement initiation (action cancelation) as measured using the stop-signal task (SST) in which subjects inhibit responding to an infrequent stop signal with onset following a delay after the Go signal (Aron 2011; Chambers et al. 2009; Eagle et al. 2008). In the SST, the stop-signal delay changes as a function of successful stopping such that successful and failed stops occur with a 50 % probability. The shorter the delay the greater likelihood of stopping, while the longer the delay the lower likelihood of stopping. Whereas the Go/NoGo task measures discrete commission errors, the SST assesses the internal speed of stopping (stop-signal reaction time SSRT) by assessing the speed at the Go signal and the probability of stopping at the stop-signal delay (Logan et al. 1984; Verbruggen and Logan 2009). This is conceptually based on the Logan's Race model or a race competition between the process of Go and Stop in which the process that crosses the threshold first is executed. As evidence for the neurobiology underlying human and rodent SSRT and Go/NoGo tasks have been extensively discussed (Aron 2011; Chambers et al. 2009; Eagle et al. 2008), only a brief summary will be presented here.

Meta-analyses show deficits in response inhibition particularly with the SSRT and to a lesser extent with Go/NoGo in ADHD (Lipszyc and Schachar 2010) and across most but not all substance use disorders including stimulants, nicotine, alcohol, and in pathological gambling and Internet use disorder but not opioid or cannabis abuse (Smith et al. 2014). Impairments in the SST have also been shown in unaffected siblings of stimulant-dependent subjects suggesting an endophenotypic risk factor for the development of addiction (Ersche et al. 2012) and predicted adolescent alcohol and drug-related problems (Nigg et al. 2006) and the progression from heavy alcohol use in adults to alcohol dependence (Rubio et al. 2008). Similarly, both OCD subjects and unaffected family members show impairments in the SST suggesting a cognitive endophenotype (Menzies et al. 2007) underlying the development of OCD. However, unlike premature responding, rodent studies have not yet shown that this form of impulsivity predicts compulsive substance use.

2.2.1 Neural Substrates

In rodent studies, the SST appears to be influenced by NE-ergic mechanisms with a rather more limited influence from DA to 5-HT. In rodents, methylphenidate and acute amphetamine influences SST as a function of baseline (Eagle et al. 2007; Feola et al. 2000). Low-dose acute amphetamine also improved the impairment in response inhibition produced by a medial striatal lesion but high doses further impaired response inhibition (Eagle and Robbins 2003). Atomoxetine infused into the orbitofrontal cortex (OFC) and dorsal prelimbic cortex improved response inhibition on the SST (Robinson et al. 2008a; Bari et al. 2011; Bari et al. 2009), an effect blocked by guanfacine, an alpha-2a receptor agonist, but not by alpha-flupenthixol, a D1/D2 receptor antagonist (Bari et al. 2011). Similarly, an alpha-2a antagonist improved response inhibition unlike DA receptor agonists and antagonists (Bari and Robbins 2013). Further, neither D1/D2 receptor antagonists (Eagle et al. 2007) nor DA transporter inhibition (Bari et al. 2009) was shown to affect SSRT, similar to the lack of effects of 5-HT depletion and 5-HT transporter inhibition (Bari et al. 2009; Eagle et al. 2009). D1 and D2 antagonists infused in the dorsomedial striatum have been shown to improve and impair SSRT respectively.

In humans, methylphenidate improves SST performance in those with impaired SSRT such as cocaine-dependent subjects (Li et al. 2010) as well as children (DeVito et al. 2009; Tannock et al. 1989) and adults (Aron et al. 2003a) with ADHD. In healthy subjects, direct comparisons show that acute methylphenidate but not atomoxetine or citalopram improves response inhibition on the SST (Nandam et al. 2011) although a direct comparison shows efficacy of atomoxetine but not citalopram in improving response inhibition on the SST (Chamberlain et al. 2007a). However, higher doses of atomoxetine (80 mg vs. 40 mg) impaired response inhibition on a Go/NoGo task (Graf et al. 2011). Thus, converging with rodent studies, NE appears to play a role in reactive stopping in humans with a possible U-shaped dose–response relationship.

Another type of inhibitory control, which has been more extensively studied in human subjects, is that of proactive stopping in which motor control is preparatory and goal-directed (Aron 2011; Jaffard et al. 2008). This form of stopping has similarities to 'braking' and 'conflict-induced slowing' in which slowing of reaction time may occur in the context of a conflict to prevent impulsive decisions until a decision is made (Frank et al. 2007; Frank 2006). Proactive stopping can be differentiated from reactive stopping by comparing response inhibition tasks in which either the stop signal is acted upon to countermand the action (proactive condition) or the stop signal is ignored or not present (baseline condition). Proactive inhibition is associated with a decrease in motor-evoked potential, an index of cortical excitability, which is suppressed to a greater extent than at rest when anticipating a stopping response (Cai et al. 2011).

Response inhibition implicates a network across rodent and human studies including the SMA, right inferior frontal cortex (rIFC), the STN, and caudate and has been extensively reviewed elsewhere (Dalley et al. 2011; Morris et al. 2015; Aron 2011; Aron et al. 2003b) (Fig. 2). Studies of proactive inhibition implicate the same regions as reactive stopping including the SMA, rIFC, and STN (Jaffard et al. 2008; Zandbelt and Vink 2010; Obeso et al. 2013; Ballanger et al. 2009). The hyperdirect connections from the SMA and rIFC to STN are thought underlie reactive stopping, whereas fronto-striatal circuitry via the direct and indirect pathways appears to be more important for proactive stopping (Zandbelt and Vink 2010; Smittenaar et al. 2013a; Majid et al. 2013).

2.3 Delay Discounting

Animal and humans demonstrate an inherent tendency to discount or devalue future outcomes (Ainslie 1975). Impulsive choice or delay discounting is a form of impulsivity that can be measured using intertemporal choice tasks. In these tasks, subjects choose between a small immediate reward and a larger but delayed reward. Intertemporal choice tasks used in animal studies are invariably based on feedback and range in duration from seconds to minutes, whereas, in human studies, hypothetical tasks involving longer durations from days to years are generally used. However, discounting can occur in humans over relatively short delay intervals of just seconds in humans (Schweighofer et al. 2008; Gregorios-Pippas et al. 2009). The Experiential Discounting Task is an intertemporal choice task paired with real-time coin machine feedback developed to be sensitive to state changes in discounting and to model naturalistic choice context (Reynolds and Schiffbauer 2004). The devaluation of future reward can be reliably modeled by a hyperbolic function ($V_S = V_A/1 + Kd$) with steeper slopes closer to the time of reward receipt than an exponential function ($V_S = V_A e^{-kd}$) with equal slopes over delay intervals (Ainslie 1975; Mazur 1987). With such functions, the subjective value, V_S, is a modification of the actual value, V_A, by the delay (d) and a discount constant (K). K represents the steepness of the temporal discounting curve and represents a measure of impulsivity. A hyperbolic fit implies that when the smaller reward is imminently available, the subjective value of the smaller immediate reward will be greater than the subjective value of the larger delayed reward, thus resulting in a preference reversal away from the larger future outcome toward the smaller immediate outcome (Ainslie 1975).

In animal studies, pre-existing impairments in impulsive choice predispose to greater cocaine self-administration and reinstatement of cocaine-seeking behavior (Perry et al. 2005, 2008), and greater use of alcohol (Mitchell et al. 2006; Poulos et al. 1995) and nicotine (Diergaarde et al. 2008). In humans, delay discounting is a core impairment implicated in ADHD (Noreika et al. 2013) and substance addictions across multiple drug categories, pathological gambling (Bickel et al. 2013; Bickel et al. 2014), and in obesity both with and without BED (Mole et al. 2014; Voon 2015).

2.3.1 Neural Substrates

Converging evidence implicates a role for DA in delay discounting. In primates, single unit striatal recordings to reward-predictive cues show that dopaminergic neuronal activity scales with magnitude and decreases with delay, thus reflecting the subjective devaluation of the delayed reward (Kobayashi and Schultz 2008). Similarly, in rodents, DA levels in the NAcb, assessed by in vivo voltammetry, scale with reward magnitude and decrease with delay (Saddoris et al. 2015). Notably, higher DA levels correlated with choice preference at shorter delays (Saddoris et al. 2015) and optogenetically enhanced DA release in the NAcb during reward-predictive cues shifted choice preference for delay but not magnitude-related decisions (Saddoris et al. 2015).

In rodents, delay discounting has been shown to correlate with 'trait' premature or anticipatory responding (Robinson et al. 2009), the latter associated with lower D2/3 receptor density in the ventral striatum (Dalley et al. 2007a). Acute administration of low and moderate doses of amphetamine decreases impulsive choice in rodent studies (Floresco et al. 2008; Richards et al. 1999; Wade et al. 2000; van Gaalen et al. 2006), whereas high, chronic doses of methamphetamine and cocaine increase impulsive choice in rodents (Richards et al. 1999; Roesch et al. 2007; Simon et al. 2007).

Studies in healthy humans show that increased questionnaire-based trait impulsivity (although not specifically delay discounting) is correlated with decreased D2/3 autoreceptor midbrain availability and with increased amphetamine-induced DA release in the striatum (Buckholtz et al. 2010). Striatal DA release was associated with greater wanting or desire for the stimulant. Using mediation analysis, the relationship between D2/3 receptor binding and impulsivity was in part mediated by striatal DA release. Lower D2/3 receptor availability in the ventral striatum appears to correlate more specifically with greater delay discounting in pathological gamblers (Joutsa et al. 2015), methamphetamine dependence (Ballard et al. 2015), and alcohol use disorders (Oberlin et al. 2015). Levodopa, a precursor to DA in healthy humans, increases impulsive choice (Pine et al. 2010) and increases delay aversion in patients with PD (Cools et al. 2003). Prefrontal cortical DA has also been implicated in impulsive choice. Thus, genetic polymorphisms associated with catechol-o-methyltransferase (COMT), an enzyme found in the PFC and responsible for DA breakdown, are associated with a U-shaped relationship between prefrontal dopaminergic function and impulsive choice (Kayser et al. 2012).

Norepinephrinergic and serotonergic mechanisms have also been implicated in impulsive choice. Atomoxetine decreases impulsive choice in rodents (Robinson et al. 2008a; Sun et al. 2012). By contrast, central 5-HT depletion increases impulsive choice in rodents (Mobini et al. 2000) and attenuates the decrease in impulsive choice induced by low-to-moderate doses of d-amphetamine (Winstanley et al. 2003). Furthermore, 5-HT1A receptor agonists similarly attenuate the effects of d-amphetamine but have no effect in rats depleted of DA in the NAcb (Winstanley et al. 2005). Interestingly, during the waiting period for primary and conditioned rewards, serotonergic activity is tonically increased in rats (Miyazaki et al. 2011) with optogenetic activation of dorsal raphe serotonergic neurons increasing the ability of rats to withhold responding to delayed rewards (Fonseca et al. 2015). In humans, the role of 5-HT is less clear since tryptophan depletion does not influence delay discounting in healthy controls (Worbe et al. 2014), with and without a family history of alcohol dependence (Crean et al. 2002) or with simulated binging of alcohol (Dougherty et al. 2015).

Rodent lesion studies implicate the NAcb core, OFC, amygdala, and hippocampus in delay discounting (Cardinal et al. 2004; Cardinal et al. 2001; Winstanley et al. 2004b). Human imaging studies also implicate the ventral striatum, OFC, lateral prefrontal cortex (lPFC), insula, amygdala, posterior cingulate, and parietal cortex in delay discounting for secondary rewards (Ballard and Knutson 2009; Tanaka et al. 2004; Kable and Glimcher 2007; McClure et al. 2004) and primary rewards (McClure et al. 2007) (Fig. 2).

More specifically, subregions of the OFC have been implicated in delay discounting with lesions of the medial OFC increasing delay discounting in rats and lesions of the lateral OFC decreasing delay discounting (Mar et al. 2011). These lesion studies converge with studies of single neuron activity in the OFC showing higher activity with time-discounted rewards after a short delay and lower activity after a long delay, independent of the encoding for absolute reward magnitude (Roesch et al. 2006). Similarly, stroke-induced lesions of the medial OFC increased delay discounting in humans (Sellitto et al. 2010). Thus, the OFC appears to play a specific role in encoding time-discounted rewards beyond value encoding to guide choice behavior.

The ventral striatum is a key structure implicated in single and dual valuation theories of temporal discounting in human studies. In the dual valuation system, the beta system activates limbic systems (ventral striatum and medial PFC) and is associated with the choice of the immediate reward, whereas delta regions (lateral prefrontal and parietal cortices) are activated during all decisions (McClure et al. 2004). The beta system is hypothesized to overvalue immediate rewards while the delta system is considered to discount rewards over a constant rate with time. An alternate dual valuation system is hypothesized in which delay is coded in the lPFC and magnitude coded in the ventral striatum (Ballard and Knutson 2009). The differential involvement of cortico-basal-ganglia loops has been implicated with ventro-anterior striatum and insula being preferentially involved in immediate choices and dorso-posterior striatum and insula being preferentially involved in delayed choices (Tanaka et al. 2004). In contrast, others have argued for a single valuation system with the ventral striatum representing the subjective value of the delayed choice (Kable and Glimcher 2007).

2.4 *Reflection Impulsivity*

Reflection impulsivity is predominantly assessed in humans and describes the accumulation of evidence, evaluation of options, and rapid hypothesis testing prior to a decision (Kagan 1966). This form of impulsivity can be divided into perceptual and probabilistic decisions. Perceptual tasks include the matching familiar figures task (MFFT) in which subjects decide whether a pattern matches a series of similar patterns of which all but one differs (Kagan 1966). The impulsivity score captures the core feature of the extent of information sampling or reaction time and accuracy of the decision. Children with ADHD perform more impulsively on the MFFT, which improves with psychostimulant treatment (Brown and Sleator 1979). MDMA users but not cannabis users or alcohol-dependent subjects are also impaired on the MFFT (Quednow et al. 2007; Morgan et al. 2006; Weijers et al.

2001). This task may have overlaps with other perceptual decision tasks focusing on speed-accuracy trade-offs involving sensory discrimination (e.g., random dot motion task) (Gold and Shadlen 2007; Banca et al. 2014) and conflict-induced slowing to either probabilistic or perceptual conflict (Frank 2006; Wylie et al. 2009).

Other reflection impulsivity tasks assess probabilistic decisions more directly by measuring the extent of information sampling or evidence accumulation [e.g., the Beads-in-a-Jar task ('Beads task') (Volans 1976) and the information sampling task (IST) (Clark et al. 2006)]. In the Beads task, subjects must decide from which of two jars beads are being selected based on known probabilities of the color ratio of the beads within the jars. Participants are aware of the explicit probabilities of the alternate options with each piece of evidence accumulated associated with an expected probability of being correct. Using this task, elevated probabilistic reflection impulsivity has been observed in substance use disorders, pathological gamblers (Djamshidian et al. 2012), binge drinkers (Banca et al. 2015), and patients with PD and medication-induced behavioral addictions (Djamshidian et al. 2012). Reflection impulsivity tested using the Beads task is exacerbated by DA receptor agonists though not by Levodopa in studies of PD (Djamshidian et al. 2013). The IST is a similar information sampling paradigm that asks participants to decide which color is predominant in a 5 × 5 matrix by opening boxes to make a decision (Clark et al. 2006). Current or former amphetamine and opiate users sample less information compared to healthy volunteers (Clark et al. 2006). One study has shown an impairment in binge drinkers (Banca et al. 2015) although this was not confirmed in a second study (Townshend et al. 2014).

The IST is conceptually similar to the Beads task, yet a recent study in schizophrenia unexpectedly did not demonstrate impairments in the IST despite consistent reports of impairments in the Beads task (Huddy et al. 2013). Similarly, a binge drinking study showed an impairment with the Beads task but not the IST (Banca et al. 2015). One reason for this apparent discrepancy is that unlike the Beads task, the IST presents the available information for sampling in a more explicit manner, thus encouraging thinking ahead of all possible outcomes and overall task representation. That the generative probability distribution is more uncertain and closer to 50 % in the IST as compared to the known explicit probabilities in the Beads task will also shift subjects toward being more cautious.

2.4.1 Neural Substrates

Volumetric differences between the IST and the Beads task have also been reported (Banca et al. 2015). Greater impulsivity in the Beads task was associated with smaller dlPFC and left inferior parietal volumes (Banca et al. 2015). The mechanisms underlying evidence accumulation can be subdivided into evidence-seeking or decision-making. The Beads task is associated with parietal activity during evidence-seeking and dlPFC activity during both evidence-seeking and decision-making (Furl and Averbeck 2011). The dlPFC is important for the resolution of uncertainty (Huettel et al. 2005) and computing differences between costs and benefits (Basten et al. 2010) with the accumulated difference represented in the parietal cortex signaling the final decision and confidence (Kiani and Shadlen 2009; Stern et al. 2010). In contrast, greater impulsivity in the IST was associated with greater left dorsal cingulate and right precuneus volumes (Banca et al. 2015). Similarly, in an fMRI study investigating evidence accumulation, greater uncertainty during evidence accumulation was associated with ACC and precuneus activity, whereas greater uncertainty during decision execution was associated with greater lateral frontal and parietal activity (Stern et al. 2010). The dACC is implicated in error and conflict monitoring processes (Scheffers and Coles 2000; Botvinick et al. 2001), and in coding unexpected and unpredicted outcomes during evidence accumulation (Stern et al. 2010; Oliveira et al. 2007).

3 Compulsivity

Compulsivity is generally understood to involve excessive repetitive actions that are incongruous to a situation. OCD is the canonical disorder of compulsivity characterized by anxiety-provoking obsessions and compulsive rituals (Stein and Hollander 1995). However, compulsivity in the form of repetitive and rigid stereotyped behaviors extends to autism, Tourette's syndrome (TS), disorders of impulse control (pathological gambling, trichotillomania), eating disorders, and substance-related and addictive disorders. Various methods have been used in animals to research OCD subtypes and related impulsive–compulsive disorders. Though beyond the scope of the present article, these include genetic approaches (e.g., the Sapap3 knockout mouse showing excessive self-grooming and anxiety (Welch et al. 2007), pharmacological 'models' (e.g., quinpirole-induced compulsive checking (Szechtman et al. 2001), and behavioral models of autism (Kas et al. 2014), pathological gambling (Zeeb et al. 2009) and OCD [e.g., the signal attenuation task and marble burying behavior in rodents (Albelda and Joel 2012)]. Other compulsivity constructs include cognitive inflexibility assessed by response perseveration and impaired attentional set-shifting, dominance of S-R habits, and rigid stereotyped behavior, are reviewed below.

3.1 Cognitive Inflexibility

Adaptive goal-directed behavior requires flexible cognitive control over reinforcement learning, working memory, and attentional set-shifting. Deficits in cognitive flexibility occur in PD, OCD, autism, ADHD, Alzheimer's disease, schizophrenia, addiction, among other disorders (Nilsson et al. 2015) endorsing the view that many measurable and potentially translatable constructs in psychiatry cut across diagnostic boundaries. Cognitive flexibility is most commonly assessed by reversal learning and attentional set-shifting tasks adapted for use in rodents, non-human primates, and humans (Iversen and Mishkin 1970; Fellows and Farah 2003; Tait et al. 2014). Optimal reversal learning requires the capacity to flexibly switch responding to changing stimulus-response contingencies and usually involves a single perceptual dimension where one stimulus is rewarded and the other is not. Attention set-shifting tasks such as the Wisconsin Card Sorting Test (Grant and Berg 1948), the CANTAB intra-/extra-dimensional (IED) set-shift task (Robbins 2000) involve at least two superimposed perceptual dimensions, each containing at least two different stimuli. Set-shifting assesses the capacity to switch responding to previously irrelevant stimuli or to switch in response to changes in rules requiring attentional flexibility.

Current amphetamine and methamphetamine users show impaired set-shifting (Clark et al. 2006; Ornstein et al. 2000) which improves with prolonged abstinence (van den Hout et al. 2009; Johanson et al. 2006; Toomey et al. 2003). In alcohol dependence, impairments in set-shifting are associated with years of abuse (Tarter 1973) and with relapse (Pothiyil and Alex 2013) without any improvements with abstinence (Nowakowska et al. 2007). Impaired set-shifting is also observed in obese individuals with and without BED (Duchesne et al. 2010; Wu et al. 2014). Pathological gambling is associated with both reduced and unaffected set-shifting in the IED task (Grant et al. 2011) and the WCST task (Goudriaan et al. 2006). OCD patients and unaffected first-degree relatives exhibit deficits in ED set-shifting (Chamberlain et al. 2007b) suggesting that impaired set-shifting may be a cognitive endophenotype for OCD. Similarly, patients with TS show impaired ED set-shifting (Watkins et al. 2005).

Substance use disorders can also be associated with impaired reversal learning. Thus, reversal learning is impaired in cocaine use disorders (Camchong et al. 2011; Fernandez-Serrano et al. 2012; Ersche et al. 2008; Fillmore and Rush 2006) but not in amphetamine and opiate abusers (Ersche et al. 2008). The influence of alcohol dependence on reversal learning is less clear with deficient aversive eye-blink conditioning (Fortier et al. 2008) and evidence of slower reversal but with no increase in perseverative errors (Vanes et al. 2014). However, impaired reversal learning is present in pathological gambling with gain and loss outcomes (de Ruiter et al. 2009; Patterson et al. 2006) and in TS (Watkins et al. 2005).

3.1.1 Neural Substrates

Behavioral flexibility is widely accepted to depend on the OFC, lPFC, ACC, and caudate nucleus (Clarke et al. 2008; Cools et al. 2002; Rogers et al. 2000). Thus, damage to the OFC disrupts reversal learning in humans (Fellows and Farah 2003; Hornak et al. 2004), monkeys (Iversen and Mishkin 1970; Dias et al. 1996), and rats (Schoenbaum et al. 2002; Chudasama and Robbins 2003). Reversal learning is also impaired in rats by excitotoxic lesions and DA depletion of the dorsomedial striatum (or caudate) (Castane et al. 2010; O'Neill and Brown 2007), with lesions of this region also impairing attentional set formation (Lindgren et al. 2013). In contrast, lesions of the lPFC impair ED set-shifting in marmosets (Dias et al. 1996) while a brain imaging study in humans found dissociable activations in lOFC, lPFC, and ACC, respectively, at the point of reversal, attentional control, and when new searches were initiated (Hampshire et al. 2012).

Impaired reversal learning in rats, monkeys, and humans has been strongly linked with a reduction in brain 5-HT (Fineberg et al. 2010) and specifically within the OFC (Clarke et al. 2004). Moreover, intra-OFC 5-HT2C antagonism has been shown to improve both spatial and visual reversal learning in rats (Boulougouris and Robbins 2010; Alsio et al. 2015). In addition, a recent study found reduced markers of 5-HT function in the OFC and raphé nucleus of rats exhibiting behavioral inflexibility on a spatial serial reversal learning task (Barlow et al. 2015). Such changes were accompanied by reduced and increased expression of monoamine oxidase in the raphé nucleus and OFC, respectively. Regulation of reversal learning by 5-HT appears to be mediated cortically rather than within the caudate (Clarke et al. 2007, 2011).

In contrast, DA acting at the level of the caudate has been shown to play a key role in behavioral flexibility. In humans, reversal learning performance correlates with methylphenidate-induced DA release in this region (Clatworthy et al. 2009) while reduced D2 receptor availability predicts reduced ventral striatal activation during probabilistic reversals (Jocham et al. 2009) and is linked to OCD (Denys et al. 2004). Intriguingly, the D2 receptor agonist, bromocriptine, improved performance of a task-set-shifting task but only in individuals with genetically reduced levels of DA (van Holstein et al. 2011). This improvement was abolished by pretreatment with the D2 antagonist sulpiride. In rats and monkeys, neurochemically selective depletion of DA in the dorsomedial striatum (caudate) impairs reversal learning (O'Neill and Brown 2007; Clarke et al. 2011), consistent with evidence that systemic blockade of D2 but not D1 receptors impairs reversal learning in monkeys (Lee et al. 2007).

3.2 *Stimulus-Response (Habit) Learning*

Converging animal and human studies suggests that two different processes of learning are associated with decision-making: the acquisition of goal-directed actions involves decisions made on affective outcome and are governed by knowledge of the association between actions and the value of consequences or response-outcome (R-O) associations. In contrast, habitual choices are made on previously reinforced choices or learned stimulus-response associations (S-R) based on predictive stimuli and are divorced from the value of the outcome. Both goal-directed and habitual learning are used in parallel but with extended training shifts toward habitual control (Adams and Dickinson 1981; Dickinson and Balleine 2002). The relationship between these two forms of learning are formally assessed following training in rodent studies with testing in extinction (without the outcome) with outcome devaluation or contingency degradation.

A set of paradigms using a computational account based on reinforcement learning have been applied to goal-directed and habit learning also termed model-based and model-free control (Daw et al. 2011; Dolan and Dayan 2013). Goal-directed control is prospective and computationally demanding and is based on a learned internal model of the environment, whereas habitual control is retrospective, efficient, and based on the memory of the previously reinforced actions divorced from the predicted outcomes. The two-step task is a sequential two-choice decision task in which subjects make two choices on every trial leading to a rewarded or non-rewarded outcome (Daw et al. 2011). Choices at the first stage are associated with a likely (common) and an unlikely (uncommon) transition. Model-free habitual control is based on the repetition of a previously rewarded action regardless of this transition, whereas model-based goal-directed control takes into account the probability of state–state transitions and selects actions that will more likely lead to reward on future trials. Thus, after uncommon transitions, a reward will lead a model-free subject to choose the same first-stage stimulus on the next trial since the action values are updated based on the reward from the previous action. A model-based subject representing the task structure after receiving a reward following an uncommon transition would switch to the alternate first-stage stimulus since this would be more likely to lead to reward at the second stage. Behavior is best reflected by a hybrid model integrating both model-based and model-free learning (Daw et al. 2011). Healthy volunteers are shown to use a relative mix of both types of control. The outcomes of the two-step task have been shown to correlate with outcomes in conventional overtraining and outcome devaluation tasks (Friedel et al. 2014).

3.2.1 Neural Substrates

Lesions of the posterior dorsomedial striatum or prelimbic cortex prevent the expression of goal-directed learning leaving intact habit learning insensitive to outcome devaluation and contingency degradation (Yin et al. 2005; Balleine and Dickinson 1998). Lesions to the dorsolateral striatum leave intact goal-directed behaviors and lesions to infralimbic result in intact sensitivity to outcome devaluation despite extended training (Balleine and Dickinson 1998; Yin et al. 2004; Killcross and Coutureau 2003).

Human studies have translated tasks in animals to investigate neural correlates. The relationship between goal-directed behaviors and outcome value was assessed in a study in which subjects were moderately trained on two instrumental tasks with differing food outcomes, one of which was devalued by feeding to satiety (Valentin et al. 2007). Behaviors remained goal-directed with decreased actions for the devalued outcome in extinction associated with a decrease in OFC activity. In a study investigating overtrained habitual behaviors, subjects were trained on action–outcome contingencies in which one group that was extensively trained did not retain outcome sensitivity with testing in extinction relative to the other group that was minimally trained. Greater habitual behaviors over the course of learning were associated with increased cue-related activity in the posterior putamen (Tricomi et al. 2009).

An alternate design in humans uses a conflict procedure (de Wit et al. 2009). Subjects must first learn the contingencies between a cue (fruit) and response (left or right button) and outcomes (fruit) for points. When the cue and outcome were congruent, both goal-directed and habitual systems were recruited, whereas only the habitual system was predominantly used when the cue and outcome were incongruent as using the goal-directed system would be disadvantageous. The authors show that under conditions in which goal-directed action predominated, ventromedial prefrontal cortical activity was enhanced. Following acquisition, an instructed outcome devaluation test was then performed in which subjects were presented with two open boxes with one fruit previously associated with a left button press and another with a right button press. One fruit was shown with a cross indicating it had been devalued. An additional testing 'slips-of-action' testing phase in which subjects were instructed that two of the six different fruit outcomes were devalued or would be associated with loss of points. Subjects were then shown boxes with fruit for which points could be earned for pressing valued fruit outcomes and avoiding losing points by withholding pressing for devalued fruit outcomes. Habitual 'slips-of-action' toward the outcomes that were no longer rewarding were associated with greater white matter tract strength between the premotor cortex and the posterior putamen, whereas goal-directed actions were associated with greater tract strength in the ventromedial prefrontal cortex and caudate (de Wit et al. 2012a).

Human functional imaging studies focusing on the encoding of reward value signals relevant for action selection implicate the medial OFC extending dorsally along the medial PFC. These regions represent action–outcome associations (Daw et al. 2006) separate from stimulus-related value signals (Valentin et al. 2007). The caudate is also implicated in the online computation of action–outcome contingency to guide goal-directed learning (Tanaka et al. 2008; Liljeholm et al. 2011). Using the two-step task, both outcome prediction error (the difference between received and expected outcomes used in model-free learning) and state prediction error (the discrepancy between the observed and expected state transition used in model-based learning) converge on the ventral striatum (Daw et al. 2011). Greater model-based learning is associated with greater medial OFC and caudate volumes (Voon et al. 2015b).

State-dependent prediction error relevant to goal-directed behavior in the two-step task is also represented in the lPFC and intraparietal sulcus (Glascher et al. 2010). Transcranial magnetic stimulation to the right dlPFC impairs model-based but not model-free control; in contrast, the left dlPFC disrupts model-based choices dependent on working memory capacity (Smittenaar et al. 2013b). Model-based control can be impaired based on dual-task performance with a demanding task (Otto et al. 2013) and under conditions of stress.

A study in healthy volunteers using the three-step task, an update of the two-step task, showed that values associated goal-directed forward planning trials were associated with caudate activity, whereas habitual trials over-trained over 3 days were associated with posterior putamen activity irrespective of the final choice

(Wunderlich et al. 2012). The ventromedial PFC increased connectivity with both caudate and putamen during choice and encoded the chosen value suggesting an active role as a value comparator.

In rodent studies, DA appears to strengthen habit formation as shown by the effects of a sensitizing regimen of amphetamine (Nelson and Killcross 2006). This effect depends on D1 receptor activation (Nelson and Killcross 2013). Furthermore, selective lesions of the nigrostriatal dopaminergic system impair habit formation (Faure et al. 2005). In humans, decreasing DA function with an acute dietary intervention encourages greater habitual control in the slips-of-action task (de Wit et al. 2012b). Consistent with these findings greater presynaptic DA synthesis, as measured using F-DOPA PET, correlates with greater model-based learning (Deserno et al. 2015).

A role for 5-HT in habit learning has also been suggested. Decreasing forebrain 5-HT and systemic 5HT2C antagonism enhanced compulsive cocaine seeking in rodents, which was reversed by a 5HT2C agonist (Pelloux et al. 2012). Overexpression of 5-HT6 receptors in the rodent dorsolateral striatum was associated with decreased habit learning with reduced lever pressing under extinction (Eskenazi and Neumaier 2011). Acute tryptophan depletion in healthy humans produces a shift toward habitual responding on the slips-of-action task (Worbe et al. 2015a). Similarly, this intervention impaired model-based goal-directed behaviors to reward outcomes but enhanced model-free habitual behaviors to loss outcomes in the two-step task. One possible mechanism whereby tonic 5-HT might enhance goal-directed behaviors is changing the long-run average reward representation by providing a positive or negative signal of the 'goodness' or 'badness' of the environment (Daw et al. 2002). Indeed, 5-HT signaling may signify the cost associated with deliberation (Niv et al. 2007; Keramati et al. 2011).

OCD subjects also show impaired goal-directed knowledge in the instructed outcome devaluation test with greater responding to devalued outcomes indicating greater habitual responding in the slips-of-action test (Gillan et al. 2011). Similarly, OCD subjects were impaired on model-based goal-directed learning to reward on the two-step task. This has now been independently replicated at two sites with compulsivity correlating with model-based behaviors (Voon et al. 2015b, c). The phenomenology of compulsive symptoms in OCD may be better captured by aversive avoidance which has been shown in which OCD subjects showed greater habitual responding following overtraining to a virtual devaluation of a shock outcome in an aversive shock habit task (Gillan et al. 2013). These results contrast with a study using monetary loss outcomes in the two-step task where OCD subjects had greater model-based behaviors to loss outcomes (Voon et al. 2015c). These differing results may be related to sensitivity to motivational status with decreased sensitivity to monetary rewards and enhanced sensitivity to monetary losses in the two-step task, differential behavioral responding to losses as compared to shock outcomes, or differences in task design.

Alcohol-dependent subjects show impaired goal-directed learning based on the outcome devaluation test along with decreased activity in the ventromedial PFC and anterior putamen and increased activity in the posterior putamen (Sjoerds et al.

2013). Similarly, heavy drinkers showed greater activity in the dorsal striatum to drinking cues, whereas light drinkers showed greater prefrontal and ventral striatal activity (Vollstadt-Klein et al. 2010). One study testing alcohol-dependent subjects after 2 weeks showed impaired model-free behaviors (Sebold et al. 2014), whereas another study did not show any differences from healthy volunteers but did show that abstinence improved model-based learning (Voon et al. 2015b). Subjects with methamphetamine dependence and obese subjects with BED were also shown to have impaired model-based learning on the two-step task (Voon et al. 2015b).

3.3 Motor Stereotypy

Motor stereotyped behavior such as pacing, route tracing, and repetitive orofacial movements (e.g., sniffing, rearing, licking, and gnawing) can be induced in animals by high doses of psychostimulant drugs (Divac 1972; Fog 1972; Kelly et al. 1975), D1 and D2 receptor agonists (LaHoste and Marshall 1993; Kreipke and Walker 2004), and environmental variables such as feeding times in captive and commercial animals (Lawrence and Terlouw 1993). Intense orofacial stereotypes, however, appear to require the activation of both D1 and D2 receptors (Delfs and Kelley 1990). Motor stereotypes are recognized to reflect a disruption within basal ganglia circuitry and specifically dopaminergic modulation of the direct and indirect pathways of the dorsal striatum (or caudate putamen in humans) (Arnt 1985; Langen et al. 2011). Although the release of behavior is thought to depend on the balance in activity between the D1 and D2 receptor-modulated direct and indirect pathways (DeLong and Wichmann 2015), relative activity in the striosomal (patch) and extrastriosomal (matrix) compartments of the striatum may also be important. Thus, the intensity of psychostimulant-induced stereotypy in rats has been shown to correlate with greater relative activity in striosomal neurons than matrix neurons (Canales and Graybiel 2000). Striosomal neurons receive inputs preferentially from the frontal cortical regions (e.g., ACC, prelimbic cortex) and, for the most part, express dynorphin and substance P to form the direct striatonigral pathway (Gerfen 1992; Ragsdale and Graybiel 1990). Thus, the expression of motor stereotypies appears to involve an imbalance between cortico-striatal circuits at different levels of anatomical and functional organization, which may not be mutually exclusive (Langen et al. 2011).

Motor stereotypies in humans can include behaviors such as hand flapping in autism or punding behaviors observed at peak dose in cocaine users (Rylander 1972) and in the context of parkinsonian medications (Evans et al. 2004). Punding involves excessive non-goal-oriented repetitive behaviors. Punding has been reported on high doses of cocaine which included simple motor actions such as repeatedly playing with the intravenous pole or driving motorcycles around the block in circles. Similarly, punding has been associated with high doses of Levodopa and DA agonists such as apomorphine in patients with PD (Evans et al. 2004; Miyasaki 2007) or restless legs syndrome (Voon et al. 2011) and can include simple acts such as shuffling papers, collecting buttons, rearranging handbags, to more complex behavioral sequences such as repeatedly taking apart lawn mowers or hobbies such as gardening or painting. Although systematic studies are lacking, descriptive studies suggest a link between these behaviors and previous occupations or gender stereotypes (e.g., an accountant subsequently shuffling papers or a seamstress collecting buttons) suggesting a possible role for disinhibition of previously learned motor repertoires (Evans et al. 2004; Voon 2004).

4 Synthesis and Future Perspectives

Impulsivity and compulsivity are often used interchangeably but despite being neurally and psychologically distinct can be present in the same disorder (Fineberg et al. 2010). Distinct and overlapping fronto-striatal networks implicated in various impulsivity and compulsivity subtypes are depicted in Fig. 3. We emphasize the

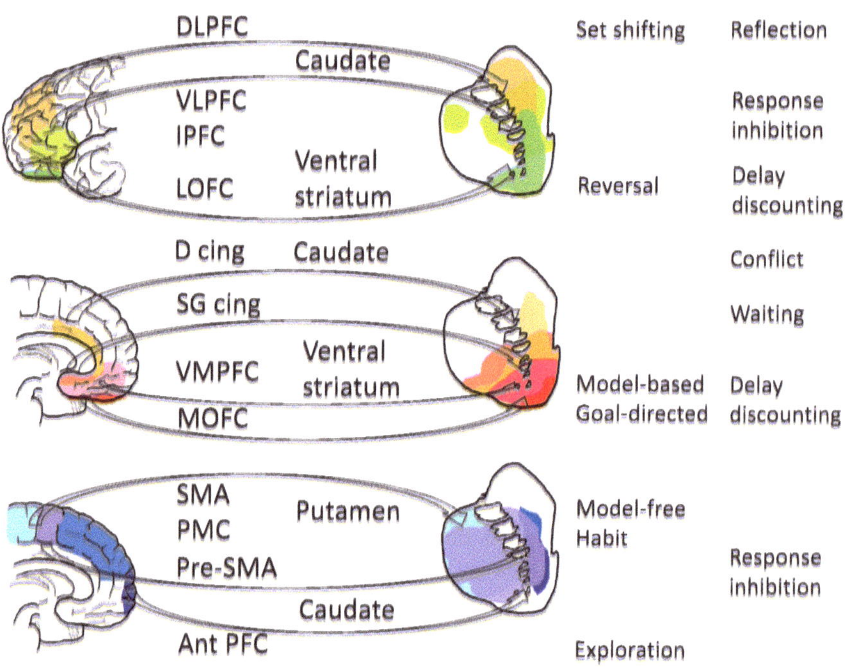

Fig. 3 Fronto-striatal substrates of impulsivity and compulsivity. The prefrontal and striatal images represent actual resting state connectivity patterns in healthy volunteers based on well-defined prefrontal functionally defined seeds (Morris et al. submitted). The columns to the *right* represent the impulsivity (*right*) and compulsivity (*left*) subtypes that map onto these fronto-striatal networks. Abbreviations: *DLPFC* dorsolateral prefrontal cortex; *VLPFC* ventrolateral prefrontal cortex; *IPFC* inferior prefrontal cortex; *LOFC* lateral orbitofrontal cortex; *D cing* dorsal cingulate; *SG cing* subgenual cingulate; *VMPFC* ventromedial prefrontal cortex; *MOFC* medial orbitofrontal cortex; *SMA* supplementary motor area; *PMC* premotor cortex; *pre-SMA* presupplementary motor area; *Ant PFC* anterior prefrontal cortex

convergent and divergent roles of both neural anatomy and neurochemistry and its relevance to impulsivity and compulsivity.

4.1 Neural Considerations: Fronto-striatal Nodes and Opponency Processes

Multiple levels of striatal anatomy and connectivity allow for convergent and divergent functional expression. The NAcb/VS and STN are critical convergent nodes in behavioral control (Fig. 2) and targets for therapeutic intervention, e.g., DBS for PD (Deuschl et al. 2006), OCD (Mallet et al. 2008), and addiction (Muller et al. 2013). Opponency processes also exist in the striatum on multiple levels providing fine-grained control and heterogeneity of function. Broadly, ventral striatal regions are more closely associated with impulsivity and dorsal striatal with compulsivity with diverging prefrontal inputs. The NAcb/VS is implicated in waiting impulsivity (Dalley et al. 2007a; Caprioli et al. 2013) and delay discounting (Cardinal et al. 2004, 2001); Winstanley et al. 2004; Ballard and Knutson 2009; Tanaka et al. 2004; Kable and Glimcher 2007; McClure et al. 2004), subtypes of impulsivity which have been shown to correlate in rodents but not in humans (Voon et al. 2014). In rodent studies, delay discounting studies are conducted with real-time feedback with short delays, whereas human studies typically use hypothetical monetary feedback with long delays. Whether waiting impulsivity correlates with delay discounting in tasks involving shorter delays with real-time feedback remains to be established. Dorsal striatal regions are implicated in proactive stopping with the putamen more specifically implicated in proactive stopping of specific motor responses (Zandbelt and Vink 2010; Smittenaar et al. 2013a; Majid et al. 2013). The dorsomedial/caudate and dorsolateral/putamen are associated with the opponent processes of goal-directed and habitual behaviors, respectively (Daw et al. 2011; Yin et al. 2005; Balleine and Dickinson 1998; Yin et al. 2004; Killcross and Coutureau 2003; Tricomi et al. 2009; de Wit et al. 2012a; Voon et al. 2015b; Wunderlich et al. 2012). The NAcb shell and dorsolateral striatal regions are integrated and linked via spiraling loops from the ventral to dorsal midbrain (Haber et al. 2000). The STN in the indirect pathway is a critical node relevant particularly to impulsivity and receives projections from fronto-striatal circuitry implicated in waiting impulsivity (Baunez and Robbins 1997; Eagle and Baunez 2010) with hyperdirect cortical connections to the STN implicated in fast reactive stopping (Aron 2011; Ballanger et al. 2009) and conflict-induced slowing (Frank et al. 2007).

Opponency processes exist on several levels in the striatum. The NAcb in rodents can be divided into a core and shell with differing afferent and efferent connections with opposing effects on waiting impulsivity of DBS and DA- and NE-based interventions (Sesia et al. 2008; Sesia et al. 2010; Economidou et al. 2012; Moreno et al. 2013). Waiting impulsivity in humans similarly implicates the ventral striatum (Morris et al. 2015) although it is unclear whether a similar subregional functional distinction exists as in rodents. Lesions of the NAcb core also impair delay discounting in rodents (Cardinal et al. 2001) with striatal DA coding for delayed rewards (Kobayashi and Schultz 2008) and optogenetic manipulation of the dopaminergic system influencing delay-related choices (Saddoris et al. 2015). The fronto-striatal direct and indirect pathways, associated with D1 and D2 receptors, respectively, provide facilitating and inhibitory functions (DeLong and Wichmann 2015). Activation and antagonism of D1 and D2 receptors also has a differential effect on stereotypies and locomotor behaviors (DeLong and Wichmann 2015). On a microstructural level, stereotypies are associated with greater engagement of striosomal (patch) compartments receiving prefrontal inputs and outputs to the direct pathway relative to extrastriosomal (matrix) compartments (Langen et al. 2011; Canales and Graybiel 2000). Thus, subtypes of impulsivity and compulsivity map onto a striatal neural organization on both macro- and microstructural levels have both convergent and divergent structural and functional organizations.

4.2 Neurochemical Considerations: Dopamine and Serotonin

DA influences multiple forms of impulsivity and compulsivity which may in part be related to differential receptor expression in the striatum in impulsivity and compulsivity implicating ventral and dorsal striatal regions, respectively. Thus, low D2/3 receptor availability in the ventral striatum, without changes in DA release, predicts high levels of premature responding in rats (Dalley et al. 2007a), and knocking down D2 receptors in the rodent putamen generates binge-like eating suggestive of compulsive behaviors (Johnson and Kenny 2010). Lower ventral striatal D2/D3 receptor availability has also been shown to correlate more specifically with greater delay discounting in pathological gamblers (Joutsa et al. 2015), methamphetamine dependence (Ballard et al. 2015), and alcohol use disorders (Oberlin et al. 2015). Further, low midbrain D2/3 receptors have been shown to correlate with questionnaire-based impulsivity in healthy volunteers, and striatal DA release (Buckholtz et al. 2010).

Impulsivity and compulsivity can be further dissociated by serotonergic function and opposing actions of the 5HT2A and 2C receptors, which may be related to opposing effects on DA function (Cunningham and Anastasio 2014; Howell and Cunningham 2015) and differential actions on prefrontal and striatal regions. In rodents, prefrontal 5-HT depletion enhances waiting impulsivity (Winstanley et al. 2004a; Robinson et al. 2008b) and central 5-HT depletion enhances delay discounting (Mobini et al. 2000) with optogenetic activation of dorsal raphe neurons enhancing the capacity to wait for delayed rewards (Fonseca et al. 2015). In humans, central 5-HT depletion similarly enhances waiting impulsivity (Worbe et al. 2014). More specifically, systemic, intra-NAcb, and prefrontal 5HT2A antagonism decrease premature responding, whereas similar 5HT2C antagonism increases premature responding (Winstanley et al. 2004a; Robinson et al. 2008b). Although one study showed effects in both prefrontal and accumbal regions, another study demonstrated specificity to accumbal and not prefrontal regions. Systemic 5HT2C and 5HT2A antagonists enhance and impair reversal learning, respectively (Boulougouris et al. 2008) with effects of 5HT2C antagonists specific to the OFC (Boulougouris and Robbins 2010). Thus, 5HT2A antagonism within the NAcb decreases premature responding and 5HT2C antagonism within the PFC improves reversal learning. In contrast, decreasing forebrain 5-HT and systemic 5HT2C antagonism enhances compulsive cocaine seeking or habitual behaviors in rodents which can be reversed by 5HT2C agonist (Pelloux et al. 2012). These findings converge with human studies in which central 5-HT depletion encourages goal-directed learning rather than habitual behaviors (Worbe et al. 2015b).

5 Conclusion

In summary, we highlight the translational and back-translational relevance of subtypes of impulsivity and compulsivity. These constructs map onto distinct fronto-striatal neural and neurochemical systems interacting both at nodal convergent points and as opponent processes highlighting both the heterogeneity and the commonalities of function. We emphasize the relevance of these constructs for understanding dimensional psychiatry.

Acknowledgements VV is a Wellcome Trust Fellow and supported by the Wellcome Trust (093705/Z/10/Z). JD acknowledges funding support from the Medical Research Council (G0701500; G0802729). The Behavioural and Clinical Neuroscience Institute (BCNI) at Cambridge University is supported by a joint award from the Medical Research Council (G1000183) and Wellcome Trust (093875/Z/10/Z). We would like to thank Laurel Morris for fronto-striatal connectivity illustrations shown in Fig. 3.

References

Adams CD, Dickinson A (1981) Instrumental responding following reinforcer devaluation. Q J Exp Psychol 33:109–122

Ainslie G (1975) Specious reward: a behavioral theory of impulsiveness and impulse control. Psychol Bull 82(4):463–496

Albelda N, Joel D (2012) Animal models of obsessive-compulsive disorder: exploring pharmacology and neural substrates. Neurosci Biobehav Rev 36(1):47–63

Aleksandrova LR, Creed MC, Fletcher PJ, Lobo DS, Hamani C, Nobrega JN (2013) Deep brain stimulation of the subthalamic nucleus increases premature responding in a rat gambling task. Behav Brain Res 245:76–82

Alsio J, Nilsson SR, Gastambide F, Wang RA, Dam SA, Mar AC et al (2015) The role of 5-HT2C receptors in touchscreen visual reversal learning in the rat: a cross-site study. Psychopharmacology 232(21–22):4017–4031

Anastasio NC, Stoffel EC, Fox RG, Bubar MJ, Rice KC, Moeller FG et al (2011) Serotonin (5-hydroxytryptamine) 5-HT(2A) receptor: association with inherent and cocaine-evoked behavioral disinhibition in rats. Behav Pharmacol 22(3):248–261

Ansquer S, Belin-Rauscent A, Dugast E, Duran T, Benatru I, Mar AC et al (2014) Atomoxetine decreases vulnerability to develop compulsivity in high impulsive rats. Biol Psychiatry 75 (10):825–832

Arnt J (1985) Antistereotypic effects of dopamine D-1 and D-2 antagonists after intrastriatal injection in rats. Pharmacological and regional specificity. Naunyn-Schmiedeberg's Arch Pharmacol 330(2):97–104

Aron AR (2011) From reactive to proactive and selective control: developing a richer model for stopping inappropriate responses. Biol Psychiatry 69(12):e55–e68

Aron AR, Dowson JH, Sahakian BJ, Robbins TW (2003a) Methylphenidate improves response inhibition in adults with attention-deficit/hyperactivity disorder. Biol Psychiatry 54(12):1465–1468

Aron AR, Fletcher PC, Bullmore ET, Sahakian BJ, Robbins TW (2003b) Stop-signal inhibition disrupted by damage to right inferior frontal gyrus in humans. Nat Neurosci 6(2):115–116

Austin JL, Bevan D (2011) Using differential reinforcement of low rates to reduce children's requests for teacher attention. J Appl Behav Anal 44(3):451–461

Baarendse PJ, Winstanley CA, Vanderschuren LJ (2013) Simultaneous blockade of dopamine and noradrenaline reuptake promotes disadvantageous decision making in a rat gambling task. Psychopharmacology 225(3):719–731

Ballanger B, van Eimeren T, Moro E, Lozano AM, Hamani C, Boulinguez P et al (2009) Stimulation of the subthalamic nucleus and impulsivity: release your horses. Ann Neurol 66 (6):817–824

Ballard K, Knutson B (2009) Dissociable neural representations of future reward magnitude and delay during temporal discounting. Neuroimage 45(1):143–150

Ballard ME, Mandelkern MA, Monterosso JR, Hsu E, Robertson CL, Ishibashi K et al (2015) Low dopamine D2/D3 receptor availability is associated with steep discounting of delayed rewards in methamphetamine dependence. Int J Neuropsychopharmacol 18(7):pyu119

Balleine BW, Dickinson A (1998) Goal-directed instrumental action: contingency and incentive learning and their cortical substrates. Neuropharmacology 37(4–5):407–419

Banca P, Vestergaard MD, Rankov V, Baek K, Mitchell S, Lapa T et al (2014) Evidence accumulation in obsessive-compulsive disorder: the role of uncertainty and monetary reward on perceptual decision-making thresholds. Neuropsychopharmacology 40(5):1192–1202

Banca P, Lange I, Worbe Y, Howell NA, Irvine M, Harrison NA et al (2015) Reflection impulsivity in binge drinking: behavioural and volumetric correlates. Addict Biol

Bari A, Robbins TW (2013) Noradrenergic versus dopaminergic modulation of impulsivity, attention and monitoring behaviour in rats performing the stop-signal task: possible relevance to ADHD. Psychopharmacology 230(1):89–111

Bari A, Eagle DM, Mar AC, Robinson ES, Robbins TW (2009) Dissociable effects of noradrenaline, dopamine, and serotonin uptake blockade on stop task performance in rats. Psychopharmacology 205(2):273–283

Bari A, Mar AC, Theobald DE, Elands SA, Oganya KC, Eagle DM et al (2011) Prefrontal and monoaminergic contributions to stop-signal task performance in rats. J Neurosci 31 (25):9254–9263

Barlow RL, Alsio J, Jupp B, Rabinovich R, Shrestha S, Roberts AC et al (2015) Markers of serotonergic function in the orbitofrontal cortex and dorsal raphe nucleus predict individual variation in spatial-discrimination serial reversal learning. Neuropsychopharmacology 40 (7):1619–1630

Basten U, Biele G, Heekeren HR, Fiebach CJ (2010) How the brain integrates costs and benefits during decision making. Proc Natl Acad Sci USA 107(50):21767–21772

Baunez C, Robbins TW (1997) Bilateral lesions of the subthalamic nucleus induce multiple deficits in an attentional task in rats. Eur J Neurosci 9(10):2086–2099

Baunez C, Nieoullon A, Amalric M (1995) In a rat model of parkinsonism, lesions of the subthalamic nucleus reverse increases of reaction time but induce a dramatic premature responding deficit. J Neurosci 15(10):6531–6541
Belin D, Mar AC, Dalley JW, Robbins TW, Everitt BJ (2008) High impulsivity predicts the switch to compulsive cocaine-taking. Science 320(5881):1352–1355
Berlin GS, Hollander E (2014) Compulsivity, impulsivity, and the DSM-5 process. CNS Spectr 19(1):62–68
Bickel WK, Koffarnus MN, Moody L, Wilson AG (2013) The behavioral- and neuro-economic process of temporal discounting: a candidate behavioral marker of addiction. Neuropharmacology 76 Pt B:518–527
Bickel WK, Koffarnus MN, Moody L, Wilson AG (2014) The behavioral- and neuro-economic process of temporal discounting: a candidate behavioral marker of addiction. Neuropharmacology. 76 Pt B:518–527
Botvinick MM, Braver TS, Barch DM, Carter CS, Cohen JD (2001) Conflict monitoring and cognitive control. Psychol Rev 108(3):624–652
Boulougouris V, Robbins TW (2010) Enhancement of spatial reversal learning by 5-HT2C receptor antagonism is neuroanatomically specific. J Neurosci 30(3):930–938
Boulougouris V, Glennon JC, Robbins TW (2008) Dissociable effects of selective 5-HT2A and 5-HT2C receptor antagonists on serial spatial reversal learning in rats. Neuropsychopharmacology 33(8):2007–2019
Brewer JA, Potenza MN (2008) The neurobiology and genetics of impulse control disorders: relationships to drug addictions. Biochem Pharmacol 75(1):63–75
Brown RT, Sleator EK (1979) Methylphenidate in hyperkinetic children: differences in dose effects on impulsive behavior. Pediatrics 64(4):408–411
Buckholtz JW, Treadway MT, Cowan RL, Woodward ND, Li R, Ansari MS et al (2010) Dopaminergic network differences in human impulsivity. Science 329(5991):532
Cai W, Oldenkamp CL, Aron AR (2011) A proactive mechanism for selective suppression of response tendencies. J Neurosci 31(16):5965–5969
Camchong J, MacDonald AW 3rd, Nelson B, Bell C, Mueller BA, Specker S et al (2011) Frontal hyperconnectivity related to discounting and reversal learning in cocaine subjects. Biol Psychiatry 69(11):1117–1123
Canales JJ, Graybiel AM (2000) Patterns of gene expression and behavior induced by chronic dopamine treatments. Ann Neurol 47(4 Suppl 1):S53–S59
Caprioli D, Sawiak SJ, Merlo E, Theobald DE, Spoelder M, Jupp B et al. (2013) Gamma aminobutyric acidergic and neuronal structural markers in the nucleus accumbens core underlie trait-like impulsive behavior. Biol Psychiatry 75:115–123
Cardinal RN, Pennicott DR, Sugathapala CL, Robbins TW, Everitt BJ (2001) Impulsive choice induced in rats by lesions of the nucleus accumbens core. Science 292(5526):2499–2501
Cardinal RN, Winstanley CA, Robbins TW, Everitt BJ (2004) Limbic corticostriatal systems and delayed reinforcement. Ann N Y Acad Sci 1021:33–50
Castane A, Theobald DE, Robbins TW (2010) Selective lesions of the dorsomedial striatum impair serial spatial reversal learning in rats. Behav Brain Res 210(1):74–83
Chamberlain SR, Del Campo N, Dowson J, Muller U, Clark L, Robbins TW et al (2007a) Atomoxetine improved response inhibition in adults with attention deficit/hyperactivity disorder. Biol Psychiatry 62(9):977–984
Chamberlain SR, Fineberg NA, Menzies LA, Blackwell AD, Bullmore ET, Robbins TW et al (2007b) Impaired cognitive flexibility and motor inhibition in unaffected first-degree relatives of patients with obsessive-compulsive disorder. Am J Psychiatry 164(2):335–338
Chambers CD, Garavan H, Bellgrove MA (2009) Insights into the neural basis of response inhibition from cognitive and clinical neuroscience. Neurosci Biobehav Rev 33(5):631–646
Chudasama Y, Robbins TW (2003) Dissociable contributions of the orbitofrontal and infralimbic cortex to pavlovian autoshaping and discrimination reversal learning: further evidence for the functional heterogeneity of the rodent frontal cortex. J Neurosci 23(25):8771–8780

Chudasama Y, Passetti F, Rhodes SE, Lopian D, Desai A, Robbins TW (2003) Dissociable aspects of performance on the 5-choice serial reaction time task following lesions of the dorsal anterior cingulate, infralimbic and orbitofrontal cortex in the rat: differential effects on selectivity, impulsivity and compulsivity. Behav Brain Res 146(1–2):105–119
Clark L, Robbins TW, Ersche KD, Sahakian BJ (2006) Reflection impulsivity in current and former substance users. Biol Psychiatry 60(5):515–522
Clarke HF, Dalley JW, Crofts HS, Robbins TW, Roberts AC (2004) Cognitive inflexibility after prefrontal serotonin depletion. Science 304(5672):878–880
Clarke HF, Walker SC, Dalley JW, Robbins TW, Roberts AC (2007) Cognitive inflexibility after prefrontal serotonin depletion is behaviorally and neurochemically specific. Cereb Cortex 17 (1):18–27
Clarke HF, Robbins TW, Roberts AC (2008) Lesions of the medial striatum in monkeys produce perseverative impairments during reversal learning similar to those produced by lesions of the orbitofrontal cortex. J Neurosci 28(43):10972–10982
Clarke HF, Hill GJ, Robbins TW, Roberts AC (2011) Dopamine, but not serotonin, regulates reversal learning in the marmoset caudate nucleus. J Neurosci 31(11):4290–4297
Clatworthy PL, Lewis SJ, Brichard L, Hong YT, Izquierdo D, Clark L et al (2009) Dopamine release in dissociable striatal subregions predicts the different effects of oral methylphenidate on reversal learning and spatial working memory. J Neurosci 29(15):4690–4696
Cole BJ, Robbins TW (1987) Amphetamine impairs the discriminative performance of rats with dorsal noradrenergic bundle lesions on a 5-choice serial reaction time task: new evidence for central dopaminergic-noradrenergic interactions. Psychopharmacology 91(4):458–466
Cole BJ, Robbins TW (1989) Effects of 6-hydroxydopamine lesions of the nucleus accumbens septi on performance of a 5-choice serial reaction time task in rats: implications for theories of selective attention and arousal. Behav Brain Res 33(2):165–179
Cools R, Clark L, Owen AM, Robbins TW (2002) Defining the neural mechanisms of probabilistic reversal learning using event-related functional magnetic resonance imaging. J Neurosci 22 (11):4563–4567
Cools R, Barker RA, Sahakian BJ, Robbins TW (2003) L-Dopa medication remediates cognitive inflexibility, but increases impulsivity in patients with Parkinson's disease. Neuropsychologia 41(11):1431–1441
Crean J, Richards JB, de Wit H (2002) Effect of tryptophan depletion on impulsive behavior in men with or without a family history of alcoholism. Behav Brain Res 136(2):349–357
Cunningham KA, Anastasio NC (2014) Serotonin at the nexus of impulsivity and cue reactivity in cocaine addiction. Neuropharmacology 6 Pt B:460–478
Dalley JW, Fryer TD, Brichard L, Robinson ES, Theobald DE, Laane K et al (2007a) Nucleus accumbens D2/3 receptors predict trait impulsivity and cocaine reinforcement. Science 315 (5816):1267–1270
Dalley JW, Laane K, Theobald DE, Pena Y, Bruce CC, Huszar AC et al (2007b) Enduring deficits in sustained visual attention during withdrawal of intravenous methylenedioxymethamphetamine self-administration in rats: results from a comparative study with d-amphetamine and methamphetamine. Neuropsychopharmacology 32(5):1195–1206
Dalley JW, Mar AC, Economidou D, Robbins TW (2008) Neurobehavioral mechanisms of impulsivity: fronto-striatal systems and functional neurochemistry. Pharmacol Biochem Behav 90(2):250–260
Dalley JW, Everitt BJ, Robbins TW (2011) Impulsivity, compulsivity, and top-down cognitive control. Neuron 69(4):680–694
Daw ND, Kakade S, Dayan P (2002) Opponent interactions between serotonin and dopamine. Neural Netw Official J Int Neural Netw Soc 15(4–6):603–616
Daw ND, O'Doherty JP, Dayan P, Seymour B, Dolan RJ (2006) Cortical substrates for exploratory decisions in humans. Nature 441(7095):876–879
Daw ND, Gershman SJ, Seymour B, Dayan P, Dolan RJ (2011) Model-based influences on humans' choices and striatal prediction errors. Neuron 69(6):1204–1215

de Ruiter MB, Veltman DJ, Goudriaan AE, Oosterlaan J, Sjoerds Z, van den Brink W (2009) Response perseveration and ventral prefrontal sensitivity to reward and punishment in male problem gamblers and smokers. Neuropsychopharmacology 34(4):1027–1038
de Wit H (2009) Impulsivity as a determinant and consequence of drug use: a review of underlying processes. Addict Biol 14(1):22–31
de Wit S, Corlett PR, Aitken MR, Dickinson A, Fletcher PC (2009) Differential engagement of the ventromedial prefrontal cortex by goal-directed and habitual behavior toward food pictures in humans. J Neurosci 29(36):11330–11338
de Wit S, Watson P, Harsay HA, Cohen MX, van de Vijver I, Ridderinkhof KR (2012a) Corticostriatal connectivity underlies individual differences in the balance between habitual and goal-directed action control. J Neurosci 32(35):12066–12075
de Wit S, Standing HR, Devito EE, Robinson OJ, Ridderinkhof KR, Robbins TW et al (2012b) Reliance on habits at the expense of goal-directed control following dopamine precursor depletion. Psychopharmacology 219(2):621–631
Delfs JM, Kelley AE (1990) The role of D1 and D2 dopamine receptors in oral stereotypy induced by dopaminergic stimulation of the ventrolateral striatum. Neuroscience 39(1):59–67
DeLong MR, Wichmann T (2015) Basal Ganglia circuits as targets for neuromodulation in parkinson disease. JAMA neurology. 2015:1–7
Denys D (2011) Obsessionality & compulsivity: a phenomenology of obsessive-compulsive disorder. Philos Ethics Humanit Med 6:3
Denys D, van der Wee N, Janssen J, De Geus F, Westenberg HG (2004) Low level of dopaminergic D2 receptor binding in obsessive-compulsive disorder. Biol Psychiatry 55 (10):1041–1045
Deserno L, Huys QJ, Boehme R, Buchert R, Heinze HJ, Grace AA et al (2015) Ventral striatal dopamine reflects behavioral and neural signatures of model-based control during sequential decision making. Proc Natl Acad Sci USA 112(5):1595–1600
Deuschl G, Schade-Brittinger C, Krack P, Volkmann J, Schafer H, Botzel K et al (2006) A randomized trial of deep-brain stimulation for Parkinson's disease. N Engl J Med 355 (9):896–908
DeVito EE, Blackwell AD, Clark L, Kent L, Dezsery AM, Turner DC et al (2009) Methylphenidate improves response inhibition but not reflection-impulsivity in children with attention deficit hyperactivity disorder (ADHD). Psychopharmacology 202(1–3):531–539
Dias R, Robbins TW, Roberts AC (1996) Dissociation in prefrontal cortex of affective and attentional shifts. Nature 380(6569):69–72
Dickinson A, Balleine BW (2002) The role of learning in the operation of motivational systems. In: Gallister CR (ed) Steven's handbook of experimental psychology: learning, motivation and emotion. vol 3, 3rd edn. Wiley , New York, pp 497–534
Diergaarde L, Pattij T, Poortvliet I, Hogenboom F, de Vries W, Schoffelmeer AN et al (2008) Impulsive choice and impulsive action predict vulnerability to distinct stages of nicotine seeking in rats. Biol Psychiatry 63(3):301–308
Diergaarde L, Pattij T, Nawijn L, Schoffelmeer AN, De Vries TJ (2009) Trait impulsivity predicts escalation of sucrose seeking and hypersensitivity to sucrose-associated stimuli. Behav Neurosci 123(4):794–803
Divac I (1972) Drug-induced syndromes in rats with large, chronic lesions in the corpus striatum. Psychopharmacologia 27(2):171–178
Djamshidian A, O'Sullivan SS, Sanotsky Y, Sharman S, Matviyenko Y, Foltynie T et al (2012) Decision making, impulsivity, and addictions: do Parkinson's disease patients jump to conclusions? Mov Disord 27(9):1137–1145
Djamshidian A, O'Sullivan SS, Foltynie T, Aviles-Olmos I, Limousin P, Noyce A et al (2013) Dopamine agonists rather than deep brain stimulation cause reflection impulsivity in Parkinson's disease. J Parkinson's Dis 3(2):139–144
Dolan RJ, Dayan P (2013) Goals and habits in the brain. Neuron 80(2):312–325
Dougherty DM, Mathias CW, Marsh DM, Moeller FG, Swann AC (2004) Suicidal behaviors and drug abuse: impulsivity and its assessment. Drug Alcohol Depend 76(Suppl):S93–S105

Dougherty DM, Mullen J, Hill-Kapturczak N, Liang Y, Karns TE, Lake SL et al (2015) Effects of tryptophan depletion and a simulated alcohol binge on impulsivity. Exp Clin Psychopharmacol 23(2):109–121

Duchesne M, Mattos P, Appolinario JC, de Freitas SR, Coutinho G, Santos C et al (2010) Assessment of executive functions in obese individuals with binge eating disorder. Rev Bras Psiquiatr 32(4):381–388

Durana JH, Barnes PA (1993) A neurodevelopmental view of impulsivity and its relationship to the superfactors of personality. In: McCown WG, Johnson JL, Shure MB (eds) The impulsive Client: Theory, Research and Treatment. American Psychological Association, Washington, D.C.

Eagle DM, Baunez C (2010) Is there an inhibitory-response-control system in the rat? Evidence from anatomical and pharmacological studies of behavioral inhibition. Neurosci Biobehav Rev 34(1):50–72

Eagle DM, Robbins TW (2003) Inhibitory control in rats performing a stop-signal reaction-time task: effects of lesions of the medial striatum and d-amphetamine. Behav Neurosci 117 (6):1302–1317

Eagle DM, Tufft MR, Goodchild HL, Robbins TW (2007) Differential effects of modafinil and methylphenidate on stop-signal reaction time task performance in the rat, and interactions with the dopamine receptor antagonist cis-flupenthixol. Psychopharmacology 192(2):193–206

Eagle DM, Bari A, Robbins TW (2008) The neuropsychopharmacology of action inhibition: cross-species translation of the stop-signal and go/no-go tasks. Psychopharmacology 199 (3):439–456

Eagle DM, Lehmann O, Theobald DE, Pena Y, Zakaria R, Ghosh R et al (2009) Serotonin depletion impairs waiting but not stop-signal reaction time in rats: implications for theories of the role of 5-HT in behavioral inhibition. Neuropsychopharmacology 34(5):1311–1321

Economidou D, Pelloux Y, Robbins TW, Dalley JW, Everitt BJ (2009) High impulsivity predicts relapse to cocaine-seeking after punishment-induced abstinence. Biol Psychiatry 65 (10):851–856

Economidou D, Theobald DE, Robbins TW, Everitt BJ, Dalley JW (2012) Norepinephrine and dopamine modulate impulsivity on the five-choice serial reaction time task through opponent actions in the shell and core sub-regions of the nucleus accumbens. Neuropsychopharmacology 37(9):2057–2066

Ersche KD, Roiser JP, Robbins TW, Sahakian BJ (2008) Chronic cocaine but not chronic amphetamine use is associated with perseverative responding in humans. Psychopharmacology 197(3):421–431

Ersche KD, Bullmore ET, Craig KJ, Shabbir SS, Abbott S, Muller U et al (2010) Influence of compulsivity of drug abuse on dopaminergic modulation of attentional bias in stimulant dependence. Arch Gen Psychiatry 67(6):632–644

Ersche KD, Jones PS, Williams GB, Turton AJ, Robbins TW, Bullmore ET (2012) Abnormal brain structure implicated in stimulant drug addiction. Science 335(6068):601–604

Eskenazi D, Neumaier JF (2011) Increased expression of 5-HT(6) receptors in dorsolateral striatum decreases habitual lever pressing, but does not affect learning acquisition of simple operant tasks in rats. Eur J Neurosci 34(2):343–351

Evans AH, Katzenschlager R, Paviour D, O'Sullivan JD, Appel S, Lawrence AD et al (2004) Punding in Parkinson's disease: its relation to the dopamine dysregulation syndrome. Mov Disord 19(4):397–405

Evenden JL (1999) Varieties of impulsivity. Psychopharmacology 146(4):348–361

Faure A, Haberland U, Conde F, El Massioui N (2005) Lesion to the nigrostriatal dopamine system disrupts stimulus-response habit formation. J Neurosci 25(11):2771–2780

Fellows LK, Farah MJ (2003) Ventromedial frontal cortex mediates affective shifting in humans: evidence from a reversal learning paradigm. Brain 126(Pt 8):1830–1837

Feola TW, de Wit H, Richards JB (2000) Effects of d-amphetamine and alcohol on a measure of behavioral inhibition in rats. Behav Neurosci 114(4):838–848

Fernandez-Serrano MJ, Perales JC, Moreno-Lopez L, Perez-Garcia M, Verdejo-Garcia A (2012) Neuropsychological profiling of impulsivity and compulsivity in cocaine dependent individuals. Psychopharmacology 219(2):673–683
Ferster CB, Skinner BF (1957) Schedules of reinforcement. Appleton-Century-Crofts, New York
Fillmore MT, Rush CR (2006) Polydrug abusers display impaired discrimination-reversal learning in a model of behavioural control. J Psychopharmacol 20(1):24–32
Fineberg NA, Potenza MN, Chamberlain SR, Berlin HA, Menzies L, Bechara A et al (2010) Probing compulsive and impulsive behaviors, from animal models to endophenotypes: a narrative review. Neuropsychopharmacology 35(3):591–604
Floresco SB, Tse MT, Ghods-Sharifi S (2008) Dopaminergic and glutamatergic regulation of effort- and delay-based decision making. Neuropsychopharmacology 33(8):1966–1979
Fog R (1972) On stereotypy and catalepsy: studies on the effect of amphetamines and neuroleptics in rats. Acta Neurol Scand Suppl 50:3–66
Fonseca MS, Murakami M, Mainen ZF (2015) Activation of dorsal raphe serotonergic neurons promotes waiting but is not reinforcing. Curr Biol 25(3):306–315
Fortier CB, Steffen EM, Lafleche G, Venne JR, Disterhoft JF, McGlinchey RE (2008) Delay discrimination and reversal eyeblink classical conditioning in abstinent chronic alcoholics. Neuropsychology 22(2):196–208
Frank MJ (2006) Hold your horses: a dynamic computational role for the subthalamic nucleus in decision making. Neural Netw Official J Int Neural Network Soc 19(8):1120–1136
Frank MJ, Samanta J, Moustafa AA, Sherman SJ (2007) Hold your horses: impulsivity, deep brain stimulation, and medication in parkinsonism. Science 318(5854):1309–1312
Friedel E, Koch SP, Wendt J, Heinz A, Deserno L, Schlagenhauf F (2014) Devaluation and sequential decisions: linking goal-directed and model-based behavior. Front Hum Neurosci 8:587
Furl N, Averbeck BB (2011) Parietal cortex and insula relate to evidence seeking relevant to reward-related decisions. J Neurosci 31(48):17572–17582
Gerfen CR (1992) The neostriatal mosaic: multiple levels of compartmental organization. J Neural Transm Suppl 36:43–59
Gillan CM, Papmeyer M, Morein-Zamir S, Sahakian BJ, Fineberg NA, Robbins TW et al (2011) Disruption in the balance between goal-directed behavior and habit learning in obsessive-compulsive disorder. Am J Psychiatry 168(7):718–726
Gillan CM, Morein-Zamir S, Urcelay GP, Sule A, Voon V, Apergis-Schoute AM et al (2013) Enhanced avoidance habits in obsessive-compulsive disorder. Biol Psychiatry
Glascher J, Daw N, Dayan P, O'Doherty JP (2010) States versus rewards: dissociable neural prediction error signals underlying model-based and model-free reinforcement learning. Neuron 66(4):585–595
Gold JI, Shadlen MN (2007) The neural basis of decision making. Ann Rev Neurosci 30:535–574
Gottesman II, Gould TD (2003) The endophenotype concept in psychiatry: etymology and strategic intentions. Am J Psychiatry 160(4):636–645
Goudriaan AE, Oosterlaan J, de Beurs E, van den Brink W (2006) Neurocognitive functions in pathological gambling: a comparison with alcohol dependence, Tourette syndrome and normal controls. Addiction 101(4):534–547
Graf H, Abler B, Freudenmann R, Beschoner P, Schaeffeler E, Spitzer M et al (2011) Neural correlates of error monitoring modulated by atomoxetine in healthy volunteers. Biol Psychiatry 69(9):890–897
Grant DA, Berg E (1948) A behavioural analysis of degree of reinforcement and ease of shifting to new responses in Weigl-type card-sorting problem. J Exp Psychol 38:404–411
Grant JE, Odlaug BL, Chamberlain SR (2011) Neurocognitive response to deep brain stimulation for obsessive-compulsive disorder: a case report. Am J Psychiatry 168(12):1338–1339
Gregorios-Pippas L, Tobler PN, Schultz W (2009) Short-term temporal discounting of reward value in human ventral striatum. J Neurophysiol 101(3):1507–1523

Gubner NR, Wilhelm CJ, Phillips TJ, Mitchell SH (2010) Strain differences in behavioral inhibition in a Go/No-go task demonstrated using 15 inbred mouse strains. Alcohol Clin Exp Res 34(8):1353–1362
Haber SN, Fudge JL, McFarland NR (2000) Striatonigrostriatal pathways in primates form an ascending spiral from the shell to the dorsolateral striatum. J Neurosci 20(6):2369–2382
Hampshire A, Chaudhry AM, Owen AM, Roberts AC (2012) Dissociable roles for lateral orbitofrontal cortex and lateral prefrontal cortex during preference driven reversal learning. NeuroImage 59(4):4102–4112
Hayes DJ, Jupp B, Sawiak SJ, Merlo E, Caprioli D, Dalley JW (2014) Brain gamma-aminobutyric acid: a neglected role in impulsivity. Eur J Neurosci 39(11):1921–1932
Higgins GA, Silenieks LB, Rossmann A, Rizos Z, Noble K, Soko AD et al (2012) The 5-HT2C receptor agonist lorcaserin reduces nicotine self-administration, discrimination, and reinstatement: relationship to feeding behavior and impulse control. Neuropsychopharmacology 37 (5):1177–1191
Hornak J, O'Doherty J, Bramham J, Rolls ET, Morris RG, Bullock PR et al (2004) Reward-related reversal learning after surgical excisions in orbito-frontal or dorsolateral prefrontal cortex in humans. J Cogn Neurosci 16(3):463–478
Howell LL, Cunningham KA (2015) Serotonin 5-HT2 receptor interactions with dopamine function: implications for therapeutics in cocaine use disorder. Pharmacol Rev 67(1):176–197
Huddy VC, Clark L, Harrison I, Ron MA, Moutoussis M, Barnes TR et al (2013) Reflection impulsivity and response inhibition in first-episode psychosis: relationship to cannabis use. Psychol Med 43(10):2097–2107
Huettel SA, Song AW, McCarthy G (2005) Decisions under uncertainty: probabilistic context influences activation of prefrontal and parietal cortices. J Neurosci 25(13):3304–3311
Insel T, Cuthbert B, Garvey M, Heinssen R, Pine DS, Quinn K et al (2010) Research domain criteria (RDoC): toward a new classification framework for research on mental disorders. Am J Psychiatry 167(7):748–751
Iversen SD, Mishkin M (1970) Perseverative interference in monkeys following selective lesions of the inferior prefrontal convexity. Exp Brain Res 11(4):376–386
Jaffard M, Longcamp M, Velay JL, Anton JL, Roth M, Nazarian B et al (2008) Proactive inhibitory control of movement assessed by event-related fMRI. Neuroimage 42(3):1196–1206
Jocham G, Klein TA, Neumann J, von Cramon DY, Reuter M, Ullsperger M (2009) Dopamine DRD2 polymorphism alters reversal learning and associated neural activity. J Neurosci 29 (12):3695–3704
Johanson CE, Frey KA, Lundahl LH, Keenan P, Lockhart N, Roll J et al (2006) Cognitive function and nigrostriatal markers in abstinent methamphetamine abusers. Psychopharmacology 185 (3):327–338
Johnson PM, Kenny PJ (2010) Dopamine D2 receptors in addiction-like reward dysfunction and compulsive eating in obese rats. Nat Neurosci 13(5):635–641
Joutsa J, Voon V, Johansson J, Niemela S, Bergman J, Kaasinen V (2015) Dopaminergic function and intertemporal choice. Transl Psychiatry 5:e491
Kable JW, Glimcher PW (2007) The neural correlates of subjective value during intertemporal choice. Nat Neurosci 10(12):1625–1633
Kagan J (1966) Reflection–impulsivity: the generality and dynamics of conceptual tempo. J Abnorm Psychol 71(1):17–24
Kaladjian A, Jeanningros R, Azorin JM, Anton JL, Mazzola-Pomietto P (2011) Impulsivity and neural correlates of response inhibition in schizophrenia. Psychol Med 41(2):291–299
Kas MJ, Glennon JC, Buitelaar J, Ey E, Biemans B, Crawley J et al (2014) Assessing behavioural and cognitive domains of autism spectrum disorders in rodents: current status and future perspectives. Psychopharmacology 231(6):1125–1146
Kayser AS, Allen DC, Navarro-Cebrian A, Mitchell JM, Fields HL (2012) Dopamine, corticostriatal connectivity, and intertemporal choice. J Neurosci 32(27):9402–9409

Kelly PH, Seviour PW, Iversen SD (1975) Amphetamine and apomorphine responses in the rat following 6-OHDA lesions of the nucleus accumbens septi and corpus striatum. Brain Res 94 (3):507–522
Keramati M, Dezfouli A, Piray P (2011) Speed/accuracy trade-off between the habitual and the goal-directed processes. PLoS Comput Biol 7(5):e1002055
Kiani R, Shadlen MN (2009) Representation of confidence associated with a decision by neurons in the parietal cortex. Science 324(5928):759–764
Killcross S, Coutureau E (2003) Coordination of actions and habits in the medial prefrontal cortex of rats. Cereb Cortex 13(4):400–408
Kobayashi S, Schultz W (2008) Influence of reward delays on responses of dopamine neurons. J Neurosci 28(31):7837–7846
Koob GF, Le Moal M (2008) Review. Neurobiological mechanisms for opponent motivational processes in addiction. Philos Trans R Soc Lond B Biol Sci 363(1507):3113–3123
Kreipke CW, Walker PD (2004) NMDA receptor blockade attenuates locomotion elicited by intrastriatal dopamine D1-receptor stimulation. Synapse 53(1):28–35
LaHoste GJ, Marshall JF (1993) New concepts in dopamine receptor plasticity. Ann N Y Acad Sci 702:183–196
Langen M, Kas MJ, Staal WG, van Engeland H, Durston S (2011) The neurobiology of repetitive behavior: of mice. Neurosci Biobehav Rev 35(3):345–355
Lawrence AB, Terlouw EM (1993) A review of behavioral factors involved in the development and continued performance of stereotypic behaviors in pigs. J Anim Sci 71(10):2815–2825
Lee B, Groman S, London ED, Jentsch JD (2007) Dopamine D2/D3 receptors play a specific role in the reversal of a learned visual discrimination in monkeys. Neuropsychopharmacology 32 (10):2125–2134
Lennox DB, Miltenberger RG, Donnelly DR (1987) Response interruption and DRL for the reduction of rapid eating. J Appl Behav Anal 20(3):279–284
Li CS, Morgan PT, Matuskey D, Abdelghany O, Luo X, Chang JL et al (2010) Biological markers of the effects of intravenous methylphenidate on improving inhibitory control in cocaine-dependent patients. Proc Natl Acad Sci USA 107(32):14455–14459
Liljeholm M, Tricomi E, O'Doherty JP, Balleine BW (2011) Neural correlates of instrumental contingency learning: differential effects of action-reward conjunction and disjunction. J Neurosci 31(7):2474–2480
Lindgren HS, Wickens R, Tait DS, Brown VJ, Dunnett SB (2013) Lesions of the dorsomedial striatum impair formation of attentional set in rats. Neuropharmacology 71:148–153
Lipszyc J, Schachar R (2010) Inhibitory control and psychopathology: a meta-analysis of studies using the stop signal task. J Int Neuropsychol Soc 16(6):1064–1076
Logan GD, Cowan WB, Davis KA (1984) On the ability to inhibit simple and choice reaction time responses: a model and a method. J Exp Psychol Hum Percept Perform 10(2):276–291
Lombardo LE, Bearden CE, Barrett J, Brumbaugh MS, Pittman B, Frangou S et al (2012) Trait impulsivity as an endophenotype for bipolar I disorder. Bipolar Disord 14(5):565–570
Majid DS, Cai W, Corey-Bloom J, Aron AR (2013) Proactive selective response suppression is implemented via the basal ganglia. J Neurosci 33(33):13259–13269
Mallet L, Polosan M, Jaafari N, Baup N, Welter ML, Fontaine D et al (2008) Subthalamic nucleus stimulation in severe obsessive-compulsive disorder. N Engl J Med 359(20):2121–2134
Mar AC, Walker AL, Theobald DE, Eagle DM, Robbins TW (2011) Dissociable effects of lesions to orbitofrontal cortex subregions on impulsive choice in the rat. J Neurosci 31(17):6398–6404
Mazur JE (1987) The effect of delayed and intervening events on reinforcement value. In: Commons ML, Mazur JE, Nevin JA, Rachlin H (eds) An adjustment procedure for studying delayed reinforcement. Erlbaum, Hillsdale, NJ
McClure SM, Laibson DI, Loewenstein G, Cohen JD (2004) Separate neural systems value immediate and delayed monetary rewards. Science 306(5695):503–507
McClure SM, Ericson KM, Laibson DI, Loewenstein G, Cohen JD (2007) Time discounting for primary rewards. J Neurosci 27(21):5796–5804

Menzies L, Achard S, Chamberlain SR, Fineberg N, Chen CH, del Campo N et al (2007) Neurocognitive endophenotypes of obsessive-compulsive disorder. Brain 130(Pt 12):3223–3236

Millan MJ, Goodwin GM, Meyer-Lindenberg A (2015) Ove Ogren S. Learning from the past and looking to the future: emerging perspectives for improving the treatment of psychiatric disorders. Eur Neuropsychopharmacol 25(5):599–656

Milstein JA, Dalley JW, Robbins TW (2010) Methylphenidate-induced impulsivity: pharmacological antagonism by beta-adrenoreceptor blockade. J Psychopharmacol 24(3):309–321

Mitchell SH, Reeves JM, Li N, Phillips TJ (2006) Delay discounting predicts behavioral sensitization to ethanol in outbred WSC mice. Alcohol Clin Exp Res 30(3):429–437

Miyasaki JM (2007) Al Hassan K, Lang AE, Voon V. Punding prevalence in Parkinson's disease. Mov Disord 22(8):1179–1181

Miyazaki K, Miyazaki KW, Doya K (2011) Activation of dorsal raphe serotonin neurons underlies waiting for delayed rewards. J Neurosci 31(2):469–479

Mobini S, Chiang TJ, Ho MY, Bradshaw CM, Szabadi E (2000) Effects of central 5-hydroxytryptamine depletion on sensitivity to delayed and probabilistic reinforcement. Psychopharmacology 152(4):390–397

Mole TB, Irvine MA, Worbe Y, Collins P, Mitchell SP, Bolton S, et al. (2014) Impulsivity in disorders of food and drug misuse. Psychol Med 2014:1–12

Moreno M, Economidou D, Mar AC, Lopez-Granero C, Caprioli D, Theobald DE et al (2013) Divergent effects of D(2)/(3) receptor activation in the nucleus accumbens core and shell on impulsivity and locomotor activity in high and low impulsive rats. Psychopharmacology 228 (1):19–30

Morgan MJ, Impallomeni LC, Pirona A, Rogers RD (2006) Elevated impulsivity and impaired decision-making in abstinent Ecstasy (MDMA) users compared to polydrug and drug-naive controls. Neuropsychopharmacology 31(7):1562–1573

Morris LS, Kundu P, Baek K, Irvine MA, Mechelmans DJ, Wood J et al (2015) Jumping the gun: mapping neural correlates of waiting impulsivity and relevance across alcohol misuse. Biol Psychiatry doi:10.1016/j.biopsych.2015.06.009

Muller UJ, Voges J, Steiner J, Galazky I, Heinze HJ, Moller M et al (2013) Deep brain stimulation of the nucleus accumbens for the treatment of addiction. Ann N Y Acad Sci 1282:119–128

Nandam LS, Hester R, Wagner J, Cummins TD, Garner K, Dean AJ et al (2011) Methylphenidate but not atomoxetine or citalopram modulates inhibitory control and response time variability. Biol Psychiatry 69(9):902–904

Navarra R, Graf R, Huang Y, Logue S, Comery T, Hughes Z et al (2008) Effects of atomoxetine and methylphenidate on attention and impulsivity in the 5-choice serial reaction time test. Prog Neuropsychopharmacol Biol Psychiatry 32(1):34–41

Nelson A, Killcross S (2006) Amphetamine exposure enhances habit formation. J Neurosci 26 (14):3805–3812

Nelson AJ, Killcross S (2013) Accelerated habit formation following amphetamine exposure is reversed by D1, but enhanced by D2, receptor antagonists. Front Neurosci 7:76

Nigg JT, Wong MM, Martel MM, Jester JM, Puttler LI, Glass JM et al (2006) Poor response inhibition as a predictor of problem drinking and illicit drug use in adolescents at risk for alcoholism and other substance use disorders. J Am Acad Child Adolesc Psychiatry 45 (4):468–475

Nilsson SR, Alsio J, Somerville EM, Clifton PG (2015) The rat's not for turning: dissociating the psychological components of cognitive inflexibility. Neurosci Biobehav Rev 56:1–14

Niv Y, Daw ND, Joel D, Dayan P (2007) Tonic dopamine: opportunity costs and the control of response vigor. Psychopharmacology 191(3):507–520

Noreika V, Falter CM, Rubia K (2013) Timing deficits in attention-deficit/hyperactivity disorder (ADHD): evidence from neurocognitive and neuroimaging studies. Neuropsychologia 51 (2):235–266

Nowakowska K, Jablkowska K, Borkowska A (2007) Cognitive dysfunctions in patients with alcohol dependence. Psychiatr Pol 41(5):693–702

Oberlin BG, Albrecht DS, Herring CM, Walters JW, Hile KL, Kareken DA et al (2015) Monetary discounting and ventral striatal dopamine receptor availability in nontreatment-seeking alcoholics and social drinkers. Psychopharmacology 232(12):2207–2216
Obeso I, Wilkinson L, Rodriguez-Oroz MC, Obeso JA, Jahanshahi M (2013) Bilateral stimulation of the subthalamic nucleus has differential effects on reactive and proactive inhibition and conflict-induced slowing in Parkinson's disease. Exp Brain Res 226(3):451–462
Oliveira FT, McDonald JJ, Goodman D (2007) Performance monitoring in the anterior cingulate is not all error related: expectancy deviation and the representation of action-outcome associations. J Cogn Neurosci 19(12):1994–2004
Oliver YP, Ripley TL, Stephens DN (2009) Ethanol effects on impulsivity in two mouse strains: similarities to diazepam and ketamine. Psychopharmacology 204(4):679–692
O'Neill M, Brown VJ (2007) The effect of striatal dopamine depletion and the adenosine A2A antagonist KW-6002 on reversal learning in rats. Neurobiol Learn Mem 88(1):75–81
Ornstein TJ, Iddon JL, Baldacchino AM, Sahakian BJ, London M, Everitt BJ et al (2000) Profiles of cognitive dysfunction in chronic amphetamine and heroin abusers. Neuropsychopharmacology 23(2):113–126
Otto AR, Gershman SJ, Markman AB, Daw ND (2013) The curse of planning: dissecting multiple reinforcement-learning systems by taxing the central executive. Psychol Sci 24(5):751–761
Parker MO, Brock AJ, Sudwarts A, Brennan CH (2014) Atomoxetine reduces anticipatory responding in a 5-choice serial reaction time task for adult zebrafish. Psychopharmacology 231 (13):2671–2679
Patterson JC 2nd, Holland J, Middleton R (2006) Neuropsychological performance, impulsivity, and comorbid psychiatric illness in patients with pathological gambling undergoing treatment at the CORE inpatient treatment center. South Med J 99(1):36–43
Pattij T, Janssen MC, Vanderschuren LJ, Schoffelmeer AN, van Gaalen MM (2007) Involvement of dopamine D1 and D2 receptors in the nucleus accumbens core and shell in inhibitory response control. Psychopharmacology 191(3):587–598
Pattij T, Schetters D, Janssen MC, Wiskerke J, Schoffelmeer AN (2009) Acute effects of morphine on distinct forms of impulsive behavior in rats. Psychopharmacology 205(3):489–502
Pelloux Y, Dilleen R, Economidou D, Theobald D, Everitt BJ (2012) Reduced forebrain serotonin transmission is causally involved in the development of compulsive cocaine seeking in rats. Neuropsychopharmacology 37(11):2505–2514
Pena-Oliver Y, Giuliano C, Economidou D, Goodlett CR, Robbins TW, Dalley JW et al (2015) Alcohol-Preferring rats show goal oriented behaviour to food incentives but are neither sign-trackers nor impulsive. PLoS ONE 10(6):e0131016
Perry JC, Korner AC (2011) Impulsive phenomena, the impulsive character (der Triebhafte Charakter) and DSM personality disorders. J Pers Disord 25(5):586–606
Perry JL, Larson EB, German JP, Madden GJ, Carroll ME (2005) Impulsivity (delay discounting) as a predictor of acquisition of IV cocaine self-administration in female rats. Psychopharmacology 178(2–3):193–201
Perry JL, Nelson SE, Carroll ME (2008) Impulsive choice as a predictor of acquisition of IV cocaine self- administration and reinstatement of cocaine-seeking behavior in male and female rats. Exp Clin Psychopharmacol 16(2):165–177
Pine A, Shiner T, Seymour B, Dolan RJ (2010) Dopamine, time, and impulsivity in humans. J Neurosci 30(26):8888–8896
Pothiyil DI, Alex J (2013) Self-regulation and set-shifting in alcohol dependence syndrome. Ind J Appl Res III(III)
Poulos CX, Le AD, Parker JL (1995) Impulsivity predicts individual susceptibility to high levels of alcohol self-administration. Behav Pharmacol 6(8):810–814
Quednow BB, Kuhn KU, Hoppe C, Westheide J, Maier W, Daum I et al (2007) Elevated impulsivity and impaired decision-making cognition in heavy users of MDMA ("Ecstasy"). Psychopharmacology 189(4):517–530

Ragsdale CW Jr, Graybiel AM (1990) A simple ordering of neocortical areas established by the compartmental organization of their striatal projections. Proc Natl Acad Sci USA 87(16):6196–6199
Reynolds B, Schiffbauer R (2004) Measuring state changes in human delay discounting: an experiential discounting task. Behav Processes 67(3):343–356
Richards JB, Sabol KE, de Wit H (1999) Effects of methamphetamine on the adjusting amount procedure, a model of impulsive behavior in rats. Psychopharmacology 146(4):432–439
Robbins TW (2000) Chemical neuromodulation of frontal-executive functions in humans and other animals. Exp Brain Res 133(1):130–138
Robbins TW (2002) The 5-choice serial reaction time task: behavioural pharmacology and functional neurochemistry. Psychopharmacology 163(3–4):362–380
Robbins TW, Gillan CM, Smith DG, de Wit S, Ersche KD (2012) Neurocognitive endophenotypes of impulsivity and compulsivity: towards dimensional psychiatry. Trends Cogn Sci. 16(1):81–91
Robinson ES, Eagle DM, Mar AC, Bari A, Banerjee G, Jiang X et al (2008a) Similar effects of the selective noradrenaline reuptake inhibitor atomoxetine on three distinct forms of impulsivity in the rat. Neuropsychopharmacology 33(5):1028–1037
Robinson ES, Dalley JW, Theobald DE, Glennon JC, Pezze MA, Murphy ER et al (2008b) Opposing roles for 5-HT2A and 5-HT2C receptors in the nucleus accumbens on inhibitory response control in the 5-choice serial reaction time task. Neuropsychopharmacology 33 (10):2398–2406
Robinson ES, Eagle DM, Economidou D, Theobald DE, Mar AC, Murphy ER et al (2009) Behavioural characterisation of high impulsivity on the 5-choice serial reaction time task: specific deficits in 'waiting' versus 'stopping'. Behav Brain Res 196(2):310–316
Roesch MR, Taylor AR, Schoenbaum G (2006) Encoding of time-discounted rewards in orbitofrontal cortex is independent of value representation. Neuron 51(4):509–520
Roesch MR, Takahashi Y, Gugsa N, Bissonette GB, Schoenbaum G (2007) Previous cocaine exposure makes rats hypersensitive to both delay and reward magnitude. J Neurosci 27 (1):245–250
Rogers RD, Andrews TC, Grasby PM, Brooks DJ, Robbins TW (2000) Contrasting cortical and subcortical activations produced by attentional-set shifting and reversal learning in humans. J Cogn Neurosci 12(1):142–162
Rubio G, Jimenez M, Rodriguez-Jimenez R, Martinez I, Avila C, Ferre F et al (2008) The role of behavioral impulsivity in the development of alcohol dependence: a 4-year follow-up study. Alcohol Clin Exp Res 32(9):1681–1687
Rylander G (1972) Psychoses and the punding and choreiform syndromes in addiction to central stimulant drugs. Psychiatr Neurol Neurochir 75(3):203–212
Saddoris MP, Sugam JA, Stuber GD, Witten IB, Deisseroth K, Carelli RM (2015) Mesolimbic dopamine dynamically tracks, and is causally linked to, discrete aspects of value-based decision making. Biol Psychiatry 77(10):903–911
Sanchez-Roige S, Baro V, Trick L, Pena-Oliver Y, Stephens DN, Duka T (2014) Exaggerated waiting impulsivity associated with human binge drinking, and high alcohol consumption in mice. Neuropsychopharmacology 39:2919–2927
Sanchez-Roige S, Baro V, Trick L, Pena-Oliver Y, Stephens DN, Duka T (2014b) Exaggerated waiting impulsivity associated with human binge drinking, and high alcohol consumption in mice. Neuropsychopharmacology 39(13):2919–2927
Scheffers MK, Coles MG (2000) Performance monitoring in a confusing world: error-related brain activity, judgments of response accuracy, and types of errors. J Exp Psychol Hum Percept Perform 26(1):141–151
Schel MA, Kuhn S, Brass M, Haggard P, Ridderinkhof KR, Crone EA (2014) Neural correlates of intentional and stimulus-driven inhibition: a comparison. Front Hum Neurosci 8:27
Schoenbaum G, Nugent SL, Saddoris MP, Setlow B (2002) Orbitofrontal lesions in rats impair reversal but not acquisition of go, no-go odor discriminations. NeuroReport 13(6):885–890

Schweighofer N, Bertin M, Shishida K, Okamoto Y, Tanaka SC, Yamawaki S et al (2008) Low-serotonin levels increase delayed reward discounting in humans. J Neurosci 28(17):4528–4532

Sebold M, Deserno L, Nebe S, Schad DJ, Garbusow M, Hagele C et al (2014) Model-based and model-free decisions in alcohol dependence. Neuropsychobiology 70(2):122–131

Sellitto M, Ciaramelli E, di Pellegrino G (2010) Myopic discounting of future rewards after medial orbitofrontal damage in humans. J Neurosci 30(49):16429–16436

Sesia T, Temel Y, Lim LW, Blokland A, Steinbusch HW, Visser-Vandewalle V (2008) Deep brain stimulation of the nucleus accumbens core and shell: opposite effects on impulsive action. Exp Neurol 214(1):135–139

Sesia T, Bulthuis V, Tan S, Lim LW, Vlamings R, Blokland A et al (2010) Deep brain stimulation of the nucleus accumbens shell increases impulsive behavior and tissue levels of dopamine and serotonin. Exp Neurol 225(2):302–309

Simon NW, Mendez IA, Setlow B (2007) Cocaine exposure causes long-term increases in impulsive choice. Behav Neurosci 121(3):543–549

Sjoerds Z, de Wit S, van den Brink W, Robbins TW, Beekman AT, Penninx BW et al (2013) Behavioral and neuroimaging evidence for overreliance on habit learning in alcohol-dependent patients. Trans Psychiatry 3:e337

Smith JL, Mattick RP, Jamadar SD, Iredale JM (2014) Deficits in behavioural inhibition in substance abuse and addiction: a meta-analysis. Drug Alcohol Depend 145:1–33

Smittenaar P, Guitart-Masip M, Lutti A, Dolan RJ (2013a) Preparing for selective inhibition within frontostriatal loops. J Neurosci 33(46):18087–18097

Smittenaar P, FitzGerald TH, Romei V, Wright ND, Dolan RJ (2013b) Disruption of dorsolateral prefrontal cortex decreases model-based in favor of model-free control in humans. Neuron 80 (4):914–919

Sonuga-Barke EJ, Taylor E, Sembi S, Smith J (1992) Hyperactivity and delay aversion-I. The effect of delay on choice. J Child Psychol Psychiatry 33(2):387–398

Stein DJ, Hollander E (1995) Obsessive-compulsive spectrum disorders. J Clin Psychiatry 56 (6):265–266

Stern ER, Gonzalez R, Welsh RC, Taylor SF (2010) Updating beliefs for a decision: neural correlates of uncertainty and underconfidence. J Neurosci 30(23):8032–8041

Sun H, Cocker PJ, Zeeb FD, Winstanley CA (2012) Chronic atomoxetine treatment during adolescence decreases impulsive choice, but not impulsive action, in adult rats and alters markers of synaptic plasticity in the orbitofrontal cortex. Psychopharmacology 219(2):285–301

Szechtman H, Eckert MJ, Tse WS, Boersma JT, Bonura CA, McClelland JZ et al (2001) Compulsive checking behavior of quinpirole-sensitized rats as an animal model of Obsessive-Compulsive Disorder(OCD): form and control. BMC Neurosci 2:4

Tait DS, Chase EA, Brown VJ (2014) Attentional set-shifting in rodents: a review of behavioural methods and pharmacological results. Curr Pharm Des 20(31):5046–5059

Tanaka SC, Doya K, Okada G, Ueda K, Okamoto Y, Yamawaki S (2004) Prediction of immediate and future rewards differentially recruits cortico-basal ganglia loops. Nat Neurosci 7 (8):887–893

Tanaka SC, Balleine BW, O'Doherty JP (2008) Calculating consequences: brain systems that encode the causal effects of actions. J Neurosci 28(26):6750–6755

Tannock R, Schachar RJ, Carr RP, Chajczyk D, Logan GD (1989) Effects of methylphenidate on inhibitory control in hyperactive children. J Abnorm Child Psychol 17(5):473–491

Tarter RE (1973) An analysis of cognitive deficits in chronic alcoholics. J Nerv Ment Dis 157 (2):138–147

Tomlinson A, Grayson B, Marsh S, Harte MK, Barnes SA, Marshall KM et al (2014) Pay attention to impulsivity: modelling low attentive and high impulsive subtypes of adult ADHD in the 5-choice continuous performance task (5C-CPT) in female rats. Eur Neuropsychopharmacol 24 (8):1371–1380

Toomey R, Lyons MJ, Eisen SA, Xian H, Chantarujikapong S, Seidman LJ et al (2003) A twin study of the neuropsychological consequences of stimulant abuse. Arch Gen Psychiatry 60 (3):303–310

Townshend JM, Kambouropoulos N, Griffin A, Hunt FJ, Milani RM (2014) Binge drinking, reflection impulsivity, and unplanned sexual behavior: impaired decision-making in young social drinkers. Alcohol Clin Exp Res 38(4):1143–1150

Tricomi E, Balleine BW, O'Doherty JP (2009) A specific role for posterior dorsolateral striatum in human habit learning. Eur J Neurosci 29(11):2225–2232

Valentin VV, Dickinson A, O'Doherty JP (2007) Determining the neural substrates of goal-directed learning in the human brain. J Neurosci 27(15):4019–4026

van den Hout MA, Engelhard IM, Smeets M, Dek EC, Turksma K, Saric R (2009) Uncertainty about perception and dissociation after compulsive-like staring: time course of effects. Behav Res Ther 47(6):535–539

van Gaalen MM, van Koten R, Schoffelmeer AN, Vanderschuren LJ (2006) Critical involvement of dopaminergic neurotransmission in impulsive decision making. Biol Psychiatry 60(1):66–73

van Holstein M, Aarts E, van der Schaaf ME, Geurts DE, Verkes RJ, Franke B et al (2011) Human cognitive flexibility depends on dopamine D2 receptor signaling. Psychopharmacology 218 (3):567–578

Vanes LD, van Holst RJ, Jansen JM, van den Brink W, Oosterlaan J, Goudriaan AE (2014) Contingency learning in alcohol dependence and pathological gambling: learning and unlearning reward contingencies. Alcohol Clin Exp Res 38(6):1602–1610

Verbruggen F, Logan GD (2009) Models of response inhibition in the stop-signal and stop-change paradigms. Neurosci Biobehav Rev 33(5):647–661

Verdejo-Garcia A, Lawrence AJ, Clark L (2008) Impulsivity as a vulnerability marker for substance-use disorders: review of findings from high-risk research, problem gamblers and genetic association studies. Neurosci Biobehav Rev 32(4):777–810

Volans PJ (1976) Styles of decision-making and probability appraisal in selected obsessional and phobic patients. Br J Soc Clin Psychol 15(3):305–317

Vollstadt-Klein S, Wichert S, Rabinstein J, Buhler M, Klein O, Ende G et al (2010) Initial, habitual and compulsive alcohol use is characterized by a shift of cue processing from ventral to dorsal striatum. Addiction 105(10):1741–1749

Voon V (2004) Repetition, repetition, and repetition: compulsive and punding behaviors in Parkinson's disease. Mov Disord 19(4):367–370

Voon V (2014) Models of impulsivity with a focus on waiting impulsivity: translational potential for neuropsychiatric disorders. Current Addict Rep 1(4):281–288

Voon V (2015) Cognitive biases in binge eating disorder: the hijacking of decision making. CNS Spectrums

Voon V, Schoerling A, Wenzel S, Ekanayake V, Reiff J, Trenkwalder C et al (2011) Frequency of impulse control behaviours associated with dopaminergic therapy in restless legs syndrome. BMC Neurol 11:117

Voon V, Irvine MA, Derbyshire K, Worbe Y, Lange I, Abbott S et al (2014) Measuring "waiting" impulsivity in substance addictions and binge eating disorder in a novel analogue of rodent serial reaction time task. Biol Psychiatry 75(2):148–155

Voon V, Chang-Webb YC, Morris LS, Cooper E, Sethi A, Baek K, et al (2015) Waiting impulsivity: the influence of acute methylphenidate and feedback. Int J Neuropsychopharmacol doi:10.1093/ijnp/pyv074

Voon V, Derbyshire K, Ruck C, Irvine MA, Worbe Y, Enander J et al (2015b) Disorders of compulsivity: a common bias towards learning habits. Mol Psychiatry 20(3):345–352

Voon V, Baek K, Enander J, Worbe Y, Morris LS, Harrison NA, et al (2015) Motivation and value influences in the relative balance of goal-directed and habitual behaviours in obsessive-compulsive disorder. Trans Psychiatry

Wade TR, de Wit H, Richards JB (2000) Effects of dopaminergic drugs on delayed reward as a measure of impulsive behavior in rats. Psychopharmacology 150(1):90–101

Walker SE, Pena-Oliver Y, Stephens DN (2011) Learning not to be impulsive: disruption by experience of alcohol withdrawal. Psychopharmacology 217(3):433–442

Watkins LH, Sahakian BJ, Robertson MM, Veale DM, Rogers RD, Pickard KM et al (2005) Executive function in Tourette's syndrome and obsessive-compulsive disorder. Psychol Med 35(4):571–582

Weijers H-G, Wiesbeck GA, Böning J (2001) Reflection-impulsivity, personality and performance: a psychometric and validity study of the Matching Familiar Figures Test in detoxified alcoholics. Pers Individ Differ 31(5):731–754

Welch JM, Lu J, Rodriguiz RM, Trotta NC, Peca J, Ding JD et al (2007) Cortico-striatal synaptic defects and OCD-like behaviours in Sapap3-mutant mice. Nature 448(7156):894–900

Winstanley CA, Dalley JW, Theobald DE, Robbins TW (2003) Global 5-HT depletion attenuates the ability of amphetamine to decrease impulsive choice on a delay-discounting task in rats. Psychopharmacology 170(3):320–331

Winstanley CA, Theobald DE, Dalley JW, Glennon JC, Robbins TW (2004a) 5-HT2A and 5-HT2C receptor antagonists have opposing effects on a measure of impulsivity: interactions with global 5-HT depletion. Psychopharmacology 176(3–4):376–385

Winstanley CA, Theobald DE, Cardinal RN, Robbins TW (2004b) Contrasting roles of basolateral amygdala and orbitofrontal cortex in impulsive choice. J Neurosci 24(20):4718–4722

Winstanley CA, Theobald DE, Dalley JW, Robbins TW (2005) Interactions between serotonin and dopamine in the control of impulsive choice in rats: therapeutic implications for impulse control disorders. Neuropsychopharmacology. 30(4):669–682

Winstanley CA, Eagle DM, Robbins TW (2006) Behavioral models of impulsivity in relation to ADHD: translation between clinical and preclinical studies. Clin Psychol Rev 26(4):379–395

Wiskerke J, Schetters D, van Es IE, van Mourik Y, den Hollander BR, Schoffelmeer AN et al (2011) mu-Opioid receptors in the nucleus accumbens shell region mediate the effects of amphetamine on inhibitory control but not impulsive choice. J Neurosci 31(1):262–272

Worbe Y, Savulich G, Voon V, Fernandez-Egea E, Robbins TW (2014) Serotonin depletion induces 'waiting impulsivity' on the human four-choice serial reaction time task: cross-species translational significance. Neuropsychopharmacology 39(6):1519–1526

Worbe Y, Savulich G, de Wit S, Fernandez-Egea E, Robbins TW (2015) Tryptophan depletion promotes habitual over goal-directed control of appetitive responding in humans. Int J Neuropsychopharmacol 18(10):pyv013

Worbe Y, Palminteri S, Savulich G, Daw ND, Fernandez-Egea E, Robbins TW et al (2015) Valence-dependent influence of serotonin depletion on model-based choice strategy. Mol Psychiatry

Wu M, Brockmeyer T, Hartmann M, Skunde M, Herzog W, Friederich HC (2014) Set-shifting ability across the spectrum of eating disorders and in overweight and obesity: a systematic review and meta-analysis. Psychol Med 44(16):3365–3385

Wunderlich K, Dayan P, Dolan RJ (2012) Mapping value based planning and extensively trained choice in the human brain. Nat Neurosci 15(5):786–791

Wylie SA, van den Wildenberg WP, Ridderinkhof KR, Bashore TR, Powell VD, Manning CA et al (2009) The effect of speed-accuracy strategy on response interference control in Parkinson's disease. Neuropsychologia 47(8–9):1844–1853

Yin HH, Knowlton BJ, Balleine BW (2004) Lesions of dorsolateral striatum preserve outcome expectancy but disrupt habit formation in instrumental learning. Eur J Neurosci 19(1):181–189

Yin HH, Ostlund SB, Knowlton BJ, Balleine BW (2005) The role of the dorsomedial striatum in instrumental conditioning. Eur J Neurosci 22(2):513–523

Zandbelt BB, Vink M (2010) On the role of the striatum in response inhibition. PLoS ONE 5(11): e13848

Zeeb FD, Robbins TW, Winstanley CA (2009) Serotonergic and dopaminergic modulation of gambling behavior as assessed using a novel rat gambling task. Neuropsychopharmacology 34 (10):2329–2343

Translational Models of Gambling-Related Decision Making

Catharine A. Winstanley and Luke Clark

Abstract Gambling is a harmless, recreational pastime that is ubiquitous across cultures. However, for some, gambling becomes maladaptive and compulsive, and this syndrome is conceptualized as a behavioural addiction. Laboratory models that capture the key cognitive processes involved in gambling behaviour, and that can be translated across species, have the potential to make an important contribution to both decision neuroscience and the study of addictive disorders. The Iowa gambling task has been widely used to assess human decision-making under uncertainty, and this paradigm can be successfully modelled in rodents. Similar neurobiological processes underpin choice behaviour in humans and rats, and thus, a preference for the disadvantageous "high-risk, high-reward" options may reflect meaningful vulnerability for mental health problems. However, the choice behaviour operationalized by these tasks does not necessarily approximate the vulnerability to gambling disorder (GD) per se. We consider a number of psychological challenges that apply to modelling gambling in a translational way, and evaluate the success of the existing models. Heterogeneity in the structure of gambling games, as well as in the motivations of individuals with GD, is highlighted. The potential issues with extrapolating too directly from established animal models of drug dependency are discussed, as are the inherent difficulties in validating animal models of GD in the absence of any approved treatments for GD. Further advances in modelling the cognitive biases endemic in human decision-making, which appear to be exacerbated in GD, may be a promising line of research.

C.A. Winstanley (✉) · L. Clark
Department of Psychology, University of British Columbia, Vancouver, Canada
e-mail: cwinstanley@psych.ubc.ca

L. Clark
e-mail: luke.clark@psych.ubc.ca

C.A. Winstanley
Djavad Mowafaghian Centre for Brain Health, University of British Columbia, Vancouver, Canada

L. Clark
Centre for Gambling Research, University of British Columbia, Vancouver, Canada

Curr Topics Behav Neurosci (2016) 28: 93–120
DOI 10.1007/7854_2015_5014

Keywords Gambling · Risk · Addiction · Dopamine · Near-miss effect · Commitment · Escalation

Contents

1 Introduction 94
1.1 The Iowa Gambling Task 95
2 Challenges to Translational Research into Gambling 99
2.1 Multiple Types of Gambling 99
2.2 Reliance upon Models of Drug Addiction 100
2.3 The Puzzle of Dopamine and GD 102
2.4 Comorbidity of GD with Other Psychiatric Conditions 104
2.5 Subjective Biases in Decision-Making Under Uncertainty 105
3 Escalation of Commitment 107
3.1 Loss-Chasing 109
4 The Near-miss Effect 110
5 Conclusion 113
References 114

1 Introduction

Gambling has long been part of human society and is a culturally ubiquitous form of entertainment that constitutes a harmless pastime for many healthy individuals. Worldwide, the gambling industry is worth billions of dollars and makes a substantial contribution to government revenue streams. At the heart of every gamble is the decision to risk losing something of value for the chance to obtain a larger prize. Given the inherent predictive complexity and unpredictability of the natural world, gambling shares many features with the kinds of decision-making processes that humans and non-human animals engage in on a daily basis; almost every meaningful choice we make can be conceptualized as a cost/benefit trade-off with elements of uncertainty and ambiguity. Understanding the processes involved in gambling therefore overlaps significantly with research into the basic tenets of decision-making itself (Clark et al. 2013).

Although gambling may tap into decision-making processes typically engaged in everyday choices, gambling scenarios are games of chance that occupy a distinct decision space, where the application of rules or heuristics that are evolutionarily optimized for engagement with the wider world may actually work against us. Clearly, gambling has a "dark side". In a subset of consumers, the act of gambling becomes excessive, compulsive, and maladaptive. The similarities between this clinical construct and drug addiction have been recognized recently in the DSM-5 (APA 2013), in the reclassification of "Pathological Gambling" to gambling disorder (GD) in a new category termed "Substance-Related and Addictive Disorders". Understanding the decision-making processes that underlie the general appeal of gambling, and its capacity of becoming excessive in some people, is therefore of interest to economists, neuroscientists and health practitioners.

In seeking to understand the neurobiological mechanisms underpinning cognition and behaviour, laboratory-based tasks that accurately capture relevant psychological processes have proved an invaluable tool. The development of appropriate animal analogues of these tasks with strong face, construct and predictive validity also allows for more direct tests of hypotheses regarding the causal involvement of neural circuits or neurotransmitter systems. Extrapolating from the contributions that such translational research has made to our understanding of drug addiction (Shaham et al. 2003; Le Moal and Koob 2007; Lerman et al. 2007), the potential insights to be gained into gambling and gambling disorder could be substantial (Potenza 2009). However, there are also some unique challenges to be considered in harnessing the translational power of animal models of gambling behaviour. Recent progress in overcoming these challenges is the focus of the present review.

We will begin by discussing one of the most widely used paradigms for assess decision-making under uncertainty—the Iowa gambling task (IGT) (Bechara et al. 1994). Despite its name, the IGT was developed to model "real-world" risky decision-making rather than gambling per se; for example, there is not a conventional wager in the task, and participants typically play for hypothetical monetary amounts. Nevertheless, the task has become synonymous with gambling research due to a surface similarity to the gambling process. With translational models in mind, the development of a rodent analogue of the IGT was a logical first step in exploring the neurobiological basis of gambling-related decision-making and also serves to exemplify many of the pitfalls and challenges with modelling disordered gambling with laboratory tasks.

1.1 The Iowa Gambling Task

The development of this task was inspired by the neurological observation that a series of cases with acquired brain injury to the ventromedial portion of the prefrontal cortex (vmPFC) showed evident disruption in their day-to-day judgment, risk perception and emotional decision-making, but performed well on existing neuropsychological probes of executive function, such as the tests of the Wisconsin Card Sort Test (Damasio 1994). Damasio and Bechara designed a novel task to capture the kinds of choice situations in which these patients struggled (Bechara et al. 1994). The participant makes a series of card choices between 4 decks, where each choice can win or lose varying amounts of pretend money. Unbeknownst to the participants, 2 decks are risky, generating large occasional penalties that result in net loss over time; the other 2 decks are "safe", resulting in gradual accumulation of profit. Patients with vmPFC damage were unable to learn the advantageous strategy, persisting with choices from the superficially attractive risky decks. Of course, it is unlikely that this brain area alone would be responsible for the complex cognitive operations required by this task. In subsequent work, a network of areas has been described, involved in the coding and retrieval of "somatic markers": emotional tags that are initially coded by the amygdala (Bechara et al. 1999) and contain a visceral, physiological component that is represented in the somatosensory cortex and insula (Bar-On et al. 2003). In subsequent decisions that involve stimuli with these emotional associations as options, the somatic markers are reactivated by the ventromedial PFC and can thereby bias (or entirely short-cut) a more methodical, cost–benefit mode approach to decision-making (Bechara et al. 2005).

Given the relevance of decision-making and risk-taking to many forms of mental illness, the IGT has understandably become highly influential within psychiatry, with impairments associated with bipolar mania (Clark et al. 2001), psychopathy (van Honk et al. 2002), substance use disorders (Bechara and Damasio 2002) and the presence of suicidal features (Jollant et al. 2005), to name just a few examples. The task is also highly sensitive to gambling disorder. In the first demonstration of this effect, 20 treatment-seeking pathological gamblers were impaired relative to healthy controls on the IGT, and in support of their resemblance to the vmPFC lesion syndrome, the pathological gamblers displayed an overt preference for the risky decks, and were unimpaired on the Wisconsin Card Sort Test (Cavedini et al. 2002). Numerous studies replicated these effects; for example, the IGT provides independent discriminatory power over simpler tasks of response inhibition (Kertzman et al. 2011) and has predictive value in relation to treatment dropout (Alvarez-Moya et al. 2011). Functional neuroimaging experiments indicate that these changes in performance are mediated by vmPFC (Tanabe et al. 2007; Power et al. 2012).

Several versions of a rodent analogue of the IGT have been described in recent years, and these tasks vary in the number of choice options, and the nature and schedule of reinforcement (see (de Visser et al. 2011b) for review). One critical factor concerns the representation of "loss". When people gamble in real life, a (monetary) bet is placed, and this bet can be offset by winning outcomes. In the human IGT, the loss of the wager is represented by the deduction of money or points from the individual's current tally, which fluctuates across the session. In both scenarios, unsuccessful trials exact a penalty such that the individual ends up at a disadvantage compared to their initial state. Traditional models of rodent conditioning involve simple gain or no gain outcomes—the animal either receives sugar pellets, or it does not. Approximating loss as the withholding of a potential reward is unsatisfying; there is no "step backward". Moreover, as sugar pellets are consumed immediately rather than cached for future consumption, it is not possible to deduct rewards from prior trials.

In modelling gambling-like behaviour in rodents, researchers have sought to approximate financial loss either by delivering an aversive stimulus such as mild electric shocks (Simon et al. 2009) or bitter-tasting, inedible food pellets (van den Bos et al. 2006), or by imposing timeout penalties to explicitly signal opportunity costs in time-limited paradigms (Zeeb et al. 2009; Rogers et al. 2013). While each of these approaches only approximates the physical removal of a valuable item, such tasks are dissociable from those involving simple reward omission. For example, whereas amphetamine promotes choice of larger, probabilistic rewards when the negative outcome is reward omission (St. Onge and Floresco 2009; Cocker et al. 2012), the drug decreases preference for larger, riskier reward options on tasks with explicit penalties, either in the form of electric shocks (Simon et al. 2009; Mitchell et al. 2011) or timeout periods (Zeeb et al. 2009, 2013; Silveira et al. 2015).

Of the different translational versions of the IGT, the rodent gambling task (rGT; Zeeb et al. 2009) has been the most widely adopted to date, and thus face, construct and predictive validity can be examined using data from multiple laboratories. In the rGT, the rat has a limited amount of time (30 min) to maximize their sugar pellet profits by sampling between four options (see Fig. 1). Each option is associated with a distinct reinforcement schedule that differs in the magnitude and probability of rewards and penalty timeout punishments. The optimal strategy is to favour the options associated with lower per trial gain and avoid the tempting high-reward choices that are paired with disproportionately larger and more frequent timeout punishments. Persistence of such a risky strategy results in a lower net reward per session, akin to the human version.

This task is typically run repeatedly over multiple sessions in order to obtain stable baseline behaviour on which to assess the impact of discrete manipulations. This enables within-subjects experimental designs that reduce total animals used, and offers a further advantage in characterizing baseline risk preferences for examination of individual differences. Consistent with the human lesion data, experimental lesions to the orbitofrontal cortex and basolateral amygdala both retard adoption of the optimal strategy when the brain damage is inflicted prior to task acquisition, and a functional disconnection study demonstrated these two areas work together to optimize adoption of the most advantageous choice strategy (Zeeb and Winstanley 2011, 2013). In contrast, neither lesions to, nor inactivations of, the OFC disrupt choice once a stable baseline has been established (Zeeb and Winstanley 2011; Zeeb et al. 2015), and disconnection of the BLA and OFC did not prevent animals from eventually exhibiting advantageous choice (Zeeb and Winstanley 2013). However, this latter group failed to adjust their choice preference in response to reward devaluation, suggestive of cognitive inflexibility. Interestingly only post-acquisition lesions to the BLA, but not OFC, increased preference for the disadvantageous options (Zeeb and Winstanley 2011), positing a specific, ongoing role for the BLA in maintaining an optimal decision-making strategy.

Besides simply corroborating the human data, the rGT observations speak to some unresolved questions regarding the human cognitive neuroscience literature. Some data indicate that the vmPFC impairment in human lesion cases reflects a deficit in reversal learning rather than altered decision-making under uncertainty per se (Fellows and Farah 2005; Lawrence et al. 2009). The rodent data suggest that the OFC is recruited into analysis of risk-based decision-making when learning demands are high, and the subject must resolve which out of many potentially promising options is the most advantageous. However, once the strategy is set, the OFC plays a less prominent role unless task contingencies change. In contrast,

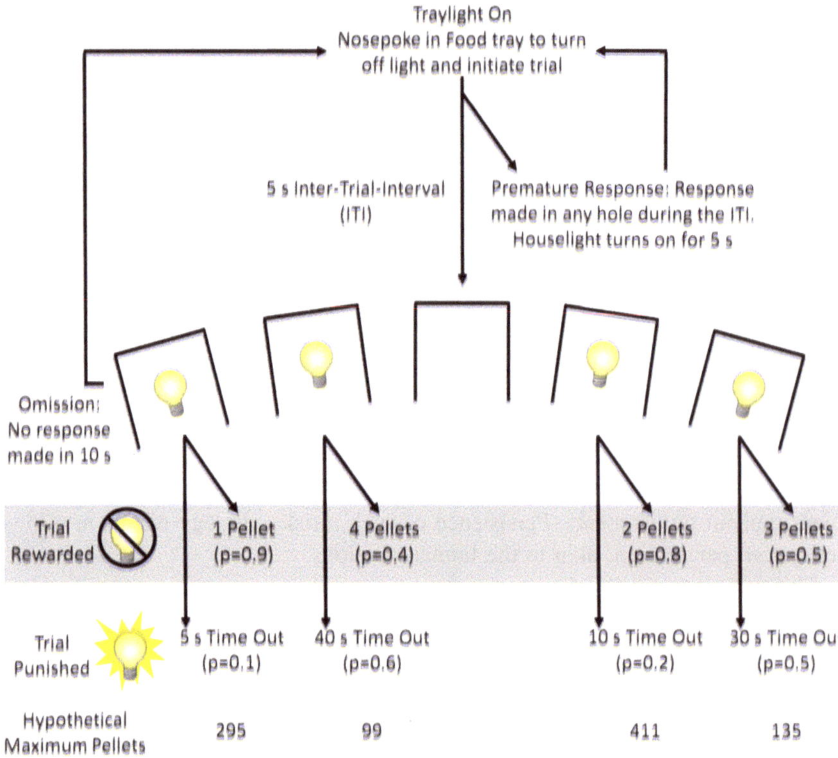

Fig. 1 Task schematic of the rGT paradigm. The task begins with illumination of the tray light. A nosepoke response in the food tray extinguishes the tray light and initiates a new trial. After an inter-trial interval (ITI) of 5 s, four stimulus lights are turned on in holes 1, 2, 4, and 5, each of which is associated with a different number of sugar pellets (P1–P4). The animal is required to respond in one of these holes within 10 s. This response is then rewarded or punished depending on the reinforcement schedule for that option (indicated by the probability of a win or loss in *brackets*). If the animal is rewarded, the stimulus lights are extinguished and the animal receives the corresponding number of pellets in the now-illuminated food tray. If the animal is punished, the stimulus light in the corresponding hole flashes at a frequency of 0.5 Hz for the duration of the punishing timeout, and all other lights are extinguished. At the end of the punishment period, the tray light is turned on and the animal can initiate a new trial. Failure to respond at the illuminated holes results in an omission, whereas a response during the ITI is classified as a premature response and punished by a 5 s timeout during which the houselight is turned on. The order of the options from *left* to *right* is counterbalanced within each cohort (version A as shown, version B: 4, 1, 3, 2). The maximum number of pellets that could be theoretically obtained if the option was chosen exclusively in a 30-min session (not allowing time for choice/food consumption for consistency) is given, hence providing an objective value for each option. Figure reproduced from (Barrus et al. 2015)

output from the basolateral amygdala helps subjects to maintain an optimal decision-making strategy, potentially through representation of the negative outcomes associated with loss (Tremblay et al. 2014), leading to an ongoing suppression of risky choice.

More recent data also highlight an important role for the agranular insular and medial prefrontal regions, particularly the infralimbic area, in mediating optimal choice on the rGT. Lesions or inactivations to these areas result in a reduction in choice of the best option (P2), but an increase in choice of the options associated with the lowest per trial gain but also the shortest and most infrequent punishments (Paine et al. 2013; Pushparaj et al. 2015; Zeeb et al. 2015). Although not equivalent to an increase in risky choice per se, this profile is clearly suboptimal. Human imaging and lesion studies have indicated some involvement of dorsolateral PFC in IGT performance [e.g. (Fellows and Farah 2005; Lawrence et al. 2009)], and such functions are likely localized to areas of medial PFC in the rat (Chudasama 2011). These regions are also implicated in alternative rodent analogues of the IGT (Rivalan et al. 2011; van den Bos et al. 2014).

2 Challenges to Translational Research into Gambling

Having established the translational potential of gambling tasks from human to rodent, and the convergence in core underlying neural circuitry, the question then arises as to what extent such procedures are really representative of disordered gambling. We will now consider some of the important features of this argument.

2.1 *Multiple Types of Gambling*

Gambling is an umbrella term for a variety of distinct games that vary in a number of psychologically important dimensions. Many popular forms of gambling represent games of pure chance (lotteries, roulette, slot machines). Others involve an overt element of skill (e.g. poker), while some require a more opaque degree of skill (e.g. sports betting) [e.g. (Cantinotti et al. 2004)]. Some games are continuous forms (e.g. slot machines), with short event durations and the potential to play repeatedly, whereas others (e.g. lotteries) impose lengthy anticipatory periods and more limited potential for repeat play (Griffiths and Auer 2012). The computerized format of modern electronic gambling machines (EGMs) has enabled a spiralling complexity of psychological features. Laboratory studies—whether in humans or animals—naturally seek to characterize single manipulations while holding other features constant, but this reverse engineering is increasingly difficult in the case of EGMs. We also recognize heterogeneity among problem gamblers (Blaszczynski and Nower 2002; Petry 2003), and different forms of gambling may selectively appeal to players with distinct motives, such as gambling to relieve aversive emotional states versus gambling for excitement or thrill seeking (Griffiths 1995; Stewart et al. 2008).

With respect to the IGT, the task represents a generic measure of a participant's ability to discriminate choice options that vary in their probabilistic reinforcement contingencies, but the task does not resemble any actual commercial form of gambling. A participant's capacity to withhold responses to superficially attractive options that have the potential for large penalties may arguably capture the ability to resist temptations when gambling, but the IGT task structure is greatly simplified compared to modern forms of gambling and lacks many of the psychological features that are linked to disordered play, such as near-misses (Clark et al. 2009) or high sensory stimulation (Loba et al. 2001; Dixon et al. 2014). Judging by the variety and complexity in gambling opportunities, arguably one single laboratory task will not be able to model all pertinent aspects of gambling disorder. The development of a battery of tasks, each capable of tapping into one or more key cognitive processes, may hold more promise in this regard.

2.2 *Reliance upon Models of Drug Addiction*

The medical concept of behavioural addiction is reasonably novel, contrasting with historical notions of addiction that are linked to the pharmacological action of exogenous chemicals (Potenza 2006). A number of question marks remain over the features that are necessary or sufficient for defining addiction, and what other non-chemical addictions might join gambling disorder in future iterations (Clark 2014). One approach for improving upon the IGT/rGT model of gambling is to focus on specific features of the clinical manifestation of gambling disorder. The DSM-5 diagnosis is based upon 9 criteria (with a threshold of 4 for diagnosis). These criteria include some characteristic "hallmarks" of an addiction syndrome including preoccupation with gambling, wagering larger sums of money over time (which is assumed to reflect tolerance) and experiencing restlessness or irritability when stopping gambling (akin to withdrawal) (Shaffer et al. 2004). Other features pertain to the negative consequences of excessive gambling in order to confirm functional decline, and include an impact on one's job or family life, borrowing money and lying about gambling losses. These features were all adapted from the criteria for substance dependence dating back to the DSM-III, and as such, translational models of these features will inevitably overlap with models of drug addiction. The obvious question is, therefore, how much the fledgling field of modelling gambling and gambling disorder can learn from the study of substance use disorder.

One clear message is that the act of engaging in addictive behaviour does not equate to the manifestation of an addiction (Ahmed 2012). Moreover, pharmacotherapies developed to attenuate self-administration in preclinical models have failed to translate into treatments for addiction (e.g. D_1 receptor antagonists; (Haney and Spealman 2008; Pierce et al. 2012). Working from the DSM-IV and DSM-5 criteria for substance abuse disorder, researchers have begun to develop models to capture specific symptoms that separate drug addiction from simple drug-taking, and thus represent targets for treatment. However, translating these "addiction-like" methodological approaches into models of gambling-related choice is not trivial.

Take, for example, extending the length of drug self-administration sessions from shorter (2–3 h) to longer (6–8 h) access periods; this produces an escalation in drug intake that is reminiscent of the bingeing and loss of control central to the addicted state (Ahmed and Koob 1998). In contrast, "binge" gambling is unlikely to be maintained when sugar pellets are used as a reinforcer due to inevitable satiation effects. Maintained drug use in the face of negative consequences, another key DSM criterion for SUD, has been modelled in a subset of animals that will continue to respond for cocaine when such self-administration is accompanied by electric shock delivery (Belin et al. 2008). Gambling models such as the rGT already use penalties to convey the negative consequences of risky choice, so the rationale for additional electric shock penalties would be unclear.

Persistent responding for addictive drugs in extinction may also model the inability to refrain from drug-seeking and is also indicative of habitual rather than goal-directed behaviour (Belin et al. 2008). However, while habitual responding is generally easy to reproduce with overtraining on simple lever-press schedules of reinforcement (Dickinson 1985), it is not clear that complex cognitive tasks such as the rGT/IGT could ever be performed in such a manner or that risky choice is an automatic "habit" in GD. Furthermore, when tested during extinction on the rGT, rats decreased choice of P2 and distributed their responses more equally among all other options, both risk averse and risky, presumably reflecting a return to a strategy of exploration in the face of non-reward (Zeeb and Winstanley 2013). Devaluing the specific reinforcer used in the rGT—which should also alter choice preferences if decision-making is still goal-directed—actually results in a shift away from the best option, P2, and towards P1, presumably because the value of the reward no longer offsets the penalty timeouts (Zeeb and Winstanley 2013). While these results are interesting in the context of decision neuroscience, they do not seem particularly informative with respect to modelling GD.

Other drug addiction models have focused on relapse by looking at environmental factors that predict reinstatement after the extinction of the drug-seeking response. The fact that many addicts relapse at timepoints distal to cessation of use, indicative of an incubation of craving for drug, has also been successfully modelled in both rodent and human laboratory-based tests (Grimm et al. 2001; Bedi et al. 2011; Wang et al. 2013). Similar to the methods reviewed above, there are logical problems in adapting reinstatement models of relapse to GD: if sugar pellet rewards were omitted in the rGT and then reinstated 2 weeks later, we would expect the animals to resume responding—sugar pellets are a primary reward and there is nothing pathological about such food-seeking. In contrast, the fact that an addictive substance like cocaine is an exogenous reward, with no biological value, makes the resumption of drug-seeking more clearly reminiscent of a maladaptive behaviour.

Beyond the reliance upon non-drug (and primary) rewards, many of the difficulties in translation between drug addiction and gambling models come down to the fact that the poor cost/benefit decision-making that is captured on the IGT/rGT is not itself a pathological state exclusively linked to GD, but a more general measure of poor executive functioning that is associated with many forms of psychopathology. By direct contrast, self-administration models of drug addiction are not considered a "proxy" for any other forms of psychiatric illness besides drug addiction. Seeking to develop a model of GD by adapting models of gambling-related decision-making to mimic animal models of drug addiction may therefore be of limited utility.

2.3 *The Puzzle of Dopamine and GD*

One observation that has held remarkably true for all manner of pharmacologically diverse addictive substances is their powerful ability to potentiate dopamine release, particularly within the mesolimbic projections from the ventral tegmental area to the nucleus accumbens (Wise 1996). Direct electrical stimulation of this pathway itself is highly reinforcing (Olds and Milner 1954). If gambling also potentiates striatal dopamine release in individuals with gambling disorder, might this serve as a neurobiological marker for animal models? The measurement of dopamine transmission in the human brain is deceptively challenging. Early studies measuring changes in plasma/cerebral spinal fluid dopamine and dopamine metabolites did detect changes in groups of pathological gamblers (Roy et al. 1988; Bergh et al. 1997; Meyer et al. 2004), but the direction of effect was not consistent, and peripheral measures correlate only weakly with central dopamine markers. Genetic studies also point to altered rates of certain polymorphisms affecting dopamine function; the TaqA1 DRD2 gene has been implicated in a number of studies (Comings and Blum 2000), and a recent study in 400 cases with gambling disorder identified changes in the DRD3 gene that was corroborated in a rodent translational arm by correlations between DRD3 mRNA expression and performance on the rGT (Lobo et al. 2014).

Arguably, the clearest sign of a key role for dopamine in gambling behaviour is that chronic DA agonist therapy in parkinsonian patients can induce GD de novo, as well as other reward-driven, impulsive behaviours (hypersexuality and compulsive shopping), and this normalizes once drug administration ceases (Weintraub and Potenza 2006; Weintraub et al. 2006, 2010; Leeman and Potenza 2011; Voon et al. 2011). This syndrome is most linked to pramipexole, ropinirole and cabergoline, agents that all have notably high affinities for the D3 receptor subtype. Electrophysiological data obtained from non-human primates indicate that CSs that predict reward with maximal uncertainty (50 %) lead to the greatest activation of dopaminergic neurons (Fiorillo et al. 2003). In recent work, repeated exposure to such unpredictable CSs was seen to sensitise animals' locomotor responses to amphetamine. This is a provocative indication that non-drug-related behaviours not only excite the DA system but that this activity potentiates over time, much like addictive drugs (Singer et al. 2012; Zack et al. 2014). By inference, repeated risk-taking within operant behavioural testing could itself alter the neurobiological response to uncertain outcomes, although this has yet to be determined.

One of the most well-documented clinical findings in drug addiction is of reduced $D_{2/3}$ receptor availability in the striatum, an effect that has been considered a biological endophenotype for addiction vulnerability (Volkow et al. 2004, 2007;

Wang et al. 2004; Dalley et al. 2007; de Weijer et al. 2011; Cocker et al. 2012). Notably, a corresponding reduction in $D_{2/3}$ receptor density was not observed in GD, across four independent studies (Linnet et al. 2011; Clark et al. 2012; Joutsa et al. 2012; Boileau et al. 2013b). Some individual differences were reported, in relation to trait impulsivity (Clark et al. 2012) and gambling severity (Boileau et al. 2013b) that do support a more nuanced account of dopamine dysregulation. Other clinical PET studies have begun to investigate dopamine release, by measuring tracer displacement in response to either stimulant challenge or an actual gambling task. Using amphetamine administration with the alternative radiotracer [11C] PNHO (a dopamine D3 preferent ligand), individuals with gambling disorder showed heightened dopamine release in the dorsal striatum and correlations with gambling severity in the ventral striatum (Boileau et al. 2013a). Two studies with [11C]raclopride also described significant positive correlations between gambling-induced dopamine release and subjective excitement on the Iowa gambling task (Linnet et al. 2011) and gambling severity using a slot machine game (Joutsa et al. 2012).

To add to the confusion, D_2 receptors are inhibitory and can exist not only presynaptically as autoreceptors on DA afferents but also post-synaptically as heteroreceptors on both inhibitory interneurons and the medium spiny neurons—the major projection neurons—of the striatum (Seamans and Yang 2004; Ford 2014). Trying to determine the impact of broadscale increases or decreases in receptor number, or administration of a D_2 receptor agonist/antagonist, on activity in a region or its response to DA is therefore challenging. Given the heterogeneity of gambling behaviours as discussed above (Blaszczynski and Nower 2002; Petry 2003), it is also possible that a hyper/hypodopaminergic state may only be prominent in mediating some types of gambling. Alternatively, it is well known that the relationship between performance and DA level can follow an inverted U-shaped curve, such that both too much and too little DA can result in impairments (Cools and D'Esposito 2011).

In sum, using sensitivity to dopaminergic manipulations as a benchmark against which to test the validity of animal models of GD is fraught with complications. While it is still worthwhile to qualify the impact of dopaminergic manipulations on gambling-related behaviour in non-human animals due to the obvious parallels to drug addiction, it is unclear as to what kind of DA dysfunction, if any, is critical for the manifestation of GD (Potenza 2013). With that being said, chronic administration of $D_{2/3}$ receptor agonists can increase choice of uncertain options in some animal models, particularly those in which loss has not been explicitly signalled, but does not shift choice behaviour in the rGT (Rokosik and Napier 2012; Tremblay et al. 2013).

In parallel to the discrepant findings concerning the role of the dopamine system in gambling and GD, there are similarly inconclusive data regarding the role of other neurotransmitter systems in mediating this condition (Williams et al. 2008). As such, it is unsurprising that there are no recognized effective pharmacological treatments for GD (Lupi et al. 2014). Studies attempting to verify whether antidepressant drugs, such as serotonin reuptake inhibitors, are effective in treating GD have been inconclusive (Grant and Kim 2006). Furthermore, it has been suggested that the efficacy of certain pharmacological treatments, such as opiate antagonists, in treating GD may depend on the presence or absence of certain features such as craving or impulse control deficits (Grant and Kim 2002; Grant et al. 2006). While such studies indicate that a personalized medicine approach may be a more effective strategy in developing GD treatments than the traditional "one medicine fits all" pharmacotherapy strategy, a clear consensus as to what drugs may be effective in which populations of gambling addicts is still lacking. Besides the negative impact this may have on treatment-seeking patients, this also makes the establishment of predictive validity in animal models of GD an almost impossible task. Given that demonstrating efficacy of known treatments for a disorder is normally the key metric by which success or failure of the model is judged, this inevitably means that models of GD are currently hard to truly validate (Dawson et al. 1992).

2.4 *Comorbidity of GD with Other Psychiatric Conditions*

As with most other areas of mental illness, gambling disorder rarely exists in the absence of other mental health problems. Comorbidities with mood disorders, substance use disorders and personality disorders are widely reported (Lorains et al. 2011). It is unclear how these issues of clinical overlap should be approached in animal models. Application of the research domain criteria (RDoC) approach may, in some respects, render this debate somewhat moot in that the focus is to understand neurobiological regulation of specific behavioural endophenotypes, irrespective of clinical presentation or comorbidity (see (Casey et al. 2013) for discussion). However, co-occurrence of psychiatric disorders will potentially complicate pharmacotherapy, in that some drug combinations are contraindicated and even alter the response to medical intervention.

Despite these potential problems with comorbidity, there may be potential benefits to recognising the co-occurrence of GD with other psychiatric disorders with respect to animal modelling. More specifically, in the absence of definitive pharmacological tests to demonstrate validity of animal models of GD, it may be possible to use co-occurrence of behavioural traits synonymous with GD as at least indirect support of a GD endophenotype. High impulsivity has been repeatedly and reliably documented in GD populations and has been found to contribute to severity of the disorder (Verdejo-Garcia et al. 2008). Furthermore, high impulsivity may endow vulnerability to multiple addictions, including drug addiction (Jentsch and Taylor 1999; Winstanley et al. 2010), and may even dissociate the controversial diagnosis of food addiction from other forms of pathological engagement with food (Ziauddeen and Fletcher 2013). Demonstrating that a certain pattern of behaviour in animal models of gambling-related cognitions could therefore help to substantiate them as indicative of vulnerability to GD.

One potential complication that should be acknowledged in this regard is that impulsivity is itself a multifaceted construct, and measures of impulsive choice and impulsive action do not typically correlate with populations of healthy volunteers or normal animals (see (Moeller et al. 2001; Winstanley et al. 2006) for review). Nevertheless, multiple forms of impulsivity are elevated in clinical conditions, suggesting at the extreme end of the spectrum, these behaviours do coalesce. One advantage of the rGT discussed above is that it enables the concurrent measurement of a form of motor impulsivity, as indexed by premature responding at the stimulus array prior to illumination of the apertures (Zeeb et al. 2009). This measure of impulse control has been extensively validated in the five-choice serial reaction time task in rats as being indicative of impulsive responding in the continuous performance task in human subjects and has also recently been successfully back-translated to demonstrate elevated impulsivity in substance-dependent individuals, binge drinkers and those with binge-eating disorder (Sanchez-Roige et al. 2014; Voon et al. 2014). Despite the lack of any obvious relationship between premature responding and risky choice on the rGT in individual cohorts (n = 16–24), a meta-analysis of data from over 200 animals clearly indicates a positive correlation between motor impulsivity and preference for the disadvantageous options, as well as increased speed of decision-making (Barrus et al. 2015). Furthermore, other rodent analogues of the IGT indicate a relationship between maladaptive preference for risky options and anxiety (de Visser et al. 2011a) that has been observed in some (Miu et al. 2008) but not all human studies (Drost et al. 2014).

While the demonstration that poor choice under uncertainty correlates with high impulsivity or high anxiety in rodents may capture the comorbidity of these traits in clinical populations, neither of these relationships is observed exclusively in GD. Again, such studies seem to confirm that the IGT and rodent analogues can be highly informative in elucidating the phenomenological and neurobiological basis of decision-making impairments, and deficits in choice do appear to result from the same cognitive or emotional processes that are impaired in clinically important conditions, but these findings do not speak directly or specifically to GD.

2.5 Subjective Biases in Decision-Making Under Uncertainty

A final consideration with regard to modelling gambling, and vulnerability to GD, in non-human subjects is that healthy human decision-making is remarkably "non-normative", as described in a substantial programme of research inspired by Kahneman and Tversky's work on heuristics and biases (Kahneman and Tversky 1979). Indeed, such as the severity of these psychological distortions, certain forms of psychopathology have been associated with more "rational" behaviour, e.g. depressive realism (Alloy and Abramson 1979), or jumping to conclusions in psychosis (Garety and Freeman 1999). The heuristics approach argues that the decision space for many "real-world" choices is simply too large and too ambiguous for choices to be made based on a thorough and algorithmic "cost–benefit" analysis, and instead, we rely on cognitive shortcuts that provide rapid and effective solutions in most circumstances (see (Kahneman 2011) for review and discussion).

In the specific case of gambling, we see a wide variety of faulty beliefs during gambling play, and these cognitive distortions create an inappropriately high expectancy of winning (Ladouceur and Walker 1996). One set of erroneous beliefs apply to predictions over successive gambles with a random outcomes; in the classic *gambler's fallacy*, the player is less likely to choose a recent outcomes, as they expect this sequence to balance out in the long term (Oskarsson et al. 2009). In a complementary effect, the hot hand players expect recent winning streaks to continue (Gilovich et al. 1985). Another set of beliefs concern the players' sense that they possess some control or skill over a randomly determined outcome; various features of gambling games promote this illusion of control (Langer 1975), for example the ability to choose one's lottery numbers or throw a ball or dice. The gambler's fallacy and the illusion of control are extensively described in the healthy population and often in large-scale economic data sets such as the stock market or Internet gambling data (Clotfelter and Cook 1993; Smith et al. 2009). However, there is increasing evidence that the frequency (or strength) of these distorted thoughts is elevated in people with gambling disorder. An important ingredient of cognitive behavioural therapy for problem gambling is to restructure these faulty cognitions.

It is a fascinating question to what extent these non-normative biases exist in non-human species. Within behavioural economics, a recent paper describes similar non-linearities in processing probability information in macaques to the classic effects described in humans in prospect theory; the macaques overestimate low probability events (c.f. lottery wins) and underestimate high probability events (Stauffer et al. 2015). Using a human version of this task in pathological gamblers, an overall increase in subjective likelihood was observed across the full range of probabilities from 0 to 1 (Ligneul et al. 2013). Rhesus monkeys also show evidence of a hot hand effect: in a two-choice discrimination task where the alternatives were equally reinforced (i.e. $p = 0.5$), the monkeys implemented a win-stay, lose-shift strategy that persisted across extensive training (Blanchard et al. 2014).

Such findings open up the possibility that, just as in human subjects, laboratory animals which show higher levels of these choice biases may approximate individuals vulnerable to GD. The adoption of such an approach to modelling gambling and GD sidesteps many of the criticisms discussed above with respect to the rGT, namely the degree to which behaviour is selective for GD or represents more widespread executive dysfunction or addiction vulnerability. While research with non-human primates offers certain key advantages in investigating cognitive processes comparable to the human condition, demonstrating these choice biases in rodents would significantly advance the utility of laboratory models of GD and enable a broader range and size of experiments with faster throughput for drug discovery (see Cocker and Winstanley 2015). In the final section, we will consider three specific examples of gambling-related cognitive distortions that have been targeted in translational models in rodents.

3 Escalation of Commitment

Imagine a 50–50 gamble (Gamble A) where you can win \$10 or lose \$5; many people are inclined to accept such gambles. Now, simply scale up the outcomes (Gamble B): a 50–50 gamble on winning \$10,000 or losing \$5000. Many people who would accept Gamble A, decline Gamble B, despite the equivalent odds. As the stakes get high, people become more cautious; this effect is termed "escalation of commitment" (Staw 1981). A similar effect is evident in rodents, who shift their preference towards a guaranteed reward over a larger probabilistic ($p = 0.5$) reward, as the size of the reward increases (Fig. 2), (Cocker et al. 2012). This bias persists over extensive training and also shows considerable individual variation; whereas some animals show a marked reduction in preference for uncertainty as the stake

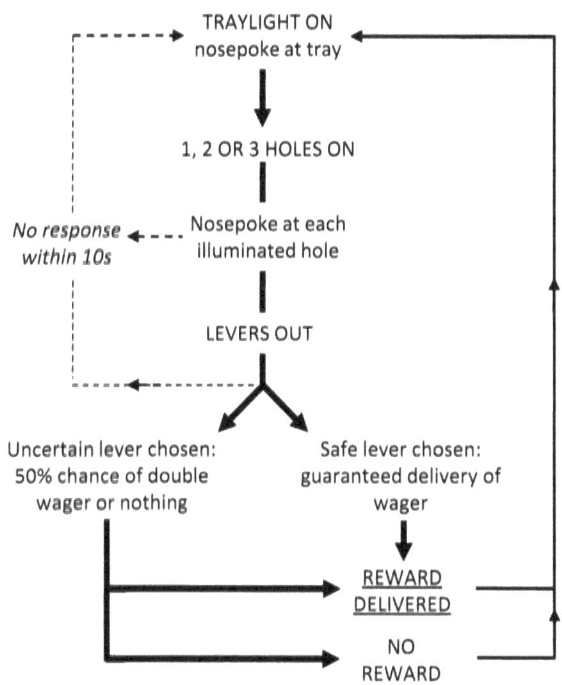

Fig. 2 Task schematic of the betting paradigm. The rat initiates each trial by making a nosepoke response at the illuminated food tray. The tray light was then extinguished, and 1–3 response holes were illuminated, signalling the size of the bet or wager (1–3 sugar pellets). A nosepoke response at an illuminated aperture turned off the light inside it. Once all the aperture lights had been extinguished in this manner, 2 levers were presented to the rat. Selection of the uncertain lever resulted in a 50:50 chance of receiving either double the wager or nothing, whereas selection of the safe lever always leads to delivery of the wager. The trial was scored as a choice omission if the rat failed to choose one of the levers within 10 s. Likewise, if the rat failed to respond at each illuminated response hole within 10 s, the trial was scored as a hole omission. Adapted from Cocker et al. (2012)

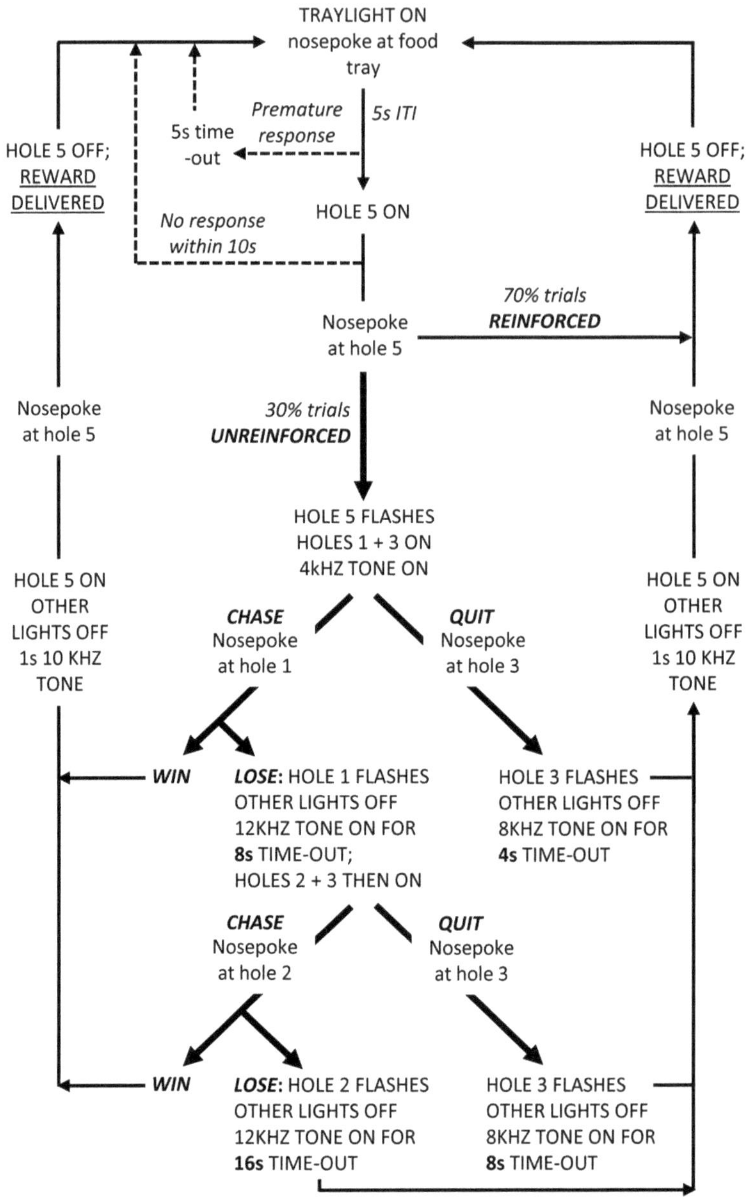
TRAYLIGHT ON
nosepoke at food tray
5s ITI
Premature response
5s time -out
HOLE 5 ON
No response within 10s
HOLE 5 OFF; REWARD DELIVERED
HOLE 5 OFF; REWARD DELIVERED
Nosepoke at hole 5
70% trials REINFORCED
30% trials UNREINFORCED
Nosepoke at hole 5
Nosepoke at hole 5
HOLE 5 FLASHES
HOLES 1 + 3 ON
4kHZ TONE ON
HOLE 5 ON OTHER LIGHTS OFF 1s 10 KHZ TONE
HOLE 5 ON OTHER LIGHTS OFF 1s 10 KHZ TONE
CHASE Nosepoke at hole 1
QUIT Nosepoke at hole 3
WIN
LOSE: HOLE 1 FLASHES OTHER LIGHTS OFF 12KHZ TONE ON FOR 8s TIME-OUT; HOLES 2 + 3 THEN ON
HOLE 3 FLASHES OTHER LIGHTS OFF 8KHZ TONE ON FOR 4s TIME-OUT
CHASE Nosepoke at hole 2
QUIT Nosepoke at hole 3
WIN
LOSE: HOLE 2 FLASHES OTHER LIGHTS OFF 12KHZ TONE ON FOR 16s TIME-OUT
HOLE 3 FLASHES OTHER LIGHTS OFF 8KHZ TONE ON FOR 8s TIME-OUT

◀ **Fig. 3** Task schematic of the loss-chasing paradigm. Each trial begins when the rat makes a nosepoke response at the illuminated food tray. The play hole is then turned on. A trial is aborted if the rat fails to respond at the play hole within 10 s (scored as an omission), or responds prematurely at any hole prior to illumination of the play hole. Responding at the illuminated play hole leads to immediate delivery of food reward on win trials. However, on 30 % of trials, this response triggers a loss, and subjects are required to choose between a guaranteed time penalty of 4 s ("quitting") and a 50:50 chance of double that time penalty or evading the timeout ("chasing"). The rat signals its preference by responding at either of the illuminated quit and chase1 holes, respectively. If the rat chases and wins, the play hole is instantly reilluminated and the rat can again earn reward, whereas quitting delays illumination of the play hole by 4 s. After the 8 s timeout penalty incurred if the rat chased and lost, the options to chase or quit are presented a second time, signalled by illumination of the chase2 and quit holes. The penalties at this second choice point are double that at the first. Winning the chase still results in immediate reillumination of the play hole, whereas quitting or losing the chase delays this opportunity to earn reward by 8 and 16 s, respectively. Reproduced from (Rogers et al. 2013)

increases, others maintain a more mathematically normative indifference between the two options regardless of the size of the reward in play.

MicroPET imaging using [11C] raclopride indicated that the degree of wager sensitivity was inversely correlated with the level of dopamine $D_{2/3}$ receptor density in the dorsal striatum, a finding that was subsequently confirmed using ex vivo dopamine receptor autoradiography (Cocker et al. 2012). As such, the degree to which rats exhibited an irrational choice bias under uncertainty was associated with this putative marker of addiction vulnerability. The OFC also appears to be a key neural locus mediating the subjective preference for the uncertain vs safe outcomes in this paradigm, in that local inactivation ameliorated the impact of bet size on choice preference in wager-sensitive rats, with null effects observed following lesions or inactivations of the BLA (Tremblay et al. 2014), prelimbic and infralimbic cortices (Barrus and Winstanley, in press). Interestingly, chronic administration of the $D_{2/3}$ receptor agonist ropinirole significantly increased preference for the uncertain options across all bet sizes in a subgroup of rats, independent of their baseline choice pattern (Tremblay et al. 2013). The same drug regimen did not alter behaviour on the rGT, suggesting that this task may capture something of the bias towards risky outcomes that can be induced in a subset of Parkinson's patients given such medications.

Collectively, these data suggest that behaviour on this task is representative of a subjective bias modulating preference for uncertain outcomes via value judgements in the OFC and dopaminergic innervation to the striatum. While these data are promising in terms of capturing biases in decision-making relevant to GD, the assumption would be that treatments which increase choice of the guaranteed options may hold promise in reducing the drive for uncertainty in either iatrogenic or idiopathic GD. As yet, no such ligands have been reported, but this hypothesis remains open to future verification.

3.1 Loss-Chasing

One key symptom of GD that does not have a clear analogy in drug addiction is loss-chasing: the desperate continuation of gambling, or returning to gamble another day, in an effort to recoup previous losses. Loss-chasing is often regarded as the single feature that best distinguishes a casual or recreational gambler from a problem gambler, and in epidemiological data sets, it is the most commonly endorsed symptom (Toce-Gerstein et al. 2003), particularly so among younger problem gamblers (Strong and Kahler 2007).

A rodent loss-chasing task has been developed, based on a laboratory-based paradigm used to assess chasing behaviour in healthy volunteers (Fig. 3, (Campbell-Meiklejohn et al. 2007). In essence, the subject has a limited amount of time in which to earn sugar reward. Periodically, the animal must choose between accepting a set timeout penalty, or gambling in order to try and evade any such penalty at the risk of receiving double the punishing timeout with 50:50 odds (Rogers et al. 2013). Similar to human subjects, rats exhibited a strong tendency to chase their losses, gambling to avoid the timeout on over 80 % of trials. Lesions to the BLA significantly attenuated this chasing bias, suggesting that this region is important in representing the aversive nature of loss, as in the rGT (Tremblay et al. 2014). Chasing could also be attenuated through administration of a D_2 receptor antagonist and an agonist at the 5-HT_{1A} receptor (Rogers et al. 2013). While this task has considerable face validity in terms of modelling a choice bias that is known to be problematic in human gamblers, and in a manner that allows for ready detection of improvements in this choice behaviour, the high level of loss-chasing in healthy animals also makes it difficult to determine what a "pathological" level of engagement in this behaviour would look like. Future studies using either repeated administration of dopamine agonists may help in this regard. Likewise, the modification of the task parameters to draw out a greater degree of individual differences in the tendency to chase could allow for the ready identification of a subgroup of rats that chase persistently despite, for example, negative consequences.

4 The Near-miss Effect

A number of groups have demonstrated that rats and pigeons may be vulnerable to the effects of near-miss outcomes (Peters et al. 2010; Scarf et al. 2011), which are known to be effective stimuli in motivating gambling tendencies and persistent play in human gamblers (Kassinove and Schare 2001; Clark et al. 2009). In the rat slot machine task (rSMT), the animal must respond at three successive apertures (equivalent to three slot machine reels), each of which contains a flashing stimulus light (Fig. 4), (Winstanley et al. 2011). Once a nosepoke is detected at each hole, the light inside sets to on or off, and the animal must then use the pattern of lights to determine whether it should respond on the collect lever and receive any available reward, or to start a new trial by hitting the roll lever instead. A collect response

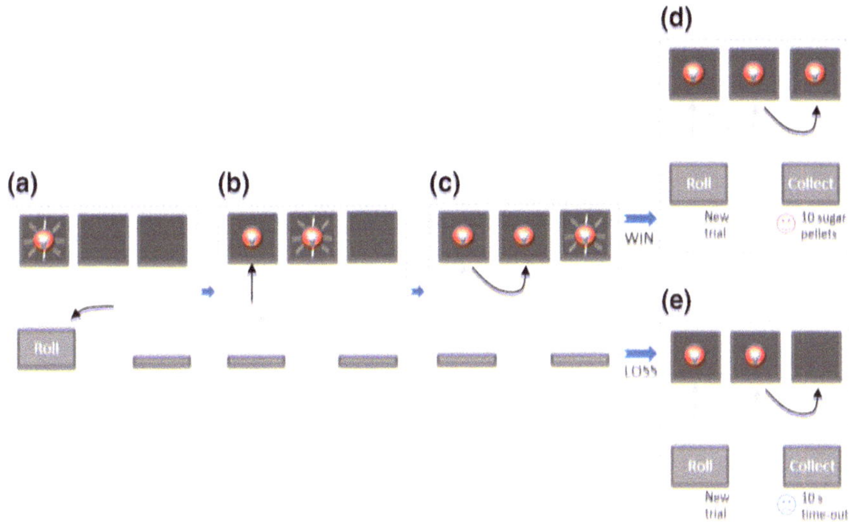

Fig. 4 Schematic diagram showing the trial structure for the slot machine task. A response on the roll lever starts the first light flashing (**a**). Once the animal responds in each flashing aperture, the light inside sets to on or off and the neighbouring hole starts to flash (**b–c**). Once all three lights have been set, the rat has the choice to start a new trial, by responding on the roll lever, or responding on the collect lever. On win trials, where all the lights have set to on, a collect response delivers 10 sugar pellets (**d**). If any of the lights have set to off, a response on the collect lever instead results in a 10-s timeout period (**e**). Figure adapted from (Winstanley et al. 2011)

only results in reward delivery on win trials, on which all 3 lights are illuminated, but a 10-s timeout period on any other trial type. As with the rGT and loss-chasing tasks, animals only have a limited amount of time to earn the maximal amount of reward possible; therefore, these timeouts are frustrating and negatively impact the number of sugar pellets earned.

While rats find it relatively easy to discriminate wins and losses in which zero or only 1 light is illuminated, erroneous collect responses are frequent on 2-light trials, with many rats responding on the collect lever on 70–80 % of these trials. It would therefore appear that animals are responding to this pattern of lights as being more indicative of a win than a loss, leading to the anticipation of available reward. As such, this response pattern shares fundamental similarities with the near-miss effect. The high incidence of erroneous responding on these putative near-miss trials could not be attributed to simple perceptual discrimination, or the animals' failure to learn the task contingencies (see (Winstanley et al. 2011; Cocker and Winstanley 2015) for further discussion). Furthermore, animals were consistently faster to respond at the subsequent hole if the preceding light set to on, implying animals remain sensitive to the positive or negative feedback provided by the visual stimuli.

Given the structure of the rSMT, it was possible to implement an extinction and reinstatement like manipulation similar to that performed in studies of drug addiction (see Fig. 5). Hence, if near-miss events contributed to the addictive nature

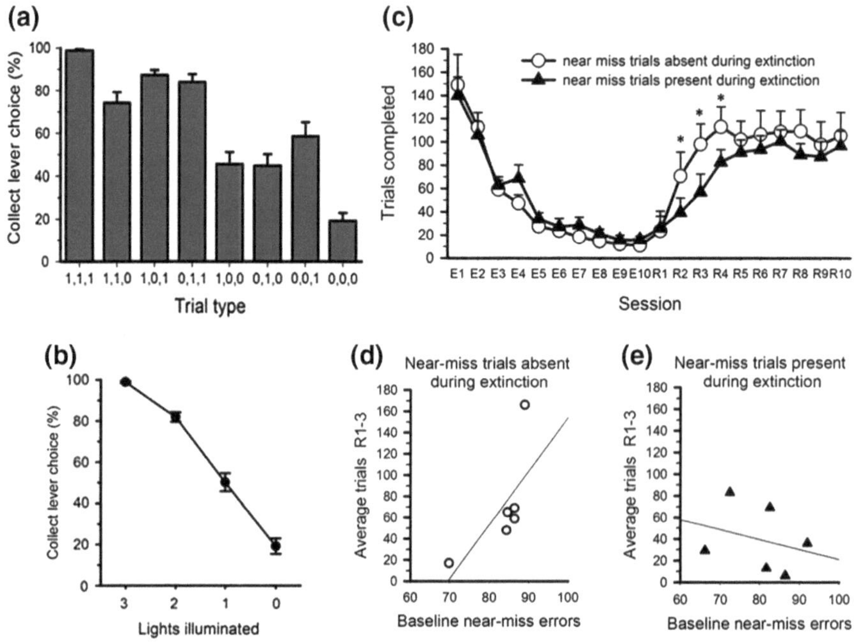

Fig. 5 Performance of the slot machine task at baseline, and the effect of removing near-miss trials during extinction on both the rate of extinction and subsequent reinstatement of task performance. On win trials, when all 3 lights had set to on ([1,1,1]), animals chose the collect lever 100 % of the time (**a**, **b**). As the number of lights illuminated decreased, so did choice of the collect lever (**b**). Animals consistently showed a strong preference for the collect lever on 2-light losses, or near-miss trials. The proportion of collect responses made on both 2-light and 1-light losses also varied according to the precise pattern of lights illuminated (**a**). Animals clearly discriminate wins from losses on which all three lights set to off. The presence or absence of near-miss trials did not affect the rate of extinction as indicated by the number of trials completed per session (**c**). During reinstatement, near-miss trials were again present for both groups. Rats that had not experienced near-miss trials during extinction were faster to pick up the task again once win trials were rewarded (**c**; sessions R1-3: session x group $F_{2,20} = 4.31$, $p = 0.028$). In this group, near-miss errors at baseline were significantly correlated with the degree of reinstatement; the more near-miss errors animals made, the greater the reengagement with the task (**d**; $r^2 = 0.48$, $p < 0.049$). In contrast, if near-miss trials had been explicitly devalued during extinction by pairing them with non-reinforced wins, reinstatement was slower and there was no longer a relationship between near-miss errors at baseline and task re-engagement (**e**; $r^2 = 0.08$, NS) $N = 6$ per group. Data shown are mean ± SEM. * indicates significant ($p < 0.05$) group differences. Data adapted from (Winstanley et al. 2011)

of slot machine play, then the presence or absence of such trials should influence extinction and/or reinstatement. The number of trials completed dropped dramatically when wins were no longer rewarded, but this rate of extinction did not vary depending on whether near-miss trials were present (Winstanley et al. 2011). Hence, near-misses could not sustain slot machine play in the absence of reinforced win trials. However, when win trials were once again rewarded, animals which had *not* experienced near-miss trials during the extinction test were quicker to re-engage with the task and completed more trials in the first few days of play. Furthermore, the rate of reinstatement in this group was positively correlated with the number of near-miss errors rats made at baseline (Fig. 5d). Such a relationship matches clinical data showing that the magnitude of the near-miss effect correlates with the severity of PG, (Chase and Clark 2010; Habib and Dixon 2010) further validating the rSMT as a useful model. If near-miss stimuli were explicitly paired with non-reinforced win trials during extinction, reinstatement was slower and the degree to which animals previously responded to these stimuli as if they were wins did not influence reinstatement of game play. These data suggest that the ability of near-miss errors to drive reinstatement stems from a false association between near-miss stimuli and winning outcomes (Winstanley et al. 2011).

These data also suggest that the incentive value of near-miss trials is represented independently from the incentive value of wins and that a change in the latter does not automatically result in a change in the former. Thus, unless near-miss trials are explicitly paired with non-reinforced win trials, they retain the ability to invigorate behaviour (see (Balleine and Dickinson 1992) for a discussion of such incentive shifts). This disconnection is reminiscent of the notorious ability of drug-paired cues to continue to drive drug-seeking even though the drug itself has lost much of its rewarding properties (Robinson and Berridge 1993).

This particular type of gambling-related decision-making also appears very sensitive to modulation by the dopamine system and, in particular, by D_4 receptor ligands. Administration of a D_4 agonist can increase erroneous collect errors on non-win trials, whereas a D_4 receptor antagonist improved discriminative performance and attenuated the marked increased in collect errors made the following administration of a$D_{2/3}$ agonist (Cocker et al. 2013). Cognitive strategies designed to help players correctly identify near-miss trials as losses, rather than predictors of imminent wins, have shown promise in terms of reducing the time spent playing slot machines (Dixon et al. 2009). Hence, D_4 receptor antagonists may be useful in the treatment of pathological slot machine play by decreasing the ability of near-misses to evoke the expectation of reward. Recent data also implicate the anterior cingulate as a key area in which D_4 receptor agents may exert beneficial effects (Cocker et al. 2016). Ongoing studies are verifying whether similar neural circuitry is involved in mediating the near-miss effect in rats as would be expected based on neuroimaging studies. Current data, however, suggest this task offers considerable promise in tapping into a cognitive process relevant for the study of GD.

5 Conclusion

Given the challenges that we have described in developing valid models of gambling-related decision-making, with a view to ultimately provide insight into GD, it becomes clear that a single paradigm will never be able to capture all the features and cognitions subsumed by human gambling. When considering how performance of tasks such as the IGT/rGT may be used to inform our understanding of behavioural addictions, there is a real danger of operationalizing risky decision-making as representative of GD, when in fact this cognitive trait may instead represent an endophenotype indicative of vulnerability to numerous addictions as well as non-addiction related pathologies. Attempts to use methodological advances in preclinical models of drug addiction to inform the development of models of GD are largely confounded due to the lack of a specific preclinical marker for pathological engagement in gambling-like behaviour, and the use of a primary reward as a reinforcer. A lack of knowledge regarding specific neurobiological features associated exclusively with GD, or the reliable identification of any pharmaceutical treatments with proven efficacy in the remediation of GD, further complicates the validation of an animal model of GD.

Notwithstanding these obstacles in the field, recent data suggest that the decision-making of non-human animals may be subject to similar heuristics and biases that hallmark human choice under uncertainty. In describing these biases, we have focussed on animal models of escalating commitment, loss-chasing, and near-misses. It is increasingly recognized that the degree to which gamblers subscribe to certain irrational beliefs regarding the relationship between probabilistic events and outcomes can contribute to the trajectory of GD. As such, rodent models that capture individual differences in sensitivity to these biases may offer considerable future promise as the foundation to a meaningful model of vulnerability to GD.

References

Ahmed SH (2012) The science of making drug-addicted animals. Neuroscience 211:107–125

Ahmed SH, Koob GF (1998) Transition from moderate to excessive drug intake: change in hedonic set point. Science 282:298–300

Alloy L, Abramson L (1979) Judgement of contingency in depressed and nondepressed students: sadder but wiser? J Exp Psychol Gen 108:441–485

Alvarez-Moya EM, Ochoa C, Jimenez-Murcia S, Aymami MN, Gomez-Pena M, Fernandez-Aranda F, Santamaria J, Moragas L, Bove F, Menchon JM (2011) Effect of executive functioning, decision-making and self-reported impulsivity on the treatment outcome of pathologic gambling. J Psychiatry Neurosci 36:165–175

APA (2013) Diagnostic and statistical manual 5, 5th edn. American Psychiatric Press, Washington, D.C.

Balleine B, Dickinson A (1992) Signalling and incentive processes in instrumental reinforcer devaluation. Q J Exp Psychol B 45:285–301

Bar-On R, Tranel D, Denburg NL, Bechara A (2003) Exploring the neurological substrate of emotional and social intelligence. Brain 126:1790–1800

Barrus MM, Winstanley CA (in press) Inactivation of the orbitofrontal cortex reduces irrational choice on a rodent betting task. Neuroscience

Barrus MM, Hosking JG, Zeeb FD, Tremblay M, Winstanley CA (2015) Disadvantageous decision-making is associated with increased motor impulsivity at the population level on a rodent gambling task. J Psychiatry Neurosci 40:108–117

Bechara A, Damasio H (2002) Decision-making and addiction (part I): impaired activation of somatic states in substance dependent individuals when pondering decisions with negative future consequences. Neuropsychologia 40:1675–1689
Bechara A, Damasio AR, Damasio H, Anderson SW (1994) Insensitivity to future consequences following damage to human prefrontal cortex. Cognition 50:7–15
Bechara A, Damasio H, Damasio AR, Lee GP (1999) Different contributions of the human amygdala and ventromedial prefrontal cortex to decision-making. J Neurosci 19:5473–5481
Bechara A, Damasio H, Tranel D, Damasio AR (2005) The Iowa Gambling Task and the somatic marker hypothesis: some questions and answers. Trends Cogn Sci 9:159–162 (Discussion 162–154)
Bedi G, Preston KL, Epstein DH, Heishman SJ, Marrone GF, Shaham Y, de Wit H (2011) Incubation of cue-induced cigarette craving during abstinence in human smokers. Biol Psychiatry 69:708–711
Belin D, Mar AC, Dalley JW, Robbins TW, Everitt BJ (2008) High impulsivity predicts the switch to compulsive cocaine-taking. Science 320:1352–1355
Bergh C, Eklund T, Sodersten P, Nordin C (1997) Altered dopamine function in pathological gambling. Psychol Med 27:473–475
Blanchard TC, Wilke A, Hayden BY (2014) Hot-hand bias in rhesus monkeys. J Exp Psychol Anim Learn Cogn 40:280–286
Blaszczynski A, Nower L (2002) A pathways model of problem and pathological gambling. Addiction 97:487–499
Boileau I, Payer D, Chugani B, Lobo DS, Houle S, Wilson AA, Warsh J, Kish SJ, Zack M (2013a) In vivo evidence for greater amphetamine-induced dopamine release in pathological gambling: a positron emission tomography study with [C]-(+)-PHNO. Mol Psychiatry 19(12):1305–1313
Boileau I, Payer D, Chugani B, Lobo D, Behzadi A, Rusjan PM, Houle S, Wilson AA, Warsh J, Kish SJ, Zack M (2013b) The D2/3 dopamine receptor in pathological gambling: a positron emission tomography study with [11C]-(+)-propyl-hexahydro-naphtho-oxazin and [11C] raclopride. Addiction 108:953–963
Campbell-Meiklejohn DK, Woolrich MW, Passingham RE, Rogers RD (2007) Knowing when to stop: the brain mechanisms of chasing losses. Biol Psychiatry 63:293–300
Cantinotti M, Ladouceur R, Jacques C (2004) Sports betting: can gamblers beat randomness? Psychol Addict Behav 18:143–147
Casey BJ, Craddock N, Cuthbert BN, Hyman SE, Lee FS, Ressler KJ (2013) DSM-5 and RDoC: progress in psychiatry research? Nat Rev Neurosci 14:810–814
Cavedini P, Riboldi G, Keller R, D'Annucci A, Bellodi L (2002) Frontal lobe dysfunction in pathological gambling patients. Biol Psychiatry 51:334–341
Chase HW, Clark L (2010) Gambling severity predicts midbrain response to near-miss outcomes. J Neurosci 30:6180–6187
Chudasama Y (2011) Animal models of prefrontal-executive function. Behav Neurosci 125:327–343
Clark L (2014) Disordered gambling: the evolving concept of behavioral addiction. Ann N Y Acad Sci 1327:46–61
Clark L, Iversen SD, Goodwin GM (2001) A neuropsychological investigation of prefrontal cortex involvement in acute mania. Am J Psychiatry 158:1605–1611
Clark L, Lawrence AJ, Astley-Jones F, Gray N (2009) Gambling near-misses enhance motivation to gamble and recruit win-related brain circuitry. Neuron 61:481–490
Clark L, Averbeck B, Payer D, Sescousse G, Winstanley CA, Xue G (2013) Pathological choice: the neuroscience of gambling and gambling addiction. J Neurosci 33:17617–17623
Clark L, Stokes PR, Wu K, Michalczuk R, Benecke A, Watson BJ, Egerton A, Piccini P, Nutt DJ, Bowden-Jones H, Lingford-Hughes AR (2012) Striatal dopamine D(2)/D(3) receptor binding in pathological gambling is correlated with mood-related impulsivity. Neuroimage 63:40–46
Clotfelter CT, Cook PJ (1993) Notes: The "Gambler's Fallacy" in lottery play. Manage Sci 39:1521–1525
Cocker PJ, Winstanley CA (2015) Irrational beliefs, biases and gambling: exploring the role of animal models in elucidating vulnerabilities for the development of pathological gambling. Behav Brain Res 279:259–273

Cocker PJ, Hosking JG, Winstanley CA (2016) Activation of dopamine D4 receptors within the anterior cingulate cortex enhances the erroneous expectation of reward on a rat slot machine task. Neuropsychopharmacology, Jan 15th [EPub ahead of print]
Cocker PJ, Le Foll B, Rogers RD, Winstanley CA (2013) A selective role for dopamine D receptors in modulating reward expectancy in a rodent slot machine task. Biol Psychiatry
Cocker PJ, Dinelle K, Kornelson R, Sossi V, Winstanley CA (2012) Irrational choice under uncertainty correlates with lower striatal D(2/3) receptor binding in rats. J Neurosci 32:15450–15457
Comings DE, Blum K (2000) Reward deficiency syndrome: genetic aspects of behavioral disorders. Prog Brain Res 126:325–341
Cools R, D'Esposito M (2011) Inverted-U Shaped dopamine actions on human working memory and cognitive control. Biol Psychiatr 69:e113–e125
Dalley JW, Fryer TD, Brichard L, Robinson ES, Theobald DE, Laane K, Pena Y, Murphy ER, Shah Y, Probst K, Abakumova I, Aigbirhio FI, Richards HK, Hong Y, Baron JC, Everitt BJ, Robbins TW (2007) Nucleus accumbens D2/3 receptors predict trait impulsivity and cocaine reinforcement. Science 315:1267–1270
Damasio AR (1994) Descartes' error. Putman, New York
Dawson GR, Heyes CM, Iversen SD (1992) Pharmacological mechanisms and animal models of cognition. Behav Pharmacol 3:285–297
de Visser L, Baars AM, Lavrijsen M, van der Weerd CM, van den Bos R (2011a) Decision-making performance is related to levels of anxiety and differential recruitment of frontostriatal areas in male rats. Neuroscience 184:97–106
de Visser L, Homberg JR, Mitsogiannis M, Zeeb FD, Rivalan M, Fitoussi A, Galhardo V, van den Bos R, Winstanley CA, Dellu-Hagedorn F (2011b) Rodent versions of the iowa gambling task: opportunities and challenges for the understanding of decision-making. Front Neurosci 5:109
de Weijer BA, van de Giessen E, van Amelsvoort TA, Boot E, Braak B, Janssen IM, van de Laar A, Fliers E, Serlie MJ, Booij J (2011) Lower striatal dopamine D2/3 receptor availability in obese compared with non-obese subjects. EJNMMI Res 1:37
Dickinson A (1985) Actions and habits—the development of behavioral autonomy. Phil Transact Roy Soc London B—Biol Sci 308:67–78
Dixon MJ, Harrigan KA, Santesso DL, Graydon C, Fugelsang JA, Collins K (2014) The impact of sound in modern multiline video slot machine play. J Gambl Stud 30:913–929
Dixon MR, Nastally BL, Jackson JE, Habib R (2009) Altering the near-miss effect in slot machine gamblers. J Appl Behav Anal 42:913–918
Drost J, Spinhoven P, Kruijt AW, Van der Does W (2014) The influence of worry and avoidance on the Iowa Gambling Task. J Behav Ther Exp Psychiatry 45:74–80
Fellows LK, Farah MJ (2005) Different underlying impairments in decision-making following ventromedial and dorsolateral frontal lobe damage in humans. Cereb Cortex 15:58–63
Fiorillo CD, Tobler PN, Schultz W (2003) Discrete coding of reward probability and uncertainty by dopamine neurons. Science 299:1898–1902
Ford CP (2014) The role of D2-autoreceptors in regulating dopamine neuron activity and transmission. Neuroscience 282C:13–22
Garety PA, Freeman D (1999) Cognitive approaches to delusions: a critical review of theories and evidence. Br J Clin Psychol 38(Pt 2):113–154
Gilovich T, Vallone R, Tversky A (1985) The hot hand in basketball: on the misperception of random sequences. Cogn Psychol 17:295–314
Grant JE, Kim SW (2002) Effectiveness of pharmacotherapy for pathological gambling: a chart review. Ann Clin Psychiatry 14:155–161
Grant JE, Kim SW (2006) Medication management of pathological gambling. Minn Med 89:44–48
Grant JE, Potenza MN, Hollander E, Cunningham-Williams R, Nurminen T, Smits G, Kallio A (2006) Multicenter investigation of the opioid antagonist nalmefene in the treatment of pathological gambling. Am J Psychiatry 163:303–312
Griffiths M (1995) The role of subjective mood states in the maintenance of fruit machine gambling behaviour. J Gambl Stud 11:123–135

Griffiths MD, Auer M (2012) The irrelevancy of game-type in the acquisition, development, and maintenance of problem gambling. Front Psychol 3:621
Grimm JW, Hope BT, Wise RA, Shaham Y (2001) Neuroadaptation. Incubation of cocaine craving after withdrawal. Nature 412:141–142
Habib R, Dixon MR (2010) Neurobehavioral evidence for the "near-miss" effect in pathological gamblers. J Exp Anal Behav 93:313–328
Haney M, Spealman R (2008) Controversies in translational research: drug self-administration. Psychopharmacology 199:403–419
Jentsch JD, Taylor JR (1999) Impulsivity resulting from frontostriatal dysfunction in drug abuse: implications for the control of behavior by reward-related stimuli. Psychopharmacology 146:373–390
Jollant F, Bellivier F, Leboyer M, Astruc B, Torres S, Verdier R, Castelnau D, Malafosse A, Courtet P (2005) Impaired decision making in suicide attempters. Am J Psychiatry 162:304–310
Joutsa J, Johansson J, Niemela S, Ollikainen A, Hirvonen MM, Piepponen P, Arponen E, Alho H, Voon V, Rinne JO, Hietala J, Kaasinen V (2012) Mesolimbic dopamine release is linked to symptom severity in pathological gambling. Neuroimage 60:1992–1999
Kahneman D (2011) Thinking fast and slow. Penguin, London
Kahneman D, Tversky A (1979) Prospect theory: an analysis of decision under risk. Econometrica 47:263–292
Kassinove JI, Schare ML (2001) Effects of the "near miss" and the "big win" on persistence at slot machine gambling. Psychol Addict Behav 15:155–158
Kertzman S, Lidogoster H, Aizer A, Kotler M, Dannon PN (2011) Risk-taking decisions in pathological gamblers is not a result of their impaired inhibition ability. Psychiatry Res 188:71–77
Ladouceur R, Walker M (1996) A cognitive perspective on gambling. In: Salkovskis PM (ed) Trends in cognitive therapy. Wiley, Oxford, pp 89–120
Langer EJ (1975) The illusion of control. J Pers Soc Psychol 32:311–328
Lawrence NS, Jollant F, O'Daly O, Zelaya F, Phillips ML (2009) Distinct roles of prefrontal cortical subregions in the Iowa Gambling Task. Cereb Cortex 19:1134–1143
Le Moal M, Koob GF (2007) Drug addiction: pathways to the disease and pathophysiological perspectives. Eur Neuropsychopharmacol 17:377–393
Leeman RF, Potenza MN (2011) Impulse control disorders in Parkinson's disease: clinical characteristics and implications. Neuropsychiatry 1:133–147
Lerman C, LeSage MG, Perkins KA, O'Malley SS, Siegel SJ, Benowitz NL, Corrigall WA (2007) Translational research in medication development for nicotine dependence. Nat Rev 6:746–762
Ligneul R, Sescousse G, Barbalat G, Domenech P, Dreher JC (2013) Shifted risk preferences in pathological gambling. Psychol Med 43:1059–1068
Linnet J, Moller A, Peterson E, Gjedde A, Doudet D (2011) Dopamine release in ventral striatum during Iowa Gambling Task performance is associated with increased excitement levels in pathological gambling. Addiction 106:383–390
Loba P, Stewart SH, Klein RM, Blackburn JR (2001) Manipulations of the features of standard video lottery terminal (VLT) games: effects in pathological and non-pathological gamblers. J Gambl Stud 17:297–320
Lobo DS, Aleksandrova L, Knight J, Casey DM, El-Guebaly N, Nobrega JN, Kennedy JL (2014) Addiction-related genes in gambling disorders: new insights from parallel human and pre-clinical models. Mol Psychiatry
Lorains FK, Cowlishaw S, Thomas SA (2011) Prevalence of comorbid disorders in problem and pathological gambling: systematic review and meta-analysis of population surveys. Addiction 106:490–498
Lupi M, Martinotti G, Acciavatti T, Pettorruso M, Brunetti M, Santacroce R, Cinosi E, Di Iorio G, Di Nicola M, Di Giannantonio M (2014) Pharmacological treatments in gambling disorder: a qualitative review. BioMed Res Int 2014:537306

Meyer G, Schwertfeger J, Exton MS, Janssen OE, Knapp W, Stadler MA, Schedlowski M, Kruger TH (2004) Neuroendocrine response to casino gambling in problem gamblers. Psychoneuroendocrinology 29:1272–1280
Mitchell MR, Vokes CM, Blankenship AL, Simon NW, Setlow B (2011) Effects of acute administration of nicotine, amphetamine, diazepam, morphine, and ethanol on risky decision-making in rats. Psychopharmacology 218:703–712
Miu AC, Heilman RM, Houser D (2008) Anxiety impairs decision-making: psychophysiological evidence from an Iowa Gambling Task. Biol Psychol 77:353–358
Moeller FG, Barratt ES, Dougherty DM, Schmitz JM, Swann AC (2001) Psychiatric aspects of impulsivity. Am J Psychiatry 158:1783–1793
Olds J, Milner P (1954) Positive reinforcement produced by electrical stimulation of septal area and other regions of rat brain. J Comp Physiol Psychol 47:419–427
Oskarsson AT, Van Boven L, McClelland GH, Hastie R (2009) What's next? Judging sequences of binary events. Psychol Bull 135:262–285
Paine TA, Asinof SK, Diehl GW, Frackman A, Leffler J (2013) Medial prefrontal cortex lesions impair decision-making on a rodent gambling task: reversal by D1 receptor antagonist administration. Behav Brain Res 243:247–254
Peters H, Hunt M, Harper D (2010) An Animal Model of Slot Machine Gambling: The Effect of Structural Characteristics on Response Latency and Persistence. J Gambl Stud:Mar 9 [EPub ahead of print]
Petry NM (2003) A comparison of treatment-seeking pathological gamblers based on preferred gambling activity. Addiction 98:645–655
Pierce RC, O'Brien CP, Kenny PJ, Vanderschuren LJ (2012) Rational development of addiction pharmacotherapies: successes, failures, and prospects. Cold Spring Harb Perspect Med 2: a012880
Potenza MN (2006) Should addictive disorders include non-substance-related conditions? Addiction 101(Suppl 1):142–151
Potenza MN (2009) The importance of animal models of decision making, gambling, and related behaviors: implications for translational research in addiction. Neuropsychopharmacology 34:2623–2624
Potenza MN (2013) How central is dopamine to pathological gambling or gambling disorder? Frontiers in behavioral neuroscience 7:206
Power Y, Goodyear B, Crockford D (2012) Neural correlates of pathological gamblers preference for immediate rewards during the iowa gambling task: an fMRI study. J Gambl Stud 28:623–636
Pushparaj A, Kim AS, Musiol M, Zangen A, Daskalakis ZJ, Zack M, Winstanley CA, Le Foll B (2015) Differential involvement of the Agranular vs granular insular cortex in the acquisition and performance of choice behavior in a Rodent Gambling Task. Neuropsychopharmacology
Rivalan M, Coutureau E, Fitoussi A, Dellu-Hagedorn F (2011) Inter-individual decision-making differences in the effects of cingulate, orbitofrontal, and prelimbic cortex lesions in a rat gambling task. Front Behav Neurosci 5:22
Robinson TE, Berridge KC (1993) The neural basis of drug craving: an incentive-sensitization theory of addiction. Brain Res Brain Res Rev 18:247–291
Rogers RD, Wong A, McKinnon C, Winstanley CA (2013) Systemic Administration of 8-OH-DPAT and Eticlopride, but not SCH23390, Alters Loss-Chasing Behavior in the Rat. Neuropsychopharmacology
Rokosik SL, Napier TC (2012) Pramipexole-induced increased probabilistic discounting: comparison between a rodent model of Parkinson's disease and controls. Neuropsychopharmacology 37:1397–1408
Roy A, Adinoff B, Roehrich L, Lamparski D, Custer R, Lorenz V, Barbaccia M, Guidotti A, Costa E, Linnoila M (1988) Pathological gambling.A psychobiological study. Arch Gen Psychiatry 45:369–373
Sanchez-Roige S, Baro V, Trick L, Pena-Oliver Y, Stephens DN, Duka T (2014) Exaggerated waiting impulsivity associated with human binge drinking, and high alcohol consumption in mice. Neuropsychopharmacology 39:2919–2927

Scarf D, Miles K, Sloan A, Goulter N, Hegan M, Seid-Fatemi A, Harper D, Colombo M (2011) Brain cells in the avian 'prefrontal cortex' code for features of slot-machine-like gambling. PLoS ONE 6:e14589
Seamans JK, Yang CR (2004) The principal features and mechanisms of dopamine modulation in the prefrontal cortex. Prog Neurobiol 74:1–58
Shaffer HJ, LaPlante DA, LaBrie RA, Kidman RC, Donato AN, Stanton MV (2004) Toward a syndrome model of addiction: multiple expressions, common etiology. Harv Rev Psychiatry 12:367–374
Shaham Y, Shalev U, Lu L, De Wit H, Stewart J (2003) The reinstatement model of drug relapse: history, methodology and major findings. Psychopharmacology 168:3–20
Silveira MM, Malcolm E, Shoaib M, Winstanley CA (2015) Scopolamine and amphetamine produce similar decision-making deficits on a rat gambling task via independent pathways. Behav Brain Res 281:86–95
Simon NW, Gilbert RJ, Mayse JD, Bizon JL, Setlow B (2009) Balancing risk and reward: a rat model of risky decision making. Neuropsychopharmacology 34:2208–2217
Singer BF, Scott-Railton J, Vezina P (2012) Unpredictable saccharin reinforcement enhances locomotor responding to amphetamine. Behav Brain Res 226:340–344
Smith G, Levere M, Kurtzmann R (2009) Poker player behavior after big wins and big losses. Manage Sci 55:1547–1555
St. Onge JR, Floresco SB (2009) Dopaminergic regulation of risk-based decision-making. Neuropsychopharmacology 34:681–697
Stauffer WR, Lak A, Bossaerts P, Schultz W (2015) Economic choices reveal probability distortion in macaque monkeys. J Neurosci 35:3146–3154
Staw BJ (1981) The escalation of commitment to a course of action. Acad Manag Rev 6:577–587
Stewart SH, Zack M, Collins P, Klein RM (2008) Subtyping pathological gamblers on the basis of affective motivations for gambling: relations to gambling problems, drinking problems, and affective motivations for drinking. Psychol Addict Behav 22:257–268
Strong DR, Kahler CW (2007) Evaluation of the continuum of gambling problems using the DSM-IV. Addiction 102:713–721
Tanabe J, Thompson L, Claus E, Dalwani M, Hutchison K, Banich MT (2007) Prefrontal cortex activity is reduced in gambling and nongambling substance users during decision-making. Hum Brain Mapp 28:1276–1286
Toce-Gerstein M, Gerstein DR, Volberg RA (2003) A hierarchy of gambling disorders in the community. Addiction 98:1661–1672
Tremblay M, Hosking JG, Winstanley CA (2013) Effects of chronic D-2/3 agonist ropinirole medication on rodent models of gambling behaviour. Movement Disord 28:S218–S218
Tremblay M, Cocker PJ, Hosking JG, Zeeb FD, Rogers RD, Winstanley CA (2014) Dissociable effects of basolateral amygdala lesions on decision-making biases in rats when loss or gain is emphasized. Cogn Affect Behav Neurosci 14:1184–1195
van den Bos R, Koot S, de Visser L (2014) A rodent version of the Iowa Gambling Task: 7 years of progress. Front Psychol 5:203
van den Bos R, Lasthuis W, den Heijer E, van der Harst J, Spruijt B (2006) Toward a rodent model of the Iowa gambling task. Behav Res Methods 38:470–478
van Honk J, Hermans EJ, Putman P, Montagne B, Schutter DJ (2002) Defective somatic markers in sub-clinical psychopathy. NeuroReport 13:1025–1027
Verdejo-Garcia A, Lawrence AJ, Clark L (2008) Impulsivity as a vulnerability marker for substance-use disorders: review of findings from high-risk research, problem gamblers and genetic association studies. Neurosci Biobehav Rev 32:777–810
Volkow ND, Fowler JS, Wang G-J (2004) The addicted human brain viewed in the light of imaging studies: brain circuits and treatment strategies. Neuropharmacology 47:3–13
Volkow ND, Fowler JS, Wang GJ, Swanson JM, Telang F (2007) Dopamine in drug abuse and addiction: results of imaging studies and treatment implications. Arch Neurol 64:1575–1579
Voon V, Mehta AR, Hallett M (2011) Impulse control disorders in Parkinson's disease: recent advances. Curr Opin Neurol 24:324–330

Voon V, Irvine MA, Derbyshire K, Worbe Y, Lange I, Abbott S, Morein-Zamir S, Dudley R, Caprioli D, Harrison NA, Wood J, Dalley JW, Bullmore ET, Grant JE, Robbins TW (2014) Measuring "waiting" impulsivity in substance addictions and binge eating disorder in a novel analogue of rodent serial reaction time task. Biol Psychiatry 75:148–155
Wang G, Shi J, Chen N, Xu L, Li J, Li P, Sun Y, Lu L (2013) Effects of length of abstinence on decision-making and craving in methamphetamine abusers. PLoS ONE 8:e68791
Wang GJ, Volkow ND, Thanos PK, Fowler JS (2004) Similarity between obesity and drug addiction as assessed by neurofunctional imaging: a concept review. J Addict Dis 23:39–53
Weintraub D, Potenza MN (2006) Impulse control disorders in Parkinson's disease. Curr Neurol Neurosci Rep 6:302–306
Weintraub D, Siderowf AD, Potenza MN, Goveas J, Morales KH, Duda JE, Moberg PJ, Stern MB (2006) Association of dopamine agonist use with impulse control disorders in Parkinson disease. Arch Neurol 63:969–973
Weintraub D, Koester J, Potenza MN, Siderowf AD, Stacy M, Voon V, Whetteckey J, Wunderlich GR, Lang AE (2010) Impulse control disorders in Parkinson disease: a cross-sectional study of 3090 patients. Arch Neurol 67:589–595
Williams WA, Grant JE, Winstanley CA, Potenza MN (2008) Current concepts in the classification, treatment and modeling of pathological gambling and other impulse control disorders. In: McArthur RA, Borsini F (eds) Reward deficit disorders. Elsevier, Burlington, MA, pp 317–357
Winstanley CA, Eagle DM, Robbins TW (2006) Behavioral models of impulsivity in relation to ADHD: translation between clinical and preclinical studies. Clin Psychol Rev 26:379–395
Winstanley CA, Cocker PJ, Rogers RD (2011) Dopamine modulates reward expectancy during performance of a slot machine task in rats: evidence for a 'near-miss' effect. Neuropsychopharmacology 36:913–925
Winstanley CA, Olausson P, Taylor JR, Jentsch JD (2010) Insight into the relationship between impulsivity and substance abuse from studies using animal models. Alcohol Clin Exp Res 34:1306–1318
Wise RA (1996) Neurobiology of addiction. Curr Opin Neurobiol 6:243–251
Zack M, Featherstone RE, Mathewson S, Fletcher PJ (2014) Chronic exposure to a gambling-like schedule of reward predictive stimuli can promote sensitization to amphetamine in rats. Front Behav Neurosci 8:36
Zeeb FD, Winstanley CA (2011) Lesions of the basolateral amygdala and orbitofrontal cortex differentially affect acquisition and performance of a rodent gambling task. J Neurosci 31:2197–2204
Zeeb FD, Winstanley CA (2013) Functional disconnection of the orbitofrontal cortex and basolateral amygdala impairs acquisition of a rat gambling task and disrupts animals' ability to alter decision-making behavior after reinforcer devaluation. J Neurosci 33:6434–6443
Zeeb FD, Robbins TW, Winstanley CA (2009) Serotonergic and dopaminergic modulation of gambling behavior as assessed using a novel rat gambling task. Neuropsychopharmacology 34:2329–2343
Zeeb FD, Wong AC, Winstanley CA (2013) Differential effects of environmental enrichment, social-housing, and isolation-rearing on a rat gambling task: dissociations between impulsive action and risky decision-making. Psychopharmacology 225:381–395
Zeeb FD, Baarendse PJ, Vanderschuren LJM, Winstanley CA (2015) Functional integrity of the medial prefrontal cortex, but not orbitofrontal or anterior cingulate cortex, is required for optimal rat gambling task performance. Psychopharmacology (Berl)
Ziauddeen H, Fletcher PC (2013) Is food addiction a valid and useful concept? Obes Rev 14:19–28

Translational Research on Nicotine Dependence

Mary Falcone, Bridgin Lee, Caryn Lerman and Julie A. Blendy

Abstract Nicotine dependence is a chronic, relapsing disorder with complex biological mechanisms underlying the motivational basis for this behavior. Although more than 70 % of current smokers express a desire to quit, most relapse within one year, underscoring the need for novel treatments. A key focus of translational research models addressing nicotine dependence has been on cross-validation of human and animal models in order to improve the predictive value of medication screening paradigms. In this chapter, we review several lines of research highlighting the utility of cross-validation models in elucidating the biological underpinnings of nicotine reward and reinforcement, identifying factors which may influence individual response to treatment, and facilitating rapid translation of findings to practice.

Keywords Nicotine dependence · Nicotine reinforcement · Nicotine withdrawal · Smoking cessation

Contents

1 Introduction and Overarching Model 122
2 Role of the Endogenous Opioid System in Nicotine Reward and Relapse 123
2.1 Biological Mechanisms 124
2.2 Future Directions for Treatment 127
3 Interactions Between Smoking and Overeating 128
3.1 Post-cessation Weight Gain 128
3.2 Altered Nicotine Reward in Obesity 131

M. Falcone · C. Lerman (✉)
Center for Interdisciplinary Research on Nicotine Addiction, Department of Psychiatry, University of Pennsylvania, Philadelphia, PA 19104, USA
e-mail: clerman@mail.med.upenn.edu

B. Lee · J.A. Blendy
Department of Pharmacology, University of Pennsylvania, Philadelphia, PA 19104, USA
e-mail: blendy@mail.med.upenn.edu

Curr Topics Behav Neurosci (2016) 28: 121–150
DOI 10.1007/7854_2015_5005

3.3 Future Directions for Treatment ... 132
4 Nicotine Withdrawal: Cognition and Affect ... 133
4.1 Withdrawal-Related Cognitive Changes ... 133
4.2 Withdrawal-Related Affective Changes ... 136
4.3 Future Directions for Treatment ... 139
5 Summary and Conclusion ... 141
References ... 141

1 Introduction and Overarching Model

Tobacco smoking is the leading cause of preventable death worldwide and presents a major public health burden (World Health Organization 2012, 2013). In the USA alone, tobacco smoking accounts for over 400,000 deaths per year, yet 17 % of adults continue to smoke (Jamal et al. 2014; US Department of Health and Human Services 2014). There are currently only three smoking cessation treatments approved by the US Food and Drug Administration: nicotine replacement therapy, bupropion, and varenicline. Even with treatment, the majority of those who try to quit relapse within one month and only 5–10 % achieve long-term abstinence (Benowitz 2010; Centers for Disease Control and Prevention 2011). These outcomes highlight the urgent need for novel therapeutics to treat nicotine dependence.

Drug dependence, including nicotine dependence, is a chronic, relapsing disorder characterized by a compulsive urge to take the drug that persists in the face of negative social consequences; loss of control over drug intake; and emergence of a withdrawal syndrome consisting of negative physical and affective symptoms in the absence of the drug (Koob and Volkow 2010). Factors promoting dependence are thought to relate to both the positive reinforcement and rewarding properties of the drug as well as the negative reinforcement of the withdrawal syndrome, and efforts to develop novel treatments for nicotine dependence have traditionally focused on one or both of these aspects (Lerman et al. 2007). We propose a conceptual framework to improve treatment for nicotine dependence which integrates related lines of research into an overarching translational model: (a) preclinical research in neuroscience, pharmacology, and genetics to improve understanding of the biological substrates of nicotine dependence and identify novel targets for new medications; (b) preclinical and human laboratory cross-evaluation of novel medications; and c) investigations of genetic variation and other individual differences that may influence treatment response (Fig. 1) (Phillips et al. 2007).

A key focus of translational research efforts has been on cross-validation of animal and human models in order to better understand the biological underpinnings of nicotine addiction and improve the predictive value of medication screening paradigms. There is also a need to leverage basic science to identify potential factors influencing individual response to nicotine dependence treatment, and to facilitate rapid translation of findings to practice by repurposing existing medications toward the treatment of nicotine dependence. In this chapter, we

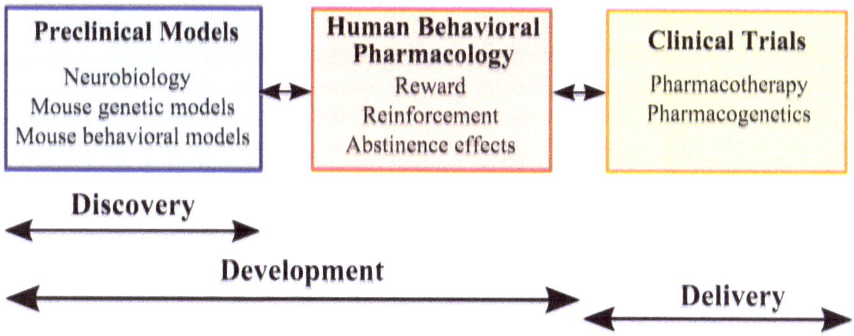

Fig. 1 Translational model of nicotine dependence (adapted from Phillips et al. 2007). This figure illustrates a conceptual framework to improve treatment for nicotine dependence by addressing discovery, development, and delivery

summarize lines of research highlighting the utility of cross-validation models in three areas: (a) elucidating the role of the endogenous opioid system in nicotine reward and dependence, (b) clarifying interactions between smoking and obesity, and c) understanding nicotine withdrawal effects on cognition and affect. Lastly, we discuss future directions for novel treatments.

2 Role of the Endogenous Opioid System in Nicotine Reward and Relapse

Although nicotine is widely accepted to be the primary addictive component of tobacco smoke, the biological mechanisms underlying its rewarding and reinforcing properties are complex. Nicotine binds to nicotinic cholinergic receptors in the brain, stimulating the release of a number of neurotransmitters including dopamine, serotonin, GABA, and glutamate (Benowitz 2010; Li et al. 2014). Like other drugs of abuse, the primary rewarding and reinforcing properties of nicotine are thought to result from stimulation of dopaminergic neurons in the ventral tegmental area (VTA) leading to increased dopamine concentrations in the nucleus accumbens (NAc) (Balfour 2015; De Biasi and Dani 2011). Nicotine also indirectly modulates the release of endogenous opioids (Berrendero et al. 2010; Boyadjieva and Sarkar 1997; Gudehithlu et al. 2012; Hadjiconstantinou and Neff 2011; Tanda and Di Chiara 1998). Mice lacking the mu-opioid receptor or its endogenous substrate, beta-endorphin, exhibit reduced responses to nicotine, including decreased antinociceptive and anxiogenic effects and reduced nicotine reward in a conditioned place preference (CPP) paradigm (Berrendero et al. 2002; Trigo et al. 2009); furthermore, nicotine-induced dopamine release in the NAc can be blocked by administration of the mu-opioid receptor (MOR) antagonist naloxanazine (Tanda and Di Chiara 1998), suggesting a role for the endogenous opioid system in nicotine reward (Berrendero et al. 2010; Hadjiconstantinou and Neff 2011).

2.1 Biological Mechanisms

To clarify the role of endogenous opioids in nicotine reward, we explored the effects of pharmacological manipulation of MOR function on nicotine reinforcement and investigated underlying molecular mechanisms in animal models (Berrendero et al. 2002; Walters et al. 2005). Mice were tested for the behavioral expression of nicotine reward using the CPP paradigm, in which nicotine, through Pavlovian conditioning, becomes associated with specific contextual cues in the environment in which it was administered (Fig. 2a). Using this paradigm, the rewarding properties of a drug are measured by the amount of time spent on the side of the chamber associated with drug administration. In addition, this study assessed brain levels of the transcription factor CREB, which has been shown to play a role in the rewarding effects of drugs of abuse (Bilbao et al. 2014; Dinieri et al. 2009; Guitart et al. 1992; Kano et al. 1995; Lane-Ladd et al. 1997; Larson et al. 2011; Nestler 2012). Results of the study showed that nicotine administration at a dose of 1.0 mg/kg increased the amount of time spent in the nicotine-paired side of the testing chamber relative to the time spent in the saline-paired chamber and also increased the amount of phosphorylated CREB in the VTA and the NAc of treated animals. In contrast, a higher dose of nicotine (2.0 mg/kg) was aversive in the CPP behavioral paradigm and did not affect pCREB levels. Pretreatment with the MOR antagonist naloxone blocked both the behavioral and molecular effects of nicotine at 1.0 mg/kg but did not impair the aversive effects of the 2.0 mg/kg dose. These data suggested that the activation of endogenous opioid receptors was specifically required for nicotine reward (Walters et al. 2005), consistent with more recent studies using genetically modified mice (Berrendero et al. 2005; Trigo et al. 2009) and MOR antagonists in rats (Goktalay et al. 2006; Ismayilova and Shoaib 2010; Liu and Jernigan 2011).

These initial studies supported a role for the endogenous opioid system in the reinforcing and rewarding properties of nicotine. To extend these findings, we decided to explore relationships between the mu-opioid system and smoking cessation in a clinical sample. The mu-opioid receptor is encoded by the human *OPRM1* gene, which includes a common Exon 1 functional Asn40Asp (A118G) single nucleotide polymorphism (SNP); the minor G allele is present in up to 30 % of individuals of European ancestry (Gelernter et al. 1999). The G allele has been associated with greater binding affinity for beta-endorphin, but also with reduced receptor expression and reduced signaling in the presence of mu-opioid agonists (Bond et al. 1998; Kroslak et al. 2007; Oertel et al. 2009; Zhang et al. 2005), suggesting an overall profile of reduced receptor function (see also discussion of more recent studies below). Smokers in an open-label randomized trial comparing transdermal nicotine to nicotine nasal spray for smoking cessation were genotyped

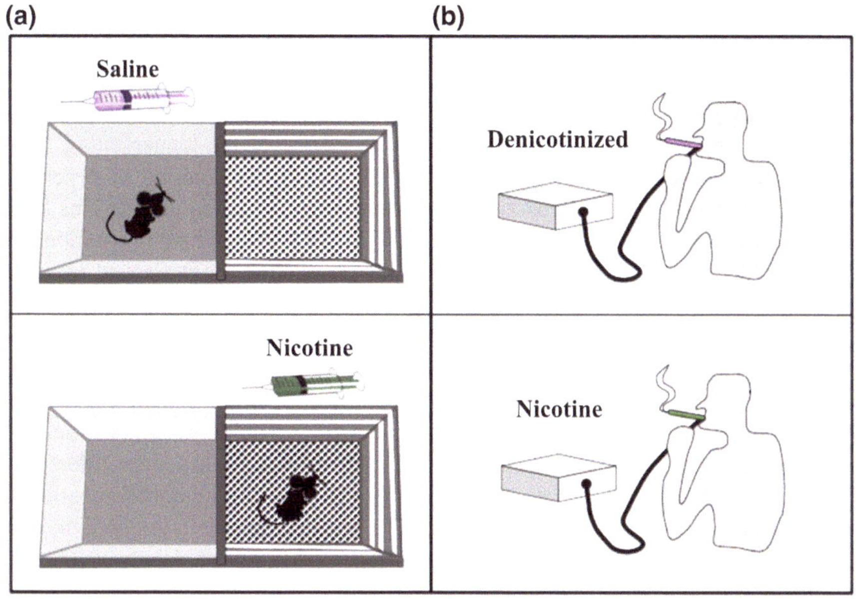

Fig. 2 Conditioned reward models in animals and humans. Conditioned nicotine associations can be established in both animals and humans. **a** Animals are exposed to conditioning boxes that consist of two distinct sides and are conditioned with saline on one side and nicotine on the other side. To test preference for nicotine, animals are allowed to roam freely between the two sides. Time spent on each side is recorded, and data are expressed as time spent on drug-paired side minus time spent on saline-paired side. A positive number indicates a preference for the drug-paired side, while a negative number indicates an aversion to the drug-paired side. A number at or near zero indicates no preference for either side. **b** Humans are exposed to two research cigarettes that are color-coded—one containing 0.6 mg nicotine and the other essentially nicotine-free (0.05 mg). Participants are instructed to take four puffs of each cigarette using a smoking topography unit to standardize puffing over time. To test preference for nicotine, participants are asked to choose puffs from any combination of the two color-coded cigarettes. The total number of times out of 16 that participants selected the 0.6 mg over the 0.05 mg nicotine cigarette during the four choice trials indicates a preference for nicotine

for *OPRM1* A118G, and differences by genotype in biochemically verified smoking status were examined (Lerman et al. 2004c). This study found that smokers carrying the G allele were significantly more likely to be abstinent at the end of treatment (8 weeks following the target quit date) than A/A homozygotes. This effect was most pronounced in the transdermal nicotine group, where 52 % of smokers who carried at least one copy of the G allele remained abstinent at the end of treatment, compared to 33 % of those homozygous for the wild-type A allele. Those carrying the G allele who received transdermal nicotine treatment also demonstrated significantly higher rates of recovery from relapse and reported greater reductions in negative affect during cessation treatment compared to A/A individuals (Lerman et al. 2004c).

Although these data provided a starting point toward identifying subgroups of smokers predisposed to relapse in a smoking cessation program, they did not provide insight into the specific biobehavioral mechanism responsible for the reduced risk of relapse associated with the *OPRM1* 118G variant. Initially, we conducted a human behavioral pharmacology study to examine the effects of the longer-acting MOR antagonist naltrexone on the relative reinforcing value of nicotine in dependent smokers (Rukstalis et al. 2005b). Utilizing a nicotine choice paradigm (Perkins 1999; Perkins et al. 1996, 1997) (Fig. 2b), this study found that acute administration of a MOR antagonist reduced the rewarding effects of nicotine, as evidenced by fewer puffs taken from the nicotine-containing cigarette following acute naltrexone administration versus placebo (Rukstalis et al. 2005b). A second human laboratory study hypothesized that carriers of the *OPRM1*-reduced activity G allele would exhibit less behavioral preference for nicotine in the nicotine choice paradigm and that administration of the MOR antagonist naltrexone would reduce nicotine preference regardless of genotype. This study also addressed possible gender differences; based on preclinical and clinical data suggesting that females are more susceptible to pharmacogenetic manipulation of the opioid system than males, it was hypothesized that the effects of *OPRM1* genotype and naltrexone on nicotine choice would be more pronounced among women smokers compared to men. A total of 60 smokers (half homozygous A/A at *OPRM1* and half carrying at least one copy of the G allele, */G) completed two counterbalanced study medication phases (naltrexone vs. placebo) with a 5- to 7-day washout period between phases (Ray et al. 2006). In contrast to a previous study (which employed acute dosing of naltrexone) (Rukstalis et al. 2005b), naltrexone was administered once daily for four days in a dose run-up fashion (12.5, 25, 50, 50 mg). On the fourth day of each medication period, subjects completed the nicotine choice paradigm. The results showed a significant interaction of *OPRM1* genotype by sex, such that women with at least one copy of the reduced-function G allele chose significantly fewer puffs from the nicotine-containing cigarette, and reported less difference between the nicotine-containing and denicotinized cigarettes with respect to satisfaction and strength, than those homozygous for the wild-type allele (A/A). There was no association between *OPRM1* genotype and nicotine choice among men. In contrast to the first study (Rukstalis et al. 2005b), there was no effect of naltrexone on nicotine preference regardless of gender or genotype, possibly indicating a difference in medication effects following short-term daily dosing compared to a single acute dose.

To explore the molecular basis of these genetic effects, we generated a mouse possessing the equivalent SNP (A112G), which corresponds to an amino acid (N38D) substitution similar to that found in humans due to high homology between mouse and human sequence at both the nucleotide (86.9 %) and amino acid level (92.3 %) (Mague et al. 2009). Mice harboring this SNP (A112G) demonstrated several phenotypic similarities to humans carrying the A118G SNP, including reduced mRNA, decreased pain threshold, and reduced morphine-mediated antinociception (Mague et al. 2009). Additional phenotypes associated with this SNP included sex-specific alterations in conditioned reward, wherein only female mice homozygous for the G allele (G/G) show reduced conditioned place preference for morphine (Mague et al.

2009), but not cocaine or nicotine (Mague, Hilario and Blendy, unpublished data). In self-administration studies, both male and female G/G mice respond for heroin significantly more (and had greater intake) than mice homozygous for the A allele (A/A). Furthermore, heroin-induced increases in striatal dopamine levels are higher in the G/G mice than in the A/A mice (Zhang et al. 2014). A complementary approach was used to generate a different mouse line for this SNP that expressed humanized receptors with and without the A118G variant (Ramchandani et al. 2011). These mice also show enhanced DA responses, in this case to alcohol (Ramchandani et al. 2011).

While additional studies are needed to explore the neurochemical and molecular changes associated with nicotine self-administration or chronic nicotine exposure and withdrawal, findings from the A112G mouse model suggested an avenue for exploration in a translational human study. Twenty-two smokers were recruited after pre-screening for *OPRM1* genotype (12 A/A, 10 */G) and completed two positron emission tomography (PET) imaging sessions following overnight abstinence (Ray et al. 2011); to provide a control group, twenty age- and sex-matched non-smokers (10 A/A and 10 */G) completed a single PET imaging session. The radiotracer used during the PET sessions was [^{11}C]carfentanil, a radioligand specific for the mu-opioid receptor; this allowed the researchers to estimate mu-opioid receptor availability in the brain. Just prior to the PET scan, smokers were asked to smoke a cigarette. During one session, this cigarette contained normal amounts of nicotine, but during the other session, a denicotinized placebo cigarette was used. The order of sessions was single-blind and counterbalanced. Smokers who carried the G allele were found to exhibit reduced MOR availability in the bilateral amygdala, left thalamus, and left anterior cingulate cortex during both sessions compared to A/A homozygotes; furthermore, the difference in receptor availability after smoking a nicotine-containing cigarette compared to a denicotinized cigarette was significantly associated with differences in subjective reward reported by G allele carriers, but not A/A homozygotes. There was no difference in receptor availability by genotype among non-smokers. These results suggest that reduced MOR availability may underlie the reduced nicotine reward demonstrated by G allele carriers. However, another study found that carriers of the G allele demonstrated greater striatal dopamine release after smoking a cigarette (Domino et al. 2012), and others observed that although greater endogenous opioid release in frontal regions (as measured by a decrease in MOR availability) after smoking a nicotine cigarette was associated with subjective reward measures, greater baseline MOR availability in the superior temporal lobe was associated with lower nicotine dependence scores (Kuwabara et al. 2014). Additional research is necessary to fully clarify the neurochemical basis for differences in nicotine reward associated with *OPRM1* genotype.

2.2 *Future Directions for Treatment*

Currently, all three medications that are efficacious and available to treat tobacco dependence (nicotine replacement therapy, varenicline, and bupropion) affect nicotinic receptor function (Coe et al. 2005; Slemmer et al. 2000). However, these pharmacological approaches to smoking cessation are only partially effective (Centers for Disease Control and Prevention 2008a; Fiore 2000; Hughes et al. 2004; US Department of Health and Human Services 2014), and there is a great need for improved treatments. The endogenous opioid system clearly has a role in maintaining a "reward pathway" in non-opiate-dependent animals. Many studies have documented the nicotine-stimulated release of endogenous opioids in various brain regions involved in reward and reinforcement. We and others have shown that nicotine acts to augment these endogenous opioids. Thus, a shift of focus from dopamine to endogenous opioids may allow for the development of more effective therapies in combating nicotine dependence. Previous studies have attempted to demonstrate opioid modulation of nicotine reinforcement in both human and animal studies with mixed results. In particular, studies reporting a negative effect of opiate antagonists on smoking behavior in humans have administered naltrexone either chronically (Wong et al. 1999), or 1–24 h prior to smoking a cigarette (Nemeth-Coslett and Griffiths 1986; Sutherland et al. 1995) and subsequent smoking behavior was evaluated in a hospital or laboratory setting. The rationale for taking a second look at opioid receptor modulation and nicotine reward lies in the ability to extract novel mechanistic information via animal models that can recapitulate the *OPRM1* SNP genotypes and the ability to directly translate the evaluation of these SNPs to the human equivalent SNP populations.

3 Interactions Between Smoking and Overeating

3.1 Post-cessation Weight Gain

Tobacco use and obesity are the two leading causes of preventable death in the USA (US Burden of Disease Collaborators 2013; Danaei et al. 2009; Mokdad et al. 2004) and take a significant economic toll (Centers for Disease Control and Prevention 2008b; Wang et al. 2011). Estimates suggest that up to 20 % of current smokers are obese (Wee et al. 2001), and the combination of these health risks may act synergistically to increase morbidity and mortality (Freedman et al. 2006; Perkins 1989). Yet, these behaviors are resistant to change; despite widespread knowledge of the risks, most people revert to their former practices of cigarette smoking or overeating (Benowitz 2010; Wing and Phelan 2005). Furthermore, tobacco use and obesity share underlying reward mechanisms, and reductions in one behavior are often accompanied by increases in the other. Following withdrawal from chronic nicotine exposure, animals and human smokers increase their daily caloric intake significantly, an effect observable on day one and lasting for weeks or months, resulting in significant weight gain (Hatsukami et al. 1984; Hughes and Hatsukami 1986; LeSage et al. 2006; Perkins 1992a; Perkins et al. 1990a, b; Robinson and York 1986; Spring et al. 1991). Indeed, cohort studies and clinical trials indicate that roughly 80 % of successful quitters gain weight (Klesges et al. 1997; Lycett et al. 2011; US Department of Health 1990); within one year, ~50 % of smokers gain > 11 lbs and ~15 % gain > 22lbs (Aubin et al. 2012). As such, post-cessation weight gain (PCWG) is often cited as a primary reason for returning to smoking (Audrain-McGovern and Benowitz 2011; Klesges et al. 1988; Meyers et al. 1997). Although some treatments have been found to reduce PCWG (Schnoll et al. 2012), many of the interventions tested tend to be ineffective or to have short-lived benefits (Farley et al. 2012; Spring et al. 2009). A recent Cochrane review found that weight management interventions can actually impede smoking cessation (Farley et al. 2012; Hall et al. 1992; Pirie et al. 1992). Given the significance of this problem for public health, novel therapeutic approaches are urgently needed. This requires identifying new intervention targets based on a better understanding of the mechanisms underlying PCWG, a topic well suited for translational research designs.

Initial investigations into PCWG focused on metabolic changes which occur during smoking cessation (Dallosso and James 1984; Perkins 1992b; Perkins et al. 1990a; Stamford et al. 1986). However, although nicotine can suppress appetite and increase resting metabolic rate, these effects tend to be very small (<5 % change) and brief (<30 min) and cannot account for the sizeable increases in caloric intake and weight gain (Klesges et al. 1997; Spring et al. 2003). On the other hand, tobacco dependence and obesity are associated with alterations in similar brain reward circuits (Volkow et al. 2012). High-calorie foods activate brain regions in reward and salience pathways that overlap with those activated by nicotine and other drugs of abuse (Tomasi and Volkow 2013; Volkow et al. 2008), and sensory cues related to highly palatable and high-calorie foods induce brain responses similar to those evoked by smoking cues (Tang et al. 2012). Neural responses to food cues have been shown to predict immediate food consumption and long-term weight gain in healthy individuals, as well as predicting outcomes of weight loss programs for overweight and obese individuals (Demos et al. 2012; Lawrence et al. 2012; Mehta et al. 2012; Murdaugh et al. 2012). Nicotine exposure is known to alter reward thresholds and food cue reactivity in mice and in humans (Hilario et al. 2012; Kroemer et al. 2013), and greater food reward is associated with increased food intake in smokers (Epstein et al. 2004). Furthermore, nicotine withdrawal is associated with changes in food reward which are associated with weight gain during smoking cessation treatment (Lerman et al. 2004a; Spring et al. 2003). It is possible that nicotine withdrawal may alter neural responses to food cues and food reward as part of a pattern of reward dysregulation, leading to increases in intake of high-calorie foods following smoking cessation.

Evidence suggests that PCWG may be modulated by the endogenous opioid system in a similar fashion to acute nicotine reward. It has been well established that endogenous opioids contribute to the regulation of ingestive behavior and food hedonics (Olszewski et al. 2011; Pecina and Smith 2010; Yeomans and Gray 2002). Morphine, the mu-opioid endogenous agonist beta-endorphin, and synthetic agonist DAMGO induce eating and increase saccharin intake in animal studies (Grandison and Guidotti 1977; Marks-Kaufman 1982; Noel and Wise 1995; Sanger and

McCarthy 1980). In contrast, opioid receptor antagonists including naltrexone and naloxone suppress food and water intake and decrease meal size in rats (Mandenoff et al. 1984; Marks-Kaufman et al. 1984; Thornhill et al. 1982). Preclinical studies suggest that the anorectic effects of opioid receptor antagonists are sensitized following a palatable diet. In contrast to the high doses of naltrexone needed to suppress normal chow food intake, only low doses of this opioid antagonist are needed to suppress hyperphagia associated with a cafeteria diet (Mandenoff et al. 1984) and cookie consumption (Levine et al. 1995) in rats. Treatment with naloxone also decreases intake of sucrose solutions and high-fat or high-sugar diets (Marks-Kaufman and Kanarek 1990), and central administration potently reduces intake of preferred, high-fat diet relative to non-preferred normal chow diets (Glass et al. 1996). This effect does not appear to affect taste preference in some studies, as naloxone fails to alter an operant discrimination task in which increasing concentrations of sucrose are used to gain food (O'Hare et al. 1997). In contrast, naloxone does decrease motivation to respond to sucrose using a progressive reinforcement schedule (Cleary et al. 1996) and these effects are enhanced in animals maintained on a sucrose diet (Rudski et al. 1997). Similar studies examining high-fat foods have not been examined.

Opioid receptor density, availability, and location within the brain all contribute to mediating the effect of the opioid system on feeding behavior. Rats that receive a palatable diet containing Nestlé condensed milk and sucrose for 17 weeks show significantly increased amounts of mu-opioid receptor binding in regions critically associated with the reward circuitry of the brain (Smith et al. 2002). In addition to increased binding, increases in protein levels of mu-opioid receptors are observed in the brains of rats susceptible to diet-induced obesity compared to rats resistant to diet-induced obesity following a high-fat diet (Barnes et al. 2006). Although an increase in mu-opioid receptors is associated with increased food intake and obesity in rodents, mice deficient in the mu-opioid receptor (MOR-/-) show decreased motivation to work for food pellets compared to wild-type mice (Papaleo et al. 2007) and decreased susceptibility to diet-induced obesity (Zuberi et al. 2008).

In humans, individuals who carried the reduced activity G allele of *OPRM1* were observed to gain less weight after quitting smoking with nicotine replacement therapy than those homozygous for the wild-type A allele (Lerman et al. 2004c), suggesting that opioid-based regulation plays a clinical role in PCWG. Administration of the opioid antagonist naltrexone decreases food intake and reduces subjective ratings of food pleasantness in healthy, normal weight volunteers (Yeomans and Gray 1996, 1997). Naltrexone showed little efficacy for smoking cessation when used alone, yet showed some promise for enhancing quit rates and reducing weight gain when used in conjunction with nicotine replacement therapy (Krishnan-Sarin et al. 1999; O'Malley et al. 2006). Building on these findings led to a larger clinical trial by O'Malley and colleagues at Yale which randomly assigned 400 treatment-seeking smokers to one of four treatment groups: 0, 25, 50, or 100 mg/day of naltrexone hydrochloride in combination with the 21 mg nicotine patch for six weeks (O'Malley et al. 2006). Although data from this sample demonstrated a trend toward improved cessation outcomes (assessed as continuous abstinence for the last four weeks of treatment) among the highest dose group compared to placebo, the effect did not reach statistical significance. However, among those who were abstinent at the end of treatment, the 25 mg/day dose of naltrexone was associated with reduced weight gain compared to placebo (O'Malley et al. 2006).

Based on the observation of reduced weight gain in the lowest dose group above, a subsequent clinical trial was designed to examine whether augmenting nicotine replacement therapy with low-dose naltrexone could improve outcomes among smokers who are particularly concerned about gaining weight after quitting (Toll et al. 2010). This trial included 172 smokers who were randomized to receive either 25 mg/day naltrexone or placebo in combination with the nicotine patch for 27 weeks (1 week prior and 26 weeks after the target quit date). Weight gain and 7-day point prevalence abstinence were assessed at 26 weeks post-quit. Unfortunately, this study observed a relatively small decrease in weight gain combined with a slight reduction in quit rates among smokers receiving 25 mg/day naltrexone compared to placebo; neither difference was statistically significant (Toll et al. 2010). Results from a subsequent trial suggest that the effects of naltrexone on smoking cessation and PCWG gain may be gender-dependent: men may show greater reductions in smoking behavior, whereas women show greater reductions in weight gain when treated with naltrexone combination therapy compared to placebo (King et al. 2012). A secondary analysis combined the samples from two clinical trials (King et al. 2012; O'Malley et al. 2006) in order to obtain a larger sample size of successful quitters at 6- and 12-months post-quit; this analysis showed that naltrexone combination treatment was associated with long-term reductions in weight gain among women who successfully quit smoking, but not among men (King et al. 2013). Further research clarifying the mechanisms underlying gender differences in response to low-dose naltrexone augmentation of smoking cessation treatment is necessary to determine whether this novel treatment may be translated into practice.

3.2 Altered Nicotine Reward in Obesity

Progress has been made in assessing and treating the effects of smoking cessation on body weight; however, few studies have examined the converse relationship—i.e., differences in smoking behavior and cessation among obese smokers compared to normal weight smokers. We examined differences in smoking cessation treatment response among participants in our prior clinical trial described above (Lerman et al. 2004b). Non-obese individuals (Body mass index; BMI < 30) were more likely to quit when treated with transdermal nicotine, which provides a slow onset and steady delivery of nicotine, whereas obese individuals (BMI ≥ 30) were more likely to benefit from nicotine nasal spray, which provides more rapid nicotine delivery. These data suggested that obese smokers might find nicotine more rewarding than non-obese smokers.

To test this hypothesis, a set of translational research experiments was conducted in mice and human subjects (Blendy et al. 2005). The human laboratory study was designed utilizing the nicotine choice paradigm described previously to assess whether the reinforcing properties of nicotine varied in obese and non-obese smokers. Contrary to the hypothesis, obese subjects took fewer puffs from the nicotine-containing cigarette than non-obese subjects and reported less difference in hedonic effects (such as satisfaction and liking) between the nicotine-containing and denicotinized cigarettes. These findings suggested that obese smokers may find the non-nicotine rewarding aspects of smoking (such as conditioned associations and sensory effects of smoking) at least as rewarding as the pharmacological effects of nicotine. In parallel, a mouse study was employed to allow for experimental manipulation of subject body weight and neurobiological assays that would be impossible in human subjects. One group of mice was fed a high-fat diet for 15 weeks to induce obesity, while a second group was maintained on standard low-fat rodent chow. After the 15-week diet manipulation, nicotine reward was assessed in the conditioned place preference paradigm, and concentrations of mRNA for the MOR and leptin (a peptide hormone involved in the control of energy expenditure and appetite which may alter brain processes related to reward) in the brain were examined. Contrary to normal chow-fed mice, obese mice failed to demonstrate a preference for nicotine in the CPP paradigm. This suggests a reduced nicotine reward in obese mice, aligning this preclinical finding with the results of the human study (Fig. 3a, b). Furthermore, mRNA levels for both the MOR and leptin receptors were significantly downregulated in the ventral tegmental area of obese mice compared to lean mice, suggesting a possible mechanism underlying reduced nicotine reward in obese individuals (Blendy et al. 2005). Taken together, the results of the translational studies do not support the hypothesis that different responses in obese smokers compared to non-obese smokers observed during the clinical trial (Lerman et al. 2004b) are due to enhanced nicotine reward in obese smokers; further research is necessary in order to replicate and clarify this finding.

3.3 *Future Directions for Treatment*

Post-cessation weight gain is a risk factor for cancer, diabetes, and heart disease. In addition to increased body weight gain, smoking cessation is associated with an increased risk of metabolic syndrome (Cena et al. 2013). A recent study reported increased diabetes and elevated fasting glucose levels three years following a quit attempt (Stein et al. 2014), and insulin resistance caused by smoking persists for two years following cessation (Chang 2012). Therefore, therapy that addresses the weight changes seen during nicotine withdrawal would not only provide improved treatments for nicotine addiction, but may also address two other leading causes of preventable diseases, i.e., obesity and type 2 diabetes. Identifying mechanisms underlying PCWG will result in a more rational approach to therapy. Animal models are a tractable system to interrogate mechanisms underlying PCWG. Like

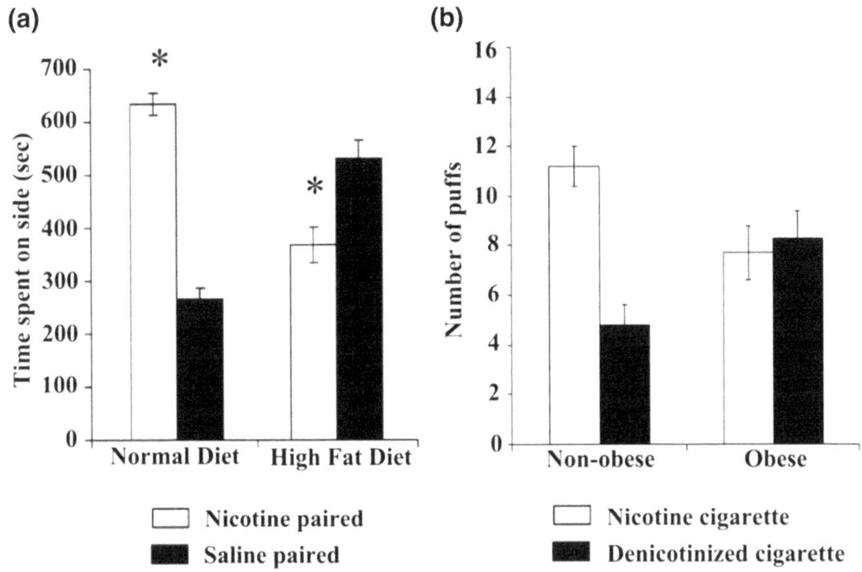

Fig. 3 Nicotine reward in obesity. In mice as well as human smokers, obese subjects display reduced preference for nicotine. **a** In a conditioned place paradigm, mice fed a normal diet showed greater preference for the side of the chamber paired with nicotine, whereas mice were fed a high-fat diet to induce obesity showed greater preference for the unpaired chamber. **b** Non-obese humans (BMI < 30) took more puffs from a nicotine-containing cigarette than a denicotinized cigarette in a nicotine choice paradigm, whereas obese smokers (BMI ≥ 30) did not discriminate between the two cigarettes. Adapted from (Blendy et al. 2005)

humans, rodents treated with nicotine exhibit blunted weight gain (Hussmann et al. 2014; Mangubat et al. 2012), decreased caloric intake (Wellman et al. 2005), and decreased consumption of palatable foods (Grunberg et al. 1988). In contrast, accelerated weight gain and increased food intake are observed during nicotine withdrawal (Dandekar et al. 2011; Mangubat et al. 2012). In rodents, females are more sensitive to these effects than males, consistent with some human studies (Grunberg et al. 1985, 1988). Research directed at the interface of basic and clinical science has a greater chance of identifying novel targets for therapies treating PCWG.

4 Nicotine Withdrawal: Cognition and Affect

4.1 Withdrawal-Related Cognitive Changes

In addition to dysregulation of reward, nicotine withdrawal is associated with deficits in executive cognitive functions such as working memory, attention, and response inhibition (Ashare et al. 2014; Hughes 2007). These deficits can be measured objectively in animals and in humans and are reversed by nicotine re-exposure, suggesting that relapse to smoking may occur as an attempt to ameliorate these deficits. Importantly, cognitive deficits arising during early abstinence predict relapse in human smokers (Culhane et al. 2008; Kassel et al. 2007; Krishnan-Sarin et al. 2007). These withdrawal-related cognitive deficits were the subject of a series of translational research studies designed to clarify their molecular and neural basis in order to identify potential targets for smoking cessation treatment.

Effects of nicotine withdrawal on learning and memory were examined in mice using the contextual fear conditioning paradigm (Davis et al. 2005). In this paradigm, mice are trained in a specific context by pairing an auditory conditioned stimulus (CS) with an aversive unconditioned stimulus (US) such as a footshock. Associations formed between the context and the US are hippocampal-dependent and result in fear behavior (i.e., freezing) upon exposure to the context without the CS (known as contextual fear conditioning); associations formed between the CS and US are not hippocampal-dependent and result in freezing in response to the CS in a different context (cued fear conditioning). To examine effects of nicotine withdrawal, mice were implanted with micro-osmotic mini pumps that administered 6.3 mg/kg/day nicotine or saline for 12 days, after which the pumps were removed. Training in the fear conditioning paradigm occurred on day 13 (after 24 h of withdrawal), and mice were tested for freezing to the context and CS on day 14. Mice withdrawn from chronic nicotine displayed less contextual fear conditioning than saline-treated mice; there was no difference between the groups in cued fear conditioning. Another group of mice underwent identical chronic exposure and withdrawal procedures, but received acute injections of nicotine or saline 5 min prior to training and testing. In this experiment, acute nicotine not only reversed the deficit in contextual fear learning in mice withdrawn from chronic nicotine, but in fact increased learning in these mice to a level similar to that seen in saline-treated mice who received acute nicotine (Davis et al. 2005). Follow-up experiments utilizing the same paradigm demonstrated that the learning deficits seen during nicotine withdrawal were mediated by nicotinic acetylcholine receptors containing the β2 subunit and could be reversed by administration of bupropion or varenicline, two effective smoking cessation medications (Portugal and Gould 2007; Portugal et al. 2008; Raybuck et al. 2008).

To translate these findings into human smokers, sixty-seven treatment-seeking smokers completed a human laboratory study investigating the cognitive effects of nicotine withdrawal. Each participant completed two identical medication periods (varenicline vs. placebo) in a double-blind, within-subject crossover design (Patterson et al. 2009). Each period consisted of a 10-day medication run-up followed by a mandatory 3-day smoking abstinence period (days 11–13), during which cognitive performance was assessed using the letter n-back task (a measure of working memory) and the continuous performance task (a measure of sustained attention). Subjects were exposed to a scheduled smoking lapse on day 14 and were then monitored for smoking lapses during a 7-day "practice quit attempt" (days 15–21). Subjects in this study displayed faster response times on the letter n-back and increased accuracy on the continuous performance task during the varenicline period compared to placebo (Patterson et al. 2010). These findings were consistent with the results of the mouse studies in demonstrating that varenicline reverses withdrawal-related cognitive deficits. Furthermore, response times during the letter n-back task were significantly associated with days to relapse during the placebo period, but not during the varenicline period; in the placebo condition, slower response times on the letter n-back task predicted faster relapse during the 7-day practice quit attempt (Patterson et al. 2010), suggesting that working memory deficits during early abstinence could represent a risk factor for relapse. However, further understanding of the mechanisms underlying these deficits was necessary in order to facilitate translation into treatment.

To elucidate the neural underpinnings of the observed working memory deficits, twenty-two smokers were recruited for a functional magnetic resonance imaging (fMRI) study. Using a similar within-subject crossover design, subjects completed two medication periods (varenicline vs. placebo) consisting of a 10-day medication run-up followed by a mandatory 3-day abstinence period. On the third day of abstinence, subjects completed a blood-oxygen-level-dependent (BOLD) fMRI scan during performance of an n-back working memory task. Results of this study showed a significant effect of treatment on mean percent BOLD signal change (a measure of brain activation): varenicline significantly increased activation in the dorsal anterior cingulate/medial frontal cortex and the left and right dorsolateral prefrontal cortex compared to placebo (Loughead et al. 2010). These brain regions are part of the executive control network and are robustly activated by working memory tasks in a load-dependent fashion; as memory load increases, activation in these regions increases (Owen et al. 2005). Using a cross-region model, there was a significant treatment by memory–load interaction effect, indicating that the effects of varenicline on brain activation were strongest at the highest memory loads (Loughead et al. 2010).

These results, combined with the laboratory data showing that working memory deficits predicted relapse, suggested that working memory-related brain activation during early abstinence could potentially serve as a biomarker for relapse risk in smokers who wanted to quit. This hypothesis was tested in a subsequent clinical trial (Falcone et al. 2014; Lerman et al. 2014; Loughead et al. 2014). In this trial, 80 treatment-seeking smokers completed two BOLD fMRI scanning sessions (one while smoking as usual and one following 24 h of smoking abstinence). During the scanning sessions, BOLD data were acquired at rest and while subjects performed an n-back working memory task similar to the one used in the prior experiment. After completing both scanning sessions, subjects received smoking cessation counseling and attempted to quit smoking without medication. Short-term smoking relapse was assessed at a follow-up visit seven days after the target quit date, where smoking status was biochemically verified by measuring levels of cotinine (the primary metabolite of nicotine) in urine. Subjects demonstrated slower response times and reduced BOLD signal in the dorsolateral prefrontal cortex (DLPFC) and medial frontal/cingulate cortex during performance of the n-back task during abstinence compared to smoking as usual (Falcone et al. 2014). Increased activation during abstinence was noted in the posterior cingulate cortex (PCC), part of the default mode network. The default mode network is normally active at rest and suppressed during task performance; greater activation during abstinence could indicate inability to suppress task-irrelevant functions. Resting state functional connectivity was examined in a subset ($n = 37$) of these subjects; these data indicated that reduced functional connectivity between the default mode network and the salience network during abstinence predicted failure to suppress default mode activity during performance of the n-back task. Furthermore, reduced functional connectivity between the executive control, salience, and default mode networks during abstinence predicted withdrawal-related cravings to smoke (Lerman et al. 2014).

Finally, brain activation during working memory performance along with other clinical measures was used to model smoking relapse during the seven-day follow-up period. Abstinence-induced decreases in left DLPFC activation, along with failure to suppress activation in the PCC, were found to predict relapse during the first week of a quit attempt above and beyond standard clinical measures (Fig. 4a–c) (Loughead et al. 2014). Alterations in working memory-related activation may therefore represent a useful biomarker in predicting relapse risk. Although broad implementation of neuroimaging in clinical practice is not currently feasible, it may provide a useful biomarker for early efficacy assessment of novel smoking cessation treatments. These data also support targeting treatments for withdrawal-related cognitive deficits to improve smoking cessation outcomes.

4.2 *Withdrawal-Related Affective Changes*

In addition to reward deficits and attentional dysregulation (Rukstalis et al. 2005a), nicotine withdrawal is associated with affective state, which has been shown to significantly impact relapse rate (Dani and Harris 2005; DiMatteo et al. 2000). Many smokers experience anxiety symptoms during acute abstinence, which may contribute to relapse (Dani and Harris 2005). Not surprisingly, adherence to treatment for tobacco dependence has been shown to predict abstinence rates (Hays et al. 2010) and affective state contributes to poor adherence to treatment regimens (DiMatteo et al. 2000). Furthermore, clinical and epidemiological studies have shown that depression and anxiety are both highly comorbid with nicotine dependence (Bertrand 2005; Lasser et al. 2000; Paperwalla et al. 2004; Zvolensky et al. 2008). An especially striking finding by Lasser et al. concluded that nearly one-half of all cigarettes smoked in the USA are consumed by persons with mental illness, suggesting nicotine usage may be an attempt at self-medication in these conditions (Lasser et al. 2000).

Varenicline, the most effective medication for smoking cessation currently available (Chantix; Pfizer), is selective for $\alpha4\beta2$ nAChRs (Coe et al. 2005) and has been shown to function as a partial agonist with 40–60 % efficacy compared to nicotine (Rollema et al. 2007a). However, in addition to the $\alpha4\beta2$ nAChRs, varenicline can also bind to α4β3 and α7 subtypes of nAChRs (Rollema et al. 2007b). In contrast, a novel α4β2 nAChR, sazetidine-A, is much more selective for this subtype. Both sazetidine-A and varenicline ameliorate cognitive withdrawal symptoms (Raybuck et al. 2008; Rezvani et al. 2010) and reduce nicotine self-administration (Levin et al. 2010; O'Connor et al. 2010; Rezvani et al. 2010). Furthermore, both varenicline and sazetidine-A have shown antidepressant efficacy in the forced swim test (Caldarone et al. 2011; Kozikowski et al. 2009; Mineur et al. 2011; Rollema et al. 2009; Turner et al. 2010). However, only infusions of sazetidine-A (but not varenicline) into the ventral hippocampus are efficacious in reducing nicotine withdrawal-related anxiety-like phenotypes (Turner et al. 2013).

A human behavioral pharmacology study with varenicline showed that this medication attenuated abstinence-induced reductions in positive affect and increases in negative affect, while restoring abstinence-induced cognitive impairments (Patterson et al. 2009). In addition, several studies support a positive or no effect of varenicline on depressive symptoms (Avery et al. 2014; Foulds et al. 2013; Hong et al. 2014), despite an earlier black box warning for adverse neuropsychiatric events. Varenicline increased smoking cessation in smokers with a history of depression as well as in those that were stably treated without worsening anxiety or depression (Anthenelli et al. 2013). Additionally, coadministration of varenicline was shown to positively enhance the effects of antidepressants in depressed smokers in a small open-label study (Philip et al. 2009).

Another first-line treatment option for smoking cessation is bupropion, a dopamine and norepinephrine reuptake inhibitor (Li et al. 2002), and nicotinic acetylcholine receptor (nAChR) antagonist (Fryer and Lukas 1999; Slemmer et al. 2000). When compared to placebo or nicotine patch treatment, bupropion produced greater long-term rates of smoking cessation (Hurt et al. 1997; Jorenby et al. 1999). In another study, abstinence rates were not significantly different between bupropion and nicotine replacement therapy, or combination treatment; however, there was some evidence that bupropion was more beneficial than nicotine replacement in those smokers with depression history (Stapleton et al. 2013).

In addition, nortriptyline, a tricyclic antidepressant which acts as a norepinephrine and serotonin reuptake inhibitor, has been recommended as a second-line pharmacotherapy for smoking cessation (Lerman et al. 2007). A review of published studies to date concluded that nortriptyline approximately doubled smoking cessation rates compared to placebo (Herman and Sofuoglu 2010). However, though these drugs presumably influence affective state as a function of their antidepressant effects, clinical studies evaluating their utility in individuals comorbid for anxiety disorders found no significant effect on maintained abstinence (Piper et al. 2011). This suggests that though these drugs may positively impact affective withdrawal state, they may not significantly attenuate other withdrawal symptoms. Alternatively, perhaps individuals comorbid for affective disorders and nicotine dependence comprise the "antidepressant treatment resistant" population, hence their dependence on nicotine in the first place and the lack of effect of antidepressants on other behaviors associated with smoking cessation. Though it is necessary to further evaluate the efficacy of all smoking cessation treatments in

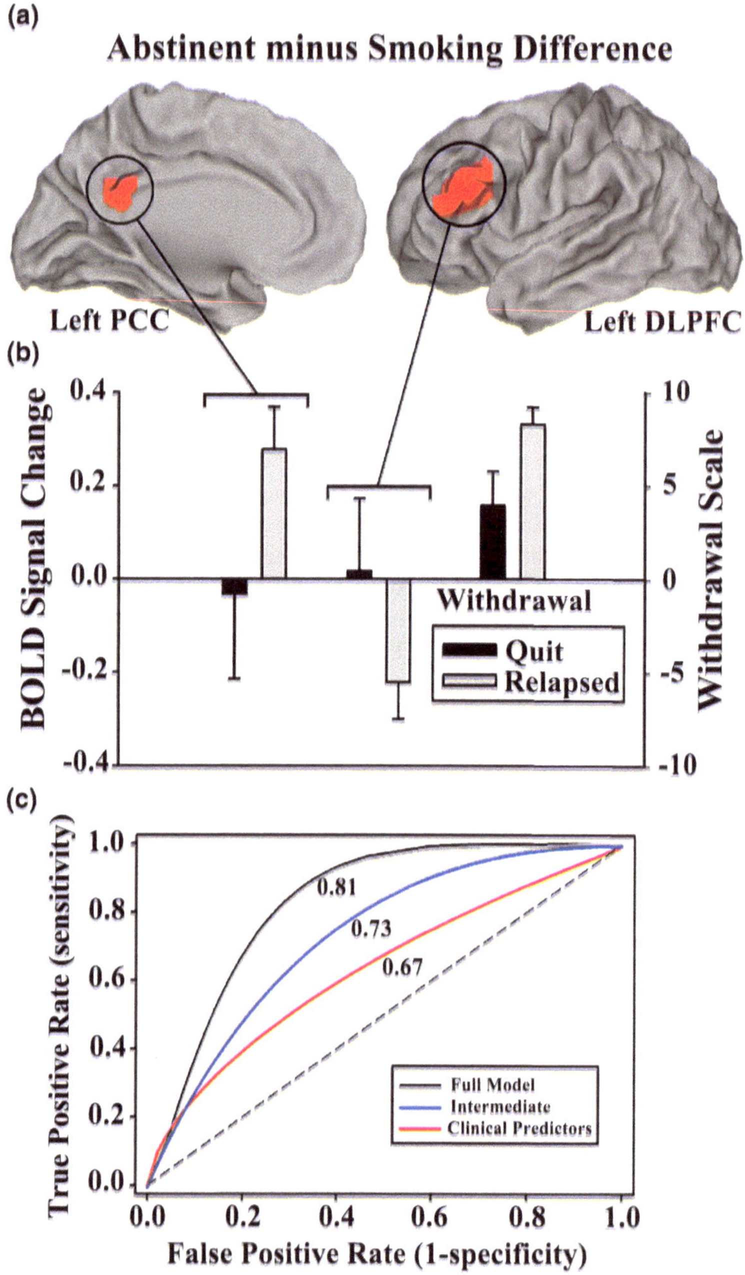
(a)
Abstinent minus Smoking Difference
Left PCC
Left DLPFC
(b)
BOLD Signal Change
0.4
0.2
0.0
-0.2
-0.4
Withdrawal
Quit
Relapsed
Withdrawal Scale
10
5
0
-5
-10
(c)
True Positive Rate (sensitivity)
1.0
0.8
0.6
0.4
0.2
0.0
0.81
0.73
0.67
Full Model
Intermediate
Clinical Predictors
False Positive Rate (1-specificity)

◂ **Fig. 4** Working memory-related BOLD signal change predicts short-term cessation. **a** Functionally defined brain regions responsive to the n-back task and associated with 7-day quit status: left dorsolateral prefrontal cortex (DLPFC) and posterior cingulate cortex (PCC). **b** The left DLPFC, left PCC, and nicotine withdrawal scores were sensitive to 24-hour abstinence challenge (versus smoking as usual); greater abstinence-induced change in withdrawal (increase), left DLPFC percent signal change (reduced activation), and PCC percent signal change (less deactivation) were predictive of relapse. **c** Receiver operating characteristic curves for three predictive models of 7-day quit status. The full model (*black*) includes clinical, withdrawal, and brain variables and significantly improved prediction above and beyond clinical predictors alone (*red*) and an intermediate model utilizing clinical and withdrawal scores (*blue*). Adapted from (Loughead et al. 2014)

comorbid populations, more in-depth investigation of these compounds in preclinical affective behavioral paradigms may help to inform clinical experimental design.

4.3 *Future Directions for Treatment*

Translational designs offer an optimal framework for testing novel treatments targeting the cognitive and affective symptoms of nicotine withdrawal associated with smoking relapse. A few initial studies which employ translational paradigms to examine procognitive treatments for nicotine dependence have shown some promise. For example, donepezil and galantamine are acetylcholinesterase inhibitors (AChEIs) currently approved by the US Food and Drug Administration for treating the cognitive symptoms of Alzheimer's disease (Terry and Buccafusco 2003). Acetylcholinesterase is an enzyme responsible for metabolizing acetylcholine in the synapse; AChEIs enhance cholinergic transmission in the brain by increasing extracellular concentrations of acetylcholine. Both donepezil and galantamine were shown to reduce nicotine self-administration in rats (Hopkins et al. 2012; Kimmey et al. 2014), and in mice, donepezil may reverse withdrawal induced deficits in contextual learning (Poole et al. 2014). Donepezil was chosen for follow-up in humans because it was the most commonly prescribed AChEI for treating cognitive deficits related to Alzheimer's disease. A human laboratory study demonstrated improved working memory in non-abstinent smokers after four weeks of treatment with donepezil (5 mg/day) compared to placebo in a double-blind, between-subject designs (Ashare et al. 2012). Cognitive performance was assessed using an n-back working memory task administered at baseline and week four. Participants receiving donepezil showed greater improvements in performance on the task (measured by number of correct responses) at week 4 compared to baseline than participants in the placebo group, especially at the highest memory load. Although there was no evidence that donepezil spontaneously reduced smoking behavior in this sample, of importance the participants were not actively attempting to quit smoking or reducing their smoking quantity (Ashare et al. 2012).

Guanfacine, an alpha2-adrenergic agonist, is another potential cognitive enhancer examined for use in smoking cessation. Noradrenergic transmission plays a critical role in executive function, and guanfacine has been shown to mitigate stress-induced nicotine reinstatement (Arnsten 2009; Yamada and Bruijnzeel 2011). A translational study at Yale University led by Sherry McKee investigated whether chronic treatment with guanfacine (titrated to steady state at 3 mg/day over 21 days) could improve cognitive control and reduce stress-precipitated smoking in a human laboratory paradigm. Thirty-three subjects completed the medication run-up period, 17 received guanfacine and 16 received placebo in a double-blind, between-subject design. After medication dosing had reached steady state, participants completed two laboratory sessions after overnight abstinence. Each session utilized the smoking lapse paradigm, a validated laboratory model of smoking relapse (McKee et al. 2012). In this paradigm, abstinent smokers are offered the opportunity to smoke a cigarette, or to receive money for not smoking. For each five-minute block of time that participants choose not to smoke, they earn $1, up to a maximum of $10 for 50 min of not smoking. Participants listened to one of two scripts at the start of each session (a personalized stress imagery script or a neutral/relaxation script) and then underwent the smoking lapse paradigm. Participants in the placebo group demonstrated a faster latency to smoke during the stress condition compared to the neutral condition, an effect that was significantly reduced in the guanfacine group. Subsets of these participants completed a separate fMRI session and/or continued into a 4-week smoking cessation treatment phase. Compared to placebo, guanfacine altered prefrontal activity during a cognitive control task and reduced cigarette use during the treatment phase (but did not affect complete abstinence) (McKee et al. 2014). Additional research is necessary to replicate these findings in a larger sample; however, in conjunction with the results for AChEIs described above, they support further translational investigation into cognitive enhancers as novel treatments for nicotine addiction.

While reinforcement is still a hallmark attribute of abused drugs, a number of laboratories have become increasingly interested in evaluating drug *dependence*. Drug dependence is classically defined as occurring when removal of drug results in withdrawal (DSM-IV 2000). Overt physical signs associated with nicotine withdrawal can be quantified (Damaj et al. 2003; Grabus et al. 2005); however, other measures of abstinence, such as cognitive impairment and negative affect, may be more sensitive measures of nicotine dependence. Furthermore, in humans, nicotine does not elicit severe somatic withdrawal signs and symptoms such as those observed during opiate withdrawal. Moreover, a number of studies suggest that traditional measures of withdrawal that focus on physical symptoms are relatively weaker predictors of relapse compared to affective and cognitive measures (Baker et al. 2004; Kenford et al. 2002; Loughead et al. 2014). Understanding the effects of nicotine abstinence on affect, including which measures of affect are most susceptible and what pharmacotherapies are most effective in ameliorating these deficits, will aid in treating nicotine addiction and developing new medications.

5 Summary and Conclusion

Despite the progress that has been made in understanding the neurobiology of nicotine addiction, only a fraction of smokers who attempt to quit using currently available treatments are successful. Current models used for the development of medications for the treatment of drug dependence evaluate the efficacy of potential compounds in animal models of drug dependence, such as intravenous drug self-administration, or place preference conditioning paradigms. These models have good face validity and predictability, allowing the evaluation of drug-taking behavior and the motivation to take drugs under various experimental conditions. However, this approach ignores some of the major deterrents to smoking cessation such as cognitive impairment, negative affect, and PCWG. Therefore, efforts should be made to invest in cross-species animal and human studies to evaluate novel targets causally linked to these factors. This approach could provide further insight into the mechanisms of nicotine withdrawal and ultimately improve the identification of novel, more effective therapeutics for smoking cessation.

References

Anthenelli RM, Morris C, Ramey TS, Dubrava SJ, Tsilkos K, Russ C et al (2013) Effects of varenicline on smoking cessation in adults with stably treated current or past major depression: a randomized trial. Ann Intern Med 159(6):390–400

Arnsten AF (2009) Stress signalling pathways that impair prefrontal cortex structure and function. Nat Rev Neurosci 10(6):410–422

Ashare RL, Falcone M, Lerman C (2014) Cognitive function during nicotine withdrawal: implications for nicotine dependence treatment. Neuropharmacology 76 Pt B: 581–591

Ashare RL, Ray R, Lerman C, Strasser AA (2012) Cognitive effects of the acetylcholinesterase inhibitor, donepezil, in healthy, non-treatment seeking smokers: a pilot feasibility study. Drug Alcohol Depend 126(1–2):263–267

Aubin HJ, Farley A, Lycett D, Lahmek P, Aveyard P (2012) Weight gain in smokers after quitting cigarettes: meta-analysis. BMJ 345:e4439

Audrain-McGovern J, Benowitz NL (2011) Cigarette smoking, nicotine, and body weight. Clin Pharmacol Ther 90(1):164–168

Avery N, Kenny AM, Kleppinger A, Brindisi J, Litt MD, Oncken CA (2014) Effects of varenicline, nicotine or placebo on depressive symptoms in postmenopausal smokers. Am J Addict 23(5):459–465

Baker TB, Brandon TH, Chassin L (2004) Motivational influences on cigarette smoking. Annu Rev Psychol 55:463–491

Balfour DJ (2015) The role of mesoaccumbens dopamine in nicotine dependence. Curr Top Behav Neurosci 24:55–98

Barnes MJ, Holmes G, Primeaux SD, York DA, Bray GA (2006) Increased expression of mu opioid receptors in animals susceptible to diet-induced obesity. Peptides 27(12):3292–3298

Benowitz NL (2010) Nicotine addiction. N Engl J Med 362(24):2295–2303

Berrendero F, Kieffer BL, Maldonado R (2002) Attenuation of nicotine-induced antinociception, rewarding effects, and dependence in mu-opioid receptor knock-out mice. J Neurosci 22(24):10935–10940

Berrendero F, Mendizabal V, Robledo P, Galeote L, Bilkei-Gorzo A, Zimmer A et al (2005) Nicotine-induced antinociception, rewarding effects, and physical dependence are decreased in mice lacking the preproenkephalin gene. J Neurosci 25(5):1103–1112

Berrendero F, Robledo P, Trigo JM, Martin-Garcia E, Maldonado R (2010) Neurobiological mechanisms involved in nicotine dependence and reward: participation of the endogenous opioid system. Neurosci Biobehav Rev 35(2):220–231

Bertrand D (2005) The possible contribution of neuronal nicotinic acetylcholine receptors in depression. Dialogues Clin Neurosci 7(3):207–216

Bilbao A, Rieker C, Cannella N, Parlato R, Golda S, Piechota M et al (2014) CREB activity in dopamine D1 receptor expressing neurons regulates cocaine-induced behavioral effects. Front Behav Neurosci 8:212

Blendy JA, Strasser A, Walters CL, Perkins KA, Patterson F, Berkowitz R et al (2005) Reduced nicotine reward in obesity: cross-comparison in human and mouse. Psychopharmacology 180(2):306–315

Bond C, LaForge KS, Tian M, Melia D, Zhang S, Borg L et al (1998) Single-nucleotide polymorphism in the human mu opioid receptor gene alters beta-endorphin binding and activity: possible implications for opiate addiction. Proc Natl Acad Sci USA 95(16):9608–9613

Boyadjieva NI, Sarkar DK (1997) The secretory response of hypothalamic beta-endorphin neurons to acute and chronic nicotine treatments and following nicotine withdrawal. Life Sci 61(6): PL59–66

Caldarone BJ, Wang D, Paterson NE, Manzano M, Fedolak A, Cavino K et al (2011) Dissociation between duration of action in the forced swim test in mice and nicotinic acetylcholine receptor occupancy with sazetidine, varenicline, and 5-I-A85380. Psychopharmacology 217(2):199–210

Centers for Disease Control and Prevention (2008a) Cigarette smoking among adults—United States, 2007. MMWR 57(47):1281

Centers for Disease Control and Prevention (2008b) Smoking attributable mortality, years of potential life lost, and productivity losses. MMWR Morb Mortal Wkly Rep 57(45):1226–1228

Centers for Disease Control and Prevention (2011) Quitting smoking among adults—United States, 2001–2010. MMWR Morb Mortal Wkly Rep 60(44):1513–1519

Cena H, Tesone A, Niniano R, Cerveri I, Roggi C, Turconi G (2013) Prevalence rate of metabolic syndrome in a group of light and heavy smokers. Diabetol Metab Syndr 5(1):28

Chang SA (2012) Smoking and type 2 diabetes mellitus. Diabetes Metab J 36(6):399–403

Cleary J, Weldon DT, O'Hare E, Billington C, Levine AS (1996) Naloxone effects on sucrose-motivated behavior. Psychopharmacology (Berl) 126(2):110–114

Coe JW, Brooks PR, Vetelino MG, Wirtz MC, Arnold EP, Huang J et al (2005) Varenicline: an alpha4beta2 nicotinic receptor partial agonist for smoking cessation. J Med Chem 48(10):3474–3477

Culhane MA, Schoenfeld DA, Barr RS, Cather C, Deckersbach T, Freudenreich O et al (2008) Predictors of early abstinence in smokers with schizophrenia. J Clin Psychiatry 69(11):1743–1750

Dallosso HM, James WP (1984) The role of smoking in the regulation of energy balance. Int J Obes 8(4):365–375

Damaj MI, Kao W, Martin BR (2003) Characterization of spontaneous and precipitated nicotine withdrawal in the mouse. J Pharmacol Exp Ther 307(2):526–534

Danaei G, Ding EL, Mozaffarian D, Taylor B, Rehm J, Murray CJ et al (2009) The preventable causes of death in the United States: comparative risk assessment of dietary, lifestyle, and metabolic risk factors. PLoS Med 6(4):e1000058

Dandekar MP, Nakhate KT, Kokare DM, Subhedar NK (2011) Effect of nicotine on feeding and body weight in rats: involvement of cocaine- and amphetamine-regulated transcript peptide. Behav Brain Res 219(1):31–38

Dani JA, Harris RA (2005) Nicotine addiction and comorbidity with alcohol abuse and mental illness. Nat Neurosci 8(11):1465–1470

Davis JA, James JR, Siegel SJ, Gould TJ (2005) Withdrawal from chronic nicotine administration impairs contextual fear conditioning in C57BL/6 mice. J Neurosci 25(38):8708–8713

De Biasi M, Dani JA (2011) Reward, addiction, withdrawal to nicotine. Annu Rev Neurosci 34:105–130

Demos KE, Heatherton TF, Kelley WM (2012) Individual differences in nucleus accumbens activity to food and sexual images predict weight gain and sexual behavior. J Neurosci 32(16):5549–5552

DiMatteo MR, Lepper HS, Croghan TW (2000) Depression is a risk factor for noncompliance with medical treatment: meta-analysis of the effects of anxiety and depression on patient adherence. Arch Intern Med 160(14):2101–2107

Dinieri JA, Nemeth CL, Parsegian A, Carle T, Gurevich VV, Gurevich E et al (2009) Altered sensitivity to rewarding and aversive drugs in mice with inducible disruption of cAMP response element-binding protein function within the nucleus accumbens. J Neurosci 29(6):1855–1859

Domino EF, Evans CL, Ni L, Guthrie SK, Koeppe RA, Zubieta JK (2012) Tobacco smoking produces greater striatal dopamine release in G-allele carriers with mu opioid receptor A118G polymorphism. Prog Neuropsychopharmacol Biol Psychiatry 38(2):236–240

DSM-IV (2000) American Psychiatric Association (2000) Diagnostic and statistical manual of mental disorders, 4th edn. Washington, DC

Epstein LH, Wright SM, Paluch RA, Leddy J, Hawk LW Jr, Jaroni JL et al (2004) Food hedonics and reinforcement as determinants of laboratory food intake in smokers. Physiol Behav 81(3):511–517

Falcone M, Wileyto EP, Ruparel K, Gerraty RT, LaPrate L, Detre JA et al (2014) Age-related differences in working memory deficits during nicotine withdrawal. Addict Biol 19(5):907–917

Farley AC, Hajek P, Lycett D, Aveyard P (2012) Interventions for preventing weight gain after smoking cessation. Cochrane Database Syst Rev 1:CD006219

Fiore MC (2000) US public health service clinical practice guideline: treating tobacco use and dependence. Respir Care 45(10):1200–1262

Foulds J, Russ C, Yu CR, Zou KH, Galaznik A, Franzon M et al (2013) Effect of varenicline on individual nicotine withdrawal symptoms: a combined analysis of eight randomized, placebo-controlled trials. Nicotine Tob Res 15(11):1849–1857

Freedman DM, Sigurdson AJ, Rajaraman P, Doody MM, Linet MS, Ron E (2006) The mortality risk of smoking and obesity combined. Am J Prev Med 31(5):355–362

Fryer JD, Lukas RJ (1999) Antidepressants noncompetitively inhibit nicotinic acetylcholine receptor function. J Neurochem 72(3):1117–1124

Gelernter J, Kranzler H, Cubells J (1999) Genetics of two mu opioid receptor gene (OPRM1) exon I polymorphisms: population studies, and allele frequencies in alcohol- and drug-dependent subjects. Mol Psychiatry 4(5):476–483

Glass MJ, Grace M, Cleary JP, Billington CJ, Levine AS (1996) Potency of naloxone's anorectic effect in rats is dependent on diet preference. Am J Physiol 271(1 Pt 2):R217–221

Goktalay G, Cavun S, Levendusky MC, Hamilton JR, Millington WR (2006) Glycyl-glutamine inhibits nicotine conditioned place preference and withdrawal. Eur J Pharmacol 530(1–2):95–102

Grabus SD, Martin BR, Batman AM, Tyndale RF, Sellers E, Damaj MI (2005) Nicotine physical dependence and tolerance in the mouse following chronic oral administration. Psychopharmacology 178(2–3):183–192

Grandison L, Guidotti A (1977) Stimulation of food intake by muscimol and beta endorphin. Neuropharmacology 16(7–8):533–536

Grunberg NE, Bowen DJ, Maycock VA, Nespor SM (1985) The importance of sweet taste and caloric content in the effects of nicotine on specific food consumption. Psychopharmacology 87(2):198–203

Grunberg NE, Popp KA, Winders SE (1988) Effects of nicotine on body weight in rats with access to "junk" foods. Psychopharmacology 94(4):536–539

Gudehithlu KP, Duchemin AM, Tejwani GA, Neff NH, Hadjiconstantinou M (2012) Nicotine-induced changes of brain beta-endorphin. Neuropeptides 46(3):125–131

Guitart X, Thompson MA, Mirante CK, Greenberg ME, Nestler EJ (1992) Regulation of cyclic AMP response element-binding protein (CREB) phosphorylation by acute and chronic morphine in the rat locus coeruleus. J Neurochem 58(3):1168–1171
Hadjiconstantinou M, Neff NH (2011) Nicotine and endogenous opioids: neurochemical and pharmacological evidence. Neuropharmacology 60(7–8):1209–1220
Hall SM, Tunstall CD, Vila KL, Duffy J (1992) Weight gain prevention and smoking cessation: cautionary findings. Am J Public Health 82(6):799–803
Hatsukami DK, Hughes JR, Pickens RW, Svikis D (1984) Tobacco withdrawal symptoms: an experimental analysis. Psychopharmacology 84(2):231–236
Hays JT, Leischow SJ, Lawrence D, Lee TC (2010) Adherence to treatment for tobacco dependence: association with smoking abstinence and predictors of adherence. Nicotine Tob Res 12(6):574–581
Herman AI, Sofuoglu M (2010) Comparison of available treatments for tobacco addiction. Curr Psychiatry Rep 12(5):433–440
Hilario MR, Turner JR, Blendy JA (2012) Reward sensitization: effects of repeated nicotine exposure and withdrawal in mice. Neuropsychopharmacology 37(12):2661–2670
Hong AS, Elrashidi MY, Schroeder DR, Ebbert JO (2014) Depressive symptoms among patients receiving varenicline and bupropion for smoking cessation. J Subst Abuse Treat
Hopkins TJ, Rupprecht LE, Hayes MR, Blendy JA, Schmidt HD (2012) Galantamine, an acetylcholinesterase inhibitor and positive allosteric modulator of nicotinic acetylcholine receptors, attenuates nicotine taking and seeking in rats. Neuropsychopharmacology 37(10):2310–2321
Hughes JR (2007) Effects of abstinence from tobacco: valid symptoms and time course. Nicotine Tob Res 9(3):315–327
Hughes JR, Hatsukami D (1986) Signs and symptoms of tobacco withdrawal. Arch Gen Psychiatry 43(3):289–294
Hughes JR, Keely J, Naud S (2004) Shape of the relapse curve and long-term abstinence among untreated smokers. Addiction 99(1):29–38
Hurt RD, Sachs DP, Glover ED, Offord KP, Johnston JA, Dale LC et al (1997) A comparison of sustained-release bupropion and placebo for smoking cessation. N Engl J Med 337(17):1195–1202
Hussmann GP, DeDominicis KE, Turner JR, Yasuda RP, Klehm J, Forcelli PA et al (2014) Chronic sazetidine-A maintains anxiolytic effects and slower weight gain following chronic nicotine without maintaining increased density of nicotinic receptors in rodent brain. J Neurochem 129(4):721–731
Ismayilova N, Shoaib M (2010) Alteration of intravenous nicotine self-administration by opioid receptor agonist and antagonists in rats. Psychopharmacology 210(2):211–220
Jamal A, Agaku IT, O'Connor E, King BA, Kenemer JB, Neff L (2014) Current cigarette smoking among adults–United States, 2005–2013. MMWR Morb Mortal Wkly Rep 63(47):1108–1112
Jorenby DE, Leischow SJ, Nides MA, Rennard SI, Johnston JA, Hughes AR et al (1999) A controlled trial of sustained-release bupropion, a nicotine patch, or both for smoking cessation. N Engl J Med 340(9):685–691
Kano T, Suzuki Y, Shibuya M, Kiuchi K, Hagiwara M (1995) Cocaine-induced CREB phosphorylation and c-Fos expression are suppressed in Parkinsonism model mice. NeuroReport 6(16):2197–2200
Kassel JD, Yates M, Brown RA (2007) Baseline reaction time predicts 12-month smoking cessation outcome in formerly depressed smokers. Psychol Addict Behav 21(3):415–419
Kenford SL, Smith SS, Wetter DW, Jorenby DE, Fiore MC, Baker TB (2002) Predicting relapse back to smoking: contrasting affective and physical models of dependence. J Consult Clin Psychol 70(1):216–227
Kimmey BA, Rupprecht LE, Hayes MR, Schmidt HD (2014) Donepezil, an acetylcholinesterase inhibitor, attenuates nicotine self-administration and reinstatement of nicotine seeking in rats. Addict Biol 19(4):539–551

King AC, Cao D, O'Malley SS, Kranzler HR, Cai X, deWit H et al (2012) Effects of naltrexone on smoking cessation outcomes and weight gain in nicotine-dependent men and women. J Clin Psychopharmacol 32(5):630–636
King AC, Cao D, Zhang L, O'Malley SS (2013) Naltrexone reduction of long-term smoking cessation weight gain in women but not men: a randomized controlled trial. Biol Psychiatry 73(9):924–930
Klesges RC, Brown K, Pascale RW, Murphy M, Williams E, Cigrang JA (1988) Factors associated with participation, attrition, and outcome in a smoking cessation program at the workplace. Health Psychol 7(6):575–589
Klesges RC, Winders SE, Meyers AW, Eck LH, Ward KD, Hultquist CM et al (1997) How much weight gain occurs following smoking cessation? A comparison of weight gain using both continuous and point prevalence abstinence. J Consult Clin Psychol 65(2):286–291
Koob GF, Volkow ND (2010) Neurocircuitry of addiction. Neuropsychopharmacology 35(1):217–238
Kozikowski AP, Eaton JB, Bajjuri KM, Chellappan SK, Chen Y, Karadi S et al (2009) Chemistry and pharmacology of nicotinic ligands based on 6-[5-(azetidin-2-ylmethoxy)pyridin-3-yl] hex-5-yn-1-ol (AMOP-H-OH) for possible use in depression. ChemMedChem 4(8):1279–1291
Krishnan-Sarin S, Reynolds B, Duhig AM, Smith A, Liss T, McFetridge A et al (2007) Behavioral impulsivity predicts treatment outcome in a smoking cessation program for adolescent smokers. Drug Alcohol Depend 88(1):79–82
Krishnan-Sarin S, Rosen MI, O'Malley SS (1999) Naloxone challenge in smokers. Preliminary evidence of an opioid component in nicotine dependence. Arch Gen Psychiatry 56(7):663–668
Kroemer NB, Guevara A, Vollstadt-Klein S, Smolka MN (2013) Nicotine alters food-cue reactivity via networks extending from the hypothalamus. Neuropsychopharmacology 38(11):2307–2314
Kroslak T, Laforge KS, Gianotti RJ, Ho A, Nielsen DA, Kreek MJ (2007) The single nucleotide polymorphism A118G alters functional properties of the human mu opioid receptor. J Neurochem 103(1):77–87
Kuwabara H, Heishman SJ, Brasic JR, Contoreggi C, Cascella N, Mackowick KM et al (2014) Mu opioid receptor binding correlates with nicotine dependence and reward in smokers. PLoS ONE 9(12):e113694
Lane-Ladd SB, Pineda J, Boundy VA, Pfeuffer T, Krupinski J, Aghajanian GK et al (1997) CREB (cAMP response element-binding protein) in the locus coeruleus: biochemical, physiological, and behavioral evidence for a role in opiate dependence. J Neurosci 17(20):7890–7901
Larson EB, Graham DL, Arzaga RR, Buzin N, Webb J, Green TA et al (2011) Overexpression of CREB in the nucleus accumbens shell increases cocaine reinforcement in self-administering rats. J Neurosci 31(45):16447–16457
Lasser K, Boyd JW, Woolhandler S, Himmelstein DU, McCormick D, Bor DH (2000) Smoking and mental illness: a population-based prevalence study. JAMA 284(20):2606–2610
Lawrence NS, Hinton EC, Parkinson JA, Lawrence AD (2012) Nucleus accumbens response to food cues predicts subsequent snack consumption in women and increased body mass index in those with reduced self-control. NeuroImage 63(1):415–422
Lerman C, Berrettini W, Pinto A, Patterson F, Crystal-Mansour S, Wileyto EP et al (2004a) Changes in food reward following smoking cessation: a pharmacogenetic investigation. Psychopharmacology 174(4):571–577
Lerman C, Gu H, Loughead J, Ruparel K, Yang Y, Stein EA (2014) Large scale brain network coupling predicts acute nicotine abstinence effects on craving and cognitive function. JAMA Psych (Epub ahead of print)
Lerman C, Kaufmann V, Rukstalis M, Patterson F, Perkins K, Audrain-McGovern J et al (2004b) Individualizing nicotine replacement therapy for the treatment of tobacco dependence: a randomized trial. Ann Intern Med 140(6):426–433
Lerman C, LeSage MG, Perkins KA, O'Malley SS, Siegel SJ, Benowitz NL et al (2007) Translational research in medication development for nicotine dependence. Nat Rev Drug Discov 6(9):746–762

Lerman C, Wileyto EP, Patterson F, Rukstalis M, Audrain-McGovern J, Restine S et al (2004c) The functional mu opioid receptor (OPRM1) Asn40Asp variant predicts short-term response to nicotine replacement therapy in a clinical trial. Pharmacogenomics J 4(3):184–192

LeSage MG, Burroughs D, Pentel PR (2006) Effects of nicotine withdrawal on performance under a progressive-ratio schedule of sucrose pellet delivery in rats. Pharmacol Biochem Behav 83(4):585–591

Levin ED, Rezvani AH, Xiao Y, Slade S, Cauley M, Wells C et al (2010) Sazetidine-A, a selective alpha4beta2 nicotinic receptor desensitizing agent and partial agonist, reduces nicotine self-administration in rats. J Pharmacol Exp Ther 332(3):933–939

Levine AS, Weldon DT, Grace M, Cleary JP, Billington CJ (1995) Naloxone blocks that portion of feeding driven by sweet taste in food-restricted rats. Am J Physiol 268(1 Pt 2):R248–252

Li SX, Perry KW, Wong DT (2002) Influence of fluoxetine on the ability of bupropion to modulate extracellular dopamine and norepinephrine concentrations in three mesocorticolimbic areas of rats. Neuropharmacology 42(2):181–190

Li X, Semenova S, D'Souza MS, Stoker AK, Markou A (2014) Involvement of glutamatergic and GABAergic systems in nicotine dependence: Implications for novel pharmacotherapies for smoking cessation. Neuropharmacology 76 Pt B: 554–565

Liu X, Jernigan C (2011) Activation of the opioid mu1, but not delta or kappa, receptors is required for nicotine reinforcement in a rat model of drug self-administration. Prog Neuropsychopharmacol Biol Psychiatry 35(1):146–153

Loughead J, Ray R, Wileyto EP, Ruparel K, Sanborn P, Siegel S et al (2010) Effects of the alpha4beta2 partial agonist varenicline on brain activity and working memory in abstinent smokers. Biol Psychiatry 67(8):715–721

Loughead J, Wileyto EP, Ruparel K, Falcone M, Hopson R, Gur R et al (2014) Working memory-related neural activity predicts future smoking relapse. Neuropsychopharmacology

Lycett D, Munafo M, Johnstone E, Murphy M, Aveyard P (2011) Associations between weight change over 8 years and baseline body mass index in a cohort of continuing and quitting smokers. Addiction 106(1):188–196

Mague SD, Isiegas C, Huang P, Liu-Chen LY, Lerman C, Blendy JA (2009) Mouse model of OPRM1 (A118G) polymorphism has sex-specific effects on drug-mediated behavior. Proc Natl Acad Sci USA 106(26):10847–10852

Mandenoff A, Bertiere MC, Betoulle D, Apfelbaum M (1984) Difference in sensitivity of the mu and kappa systems in cafeteria rats. Neuropeptides 5(1–3):265–268

Mangubat M, Lutfy K, Lee ML, Pulido L, Stout D, Davis R et al (2012) Effect of nicotine on body composition in mice. J Endocrinol 212(3):317–326

Marks-Kaufman R (1982) Increased fat consumption induced by morphine administration in rats. Pharmacol Biochem Behav 16(6):949–955

Marks-Kaufman R, Balmagiya T, Gross E (1984) Modifications in food intake and energy metabolism in rats as a function of chronic naltrexone infusions. Pharmacol Biochem Behav 20(6):911–916

Marks-Kaufman R, Kanarek RB (1990) Diet selection following a chronic morphine and naloxone regimen. Pharmacol Biochem Behav 35(3):665–669

McKee SA, Potenza MN, Kober H, Sofuoglu M, Arnsten AF, Picciotto MR et al (2014) A translational investigation targeting stress-reactivity and prefrontal cognitive control with guanfacine for smoking cessation. J Psychopharmacol

McKee SA, Weinberger AH, Shi J, Tetrault J, Coppola S (2012) Developing and validating a human laboratory model to screen medications for smoking cessation. Nicotine Tob Res Official J Soc Res Nicotine Tob 14(11):1362–1371

Mehta S, Melhorn SJ, Smeraglio A, Tyagi V, Grabowski T, Schwartz MW et al (2012) Regional brain response to visual food cues is a marker of satiety that predicts food choice. Am J Clin Nutr 96(5):989–999

Meyers AW, Klesges RC, Winders SE, Ward KD, Peterson BA, Eck LH (1997) Are weight concerns predictive of smoking cessation? A prospective analysis. J Consult Clin Psychol 65(3):448–452

Mineur YS, Einstein EB, Seymour PA, Coe JW, O'Neill BT, Rollema H et al (2011) alpha4beta2 nicotinic acetylcholine receptor partial agonists with low intrinsic efficacy have antidepressant-like properties. Behav Pharmacol 22(4):291–299
Mokdad AH, Marks JS, Stroup DF, Gerberding JL (2004) Actual causes of death in the United States, 2000. JAMA 291(10):1238–1245
Murdaugh DL, Cox JE, Cook EW 3rd, Weller RE (2012) fMRI reactivity to high-calorie food pictures predicts short- and long-term outcome in a weight-loss program. NeuroImage 59(3):2709–2721
Nemeth-Coslett R, Griffiths RR (1986) Naloxone does not affect cigarette smoking. Psychopharmacology 89(3):261–264
Nestler EJ (2012) Transcriptional mechanisms of drug addiction. Clin Psychopharmacol Neurosci 10(3):136–143
Noel MB, Wise RA (1995) Ventral tegmental injections of a selective mu or delta opioid enhance feeding in food-deprived rats. Brain Res 673(2):304–312
O'Connor EC, Parker D, Rollema H, Mead AN (2010) The alpha4beta2 nicotinic acetylcholine-receptor partial agonist varenicline inhibits both nicotine self-administration following repeated dosing and reinstatement of nicotine seeking in rats. Psychopharmacology 208(3):365–376
O'Hare EO, Cleary J, Bartz PJ, Weldon DT, Billington CJ, Levine AS (1997) Naloxone administration following operant training of sucrose/water discrimination in the rat. Psychopharmacology (Berl) 129:289–294
O'Malley SS, Cooney JL, Krishnan-Sarin S, Dubin JA, McKee SA, Cooney NL et al (2006) A controlled trial of naltrexone augmentation of nicotine replacement therapy for smoking cessation. Arch Intern Med 166(6):667–674
Oertel BG, Kettner M, Scholich K, Renne C, Roskam B, Geisslinger G et al (2009) A common human micro-opioid receptor genetic variant diminishes the receptor signaling efficacy in brain regions processing the sensory information of pain. J Biol Chem 284(10):6530–6535
Olszewski PK, Alsio J, Schioth HB, Levine AS (2011) Opioids as facilitators of feeding: can any food be rewarding? Physiol Behav 104(1):105–110
Owen AM, McMillan KM, Laird AR, Bullmore E (2005) N-back working memory paradigm: a meta-analysis of normative functional neuroimaging studies. Hum Brain Mapp 25(1):46–59
Papaleo F, Kieffer BL, Tabarin A, Contarino A (2007) Decreased motivation to eat in mu-opioid receptor-deficient mice. Eur J Neurosci 25(11):3398–3405
Paperwalla KN, Levin TT, Weiner J, Saravay SM (2004) Smoking and depression. Med Clin North Am 88(6):1483–1494, x–xi
Patterson F, Jepson C, Loughead J, Perkins K, Strasser AA, Siegel S et al (2010) Working memory deficits predict short-term smoking resumption following brief abstinence. Drug Alcohol Depend 106(1):61–64
Patterson F, Jepson C, Strasser AA, Loughead J, Perkins KA, Gur RC et al (2009) Varenicline improves mood and cognition during smoking abstinence. Biol Psychiatry 65(2):144–149
Pečina S, Smith KS (2010) Hedonic and motivational roles of opioids in food reward: implications for overeating disorders. Pharmacol Biochem Behav 97(1):34–46
Perkins KA (1989) Interactions among coronary heart disease risk factors. Ann Behav Med 11:3–11
Perkins KA (1992a) Effects of tobacco smoking on caloric intake. Br J Addict 87(2):193–205
Perkins KA (1992b) Metabolic effects of cigarette smoking. J Appl Physiol 72(2):401–409
Perkins KA (1999) Nicotine self-administration. Nicotine Tob Res 1(Suppl 2): S133–137, discussion S139–140
Perkins KA, Epstein LH, Pastor S (1990a) Changes in energy balance following smoking cessation and resumption of smoking in women. J Consult Clin Psychol 58(1):121–125
Perkins KA, Epstein LH, Sexton JE, Pastor S (1990b) Effects of smoking cessation on consumption of alcohol and sweet, high-fat foods. J Subst Abuse 2(3):287–297
Perkins KA, Grobe JE, Caggiula A, Wilson AS, Stiller RL (1997) Acute reinforcing effects of low-dose nicotine nasal spray in humans. Pharmacol Biochem Behav 56(2):235–241

Perkins KA, Grobe JE, Weiss D, Fonte C, Caggiula A (1996) Nicotine preference in smokers as a function of smoking abstinence. Pharmacol Biochem Behav 55(2):257–263

Philip NS, Carpenter LL, Tyrka AR, Whiteley LB, Price LH (2009) Varenicline augmentation in depressed smokers: an 8-week, open-label study. J Clin Psychiatry 70(7):1026–1031

Phillips JM, Siegel SJ, Shields AE, Patterson F, Gould TJ, Strasser AA et al (2007) Translating basic science to improve pharmacotherapy for nicotine dependence. Nicotine Tob Res 9(Suppl 4):S583–598

Piper ME, Cook JW, Schlam TR, Jorenby DE, Baker TB (2011) Anxiety diagnoses in smokers seeking cessation treatment: relations with tobacco dependence, withdrawal, outcome and response to treatment. Addiction 106(2):418–427

Pirie PL, McBride CM, Hellerstedt W, Jeffery RW, Hatsukami D, Allen S et al (1992) Smoking cessation in women concerned about weight. Am J Public Health 82(9):1238–1243

Poole RL, Connor DA, Gould TJ (2014) Donepezil reverses nicotine withdrawal-induced deficits in contextual fear conditioning in C57BL/6J mice. Behav Neurosci 128(5):588–593

Portugal GS, Gould TJ (2007) Bupropion dose-dependently reverses nicotine withdrawal deficits in contextual fear conditioning. Pharmacol Biochem Behav 88(2):179–187

Portugal GS, Kenney JW, Gould TJ (2008) Beta2 subunit containing acetylcholine receptors mediate nicotine withdrawal deficits in the acquisition of contextual fear conditioning. Neurobiol Learn Mem 89(2):106–113

Ramchandani VA, Umhau J, Pavon FJ, Ruiz-Velasco V, Margas W, Sun H et al (2011) A genetic determinant of the striatal dopamine response to alcohol in men. Mol Psychiatry 16(8):809–817

Ray R, Jepson C, Patterson F, Strasser A, Rukstalis M, Perkins K et al (2006) Association of OPRM1 A118G variant with the relative reinforcing value of nicotine. Psychopharmacology 188(3):355–363

Ray R, Ruparel K, Newberg A, Wileyto EP, Loughead JW, Divgi C et al (2011) Human Mu Opioid Receptor (OPRM1 A118G) polymorphism is associated with brain mu-opioid receptor binding potential in smokers. Proc Natl Acad Sci USA 108(22):9268–9273

Raybuck JD, Portugal GS, Lerman C, Gould TJ (2008) Varenicline ameliorates nicotine withdrawal-induced learning deficits in C57BL/6 mice. Behav Neurosci 122(5):1166–1171

Rezvani AH, Slade S, Wells C, Petro A, Lumeng L, Li TK et al (2010) Effects of sazetidine-A, a selective alpha4beta2 nicotinic acetylcholine receptor desensitizing agent on alcohol and nicotine self-administration in selectively bred alcohol-preferring (P) rats. Psychopharmacology 211(2):161–174

Robinson S, York DA (1986) The effect of cigarette smoking on the thermic response to feeding. Int J Obes 10(5):407–417

Rollema H, Chambers LK, Coe JW, Glowa J, Hurst RS, Lebel LA et al (2007a) Pharmacological profile of the alpha4beta2 nicotinic acetylcholine receptor partial agonist varenicline, an effective smoking cessation aid. Neuropharmacology 52(3):985–994

Rollema H, Coe JW, Chambers LK, Hurst RS, Stahl SM, Williams KE (2007b) Rationale, pharmacology and clinical efficacy of partial agonists of alpha4beta2 nACh receptors for smoking cessation. Trends Pharmacol Sci 28(7):316–325

Rollema H, Guanowsky V, Mineur YS, Shrikhande A, Coe JW, Seymour PA et al (2009) Varenicline has antidepressant-like activity in the forced swim test and augments sertraline's effect. Eur J Pharmacol 605(1–3):114–116

Rudski JM, Billington CJ, Levine AS (1997) A sucrose-based maintenance diet increases sensitivity to appetite suppressant effects of naloxone. Pharmacol Biochem Behav 58(3):679–682

Rukstalis M, Jepson C, Patterson F, Lerman C (2005a) Increases in hyperactive-impulsive symptoms predict relapse among smokers in nicotine replacement therapy. J Subst Abuse Treat 28(4):297–304

Rukstalis M, Jepson C, Strasser A, Lynch KG, Perkins K, Patterson F et al (2005b) Naltrexone reduces the relative reinforcing value of nicotine in a cigarette smoking choice paradigm. Psychopharmacology 180(1):41–48

Sanger DJ, McCarthy PS (1980) Differential effects of morphine on food and water intake in food deprived and freely-feeding rats. Psychopharmacology 72(1):103–106
Schnoll RA, Wileyto EP, Lerman C (2012) Extended duration therapy with transdermal nicotine may attenuate weight gain following smoking cessation. Addict Behav 37(4):565–568
Slemmer JE, Martin BR, Damaj MI (2000) Bupropion is a nicotinic antagonist. J Pharmacol Exp Ther 295(1):321–327
Smith SL, Harrold JA, Williams G (2002) Diet-induced obesity increases mu opioid receptor binding in specific regions of the rat brain. Brain Res 953(1–2):215–222
Spring B, Howe D, Berendsen M, McFadden HG, Hitchcock K, Rademaker AW et al (2009) Behavioral intervention to promote smoking cessation and prevent weight gain: a systematic review and meta-analysis. Addiction 104(9):1472–1486
Spring B, Pagoto S, McChargue D, Hedeker D, Werth J (2003) Altered reward value of carbohydrate snacks for female smokers withdrawn from nicotine. Pharmacol Biochem Behav 76(2):351–360
Spring B, Wurtman J, Gleason R, Wurtman R, Kessler K (1991) Weight gain and withdrawal symptoms after smoking cessation: a preventive intervention using d-fenfluramine. Health Psychol 10(3):216–223
Stamford BA, Matter S, Fell RD, Papanek P (1986) Effects of smoking cessation on weight gain, metabolic rate, caloric consumption, and blood lipids. Am J Clin Nutr 43(4):486–494
Stapleton J, West R, Hajek P, Wheeler J, Vangeli E, Abdi Z et al (2013) Randomized trial of nicotine replacement therapy (NRT), bupropion and NRT plus bupropion for smoking cessation: effectiveness in clinical practice. Addiction 108(12):2193–2201
Stein JH, Asthana A, Smith SS, Piper ME, Loh WY, Fiore MC et al (2014) Smoking cessation and the risk of diabetes mellitus and impaired fasting glucose: three-year outcomes after a quit attempt. PLoS ONE 9(6):e98278
Sutherland G, Stapleton JA, Russell MA, Feyerabend C (1995) Naltrexone, smoking behaviour and cigarette withdrawal. Psychopharmacology 120(4):418–425
Tanda G, Di Chiara G (1998) A dopamine-mu1 opioid link in the rat ventral tegmentum shared by palatable food (Fonzies) and non-psychostimulant drugs of abuse. Eur J Neurosci 10(3):1179–1187
Tang DW, Fellows LK, Small DM, Dagher A (2012) Food and drug cues activate similar brain regions: a meta-analysis of functional MRI studies. Physiol Behav 106(3):317–324
Terry AV Jr, Buccafusco JJ (2003) The cholinergic hypothesis of age and Alzheimer's disease-related cognitive deficits: recent challenges and their implications for novel drug development. J Pharmacol Exp Ther 306(3):821–827
Thornhill JA, Taylor B, Marshall W, Parent K (1982) Central, as well as peripheral naloxone administration suppresses feeding in food-deprived Sprague-Dawley and genetically obese (Zucker) rats. Physiol Behav 29(5):841–846
Toll BA, White M, Wu R, Meandzija B, Jatlow P, Makuch R et al (2010) Low-dose naltrexone augmentation of nicotine replacement for smoking cessation with reduced weight gain: a randomized trial. Drug Alcohol Depend 111(3):200–206
Tomasi D, Volkow ND (2013) Striatocortical pathway dysfunction in addiction and obesity: differences and similarities. Crit Rev Biochem Mol Biol 48(1):1–19
Trigo JM, Zimmer A, Maldonado R (2009) Nicotine anxiogenic and rewarding effects are decreased in mice lacking beta-endorphin. Neuropharmacology 56(8):1147–1153
Turner JR, Castellano LM, Blendy JA (2010) Nicotinic partial agonists varenicline and sazetidine-A have differential effects on affective behavior. J Pharmacol Exp Ther 334(2):665–672
Turner JR, Wilkinson DS, Poole RL, Gould TJ, Carlson GC, Blendy JA (2013) Divergent functional effects of sazetidine-a and varenicline during nicotine withdrawal. Neuropsychopharmacology 38(10):2035–2047
US Burden of Disease Collaborators (2013) The state of US health, 1990–2010: burden of diseases, injuries, and risk factors. JAMA 310(6):591–608

US Department of Health and Human Services (2014) The health consequences of smoking—50 years of progress: a report of the surgeon general. In: U.S. Department of Health and Human Services, Office on Smoking and Health (ed). U.S. Department of Health and Human Services, Atlanta

US Department of Health and Human Services (1990) (Public Health Service). Healthy people 2000: national health promotion and disease prevention objectives. US Department of Health and Human Services, Public Health Service, Washington, DC. DHHS publication no. (PHS) 90–50212

Volkow ND, Wang GJ, Fowler JS, Telang F (2008) Overlapping neuronal circuits in addiction and obesity: evidence of systems pathology. Philos Trans R Soc Lond B Biol Sci 363(1507):3191–3200

Volkow ND, Wang GJ, Fowler JS, Tomasi D, Baler R (2012) Food and drug reward: overlapping circuits in human obesity and addiction. Curr Top Behav Neurosci 11:1–24

Walters CL, Cleck JN, Kuo YC, Blendy JA (2005) Mu-opioid receptor and CREB activation are required for nicotine reward. Neuron 46(6):933–943

Wang YC, McPherson K, Marsh T, Gortmaker SL, Brown M (2011) Health and economic burden of the projected obesity trends in the USA and the UK. Lancet 378(9793):815–825

Wee CC, Rigotti NA, Davis RB, Phillips RS (2001) Relationship between smoking and weight control efforts among adults in the United States. Arch Intern Med 161(4):546–550

Wellman PJ, Bellinger LL, Cepeda-Benito A, Susabda A, Ho DH, Davis KW (2005) Meal patterns and body weight after nicotine in male rats as a function of chow or high-fat diet. Pharmacol Biochem Behav 82(4):627–634

World Health Organization (2012) WHO global report: mortality attributable to tobacco

World Health Organization (2013) WHO report on the global tobacco epidemic 2013

Wing RR, Phelan S (2005) Long-term weight loss maintenance. Am J Clin Nutr 82(1 Suppl): 222S–225S

Wong GY, Wolter TD, Croghan GA, Croghan IT, Offord KP, Hurt RD (1999) A randomized trial of naltrexone for smoking cessation. Addiction 94(8):1227–1237

Yamada H, Bruijnzeel AW (2011) Stimulation of alpha2-adrenergic receptors in the central nucleus of the amygdala attenuates stress-induced reinstatement of nicotine seeking in rats. Neuropharmacology 60(2–3):303–311

Yeomans MR, Gray RW (1996) Selective effects of naltrexone on food pleasantness and intake. Physiol Behav 60(2):439–446

Yeomans MR, Gray RW (1997) Effects of naltrexone on food intake and changes in subjective appetite during eating: evidence for opioid involvement in the appetizer effect. Physiol Behav 62(1):15–21

Yeomans MR, Gray RW (2002) Opioid peptides and the control of human ingestive behaviour. Neurosci Biobehav Rev 26(6):713–728

Zhang Y, Picetti R, Butelman ER, Ho A, Blendy JA, Kreek MJ (2014) Mouse model of the OPRM1 (A118G) polymorphism: differential heroin self-administration behavior compared with wild-type mice. Neuropsychopharmacology

Zhang Y, Wang D, Johnson AD, Papp AC, Sadee W (2005) Allelic expression imbalance of human mu opioid receptor (OPRM1) caused by variant A118G. J Biol Chem 280(38):32618–32624

Zuberi AR, Townsend L, Patterson L, Zheng H, Berthoud HR (2008) Increased adiposity on normal diet, but decreased susceptibility to diet-induced obesity in mu-opioid receptor-deficient mice. Eur J Pharmacol 585(1):14–23

Zvolensky MJ, Gonzalez A, Bonn-Miller MO, Bernstein A, Goodwin RD (2008) Negative reinforcement/negative affect reduction cigarette smoking outcome expectancies: incremental validity for anxiety focused on bodily sensations and panic attack symptoms among daily smokers. Exp Clin Psychopharmacol 16(1):66–76

The Need for Treatment Responsive Translational Biomarkers in Alcoholism Research

Markus Heilig, Wolfgang H. Sommer and Rainer Spanagel

Abstract Over the past two decades, major advances have been made in the basic neuroscience of alcohol addiction. However, few of these have been translated into clinically useful treatments, which remain limited. In the past decade, psychiatric drug development in general has been stalled, with many preclinically validated mechanisms failing in clinical development. Despite the existence of appealing preclinical models in the area of addictive disorders, drug development for these conditions has been impacted by the exodus of major pharma from psychiatric neuroscience. Here, we discuss translational biomarker strategies that may help turn this tide. Following an approach patterned on an endophenotype approach to complex behavioral traits, we hypothesize that relatively simple biological measures should be sought that can be obtained both in experimental animals and in humans, and that may be responsive to alcoholism medications. These biomarkers have to be tailored to the specific mechanism targeted by candidate medications and may in fact also help identify biologically more homogeneous subpopulations of patients. We introduce as examples alcohol-induced dopamine (DA) release, measures of central glutamate levels, and network connectivity, and discuss our experience to date with these biomarker strategies.

Keywords Alcohol endophenotype · Biomarker genetics glutamate · CRF naltrexone dopamine · Research domain criteria

M. Heilig (✉)
Center for Social and Affective Neuroscience, Department of Clinical and Experimental Medicine, Linköping University, SE-58183 Linköping, Sweden
e-mail: mheilig@mail.nih.gov

W.H. Sommer · R. Spanagel
Institute of Psychopharmacology, Central Institute of Mental Health, Faculty of Medicine Mannheim, University of Heidelberg, Heidelberg, Germany

Curr Topics Behav Neurosci (2016) 28: 151–171
DOI 10.1007/7854_2015_5006

Contents

1 Alcohol Addiction: An Area of Major Unmet Medical Needs 122
2 The Aspirations of Translational Research for Alcohol Addiction 124
3 The Promise of Translational Biomarkers 126
3.1 Alcohol-Induced Dopamine (DA) Activation as Translational Biomarker 127
3.2 Measures of Glutamate Activity as Translational Biomarkers 131
4 Future Directions: The Whole—More than the Sum of Its Parts? 134
5 Conclusion 135
References 136

1 Alcohol Addiction: An Area of Major Unmet Medical Needs

Alcohol use accounts for a substantial proportion of global health burden. In industrialized countries, about 10 % of all disability-adjusted life years (DALYs) lost are due to the consumption of alcohol (Whiteford et al. 2013). An expert evaluation in the United Kingdom concluded that in aggregate, the harm to self and others inflicted by alcohol exceeds that caused by heroin or cocaine (Nutt et al. 2010). Reducing alcohol-related harm is therefore a major public health priority. Policies that determine availability, price, and attitudes interact with individual susceptibility factors to result in harmful consumption. In the end, alcohol-related harm is closely correlated with the individual level of consumption (Rehm et al. 2003). Accordingly, both preventative measures aimed at reducing overall levels of alcohol use in society and treatments that can help individuals with established alcohol problems are important elements in a tool-kit of evidence-based strategies to reduce alcohol-related harm (Anderson et al. 2009).

A modern, disease-oriented view of alcohol problems emerged only with the beginnings of the Alcoholics Anonymous (AA) movement in the USA, in large part as a reaction to views of drinking as a moral failing. The construct of an "alcohol dependence syndrome" was introduced almost half a century later (Edwards and Gross 1976). Epidemiological studies found that "alcohol dependence" as subsequently defined by the criteria of the fourth edition of the American Diagnostic and Statistical Manual of Mental Disorders (DSM-IV; American Psychiatric Association 2000) affected about 12 % of Americans at some point in their lives (Grant et al. 2004). Although alcoholics lose more than 20 years in average life expectancy compared to the normal population (John et al. 2013; Westman et al. 2014), only a minority of individuals identified by these criteria ever seek treatment.

Alcohol dependence is highly heterogeneous, and this also applies to the large population of people in the community who meet formal criteria for this condition but do not seek treatment. Many of these individuals do not seek or receive treatment despite suffering obvious negative consequences of their alcohol use. They are caught in a vicious cycle of addictive behavior, do not have easy access to treatments they find appealing, and have learned that for a few hours, resuming alcohol use is the easiest way to escape the misery of alcohol addiction. The resulting treatment gap is greater in alcoholism than in other psychiatric conditions such as depression, where arrival of well-tolerated therapies resulted in progressively decreased stigma and more patients being reached by treatment. Developing treatments that can attract and effectively treat the large population of alcohol-dependent people who could benefit from them is clearly a major public health priority.

Others, on the other hand, may fulfill diagnostic criteria on an interview, but do not seem to experience measurable negative consequences. The diagnostic constructs enshrined in the DMS criteria may not be particularly useful when attempting to delineate these populations, or identify patients responsive to specific treatments. For instance, when a life-time diagnosis of DSM-IV of "alcohol dependence" was established in people in the general population, and these people were reassessed five years later, two-thirds of the women and one-third of the men no longer received the same diagnosis. People with clinically significant alcohol problems seem however to know something about themselves that the diagnostic criteria do not capture. In the same study, among individuals who had ever sought treatment, diagnosis was highly stable (Culverhouse et al. 2005). The "alcohol use disorder" construct recently introduced in the fifth edition of the DSM has unfortunately further diluted the diagnostic category. Recent epidemiological studies have shown that "moderate" to "severe" alcohol use disorder corresponds to what was previously captured by "alcohol dependence"; the addition of a "mild" alcohol use disorder category represents an expansion into unknown territory (Compton et al. 2013).

In seeking to develop novel treatments, it may therefore be helpful to look beyond the ever-changing diagnostic labels. A key characteristic of people with excessive alcohol use who seek treatment is that their consumption continues despite knowledge of adverse consequences. This pattern reflects what is increasingly held to be a core feature of addictive disorders (Piazza and Deroche-Gamonet 2013). Our discussion will therefore focus on people with this type of "aversion-resistant alcohol use," and interchangeably refer to their condition as "alcohol addiction" or simply "alcoholism," the term preferred by the largest client organization, AA. Even within this population, different pathophysiological mechanisms are at play in different individuals, or in the same individual at different time points of their addictive career. Translational medications development efforts must recognize this heterogeneity to succeed. A key determinant of success will be the ability to inform the selection of individuals and outcomes in early clinical trials, which allows these studies to enrich efficacy signals that may be present in more homogenous subpopulations of patients, but wash out when studies enroll all-comers.

Whatever the label and defining criteria used, unmet medical needs related to alcohol addiction are enormous. For many countries, reliable epidemiological data to assess these unmet needs are outdated or missing. It will be an important task for epidemiologists to provide those data for better policy making. However, even among people who do seek and receive help for their alcoholism, few receive evidence-based treatments (Hester and Miller 1995). The opioid antagonists naltrexone and nalmefene, as well acamprosate, a functional glutamate antagonist (see also Spanagel et al. 2014), are approved for clinical use and have documented efficacy, but their effect sizes are limited, treatment responses are highly heterogeneous (Heilig and Egli 2006; Heilig et al. 2011; Jonas et al. 2014), and prescription rates are low (Mark et al. 2003, 2009). It is therefore clear that expanding the range of options for alcoholism treatment and developing personalized approaches to treatment should continue to be a major public health priority for a long time to come.

2 The Aspirations of Translational Research for Alcohol Addiction

In the early years of the millennium, there was considerable enthusiasm that advances in the neuroscience of alcohol addiction would soon translate into mechanistically novel alcoholism therapies, expanding the range of treatment options, and addressing unmet medical needs of patients (see, e.g., Heilig and Egli 2006). In particular, there was hope that cleverly designed experimental paradigms in rodents and humans would make it possible to accelerate the process of translation from preclinical target discovery and validation into clinical development. Since then, some advances have indeed occurred. For instance, nalmefene given as needed (rather than continuously) was recently brought to the market (Mann et al. 2013), and some medications already approved for other indications may be possible to repurpose for treatment of alcohol addiction. However, several novel mechanisms that appeared to hold considerable promise based on preclinical data failed to be translated to the human condition.

Perhaps the greatest and most surprising disappointment has been the failure of corticotropin-releasing hormone (CRH1) receptor antagonism (Kwako et al. 2014). The target validation that provided the rationale for clinical trials with CRH1 antagonists in alcoholism came from several animal models of excessive drinking (Heilig and Koob 2007; Zorrilla et al. 2013). In hindsight, there were also indications that CRH1 antagonism may not consistently reduce excessive alcohol consumption. In fact, global CRH1 receptor deletion was found to result in a prolonged increase (rather than decrease) of alcohol consumption following stress exposure, and it appears that central and pituitary CRH1 receptors have opposing influences on stress-induced alcohol drinking (Sillaber et al. 2002; Molander et al. 2012).

The failure of CRH1 antagonism in alcoholism follows failures of several CRH1 blockers in other stress-related indications, such as major depression and generalized anxiety disorder (Binneman et al. 2008; Coric et al. 2010). It has been shown that blockade of CRH1 receptors can result in opposing effects on stress responses depending on the neuronal population targeted (Refojo et al. 2011). Because there are major species differences in CRH1 receptor expression, this may result in major differences in the net effect of a global CRH1 blockade. The emerging clinical observations that global CRH1 receptor blockade is not effective may in retrospect be less surprising than they appear. Nevertheless, they have had a discouraging effect on drug development efforts. Conceptually, strategies selectively targeting specific populations of CRH1 receptor expressing neurons could potentially overcome these challenges, but it is unlikely that this mechanism will be revisited by drug companies any time soon.

The loss of confidence in commonly used preclinical target validation assays for behavioral disorders is shared between disease areas. We continue to believe that current preclinical approaches have better chances of success in the area of addiction than other areas of psychiatry. In general however, it is clear that enthusiasm for psychiatric drug development has waned. While the cost of clinical development in behavioral disorders has skyrocketed, mechanism after mechanism that appeared promising in preclinical studies has simply failed to show clinical efficacy and been abandoned, leading to an exodus of major pharmaceutical companies (Miller 2010). Rightly or wrongly, addictive disorders have also been swept up in this wave. Dr. Steven Hyman, the former director of the National Institute on Mental Health, summed up the experience in a provocative piece that stated (Hyman 2012):

> current animal-based assays have failed to identify efficacious drugs with new molecular mechanisms, and given scant understanding of the pathophysiology of common psychiatric disorders, it is difficult to develop better models. Furthermore, objective diagnostic tests and treatment responsive biomarkers are lacking. Without the latter, clinical trials of psychiatric treatments are dependent on disease definitions grounded in the descriptive psychiatry of the 1960s and 1970s

Although this statement primarily referred to therapeutic areas within general psychiatry, such as schizophrenia or major depression, giving it some thought is important for addiction neuroscience as well. To positively impact the lives of patients with alcoholism, tools and strategies are needed that can persuade pharmaceutical companies to pursue discovery of novel, innovative treatments. Given the exceedingly high cost of drug development, pharmaceutical companies will not make that investment unless they believe preclinical tools exist that can help predict clinical success, and inform the design of initial clinical studies.

Target validation in alcoholism has typically relied on behavioral assays centered around drug-seeking and self-administration, often thought to reflect the clinical construct of craving (Egli 2005). But craving is not uniformly experienced by patients. Other behavioral domains, such as impulsivity or attention bias, may also offer strategies for preclinical target validation. These domains will, however, need to be further deconstructed, mapped to clinical symptoms, and evaluated for predictive validity. For example, two drugs that both have documented clinical efficacy to reduce impulsivity in patients with ADHD have very different signatures in rodents: Amphetamine improves performance in delay discounting tests but worsens premature responding in a stop signal task, while atomoxetine has the opposite effects (Broos et al. 2012). Furthermore, when impulsivity measures were improved in alcoholics as a result of modafinil treatment, clinical drinking outcomes were dependent on baseline patient characteristics: Patients with initially poor inhibitory control showed improvement in abstinence and drinking rates, but those with better baseline impulsivity scores worsened in response to the same treatment (Joos et al. 2013).

We continue to work toward and hope for increasingly refined behavioral assays for target validation and translation, but also note that simpler, biochemical measures may have advantageous measurement properties and translational utility. In the remainder of this chapter, we discuss some of these more biochemically oriented translational strategies.

3 The Promise of Translational Biomarkers

In search of strategies to better guide medications development in alcoholism, some useful lessons may be learned from advances in the genetics of complex traits. Despite well-established heritability of many complex behavioral traits, attempts to identify their genetic underpinnings have over the years been fraught with frustration and seemingly spectacular findings that have then failed to replicate. One successful strategy to address this challenge has been to expand sample sizes far beyond what was once thought necessary. A recent showcase project in the field of schizophrenia recently obtained 128 genome-wide significant findings in a multi-stage schizophrenia GWAS including 36,989 cases and 113,075 controls (Schizophrenia Working Group of the Psychiatric Genomics 2014). These kinds of sample sizes are, however, not realistic in clinical development.

Another, complementary strategy from complex behavioral genetics may have a greater appeal. The influential concept of endophenotypes as empirically tractable constructs with advantageous measurement properties was originally introduced largely on theoretical grounds (Gottesman and Gould 2003), but its practical value has since been proven in numerous instances (see, e.g., Munafo et al. 2008), where it has helped inform findings obtained with more complex, distal phenotypes such as depression (Caspi et al. 2010). The basic contention of this approach is that relatively simple traits that are not visible to the unaided eye and are closer to the biology sometimes offer advantages over attempts to use the distal, complex traits themselves as readouts.

An important example of this concept are the Research Domain Criteria (RDoC). These attempt to describe psychiatric conditions within a matrix of functional dimensions such as positive or negative valence systems (e.g., reward learning or acute/potential threat, respectively) or cognitive control (e.g., inhibition/suppression) that are investigated across a number of variables ranging from genetics and circuit activity to psychology and behavior (Insel and Cuthbert 2009; Cuthbert 2014). The RDoC approach studies dysfunction in basic systems aiming to understand sets of symptoms that may cut across multiple disorders. RDoC provide a multilevel framework and database for integrating a large variety of studies from many sites.

Capturing targetable feature of alcohol addiction through a set of intermediate phenotypes directly implies that these responses may be utilized as translational biomarkers for medication development. Along these lines, "a translational biomarker strategy" can be applied to medication development in alcohol addiction. In this approach, preclinical target validation does not have to rely on attempts to model complex behavioral phenotypes thought to be more or less homologous to behavioral elements of alcohol addiction. Instead, a biomarker strategy that focuses on simpler biological measures is chosen because these have some mechanistic relation to alcohol effects on the nervous system, have advantageous measurement properties, can be obtained both in experimental animals and in humans, and are predictably responsive to pharmacological interventions. While the interpretation of any behavioral response in terms of underlying mental processes is difficult in both humans and animals, a more direct approach is provided by neuroimaging methods that allow to objectively ascertaining brain responses at various levels, including molecular, structural, and functional levels, and thus may comprise a translational tool for the proposed biomarker strategy. The remainder of this chapter will focus on examples that illustrate this strategy, and discuss our experience with these approaches to date.

3.1 Alcohol-Induced Dopamine (DA) Activation as Translational Biomarker

An ability to activate mesolimbic alcohol-induced dopamine (DA) transmission is a shared feature of most addictive drugs and has long been thought to play a key role in addictive processes (Wise and Bozarth 1987; Di Chiara and Imperato 1988). It has since become clear that this is only one of numerous processes involved in the development and maintenance of addiction (e.g., Koob and Le Moal 2008), but drug-induced DA-activation fulfills several criteria outlined above: When it occurs, it is simpler than the complex processes of drug-seeking and can be measured with reasonable reliability and precision in both experimental animals and humans. Drug-induced mesolimbic DA-release was originally established using animal models, pharmacological manipulations, and brain microdialysis, but was subsequently found to translate to the human situation (reviewed, e.g., in Volkow et al. 2004). Similar to other addictive drugs, mesolimbic DA-activation is also seen in response to alcohol, although alcohol-induced activation is less pronounced and more variable than that resulting from, e.g., cocaine or stimulants. Alcohol-induced DA-release was originally demonstrated in experimental animals using passive, non-contingent alcohol administration (Di Chiara and Imperato 1988; Tanda and Di Chiara 1998), but was subsequently also observed as a result of oral alcohol self-administration in rats (Weiss et al. 1993). An appealing feature of the latter observation was that DA-activation in Wistar-derived rats with high-alcohol preference as a result of genetic selection was markedly higher than that of non-selected Wistar rats. In a recent meta-analysis of in vivo microdialysis DA datasets, derived from more than 7400 rats, ethanol dose-dependently, globally, and independently of brain site increased DA basal concentrations up to 270 % (Brand et al. 2013).

Detection of alcohol-induced mesolimbic DA activity in humans must necessarily rely on more indirect measures than microdialysis in animals. Despite this challenge, mesolimbic DA-activation by alcohol has been demonstrated. Key observations have been made using PET, and alcohol-induced reduction in binding potential, or ΔBP for short, for the D2/3 ligand [^{11}C]-raclopride (Boileau et al. 2003; Martinez et al. 2005; Urban et al. 2010; Ramchandani et al. 2011). Although this PET-based approach to measuring human DA-responses to addictive drugs is indirect, experiments in non-human primates have established that it reliably detects the effects of DA-releasing reference drugs (Dewey et al. 1993) and yields measures that have an excellent correlation with those obtained directly using microdialysis (Endres et al. 1997).

Drug-induced changes in raclopride BP are commonly called "displacement," implying a competitive mechanism for the reduction in BP, but it may be worth noting that non-competitive mechanisms, such as ligand-induced receptor internalization, could also be involved. These processes may be affected by disease pathophysiology in some conditions, such as schizophrenia, where they would make interpretation of data more complex (Ginovart 2005). No data suggest, however, that changes in receptor internalization would confound PET measures of DA-release in alcohol research. Drug-induced ΔBP for raclopride is a neurochemically specific measure of resulting endogenous DA-release, and a gold standard in human studies. While fundamentally reflecting the same approach, a more recent DA-D2/3 ligand, [^{11}C]-PHNO, may have potential to offer an improvement over the modest sensitivity offered by raclopride, because PHNO is an agonist that selectively labels DA-receptors in their high-affinity state (Willeit et al. 2008; Shotbolt et al. 2012).

PET-based measures of alcohol-induced DA-activation are appealing, because they have high neurochemical specificity, strong validation against direct measures of DA-release, and excellent measurement properties, while advances in ligand development may offer further improvements in sensitivity. However, PET-based measures also have distinct disadvantages as drug development tools. Most importantly, they are costly to obtain, require on-site radiosynthesis capacity, and have a low temporal resolution. Functional magnetic imaging (fMRI) lacks the neurochemical specificity of PET and is considerably noisier. However, the BOLD fMRI signal from the ventral striatum largely reflects DA-transmission in this structure (Knutson and Gibbs 2007), has small marginal cost, and allows a temporal resolution that is superior to that offered by PET. Using this approach, brain alcohol exposure closely controlled for pharmacokinetic variation has been shown to generate a robust signal from the ventral striatum of social drinkers, and the magnitude of this signal correlated strongly with subjective intoxication (Gilman et al. 2008). These data suggest that fMRI-based measures of alcohol-induced ventro-striatal activity may offer a useful complement to PET-based measures, although they are likely to require larger group sizes because of their inherently higher variance.

A consistent line of evidence supports the notion that alcohol-induced mesolimbic DA-activation can serve as a treatment responsive translational biomarker in alcoholism studies. The opioid antagonist naltrexone, an approved and effective alcoholism therapy (Jonas et al. 2014), was originally discovered in the absence of in-depth mechanistic understanding (Altshuler et al. 1980; O'Malley et al. 1992; Volpicelli et al. 1992). However, subsequent work showed that naltrexone and other, more mu-selective opioid antagonists, suppress alcohol-induced DA-activation in experimental animals (Gonzales and Weiss 1998; Tanda and Di Chiara 1998). This predicted that naltrexone would block alcohol-induced DA-release in humans. Surprisingly, this prediction has to our knowledge not been directly evaluated to date. However, data from a natural experiment provide strong indirect support for this notion.

To obtain these data, we capitalized on naturally occurring functional genetic variation at the locus encoding the target for naltrexone, an *OPRM1* A118G single nucleotide polymorphism (SNP) that encodes an amino acid substitution in the *N*-terminal extracellular loop of the receptor, and mutates out a glycosylation site (Bond et al. 1998). Using PET and raclopride displacement, we found that the mu-opioid receptor variant encoded by the major 118A allele at this locus is associated with markedly lower mesolimbic DA-response to alcohol than the minor G-allele, essentially mimicking the functional consequences of mu-opioid receptor antagonism. As a reverse-translational tool, we then generated two humanized *OPRM1* mouse lines, identical throughout the genome with the exception of the *OPRM1* polymorphism. In agreement with the human PET findings, alcohol-induced DA-release in the ventral striatum of mice homozygous for the A-allele was dramatically lower than that found in GG-mice (Ramchandani et al. 2011).

Subsequent experiments in these humanized mice have provided data supporting the potential of alcohol-induced DA-release as a treatment responsive translational biomarker. Using a classical model of drug reward, suppression of intracranial self-administration (ICSS) thresholds, we found robust rewarding effects of alcohol in the GG-mice, while these effects were markedly attenuated or absent in the AA-mice. ICSS measures of alcohol reward were blocked in the GG-mice by naltrexone, while no effects of naltrexone were seen in AA-mice. Finally, in agreement with these data, naltrexone as well as another opioid antagonist, nalmefene, was markedly more effective in its ability to suppress alcohol self-administration in GG-mice compared to AA-animals (Bilbao et al. 2014). The latter data are in agreement with the original proposition, based on a secondary analysis of clinical trial data, that alcohol-addicted patients carrying the *OPRM1* 118G allele are particularly responsive to naltrexone treatment (Oslin et al. 2003). More recent secondary analyses, based on larger patient samples, have provided further support for this observation (Chamorro et al. 2012, Garbutt et al. 2014). A subsequent, prospectively genotyped study failed to detect a moderating effect of *OPRM1* A118G variation on naltrexone efficacy (Oslin et al. in press), but that study also failed to detect overall efficacy of naltrexone as such, and can therefore not really inform the question of genetic moderation. It is simply a failed trial, as are indeed for unknown reasons about half of psychiatric trials that include a medication with documented efficacy (Khin et al. 2011).

Alcohol-induced DA-activation is not uniformly seen in all individuals. We have already noted that genetic variation at the *OPRM1* locus is a potent determinant of this response, but other factors that influence this response have also been established. In a provocative PET-study, it was shown that the mesolimbic DA-response to alcohol in females is negligible compared to males (Urban et al. 2010). This observation is consistent with our findings in non-human primates, where only males responded with psychomotor stimulation to an alcohol challenge (Barr et al. 2007). Furthermore, several studies suggest that activation of mesolimbic DA by alcohol declines with progression into alcoholism. This has been observed in the ventral striatum both using [^{11}C]-raclopride displacement (Martinez et al. 2005) and fMRI (Gilman et al. 2012; Spagnolo et al. 2014). Rather than reflecting limitations of the biomarker, these data highlight its strength. If the signature of a novel therapeutic mechanism is to attenuate alcohol-induced DA-release in experimental animals, then the absence or low magnitude of a DA-response to alcohol will help identify patients who are less likely to respond to therapeutics targeting that mechanism. Such individuals should then not be included in clinical efficacy trials, where they would only dilute efficacy signals from responsive patients. Accordingly, although available data may have some methodological limitations, being male appears to be a predictor of naltrexone response (Garbutt et al. 2014).

In summary, both PET and fMRI-based measures of alcohol-induced mesolimbic DA-activation should have considerable potential as treatment responsive translational biomarkers in developing novel alcoholism pharmacotherapies. It is important to point out that a narrow use of DA-responses as translational biomarkers of drug effects does not rest on any specific hypotheses about the role of the DA system in addiction pathophysiology. Several novel mechanisms of potential interest as alcoholism therapies have shown a signature that includes an ability to inhibit alcohol-induced mesolimbic DA-activation in preclinical studies. Among these, blockade of receptors for the appetite regulating hormone ghrelin appears particularly promising (Jerlhag et al. 2006, 2009; Landgren et al. 2012). Initial translation of this mechanism is now underway by Leggio and colleagues at the National Institutes of Health using the non-peptide ghrelin-1a receptor antagonist PF-05190457 (NCT02039349). If determined to be safe, this mechanism will be able to benefit from the biomarker strategy described here. Other candidate mechanisms with an ability to inhibit measures of alcohol-induced DA-activation are also in the pipeline, such as antagonism of the appetite regulating neuropeptides melanin-concentrating hormone type-1 receptors (MCH1-R) (Cippitelli et al. 2010). Successful translation of these promising preclinical findings will benefit from the application of a biomarker strategy outlined above.

On the other hand, observations that females and patients in later stages of alcohol addiction do not show robust alcohol-induced DA-activation suggest considerable heterogeneity among "alcoholics." Clearly, a "smorgasbord" of biomarker strategies will be needed to facilitate translation across these heterogeneous populations.

3.2 *Measures of Glutamate Activity as Translational Biomarkers*

It is increasingly recognized that alcohol addiction is an evolving process, characterized by progressive and widespread recruitment of neuroadaptive changes in the central nervous system over time. At a behavioral level, this evolving process has been characterized as a shift from positively reinforced alcohol use to consumption that is increasingly driven by negative reinforcement, or from "reward craving" to "relief craving." Neuroadaptive changes are initially triggered acutely during states of acute withdrawal, but ultimately persist into what can be called "protracted abstinence," at which time they generate powerful incentives to resume alcohol-seeking and use (Heilig et al. 2010; Glockner-Rist et al. 2013; Meinhardt and Sommer 2015). Of course, in reality this process is not nearly as uniform and neat as described here. Based on the presence or absence of pre-existing genetic susceptibility factors and exposure to environmental influences, such as drug consumption itself or life stressors, people arrive at "neuroadapted" alcoholism through very different trajectories (Heilig et al. 2011). Irrespective of trajectory, however, once these neuroadaptations are in place, they are likely to offer additional translational biomarkers.

Among a multitude of neuroadaptive changes reported in alcoholism, adaptations within the glutamatergic system appear to be prominent, and offer a rich pharmacology that holds the promise of yielding novel alcoholism treatments (Spanagel and Kiefer 2008; Spanagel 2009; Holmes et al. 2013). Chronic excessive use of alcohol ultimately results in a hyperglutamatergic state, characterized by elevated extracellular glutamate and altered glutamate receptors and transporters. Pharmacologically manipulating glutamatergic neurotransmission alters a wide range of alcohol-related behaviors, such as acute intoxication and withdrawal, but also alcohol-seeking and consumption, in both rodents and humans. Accordingly, several elements of glutamatergic neurotransmission have been proposed as attractive targets for novel alcoholism treatments. For instance, blockade of NMDA and AMPA receptors reduces alcohol consumption in rats and mice. However, side effects are likely to limit the therapeutic potential of drugs that block ionotropic glutamate receptors, and experience with this strategy in stroke has not been encouraging (Gladstone et al. 2002). Targeting metabotropic glutamate receptors (mGluRs) may offer a better tolerated approach, in particular if pursued using allosteric modulators. Indeed, blocking mGluR5 potently affects various alcohol-related behaviors in rodents, and mGluR2/3 agonism or mGluR2 positive allosteric modulation also suppresses alcohol consumption (Spanagel 2010).

Finally, glutamate transporter upregulation may mitigate behavioral and neurotoxic sequelae of excess glutamate caused by alcohol, and attenuate alcohol drinking.

The possibility that targeting the glutamate system holds promise for developing new alcoholism treatments makes it a priority to establish translational biomarkers responsive to glutamatergic medications. Once again, in establishing biomarkers of glutamate function, one can remain fairly agnostic regarding underlying pathophysiology or synaptic function. The measures that will be best able to aid translational efforts are those that are relatively simple, robust, and can be obtained in experimental animals as well as in humans with alcoholism, and are similarly responsive to drugs in animal and human models.

Early rat studies showed that extracellular glutamate levels, measured in rat striatum using brain microdialysis, increase during acute alcohol withdrawal (Rossetti and Carboni 1995). Benzodiazepines, the standard clinical treatment for acute alcohol withdrawal, blocked the behavioral withdrawal signs in that study, but did not prevent the glutamate surge. In contrast, the non-competitive NMDA antagonist MK-801 blocked both the behavioral and the neurochemical withdrawal symptoms. It was subsequently shown that within the ventral striatum, withdrawal-induced glutamate elevations increase with repeated cycles of withdrawal (Dahchour et al. 1998). A different pattern was found in the hippocampus, where glutamate elevations were found after a single withdrawal episode, but dissipated over subsequent cycles (Dahchour and De Witte 1999).

In a recent meta-analysis of in vivo microdialysis datasets, derived from 104 alcohol-dependent rats, consistent evidence was obtained for elevated extracellular glutamate levels in various brain sites that correlated with the intensity of the withdrawal response (Fliegel et al. 2013). Recently, Sommer and colleagues were able to detect withdrawal-induced increases in glutamate levels in rats using proton MR-spectroscopy (MRS) at high field, 9.4T (Hermann et al. 2012). These data represent a major advance because they tie together the direct, microdialysis-based measures, with those detected by MRS, showing that the latter to some extent reflect the extracellular glutamate pool that originates from synaptic transmission, and emphasizing the feasibility of a translational neuroimaging approach

Perhaps the best evidence for a potential of glutamate MRS as a translational biomarker in alcohol studies comes from data obtained with the glutamatergic modulator acamprosate, a clinically approved therapy for alcoholism. In mice with a deletion of the clock gene *Per2,* escalated voluntary alcohol consumption was observed by the Spanagel laboratory and was tied to decreased clearance of glutamate by the glial Glutamate Aspartate Transporter (GLAST; also Excitatory Amino Acid Transporter 1, EAAT1, or SLC1A3). Acamprosate rescued both the escalated alcohol self-administration and the striatal elevations of extracellular glutamate, measured directly by microdialysis, in the *Per2* null-mutants (Spanagel et al. 2005). These data provided some of the most important support for a causal role of a hyperglutamatergic state in driving excessive alcohol consumption. Because MRS is able to tap into a signal that represents elevated overflow of synaptic glutamate, we hypothesized that MRS-based measures may offer useful biomarkers in drug development for alcoholism.

To address this hypothesis, we carried out an experimental medicine study and attempted to establish whether MRS at 3T would be sensitive and specific enough to detect a reduction in central glutamate resulting from clinical acamprosate treatment. A challenge for MRS studies at 3T is that glutamate and its precursor glutamine overlap significantly in the 1H resonance spectrum. Higher magnetic field strength makes it possible to resolve their respective resonances, but is not yet widely available for human studies. At 3T, the overlapping glutamate and glutamine peaks are frequently combined into a "GluX" peak, but because of the glutamine–glutamate cycle (Bak et al. 2006), this approach does not provide sufficiently detailed information about the functional state of glutamatergic transmission. An echo-time-averaged, point-resolved technique (TE-averaged PRESS) has been shown to detect an unobstructed glutamate signal at 3T that is resolved from glutamine at 2.35 ppm. TE-averaged PRESS therefore provides an unambiguous measurement of glutamate at 3T (Hurd et al. 2004; Srinivasan et al. 2006) and holds potential as a biomarker.

We used TE-averaged PRESS and scanned treatment-seeking alcohol-addicted patients randomized to acamprosate or placebo, as well as healthy volunteers who were scanned for comparison. Our first scan awaited steady state for acamprosate to be reached and was therefore carried out after acute withdrawal had subsided. At this point, there was no elevation of glutamate within a voxel placed in the anterior cingulate cortex (ACC) compared to controls. When these patients were rescanned three weeks later, however, the placebo-treated group showed significantly elevated glutamate levels, while levels in the acamprosate-treated group had, if anything, declined; the two groups were clearly separated at that time (Umhau et al. 2010).

In one important aspect, our data are complementary to those obtained by Sommer and colleagues (Hermann et al. 2012). Piecing together a time course from these two studies, it appears that acute alcohol withdrawal may be associated with a transient elevation of central glutamate, and that, in alcohol-dependent patients, levels start creeping up again in protracted abstinence, when relapse most frequently occurs. The suppression of that delayed elevation by acamprosate can be detected by MRS at 3T using TE-averaged PRESS technology. Of note, measures of glutamate in the cerebrospinal fluid (CSF) do not offer an alternative to the MRS-based biomarker; CSF glutamate appears to have a different origin (Umhau et al. 2010).

These approaches hold considerable potential to facilitate alcoholism therapies targeting mGluR2 receptors, a key player in controlling glutamate homeostasis identified as a promising target by converging lines of evidence. In rats, chronic intermittent alcohol intoxication results in a post-dependent state characterized by escalation of subsequent voluntary alcohol intake (Heilig et al. 2010; Meinhardt and Sommer 2015). Post-dependent animals also show long-lasting deficits in executive control, e.g., attention and response selection (Trantham-Davidson et al. 2014), functions dependent on the medial prefrontal cortex (mPFC). We have shown that the post-dependent state is associated with a persistent insult to the infralimbic mPFC, where a lasting suppression of mGluR2 receptor expression occurs (Meinhardt et al. 2013). The translational value of these findings is supported by human postmortem data showing a reduction in mGluR2 expression in the corresponding PFC region in alcoholics. A mechanistic role for the loss of infralimbic mGluR2 expression is demonstrated by the finding that viral vector-mediated rescue of mGluR2 expression in the infralimbic mPFC of post-dependent rats results in a rescue of their escalated self-administration.

Convergent support for a key role of mGluR2 receptors in control of alcohol-seeking and consumption comes from a recent study in P rats, a line selected for high innate alcohol preference. This work identified a premature stop codon in *Grm2*, the gene encoding mGluR2 receptors, which contributes about 25 % of the increased alcohol consumption in these animals (Zhou et al. 2013) Collectively, these results suggest that mGluR2 loss in rodent and human neural circuits, which provide cortical control over deeper brain structures involved in motivational and emotional regulation, may be a major consequence of alcohol dependence and a key pathophysiological mechanism for the increased propensity to relapse. Observations that PFC control over drug-seeking behavior can be restored by rescuing or enhancing the function of metabotropic glutamate receptors (Meinhardt et al. 2013; Gass et al. 2014) provides a strong rationale for translational studies aimed at assessing these processes in alcoholic patients.

Although glutamate MRS is an appealing translational biomarker for these studies, fMRI-based approaches may once again have utility. This is illustrated by a recent animal study that used pharmacological fMRI in awake rats, and demonstrated a modulation of ketamine-induced BOLD response by an mGluR2/3 agonist (Chin et al. 2011). Several mGluR2/3 agonists and mGluR2 positive allosteric modulators (PAMs) have cleared human safety studies. Both MRS and pharmacofunctional MRI-based biomarkers can likely be used to probe the functional state of the glutamate systems and provide a window on the molecular and neuronal basis of executive function impairment seen in many alcoholics. These probes should offer attractive translational biomarkers for medication development that targets dysregulated glutamate transmission in order to improve executive control in these patients.

4 Future Directions: The Whole—More than the Sum of Its Parts?

Specific neurochemical systems contribute to alcohol addiction, but the disorder is a systems level problem. Neuroimaging can provide an unbiased view of brain function at the systems level that allows ascertainment of network activity, but so far, only a few studies have used these tools in alcoholism (Camchong et al. 2013; Muller-Oehring et al. 2014). In general, functional connectivity analysis in abstinent alcoholics has shown an overall integrity of large-scale functional networks, but has found specific pathology in distinct sub-networks, in particular expanded connectivity in attention and visual input networks, supporting the concept of network expansion as a neural mechanism for functional compensation.

Expanding this approach could help identify a "disease-network" for alcohol use disorders. Because aberrant network states can be shifted by pharmacological interventions (Schmaal et al. 2013), "disease-network" states could potentially function as biomarkers for treatment development, even in the absence of a distinct target mechanism. The biomarker would then be an ability to force network activity in the direction of a "normal" state. Imaging studies have shown that spatial and functional characteristics of intrinsic brain network architecture are conserved across species, including rodents and humans. Thus, network analysis may emerge as a translational tool for medication development. In an attempt to explore the potential of brain network analysis in rats for studying alcoholism-related brain network states, Dudek and colleagues (Dudek et al. 2014) used manganese-enhanced MRI to investigate network activity after alcohol drinking and abstinence. Many of the activated brain regions have previously been implicated in alcohol reward, but in the prelimbic cortex, ventral hippocampus, and subthalamic nucleus, activation persisted into abstinence, supporting the idea of long-term changes comprising a "relapse-prone" network state. Future studies will address to what extent "relapse-prone" networks overlap between animals and humans, and can offer translational biomarkers.

5 Conclusion

Over the past two decades, major advances have been made in the neuroscience of alcohol addiction, but few of these have directly translated into improved treatments for patients with alcoholism. During this time, the field has devoted considerable attention to developing and debating behavioral animal assays that might hold promise of predictive validity, and that might facilitate translation (Egli 2005; Heilig and Egli 2006; Litten et al. 2012). To bridge the proverbial "translational valley of death," other types of tools may have to be deployed. Treatment responsive translational biomarkers top the list of such tools, and our present paper describes some of those already available.

Tools that follow the same principles as those described, but can index alcohol and drug effects on other systems need to be added to the toolkit. A notable example in this category would be a displaceable ligand for the endocannabinoid CB1 receptor. Successful development and deployment of translational biomarkers will to a large extent have to rely on the ability of academia and pharma to work together, in turn related to the ability of governments to provide incentives and regulatory space for that kind of efforts (Hudson and Khazragui 2013). Although the pullout of major pharmaceutical companies from the CNS area confronts us with a rather gloomy picture compared to the hopes and expectations of only a decade ago, there may be a silver lining. As major pharmaceutical companies discontinue or scale down their psychiatry and addiction programs, many interesting molecules become available for licensing by consortia of biotech and academia. Those consortia may have to rely on venture capital or other sources, because few public funding mechanisms are adequate to support the costs of clinical development. Data obtained using translational biomarkers will be an important part of any portfolio that can attract investment, and ultimately benefit patients through development of novel therapeutics.

References

Altshuler HL, Phillips PE, Feinhandler DA (1980) Alteration of ethanol self-administration by naltrexone. Life Sci 26:679–688

American Psychiatric Association (2000) Diagnostic criteria from DSM-IV-TR. American Psychiatric Association, Washington, D.C.

Anderson P, Chisholm D, Fuhr DC (2009) Effectiveness and cost-effectiveness of policies and programmes to reduce the harm caused by alcohol. Lancet 373:2234–2246

Bak LK, Schousboe A, Waagepetersen HS (2006) The glutamate/GABA-glutamine cycle: aspects of transport, neurotransmitter homeostasis and ammonia transfer. J Neurochem 98:641–653

Barr CS, Schwandt M, Lindell SG, Chen SA, Goldman D, Suomi SJ, Higley JD, Heilig M (2007) Association of a functional polymorphism in the mu-opioid receptor gene with alcohol response and consumption in male rhesus macaques. Arch Gen Psychiatry 64:369–376

Bilbao A, Robinson JE, Heilig M, Malanga CJ, Spanagel R, Sommer WH, Thorsell A (2014) A pharmacogenetic determinant of mu-opioid receptor antagonist effects on alcohol reward and consumption: evidence from humanized mice. Biol Psychiatry

Binneman B, Feltner D, Kolluri S, Shi Y, Qiu R, Stiger T (2008) A 6-week randomized, placebo-controlled trial of CP-316,311 (a selective CRH1 antagonist) in the treatment of major depression. AJ Psychiatry 165:617–620

Boileau I, Assaad JM, Pihl RO, Benkelfat C, Leyton M, Diksic M, Tremblay RE, Dagher A (2003) Alcohol promotes dopamine release in the human nucleus accumbens. Synapse 49:226–231

Bond C, LaForge KS, Tian M, Melia D, Zhang S, Borg L, Gong J, Schluger J, Strong JA, Leal SM, Tischfield JA, Kreek MJ, Yu L (1998) Single-nucleotide polymorphism in the human mu opioid receptor gene alters beta-endorphin binding and activity: possible implications for opiate addiction. Proc Natl Acad Sci U S A 95:9608–9613

Brand I, Fliegel S, Spanagel R, Noori HR (2013) Global ethanol-induced enhancements of monoaminergic neurotransmission: a meta-analysis study. Alcohol Clin Exp Res 37:2048–2057

Broos N, Schmaal L, Wiskerke J, Kostelijk L, Lam T, Stoop N, Weierink L, Ham J, de Geus EJ, Schoffelmeer AN, van den Brink W, Veltman DJ, de Vries TJ, Pattij T, Goudriaan AE (2012) The relationship between impulsive choice and impulsive action: a cross-species translational study. PLoS ONE 7:e36781

Camchong J, Stenger VA, Fein G (2013) Resting-state synchrony in short-term versus long-term abstinent alcoholics. Alcohol Clin Exp Res 37:794–803

Caspi A, Hariri AR, Holmes A, Uher R, Moffitt TE (2010) Genetic sensitivity to the environment: the case of the serotonin transporter gene and its implications for studying complex diseases and traits. AJ Psychiatry 167:509–527

Chamorro AJ, Marcos M, Miron-Canelo JA, Pastor I, Gonzalez-Sarmiento R, Laso FJ (2012) Association of mu-opioid receptor (OPRM1) gene polymorphism with response to naltrexone in alcohol dependence: a systematic review and meta-analysis. Addict Biol 17:505–512

Chin CL, Upadhyay J, Marek GJ, Baker SJ, Zhang M, Mezler M, Fox GB, Day M (2011) Awake rat pharmacological magnetic resonance imaging as a translational pharmacodynamic biomarker: metabotropic glutamate 2/3 agonist modulation of ketamine-induced blood oxygenation level dependence signals. J Pharmacol Exp Ther 336:709–715

Cippitelli A, Karlsson C, Shaw JL, Thorsell A, Gehlert DR, Heilig M (2010) Suppression of alcohol self-administration and reinstatement of alcohol seeking by melanin-concentrating hormone receptor 1 (MCH1-R) antagonism in Wistar rats. Psychopharmacology 211:367–375

Compton WM, Dawson DA, Goldstein RB, Grant BF (2013) Crosswalk between DSM-IV dependence and DSM-5 substance use disorders for opioids, cannabis, cocaine and alcohol. Drug Alcohol Depend 132(1):387–390

Coric V, Feldman HH, Oren DA, Shekhar A, Pultz J, Dockens RC, Wu X, Gentile KA, Huang SP, Emison E, Delmonte T, D'Souza BB, Zimbroff DL, Grebb JA, Goddard AW, Stock EG (2010) Multicenter, randomized, double-blind, active comparator and placebo-controlled trial of a corticotropin-releasing factor receptor-1 antagonist in generalized anxiety disorder. Depress Anxiety 27:417–425

Culverhouse R, Bucholz KK, Crowe RR, Hesselbrock V, Nurnberger JI Jr, Porjesz B, Schuckit MA, Reich T, Bierut LJ (2005) Long-term stability of alcohol and other substance dependence diagnoses and habitual smoking: an evaluation after 5 years. Arch Gen Psychiatry 62:753–760

Cuthbert BN (2014) Translating intermediate phenotypes to psychopathology: the NIMH research domain criteria. Psychophysiology 51:1205–1206

Dahchour A, De Witte P (1999) Effect of repeated ethanol withdrawal on glutamate microdialysate in the hippocampus. Alcohol Clin Exp Res 23:1698–1703

Dahchour A, De Witte P, Bolo N, Nedelec JF, Muzet M, Durbin P, Macher JP (1998) Central effects of acamprosate: part 1. Acamprosate blocks the glutamate increase in the nucleus accumbens microdialysate in ethanol withdrawn rats. Psychiatry Res 82:107–114

Dewey SL, Smith GS, Logan J, Brodie JD, Fowler JS, Wolf AP (1993) Striatal binding of the PET ligand 11C-raclopride is altered by drugs that modify synaptic dopamine levels. Synapse 13:350–356

Di Chiara G, Imperato A (1988) Drugs abused by humans preferentially increase synaptic dopamine concentrations in the mesolimbic system of freely moving rats. Proc Natl Acad Sci U S A 85:5274–5278

Dudek M, Abo-Ramadan U, Hermann D, Brown M, Canals S, Sommer WH, Hyytia P (2014) Brain activation induced by voluntary alcohol and saccharin drinking in rats assessed with manganese-enhanced magnetic resonance imaging. Addiction biology

Edwards G, Gross MM (1976) Alcohol dependence: provisional description of a clinical syndrome. Br Med J 1:1058–1061

Egli M (2005) Can experimental paradigms and animal models be used to discover clinically effective medications for alcoholism? Addict Biol 10:309–319

Endres CJ, Kolachana BS, Saunders RC, Su T, Weinberger D, Breier A, Eckelman WC, Carson RE (1997) Kinetic modeling of [11C]raclopride: combined PET-microdialysis studies. J Cereb Blood Flow Metab 17:932–942

Fliegel S, Brand I, Spanagel R, Noori HR (2013) Ethanol-induced alterations of amino acids measured by in vivo microdialysis in rats: a meta-analysis. Silico Pharmacology 1:7

Garbutt JC, Greenblatt AM, West SL, Morgan LC, Kampov-Polevoy A, Jordan HS, Bobashev GV (2014) Clinical and biological moderators of response to naltrexone in alcohol dependence: a systematic review of the evidence. Addiction

Gass JT, Trantham-Davidson H, Kassab AS, Glen WB Jr, Olive MF, Chandler LJ (2014) Enhancement of extinction learning attenuates ethanol-seeking behavior and alters plasticity in the prefrontal cortex. J Neurosci 34:7562–7574

Gilman JM, Ramchandani VA, Crouss T, Hommer DW (2012) Subjective and neural responses to intravenous alcohol in young adults with light and heavy drinking patterns. Neuropsychopharmacology 37:467–477

Gilman JM, Ramchandani VA, Davis MB, Bjork JM, Hommer DW (2008) Why we like to drink: a functional magnetic resonance imaging study of the rewarding and anxiolytic effects of alcohol. J Neurosci 28:4583–4591

Ginovart N (2005) Imaging the dopamine system with in vivo [11C]raclopride displacement studies: understanding the true mechanism. Molecular imaging and biology: MIB: the official publication of the Academy of Molecular Imaging 7:45–52
Gladstone DJ, Black SE, Hakim AM, Heart, Stroke Foundation of Ontario Centre of Excellence in Stroke R (2002) Toward wisdom from failure: lessons from neuroprotective stroke trials and new therapeutic directions. Stroke 33:2123–2136
Glockner-Rist A, Lemenager T, Mann K, Group PSR (2013) Reward and relief craving tendencies in patients with alcohol use disorders: results from the PREDICT study. Addict Behav 38:1532–1540
Gonzales RA, Weiss F (1998) Suppression of ethanol-reinforced behavior by naltrexone is associated with attenuation of the ethanol-induced increase in dialysate dopamine levels in the nucleus accumbens. J Neurosci 18:10663–10671
Gottesman II, Gould TD (2003) The endophenotype concept in psychiatry: etymology and strategic intentions. AJ Psychiatry 160:636–645
Grant BF, Stinson FS, Dawson DA, Chou SP, Dufour MC, Compton W, Pickering RP, Kaplan K (2004) Prevalence and co-occurrence of substance use disorders and independent mood and anxiety disorders: results from the national epidemiologic survey on alcohol and related conditions. Arch Gen Psychiatry 61:807–816
Heilig M, Egli M (2006) Pharmacological treatment of alcohol dependence: target symptoms and target mechanisms. Pharmacol Ther 111:855–876
Heilig M, Egli M, Crabbe JC, Becker HC (2010) Acute withdrawal, protracted abstinence and negative affect in alcoholism: are they linked? Addict Biol 15:169–184
Heilig M, Goldman D, Berrettini W, O'Brien CP (2011) Pharmacogenetic approaches to the treatment of alcohol addiction. Nat Rev Neurosci 12:670–684
Heilig M, Koob GF (2007) A key role for corticotropin-releasing factor in alcohol dependence. Trends Neurosci 30:399–406
Hermann D, Weber-Fahr W, Sartorius A, Hoerst M, Frischknecht U, Tunc-Skarka N, Perreau-Lenz S, Hansson AC, Krumm B, Kiefer F, Spanagel R, Mann K, Ende G, Sommer WH (2012) Translational magnetic resonance spectroscopy reveals excessive central glutamate levels during alcohol withdrawal in humans and rats. Biol Psychiatry 71:1015–1021
Hester RK, Miller WR (1995) Handbook of alcoholism treatment approaches: effective alternatives. Allyn & Bacon, Boston
Holmes A, Spanagel R, Krystal JH (2013) Glutamatergic targets for new alcohol medications. Psychopharmacology 229:539–554
Hudson J, Khazragui HF (2013) Into the valley of death: research to innovation. Drug Discovery Today 18:610–613
Hurd R, Sailasuta N, Srinivasan R, Vigneron DB, Pelletier D, Nelson SJ (2004) Measurement of brain glutamate using TE-averaged PRESS at 3T. Magn Reson Med 51:435–440
Hyman SE (2012) Revolution stalled. Sci Trans Med 4:155cm111
Insel TR, Cuthbert BN (2009) Endophenotypes: bridging genomic complexity and disorder heterogeneity. Biol Psychiatry 66:988–989
Jerlhag E, Egecioglu E, Dickson SL, Andersson M, Svensson L, Engel JA (2006) Ghrelin stimulates locomotor activity and accumbal dopamine-overflow via central cholinergic systems in mice: implications for its involvement in brain reward. Addict Biol 11:45–54
Jerlhag E, Egecioglu E, Landgren S, Salome N, Heilig M, Moechars D, Datta R, Perrissoud D, Dickson SL, Engel JA (2009) Requirement of central ghrelin signaling for alcohol reward. Proc Natl Acad Sci U S A 106:11318–11323
John U, Rumpf HJ, Bischof G, Hapke U, Hanke M, Meyer C (2013) Excess mortality of alcohol-dependent individuals after 14 years and mortality predictors based on treatment participation and severity of alcohol dependence. Alcohol Clin Exp Res 37:156–163
Jonas DE, Amick HR, Feltner C, Bobashev G, Thomas K, Wines R, Kim MM, Shanahan E, Gass CE, Rowe CJ, Garbutt JC (2014) Pharmacotherapy for adults with alcohol use disorders in outpatient settings a systematic review and meta-analysis. Jama-J Am Med Assoc 311:1889–1900

Joos L, Goudriaan AE, Schmaal L, Fransen E, van den Brink W, Sabbe BG, Dom G (2013) Effect of modafinil on impulsivity and relapse in alcohol dependent patients: a randomized, placebo-controlled trial. Eur Neuropsychopharmacol 23:948–955
Khin NA, Chen YF, Yang Y, Yang P, Laughren TP (2011) Exploratory analyses of efficacy data from major depressive disorder trials submitted to the US food and drug administration in support of new drug applications. J Clin Psychiatry 72:464–472
Knutson B, Gibbs SE (2007) Linking nucleus accumbens dopamine and blood oxygenation. Psychopharmacology 191:813–822
Koob GF, Le Moal M (2008) Addiction and the brain antireward system. Annu Rev Psychol 59:29–53
Kwako LE, Spagnolo PA, Schwandt ML, Thorsell A, George DT, Momenan R, Rio DE, Huestis M, Anizan S, Concheiro M, Sinha R, Heilig M (2014) The corticotropin releasing hormone-1 (CRH1) receptor antagonist pexacerfont in alcohol dependence: a randomized controlled experimental medicine study. Neuropsychopharmacology
Landgren S, Simms JA, Hyytia P, Engel JA, Bartlett SE, Jerlhag E (2012) Ghrelin receptor (GHS-R1A) antagonism suppresses both operant alcohol self-administration and high alcohol consumption in rats. Addict Biol 17:86–94
Litten RZ, Egli M, Heilig M, Cui C, Fertig JB, Ryan ML, Falk DE, Moss H, Huebner R, Noronha A (2012) Medications development to treat alcohol dependence: a vision for the next decade. Addict Biol 17:513–527
Mann K, Bladstrom A, Torup L, Gual A, van den Brink W (2013) Extending the treatment options in alcohol dependence: a randomized controlled study of as-needed nalmefene. Biol Psychiatry 73:706–713
Mark TL, Kassed CA, Vandivort-Warren R, Levit KR, Kranzler HR (2009) Alcohol and opioid dependence medications: prescription trends, overall and by physician specialty. Drug Alcohol Depend 99:345–349
Mark TL, Kranzler HR, Song X (2003) Understanding US addiction physicians' low rate of naltrexone prescription. Drug Alcohol Depend 71:219–228
Martinez D, Gil R, Slifstein M, Hwang DR, Huang Y, Perez A, Kegeles L, Talbot P, Evans S, Krystal J, Laruelle M, Abi-Dargham A (2005) Alcohol dependence is associated with blunted dopamine transmission in the ventral striatum. Biol Psychiatry 58:779–786
Meinhardt MW, Hansson AC, Perreau-Lenz S, Bauder-Wenz C, Stahlin O, Heilig M, Harper C, Drescher KU, Spanagel R, Sommer WH (2013) Rescue of infralimbic mGluR2 deficit restores control over drug-seeking behavior in alcohol dependence. J Neurosci 33:2794–2806
Meinhardt MW, Sommer WH (2015) Postdependent state in rats as a model for medication development in alcoholism. Addict Biol 20:1–21
Miller G (2010) Is pharma running out of brainy ideas? Science 329:502–504
Molander A, Vengeliene V, Heilig M, Wurst W, Deussing JM, Spanagel R (2012) Brain-specific inactivation of the Crhr1 gene inhibits post-dependent and stress-induced alcohol intake, but does not affect relapse-like drinking. Neuropsychopharmacology 37:1047–1056
Muller-Oehring EM, Jung YC, Pfefferbaum A, Sullivan EV, Schulte T (2014) The resting brain of alcoholics. Cereb Cortex
Munafo MR, Brown SM, Hariri AR (2008) Serotonin transporter (5-HTTLPR) genotype and amygdala activation: a meta-analysis. Biol Psychiatry 63:852–857
Nutt DJ, King LA, Phillips LD, Independent Scientific Committee on D (2010) Drug harms in the UK: a multicriteria decision analysis. Lancet 376:1558–1565
O'Malley SS, Jaffe AJ, Chang G, Schottenfeld RS, Meyer RE, Rounsaville B (1992) Naltrexone and coping skills therapy for alcohol dependence. A controlled study. Arch Gen Psychiatry 49:881–887
Oslin DW, Berrettini W, Kranzler HR, Pettinati H, Gelernter J, Volpicelli JR, O'Brien CP (2003) A functional polymorphism of the mu-opioid receptor gene is associated with naltrexone response in alcohol-dependent patients. Neuropsychopharmacology 28:1546–1552
Piazza PV, Deroche-Gamonet V (2013) A multistep general theory of transition to addiction. Psychopharmacology 229:387–413

Ramchandani VA, Umhau J, Pavon FJ, Ruiz-Velasco V, Margas W, Sun H, Damadzic R, Eskay R, Schoor M, Thorsell A, Schwandt ML, Sommer WH, George DT, Parsons LH, Herscovitch P, Hommer D, Heilig M (2011) A genetic determinant of the striatal dopamine response to alcohol in men. Mol Psychiatry 16:809–817

Refojo D, Schweizer M, Kuehne C, Ehrenberg S, Thoeringer C, Vogl AM, Dedic N, Schumacher M, von Wolff G, Avrabos C, Touma C, Engblom D, Schutz G, Nave KA, Eder M, Wotjak CT, Sillaber I, Holsboer F, Wurst W, Deussing JM (2011) Glutamatergic and dopaminergic neurons mediate anxiogenic and anxiolytic effects of CRHR1. Science 333:1903–1907

Rehm J, Room R, Graham K, Monteiro M, Gmel G, Sempos CT (2003) The relationship of average volume of alcohol consumption and patterns of drinking to burden of disease: an overview. Addiction 98:1209–1228

Rossetti ZL, Carboni S (1995) Ethanol withdrawal is associated with increased extracellular glutamate in the rat striatum. Eur J Pharmacol 283:177–183

Schizophrenia Working Group of the Psychiatric Genomics C (2014) Biological insights from 108 schizophrenia-associated genetic loci. Nature 511:421–427

Schmaal L, Goudriaan AE, Joos L, Kruse AM, Dom G, van den Brink W, Veltman DJ (2013) Modafinil modulates resting-state functional network connectivity and cognitive control in alcohol-dependent patients. Biol Psychiatry 73:789–795

Shotbolt P, Tziortzi AC, Searle GE, Colasanti A, van der Aart J, Abanades S, Plisson C, Miller SR, Huiban M, Beaver JD, Gunn RN, Laruelle M, Rabiner EA (2012) Within-subject comparison of [(11)C]-(+)-PHNO and [(11)C]raclopride sensitivity to acute amphetamine challenge in healthy humans. J Cereb Blood Flow Metab 32:127–136

Sillaber I, Rammes G, Zimmermann S, Mahal B, Zieglgansberger W, Wurst W, Holsboer F, Spanagel R (2002) Enhanced and delayed stress-induced alcohol drinking in mice lacking functional CRH1 receptors. Science 296:931–933

Spagnolo PA, Ramchandani VA, Schwandt ML, Zhang L, Blaine SK, Usala JM, Diamond KA, Phillips MJ, George DT, Momenan R, Heilig M (2014) Effects of naltrexone on neural and subjective response to alcohol in treatment-seeking alcohol-dependent patients. Alcohol Clin Exp Res 38:3024–3032

Spanagel R (2009) Alcoholism: a systems approach from molecular physiology to addictive behavior. Physiol Rev 89:649–705

Spanagel R (2010) Are metabotropic glutamate receptors promising targets for the treatment of alcoholism? Biol Psychiatry 67:798–799

Spanagel R, Kiefer F (2008) Drugs for relapse prevention of alcoholism: ten years of progress. Trends Pharmacol Sci 29:109–115

Spanagel R, Pendyala G, Abarca C, Zghoul T, Sanchis-Segura C, Magnone MC, Lascorz J, Depner M, Holzberg D, Soyka M, Schreiber S, Matsuda F, Lathrop M, Schumann G, Albrecht U (2005) The clock gene Per2 influences the glutamatergic system and modulates alcohol consumption. Nat Med 11:35–42

Spanagel R, Vengeliene V, Jandeleit B, Fischer WN, Grindstaff K, Zhang X, Gallop MA, Krstew EV, Lawrence AJ, Kiefer F (2014) Acamprosate produces its anti-relapse effects via calcium. Neuropsychopharmacology 39:783–791

Srinivasan R, Cunningham C, Chen A, Vigneron D, Hurd R, Nelson S, Pelletier D (2006) TE-averaged two-dimensional proton spectroscopic imaging of glutamate at 3 T. Neuroimage 30:1171–1178

Tanda G, Di Chiara G (1998) A dopamine-mu1 opioid link in the rat ventral tegmentum shared by palatable food (Fonzies) and non-psychostimulant drugs of abuse. Eur J Neurosci 10:1179–1187

Trantham-Davidson H, Burnett EJ, Gass JT, Lopez MF, Mulholland PJ, Centanni SW, Floresco SB, Chandler LJ (2014) Chronic alcohol disrupts dopamine receptor activity and the cognitive function of the medial prefrontal cortex. J Neurosci 34:3706–3718

Umhau JC, Momenan R, Schwandt ML, Singley E, Lifshitz M, Doty L, Adams LJ, Vengeliene V, Spanagel R, Zhang Y, Shen J, George DT, Hommer D, Heilig M (2010) Effect of acamprosate on magnetic resonance spectroscopy measures of central glutamate in detoxified alcohol-dependent individuals: a randomized controlled experimental medicine study. Arch Gen Psychiatry 67:1069–1077

Urban NB, Kegeles LS, Slifstein M, Xu X, Martinez D, Sakr E, Castillo F, Moadel T, O'Malley SS, Krystal JH, Abi-Dargham A (2010) Sex differences in striatal dopamine release in young adults after oral alcohol challenge: a positron emission tomography imaging study with [(11)C] raclopride. Biol Psychiatry 68:689–696

Volkow ND, Fowler JS, Wang GJ, Swanson JM (2004) Dopamine in drug abuse and addiction: results from imaging studies and treatment implications. Mol Psychiatry 9:557–569

Volpicelli JR, Alterman AI, Hayashida M, O'Brien CP (1992) Naltrexone in the treatment of alcohol dependence. Arch Gen Psychiatry 49:876–880

Weiss F, Lorang MT, Bloom FE, Koob GF (1993) Oral alcohol self-administration stimulates dopamine release in the rat nucleus accumbens: genetic and motivational determinants. J Pharmacol Exp Ther 267:250–258

Westman J, Wahlbeck K, Laursen TM, Gissler M, Nordentoft M, Hallgren J, Arffman M, Osby U (2014) Mortality and life expectancy of people with alcohol use disorder in Denmark, Finland and Sweden. Acta Psychiatr Scand

Whiteford HA, Degenhardt L, Rehm J, Baxter AJ, Ferrari AJ, Erskine HE, Charlson FJ, Norman RE, Flaxman AD, Johns N, Burstein R, Murray CJ, Vos T (2013) Global burden of disease attributable to mental and substance use disorders: findings from the Global Burden of Disease Study 2010. Lancet 382:1575–1586

Willeit M, Ginovart N, Graff A, Rusjan P, Vitcu I, Houle S, Seeman P, Wilson AA, Kapur S (2008) First human evidence of d-amphetamine induced displacement of a D2/3 agonist radioligand: A [11C]-(+)-PHNO positron emission tomography study. Neuropsychopharmacology 33:279–289

Wise RA, Bozarth MA (1987) A psychomotor stimulant theory of addiction. Psychol Rev 94:469–492

Zhou Z, Karlsson C, Liang T, Xiong W, Kimura M, Tapocik JD, Yuan Q, Barbier E, Feng A, Flanigan M, Augier E, Enoch MA, Hodgkinson CA, Shen PH, Lovinger DM, Edenberg HJ, Heilig M, Goldman D (2013) Loss of metabotropic glutamate receptor 2 escalates alcohol consumption. Proc Natl Acad Sci U S A 110:16963–16968

Zorrilla EP, Heilig M, de Wit H, Shaham Y (2013) Behavioral, biological, and chemical perspectives on targeting CRF(1) receptor antagonists to treat alcoholism. Drug Alcohol Depend 128:175–186

On the Road to Translation for PTSD Treatment: Theoretical and Practical Considerations of the Use of Human Models of Conditioned Fear for Drug Development

Victoria B. Risbrough, Daniel E. Glenn and Dewleen G. Baker

Abstract The use of quantitative, laboratory-based measures of threat in humans for proof-of-concept studies and target development for novel drug discovery has grown tremendously in the last 2 decades. In particular, in the field of posttraumatic stress disorder (PTSD), human models of fear conditioning have been critical in shaping our theoretical understanding of fear processes and importantly, validating findings from animal models of the neural substrates and signaling pathways required for these complex processes. Here, we will review the use of laboratory-based measures of fear processes in humans including cued and contextual conditioning, generalization, extinction, reconsolidation, and reinstatement to develop novel drug treatments for PTSD. We will primarily focus on recent advances in using behavioral and physiological measures of fear, discussing their sensitivity as biobehavioral markers of PTSD symptoms, their response to known and novel PTSD treatments, and in the case of d-cycloserine, how well these findings have translated to outcomes in clinical trials. We will highlight some gaps in the literature and needs for future research, discuss benefits and limitations of these outcome measures in designing proof-of-concept trials, and offer practical guidelines on design and interpretation when using these fear models for drug discovery.

Keywords Posttraumatic stress disorder · Fear · Anxiety · Panic disorder · D-cycloserine · Extinction · Exposure · Consolidation · Norpepinephrine

V.B. Risbrough · D.E. Glenn · D.G. Baker
Center of Excellence for Stress and Mental Health, San Diego, VA, USA

V.B. Risbrough (✉) · D.G. Baker
Department of Psychiatry, University of California San Diego,
9500 Gilman Dr. MC0804, La Jolla, San Diego, CA 92093, USA
e-mail: vrisbrough@ucsd.edu

Curr Topics Behav Neurosci (2016) 28: 173–196
DOI 10.1007/7854_2015_5010

Contents

1 Introduction ... 174
1.1 Posttraumatic Stress Disorder Prevalence and Treatment Options ... 174
1.2 Considerations of Benefits and Limitations of Laboratory-Based Measures of Behavior for Drug Discovery ... 176
2 Learned Fear Processes ... 177
2.1 Fear Conditioning and Cued Recall ... 177
2.2 Fear Extinction, Reconsolidation, and Reinstatement ... 180
2.3 Reconsolidation and Reinstatement ... 183
2.4 Contextual Modification and Generalization of Learned Fear and Extinction ... 184
2.5 Practical Considerations When Using Learned Fear Processes as a Marker of Drug Efficacy ... 188
3 Summary ... 189
References ... 189

1 Introduction

1.1 Posttraumatic Stress Disorder Prevalence and Treatment Options

Posttraumatic stress disorder (PTSD) affects 7–8 % of the general US population and is higher in recently deployed combat veterans (up to 20 %) (Thomas et al. 2010). Mental disorders, in particular PTSD, are associated with higher rates of physical symptoms, chronic physical illness, and overall mortality (for review see Baker et al. 2009). Research shows that this increased liability of physical disease translates into greater non-mental health medical service utilization (e.g., O'Donnell et al. 2013), creating substantial burdens for the patients, families, and societal resources. Best evidence treatment for PTSD includes cognitive behavioral therapies, i.e., cognitive processing therapy (CPT) and prolonged exposure (PE), and psychotropic medications (Institute of Medicine 2014). Although cognitive behavioral approaches have proven efficacy for PTSD, non-response can be as high as 50 %, leaving unresponsive or partially responsive patients with PTSD reliant upon pharmacotherapy (Baker et al. 2009; Institute of Medicine 2014; Berger et al. 2009). As with many psychiatric disorders, the pharmacological tool kit for PTSD treatment is relatively small, predominantly selective serotonin or norepinephrine reuptake inhibitors (SSRI/SNRI) and adjunctive treatments such as prazosin, a sympatholytic drug with alpha-1 receptor blocking activity (Baker et al. 2009; Steckler and Risbrough 2012). These medications also have high non-response rates as well as side effects (Baker et al. 2009; Steckler and Risbrough 2012). There is an unquestionable need to advance development of new treatments for PTSD, with part of this effort lying in developing innovative approaches to drug development in clinical populations.

One of the difficulties of identifying biological mechanisms for PTSD, and thus in turn developing beneficial treatments, is the heterogeneous patient population and wide spectrum of potential symptoms. According to the DSM-5 (American Psychiatric Association 2013), PTSD now comprises 20 individual symptoms. These symptoms are grouped into four symptom clusters: persistent intrusive memories of the trauma, hyperarousal and reactivity, avoidance of stimuli related to the trauma event, and negative alterations in cognitions and mood. Thus, there is a wide range of symptoms that can be endorsed to comprise a PTSD diagnosis, with many possible patterns of symptom type and severity across these clusters (Galatzer-Levy and Bryant 2013). This heterogeneity suggests that several potential biological mechanisms could drive the development of PTSD. This multiplicity of potential biological mechanisms will induce substantial variance in how any given treatment will affect a patient's treatment response.

As such, the potential for numerous different underlying pathologies in patient groups makes identification of specific mechanisms across the population very difficult. One approach to this problem is to identify biological or behavioral phenotypes that are highly represented in the diagnostic class compared to specific symptoms so as to target a "core" biological pathway that is disrupted in most patients. This approach assumes that the heterogeneity is due to noise in the self-report measurements of symptoms and how they are experienced and/or articulated, but perhaps only a few biological mechanisms actually drive clinical dysfunction. The second potential approach is to identify phenotypes that are relevant to particular symptom classes that are most severe in a given individual. This approach assumes that certain discrete phenotypes may better classify dimensions of specific symptoms experienced by subpopulations within the diagnostic group as a whole, each with potentially differing biological mechanisms (Schmidt 2015).

Development of laboratory-based behavioral measures of disease-related processes is a critical component of the evolution of translational research (Bowers and Ressler 2015). These tasks can bridge complex clinical presentations with discrete biological mechanisms (Braff 2015; Gottesman and Gould 2003; Rasetti and Weinberger 2011; Risbrough 2010). This strategy is now endorsed by the Research Domain Criteria (RDoC) project by the National Institute for Mental Health (Cuthbert and Insel 2013). Similarly, industry and academia have now increasingly turned to biological and behavioral markers in initial proof-of-concept studies to identify efficacy across specific emotional and cognitive constructs of PTSD to guide future phase II clinical trial designs. Here, we will discuss the promise and pitfalls of commonly used laboratory-based measures of conditioned fear processes to support novel drug development for PTSD.

1.2 Considerations of Benefits and Limitations of Laboratory-Based Measures of Behavior for Drug Discovery

1.2.1 Benefits of Validated Behavioral Phenotypes to Complement Symptom Assessments

(1) Objective, quantifiable assessments of function compared to self-report.
(2) Often have well characterized biological mechanism(s) and neural circuit(s).
(3) Responses are predictably controlled by specific experimental parameters in keeping with their use as an operational measure of a defined construct (e.g., anxiety, fear, arousal).
(4) Observable behaviors enable cross-species translation to lower order organisms for direct mechanistic studies and drug development (Donaldson and Hen 2015).
(5) Compared to symptoms, laboratory-based measures are observable across healthy controls and clinical populations, supporting efforts to disentangle mechanisms that cause risk versus mechanisms related to symptom onset and severity. This point is particularly important for informing treatment approaches, e.g., prophylactic versus therapeutically.
(6) Unlike symptoms, behaviors can be measured in unaffected relatives to aid in identification of genetic risk factors [e.g., behavioral endophenotypes or intermediate phenotypes (Lenzenweger 2013)].
(7) Because they are typically based on continuous measures, they offer more statistical power than dichotomous diagnostic classes.
(8) Most importantly for drug discovery, they may probe a more specific conceptual target for pharmacotherapy indicated by preclinical studies (e.g., effective for enhancing fear extinction). This last point is the primary reason behavioral tests are being used more frequently, as they may offer a greater ability to translate drug effects that are based on specific circuit actions and behavioral effects in preclinical models.

1.2.2 Limitations

(1) Lack of specificity: It is often the case that some individuals with disrupted performance in a behavioral task may not show overt functional deficits or clinical presentation. For example, menstrual cycle phases are associated with reductions in fear extinction in healthy women (Glover et al. 2015; Milad et al. 2006).
(2) In the context of genetic studies, even relatively "simple" or discrete laboratory-based behaviors do not guarantee greater heritability or simpler genetic architecture than the disorder (Greenwood et al. 2007), as would be

hoped from an intermediate phenotype or endophenotype. For example, even a behavior as simple as the startle reflex may be modulated by a huge array of biological pathways (Zhang et al. 2011).

(3) Behaviors that initially seemed relatively simple in terms of core neural circuit, e.g., extinction requiring prefrontal cortex activation of inhibitory circuits in the amygdala, can have extensive modulatory circuits that may play a stronger role in how this phenotype is altered in a given disorder compared to the "core neural circuit" (Acheson et al. 2015c; Maren and Holmes 2015; Milad et al. 2013). Thus, using behavioral performance as a proxy for the function of a specific neural circuit or brain region is limited unless it is accompanied by other information such as functional imaging.

Here, we will review the state of the art in laboratory-based measures of fear response in assessing symptom state and response to treatment in healthy controls and PTSD patients within the fear learning domains. We will also offer some practical considerations for study design and interpretation pitfalls for future planning of drug efficacy using these measures.

2 Learned Fear Processes

One of the predominant features of PTSD symptoms is robust, uncontrollable memories of the traumatic event, i.e., re-experiencing. Secondly, external or internal cues that act as trauma reminders induce re-experiencing with flashbacks and dissociation at the most severe end of the spectrum, as well as strong emotional and physiological fear responses including intense anxiety and panic. Unsurprisingly, the disorder is associated with implicit and explicit strategies for cue avoidance, which can be disruptive to daily function and interfere with long-term recovery. Thus, PTSD may be caused at least in part by disruption in one or more elements of the learned fear process (Lissek and van Meurs 2014). Here, we will describe common laboratory-based measures of these processes, their relationship to symptom clusters and predictive validity for subsequent clinical trials if available, response to pharmacological treatment in both controls and PTSD patients, and considerations of their use in drug development studies.

2.1 Fear Conditioning and Cued Recall

Laboratory-based tasks to elicit Pavlovian fear conditioning in humans induce learned fear typically by presenting a visual conditioned stimuli (CS), such as simple shapes or images in combination with an aversive unconditioned stimulus (US) such as shock to the wrist or air puff to the throat. Operational measurement of fear responding to the CS+ (CS associated with US) is derived by comparing behavior or physiological responses to the CS+ compared to CSs that are not presented with the US (i.e., safety signal, CS−) or when no cues are presented. Variations include examining responses to "contextual" versus discrete CS+ [to examine phasic versus sustained fear responses (Garfinkel et al. 2014; Glenn et al. 2014; Grillon et al. 2006)].

2.1.1 Do PTSD Patients Exhibit Increased Fear Learning/Expression? Is Fear Learning/Expression Related to Specific Symptom Clusters?

The short answer is it depends on the measure. PTSD patients exhibit increased potentiated startle responses to discrete fear cues (Briscione et al. 2014; Norrholm et al. 2011) and contextual fear cues (Grillon et al. 2009b); however, increased fear is not consistently detected using other behavioral or physiological measures such as self-report or skin conductance response (SCR) (Glover et al. 2011; Milad et al. 2008). This difference may be related to specific fear circuitry that is being probed by these behavioral measures, as startle reactivity is thought to reflect "automatic" fear conditioning processes that do not rely on contingency awareness, while SCR and self-report reflect fear processes that require contingency awareness (Jovanovic et al. 2006; Tabbert et al. 2006). Given that increased startle reactivity is commonly described by patients (DSM-IV, DSM-5), startle measures of fear may specifically probe abnormal circuits and mechanisms in PTSD that drive "automatic" fear responses (Grillon 2009). As might be expected, increased fear-potentiated startle is associated with high levels of re-experiencing symptoms in PTSD patients (Norrholm et al. 2011) and attentional bias to threat (Fani et al. 2012). However, in a study that directly compared fear acquisition across subjects with PTSD, general anxiety, or depression symptoms, increased fear expression was significantly higher in individuals with general cognitive and somatic anxiety symptoms rather than PTSD or depression symptoms (Acheson et al. 2015b). Greater conditioned fear expression has also been reported in other anxiety disorders, such as panic disorder (Grillon et al. 2008) and bipolar disorder (Acheson et al. 2015c). Thus, increased fear expression may reflect a biological abnormality in subpopulations of anxiety and mood disorder patients, crossing diagnostic classifications.

2.1.2 Is Conditioned Fear Responding Sensitive to Drugs that Are Effective for PTSD?

A reasonable question when considering a laboratory-based task for drug discovery is whether the task shows predictive validity for known therapeutic compounds. Unfortunately, there is disappointingly little work in this area. In healthy controls, fear-potentiated startle responses to cues with moderate contingency prediction which are thought to elicit sustained anxiety are attenuated by sub-chronic (2 week) SSRI treatment and acute benzodiazepine treatment, while cues with 100 %

contingency for the aversive US are not (Acheson et al. 2012b; Grillon et al. 2006, 2009a). Fear conditioning as assessed by skin conductance is unaffected by sub-chronic SSRI treatment (Bui et al. 2013). These data suggest that fear-potentiated startle has predictive validity as a laboratory-based measure of fear acquisition/expression for PTSD under certain conditions, particularly when cues elicit more prolonged anxiety-like responses which may be activating differential neural circuits [e.g., bed nucleus stria terminalis, for review see Avery et al. (2015) and Burghardt and Bauer (2013)]. Does this mean discrete fear conditioning tasks are not predictive for PTSD therapeutics? Perhaps, but an alternative explanation is that current treatments, which work in 50 % or less of the population (Berger et al. 2009), are unable to treat this particular facet of the disorder and thus are not useful positive controls. Further evidence for predictive validity for SSRI effects in patients is that acute SSRI treatment potentiates fear expression in conditioned fear models, similar to accounts of increased anxiety symptoms in patients in the initial phase of SSRI treatment (Garcia-Leal et al. 2010; Grillon et al. 2007; Silva et al. 2001). Effects of prazosin, used for treating nightmares in PTSD patients and which has some efficacy in animal models of conditioned fear responding, have not been studied yet in these human models (Do Monte et al. 2013; Raskind et al. 2013). This lack of data is partly due to the requirement for incremental dosing increases over weeks to reach therapeutic levels necessary for efficacy for treatment of nightmares in PTSD, reducing the feasibility of using this compound for validation studies. Effects of behavioral therapy on conditioned fear are also relatively untested. One small study found no significant reductions in potentiated startle to trauma-related cues after exposure therapy despite >50 % reduction in symptoms (Robison-Andrew et al. 2014); however, another larger study did find that exposure therapy reduced trauma-potentiated startle (Rothbaum et al. 2014). Overall, the evidence for predictive validity in terms of sensitivity to SSRI treatment is suggestive, but there are clear nuances to the parameters and dosing strategy that must be considered if these models are to be used.

2.1.3 Does Fear Conditioning Predict Treatment Response?

Again, there is very little work in this area. One small pilot study ($n = 9 - 10$/group) showed that only patients that show discrimination in SCRs between the CS+ and CS− respond to SSRI treatment (Aikins et al. 2011). These data support the speculation that cue discrimination may probe neural circuits that are responsive to SSRI treatment, but more research is needed to confirm this preliminary finding.

2.1.4 Is There Evidence for Fear Conditioning to Be an "Intermediate Phenotype" Associated with Genes that Confer Risk for PTSD?

There is some suggestion that genes that confer risk for PTSD are also associated either with heightened fear conditioning or with disruption in ability to inhibit conditioned fear in humans [see next section below and see Skelton et al. (2012) for review of genetic approaches to fear learning phenotypes]. Examples are genes involved in noradrenergic (ADRA2B), serotonergic (SLC6A4), and catecholamine signaling (COMT), in cellular signaling pathways that support neural plasticity [PRKCA and WWC1; for review see Wilker et al. (2014)], and in genes involved in the neuroendocrine stress response [PACAP/PAC1, Ressler et al. (2011)] and opioid signaling (Andero et al. 2013). Thus far, however, only candidate gene studies have been conducted on fear acquisition and expression phenotypes, no genome-wide association studies have been published yet.

2.2 *Fear Extinction, Reconsolidation, and Reinstatement*

Fear conditioning is vital for survival, enabling threat prediction and consequent behavioral responses to avoid harm. As cues become less predictive of aversive stimuli, however, organisms adapt to this change with reduced conditioned responding termed extinction. The process of fear extinction is subserved by a hippocampal–amygdala–prefrontal cortex circuit, with the prefrontal cortex activation of inhibitory circuits in the amygdala resulting in reduced fear responses to previously learned fear cues (for review see Milad and Quirk 2012). Extinction does not modify or "erase" the original CS–US association, but instead represents new inhibitory learning that actively competes with the original excitatory CS–US associative memory (Bouton 1993). This hypothesis is supported by a number of return of fear phenomena including reinstatement of conditioned fear, in which following fear extinction a brief re-exposure to an unpaired US induces full recovery of the original conditioned fear response (Haaker et al. 2014; Myers and Davis 2002). Modification of the original fear memory can occur, however, via reconsolidation, a period in which a memory is activated and is thus transiently labile, thought to subserve an "updating" function [see following sections below for further details (Nader 2015)].

2.2.1 Do PTSD Patients Exhibit Changes in Fear Extinction Processes?

PTSD has been described as a disorder characterized by a failure in extinction. Most trauma survivors exhibit PTSD symptoms initially after the traumatic experience; however, over time most survivors (80–90 %) will return to normal functioning, while a small subset continues to exhibit robust, debilitating trauma memories that interfere with normal functioning (Rothbaum et al. 1992; Rothbaum and Davis 2003). Extinction is a critical component to the efficacy of exposure therapy for PTSD, which exposes the patient to trauma-related memories and/or cues both in the clinic and in vivo (Craske et al. 2014).

PTSD patients exhibit reduced fear extinction learning and retention in the laboratory, indicating that poor extinction of fear responses to trauma-related cues may be a mechanism underlying PTSD (Acheson et al. 2015b; Milad et al. 2008; Norrholm et al. 2011). In a recent comparative study across subjects reporting primarily PTSD, general anxiety, or depression symptoms, extinction deficits were only observed in subjects with PTSD (Acheson et al. 2015b), suggesting that poor extinction is specifically related to trauma-related symptoms as opposed to general symptoms of low mood or ruminative anxiety. PTSD patients also exhibit functional and structural abnormalities in the fear extinction network including the hippocampus, amygdala, and frontal cortex [for review see Acheson et al. (2012a), Shvil et al. (2013)]. During extinction learning, PTSD is associated with reduced activation of the ventral medial prefrontal cortex and increased activation of the amygdala and dorsal anterior cingulate, suggesting reduced inhibitory modulation by cortical inputs to fear circuits (Shvil et al. 2013). Twin studies suggest that poor extinction observed in PTSD is associated with symptom state, rather than a vulnerability trait for PTSD (but see Lommen et al. 2013; Milad et al. 2008), suggesting it could play a role in maintenance of PTSD symptoms once they emerge. Hence, pharmacological enhancement of the neuroplasticity of this circuit is of particular interest for novel therapeutic approaches to PTSD, particularly in conjunction with exposure therapy.

2.2.2 Pharmacological Approaches for Fear Extinction in PTSD

There has been an explosion of basic and clinical research on mechanisms of fear extinction, with a large literature on the cell signaling mechanisms that mediate and modulate fear extinction learning and recall. This literature has recently been comprehensively reviewed (Maren and Holmes 2015; Singewald et al. 2015); thus, here, we will focus on a brief synopsis of the use of d-cycloserine (DCS), as this treatment is the most advanced, providing a primer in the successes and difficulties of translating animal and preclinical findings in fear behavior to clinical treatment strategies.

The concept of developing adjunctive pharmacotherapies for cognitive or exposure-based therapies was largely driven by the work of Michael Davis and Kerry Ressler. They first showed that DCS, a partial NMDA receptor agonist, administered during extinction training resulted in enhanced fear extinction recall in animals. Subsequently, they showed that DCS administered during virtual reality-based exposure therapy for fear of heights significantly increased the therapy's efficacy in reducing phobia symptoms (Ressler et al. 2004; Walker et al. 2002). These seminal papers more than a decade ago led to a burst of activity across a number of disorders, showing initial increased efficacy of DCS treatment for exposure therapies for phobias, panic disorder, and obsessive compulsive disorder which has been confirmed by two meta-analyses (Bontempo et al. 2012; Norberg et al. 2008). "High-throughput" clinical trials have been developed to test efficacy of drugs for enhancement of exposure-based therapy (Rodebaugh and Lenze 2013; Rodebaugh et al. 2013). However, the translation to exposure therapy effects in

PTSD patients is less compelling. Four studies have examined DCS enhancement of exposure therapy, with either positive effects (Difede et al. 2014), equivocal, or marginal effects (de Kleine et al. 2012; Rothbaum et al. 2014), negative effects (Scheeringa and Weems 2014), or even deleterious effects (Litvin et al. 2007). These mixed results have suggested a number of potential issues that need consideration when designing treatment trials for DCS (and other putative extinction enhancing treatments): (1) are the effects of DCS more on *speed* of response rather than *magnitude* of response to exposure, two differing hypotheses that will require different experimental designs/analysis to probe efficacy; (2) what is the correct dosing/timing of treatment; (3) does DCS's cognitive enhancement promote inhibitory learning to the extinction context, which might subsequently contribute to contextual renewal of fear (Vervliet 2008); and (4) does DCS need to be targeted toward only the successful therapy sessions [for a detailed review, see Hofmann et al. (2015)]. This latter issue is because DCS is a broad cognitive enhancer, it can enhance both fear learning and extinction learning (Lee et al. 2006); thus, if the exposure session is unsuccessful in promoting extinction, it could instead promote reconsolidation (i.e., strengthening of conditioned fear to trauma memories and cues) that is then increased by DCS treatment. Thus far, however, predicting a "successful" session versus an unsuccessful one has been elusive. Alternatively, other groups are working to identify prescriptive variables that predict which subjects would most benefit from treatment, i.e., those with the most severe PTSD, specific symptom classes, or other traits (de Kleine et al. 2012, 2014).

It is worth noting that in humans, DCS has generally been found to be more efficacious in adjunct trials with exposure therapy in patient populations, compared to enhancing extinction of conditioned fear produced in the laboratory in healthy controls. One study (Kuriyama et al. 2011) out of 3 found DCS (and valproic acid) to enhance extinction. This study was the only one to utilize a reinstatement component, with DCS during extinction training affecting not within-session learning or recall, but instead suppressing reinstatement. DCS was ineffective in studies that limited their design to testing extinction acquisition and 24-h recall (Guastella et al. 2007; Klumpers et al. 2012). It has been suggested that this lack of translation of DCS effects on extinction in animals to extinction in healthy human subjects may be because extinction protocols in the laboratory are not probing "automatic" learned fear and extinction processes, but are instead governed by top-down executive functions (Grillon 2009). More recent studies, however, suggest that extinction in healthy controls is sensitive to putative extinction enhancing drugs such as cannabinoid receptor agonists and oxytocin (Acheson et al. 2013; Das et al. 2013; Eckstein et al. 2014; Rabinak et al. 2013), which suggests that these tests are "translational" in that they are sensitive to drugs that have shown efficacy in animal extinction studies (Singewald et al. 2015). Whether these drugs can then also make the leap to enhancement of exposure therapy or PTSD treatment is thus far mixed. Efficacy of cannabinoid receptor agonists for treating PTSD symptoms is promising (Cameron et al. 2014; Roitman et al. 2014), while oxytocin effects on exposure therapy are less clear (Acheson et al. 2013, 2015a; Guastella et al. 2009; Acheson and Risbrough 2015).

2.2.3 Is Fear Extinction Sensitive to Drugs that Are Effective for PTSD?

Although the bulk of pharmacology directed at extinction processes has been of drugs that are hypothesized to specifically act on this mechanism, it is fair to ask whether extinction is sensitive to current treatments. Chronic fluoxetine in rodents facilitates extinction learning and extinction memory recall, particularly in females (Deschaux et al. 2011; Fitzgerald et al. 2014; Lebron-Milad et al. 2013), and escitalopram enhances extinction in healthy humans (Bui et al. 2013), suggesting that examining effects of a drug on extinction may predict efficacy as an overall treatment beyond use as an adjunctive treatment with therapy. Paroxetine transiently enhanced effects of exposure therapy (Schneier et al. 2012); however, other studies show no efficacy of SSRIs to enhance exposure therapy in PTSD (Foa et al. 2005; Hetrick et al. 2010). It should be noted that when undergoing exposure therapy, many opportunities for exposure are outside of the therapist's office via "homework" developed to promote in vivo exposure in the patient's environment [in addition to imaginal exposure in prolonged exposure]; thus, a drug that can be given chronically may actually be more effective than a drug limited to exposure session treatments. Based on lessons learned from DCS in terms of potential unintentional enhancement of fear learning/reconsolidation, chronic treatment will depend on how selectively the drug acts on fear extinction mechanisms versus broader mechanisms of neural plasticity. (Besides its non-selective effects on extinction, DCS cannot be given chronically due to rapid tolerance.) An example of a potential target with more selective effects on extinction enhancement are agonists of the cannabinoid 1 receptor, in particular drugs that enhance endogenous ligand availability via inhibition of degradation (Steckler and Risbrough 2012).

2.2.4 Does fear extinction performance predict treatment response?

Currently, it is unknown whether extinction performance or other markers of extinction (e.g., ventral medial frontal cortex activation during recall) predict what type of treatment (e.g., pharmacology versus exposure therapy) or how much treatment (e.g., how many exposure sessions) might be most beneficial for patients. This question is of great interest in terms of supporting personalized medicine approaches and is actively being pursued by a number of research groups.

2.3 Reconsolidation and Reinstatement

Reconsolidation occurs when a memory is reactivated resulting in a period of transient lability of the underlying neuroplastic mechanisms supporting the memory. During reconsolidation, old memories can be strengthened or disrupted by drugs that modulate consolidation mechanisms. The best characterized manipulation of reconsolidation of conditioned fear is via noradrenergic manipulations, with propranolol, a beta-adrenergic receptor antagonist, disrupting reconsolidation and subsequent conditioned fear responses in both animals and humans [for review see Otis et al. (2015)]. A recent meta-analysis indicates that propranolol is effective for blocking both consolidation and reconsolidation of fear memories in healthy humans (Lonergan et al. 2013). Recent studies however suggest that experimental design may be critical, with efficacy of propranolol given before memory reactivation having limited effect (Wood et al. 2015). Sevenster and colleagues showed that propranolol effects were only observable in conditions in which reconsolidation occurred under prediction uncertainty (i.e., the CS+ may or may not be followed by the US), suggesting that reconsolidation only occurs if the memory is actively being updated with new information (Sevenster et al. 2012). This group also cleverly showed that reconsolidation can be triggered not just by the specific CS+, but also by a semantically similar stimulus. Memory reactivation by semantically similar stimuli was sensitive to propranolol disruption (Soeter and Kindt 2015). This finding supports the feasibility of reconsolidation-based therapy, given the difficulty in accurately reconstructing trauma specific cues.

Reinstatement is when previously extinguished conditioned responding is "re-instated" after re-exposure to a US (Rescorla and Heth 1975). This phenomenon supports the now established view that extinction training does not "erase" the fear memory, but instead creates a competing CS–"No US" association with the original CS–US association. This CS–"No-US" association is further complicated by its dependence upon the extinction training context (Bouton 2014; Bouton and Todd 2014.) Studies of fear reinstatement in humans are relatively new and thus far primarily in healthy human controls (Dirikx et al. 2007; Hermans et al. 2005; Neumann 2008; Sokol and Lovibond 2012). Preliminary evidence suggests that cannabinoid receptor agonists given during or immediately after extinction training may suppress reinstatement (Das et al. 2013). There is an excellent review of current findings, methodology, and considerations for developing reinstatement protocols for drug development from the Lonsdorf laboratory (Haaker et al. 2014).

2.4 Contextual Modification and Generalization of Learned Fear and Extinction

Pavlovian fear conditioning occurs not only to discrete cues associated with a trauma, but also to the context in which a trauma occurs. The definition of what constitutes an associative context remains broad, but typically includes at least one of the following qualities: (1) unpredictable prediction of the US; (2) longer duration than a common discrete CS; and (3) complex, multimodal features. Contexts have been operationalized in numerous ways in laboratory tasks, including the experimental setting itself, a virtual reality setting, pictures of rooms, and simple cues with an unpredictable US association (e.g., Alvarez et al. 2011; Armony and Dolan 2001; Bouton et al. 2006; Glenn et al. 2014; Grillon 2002; Effting and Kindt 2007; Neumann et al. 2007).

2.4.1 Do PTSD Patients Have Altered Contextual Fear Learning?

There is substantial research on contextual fear learning in animal models of PTSD (e.g., Daskalakis et al. 2013), though laboratory research on contextual learning in PTSD patients remains limited. Elevated startle response to unpredictable contextual threat has been found in PTSD patients (Grillon et al. 2009a, b). This finding suggests that PTSD patients may have elevated sensitivity to unpredictable threat, which contributes to sustained tonic "anxiety" responding, associated with activity in the bed nucleus of the stria terminalis (Walker et al. 2003).

Successful fear learning about multimodal contextual features depends upon configural processing in which a single configural representation binds together numerous co-occurring contextual elements (e.g., Rudy et al. 2004). Configural representation is a hippocampus-dependent learning process supporting identification of whether a context is similar ("pattern completion") or dissimilar ("pattern separation") to a previously encountered context. Impaired configural processing of a traumatic context has been theorized to contribute to contextual overgeneralization of fear experienced in PTSD (Acheson et al. 2012a, b; Glenn et al. 2014). Few, if any, studies have directly examined configural fear learning processes in PTSD patients. A fear conditioning study using two-dimensional images of similar-looking rooms as distinct contexts found that PTSD patients demonstrated poorer differentiation than healthy controls between threat versus safe contexts in contingency ratings (Steiger et al. 2015). The authors note that the contextual stimuli used in this study were relatively simple static photographs of rooms (hallway, library) so contextual differentiation in this task may not have required configural processing. For example, it would have been possible to distinguish between contexts by attending to a single contextual element (the presence or absence of books on the walls) without considering the overall configurations, meaning that this task did not necessarily evaluate hippocampus-dependent contextual fear learning deficits in PTSD. Configural learning deficits have been found in PTSD combat veterans, and their non-trauma exposed twins relative to non-PTSD combat veterans (Gilbertson et al. 2007), though this study utilized a "cube and paper test" which did not examine contextual learning in relation to fear conditioning.

PTSD patients have been shown to exhibit deficient extinction of contextual fear (Steiger et al. 2015). There is an extensive literature on contextual modulation of extinction and return of fear in patients with anxiety disorders (e.g., Vervliet et al. 2013) and some evidence of altered contextual modulation of extinction in PTSD patients (Rougemont-Bücking et al. 2011).

2.4.2 Do PTSD Patients Have Altered Generalization of Fear?

Generalization of fear is the process whereby conditioned fear responding occurs not only to stimuli directly associated with the US, but also to stimuli similar to the CS (e.g., Dunsmoor and Paz 2015; Dymond et al. 2014). Fear generalization is a particularly relevant process for PTSD as much of the fear experienced by PTSD patients is triggered by encountering generalization stimuli (GS) which act as reminders of the trauma due to similarity to the original conditional stimuli, rather than through encountering the actual stimuli directly involved in the trauma. Laboratory assessment of fear generalization typically includes two phases: (1) a standard differential fear conditioning phase involving both a CS+ repeatedly predictive of an aversive US and a CS− never paired with the US and (2) a generalization test measuring responding to GSs with varying levels of similarity or relatedness to the CS+. The CS+ and CS− in generalization tasks commonly differ along a particular observable gradient, such as size or color (e.g., small circle/large circle, black square/white square), but there has been extensive research on non-perceptual forms of generalization as well including category-based, semantic, and symbolic fear generalization [for reviews see Dunsmoor and Paz (2015), and Dymond et al. (2014)]. Through such methodology, a generalization gradient is generated, indicating the extent to which strong conditional responding occurs only to GSs very similar to the CS+ (steep gradient) versus responding to GSs with high and low CS+ similarity (shallow gradient).

Despite a robust literature on fear generalization and a sound theoretical basis for the relevance of generalization to PTSD, laboratory research on fear generalization in PTSD patients is extremely limited. Relative to healthy controls, PTSD patients as well as panic disorder and generalized anxiety disorder patients show shallow fear generalization gradients, indicating overgeneralization of conditioned fear (Lissek et al. 2010, 2014a; Lissek and van Meurs 2014). These data are in line with findings that subjects with PTSD do not show physiological discrimination between CS+ and CS− cues, even though they report contingency awareness perfectly accurately (Acheson et al. 2015b; Jovanovic et al. 2012). This deficit in "automatic" fear discrimination between safe and threat cues appears to be specific to PTSD symptoms compared to generalized anxiety or depression symptoms (Acheson et al. 2015b). Thus, pharmacological enhancement of cue discrimination may be an effective strategy for a number of anxiety disorders, not just PTSD.

Recent neural models of fear generalization identify hippocampal substrates involved in both pattern completion (CA3 region, involved in recognizing a GS as similar to previously encountered CS+) and pattern separation (i.e., dentate gyrus, involved in recognizing a GS as dissimilar from previously encountered CS+), while subregions of the central and lateral amygdala, the bed nucleus of the stria terminalis, and the ventromedial prefrontral cortex have been implicated in expression of generalized fear (Besnard and Sahay 2015; Dunsmoor and Paz 2015; Lissek et al. 2014b). It is noteworthy that models of pattern completion and separation in fear generalization are similar to hippocampus-centered models of contextual fear learning (Kheirbek et al. 2012; Rudy et al. 2004). Configural learning is thought to encode complex, multimodal features of the trauma environment, however, while the term fear generalization is typically used in relation to discrimination across relatively simple stimulus gradients. Greater generalization of simple stimuli may be expected when configural learning of contextual information is impaired such that context learning must be learned through elemental representation, a learning process in which individual contextual elements are not bound together but independently associated with the negative outcome (Maren et al. 1997; Rudy et al. 2004).

2.4.3 Are Contextual Fear Learning and Fear Generalization Processes Sensitive to Drugs that Are Effective for PTSD?

No research to date has examined drug effects on contextual fear learning or fear generalization processes in PTSD patients, though preliminary experimental research suggests that acute glucose consumption may enhance retention of differential configural fear learning (Glenn et al. 2014). In healthy subjects, acute administration of 1 mg of the benzodiazepine alprazolam reduced sustained startle responding in both predictable and unpredictable "context" periods, but did not alter responding to discrete cues associated with predictable and unpredictable threat (Grillon et al. 2006). These findings tentatively suggest that acute benzodiazepine administration might reduce sustained contextual anxiety in PTSD patients, though they do not indicate treatment effects for sensitivity to unpredictable threat.

Findings from animal research are mixed regarding medication effects on contextual fear learning. One recent review concludes that both acute and chronic SSRI administration reduce plasticity in the hippocampus and decrease expression of contextual fear learning (Burghardt and Bauer 2013), while another review suggests that chronic antidepressant administration enhances configural learning processes through promotion of neurogenesis in the dentate gyrus (Castren and Hen 2013). Given the involvement of pattern separation and pattern completion in both fear generalization and contextual fear learning, there is reason to expect that drugs promoting neurogenesis in the dentate gyrus might be used to both improve configural learning of contextual information and decrease overgeneralization of feared stimuli in PTSD patients (Besnard and Sahay 2015; Castren and Hen 2013). No research has directly examined drug modulation of contextual fear extinction in PTSD, though it has been argued that DCS promotes contextual safety learning (Vervliet 2008; Woods and Bouton 2006). Theoretically, drugs that improve pattern completion and separation could be used prophylactically during or immediately following trauma to improve specificity of learning and prevent overgeneralization of contextual or discrete fear (Glenn et al. 2014). Conversely, such drugs may be contraindicated for use in conjunction with exposure therapy for PTSD and other anxiety disorders given concerns that greater contextual specificity of fear extinction learning increases the probability of contextually mediated renewal of fear (Bouton et al. 2006; Vervliet et al. 2013).

2.5 Practical Considerations When Using Learned Fear Processes as a Marker of Drug Efficacy

Because fear conditioning involves active learning, consolidation, and recall, treatment regimens will have critical consequences on how drug effects can be interpreted. Whether a treatment is hypothesized to block fear consolidation (i.e., potential utility as prophylactic) versus simply block fear expression (i.e., therapeutic utility) is a key component to appropriate study design. Sub-chronic or chronic dosing regimens are the norm for initial early phase studies. Animal studies of when the drug is most effective, either at blocking fear conditioning or at expression, are critical in planning interpretable fear conditioning studies across the dosing timeline (e.g., condition before or during dosing to test drug effects on expression versus conditioning, respectively). There is a similar issue for studies of extinction, with a note of caution from our own studies on oxytocin effects on extinction. To test the effects of oxytocin on extinction, we employed a common 2-day protocol; on the first day, fear conditioning was followed by drug treatment and subsequent extinction training trials, with the fear recall test 24 h later. We found a significant increase in extinction recall in the oxytocin group (i.e., less fear than placebo), suggesting a potential enhancement of extinction encoding/consolidation (Acheson et al. 2013). A recent study using fMRI with a very similar 1-day design of fear conditioning being followed by treatment and extinction training confirmed that within-session extinction could be enhanced by pretraining oxytocin (Eckstein et al. 2014). These findings supported subsequent examination of oxytocin to enhance extinction-based therapy. However, a preliminary study we conducted in spider phobia subjects indicated that oxytocin treatment has the opposite effect than expected, and it interfered with exposure therapy effects, with placebo treated subjects exhibiting better long-term reductions in phobia symptoms than the oxytocin-treated subjects (Acheson et al. 2015a). It is not clear whether this lack of translation is due to a potential design problem in the exposure therapy trial, including too short an exposure regimen (1 session), or whether our interpretation of oxytocin effects in laboratory-based tasks was erroneous. An alternate interpretation is that oxytocin treatment, administered soon after fear conditioning, could instead have disrupted consolidation of the fear memory (Acheson and Risbrough 2015). Thus, what was interpreted as effects on improving extinction training/recall may have actually been interfered with fear consolidation, and only a test design in which conditioning and extinction are separated more widely in time (i.e., 24 h) can be sure of the correct interpretation. A 3-day design, with conditioning, extinction, and recall on separate days, is of course more difficult in terms of retraining subjects; however, such a design will greatly enhance accurate interpretation.

An additional concern in terms of drugs effects on fear extinction is whether inhibitory learning processes are expedited (i.e., faster reduction in fear) or made more robust to relapse. It has recently been noted that in exposure therapy, the extent to which reductions in fear are long-lasting and resistant to relapse may be of greater clinical value than the sheer magnitude of decrease in fear (Vervliet et al. 2013).

This same consideration should be given to evaluating drugs targeting fear extinction, with designs that incorporate assessment of long-term recall and resistance to return of fear.

3 Summary

In conclusion, the use of laboratory-based measures of fear processes has offered the promise of exciting new targets for PTSD. Although the field continues to have gaps between findings in laboratory-based fear and effects in exposure-based therapy (e.g., DCS and oxytocin), parallel work in better defining DCS effects on fear processes and how these effects might both impede and facilitate exposure are currently underway. Using laboratory measures of fear learning processes to predict treatment response in patients is also potential evolution of the utility of fear-based tasks in informing treatment approaches. As discussed above, careful evaluation of study design and treatment approaches within the fear learning/extinction continuum will be critical in early-phase proof-of-concept studies. Designing studies with assessment of long-term recall/resistance to reinstatement will also be critical in evaluating drug effects either on fear consolidation (inhibitory) or on fear extinction (enhancement or improved generalization) for the chances of efficacy in the clinic.

Acknowledgements All authors are supported by the Center of Excellence for Stress and Mental Health. In addition, Dr. Risbrough is supported by a Veterans Administration Merit Award and Dr. Baker is supported by VA Cooperative Studies Program and the Department of Defense (Navy BUMED and CDMRP). All authors have no conflicts to disclose.

References

Acheson DT, Risbrough VB (2015) Oxytocin enhancement of fear extinction: a new target for facilitating exposure-based treatments? Biol Psychiatry 78:154–155

Acheson DT, Gresack JE, Risbrough VB (2012a) Hippocampal dysfunction effects on context memory: possible etiology for posttraumatic stress disorder. Neuropharmacology 62:674–685

Acheson DT, Stein MB, Paulus MP, Ravindran L, Simmons AN, Lohr JB, Risbrough VB (2012b) Effects of anxiolytic treatment on potentiated startle during aversive image anticipation. Hum Psychopharmacol 27:419–427

Acheson D, Feifel D, de Wilde S, McKinney R, Lohr J, Risbrough V (2013) The effect of intranasal oxytocin treatment on conditioned fear extinction and recall in a healthy human sample. Psychopharmacology 229:199–208

Acheson DT, Feifel D, Kamenski M, McKinney R, Risbrough VB (2015a) Intranasal oxytocin administration prior to exposure therapy for arachnophobia impedes treatment response. Depress Anxiety 32:400–407

Acheson DT, Geyer MA, Risbrough VB (2015b) Conditioned fear and extinction learning performance and its association with psychiatric symptoms in active duty marines. Psychoneuroendocrinology 51:495–505

Acheson D, Ehler L, Resovsky J, Tsan E, Risbrough V (2015c) Fear extinction memory performance in a sample of stable, euthymic patients with bipolar disorder. J Affect Disord 185:230–238

Aikins DE, Jackson ED, Christensen A, Walderhaug E, Afroz S, Neumeister A (2011) Differential conditioned fear response predicts duloxetine treatment outcome in male veterans with PTSD: a pilot study. Psychiatry Res 188:453–455

Alvarez RP, Chen G, Bodurka J, Kaplan R, Grillon C (2011) Phasic and sustained fear in humans elicits distinct patterns of brain activity. Neuroimage 55:389–400

Andero R, Brothers SP, Jovanovic T, Chen YT, Salah-Uddin H, Cameron M, Bannister TD, Almli L, Stevens JS, Bradley B, Binder EB, Wahlestedt C and Ressler KJ (2013) Amygdala-dependent fear is regulated by Oprl1 in mice and humans with PTSD. *Sci Transl Med* 5:188ra173

Armony JL, Dolan RJ (2001) Modulation of auditory neural responses by a visual context in human fear conditioning. NeuroReport 12:3407–3411

American Psychiatric Association (2013) Diagnostic and statistical manual of mental disorders (5th Ed). Washington DC

Avery SN, Clauss JA, Blackford JU (2015) The human BNST: functional role in anxiety and addiction. Neuropsychopharmacology 41;126–141

Baker DG, Nievergelt CM, Risbrough VB (2009) Post-traumatic stress disorder: emerging concepts of pharmacotherapy. Expert Opin Emerg Drugs 14:251–272

Berger W, Mendlowicz MV, Marques-Portella C, Kinrys G, Fontenelle LF, Marmar CR, Figueira I (2009) Pharmacologic alternatives to antidepressants in posttraumatic stress disorder: a systematic review. Prog Neuropsychopharmacol Biol Psychiatry 33:169–180

Besnard A, Sahay A (2015) Adult hippocampal neurogenesis, fear generalization, and stress. Neuropsychopharmacology 41:1–21

Bontempo A, Panza KE, Bloch MH (2012) D-cycloserine augmentation of behavioral therapy for the treatment of anxiety disorders: a meta-analysis. J Clin Psychiatry 73:533–537

Bouton ME (1993) Context, time and memory retrieval in the interference paradigms of Pavlovian learning. Psychol Bull 114:90–99

Bouton ME (2014) Why behavior change is difficult to sustain. Prev Med 68:29–36

Bouton ME, Todd TP (2014) A fundamental role for context in instrumental learning and extinction. Behav Process 104:13–19

Bouton ME, Westbrook RF, Corcoran KA, Maren S (2006) Contextual and temporal modulation of extinction: behavioral and biological mechanisms. Biol Psychiatry 60:352–360

Bowers ME, Ressler KJ (2015) An overview of translationally informed treatments for posttraumatic stress disorder: animal models of pavlovian fear conditioning to human clinical trials. Biol Psychiatry 78(5):E15–E27

Braff DL (2015) The importance of endophenotypes in schizophrenia research. Schizophr Res 163:1–8

Briscione MA, Jovanovic T, Norrholm SD (2014) Conditioned fear associated phenotypes as robust, translational indices of trauma-, stressor-, and anxiety-related behaviors. Front Psychiatry 5:88

Bui E, Orr SP, Jacoby RJ, Keshaviah A, LeBlanc NJ, Milad MR, Pollack MH, Simon NM (2013) Two weeks of pretreatment with escitalopram facilitates extinction learning in healthy individuals. Hum Psychopharmacol 28:447–456

Burghardt NS, Bauer EP (2013) Acute and chronic effects of selective serotonin reuptake inhibitor treatment on fear conditioning: implications for underlying fear circuits. Neuroscience 247:253–272

Cameron C, Watson D, Robinson J (2014) Use of a synthetic cannabinoid in a correctional population for posttraumatic stress disorder-related insomnia and nightmares, chronic pain, harm reduction, and other indications: a retrospective evaluation. J Clin Psychopharmacol 34:559–564

Castren E, Hen R (2013) Neuronal plasticity and antidepressant actions. Trends Neurosci 36:259–267

Craske MG, Treanor M, Conway CC, Zbozinek T, Vervliet B (2014) Maximizing exposure therapy: an inhibitory learning approach. Behav Res Ther 58:10–23
Cuthbert B, Insel T (2013) Toward the future of psychiatric diagnosis: the seven pillars of RDoC. BMC Med 11:126
Das RK, Kamboj SK, Ramadas M, Yogan K, Gupta V, Redman E, Curran HV, Morgan CJ (2013) Cannabidiol enhances consolidation of explicit fear extinction in humans. Psychopharmacology 226:781–792
Daskalakis NP, Yehuda R, Diamond DM (2013) Animal models in translational studies of PTSD. Psychoneuroendocrinology 38:1895–1911
de Kleine RA, Hendriks GJ, Kusters WJ, Broekman TG, van Minnen A (2012) A randomized placebo-controlled trial of D-cycloserine to enhance exposure therapy for posttraumatic stress disorder. Biol Psychiatry 71:962–968
de Kleine RA, Hendriks GJ, Smits JA, Broekman TG, van Minnen A (2014) Prescriptive variables for d-cycloserine augmentation of exposure therapy for posttraumatic stress disorder. J Psychiatr Res 48:40–46
Deschaux O, Spennato G, Moreau JL, Garcia R (2011) Chronic treatment with fluoxetine prevents the return of extinguished auditory-cued conditioned fear. Psychopharmacology 215:231–237
Difede J, Cukor J, Wyka K, Olden M, Hoffman H, Lee FS, Altemus M (2014) D-cycloserine augmentation of exposure therapy for post-traumatic stress disorder: a pilot randomized clinical trial. Neuropsychopharmacology 39:1052–1058
Dirikx T, Hermans D, Vansteenwegen D, Baeyens F, Eelen P (2007) Reinstatement of conditioned responses in human differential fear conditioning. J Behav Ther Exp Psychiatry 38:237–251
Do Monte FH, Souza RR, Wong TT, Carobrez Ade P (2013) Systemic or intra-prelimbic cortex infusion of prazosin impairs fear memory reconsolidation. Behav Brain Res 244:137–141
Donaldson ZR, Hen R (2015) From psychiatric disorders to animal models: a bidirectional and dimensional approach. Biol Psychiatry 77:15–21
Dunsmoor DE, Paz R (2015) Fear generalization and anxiety: behavioral and neural mechanisms. Biol Psychiatry 78:336–343
Dymond S, Dunsmoor JE, Vervliet B, Roche B, Hermans D (2014) Fear generalization in humans: systematic review and implications for anxiety disorder research. Behav Ther. doi:10.1016/j.beth.2014.10.001
Eckstein M, Becker B, Scheele D, Scholz C, Preckel K, Schlaepfer TE, Grinevich V, Kendrick KM, Maier W, Hurlemann R (2014) Oxytocin facilitates the extinction of conditioned fear in humans. Biol Psychiatry 78:194–202
Effting M, Kindt M (2007) Contextual control of human fear associations in a renewal paradigm. Behav Res Ther 45:2002–2018
Fani N, Tone EB, Phifer J, Norrholm SD, Bradley B, Ressler KJ, Kamkwalala A, Jovanovic T (2012) Attention bias toward threat is associated with exaggerated fear expression and impaired extinction in PTSD. Psychol Med 42:533–543
Fitzgerald PJ, Seemann JR, Maren S (2014) Can fear extinction be enhanced? A review of pharmacological and behavioral findings. Brain Res Bull 105:46–60
Foa EB, Hembree EA, Cahill SP, Rauch SA, Riggs DS, Feeny NC, Yadin E (2005) Randomized trial of prolonged exposure for posttraumatic stress disorder with and without cognitive restructuring: outcome at academic and community clinics. J Consult Clin Psychol 73:953–964
Garcia-Leal C, Del-Ben CM, Leal FM, Graeff FG, Guimaraes FS (2010) Escitalopram prolonged fear induced by simulated public speaking and released hypothalamic-pituitary-adrenal axis activation. J Psychopharmacol 24:683–694
Galatzer-Levy IR, Bryant RA (2013) 636,120 ways to have posttraumatic stress disorder. Perspect Psychol Sci 8(6):651–662
Garfinkel SN, Abelson JL, King AP, Sripada RK, Wang X, Gaines LM, Liberzon I (2014) Impaired contextual modulation of memories in PTSD: an fMRI and psychophysiological study of extinction retention and fear renewal. J Neurosci 34:13435–13443
Gilbertson MW, Williston SK, Paulus LA, Lasko NB, Gurvits TV, Shenton ME, Pitman RK, Orr SP (2007) Configural cue performance in identical twins discordant for posttraumatic stress

disorder: theoretical implications for the role of hippocampal function. Biol Psychiatry 62:513–520
Glenn DE, Minor TR, Vervliet B, Craske MG (2014) The effect of glucose on hippocampal-dependent contextual fear conditioning. Biol Psychiatry 75:847–854
Glover EM, Phifer JE, Crain DF, Norrholm SD, Davis M, Bradley B, Ressler KJ, Jovanovic T (2011) Tools for translational neuroscience: PTSD is associated with heightened fear responses using acoustic startle but not skin conductance measures. Depress Anxiety 28:1058–1066
Glover EM, Jovanovic T, Norrholm SD (2015) Estrogen and extinction of fear memories: implications for posttraumatic stress disorder treatment. Biol Psychiatry 78:178–185
Gottesman II, Gould TD (2003) The endophenotype concept in psychiatry: etymology and strategic intentions. Am J Psychiatry 160:636–645
Greenwood TA, Braff DL, Light GA, Cadenhead KS, Calkins ME, Dobie DJ, Freedman R, Green MF, Gur RE, Gur RC, Mintz J, Nuechterlein KH, Olincy A, Radant AD, Seidman LJ, Siever LJ, Silverman JM, Stone WS, Swerdlow NR, Tsuang DW, Tsuang MT, Turetsky BI, Schork NJ (2007) Initial heritability analyses of endophenotypic measures for schizophrenia: the consortium on the genetics of schizophrenia. Arch Gen Psychiatry 64:1242–1250
Grillon C (2002) Associative learning deficits increase symptoms of anxiety in humans. Biol Psychiatry 51:851–858
Grillon C (2009) D-Cycloserine facilitation of fear extinction and exposure-based therapy might rely on lower-level, automatic mechanisms. Biol Psychiatry 66:636–641
Grillon C, Baas JMP, Pine DS, Lissek S, Lawley M, Ellis V, Levine J (2006) The benzodiazepine alprazolam dissociates contextual fear from cued fear in humans as assessed by fear-potentiated startle. Biol Psychiatry 60:760
Grillon C, Levenson J, Pine DS (2007) A single dose of the selective serotonin reuptake inhibitor citalopram exacerbates anxiety in humans: a fear-potentiated startle study. Neuropsychopharmacology 32:225–231
Grillon C, Lissek S, Rabin S, McDowell D, Dvir S, Pine DS (2008) Increased anxiety during anticipation of unpredictable but not predictable aversive stimuli as a psychophysiologic marker of panic disorder. Am J Psychiatry: appi.ajp.2007.07101581
Grillon C, Chavis C, Covington MF, Pine DS (2009a) Two-week treatment with the selective serotonin reuptake inhibitor citalopram reduces contextual anxiety but not cued fear in healthy volunteers: a fear-potentiated startle study. Neuropsychopharmacology 34:964–971
Grillon C, Pine DS, Lissek S, Rabin S, Bonne O, Vythilingam M (2009b) Increased anxiety during anticipation of unpredictable aversive stimuli in posttraumatic stress disorder but not in generalized anxiety disorder. Biol Psychiatry 66:47–53
Guastella AJ, Lovibond PF, Dadds MR, Mitchell P, Richardson R (2007) A randomized controlled trial of the effect of D-cycloserine on extinction and fear conditioning in humans. Behav Res Ther 45:663–672
Guastella AJ, Howard AL, Dadds MR, Mitchell P, Carson DS (2009) A randomized controlled trial of intranasal oxytocin as an adjunct to exposure therapy for social anxiety disorder. Psychoneuroendocrinology 34:917–923
Haaker J, Golkar A, Hermans D, Lonsdorf TB (2014) A review on human reinstatement studies: an overview and methodological challenges. Learn Mem 21:424–440
Hermans D, Dirikx T, Vansteenwegenin D, Baeyens F, Van den Bergh O, Eelen P (2005) Reinstatement of fear responses in human aversive conditioning. Behav Res Ther 43:533–551
Hetrick SE, Purcell R, Garner B, Parslow R (2010) Combined pharmacotherapy and psychological therapies for post traumatic stress disorder (PTSD). Cochrane Database Syst Rev: CD007316
Hofmann SG, Otto MW, Pollack MH, Smits JA (2015) D-cycloserine augmentation of cognitive behavioral therapy for anxiety disorders: an update. Curr Psychiatry Rep 17:532
Institute of Medicine (2014) Treatment for posttraumatic stress disorder in military and veteran populations: final assessment. Committee on the assessment of ongoing efforts in the treatment of posttraumatic stress disorder, board on the health of select populations, Institute of Medicine, Washington (DC), National Academies Press (US), Jun 17 2014

Jovanovic T, Norrholm SD, Keyes M, Fiallos A, Jovanovic S, Myers KM, Davis M, Duncan EJ (2006) Contingency awareness and fear inhibition in a human fear-potentiated startle paradigm. Behav Neurosci 120:995–1004
Jovanovic T, Kazama A, Bachevalier J, Davis M (2012) Impaired safety signal learning may be a biomarker of PTSD. Neuropharmacology 62:695–704
Kheirbek MA, Tannenholz L, Hen R (2012) NR2B-dependent plasticity of adult-born granule cells is necessary for context discrimination. J Neurosci 32:8696–8702
Klumpers F, Denys D, Kenemans JL, Grillon C, van der Aart J, Baas JM (2012) Testing the effects of Delta9-THC and D-cycloserine on extinction of conditioned fear in humans. J Psychopharmacol 26:471–478
Kuriyama K, Honma M, Soshi T, Fujii T, Kim Y (2011) Effect of D-cycloserine and valproic acid on the extinction of reinstated fear-conditioned responses and habituation of fear conditioning in healthy humans: a randomized controlled trial. Psychopharmacology 218:589–597
Lebron-Milad K, Tsareva A, Ahmed N, Milad MR (2013) Sex differences and estrous cycle in female rats interact with the effects of fluoxetine treatment on fear extinction. Behav Brain Res 253:217–222
Lee JLC, Milton AL, Everitt BJ (2006) Reconsolidation and extinction of conditioned fear: inhibition and potentiation. J Neurosci 26:10051–10056
Lenzenweger MF (2013) Endophenotype, intermediate phenotype, biomarker: definitions, concept comparisons, clarifications. Depress Anxiety 30:185–189
Lissek S, van Meurs B (2014) Learning models of PTSD: theoretical accounts and psychobiological evidence. Int J Psychophysiol S0167-8760(14)01647-X. doi: 10.1016/j.ijpsycho.2014.11.006. [Epub ahead of print]
Lissek S, Rabin SJ, Heller RE, Luckenbaugh D, Geraci M, Pine DS, Grillon C (2010) Overgeneralization of conditioned fear as a pathogenic marker of panic disorder. Am J Psychiatr 167:47–55
Lissek S, Bradford DE, Alvarez RP, Burton P, Espensen-Sturges T, Reynolds RC, Grillon C (2014a) Neural substrates of classically conditioned fear-generalization in humans: a parametric fMRI study. Soc Cogn Affect Neurosci 9:1134–1142
Lissek S, Kaczkurkin AN, Rabin S, Geraci M, Pine DS, Grillon C (2014b) Generalized anxiety disorder is associated with overgeneralization of classically conditioned fear. Biol Psychiatry 75:909–915
Litvin Y, Pentkowski NS, Blanchard DC, Blanchard RJ (2007) CRF type 1 receptors in the dorsal periaqueductal gray modulate anxiety-induced defensive behaviors. Horm Behav 52:244–251
Lommen MJ, Engelhard IM, Sijbrandij M, van den Hout MA, Hermans D (2013) Pre-trauma individual differences in extinction learning predict posttraumatic stress. Behav Res Ther 51:63–67
Lonergan MH, Olivera-Figueroa LA, Pitman RK, Brunet A (2013) Propranolol's effects on the consolidation and reconsolidation of long-term emotional memory in healthy participants: a meta-analysis. J Psychiatry Neurosci 38:222–231
Maren S, Holmes A (2015) Stress and fear extinction. Neuropsychopharmacology 41:58–79
Maren S, Aharonov G, Fanselow MS (1997) Neurotoxic lesions of the dorsal hippocampus and Pavlovian fear conditioning in rats. Behav Brain Res 88:261–274
Milad MR, Quirk GJ (2012) Fear extinction as a model for translational neuroscience: ten years of progress. Annu Rev Psychol 63:129–151
Milad MR, Goldstein JM, Orr SP, Wedig MM, Klibanski A, Pitman RK, Rauch SL (2006) Fear conditioning and extinction: influence of sex and menstrual cycle in healthy humans. Behav Neurosci 120:1196–1203
Milad MR, Orr SP, Lasko NB, Chang Y, Rauch SL, Pitman RK (2008) Presence and acquired origin of reduced recall for fear extinction in PTSD: results of a twin study. J Psychiatr Res 42:515–520
Milad MR, Furtak SC, Greenberg JL, Keshaviah A, Im JJ, Falkenstein MJ, Jenike M, Rauch SL, Wilhelm S (2013) Deficits in conditioned fear extinction in obsessive-compulsive disorder and neurobiological changes in the fear circuit. JAMA Psychiatry 70:608–618; quiz 554

Myers KM, Davis M (2002) Behavioral and neural analysis of extinction. Neuron 36:567–584
Nader K (2015) Reconsolidation and the dynamic nature of memory. Cold Spring Harb Perspect Biol 7:a021782
Neumann DL (2008) The effects of context changes on the reinstatement of extinguished conditioned behavior in a conditioned suppression task with humans. Learn Motiv 39:114–135
Neumann DL, Lipp OV, Cory SE (2007) Conducting extinction in multiple contexts does not necessarily attenuate the renewal of shock expectancy in a fear-conditioning procedure with humans. Behav Res Ther 45:385–394
Norberg MM, Krystal JH, Tolin DF (2008) A meta-analysis of D-cycloserine and the facilitation of fear extinction and exposure therapy. Biol Psychiatry 63:1118
Norrholm SD, Jovanovic T, Olin IW, Sands LA, Karapanou I, Bradley B, Ressler KJ (2011) Fear extinction in traumatized civilians with posttraumatic stress disorder: relation to symptom severity. Biol Psychiatry 69:556–563
O'Donnell ML, Varker T, Holmes AC, Ellen S, Wade D, Creamer M, Silove D, McFarlane A, Bryant RA, Forbes D (2013) Disability after injury: the cumulative burden of physical and mental health. J Clin Psychiatry 74:e137–e143
Otis JM, Werner CT, Mueller D (2015) Noradrenergic regulation of fear and drug-associated memory reconsolidation. Neuropsychopharmacology 40:793–803
Rabinak CA, Angstadt M, Sripada CS, Abelson JL, Liberzon I, Milad MR, Phan KL (2013) Cannabinoid facilitation of fear extinction memory recall in humans. Neuropharmacology 64:396–402
Rasetti R, Weinberger DR (2011) Intermediate phenotypes in psychiatric disorders. Curr Opin Genet Dev 21:340–348
Raskind MA, Peterson K, Williams T, Hoff DJ, Hart K, Holmes H, Homas D, Hill J, Daniels C, Calohan J, Millard SP, Rohde K, O'Connell J, Pritzl D, Feiszli K, Petrie EC, Gross C, Mayer CL, Freed MC, Engel C, Peskind ER (2013) A trial of prazosin for combat trauma PTSD with nightmares in active-duty soldiers returned from Iraq and Afghanistan. Am J Psychiatry 170:1003–1010
Rescorla RA, Heth CD (1975) Reinstatement of fear to an extinguished stimulus. J Exp Psychol Anim Behav Process 1:88–96
Ressler KJ, Rothbaum BO, Tannenbaum L, Anderson P, Graap K, Zimand E, Hodges L, Davis M (2004) Cognitive enhancers as adjuncts to psychotherapy: use of D-cycloserine in phobic individuals to facilitate extinction of fear. Arch Gen Psychiatry 61:1136–1144
Ressler KJ, Mercer KB, Bradley B, Jovanovic T, Mahan A, Kerley K, Norrholm SD, Kilaru V, Smith AK, Myers AJ, Ramirez M, Engel A, Hammack SE, Toufexis D, Braas KM, Binder EB, May V (2011) Post-traumatic stress disorder is associated with PACAP and the PAC1 receptor. Nature 470:492–497
Risbrough V (2010) Behavioral correlates of anxiety. Curr Top Behav Neurosci 2:205–228
Robison-Andrew EJ, Duval ER, Nelson CB, Echiverri-Cohen A, Giardino N, Defever A, Norrholm SD, Jovanovic T, Rothbaum BO, Liberzon I, Rauch SA (2014) Changes in trauma-potentiated startle with treatment of posttraumatic stress disorder in combat Veterans. J Anxiety Disord 28:358–362
Rodebaugh TL, Lenze EJ (2013) Lessons learned from D-cycloserine: the promise and limits of drug facilitation of exposure therapy. J Clin Psychiatry 74:415–416
Rodebaugh TL, Levinson CA, Lenze EJ (2013) A high-throughput clinical assay for testing drug facilitation of exposure therapy. Depress Anxiety 30:631–637
Roitman P, Mechoulam R, Cooper-Kazaz R, Shalev A (2014) Preliminary, open-label, pilot study of add-on oral Delta9-tetrahydrocannabinol in chronic post-traumatic stress disorder. Clin Drug Investig 34:587–591
Rothbaum BO, Davis M (2003) Applying learning principles to the treatment of post-trauma reactions. Ann NY Acad Sci 1008:112–121
Rothbaum B, Foa EB, Riggs DS, Murdock T, Walsh W (1992) A prospective examination of post-traumatic stress disorder in rape victims. J Trauma Stress 5:455–475

Rothbaum BO, Price M, Jovanovic T, Norrholm SD, Gerardi M, Dunlop B, Davis M, Bradley B, Duncan EJ, Rizzo A, Ressler KJ (2014) A randomized, double-blind evaluation of D-cycloserine or alprazolam combined with virtual reality exposure therapy for posttraumatic stress disorder in Iraq and Afghanistan War veterans. Am J Psychiatry 171:640–648

Rougemont-Bücking A, Linnman C, Zeffiro TA, Zeidan MA, Lebron-Milad K, Rodriguez-Romaguera J, Rauch SL, Pitman RK, Milad MR (2011) Altered processing of contextual information during fear extinction in PTSD: an fMRI study. CNS Neurosci Ther 17:227–236

Rudy JW, Huff NC, Matus-Amat P (2004) Understanding contextual fear conditioning: insights from a two-process model. Neurosci Biohehav Rev 28:675–685

Scheeringa MS, Weems CF (2014) Randomized placebo-controlled D-cycloserine with cognitive behavior therapy for pediatric posttraumatic stress. J Child Adolesc Psychopharmacol 24:69–77

Schmidt U (2015) A plea for symptom-based research in psychiatry. Eur J Psychotraumatol 6:27660

Schneier FR, Neria Y, Pavlicova M, Hembree E, Suh EJ, Amsel L, Marshall RD (2012) Combined prolonged exposure therapy and paroxetine for PTSD related to the World Trade Center attack: a randomized controlled trial. Am J Psychiatry 169:80–88

Sevenster D, Beckers T, Kindt M (2012) Retrieval per se is not sufficient to trigger reconsolidation of human fear memory. Neurobiol Learn Mem 97:338–345

Shvil E, Rusch HL, Sullivan GM, Neria Y (2013) Neural, psychophysiological, and behavioral markers of fear processing in PTSD: a review of the literature. Curr Psychiatry Rep 15:358

Silva M, Hetem LA, Guimaraes FS, Graeff FG (2001) Opposite effects of nefazodone in two human models of anxiety. Psychopharmacology 156:454–460

Singewald N, Schmuckermair C, Whittle N, Holmes A, Ressler KJ (2015) Pharmacology of cognitive enhancers for exposure-based therapy of fear, anxiety and trauma-related disorders. Pharmacol Ther 149:150–190

Skelton K, Ressler KJ, Norrholm SD, Jovanovic T, Bradley-Davino B (2012) PTSD and gene variants: new pathways and new thinking. Neuropharmacology 62:628–637

Soeter M, Kindt M (2015) Retrieval cues that trigger reconsolidation of associative fear memory are not necessarily an exact replica of the original learning experience. Front Behav Neurosci 9:122

Sokol N, Lovibond PF (2012) Cross-US reinstatement of human conditioned fear: return of old fears or emergence of new ones? Behav Res Ther 50:313–322

Steckler T, Risbrough V (2012) Pharmacological treatment of PTSD—established and new approaches. Neuropharmacology 62:617–627

Steiger F, Nees F, Wicking M, Lang S, Flor H (2015) Behavioral and central correlates of contextual fear learning and contextual modulation of cued fear in posttraumatic stress disorder. Int J 2015 Jul 3. pii: S0167-8760(15)30002-7. doi: 10.1016/j.ijpsycho.2015.06.009. [Epub ahead of print]

Tabbert K, Stark R, Kirsch P, Vaitl D (2006) Dissociation of neural responses and skin conductance reactions during fear conditioning with and without awareness of stimulus contingencies. Neuroimage 32:761–770

Thomas JL, Wilk JE, Riviere LA, McGurk D, Castro CA, Hoge CW (2010) Prevalence of mental health problems and functional impairment among active component and national guard soldiers 3 and 12 months following combat in Iraq. Arch Gen Psychiatry 67:614–623

Vervliet B (2008) Learning and memory in conditioned fear extinction: effects of d-cycloserine. Acta Psychol 127:601–613

Vervliet B, Craske MG, Hermans D (2013) Fear extinction and relapse: state of the art. Annu Rev Clin Psychol 9:215–248

Walker DL, Ressler KJ, Lu KT, Davis M (2002) Facilitation of conditioned fear extinction by systemic administration or intra-amygdala infusions of D-cycloserine as assessed with fear-potentiated startle in rats. J Neurosci JID 8102140(22):2343–2351

Walker DL, Toufexis DJ, Davis M (2003) Role of the bed nucleus of the stria terminalis versus the amygdala in fear, stress, and anxiety. Eur J Pharmacol 463:199–216

Wilker S, Elbert T, Kolassa IT (2014) The downside of strong emotional memories: how human memory-related genes influence the risk for posttraumatic stress disorder–a selective review. Neurobiol Learn Mem 112:75–86

Wood NE, Rosasco ML, Suris AM, Spring JD, Marin M-F, Lasko NB, Goetz JM, Fischer AM, Orr SP, Pitman RK (2015) Pharmacological blockade of memory reconsolidation in posttraumatic stress disorder: three negative psychophysiological studies. Psychiatry Res 225:31–39

Woods AM, Bouton ME (2006) D-cycloserine facilitates extinction but does not eliminate renewal of the conditioned emotional response. Behav Neurosci 120:1159–1162

Zhang L, Hu XZ, Li H, Li X, Smerin S, Benedek DM, Ursano R (2011) Startle response related genes. Med Hypotheses 77:685–691

Translational Approaches Targeting Reconsolidation

Marijn C.W. Kroes, Daniela Schiller, Joseph E. LeDoux and Elizabeth A. Phelps

Abstract Maladaptive learned responses and memories contribute to psychiatric disorders that constitute a significant socio-economic burden. Primary treatment methods teach patients to inhibit maladaptive responses, but do not get rid of the memory itself, which explains why many patients experience a return of symptoms even after initially successful treatment. This highlights the need to discover more persistent and robust techniques to diminish maladaptive learned behaviours. One potentially promising approach is to alter the original memory, as opposed to inhibiting it, by targeting memory reconsolidation. Recent research shows that reactivating an old memory results in a period of memory flexibility and requires restorage, or reconsolidation, for the memory to persist. This reconsolidation period allows a window for modification of a specific old memory. Renewal of memory flexibility following reactivation holds great clinical potential as it enables targeting reconsolidation and changing of specific learned responses and memories that contribute to maladaptive mental states and behaviours. Here, we will review translational research on non-human animals, healthy human subjects, and clinical populations aimed at altering memories by targeting reconsolidation using biological treatments (electrical stimulation, noradrenergic antagonists) or behavioural interference (reactivation–extinction paradigm). Both approaches have been used successfully to modify aversive and appetitive memories, yet effectiveness in treating clinical populations has been limited. We will discuss that memory flexibility depends on the type of memory tested and the

M.C.W. Kroes · J.E. LeDoux · E.A. Phelps
Department of Psychology, Centre for Neural Science, New York University, New York, NY 10003, USA
e-mail: marijn.kroes@nyu.edu

E.A. Phelps
e-mail: liz.phelps@nyu.edu

D. Schiller (✉)
Department of Psychiatry and Department of Neuroscience, Friedman Brain Institute, Icahn School of Medicine at Mt. Sinai, New York, NY 10029, USA
e-mail: daniela.schiller@mssm.edu

J.E. LeDoux · E.A. Phelps
Nathan Kline Institute, Orangeburg, NY 10962, USA

Curr Topics Behav Neurosci (2016) 28: 197–230
DOI 10.1007/7854_2015_5008

brain regions that underlie specific types of memory. Further, when and how we can most effectively reactivate a memory and induce flexibility is largely unclear. Finally, the development of drugs that can target reconsolidation and are safe for use in humans would optimize cross-species translations. Increasing the understanding of the mechanism and limitations of memory flexibility upon reactivation should help optimize efficacy of treatments for psychiatric patients.

Keywords Memory · Emotions · Reconsolidation · Translational approaches · Aversive conditioning · Appetitive conditioning · Norepinephrine · Beta-blockers · Reactivation–extinction

Contents

1 Introduction 198
2 Discovering Reconsolidation 200
3 Criteria to Demonstrate Reconsolidation 201
4 Biological Interventions Targeting Reconsolidation 201
4.1 Electrical Stimulation 201
4.2 Noradrenergic Manipulations Targeting Reconsolidation: Non-human Animal Studies 203
4.3 Noradrenergic Manipulations Targeting Reconsolidation: Preclinical Human Studies 205
4.4 Noradrenergic Manipulations Targeting Reconsolidation: Studies in Clinical Populations 210
4.5 Noradrenergic Manipulations Targeting Reconsolidation: Conclusions 211
5 Behavioural Interventions Targeting Reconsolidation: The Reactivation–Extinction Paradigm 212
5.1 Reactivation–Extinction: Non-human Animal Studies 213
5.2 Reactivation–Extinction: Preclinical Human Studies 216
5.3 Reactivation–Extinction: Studies in Clinical Populations 218
5.4 Reactivation–Extinction: Conclusions 219
6 Conclusion, Limitations, and Future Directions 220
6.1 Limitations 220
6.2 Future Directions 222
References 223

1 Introduction

Maladaptive learned responses and memories contribute to psychiatric disorders, which rank among the leading causes of disability worldwide, significantly contributing to the global disease burden (WHO 2011). A primary treatment for anxiety, stress-related, or addiction disorders is exposure therapy, which is based on the principles of extinction learning in Pavlovian conditioning (Morrison and Ressler 2014; Vervliet et al. 2013). For example, in Pavlovian threat conditioning, when a neutral stimulus, such as a tone (the conditioned stimulus or CS), is paired with an aversive event, such as a shock (the unconditioned stimulus or US), the tone itself may come to elicit a range of defensive responses, such as freezing or autonomic arousal (the conditioned response or CR). Extinction refers to the gradual decrease of conditioned responses to a CS when it is repeatedly presented without reinforcement. As Pavlov first suggested (Pavlov 1927), extinction does not represent unlearning of the original memory, but rather is thought to result in a novel memory trace that comes to inhibit the expression of the initial memory. One limitation of extinction is that because the original threat memory is not altered, only inhibited, maladaptive defensive behaviours can return following the passage of time (spontaneous recovery), alterations in context (renewal), and stress (reinstatement). This poses a serious challenge to exposure-based therapies after which many patients experience a return of symptoms even after initially successful treatment (Morrison and Ressler 2014; Vervliet et al. 2013).

This potential for memory recovery even after extensive extinction training or exposure therapy highlights the need to discover more persistent and robust techniques to diminish maladaptive learned behaviours. One potentially promising approach is to alter the original memory, as opposed to inhibiting it, by targeting memory reconsolidation. The traditional view of memory suggests that that following an initial consolidation period, memories are stable and 'fixed' in the brain and original memory traces remain essentially unchanged (Dudai 2004; McGaugh 2000; Müller and Pilzecker 1900). Research on reconsolidation has challenged the traditional view of memory by demonstrating that reactivating previously consolidated memories can induce renewed flexibility and an opportunity to alter the original memory long after initial learning (Alberini and LeDoux 2013; Kroes and Fernández 2012; Nader et al. 2000; Schiller and Phelps 2011). This leads to the intriguing possibility that an understanding of this persistent, flexible nature of memory will enable targeting and changing specific learned responses and memories that contribute to maladaptive mental states and behaviours.

Manipulations aimed at affecting reconsolidation have used both biological (e.g. electrical stimulation of the brain, administration of pharmacological compounds) and behavioural interventions to target reconsolidation, thus persistently altering the previously consolidated original memory following memory reactivation. In this review, we will focus on these two classes of techniques, specifically highlighting approaches that emerged from non-human animal studies and that have been applied in both healthy humans and clinical populations. First, we will discuss the discovery of reconsolidation and the criteria generated from non-human animal studies to provide convincing evidence for reconsolidation. Next, we will describe studies using electrical or pharmacological manipulations to target reconsolidation, with a specific focus on beta-adrenergic blockers. We will then review behavioural interference studies aimed at altering memory reconsolidation, with specific attention for the reactivation–extinction paradigm. For each method, we will discuss findings from studies with non-human animals, followed by preclinical studies with healthy human subjects and translational studies in clinical populations.

Finally, despite the great clinical potential, attempts to target reconsolidation to treat psychiatric disorders have had limited success. We will review potential reasons for this limited clinical efficacy and highlight future research directions that may optimize translation success. Increasing our understanding of the mechanisms underlying reconsolidation may aid to resolve the difficulties facing the translations of findings from non-human animal to patient populations and allow treatments targeting reconsolidation to live up to their clinical potential.

2 Discovering Reconsolidation

The traditional view of memory suggests that upon learning memories are initially unstable and sensitive to disturbances but stabilize over time during a process known as consolidation, following which memory is no longer sensitive to disruption (Dudai 2004; Duncan 1949; Gerard 1949; Glickman 1961; Hebb 1949; McGaugh 1966, 2000; Müller and Pilzecker 1900; Ribot 1882). Since the 1960s, great progress has been made in understanding the neural and molecular mechanisms that support memory consolidation and led to the general idea that consolidation was a process that occurred only once (e.g. McGaugh 2000). From a clinical perspective, this means that following consolidation the original memory can no longer be modified and treatments have to rely on new learning to train patients to cope with maladaptive behaviours. This leaves open the possibility that old maladaptive memories can return to dominate behaviour.

In the 1960s and 1970s, the traditional view of memory consolidation was challenged by reports that the application of electroconvulsive shock following retrieval of a consolidated memory led to memory impairment or amnesia (Lewis 1969; Lewis and Maher 1965; Misanin et al. 1968). In the wake of these findings, the idea that consolidation was not a one-time event, but that memories, could either be in a 'active' state and be sensitive to modification or in an 'inactive state' and be stable (Spear 1973). However, subsequent studies showed that following induced amnesia memories could spontaneously recover and be retrieved under more optimal cueing conditions, or be reinstated by pharmacological manipulations that do not involve potential for new learning, indicating that amnesia following reactivation can be due to temporary retrieval deficits and not a loss of the memory trace per se (for a review, see Sara and Hars 2006).

For these historical reasons, the flexibility of memory following reactivation did not receive much attention for the following decades. Renewed interest in this topic started with reports that blocking N-methyl-D-aspartic acid (NMDA) receptors after reactivation of well-trained spatial memory and beta-adrenergic blockade following reactivation of memory in an inhibitory avoidance task subsequently resulted in impaired performance, which was suggested to be evidence for memory reconsolidation (Przybyslawski et al. 1999; Przybyslawski and Sara 1997).

Interest in reconsolidation exploded when Nader et al. (2000) reported that protein synthesis inhibition within the lateral amygdala following the reactivation of a consolidated threat-conditioned response (CR) impaired subsequent expression of memory. The critical advancements of this study were that protein synthesis, known to be important for neural plasticity and memory, was inhibited within the neural circuit critical to initial consolidation and storage of conditioned threat memory (LeDoux 2000). In addition, Nader et al. (2000) showed that short-term memory was intact immediately following memory reactivation and drug administration and that the impairment took time to develop, indicating the existence of a time-dependent process following reactivation (i.e. reconsolidation). Finally, the researchers were not able to recover the expression of the original threat memory, indicating that the memory impairment was not likely due to a temporary retrieval problem. The finding that disrupting protein synthesis following reactivation can impair consolidated memories was subsequently replicated in a wide variety of species and for a wide variety of mnemonic tasks (for reviews, see Alberini 2005; Alberini and LeDoux 2013; Dudai 2004; Nader and Hardt 2009). Together these findings hold great clinical promise as they suggest that reconsolidation can be targeted to modify original maladaptive memories and behavioural responses that contribute to psychiatric disorders.

3 Criteria to Demonstrate Reconsolidation

Based on this initial research from studies with non-human animal studies, specific criteria have been generated to demonstrate that reconsolidation has been disrupted: (1) a previously consolidated memory must be reactivated by a reminder cue. (2) The manipulation aimed at altering reconsolidation should ideally be provided post-reactivation, rather than pre-reactivation. (3) Memory should be affected after a time window allowing reconsolidation to take place rather than immediately (i.e. short-term memory is intact), in line with reconsolidation being a time-dependent process (Dudai and Eisenberg 2004; Nadel and Land 2000; Nader et al. 2000; Przybyslawski and Sara 1997). A more contested criterion is that (4) the impairment should not be attributable to retrieval failure or a reactivation-locked, temporary inability to access memory traces that dissipate over time (Lattal and Abel 2004). Hence, memory should not spontaneously recover after longer delay periods or under different cueing conditions (Sara and Hars 2006). The reason this latter criterion is contested is that impairing reconsolidation may result in an attenuation of memory, not a full loss, leaving open the possibility of at least partial recovery.

4 Biological Interventions Targeting Reconsolidation

4.1 Electrical Stimulation

The first studies showing disturbance of previously consolidated memories following reactivation used electrical stimulation of the brain to evoke generalized seizure activity in non-human animals (Lewis 1969; Lewis and Maher 1965; Misanin et al. 1968). Electroconvulsive stimulation is thought to disturb normal ongoing neural activity that contributes to memory formation. The effect of electroconvulsive stimulation on disrupting previously consolidated reactivated memories in humans can be studied in depressed patients receiving this as a treatment for their psychiatric condition. In spite of the success of initial studies in non-human animals, the first attempt to use electroconvulsive therapy (ECT) to disrupt memory for a previously consolidated, reactivated memory in humans was not successful for either item or paired associative episodic memory (Squire et al. 1976). This study also found limited effects of ECT on the initial consolidation of episodic memory, suggesting that the protocol may have lacked sensitivity to detect changes in memory. In addition, in this initial study, memory was tested 6–10 h after reactivation and ECT and effects on reconsolidation may take a longer time to become apparent. More recently, it has been demonstrated that ECT following episodic memory reactivation caused a time-dependent and reactivation-specific memory impairment in humans (Kroes et al. 2014), meeting critical criteria to demonstrate reconsolidation in humans. With this in mind, it is interesting to note that an early clinical report suggested that the application of ECT immediately following patients acting out their compulsions or experiencing hallucinations resulted in a long-term absence of symptoms (Rubin et al. 1969).

Intriguingly, another technique thought to manipulate electrical activity in the brain in a more localized manner has also been suggested to alter reactivated memories in humans. Stimulation of the prefrontal cortex with transcranial magnetic stimulation (TMS) was found to cause a time-dependent enhancement of reactivated memory for word lists (Sandrini et al. 2013). However, the same group also reported that transcranial direct current stimulation (tDCS) of the same region resulted in memory enhancement for both reactivated and non-reactivated words (Sandrini et al. 2014). The mechanisms underlying this enhancement effect are currently unclear, and any applications for patients remain to be tested.

Overall, evidence suggests that ECT can impair reactivated episodic memory in a time-dependent manner, and local electrical stimulation of the brain *may* enhance reactivated memory. The use of ECT to target reconsolidation is highly invasive limiting its clinical applicability, although relatively less invasive techniques, such as TMS, may hold some promise. In the following paragraphs, we will focus on studies combining memory reactivation with pharmacological manipulations or behavioural interference techniques that are more likely to have translational significance. Although several chemical manipulations have been used to disrupt reconsolidation in non-human animals, most notably protein synthesis or protein kinase inhibitors, which present safety challenges for human use, we will specifically focus on noradrenergic manipulations as these have been used in studies with non-human animals, in human preclinical studies, and in patient populations.

4.2 Noradrenergic Manipulations Targeting Reconsolidation: Non-human Animal Studies

Pharmacological manipulations of the noradrenergic system have been used to target reconsolidation in order to modify both reactivated aversive memories and appetitive memories. The effectiveness of altering threat-related responses by noradrenergic manipulations following memory reactivation appears to vary depending on the memory task (Alberini 2011). Studies using aversive cue conditioning and contextual cue conditioning have reported that blocking norepinephrine using propranolol systemically or within the amygdala following memory reactivation results in attenuated threat-related defensive responses of recent and remote memories (Dębiec et al. 2011; Debiec and Ledoux 2004; Gamache et al. 2012; Muravieva and Alberini 2010), consistent with a role for norepinephrine in the reconsolidation of aversive cue-conditioned memories. Studies using pure context conditioning, i.e. where the onset of an aversive outcome (US) is not signalled by a specific cue, found that enhancing noradrenergic functioning following memory reactivation increases subsequent freezing, but systemic administration of propranolol following reactivation had only a modest effect on attenuating threat-related contextual conditioning (Abrari et al. 2008; Gazarini et al. 2013).

Studies using aversive inhibitory avoidance tasks have produced contradictory results, with some finding that blocking norepinephrine following memory reactivation reduces avoidance behaviour (Do Monte et al. 2013; Przybyslawski et al. 1999), and others failed to find such an effect (Muravieva and Alberini 2010). The latter researchers directly compared the effect of blocking norepinephrine on aversive cue-conditioning and inhibitory avoidance tasks (Muravieva and Alberini 2010). Noradrenergic blockade following memory reactivation was found to attenuate freezing in the cue-conditioning task. Specifically, the authors found that blocking norepinephrine following memory reactivation with the CS (tone) in the original context resulted in reduced subsequent freezing in response to both the CS and the context, whereas reactivating memory by the context alone only reduced subsequent freezing to the context, but not the CS (Muravieva and Alberini 2010). Following inhibitory avoidance training, rodents exhibit both freezing and avoidance behaviour on subsequent tests. Although noradrenergic blockade reduced retrieval, it did not seem to disrupt reconsolidation of avoidance behaviour, which was intact, but it did attenuate freezing behaviour (Muravieva and Alberini 2010). Finally, administration of the alpha-1 noradrenergic antagonists prazosin systemically or within the prelimbic cortex, but not the anterior cingulate cortex, following memory reactivation was found to attenuate avoidance behaviour (Do Monte et al. 2013). Memory research over the last century has recognized that different forms of behavioural expression involve distinct neural systems that represent specific types of memory (Fuster 2009; Henke 2010; Squire 1992; Tulving 1972). Of relevance here is that conditioned freezing responses depend on the amygdala (Davis and Whalen 2001; Fanselow and Poulos 2005; Herry and Johansen 2014; LeDoux 2000), whereas place avoidance behaviour involves the hippocampus, striatum, and neocortex (Ambrogi Lorenzini et al. 1997, 1999; Fendt and Fanselow 1999; Izquierdo et al. 1997; McGaugh 2004; McIntyre et al. 2003). These findings suggest that different neural systems that give rise to memory expressed as distinct types of behaviours may vary in sensitivity to modification by noradrenergic antagonists following retrieval.

A second line of research has focused on the modification of reactivated appetitive memories. Two tasks have mainly been used: self-administration (SA), in which an animal learns to make an operant response following a cue that leads to administration of a stimulant, such as cocaine, morphine, or alcohol, and conditioned place preference (CPP) where an animal comes to prefer the spatial location where it has received a stimulant. The acquisition of these appetitive conditioned responses depends on the ventral tegmental area, striatum, amygdala, and, for contextual information, the hippocampus (Everitt 2014; Everitt and Robbins 2005; Robbins et al. 2008). Memory reactivation combined with noradrenergic antagonists has been reported to reduce self-administration behaviour of cocaine and sucrose but may be more limited for alcohol (Diergaarde et al. 2006; Milton et al. 2008; Williams and Harding 2014; Wouda et al. 2010). In addition, studies have reported an attenuation of CPP by beta-blockers following memory reactivation (Bernardi et al. 2006, 2009; Fricks-Gleason and Marshall 2008; Robinson and Franklin 2007), although effects have not always been equally strong in that memory has been found to recover over time (Fricks-Gleason and Marshall 2008; Robinson and Franklin 2007), and more remote memories may be more resistant to modification (Robinson and Franklin 2010; Robinson et al. 2011). Note that many of the memories in these tasks are somewhat more remote as training procedures often take 1 week or more. Interestingly, repeatedly reactivating memory combined with noradrenergic treatment was found to attenuate more remote memories and prevent recovery (Fricks-Gleason and Marshall 2008). Understanding the mechanism through which noradrenergic manipulations may affect reactivated appetitive memories is complicated by the fact that several studies have given noradrenergic blockers well before memory reactivation, which lack no reactivation control conditions, and most studies have used very long reactivation times. Studies in the aversive memory domain have found that short memory reactivation results in reconsolidation, but longer reactivation trials result in extinction learning (Eisenberg et al. 2003; Pedreira and Maldonado 2003; Suzuki et al. 2004), which leaves open possible alternative interpretations for the attenuation of reactivated appetitive responses by noradrenergic blockade such as enhancing extinction, as opposed to disrupting reconsolidation. In a series of experiments, noradrenergic antagonists applied systemically prior to memory reactivation were found to cause a sustained impairment of the retrieval, but not reconsolidation, of cocaine CPP (Otis et al. 2013, 2014; Otis and Mueller 2011). This sustained retrieval impairment was found to depend on noradrenergic functioning in the prelimbic cortex and hippocampus (Otis et al. 2013, 2014). The injection of propranolol within the amygdala did not affect retrieval but did impair memory on subsequent tests in line with an impairment of reconsolidation (Bernardi et al. 2009; Otis et al. 2013). Collectively these studies indicate that combining memory reactivation with noradrenergic antagonists can reduce appetitive responses. It is less clear whether alterations in memory are the result of impaired reconsolidation or whether other mnemonic processes may also contribute to the attenuation of learned responses.

In summary, rodent studies have found that noradrenergic antagonists can attenuate reactivated aversive and appetitive memories. Yet not all memory types appear equally sensitive to reactivation-dependent flexibility. The dependence of memory on specific brain regions and its change over time with systems consolidation (Alvarez and Squire 1994; Frankland and Bontempi 2005; Marr 1971; Nadel and Moscovitch 1997; Squire 1992) appear to limit renewed flexibility. Further, noradrenergic antagonists may affect other mnemonic processes that can also lead to attenuation of learned responses, such as retrieval or extinction. Noradrenergic antagonists are at best indirect effectors of protein synthesis, but a discussion of the molecular mechanisms through which noradrenergic antagonists may assert their effects on reactivated memories is outside the scope of this chapter (for review, see Otis et al. 2015).

4.3 Noradrenergic Manipulations Targeting Reconsolidation: Preclinical Human Studies

Most studies targeting reconsolidation using pharmacological manipulations in humans have used the noradrenergic antagonist propranolol, as this is one of the few drugs used in non-human animal studies targeting reconsolidation that is safe for use in humans. Mirroring findings from non-human animals, human studies using propranolol to target reconsolidation are finding that not all types of memories are affected equally. A complicating factor in understanding how noradrenergic antagonists may alter reactivated memories in humans is that in majority of studies, the drug propranolol has been given prior to memory reactivation violating the second criterion to demonstrate reconsolidation. The reason for this is that following oral administration of propranolol, it takes 60–90 min for the drug to take effect and researchers have suggested that post-reactivation administration may miss the critical window to affect memory flexibility. Because of this, the impact of the drug on retrieval processes cannot be ruled out. Moreover, because propranolol is relatively long acting, no study has assessed possible memory alterations shortly after reactivation of drug administration (i.e. intact short-term memory), because the drug would still be active. Given this, most of these studies do not meet the second and third criteria to demonstrate reconsolidation described above. Nevertheless, an increasing body of work indicates that combining reactivation with propranolol can diminish some learned threat-related responses in humans.

The first indication that noradrenergic antagonists may attenuate reactivated memories in humans came from a study using aversive threat conditioning (Miller et al. 2004). In this study, three groups of subjects were conditioned to two visual stimuli of which one (CS+) was occasionally followed by a mild electrical shock to the wrists (US) and the other visual stimulus was never followed by a shock (CS−). Skin conductance responses (SCR), an indication of autonomic arousal, served as an index of threat-related responses. A day later, memory was reactivated by a single presentation of the CS+ and CS− and either followed by administration of the noradrenergic antagonist propranolol or placebo. A third group did not receive memory reactivation but did receive propranolol. When tested after 24 h, the group that received memory reactivation and placebo showed greater SCR to the CS+ than CS− on the first trial of the recovery test, whereas the group who received memory reactivation followed by propranolol administration did not show differential responses to the CS+ and CS− on the first trial of the test session. Although this result was encouraging, there are several aspects of the data that were problematic. First, the group that did not receive memory reactivation but was administered propranolol also did not show differential responses to the CS+ and CS− on the first trial of the test session. Second, the absence of differential responses in both propranolol groups was partially driven by increased responses to the CS− at test. Third, the lack of differential responses was limited to the first trial at test and both drug groups showed greater responses to the CS+ than CS− on subsequent trials. These results indicate not a general reduction in defensive responses but a temporary inability to discriminate between two stimuli. Finally, this effect was only present in female participants. These initial results indicate that while propranolol administered post-reactivation in human may have some effect on later expression of learned defensive responses, the effect is more limited than findings with non-human animals. It has been suggested that one possible reason for the return of conditioned defensive responses in later trials of the recovery test may be due to intact episodic knowledge of the CS–US contingencies that can drive the expression of conditioned responses in humans without reinforcement (Olsson and Phelps 2007; Phelps et al. 2001; Raio et al. 2012).

In a series of publications, Kindt and colleagues showed that administering propranolol before or after memory reactivation can decrease threat-potentiated startle responses (Kindt et al. 2009; Sevenster et al. 2012, 2013, 2014; Soeter and Kindt 2010, 2011, 2012). In these studies, subjects learn that a cue is predictive of a mild electrical shock, while another cue is never followed by a shock. Throughout the task, the startle eyeblink response to loud white noise bursts is measured. As subjects learn which cue predicts the occurrence of the shock, the magnitude of their eyeblink response to the noise burst is greater when the threatening stimulus is also presented (Grillon et al. 1991). In addition, Kindt and colleagues measured SCR and had subjects rate their expectancy of the occurrence of a shock on each trail as an online measure of explicit knowledge of the contingencies. The combination of memory reactivation and propranolol administration attenuated threat-potentiated startle responses 1 day later, but did not attenuate SCR or expectancy ratings. Propranolol without memory reactivation or memory reactivation combined with placebo had no effects. Further, the researchers were unable to reinstate the threat-potentiated startle responses (Kindt et al. 2009). In a follow-up study, the response was not found to spontaneously recover after longer delays (Soeter and Kindt 2010). These results mirror findings in non-human animals, but demonstrate that not all types of threat memory expression are equally affected by beta-blockade at memory retrieval (Alberini 2011).

It is interesting to speculate on the reason why Miller and colleagues found that propranolol temporarily attenuated reactivated aversive memory using SCR, but Kindt and colleagues found that propranolol only reduced startle responses but left threat-related SCR and explicit knowledge of the aversive events intact. It has been suggested that startle is a more direct measure of amygdala-dependent memory than SCR, the latter being an index of more explicit memory (Kindt et al. 2009). However, threat-conditioned SCR can be acquired even in the complete absence of awareness of the predictive cues (Ohman and Soares 1994; Raio et al. 2012; Schultz and Helmstetter 2010), rendering this explanation unlikely in our opinion. Both threat-conditioned SCR and startle responses involve the human amygdala at acquisition (Bechara et al. 1995; Klumpers et al. 2014; LaBar et al. 1998), and expression of both is correlated with dorsomedial prefrontal cortical activity (Klumpers et al., online prepublication). An important difference is that SCR is measured as an anticipatory response in the absence of an aversive event, but startle is measured in response to an actual aversive event (loud noise) that is enhanced by the anticipation of threat. Threat-potentiated startle thus effectively measures an emotional enhancement of a response to threat itself, and propranolol at reactivation eliminates the enhancement effect. Further, the simultaneous measurement of startle and SCR using long CS presentation times might not be optimal to assess SCR. In threat-conditioning tasks, the SCR is a phasic response to the anticipation of threat that initially occurs to the onset of a predictive cue but over learning shifts in time to the offset of the cue when the shock is expected to occur. Often at the end of threat conditioning, peak of the response occurs well after the onset of the cue and would be overshadowed by the occurrence of noise bursts or shocks if those were present at the end of a trial. Yet a study optimized to measure SCR and following the procedures of Kindt et al. (2009) also found no effect of reactivation and propranolol on subsequent SCR although it lacked an appropriate placebo control condition (Spring et al. 2015). Another important difference may be that Kindt and colleagues use higher reinforcement rates than Miller and colleagues did. In human threat conditioning, partial reinforcement rates are generally used because higher reinforcement rates result in very rapid extinction learning if threat-predictive cues are not reinforced in subsequent test sessions (Capaldi et al. 1970; LaBar et al. 1998; Phelps et al. 2004). An additional difference is the explicitness of task instructions. Whereas most human conditioning studies instruct subjects that 'there is a relationship between the cues and the shocks', Kindt and colleagues instruct subjects that 'one of the cues will be followed by a shock in most of the cases, whereas the other cue would never be followed by the shock'. In the latter case, based on explicit knowledge, learning which stimulus predicts shock and which does not can be fully learned on the first reinforced trial. Such rapid learning resulting from explicit knowledge may comprise a different neural circuitry than that which supports reinforcement learning over multiple trials. Finally, probing online expectancy judgement may affect autonomic threat-related responses

(Warren et al. 2014). Explicit knowledge of contingencies gained from episodic experience or instructions affects both skin conductance and startle responses and may result in changes in the involvement of the amygdala, hippocampus, and medial prefrontal cortex in threat-related responses (Bechara 1995; Coppens et al. 2009; Funayama et al. 2001; Phelps et al. 2001; Tabbert et al. 2006). With this in mind, it is interesting that if the shock electrodes were not attached at the time of memory reactivation, propranolol administration did not reduce subsequent threat-potentiated startle responses (Sevenster et al. 2012). These findings highlight the influence of episodic memory and explicit knowledge on the reactivation of implicit sympathetic and startle responses in humans. From a clinical perspective, it is also interesting that higher trait anxiety has been reported to limit the effectiveness of memory reactivation and propranolol administration to attenuate threat-related responses, although this was not replicated in a later study (Bos et al. 2014; Soeter and Kindt 2013). Which differences between experiments contribute to the discrepancies in findings is unclear at the moment; however, even Kindt and colleagues were not able to replicate their own findings in two separate studies (Bos et al. 2014) highlighting that it is imperative to collectively investigate these issues if we wish to reach translational applications for patients.

As mentioned earlier, memory reactivation and propranolol administration can diminish some threat-related responses but leave an index of episodic memory intact (Kindt et al. 2009). It could be considered optimal only to reduce physiological threat responses in patients, but leave patients' episodic memory for the events of a traumatic experience intact. However, if episodic memories contribute to the negative symptoms experienced by patients or contribute to a potential return of physiological threat-related responses, one may also wish to alter episodic memories as well. Several studies have specifically targeted reactivated episodic memories using propranolol (Kroes et al. 2010; Schwabe et al. 2012, 2013).

The first study had participants study neutral and emotional words (Kroes et al. 2010). On day two, participants were given placebo or propranolol followed by memory reactivation by presenting the word stems (first three letters) for a subset of the emotional and neutral words and asked to recall the words. On Day 3, memory was assessed with a cued recall task in which subjects were provided word stems and asked to remember the words. Subjects who had received placebo showed a well-known emotional enhancement effect in that they were able to correctly complete more stems of emotional words than neutral words. Interestingly, propranolol abolished the emotional enhancement effect. Critically, propranolol did not diminish memory completely but specifically abolished the beneficial contribution of emotion to memory. Further, the researchers found that emotional words that were not recalled in the presence of propranolol were also unlikely to be recalled the next day. In contrast, emotional words that were correctly recalled, and thus supposedly reactivated, in the presence of propranolol were likely to still be remembered the next day. Therefore, the authors suggest that the attenuation of memory by propranolol may result from a sustained retrieval impairment and/or new learning and not necessarily be the consequence of reconsolidation disruption (Kroes et al. 2010). Two other studies studied the effect of propranolol on the reactivation of memory for neutral and emotional pictures (Schwabe et al. 2012, 2013). Both studies also found that propranolol during memory reactivation abolished the enhanced recognition of emotional pictures on day later. Propranolol without memory reactivation had no effect. Interestingly, propranolol with memory reactivation was found to specifically reduce the number of arousing pictures subjects indicated as ‘remembered’ but did not reduce the number of arousing pictures specified as ‘known’ (Schwabe et al. 2013). With ‘remember’ responses, subjects denote that they consciously recollect having seen the picture before, whereas ‘know’ responses indicate that they merely have a feeling of familiarity. Remembering has been suggested to reflect hippocampus-dependent episodic memory, whereas knowing or familiarity is proposed to rely more on parahippocampal regions (Tulving 1985). Functional magnetic resonance imaging (fMRI) data indicated amygdala and hippocampus involvement during memory reactivation in both the placebo and propranolol groups (Schwabe et al. 2012). One day later, the propranolol group displayed reduced amygdala and hippocampus responses during recognition of arousing pictures. These studies indicate that beta-blockade at reactivation may attenuate the emotional enhancement of episodic memory (Kroes et al. 2010; Schwabe et al. 2012, 2013), but it is precarious to attribute these memory impairments as evidence for reconsolidation.

In conclusion, a growing body of studies indicated that propranolol administered at the time of memory reactivation could result in a subsequent attenuation of threat-related responses and aspects of episodic memory in humans. However, given the pharmacodynamics of propranolol, it has been difficult to meet the critical criteria to demonstrate convincing evidence for reconsolidation. All preclinical studies have investigated the effect of beta-blockade on one-day-old memory. It is currently unclear whether propranolol at reactivation has an effect on more remote memories. Furthermore, the effect of propranolol on episodic memory seems to be limited to an attenuation of the emotional enhancement of episodic memory. Considering that threat-related defensive responses involve the amygdala (LeDoux 2000) and the emotional enhancement of memory is considered to result from the amygdala modulating hippocampal processing (McGaugh 2002; Phelps 2004; Richardson et al. 2004) via a beta-adrenergic mechanism (Cahill and McGaugh 1998; McGaugh 2004; Strange and Dolan 2004), it could be that the effects of propranolol are limited to amygdala-dependent memories. The exact mechanisms through which noradrenergic blockage may affect reactivated memories will be an important question for future research. It is possible that propranolol affects reconsolidation in humans, but it is also clear that retrieval and new learning can be affected. In order to develop translational approaches, it is imperative to dissociate these mnemonic processes and understand when and how each process is evoked and how to target each process to yield optimal treatment outcomes for patients. Finally, so far all preclinical human studies have focused on the effect of beta-blockade on negative arousing memories and none on appetitive memories. Nevertheless, the findings that beta-blockade can attenuate reactivated memories have encouraged researchers to start studies in clinical populations that will be discussed next.

4.4 Noradrenergic Manipulations Targeting Reconsolidation: Studies in Clinical Populations

Translational studies targeting reconsolidation of aversive and appetitive memories with beta-blockers in clinical populations have had mixed, and limited, results. First, in a series of studies, traumatic memories of post-traumatic stress disorder (PTSD) patients were reactivated by creating a narrative script of patient's own traumatic experiences and combined with propranolol administration (Brunet et al. 2008, 2011, 2014). One week later, patients were again presented with the trauma script and mentally imagined the experience. Patients who had received propranolol had lower heart rates than those who had received placebo to a level below that normally observed in PTSD patients in this task. Skin conductance levels (note a measure reflecting overall conductance not differential responses) were also reduced in patients who had received propranolol, but not below PTSD cut-off levels. Electromyography, a third measure of arousal, was reduced in both the patients who had received propranolol and placebo (Brunet et al. 2008). The following studies replicated these results when reactivation and propranolol administration were repeated over six weekly sessions. Patients also showed a reduction of PTSD symptoms over the course of treatment and at a six-month follow-up and 70–90 % of the patients no longer meeting the criteria indicative of PTSD (Brunet et al. 2011, 2014). Conducting exploratory studies in clinical patients is demanding. Probably for this reason, the studies mentioned above were open-label, lacked placebo controls, or used patients who refused treatment (and may thus constitute a particular problematic clinical group) as comparison and did not include no reactivation control groups. Furthermore, over three independent studies, the same research group was not able to replicate the reduction in sympathetic measures or clinical symptoms in PTSD patients due to memory reactivation and propranolol administration (Wood et al. 2015). Hence, studies investigating the possibility of combining memory reactivation with propranolol to treat PTSD patients have had limited success.

Several studies have also investigated the effects of memory reactivation and propranolol on appetitive-related responses in addiction patients. The first study had heroin addicts learn 10 heroin-related positive words, 10 heroin-related negative words, and 10 neutral words (Zhao et al. 2011). The next day, subjects received propranolol or placebo and memory was reactivated by free recall of the words or memory was not reactivated. On the third day, in the absence of drug, the patients who had received propranolol and memory reactivation recalled less heroin-related positive and negative words, but similar numbers of neutral words compared to the control groups. Hence, similar to preclinical human studies (Kroes et al. 2010; Schwabe et al. 2012, 2013), propranolol and memory reactivation seem to reduce the emotional enhancement of memory. Interestingly, the reduction of memory from the first to the third day was comparable between the reactivation and propranolol administration group and no reactivation group (Zhao et al. 2011). This could indicate that propranolol blocked the enhancement of memory that can result from reactivation (Inda et al. 2011). Two other studies investigated drug craving and drug-related sympathetic responses (Pachas et al. 2014; Saladin et al. 2013). In one study, memory was reactivated by presenting cocaine addict with videos of drug paraphernalia (e.g. bags of drugs, mirrors, pipes) followed by placebo or propranolol administration. One day later, the patients who had received propranolol showed reduced blood pressure and reduced cravings, but no change in heart rate and SCR to drug-related videos (Saladin et al. 2013). Furthermore, no change in drug use was observed at follow-up on week later. In another study, nicotine addicts received propranolol or placebo and memory was reactivated via personalized smoking scripts or personal non-smoking scripts (Pachas et al. 2014). One week later, subject who had received propranolol exhibited a reduction in craving but no change in sympathetic responses to smoking scripts. However, no interaction between drug and memory reactivation was found indicating a general effect of propranolol on the reduction of cravings. In sum, studies investigating the possibility of combining memory reactivation with propranolol to treat addiction patients have also had limited success.

To conclude, translational studies targeting reconsolidation of aversive and appetitive memories with propranolol in clinical populations have had mixed and limited results. Studies have targeted reconsolidation but have had difficulty implementing important controls to meet the criteria to demonstrate reconsolidation and to demonstrate clinical effectiveness. Although several studies have had success in attenuating responses, it is unclear whether this is attributable to effects on reconsolidation or other mnemonic processes, such as exposure or extinction. In concert with findings from non-human animal and preclinical human studies, not all memory measures are equally affected by reactivation and propranolol in clinical populations. Modification of arousal-related sympathetic responses has been reported but not reductions of approach- or avoidance-related symptoms. Furthermore, the effect of propranolol on episodic memory in patients may also be limited to reducing the emotional enhancement effect. Again, this suggests that the brain regions that support the specific memory assessment may influence the effectiveness of propranolol at memory reactivation in patients. Psychiatric disorders such as PTSD and addiction are multifaceted disorders, and distinct symptoms likely involve memories dependent on different neural systems that vary in sensitivity to alteration following reactivation. Furthermore, studies in patients aimed at reducing clinical symptoms are targeting very remote memories that have had time to undergo systems consolidation. Preclinical human studies have not addressed such remote memories, and this could be a significant limitation to translational efficacy.

4.5 Noradrenergic Manipulations Targeting Reconsolidation: Conclusions

In conclusion, studies with non-human animals, in healthy human subjects, and in clinical patient populations indicate that noradrenergic antagonists can attenuate reactivated aversive and appetitive memories. However, translational studies targeting reactivated memories with propranolol to treat symptoms of patients have had mixed and limited effects. Several limiting factors have been discussed. Not all memory types appear equally affected by reactivation and noradrenergic antagonist administration. The dependency of memories on specific brain regions and their change over time with systems consolidation might limit the possibility to alter memories. More specifically, it appears that amygdala-dependent memories such as cue-conditioned arousal responses may be attenuated by noradrenergic blockade following reactivation, but these effects are limited and difficult to replicate in humans. The ability to alter hippocampus-dependent memories such as contextual conditioned memories in non-human animals or episodic memory in humans might be limited to the emotional enhancement of memory that is possibly the result of modulation of the hippocampus by the amygdala. This could be either the consequence of intrinsic qualities of distinct brain regions, or the result of the mechanisms of action of propranolol, or a combination of both. Additionally, remote memories that have undergone systems consolidation may be an important limiting factor to alter reactivated memories with noradrenergic antagonists. Further, noradrenergic antagonists may affect reconsolidation but can also have sustained effects by influencing other mnemonic processes such as retrieval or new learning. Studies in humans have had particular difficulty in dissecting distinct effects on different potential mnemonic processes. It will be important for future studies to critically investigate the mechanisms and memory processes that can lead to alterations of reactivated memories by noradrenergic antagonists if we wish to develop optimal translation applications to treat patients.

5 Behavioural Interventions Targeting Reconsolidation: The Reactivation–Extinction Paradigm

Although pharmacological manipulations are the most common techniques to alter reconsolidation in animal models, the translation to humans to date has been limited due to ethical/safety reasons, and as outlined above, the translation of drugs safe for human use has had limited success. An alternative approach is to rely on behavioural techniques proposed to influence reconsolidation, the most prominent being the reactivation–extinction paradigm.

The notion that behavioural interventions can influence reconsolidation is based on the premise that a key function of reconsolidation might be to update older memories with new information available at the time of retrieval, thus supporting the dynamic nature of memory. The first demonstration of this effect was a study conducted in humans examining motor sequence learning (Walker et al. 2003). In this study, participants learned a motor sequence on the first day as indicated by decreased reaction time over trials. The following day, they learned another motor sequence that was, or was not, proceeded by reactivation of the earlier acquired motor sequence. On the third day, memory of the first motor sequence was assessed. It was found that reactivation of the first motor sequence prior to learning a second motor sequence impaired its later retrieval relative to the no reactivation group. Importantly, this study is one of the few studies in humans that meet all the criteria for targeting reconsolidation described earlier and the first to demonstrate that behavioural techniques can also be used to influence reactivated memory.

The extension of the behavioural interference of reconsolidation to threat memories resulted in the reactivation–extinction paradigm. A common way to associate threatening cues with safety is through extinction training, where the threat-conditioned cue is presented repeatedly without the aversive outcome. This learning process creates a novel association of the CS with no US, which is traditionally thought to compete for expression with the threat association. If extinction training was sufficient, subsequent encounters with the CS would not trigger defensive responses, but returning to the threatening context, stressful exposure to threat or mere passage of time can trigger the expression of the initial threat association over the extinction memory. But what if extinction training were to occur during reconsolidation? Theoretically, the safety information learned through pairing the CS with the absence of a US might be incorporated into the threat association during the reconsolidation process. This is the rationale behind the retrieval extinction paradigm, a protocol that can prevent the return of the conditioned defensive responses in animals and humans.

5.1 Reactivation–Extinction: Non-human Animal Studies

The first report of the post-retrieval extinction effect in rodents (Monfils et al. 2009) used a threat-conditioning paradigm. On Day 1, rats were trained to associate a tone (CS) with electric shock (US). A day later, the rats were exposed either to one presentation of the CS without the US (reminder trial), or to the context only (no reminder). Extinction training followed, either 10 min or 1 h after the reminder trial (during reconsolidation), or 6 or 24 h after the reminder trial (when reconsolidation was complete). Another group had extinction, but no reminder trial preceded. A day later, the rats were exposed again to the CS under conditions that typically lead to the recovery of the threat memory following standard extinction. Only rats that received a reminder trial prior to extinction but before reconsolidation was complete (10 min or 1 h) did not show the recovery of the threat-related defensive freezing response (CR). These rats also showed impaired reacquisition when exposed to additional tone–shock pairings, suggesting that the original threat association was not erased but rather changed its meaning from threat to safety.

This initial finding spurred dozens of follow-up studies attempting to replicate this finding across species as well as to adapt the paradigm to reward and instrumental memories. Reports have been mixed, with many successful replications but also null or even opposite results. The vast parametric variation that these studies brought about outlines the probable boundaries of the post-retrieval extinction phenomenon. Recognizing boundary conditions is essential for the translation of these findings to clinical applications. One of the most important mediating factors is the age of the memory. The majority of animal studies thus far examined laboratory-made memories that were one to three days old (for reviews, see Auber et al. 2013; Flavell et al. 2013), whereas anxiety disorders often involve memories that are several months or years old. One study of post-retrieval extinction of remote memories (Costanzi et al. 2011) examined a month-old contextual memory. In this study, mice learned to associate context with a foot-shock. Approximately one month later, the mice retrieved the memory when placed in the conditioning context for 3 min (no shock was delivered) and 1 h later underwent a 30-min extinction session in the same context. The memory test was conducted a day later by placing the mice back in the conditioning context and measuring their levels of freezing. There were no differences between mice that underwent post-retrieval extinction compared to mice that underwent extinction only, indicating that post-retrieval extinction failed to attenuate remote hippocampus-dependent memories that have had time to undergo systems consolidation.

A follow-up study investigated epigenetic mechanisms differentiating recent and remote memories (Graff et al. 2014). Using cued context conditioning in mice, this study showed that retrieval of recent memories induced a time-limited period of neuronal plasticity in the hippocampus, mediated in part by epigenetic modification of gene expression involving acetylation of histone proteins. By modifying chromatic compaction, histone acetylation promotes gene transcription, thereby regulating long-lasting neuronal plasticity (Levenson and Sweatt 2005). Graff and colleagues showed that retrieval of remote memories failed to generate this temporary histone acetylation-mediated neuroplasticity in the hippocampus. The pathway critical for this process is nitrosylation of histone deacetylase 2 (HDAC2) following memory retrieval, leading to dissociation of HDAC2 from the chromatin. Using HDAC inhibitors, Graff and colleagues were able to reinstate hippocampal plasticity during post-retrieval extinction of remote memories and prevent the return of the conditioned freezing responses. In the absence of memory retrieval, treatment with HDAC inhibitors had no effect, suggesting that the original memory trace might have been modified. From a clinical perspective, these findings suggest that at least for certain types of remote memories, combining pharmacological with behavioural treatment might be more beneficial than either approach alone.

These studies demonstrate that the translation of the post-retrieval extinction procedure into clinical settings would require careful consideration of timing. In addition, the age of memory, also the duration of the reminder, the time between the reminder and extinction, and the time between post-retrieval extinction and memory test might significantly influence memory attenuation. Previous studies have shown that long exposure to the CS or the conditioned context would result in extinction rather than memory reactivation (Eisenberg et al. 2003; Power et al. 2006; Suzuki et al. 2004). For example, post-retrieval extinction of context conditioning in crabs (Perez-Cuesta and Maldonado 2009) failed to prevent the return of conditioned responses when using a relatively long reminder session (15 min). As for the time between the reminder and extinction, studies utilizing Pavlovian threat conditioning found that reconsolidation lasts more than an hour, but less than 6 h, although this window may depend on the type, strength, and age of the memory (Duvarci and Nader 2004).

Another potentially important factor for clinical treatment is the social environment. A previous study that used the post-retrieval extinction effect showed memory enhancement instead of attenuation (Chan et al. 2010). There were several parametric variations in this study compared to the original paradigm (Monfils et al. 2009), such as a different frequency of the auditory CS and a different contextual modulation. One interesting difference, though, was the housing conditions of the rats (Auber et al. 2013), which is usually overlooked. The rats in the original study were housed individually, whereas Chan and colleagues housed the rats in groups of eight. Previous studies have shown that animals respond to conspecific in distress (Panksepp and Lahvis 2011). Observing a threatened cage mate might facilitate threat learning (Knapska et al. 2010) and induce robust renewal of conditioned freezing in extinction-trained mice (Nowak et al. 2013). Such social transmission might explain the memory enhancement in the study of Chan and colleagues. Vicarious modulation of post-retrieval extinction might be an intervening factor in clinical treatments, especially those involving group dynamics.

The post-retrieval extinction paradigm has been successfully adapted to appetitive learning with implications for drug addiction (Milton and Everitt 2010). Various protocols included different positive reinforcers including sucrose (Flavell et al. 2013), grain pellets (Olshavsky et al. 2013), alcoholic beer (Millan et al. 2013), morphine, and cocaine (Ma et al. 2012; Sartor and Aston-Jones 2014; Xue et al. 2012). In contrast to Pavlovian conditioning of threat, the appetitive paradigms often involve instrumental behaviour. For example, Xue et al. (2012) trained rats to self-administer cocaine or heroin using nose poking. The drug infusions were accompanied by a light cue and a buzzing tone. The reminder consisted of a 15-min exposure to the training context, where nose poking was associated with the light and tone, but no drug was delivered. Ten minutes or 6 h later, or without the reminder, all rats underwent a 180-min extinction session, conducted similar to the reminder session. The rats repeated this retrieval extinction protocol daily for about two weeks, after which their memory was reinstated using acute drug injection (non-contingent upon nose poking). Xue and colleagues found that nose poking behaviour decreased only in rats that underwent extinction sessions 10 min post-retrieval. The authors also examined drug-related Pavlovian learning using conditioned place preference to a context associated with drug and found more robust results. This suggests that Pavlovian memories are rendered labile more readily than instrumental memories. A critical difference is that instrumental memories are usually stronger due to more intense training. Although several studies have shown that strong instrumental memories did not appear to undergo reconsolidation (e.g. Hernandez and Kelley 2004), a recent study found that introducing unexpectedness during the reminder session destabilizes a well-trained instrumental memory (Exton-McGuinness et al. 2015). Similar observation was shown for Pavlovian threat memories where targeting reconsolidation is difficult because of the strength of initial learning (Díaz-Mataix et al. 2013), consistent with the notion that some limitations of targeting reconsolidation may be overcome with variations in reactivation protocols.

The neural mechanisms mediating updating through post-retrieval extinction are still largely unknown (for reviews, see (Auber et al. 2013; Flavell et al. 2013). The working model emerging from studies thus far suggests that learning induces persistently potentiated synaptic strengthening in the lateral amygdala, by synaptic surface expression of calcium-impermeable AMPA receptors (CI-AMPAr), which are more stable at the synapse compared to the less stable calcium-permeable AMPA receptors (CP-AMPAr). Memory retrieval engages NMDA receptor-induced exchange of CI-AMPAr to CP-AMPAr. This process of CI-AMPAR endocytosis followed by CP-AMPAR insertion causes an unstable state of synaptic potentiation. The newly inserted CP-AMPARs appear to contribute to memory updating following reactivation but are removed from the synapses over the course of post-retrieval extinction (Clem and Huganir 2010; Hong et al. 2013; Monfils et al. 2009; Tedesco et al. 2014). Further understanding of the neural mechanisms underlying the reactivation–extinction paradigm may help to optimize its effectiveness in human studies.

5.2 *Reactivation–Extinction: Preclinical Human Studies*

Evidence for reactivation–extinction effect was also demonstrated in humans using threat conditioning (Schiller et al. 2010). Three groups of participants learned to associate one out of two visual cues (CS+ and CS−) with an electric shock to the wrist. The index of threat was SCR. A day later, all groups underwent extinction training. One group had regular extinction with no reactivation of the memory. The two other groups were reminded of the CS prior to extinction, one group had extinction 10 min post-retrieval (during reconsolidation), and the other waited 6 h (after reconsolidation was presumably complete). The return of conditioned SCR was tested a day later. The groups that had no reminder–extinction or extinction 6 h after the reminder showed spontaneous recovery of the threat memory. Only the participants that underwent post-retrieval extinction after 10 min showed no evidence of threat memory. A follow-up session showed that the effect persisted about a year later.

The vast majority of post-retrieval extinction studies in animals were conducted by comparing between groups, as did the protocol above in humans, where only one simple threat association was studied. Real-life memories, however, are much more complex and likely include multiple memory traces. Schiller et al. (2010) also examined whether post-retrieval extinction would influence only the memory that was retrieved but not other memories formed during the same time and within the same context but were not reactivated. To study this, the participants learned to associate two out of three visual stimuli with shock. A day later, only one of the two CSs was reactivated, and 10 min later, all stimuli were presented repeatedly without the shock in an extinction session. A day later, the participants received 4 unsignaled shocks in order to reinstate the memory, and the test was conducted 10 min later by presenting the stimuli without the shock. The results showed that only the memory that was reactivated prior to extinction was not reinstated, suggesting that post-retrieval extinction is not only effective in humans but also specific to memories that return to a labile state upon retrieval.

The specificity of post-retrieval extinction is advantageous in preventing unwarranted memory modification, but it could also be a disadvantage if we wish to modify complex memories associated with traumatic events. To address this, Liu and colleagues (Liu et al. 2014) speculated that reactivating the memory using the US would modify all CSs associated with this US. The participants underwent threat conditioning where they learned to associate two visual stimuli with shock. A day later, the participants underwent US reactivation using a weaker shock, 10 min later underwent extinction training, and were tested a day later. Liu and colleagues found that US reactivation prevented the return of conditioned threat response to both CSs. They further showed that this effect persisted at least 6 months and could also be achieved by extinguishing only one of the CSs. They also demonstrated similar results with memories that were two weeks old. These findings suggest that reactivating the central or 'binding' element of a memory might have a more overarching effect on reconsolidation. The clinical implication of this study is that re-exposure to a similar but milder adverse event might be beneficial under certain circumstances (see next section for a potential real-life demonstration of this effect).

Additional studies in humans demonstrated the post-retrieval extinction effect using a different modality of CS [auditory cue instead of visual; (Oyarzun et al. 2012)], as well as 7-day-old memories (Steinfurth et al. 2014), and in adolescents (Johnson and Casey 2015). Moreover, Agren et al. (2012b) suggested that individual differences in serotonin- and dopamine-related polymorphisms influenced post-retrieval extinction. Specifically, carriers of the short allele of the serotonin transporter length polymorphism (5-HTTLPR), and val allele homozygotes in the dopamine-related COMT Val158Met polymorphism showed enhanced reacquisition only if extinction training was performed outside, but not during, the reconsolidation window. In contrast, met allele and long-allele homozygotes did not show reacquisition regardless of reconsolidation conditions, suggesting that different allele carriers might have different reconsolidation windows. To fully assess genetic variations in reconsolidation, additional studies on sufficiently large populations are required.

Although several laboratories have reported that the reactivation–extinction paradigm persistently diminished the CR, other studies using different parameters were unable to find this effect, which might outline the boundary conditions of this phenomenon in humans (for reviews, see Agren 2014; Auber et al. 2013; Schiller and Phelps 2011). Some of these conditions include the use of CSs that were not initially neutral but rather innately frightful, such as images of spiders and snakes (Soeter and Kindt 2011). Threat conditioning to innately scary cues might induce stronger conditioning but could also engage a different neural mechanism altogether. Understanding the effect of fear-relevant stimuli might have important implications for anxiety disorders, which often involve memories that are both strong and innate, such as specific phobias (Mineka and Ohman 2002). Another important factor is the use of online expectancy ratings (Warren et al. 2014), as described in the previous paragraph. Again, it is possible that attention to the CS–US contingency during conditioning engages other neural mechanisms such as prefrontal circuits, which might diminish the impact of the reconsolidation processes within the amygdala on persistently attenuating threat responses.

Two recent studies (Agren et al. 2012a; Schiller et al. 2013) examined the neural mechanisms of the reactivation–extinction effect in the human brain using functional magnetic resonance imaging (fMRI). These studies suggest that reduction in conditioned threat responses following reactivation–extinction was coupled with reduction in amygdala reactivity to the CS (Agren et al. 2012a). Following extinction, the ventromedial prefrontal cortex (vmPFC) is thought to inhibit the amygdala and the expression of the CR enabling the expression of the extinction memory. Interestingly, unlike in standard extinction, with extinction during reconsolidation, less vmPFC involvement was detected. Furthermore, the amygdala and vmPFC were coactivated during standard extinction, but not when extinction occurred during reconsolidation (Schiller et al. 2013). These studies point to the possibility that post-retrieval extinction might circumvent extinction-related prefrontal processes, which might lead to threat memory updating within the amygdala.

5.3 Reactivation–Extinction: Studies in Clinical Populations

Persistently altering threat memories with the reactivation–extinction procedure is a promising avenue for noninvasive treatments of PTSD, but currently there are no published studies proactively examining the efficacy of a retrieval extinction approach in PTSD patients. A recent retroactive study (Weems and Graham 2014) might support the ecological validity of the reactivation–extinction paradigm in humans. This study examined the course of traumatic memories in youth from New Orleans that survived both hurricane Katrina in 2005 and hurricane Gustav in 2008. One month after hurricane Gustav, some participants recalled fewer negative memories of hurricane Katrina and had a reduction in PTSD symptoms induced by hurricane Katrina (Weems and Graham 2014). Those participants had a milder exposure to hurricane Gustav, which was significantly less devastating than hurricane Katrina. The authors found parallels between these observations and the reactivation–extinction paradigm. Exposure to hurricane Gustav may have served as a reminder for Katrina memories, similar to the US reactivation that Liu and colleagues used (described above; (Liu et al. 2014), and the experience of a milder storm may have led to the update of hurricane Katrina memories. The consequence of this sequence of events is reminiscent of the hypothesis that reactivation may serve as an update mechanism. However, this interpretation is speculative. Nevertheless, perhaps forms of treatment that mimic re-exposure to the trauma in a milder form (guided imagery, narrative rescripting, etc.) can capitalize on a reactivation update mechanism. The link between existing therapies and reconsolidation remains to be scientifically validated.

As outlined earlier, reconsolidation may also be harnessed to diminish maladaptive appetitive memories that underlie addiction. Drug-associated cues could trigger conditioned responses even after long periods of abstinence. The post-retrieval extinction procedure was recently adapted for drug addiction (Xue et al. 2012). In this study, inpatient detoxified heroin addicts underwent the following procedure: On Day 1, baseline measures of cue-induced heroin craving, including heart rate, blood pressure, and the visual analogue scale, were taken. On Days 2 and 3, the participants were divided into three groups: one group was reminded of the drug memory using a 5-min presentation of videotaped heroin cues. Ten minutes later, they underwent 1 h of extinction training comprised of four consecutive sessions of repeated exposures to three different heroin-related cues. The second group underwent a similar procedure but had a 6-h break following the reminder, and the third group had the 1-h extinction training without the reminder (they were shown 5 min of a neutral video). Change in craving compared to baseline level was assessed during Day 4, as well as approximately one month and six months later. The only group that showed significant attenuation in craving that persisted at least six months was the group had extinction training 10 min after the drug cue reminder. These findings are an encouraging first step towards implementing reconsolidation update mechanisms in drug addiction interventions and prevention of relapse and are the only clinical support of the reactivation–extinction paradigm to date.

5.4 Reactivation–Extinction: Conclusions

Taken together, the evidence from non-human animal and preclinical human studies indicates that the reactivation–extinction paradigm might be a promising avenue for the development of noninvasive techniques to modify threat- and drug-related memories. Again, the strength and age of memories as well as the type of memory may be limiting factors to flexibility. But animal and human studies suggest that the chance of inducing renewed flexibility at reactivation can be increased through specific pharmacological manipulations or by reactivating the binding element of a memory and introducing a degree of unexpectedness. The mnemonic mechanism through which the reactivation–extinction paradigm exerts its effects is not entirely clear. Modification via reconsolidation is one option, but reactivation could also allow more optimal integration between an old memory and new information through new learning. Initial neuroimaging evidence indicates that the reactivation–extinction paradigm does cause a substantial alteration of functioning within the brain network that supports threat and safety memory. Further understanding of these underlying neural mechanisms could aid optimization of the effectiveness of the paradigm. It remains to be seen whether post-retrieval extinction would prove effective in modifying real-life memories, which are typically older, stronger, and multifaceted. Yet, the first study in a clinical population provides validity to further develop the paradigm for clinical use.

6 Conclusion, Limitations, and Future Directions

Targeting reconsolidation holds great clinical promise as it allows the modification of specific memories and may prevent the return of learned responses and memories that contribute to maladaptive behaviour. Here, we have discussed translational efforts aimed at altering memories by targeting reconsolidation using biological treatments (electrical stimulation, noradrenergic antagonists) or behavioural interference (reactivation–extinction paradigm). Both approaches have been used successfully to modify aversive and appetitive memories in non-human animals, in healthy human subjects, and in clinical populations. Yet not all studies have been efficacious, and the exact mnemonic mechanisms that have been affected by the different studies are not always clear. Reconsolidation depends on de novo protein synthesis (Nader and Hardt 2009), and noradrenergic antagonists, or reactivation–extinction, are indirect effectors of protein synthesis at best. We are convinced that increasing understanding of the mechanism and limitations of memory flexibility upon reactivation can help optimize efficacy of treatments for psychiatric patients. This requires translational approaches from non-human animals, to healthy human subjects and clinical populations that take into account the translational limitations across species and study populations.

6.1 Limitations

Several limitations to memory flexibility at reactivation are becoming apparent. The dependency of memories on specific brain regions and their change over time with systems consolidation might limit the possibility to alter memories. It seems that both recent and remote memories that depend on the amygdala, such as arousal-related responses, can be altered upon reactivation. Operant behaviours that involve the striatum appear less flexible. Hippocampus-dependent memories such as contextual and episodic memories may become less sensitive to reactivation-dependent flexibility as memories undergo systems consolidation and become to depend more on cortical regions. As maladaptive memories in patients are often old and hippocampus-dependent memories may play a more prominent role in humans than rodents, this could be an important limiting factor to translational efforts.

Further, the effectiveness of noradrenergic antagonists to affect reactivated episodic memories may be limited to eliminating the emotional enhancement effect that involves amygdala-dependent modulation of the hippocampus via a beta-adrenergic mechanism. At this point, it is unclear whether this is a limitation of flexibility of memory is due to intrinsic qualities of distinct brain regions, or the result of the mechanisms of action of propranolol, or a combination of both. As episodic memories contribute to the aetiology and maintenance of maladaptive behaviours in patients, it will be critical to develop methods that can overcome this limitation to translational approaches.

Not all memory retrievals result in memory reactivation and flexibility. Brief reminder exposures trigger reconsolidation, whereas longer or repeated exposures result in extinction, although this may depend on the initial learning circumstance (Eisenberg et al. 2003; Pedreira and Maldonado 2003; Suzuki et al. 2004). In addition, the reminder cue can determine which specific memory becomes flexible and which does not (Debiec et al. 2006; Muravieva and Alberini 2010; Schiller et al. 2010). More broad memory flexibility might be induced by reactivation of a 'binding element' (Liu et al. 2014), and flexibility may require the expression of a to-be-modified response at reactivation (Sevenster et al. 2012). The reactivation of a memory thus requires the reactivation cue to be similar to the learning situation. Yet several studies also suggest that there has to be some form of mismatch between the expectation and outcome at the time of reactivation for memory flexibility to be induced (Díaz-Mataix et al. 2013; Morris et al. 2006; Sevenster et al. 2013; 2014). Thus, the conditions of reactivation can determine whether memory becomes flexible and which specific behavioural responses can be modified. The identification of the reactivation conditions that lead to the most optimal treatment outcome for a specific patient will thus be critical to overcome limitations of memory flexibility.

Several interindividual differences may also be limiting factors to translational efficacy of approaches targeting memory flexibility. One report suggested that propranolol only had an effect on reactivated aversive memories in women (Miller et al. 2004). Sex may thus be a determining factor in memory flexibility, or this effect could be due to a difference in metabolic rate or dose sensitivity. Further, individual differences in genetics may be a limiting factor to memory flexibility (Agren et al. 2012b). Sex and genetics are important considerations for translational efforts considering that almost all studies in rodents are performed in males only and given that these animals are genetically nearly identical. Future studies in non-human animals will have to address these issues.

Learned maladaptive responses to innately dangerous stimuli or innately non-nocuous stimuli (e.g. arachnophobia versus anthophobia) may also affect the ability to modify reactivated memories. To date, this topic has not received much attention.

6.2 Future Directions

Future studies will have to address limiting factors to reactivation-dependent memory flexibility to develop more optimal translational approaches. Further, reconsolidation is not the only process that can be triggered at memory retrieval. Many studies have had difficulty meeting the critical criteria to demonstrate reconsolidation, especially in humans (for review, see Schiller and Phelps 2011). Not meeting the criteria to demonstrate reconsolidation leaves open the possibility of alternative explanations for memory alterations such as retrieval impairments, or new learning processes such as extinction or secondary encoding. It is important to realize that initial learning does not happen on a tabula rasa either. Memory is adaptive, and reconsolidation is one of several processes supporting memory flexibility (Kroes and Fernández 2012; McKenzie and Eichenbaum 2011). Critically assessing the mnemonic processes that support memory alterations following retrieval is imperative. Such understanding will allow developing control over these processes and developing the most effective strategies to yield optimal treatment outcomes for patients.

Here, we discussed noradrenergic antagonists as a pharmacological approach targeting reconsolidation. The development of other pharmacological approaches that are safe for use in humans may allow effectiveness to extend beyond arousal symptoms and emotional enhancement of explicit memory. Such developments include drugs targeting the cortisol-GR-BDNF system (Abrari et al. 2008; Chen et al. 2012; Pitman et al. 2011; Schwabe and Wolf 2010; Taubenfeld et al. 2009; Tronel and Alberini 2007) and might involve targeting neuron–glia interactions (Suzuki et al. 2011), methods to increase the chance of memory destabilization at reactivation using D-cycloserine (Lee et al. 2009; Wood et al. 2015) or HDAC inhibitors (Graff et al. 2014), or potentially even dietary interventions such as the use of curcumin (Monsey et al. 2015). Repeating memory reactivation and treatment over several sessions might also optimize clinical outcomes, but efficacy might be limited to a limited number of sessions (Brunet et al. 2011, 2014).

As we have discussed, treatments targeting reconsolidation seem most effective in altering amygdala-dependent learned responses reflecting hyperarousal. Maladaptive memories in psychiatric disorders are multifaceted, and beyond hyper-arousal symptoms include approach and avoidance behaviours and cognitive ruminations. Future research will have to develop translational approaches to target these behaviours as well. Finally, altering reactivated memories can contribute to the treatment of psychiatric disorders but is not a magical cure. Patients have often over many years adjusted their delay life to their disorder and can, for example, suffer from feelings such as guilt that are not solved by modifying memories. Yet altering reactivating memories holds the promise to contribute to more optimal treatment methods and help patients break the chains of maladaptive behaviours and work towards a healthier future.

References

Abrari K, Rashidy-Pour A, Semnanian S, Fathollahi Y (2008) Administration of corticosterone after memory reactivation disrupts subsequent retrieval of a contextual conditioned fear memory: dependence upon training intensity. Neurobiol Learn Mem 89(2):178–184

Agren T (2014) Human reconsolidation: a reactivation and update. Brain Res Bull 105:70–82

Agren T, Engman J, Frick A, Björkstrand J, Larsson E-M, Furmark T et al (2012a) Disruption of reconsolidation erases a fear memory trace in the human amygdala. Science 337 (6101):1550–1552

Agren T, Furmark T, Eriksson E, Fredrikson M (2012b) Human fear reconsolidation and allelic differences in serotonergic and dopaminergic genes. Transl Psychiatry 2:e76

Alberini CM (2005) Mechanisms of memory stabilization: are consolidation and reconsolidation similar or distinct processes? Trends Neurosci 28(1):51–56

Alberini CM (2011) The role of reconsolidation and the dynamic process of long-term memory formation and storage. Front Behav Neurosci 5:12

Alberini CM, LeDoux JE (2013) Memory reconsolidation. Curr Biol (CB) 23(17):R746–R750

Alvarez P, Squire LR (1994) Memory consolidation and the medial temporal lobe: a simple network model. Proc Natl Acad Sci USA 91(15):7041–7045

Ambrogi Lorenzini CG, Baldi E, Bucherelli C, Sacchetti B, Tassoni G (1997) Role of ventral hippocampus in acquisition, consolidation and retrieval of rat's passive avoidance response memory trace. Brain Res 768(1–2):242–248

Ambrogi Lorenzini CG, Baldi E, Bucherelli C, Sacchetti B, Tassoni G (1999) Neural topography and chronology of memory consolidation: a review of functional inactivation findings. Neurobiol Learn Mem 71(1):1–18

Auber A, Tedesco V, Jones CE, Monfils MH, Chiamulera C (2013) Post-retrieval extinction as reconsolidation interference: methodological issues or boundary conditions? Psychopharmacology 226(4):631–647

Bechara A (1995) Double dissociation of conditioning and declarative knowledge relative to the amygdala and hippocampus in humans. Science 269:1115

Bechara A, Tranel D, Damasio H, Adolphs R, Rockland C, Damasio AR (1995) Double dissociation of conditioning and declarative knowledge relative to the amygdala and hippocampus in humans. Science 269(5227):1115–1118

Bernardi R, Lattal KM, Berger SP (2006) Postretrieval propranolol disrupts a cocaine conditioned place preference. NeuroReport 18(17):1443–1447

Bernardi RE, Ryabinin AE, Berger SP, Lattal KM (2009) Post-retrieval disruption of a cocaine conditioned place preference by systemic and intrabasolateral amygdala β2- and α1-adrenergic antagonists. Learn Mem 16(12):777–789

Bos MGN, Beckers T, Kindt M (2014) Noradrenergic blockade of memory reconsolidation: a failure to reduce conditioned fear responding. Front Behav Neurosci 8:412

Brunet A, Orr SP, Tremblay J, Robertson K, Nader K, Pitman RK (2008) Effect of post-retrieval propranolol on psychophysiologic responding during subsequent script-driven traumatic imagery in post-traumatic stress disorder. J Psychiatr Res 42(6):503–506

Brunet A, Poundja J, Tremblay J, Bui É, Thomas É, Orr SP et al (2011) Trauma reactivation under the influence of propranolol decreases posttraumatic stress symptoms and disorder: 3 open-label trials. J Clin Psychopharmacol 31(4):547–550

Brunet A, Thomas É, Saumier D, Ashbaugh AR, Azzoug A, Pitman RK et al (2014) Trauma reactivation plus propranolol is associated with durably low physiological responding during subsequent script-driven traumatic imagery. Can J Psychiatry 59(4):228–232

Cahill L, McGaugh JL (1998) Mechanisms of emotional arousal and lasting declarative memory. Trends Neurosci 21(7):294–299

Capaldi EJ, Ziff DR, Godbout RC (1970) Extinction and the necessity or non-necessity of anticipating reward on nonrewarded trials. Psychonomic Sci 18(1):61–63

Chan WYM, Leung HT, Westbrook RF, McNally GP (2010) Effects of recent exposure to a conditioned stimulus on extinction of Pavlovian fear conditioning. Learn Mem 17(10):512–521
Chen DY, Bambah-Mukku D, Pollonini G, Alberini CM (2012) Glucocorticoid receptors recruit the CaMKII[alpha]-BDNF-CREB pathways to mediate memory consolidation. Nat Neurosci 15(12):1707–1714
Clem RL, Huganir RL (2010) Calcium-permeable AMPA receptor dynamics mediate fear memory erasure. Science 330(6007):1108–1112
Coppens E, Spruyt A, Vandenbulcke M, Van Paesschen W, Vansteenwegen D (2009) Classically conditioned fear responses are preserved following unilateral temporal lobectomy in humans when concurrent US-expectancy ratings are used. Neuropsychologia 47(12):2496–2503
Costanzi M, Cannas S, Saraulli D, Rossi-Arnaud C, Cestari V (2011) Extinction after retrieval: effects on the associative and nonassociative components of remote contextual fear memory. Learn Mem 18(8):508–518
Davis M, Whalen P (2001) The amygdala: vigilance and emotion. Mol Psychiatry 6:13–34
Debiec J, Ledoux JE (2004) Disruption of reconsolidation but not consolidation of auditory fear conditioning by noradrenergic blockade in the amygdala. Neuroscience 129(2):267–272
Debiec J, Doyere V, Nader K, LeDoux JE (2006) Directly reactivated, but not indirectly reactivated, memories undergo reconsolidation in the amygdala. Proc Natl Acad Sci USA 103 (9):3428–3433
Dębiec J, Bush DEA, LeDoux JE (2011) Noradrenergic enhancement of reconsolidation in the amygdala impairs extinction of conditioned fear in rats—a possible mechanism for the persistence of traumatic memories in PTSD. Depress Anxiety 28(3):186–193
Díaz-Mataix L, Ruiz Martinez Raquel C, Schafe Glenn E, LeDoux Joseph E, Doyère V (2013) Detection of a temporal error triggers reconsolidation of amygdala-dependent memories. Curr Biol (CB) 23(6):467–472
Diergaarde L, Schoffelmeer ANM, De Vries TJ (2006) β-adrenoceptor mediated inhibition of long-term reward-related memory reconsolidation. Behav Brain Res 170(2):333–336
Do Monte FH, Souza RR, Wong TT, Carobrez Ade P (2013) Systemic or intra-prelimbic cortex infusion of prazosin impairs fear memory reconsolidation. Behav Brain Res 244:137–141
Dudai Y (2004) The neurobiology of consolidations, or, how stable is the engram? Annu Rev Psychol 55(1):51–86
Dudai Y, Eisenberg M (2004) Rites of passage of the engram: reconsolidation and the lingering consolidation hypothesis. Neuron 44(1):93–100
Duncan C (1949) The retroactive effect of electroshock on learning. J Comp Physiol Psychol 42:32–44
Duvarci S, Nader K (2004) Characterization of fear memory reconsolidation. J Neurosci 24 (42):9269–9275
Eisenberg M, Kobilo T, Berman DE, Dudai Y (2003) Stability of retrieved memory: inverse correlation with trace dominance. Science 301(5636):1102–1104
Everitt BJ (2014) Neural and psychological mechanisms underlying compulsive drug seeking habits and drug memories—indications for novel treatments of addiction. Eur J Neurosci 40 (1):2163–2182
Everitt BJ, Robbins TW (2005) Neural systems of reinforcement for drug addiction: from actions to habits to compulsion. Nat Neurosci 8(11):1481–1489
Exton-McGuinness MTJ, Lee JLC, Reichelt AC (2015) Updating memories—the role of prediction errors in memory reconsolidation. Behav Brain Res 278:375–384
Fanselow MS, Poulos AM (2005) The neuroscience of mammalian associative learning. Annu Rev Psychol 56(1):207–234
Fendt M, Fanselow MS (1999) The neuroanatomical and neurochemical basis of conditioned fear. Neurosci Biobehav Rev 23(5):743–760
Flavell CR, Lambert EA, Winters BD, Bredy TW (2013) Mechanisms governing the reactivation-dependent destabilization of memories and their role in extinction. Front Behav Neurosci 7:214

Frankland PW, Bontempi B (2005) The organization of recent and remote memories. Nat Rev Neurosci 6(2):119–130

Fricks-Gleason AN, Marshall JF (2008) Post-retrieval β-adrenergic receptor blockade: effects on extinction and reconsolidation of cocaine-cue memories. Learn Mem 15(9):643–648

Funayama ES, Grillon C, Davis M, Phelps EA (2001) A double dissociation in the affective modulation of startle in humans: effects of unilateral temporal lobectomy. J Cogn Neurosci 13 (6):721–729

Fuster JM (2009) Cortex and memory: emergence of a new paradigm. J Cogn Neurosci 21 (11):2047–2072

Gamache K, Pitman RK, Nader K (2012) Preclinical evaluation of reconsolidation blockade by clonidine as a potential novel treatment for posttraumatic stress disorder. Neuropsychopharmacology 37(13):2789–2796

Gazarini L, Stern CAJ, Carobrez AP, Bertoglio LJ (2013) Enhanced noradrenergic activity potentiates fear memory consolidation and reconsolidation by differentially recruiting α1- and β-adrenergic receptors. Learn Mem 20(4):210–219

Gerard (1949) Physiology and psychiatry. Am J Psychiatry 106:161–173

Glickman S (1961) Perseverative neural processes and consolidation of the memory trace. Psychol Bull 28:218–233

Graff J, Joseph NF, Horn ME, Samiei A, Meng J, Seo J et al (2014) Epigenetic priming of memory updating during reconsolidation to attenuate remote fear memories. Cell 156(1–2):261–276

Grillon C, Ameli R, Woods SW, Merikangas K, Davis M (1991) Fear-potentiated startle in humans: effects of anticipatory anxiety on the acoustic blink reflex. Psychophysiology 28 (5):588–595

Hebb D (1949) The organization of behavior. Wiley, New York

Henke K (2010) A model for memory systems based on processing modes rather than consciousness. Nat Rev Neursci 11(7):523–532

Hernandez PJ, Kelley AE (2004) Long-term memory for instrumental responses does not undergo protein synthesis-dependent reconsolidation upon retrieval. Learn Mem 11(6):748–754

Herry C, Johansen JP (2014) Encoding of fear learning and memory in distributed neuronal circuits. Nat Neurosci 17(12):1644–1654

Hong I, Kim J, Kim J, Lee S, Ko H-G, Nader K et al (2013) AMPA receptor exchange underlies transient memory destabilization on retrieval. Proc Natl Acad Sci USA 110(20):8218–8223

Inda MC, Muravieva EV, Alberini CM (2011) Memory retrieval and the passage of time: from reconsolidation and strengthening to extinction. J Neurosci 31(5):1635–1643

Izquierdo I, Quillfeldt JA, Zanatta MS, Quevedo J, Schaeffer E, Schmitz PK et al (1997) Sequential role of hippocampus and amygdala, entorhinal cortex and parietal cortex in formation and retrieval of memory for inhibitory avoidance in rats. Eur J Neurosci 9 (4):786–793

Johnson DC, Casey BJ (2015) Extinction during memory reconsolidation blocks recovery of fear in adolescents. Sci Rep 5

Kindt M, Soeter M, Vervliet B (2009) Beyond extinction: erasing human fear responses and preventing the return of fear. Nat Neurosci 12(3):256–258

Klumpers F, Kroes MC, Heitland I, Everaerd D, Akkermans SEA, Oosting RS et al (online prepublication) Dorsomedial prefrontal cortex mediates the impact of serotonin transporter linked polymorphic region genotype on anticipatory threat reactions. Biol Psychiatry

Klumpers F, Morgan B, Terburg D, Stein DJ, van Honk J (2014) Impaired acquisition of classically conditioned fear-potentiated startle reflexes in humans with focal bilateral basolateral amygdala damage. Soc Cogn Affect Neurosci

Knapska E, Mikosz M, Werka T, Maren S (2010) Social modulation of learning in rats. Learn Mem 17(1):35–42

Kroes MCW, Fernández G (2012) Dynamic neural systems enable adaptive, flexible memories. Neurosci Biobehav Rev 36(7):1646–1666

Kroes MCW, Strange BA, Dolan RJ (2010) β-Adrenergic blockade during memory retrieval in humans evokes a sustained reduction of declarative emotional memory enhancement. J Neurosci 30(11):3959–3963
Kroes MC, Tendolkar I, van Wingen GA, van Waarde JA, Strange BA, Fernandez G (2014) An electroconvulsive therapy procedure impairs reconsolidation of episodic memories in humans. Nat Neurosci 17(2):204–206
LaBar KS, Gatenby JC, Gore JC, LeDoux JE, Phelps EA (1998) Human amygdala activation during conditioned fear acquisition and extinction: a mixed-trial fMRI study. Neuron 20(5):937
Lattal KM, Abel T (2004) Behavioral impairments caused by injections of the protein synthesis inhibitor anisomycin after contextual retrieval reverse with time. Proc Natl Acad Sci USA 101 (13):4667–4672
LeDoux JE (2000) Emotion circuits in the brain. Annu Rev Neurosci 23(1):155–184
Lee JLC, Gardner RJ, Butler VJ, Everitt BJ (2009) D-cycloserine potentiates the reconsolidation of cocaine-associated memories. Learn Mem 16(1):82–85
Levenson JM, Sweatt JD (2005) Epigenetic mechanisms in memory formation. Nat Rev Neurosci 6(2):108–118
Lewis DJ (1969) Sources of experimental amnesia. Psychol Rev 76:461–472
Lewis DJ, Maher B (1965) Neural consolidation and electroconvulsive shock. Psychol Rev 72:225–239
Liu J, Zhao L, Xue Y, Shi J, Suo L, Luo Y et al (2014) An unconditioned stimulus retrieval extinction procedure to prevent the return of fear memory. Biol Psychiatry 76:895–901
Ma X, Zhang JJ, Yu LC (2012) Post-retrieval extinction training enhances or hinders the extinction of morphine-induced conditioned place preference in rats dependent on the retrieval-extinction interval. Psychopharmacology 221(1):19–26
Marr D (1971) Simple memory: a theory for archicortex. Philos Trans R Soc Lond B Biol Sci 262 (841):23–81
McGaugh JL (1966) Time-dependent processes in memory storage. Science 153 (3742):1351–1358
McGaugh JL (2000) Memory—a century of consolidation. Science 287(5451):248–251
McGaugh J (2002) Amygdala modulation of memory consolidation: interaction with other brain systems. Neurobiol Learn Mem 78(3):539–552
McGaugh JL (2004) The amygdala modulates the consolidation of memories of emotionally arousing experiences. Annu Rev Neurosci 27:1–28
McIntyre CK, Power AE, Roozendaal B, McGaugh JL (2003) Role of the basolateral amygdala in memory consolidation. Ann NY Acad Sci 985(1):273–293
McKenzie S, Eichenbaum H (2011) Consolidation and reconsolidation: two lives of memories? Neuron 71(2):224–233
Millan EZ, Milligan-Saville J, McNally GP (2013) Memory retrieval, extinction, and reinstatement of alcohol seeking. Neurobiol Learn Mem 101:26–32
Miller MM, Altemus M, Debiec J, Le Doux JE, Phelps EA (2004) Propranolol impairs reconsolidation of conditioned fear in humans. In: Neuroscience 34th annual meeting, San Diego, CA
Milton AL, Everitt BJ (2010) The psychological and neurochemical mechanisms of drug memory reconsolidation: implications for the treatment of addiction. Eur J Neurosci 31(12):2308–2319
Milton AL, Lee JLC, Everitt BJ (2008) Reconsolidation of appetitive memories for both natural and drug reinforcement is dependent on β-adrenergic receptors. Learn Mem 15(2):88–92
Mineka S, Ohman A (2002) Phobias and preparedness: the selective, automatic, and encapsulated nature of fear. Biol Psychiatry 52(10):927–937
Misanin JR, Miller RR, Lewis DJ (1968) Retrograde amnesia produced by electroconvulsive shock after reactivation of a consolidated memory trace. Science 160(3827):554–555
Monfils M-H, Cowansage KK, Klann E, LeDoux JE (2009) Extinction-reconsolidation boundaries: key to persistent attenuation of fear memories. Science 314(5929):951–955

Monsey M, Gerhard D, Boyle L, Briones M, Seligsohn M, Schafe G (2015) A diet enriched with curcumin impairs newly acquired and reactivated fear memories. Neuropsychopharmacology 13(40):1278–1288
Morris RG, Inglis J, Ainge JA, Olverman HJ, Tulloch J, Dudai Y et al (2006) Memory reconsolidation: sensitivity of spatial memory to inhibition of protein synthesis in dorsal hippocampus during encoding and retrieval. Neuron 50(3):479–489
Morrison FG, Ressler KJ (2014) From the neurobiology of extinction to improved clinical treatments. Depress Anxiety 31(4):279–290
Müller GE, Pilzecker A (1900) Experimentelle beitrage zur lehre vom gedachtnis. Z Psychol Suppl 1:1–300
Muravieva EV, Alberini CM (2010) Limited efficacy of propranolol on the reconsolidation of fear memories. Learn Mem 17(6):306–313
Nadel L, Land C (2000) Commentary—reconsolidation: memory traces revisited. Nat Rev Neursci 1(3):209–212
Nadel L, Moscovitch M (1997) Memory consolidation, retrograde amnesia and the hippocampal complex. Curr Opin Neurobiol 7(2):217–227
Nader K, Hardt O (2009) A single standard for memory: the case for reconsolidation. Nat Rev Neurosci 10(3):224–234
Nader K, Schafe GE, LeDoux JE (2000) Fear memories require protein synthesis in the amygdala for reconsolidation after retrieval. Nature 406(6797):722–726
Nowak A, Werka T, Knapska E (2013) Social modulation in extinction of aversive memories. Behav Brain Res 238:200–205
Ohman A, Soares JJF (1994) "Unconscious anxiety": phobic responses to masked stimuli. J Abnorm Psychol 103:231–240
Olshavsky ME, Song BJ, Powell DJ, Jones CE, Monfils MH, Lee HJ (2013) Updating appetitive memory during reconsolidation window: critical role of cue-directed behavior and amygdala central nucleus. Front Behav Neurosci 7:186
Olsson A, Phelps EA (2007) Social learning of fear. Nat Neurosci 10(9):1095
Otis JM, Mueller D (2011) Inhibition of β-adrenergic receptors induces a persistent deficit in retrieval of a cocaine-associated memory providing protection against reinstatement. Neuropsychopharmacology 36(9):1912–1920
Otis JM, Dashew KB, Mueller D (2013) Neurobiological dissociation of retrieval and reconsolidation of cocaine-associated memory. J Neurosci 33(3):1271–1281
Otis JM, Fitzgerald MK, Mueller D (2014) Inhibition of hippocampal [beta]-adrenergic receptors impairs retrieval but not reconsolidation of cocaine-associated memory and prevents subsequent reinstatement. Neuropsychopharmacology 39(2):303–310
Otis JM, Werner CT, Mueller D (2015) Noradrenergic regulation of fear and drug-associated memory reconsolidation. Neuropsychopharmacology 40(4):793–803
Oyarzun JP, Lopez-Barroso D, Fuentemilla L, Cucurell D, Pedraza C, Rodriguez-Fornells A et al (2012) Updating fearful memories with extinction training during reconsolidation: a human study using auditory aversive stimuli. PLoS One 7(6):e38849
Pachas G, Gilman J, Orr S, Hoeppner B, Carlini S, Loebl T et al (2014) Single dose propranolol does not affect physiologic or emotional reactivity to smoking cues. Psychopharmacology (Berl) 1–10
Panksepp JB, Lahvis GP (2011) Rodent empathy and affective neuroscience. Neurosci Biobehav Rev 35(9):1864–1875
Pavlov IP (1927) Conditioned reflexes: an investigation of the physiological activity of the cerebral cortex. Oxford University Press, London
Pedreira ME, Maldonado H (2003) Protein synthesis subserves reconsolidation or extinction depending on reminder duration. Neuron 38:863
Perez-Cuesta LM, Maldonado H (2009) Memory reconsolidation and extinction in the crab: mutual exclusion or coexistence? Learn Mem 16(11):714–721
Phelps EA (2004) Human emotion and memory: interactions of the amygdala and hippocampal complex. Curr Opin Neurobiol 14(2):198–202

Phelps EA, O'Connor KJ, Gatenby JC, Gore JC, Grillon C, Davis M (2001) Activation of the left amygdala to a cognitive representation of fear. Nat Neurosci 4(4):437–441
Phelps EA, Delgado MR, Nearing KI, LeDoux JE (2004) Extinction learning in humans: role of the amygdala and vmPFC. Neuron 43(6):897–905
Pitman RK, Milad MR, Igoe SA, Vangel MG, Orr SP, Tsareva A et al (2011) Systemic mifepristone blocks reconsolidation of cue-conditioned fear; propranolol prevents this effect. Behav Neurosci 125(4):632–638
Power AE, Berlau DJ, McGaugh JL, Steward O (2006) Anisomycin infused into the hippocampus fails to block "reconsolidation" but impairs extinction: the role of re-exposure duration. Learn Mem 13(1):27–34
Przybyslawski J, Sara SJ (1997) Reconsolidation of memory after its reactivation. Behav Brain Res 84:241–246
Przybyslawski J, Roullet P, Sara SJ (1999) Attenuation of emotional and nonemotional memories after their reactivation: role of β adrenergic receptors. J Neurosci 19(15):6623–6628
Raio CM, Carmel D, Carrasco M, Phelps EA (2012) Nonconscious fear is quickly acquired but swiftly forgotten. Curr Biol 22(12):R477–479
Ribot T (1882) Diseases of memory. Appleton, New York
Richardson MP, Strange BA, Dolan RJ (2004) Encoding of emotional memories depends on amygdala and hippocampus and their interactions. Nat Neurosci 7(3):278
Robbins TW, Ersche KD, Everitt BJ (2008) Drug addiction and the memory systems of the brain. Ann NY Acad Sci 1141(1):1–21
Robinson MJF, Franklin KBJ (2007) Central but not peripheral beta-adrenergic antagonism blocks reconsolidation for a morphine place preference. Behav Brain Res 182(1):129–134
Robinson MJF, Franklin KBJ (2010) Reconsolidation of a morphine place preference: Impact of the strength and age of memory on disruption by propranolol and midazolam. Behav Brain Res 213(2):201–207
Robinson MJF, Ross EC, Franklin KBJ (2011) The effect of propranolol dose and novelty of the reactivation procedure on the reconsolidation of a morphine place preference. Behav Brain Res 216(1):281–284
Rubin R, Fried R, Franks C (1969) New application of ECT. In: Rubin R, Franks C (eds) Advances in behavioral therapy, 1968. Academic Press, New York, pp 37–44
Saladin M, Gray K, McRae-Clark A, LaRowe S, Yeatts S, Baker N et al (2013) A double blind, placebo-controlled study of the effects of post-retrieval propranolol on reconsolidation of memory for craving and cue reactivity in cocaine dependent humans. Psychopharmacology 226(4):721–737
Sandrini M, Brambilla M, Manenti R, Rosini S, Cohen LG, Cotelli M (2014) Noninvasive stimulation of prefrontal cortex strengthens existing episodic memories and reduces forgetting in the elderly. Frontiers Aging Neurosci 6:289
Sandrini M, Censor N, Mishoe J, Cohen LG (2013) Causal role of prefrontal cortex in strengthening of episodic memories through reconsolidation. Curr Biol 23(21):2181–2184
Sara SJ, Hars B (2006) In memory of consolidation, vol 13, pp 515–521
Sartor GC, Aston-Jones G (2014) Post-retrieval extinction attenuates cocaine memories. Neuropsychopharmacology 39(5):1059–1065
Schiller D, Phelps EA (2011) Does reconsolidation occur in humans? Front Behav Neurosci 5:24
Schiller D, Monfils M-H, Raio CM, Johnson DC, LeDoux JE, Phelps EA (2010) Preventing the return of fear in humans using reconsolidation update mechanisms. Nature 463(7277):49–53
Schiller D, Kanen JW, LeDoux JE, Monfils M-H, Phelps EA (2013) Extinction during reconsolidation of threat memory diminishes prefrontal cortex involvement. Proc Natl Acad Sci USA 110(50):20040–20045
Schultz DH, Helmstetter FJ (2010) Classical conditioning of autonomic fear responses is independent of contingency awareness. J Exp Psychol Anim Behav Process 36(4):495–500
Schwabe L, Wolf OT (2010) Stress impairs the reconsolidation of autobiographical memories. Neurobiol Learn Mem 94(2):153–157

Schwabe L, Nader K, Wolf OT, Beaudry T, Pruessner JC (2012) Neural signature of reconsolidation impairments by propranolol in humans. Biol Psychiatry 71(4):380–386
Schwabe L, Nader K, Pruessner JC (2013) beta-Adrenergic blockade during reactivation reduces the subjective feeling of remembering associated with emotional episodic memories. Biol Psychol 92(2):227–232
Sevenster D, Beckers T, Kindt M (2012) Retrieval per se is not sufficient to trigger reconsolidation of human fear memory. Neurobiol Learn Mem 97(3):338–345
Sevenster D, Beckers T, Kindt M (2013) Prediction error governs pharmacologically induced amnesia for learned fear. Science 339(6121):830–833
Sevenster D, Beckers T, Kindt M (2014) Prediction error demarcates the transition from retrieval, to reconsolidation, to new learning. Learn Mem 21(11):580–584
Soeter M, Kindt M (2010) Dissociating response systems: erasing fear from memory. Neurobiol Learn Mem 94(1):30–41
Soeter M, Kindt M (2011) Disrupting reconsolidation: pharmacological and behavioral manipulations. Learn Mem 18(6):357–366
Soeter M, Kindt M (2012) Stimulation of the noradrenergic system during memory formation impairs extinction learning but not the disruption of reconsolidation. Neuropsychopharmacology 37(5):1204–1215
Soeter M, Kindt M (2013) High trait anxiety: a challenge for disrupting fear memory reconsolidation. PLoS ONE 8(11):e75239
Spear NE (1973) Retrieval of memory in animals. Psychol Rev 80(3):163–193
Spring JD, Wood NE, Mueller-Pfeiffer C, Milad MR, Pitman RK, Orr SP (2015) Prereactivation propranolol fails to reduce skin conductance reactivity to prepared fear-conditioned stimuli. Psychophysiology 52(3):407–415
Squire LR (1992) Memory and the hippocampus: a synthesis from findings with rats, monkeys, and humans. Psychol Rev 99(2):195–231
Squire LR, Slater PC, Chace PM (1976) Reactivation of recent or remote memory before electroconvulsive therapy does not produce retrograde amnesia. Behav Biol 18(3):335–343
Steinfurth EC, Kanen JW, Raio CM, Clem RL, Huganir RL, Phelps EA (2014) Young and old Pavlovian fear memories can be modified with extinction training during reconsolidation in humans. Learn Mem 21(7):338–341
Strange BA, Dolan RJ (2004) β-Adrenergic modulation of emotional memory-evoked human amygdala and hippocampal responses. Proc Natl Acad Sci USA 101(31):11454–11458
Suzuki A, Josselyn SA, Frankland PW, Masushige S, Silva AJ, Kida S (2004) Memory reconsolidation and extinction have distinct temporal and biochemical signatures. J Neurosci 24(20):4787–4795
Suzuki A, Stern SA, Bozdagi O, Huntley GW, Walker RH, Magistretti PJ et al (2011) Astrocyte-neuron lactate transport is required for long-term memory formation. Cell 144 (5):810–823
Tabbert K, Stark R, Kirsch P, Vaitl D (2006) Dissociation of neural responses and skin conductance reactions during fear conditioning with and without awareness of stimulus contingencies. Neuroimage 32(2):761–770
Taubenfeld SM, Riceberg JS, New AS, Alberini CM (2009) Preclinical assessment for selectively disrupting a traumatic memory via postretrieval inhibition of glucocorticoid receptors. Biol Psychiatry 65(3):249–257
Tedesco V, Roquet RF, DeMis J, Chiamulera C, Monfils MH (2014) Extinction, applied after retrieval of auditory fear memory, selectively increases zinc-finger protein 268 and phosphorylated ribosomal protein S6 expression in prefrontal cortex and lateral amygdala. Neurobiol Learn Mem 115:78–85
Tronel S, Alberini CM (2007) Persistent disruption of a traumatic memory by postretrieval inactivation of glucocorticoid receptors in the amygdala. Biol Psychiatry 62(1):33–39
Tulving E (1972) Episodic and sementic memory. In: Tulving E, Donaldson W (eds) Organization of memory. Academic Press, London, pp 381–403
Tulving E (1985) Memory and consciousness. Can Psychol 26:1–12

Vervliet B, Craske MG, Hermans D (2013) Fear extinction and relapse: state of the art. Annu Rev Clin Psychol 9(1):215–248

Walker MP, Brakefield T, Allan Hobson J, Stickgold R (2003) Dissociable stages of human memory consolidation and reconsolidation. Nature 425(6958):616–620

Warren VT, Anderson KM, Kwon C, Bosshardt L, Jovanovic T, Bradley B et al (2014) Human fear extinction and return of fear using reconsolidation update mechanisms: the contribution of on-line expectancy ratings. Neurobiol Learn Mem 113:165–173

Weems CF, Graham RA (2014) Resilience and trajectories of posttraumatic stress among youth exposed to disaster. J Child Adolesc Psychopharmacol 24(1):2–8

WHO (2011). The world health report

Williams KL, Harding KM (2014) Repeated alcohol extinction sessions in conjunction with MK-801, but not yohimbine or propranolol, reduces subsequent alcohol cue-induced responding in rats. Pharmacol Biochem Behav 116:16–24

Wood NE, Rosasco ML, Suris AM, Spring JD, Marin M-F, Lasko NB et al (2015) Pharmacological blockade of memory reconsolidation in posttraumatic stress disorder: three negative psychophysiological studies. Psychiatry Res 225(1):31–39

Wouda JA, Diergaarde L, Riga D, Van Mourik Y, Schoffelmeer ANM, De Vries TJ (2010) Disruption of long-term alcohol-related memory reconsolidation: role of β-adrenoceptors and NMDA receptors. Front Behav Neurosci 4:179

Xue Y-X, Luo Y-X, Wu P, Shi H-S, Xue L-F, Chen C et al (2012) A memory retrieval-extinction procedure to prevent drug craving and relapse. Science 336(6078):241–245

Zhao LY, Sun LL, Shi J, Li P, Zhang Y, Lu L (2011) Effects of beta-adrenergic receptor blockade on drug-related memory reconsolidation in abstinent heroin addicts. Drug Alcohol Depend 118 (2–3):224–229

Translational Assessment of Reward and Motivational Deficits in Psychiatric Disorders

Andre Der-Avakian, Samuel A. Barnes, Athina Markou and Diego A. Pizzagalli

Abstract Deficits in reward and motivation are common symptoms characterizing several psychiatric and neurological disorders. Such deficits may include anhedonia, defined as loss of pleasure, as well as impairments in anticipatory pleasure, reward valuation, motivation/effort, and reward learning. This chapter describes recent advances in the development of behavioral tasks used to assess different aspects of reward processing in both humans and non-human animals. While earlier tasks were generally developed independently with limited cross-species correspondence, a newer generation of translational tasks has emerged that are theoretically and procedurally analogous across species and allow parallel testing, data analyses, and interpretation between human and rodent behaviors. Such enhanced conformity between cross-species tasks will facilitate investigation of the neurobiological mechanisms underlying discrete reward and motivated behaviors and is expected to improve our understanding and treatment of neuropsychiatric disorders characterized by reward and motivation deficits.

Keywords Anhedonia · Depression · Stress · Affective bias · Anticipation · Motivation · Effort · Dopamine

Contents

1 Introduction 232
2 Deficits of Reward and Motivation in Psychiatric and Neurological Disorders 233
3 Important Considerations for Cross-Species Behavioral Assessments 233
4 Translational Assessments of Reward and Motivation 236

A. Der-Avakian · S.A. Barnes · A. Markou (✉)
Department of Psychiatry, Health Sciences, University of California San Diego, La Jolla, CA, USA
e-mail: amarkou@ucsd.edu

D.A. Pizzagalli (✉)
Department of Psychiatry, Harvard Medical School, Belmont, MA, USA
e-mail: dap@mclean.harvard.edu

Curr Topics Behav Neurosci (2016) 28: 231–262
DOI 10.1007/7854_2015_5004

4.1 Pleasure ... 236
4.2 Anticipation ... 237
4.3 Reward Valuation ... 240
4.4 Motivation/Effort ... 242
4.5 Reward Learning ... 247
5 Conclusions and Future Considerations ... 252
References ... 254

1 Introduction

Feeling joy and satisfaction when engaging in social activities or accomplishing a task is an important process that promotes positive reinforcement, ensuring that events that are vital for survival and reproductive success are repeated. Conversely, inability to feel pleasure for normally pleasurable experiences can have severely debilitating effects on many aspects of life, including interpersonal relationships, work, and health. Without pleasure, experiences, and activities that promote healthy lifestyles may not be appreciated, engaged in, and repeated.

The term anhedonia was coined by the French psychologist Ribot in the late nineteenth century to describe his patients, for which "it was impossible to find the least pleasure" (Ribot 1896). Although the term anhedonia is still widely used more than a century later, the behaviors and neurobiological mechanisms that govern impaired reward processing have since evolved to expand beyond anhedonia. Indeed, pleasure is just one aspect of reward processing that contributes to positive reinforcement. As a result, the term anhedonia does not adequately capture the multifaceted reward processes that, when disrupted, may each have debilitating effects on daily functioning and health even when other reward constructs remain intact.

Besides anhedonia, or loss of pleasure, deficits in other reward processes could result in behaviors that may be interpreted as loss of pleasure. For example, several reward-related processes precede the point at which an experience or activity could be perceived as pleasurable. Individuals must first (1) anticipate or predict expected rewards that may occur in the future; (2) determine relative values of different rewards; (3) determine the cost or effort required to obtain different rewards; (4) become motivated to perform the necessary goal-directed actions to obtain worthwhile rewards; and (5) learn from previous experiences in order to repeat pleasurable goal-directed behaviors in the future. Deficits in any of these processes may preclude an individual from engaging in goal-directed actions for rewards, regardless of whether or not the reward is perceived as pleasant once obtained. Furthermore, unless carefully assessed, deficits in any of these processes may be incorrectly interpreted as anhedonia. Because each of these distinct reward processes are subserved by distinct neurobiological mechanisms (Der-Avakian and Markou 2012), understanding which processes are affected in psychiatric and neurological disorders becomes important for elucidating pathophysiology and potential treatment options. This approach is aligned with the United States

National Institute of Mental Health (NIMH) Research Domain Criteria (RDoC) initiative, which aims to classify mental disorders based on specific behavioral dimensions, such as reward-related subdomains, that can be linked to specific neurophysiological processes (Insel et al. 2010).

2 Deficits of Reward and Motivation in Psychiatric and Neurological Disorders

Several psychiatric and neurological disorders are marked by debilitating symptoms relating to reward and motivational deficits (American Psychiatric Association 2013; World Health Organization 1992). Anhedonia has been described in major depressive disorder (MDD; core symptom) (Klein 1974), bipolar disorder (Leibenluft et al. 2003), schizophrenia (Haslam 1809; Meehl 1962), substance use disorder (particularly during withdrawal) [reviewed in (Markou et al. 1998)], eating disorders (Davis and Woodside 2002), autism spectrum disorder (Chevallier et al. 2012), post-traumatic stress disorder (PTSD) (Nawijn et al. 2015), Alzheimer's disease (Starkstein et al. 2005), and Parkinson's disease (Isella et al. 2003) (see below for some caveats to these examples). Recent interest among clinical researchers has focused on investigating which specific reward processes, beyond anhedonia, are affected in the disorders described above (reviewed in Hyman and Fenton 2003; Insel et al. 2010; Leboyer et al. 1998; Meyer-Lindenberg and Weinberger 2006; Whitton et al. 2015). This approach is consistent with research in experimental animals, where broadly defined psychiatric disorders characterized by multiple symptoms cannot be modeled in non-human animals, but rather discrete behavioral processes that are often linked to circumscribed neurobiological mechanisms can be assessed (Geyer and Markou 2002; Markou et al. 2009). Detailed investigation of the precise reward processes affected in each disorder requires novel clinical assessments that reliably distinguish one reward process from another. To take advantage of basic research in animals that can facilitate treatment development, new clinical behavioral assessments must be carefully designed.

3 Important Considerations for Cross-Species Behavioral Assessments

Of the several human and animal behavioral assessments designed to measure different components of reward processing (see below and Barnes et al. 2014; Markou et al. 2013), most were developed independently between species. Thus, attempts to translate behavior across different species have traditionally been limited by several factors, including limited correspondence between human and animal tasks designed to assess the same construct. To improve correspondence between human and animal procedures and the predictive validity of data derived from animal behavioral procedures for human behaviors, the factors described below should be considered when new tasks are developed for any species.

With regard to clinical assessments, behavior is often measured using self-report questionnaires that are subjective in nature and require verbal communication between the experimenter and participant. For obvious reasons, these types of assessments cannot be implemented in animals. Animals may be observed for changes in nonverbal behavior (e.g., locomotor activity, orofacial responses) or may be trained to perform operant responses. Importantly, behavioral output in animals is generally quantifiable and data collection is (or at least should be) objective. Thus, for any clinical assessment to be successfully back-translated to animals, testing should require no verbal communication. In the examples described in the sections below (also, see Table 1), new clinical assessments that have been

Table 1 Correspondence between human and non-human animal assessments of reward processes and associated neurobiological mechanisms

Reward Processes	Human assessments	Non-human animal assessments	Correspondence	Neurobiological mechanisms
Consummatory pleasure	Self-report (e.g., SHAPS, CPAS)		Poor to mid	Nucleus accumbens Ventral pallidum Orbitofrontal cortex Opioids Endocannabinoids
	Sucrose preference	Sucrose consumption and preference		
Anticipatory pleasure	Self-report (e.g., TEPS, ACIPS)		Poor	Anterior cingulate cortex Orbitofrontal cortex Medial prefrontal cortex Basal ganglia Dopamine
		Arousal, anticipatory locomotion, approach behaviors		
		Ultrasonic vocalizations		
		Successive contrast effects		
Reward valuation	Outcome devaluation task	Outcome devaluation task	High	Medial prefrontal cortex Dorsal striatum Nucleus accumbens Basolateral amygdala Orbitofrontal cortex Dopamine, Glutamate

(continued)

Table 1 (continued)

Reward Processes	Human assessments	Non-human animal assessments	Correspondence	Neurobiological mechanisms
Motivation	Self-report (e.g., BAS, MAP-SR)		Mid to high	Ventral tegmental area Nucleus accumbens Medial prefrontal cortex Anterior cingulate cortex Lateral hypothalamus Dopamine Glutamate
	Progressive ratio task	Progressive ratio task		
	EEfRT	Effort-related choice tasks		
Reward learning	RBPRT	RBPRT	High	Anterior cingulate cortex Orbitofrontal cortex Striatum Dorsal striatum Dopamine
	PSST	PSST		

either translated from existing animal tasks or back-translated into novel animal tasks utilize a computer-based stimulus-response "game" (Anderson et al. 2012; Pizzagalli et al. 2005; Treadway et al. 2009).

With regard to animal procedures, even though assessments generally meet the criteria above with regard to objectivity and nonverbal communication, behavior is sometimes measured in a manner that cannot be replicated in humans. For example, the forced swim and tail suspension tests commonly used with rodents, which are argued to reflect behavioral despair relating to depression, cannot be implemented in humans. Because often the experimenter is forced to interpret the behavior observed in animals in terms relating to human behavior, such an anthropomorphic interpretation can be very misleading and, even worse, has failed to predict treatment efficacy in humans for novel medications (Hyman 2012; Insel et al. 2013; Markou et al. 2009; Nestler and Hyman 2010).

Lastly, with regard to both human and animal tasks, behavioral output should ideally be accompanied by some structural or physiological measure that can be compared between species. While human and animal methods for visualizing neural structure and function each have their respective advantages and limitations, comparison of hypothesized neurobiological mechanisms across species will increase confidence that the similarities in observed behavior are manifested by similar biological mechanisms. Such neurobiological concordance across species will improve the probability that putative treatments for reward deficits tested in animals will translate to the clinic.

4 Translational Assessments of Reward and Motivation

4.1 Pleasure

To disentangle the subtle differences between pleasure and other reward processes, behavioral procedures are required that can measure aspects of pleasure that are not influenced by factors like motivation and learning. The sections below summarize the procedures commonly used to assess pleasure in humans and animals.

4.1.1 Human Assessments of Pleasure

Commonly used self-report scales that exclusively probe for pleasure deficits include the Snaith–Hamilton Pleasure Scale (SHAPS) (Snaith et al. 1995), the Fawcett–Clark Pleasure Scale (Fawcett et al. 1983), the Positive and Negative Affect Schedule (PANAS) (Watson et al. 1988), and the Revised Chapman Physical Anhedonia Scale (CPAS) (Chapman et al. 1976). In addition, a subset of scores from questions relating to anhedonia from the Beck Depression Inventory (BDI) and Hamilton Rating Scale for Depression (HAM-D) may be extracted to distinguish deficits in reward processing from other symptoms of depression, although these subscores have generally not been validated. It should be noted that there are several other self-report measures of pleasure that are not listed here, but are widely used for diagnostic and research purposes.

There are important limitations related to relying on subjective observation or self-report of anhedonia that were described above. Recent adaptations of subjective anhedonia scales have begun to address some of these concerns by parsing reward deficits into discrete reward-related constructs, such as consummatory and anticipatory anhedonia (see below). Nonetheless, these new scales remain subjective and require verbal communication, limiting their usefulness as translational measures.

4.1.2 Non-human Animal Assessments of Pleasure

In animals, common procedures used to assess anhedonia include the sucrose consumption and preference tests (Willner 2005; Willner et al. 1987). The sucrose consumption test involves measuring the consumption of a palatable sucrose solution during or after exposure to an anhedonia-producing event. Similarly, the sucrose preference test is used to measure the preference for a sucrose solution when given a choice between the palatable drink and water. Decreased or no preference for the sucrose solution over water is argued to reflect anhedonia. While either test may be conducted without prior food or water restriction that is often used to increase motivation to respond during such tasks, the influence of motivation on these tasks cannot be ruled out. Exposure to various forms of stress, a precipitating factor for several psychiatric disorders, decreased sucrose preference in rodents (Willner et al. 1992) and non-human primates (Paul et al. 2000). However, the reliability of the sucrose preference test as a measure of anhedonia in animals has been questioned by several researchers who have been unable to replicate decreases in sucrose consumption or preference after chronic mild stress exposure (Forbes et al. 1996; Harris et al. 1997; Matthews et al. 1995; Reid et al. 1997).

4.1.3 Convergence of Human and Non-human Animal Assessments of Pleasure

While the human anhedonia scales cannot be translated into analogous animal tasks, the sucrose preference test has been adapted for use in humans. Interestingly, attempts to translate the animal findings of stress-induced anhedonia to humans have largely resulted in negative results. For example, patients with MDD, schizophrenia, or autism do not show deficits in the hedonic response to sucrose compared to healthy controls (Berlin et al. 1998; Damiano et al. 2014; Dichter et al. 2010). These results from human and animal studies suggest that either: (a) the sucrose preference test is not a valid assessment of pleasure; or (b) psychiatric disorders such as MDD, schizophrenia, and autism are not associated with deficits in pleasure. Notably, several lines of evidence have begun to confirm the latter point, particularly with regard to schizophrenia (Barch et al. 2015; Gard et al. 2007; Heerey and Gold 2007).

4.2 Anticipation

Whereas hedonic, or consummatory, pleasure is defined as pleasure experienced while engaged in a rewarding activity, anticipatory pleasure is a pleasure that is experienced at the thought of an event that is expected to occur in the future. Thus, deficits in anticipatory pleasure require the formulation of mental representations of future events. It has been well established that consummatory and anticipatory pleasure are mediated by distinct neural processes (Berridge and Robinson 2003; Der-Avakian and Markou 2012; Schultz 2002). Consummatory pleasure is subserved by opioid and serotonergic mechanisms, while anticipatory pleasure is mediated primarily by dopaminergic mechanisms (Barbano and Cador 2006, 2007). Thus, it is not surprising that anticipatory pleasure has been linked to motivation, another reward construct mediated by mesolimbic dopamine transmission (see below). For example, in MDD, anticipation for a rewarding event predicted the degree of an individual's motivation to produce goal-directed actions to obtain the rewarding event (Sherdell et al. 2012). Importantly, clinical studies have highlighted a dissociation between consummatory and anticipatory pleasure in psychiatric disorders. For example, schizophrenia has been associated with disrupted anticipatory pleasure, but intact consummatory pleasure (Gard et al. 2007; Mote et al. 2014). Conversely, patients with Parkinson's disease showed deficits in consummatory, but not anticipatory, pleasure (Loas et al. 2014). Thus, understanding the precise construct that is affected in different disorders has important implications for treatment development.

4.2.1 Human Assessments of Anticipation

Clinical assessments that distinguish between consummatory and anticipatory anhedonia have been developed in recent years. The Temporal Experience of Pleasure Scale (TEPS) (Gard et al. 2006) is an 18-item self-report measure of trait anticipatory and consummatory pleasure that requires participants to respond using a Likert scale to indicate their agreement with statements that reflect either enjoyment of a reward or enjoyment relating to anticipation of a reward. An example of a statement probing anticipatory pleasure is "when something exciting is coming up in my life, I really look forward to it." Scores on the anticipatory pleasure-related items of the TEPS positively correlated with scores on the Behavioral Activation Scale (BAS), a measure of motivated behavior (Carver and White 1994), further confirming the relationship between anticipatory pleasure and motivation (Gard et al. 2007).

Similar to the TEPS, the Anticipatory and Consummatory Interpersonal Pleasure Scale (ACIPS) is a 17-item self-report measure specifically designed to assess anticipatory pleasure relating to social–interpersonal interactions as well as consummatory pleasure relating to the experience of pleasure for social–interpersonal interactions when they occurred (Gooding and Pflum 2014a). The ACIPS also uses a Likert scale that measures level of agreement with statements like "I look forward to seeing people when I'm on my way to a party or get-together." Scores on the ACIPS positively correlated with scores on the TEPS, as well as with scores on the BAS, again suggesting that anticipatory pleasure is strongly related to motivated behavior (Gooding and Pflum 2014a; Gooding and Pflum 2014b).

The TEPS and ACIPS are important tools that have provided new and detailed insights into the disrupted behaviors associated with psychiatric and neurological disorders. The advent of these scales that assess anticipatory pleasure may prompt the development of additional clinical measures that can accurately distinguish consummatory and anticipatory pleasures from motivation, valuation, learning, and so on. Nonetheless, as discussed above, the required verbal communication required to respond to these scales precludes the possibility of developing animal versions of these measures.

4.2.2 Non-human Animal Assessments of Anticipation

Animal behavioral tasks developed to assess anticipatory pleasure generally measure arousal, anticipatory locomotor activity, and approach behavior prior to presentation of an expected reward. The reward is typically presented for several days at the same time of day. After training, anticipatory behavior is assessed on a test day immediately prior to the time of day when the reward is typically presented. Increased locomotor activity has been observed in rodents in anticipation of food (known as food-anticipatory activity) (Mistlberger 1994), a sweet palatable reward

(Hsu et al. 2010; Mendoza et al. 2005), a sexually receptive female (Mendelson and Pfaus 1989; Pfaus and Phillips 1991), or a drug reward (i.e., ethanol) (Buck et al. 2014a). While many of the studies on anticipatory pleasure involved rodents, food-anticipatory activity are conserved across species, including honeybees (Moore et al. 1989), fish (Weber and Spieler 1987), birds (Wenger et al. 1991), rabbits (Jilge 1992), and monkeys (Sulzman et al. 1977). In rodents, anticipatory activity was disrupted by manipulations known to produce depression-related behaviors in humans, including social stress (Kamal et al. 2010; van der Harst et al. 2005) and withdrawal from chronic amphetamine treatment (Barr et al. 1999). Moreover, blocking dopamine, but not opioid, signaling disrupted anticipatory behavior in rats (Barbano and Cador 2006), which is consistent with current knowledge of neurotransmitter systems mediating consummatory (opioid) and anticipatory (dopamine) pleasure (Barbano and Cador 2007).

Anticipatory pleasure may also be assessed by measuring short, high ultrasonic vocalizations ($\sim$50 kHz) in rats, with higher frequency of ultrasonic vocalizations reflecting greater anticipation of an expected reward (Knutson et al. 2002). Indeed, increased anticipatory motor behavior was positively correlated with frequency of high ultrasonic vocalizations (Brenes and Schwarting 2015). High ultrasonic vocalizations were emitted in anticipation of several rewarding stimuli, including food (Buck et al. 2014b; Opiol et al. 2015), a cocaine or ethanol reward (Buck et al. 2014a; Ma et al. 2010), and being reunited with a cage mate after a period of social isolation (Willey and Spear 2012). As with anticipatory activity, blocking dopamine neurotransmission attenuated high ultrasonic vocalizations in response to food (Buck et al. 2014b).

Successive contrast effects reflect adaptations of behavior in response to unexpected changes of an anticipated reward. Positive contrast effects are observed when a greater reward (e.g., 4 food pellets) is unexpectedly obtained compared to an anticipated smaller reward (e.g., 1 food pellet), eliciting a greater behavioral response (e.g., lever pressing) than if the higher value reward was only ever experienced. Conversely, negative contrast effects are observed when a smaller reward is unexpectedly obtained compared to an anticipated greater reward, which typically elicits a greater depression of behavioral responding than if the smaller value reward was only ever experienced. The latter shift (i.e., negative contrast effect) is argued to reflect a state of disappointment that arises from the anticipated reward not being delivered. Successive contrast effects are highly conserved across species, including honeybees (Couvillon and Bitterman 1984), rats (Barr and Phillips 2002), dogs (Bentosela et al. 2009), and humans (Specht and Twining 1999). As with anticipatory activity, depression-producing events, such as amphetamine withdrawal, exacerbated negative contrast effects in rats (Barr and Phillips 2002), whereas acute amphetamine administration attenuated negative contrast effects (Phelps et al. 2015). Moreover, administration of a dopamine D1 receptor antagonist worsened negative contrast effects (Phelps et al. 2015), supporting the argument that reward anticipation is mediated by dopaminergic mechanisms. The role of dopamine in successive contrast effects is perhaps not surprising, given that the behavioral procedure may also be argued to reflect positive and negative reward prediction error processing, whereby greater- or lesser-than-expected rewards, respectively, are experienced. Positive and negative reward prediction errors are thought to be mediated by increased and decreased striatal dopamine signaling, respectively (Schultz et al. 1997).

4.2.3 Convergence of Human and Non-human Animal Assessments of Anticipation

While both human and animal assessments of anticipatory pleasure have played an important part in our understanding of reward processing in psychiatric and neurological disorders and their underlying neurobiological mechanisms, assessments are limited to the species in which they were developed. As with the assessments of consummatory pleasure reviewed above, the parallel development of objective, nonverbal assessments of anticipatory pleasure across species would benefit translational research efforts aimed at this reward construct. Nonetheless, the convergence of evidence surrounding the role of dopamine in anticipatory pleasure using different human and animal assessments suggests that the different species-specific tasks are tapping into similar behavioral and neurobiological mechanisms.

4.3 Reward Valuation

Reward valuation involves assessment of the relative value of rewards that guide approach and motivated behaviors. For example, rewards of higher value are expected to produce greater anticipation of and motivation to obtain the reward compared to rewards of lower value. Prior experiences allow individuals to create representations of reward value for future stimuli. Thus, reward valuation involves some aspects of pleasure, learning, memory, and decision making. Interestingly, pleasure and valuation have been dissociated in schizophrenia, with patients showing intact capacity to experience pleasure, but deficits in properly representing the value of future rewards (Gard et al. 2007; Gold et al. 2008). After reward valuation, effort calculations, based on the work required to obtain the reward, are integrated with value calculations to construct a cost–benefit analysis to determine whether the value of the reward justifies the effort required to obtain it. Accordingly, motivated behavior may be affected by disruptions in reward valuation. Reward valuation is highly conserved across species and has been observed in mice (Crombag et al. 2010; Hilario et al. 2007), rats (Balleine and Dickinson 1992), sheep (Catanese et al. 2011), monkeys (Burke et al. 2014; West et al. 2011), and humans (Klossek et al. 2008).

4.3.1 Human Assessments of Reward Valuation

Outcome devaluation tasks have been developed for use in humans based on existing non-human animal tasks (see below). An outcome devaluation task is similar to a successive negative contrast task in that lowering the value of an expected outcome results in a decrease in behavior for that outcome. In a typical outcome devaluation task, participants are presented with two different stimuli (e.g., two food items) and perform one of two operant responses to receive either reward. One of the stimuli is then devalued, typically by overexposing the participant to that stimulus. As a result, participants tend to respond more for the stimulus that was not devalued compared to the devalued stimulus. By contrast, impaired outcome devaluation is reflected by a lack of decreased responding for the devalued stimulus.

Several manipulations have been shown to affect outcome devaluation in humans. For example, acute alcohol exposure in healthy individuals disrupted sensitivity to outcome devaluation (i.e., participants did not reduce their responding for the devalued stimulus) (Hogarth et al. 2012). Functional magnetic resonance imaging (fMRI) studies revealed that sensitivity to outcome devaluation is associated with activity of the ventromedial prefrontal cortex (de Wit et al. 2009, 2012) and orbitofrontal cortex (Valentin et al. 2007). Furthermore, consistent with neuroanatomical studies in rodents (see below), patients with Parkinson's disease, which is characterized by dopamine depletion in the dorsal striatum, showed impaired outcome devaluation (de Wit et al. 2011), suggesting a corticostriatal mechanism involved in reward valuation.

4.3.2 Non-human Animal Assessments of Reward Valuation

Outcome devaluation in animals is often assessed in an instrumental conditioning task similar to that described above for humans (Adams and Dickinson 1981). Subjects are presented with two different rewards of different values (e.g., food and sucrose pellets), each requiring a separate behavioral response to obtain the reward (e.g., in rodents, pressing a left vs. right lever for different valued rewards). As in humans, outcome devaluation can be achieved by satiating a subject that responds for a food reward or pairing the reward with a noxious stimulus. After diminishing the value of one of the rewards, the subject is given a choice to respond on either lever and typically chooses to respond more on the lever associated with the non-devalued reward compared to the lever associated with the devalued reward.

Several animal studies using an outcome devaluation task suggest that corticolimbic structures are involved in reward valuation. Lesions of the dorsomedial striatum or administration of a *N*-methyl-D-aspartate (NMDA) receptor antagonist into this region blocked the post-conditioning decrease in behavioral responding for a devalued reinforcer in rats, reflecting insensitivity to outcome devaluation (Yin et al. 2005a, b). This role of the striatum in outcome devaluation is believed to involve the core, but not the shell, of the nucleus accumbens (Corbit et al. 2001). In agreement with studies in humans described above, lesions of the medial prefrontal cortex or basolateral amygdala of rats disrupted sensitivity to outcome devaluation (Balleine and Dickinson 1998; Corbit and Balleine 2003, 2005; Killcross and Coutureau 2003). Moreover, smaller orbitofrontal cortex volume was associated with impaired outcome devaluation in monkeys (Burke et al. 2014). Overall, these results suggest an involvement of corticolimbic circuits in reward valuation. Interestingly, repeated daily administration of amphetamine, which increases cortical and striatal dopamine signaling, also disrupted sensitivity to outcome devaluation (Nelson and Killcross 2006).

4.3.3 Convergence of Human and Rodent Assessments of Reward Valuation

While few examples exist of cross-species parallel comparisons of human and animal reward valuation tasks, results from outcome devaluation tasks are strikingly consistent across species. Perhaps the best example of cross-species correspondence in reward valuation is demonstrated by studies investigating the effects of stress on sensitivity to outcome devaluation. Humans exposed to a socially evaluated cold pressor test (Schwabe and Wolf 2010) and rats exposed to chronic unpredictable stress, which includes social defeat, forced swim, and restraint (Dias-Ferreira et al. 2009), both showed decreased sensitivity to outcome devaluation. In humans, these effects of stress were associated with decreased volume of the medial prefrontal cortex and caudate (Soares et al. 2012). Indeed, several lines of evidence from both humans and animals suggest critical involvement of medial prefrontal and orbitofrontal cortices, as well as the striatum, in outcome devaluation (for review, see Balleine and O'Doherty 2010). Importantly, because the outcome devaluation task was initially developed as an operant task in animals, translation to an analogous human task was possible and produced a framework in which to investigate reward valuation across species.

4.4 *Motivation/Effort*

Motivation is the incentive or desire to act or accomplish goals. Deficits in motivation may result from deficits in other reward constructs, such as pleasure or anticipation. For example, if an individual is unable to derive pleasure from a normally rewarding activity or from anticipation of that activity, then it is unlikely that the individual will be motivated to pursue that activity (Salamone et al. 2009; Sherdell et al. 2012). Moreover, deficits in motivation may contribute to the development of other symptoms of psychiatric illness, such as social withdrawal and cognitive impairment (Brebion et al. 2009), and can be severely debilitating with regard to functional outcome and reduced quality of life in patients (Barch and Dowd 2010; Simpson et al. 2012).

Recent findings have begun to emerge highlighting the specific role of motivation in psychiatric disorders. For example, although deficits in motivation were recognized nearly a century ago in schizophrenia (Kraepelin 1921), recent evidence suggests that such deficits are dissociable from consummatory pleasure, which is intact in individuals with schizophrenia (Barch et al. 2015; Gard et al. 2007; Heerey and Gold 2007). Additionally, autism spectrum disorder is associated with anhedonia relating to social, but not other, stimuli (Chevallier et al. 2012; Damiano et al. 2014), but motivation to complete certain tasks can be greater in autism spectrum disorder compared to healthy controls (Damiano et al. 2012).

4.4.1 Human Assessments of Motivation/Effort

As with anticipatory pleasure assessments, clinical self-report questionnaires focused on motivated behavior and drive have emerged over the last two decades. The Behavioral Activation Scale (BAS) is a 13-item Likert-based self-report questionnaire probing aspects of reward relating to "reward responsiveness" (five items), "fun-seeking" (four items), and "drive" (four items) (Carver and White 1994). Low BAS scores have been associated with increased MDD risk, severity of current MDD, and poor outcome with MDD (Kasch et al. 2002; McFarland et al. 2006; Meyer et al. 1999). Similarly, the Motivation and Pleasure Scale–Self-Report (MAP-SR) is an 18-item Likert-based self-report questionnaire probing motivation related to social/interpersonal relationships and recreational/work activities (Llerena et al. 2013). In addition, experience-specific motivation scales are available to assess motivational impairment in areas such as academics (the Academic Motivation Scale) (Vallerand et al. 1992), athletics (the Sport Motivation Scale) (Pelletier et al. 1995), and fitness/health (the Exercise Motivation Scale) (Li 1999). Again, while attempts to clinically distinguish motivational impairment from other aspects of reward processing are encouraging, these self-report measures offer minimal translational value.

Because several animal procedures exist that measure motivated behavior (see below), recent attempts have been made to adapt those procedures for use in humans in a clinical laboratory setting. For example, using a progressive ratio schedule of reinforcement, operant responding (e.g., lever pressing in rodents) for a food reward is measured as the response requirement to receive each subsequent reward is exponentially increased. Eventually, the subject terminates responding when the effort required to receive a single reward becomes too great, which is interpreted as the subject's maximum level of motivation. The final ratio completed to earn a single reward is termed a break point, with decreased break points reflecting decreased motivation (Hodos 1961). Human versions of the progressive ratio task have been developed to assess motivated behavior (Roane et al. 2001). Similar to rodents, human participants are instructed to perform an operant response (e.g., click a computer mouse or press a key on a keyboard) to earn a reward. Obtaining subsequent rewards then requires exponentially more clicks or key presses. Participants continue to perform the operant response to obtain a reward before ultimately giving up, presumably because the cumulative effort required to obtain the reward eventually outweighs the perceived value of the reward. Importantly, several studies have validated the human procedures, in which more preferred rewards elicited greater break points than less preferred rewards (Glover et al. 2008; Penrod et al. 2008; Roane et al. 2001; Trosclair-Lasserre et al. 2008). Most studies utilizing progressive ratio reinforcement schedules in humans have done so using drugs of abuse as reinforcers. As expected, dependent individuals showed high levels of motivation (i.e., break points) for drug administration (Stoops 2008). Interestingly, using money as a reinforcer, patients with schizophrenia showed decreased break points in a computerized progressive ratio task (Wolf et al. 2014).

The effort-related choice tasks represent particularly intriguing examples of translation from rodent to human behavioral testing. In rodents, effort-related choice refers to the choice an animal makes to either exert effort for a high-value reward or opt for a low-value reward that is freely available (see below) (Salamone et al. 1991). The Effort-Expenditure for Rewards Task (EEfRT; Fig. 1a) is a computer-based game that was recently developed as a human analogue of the effort-related choice task in rodents (Treadway et al. 2009). With EEfRT, participants may choose between performing a hard or easy task (i.e., completed number of key presses within a short or long period of time, respectively). Successful completion of the hard and easy tasks is subsequently rewarded with relatively high- and low-monetary rewards, respectively, and the probability of receiving a reward varies with each trial and is indicated prior to choosing task difficulty. Given a medium-to-high probability of receiving a reward, the proportion of choosing the hard task was inversely correlated with self-reported anhedonia (Treadway et al. 2009). With regard to psychiatric disorders, patients with MDD were less likely to choose the hard task compared to healthy controls (Treadway et al. 2012a), suggesting that willingness to exert effort to obtain rewards is diminished in MDD. Similarly, effort-related choice was disrupted in patients with schizophrenia using EEfRT (Barch et al. 2014; Fervaha et al. 2013; Treadway et al. 2015) as well as other similar computer-based effort-related choice tasks (Gold et al. 2013). Interestingly, using EEfRT, adults with autism spectrum disorder were more likely to choose the hard task compared to healthy controls, but were less influenced by reward probabilities, supporting the argument that people with autism spectrum disorder tend to be highly motivated, but only for very selective tasks (Damiano et al. 2012).

4.4.2 Non-human Animal Assessments of Motivation/Effort

As described above, progressive ratio tests are often used to assess the motivational properties of natural and drug reinforcers in animals (Hodos 1961; Hodos and Kalman 1963). While decreased break points are generally thought to reflect decreased motivation, alternate interpretations must be carefully considered. For example, motor impairment or satiety may impede sustained responding required to

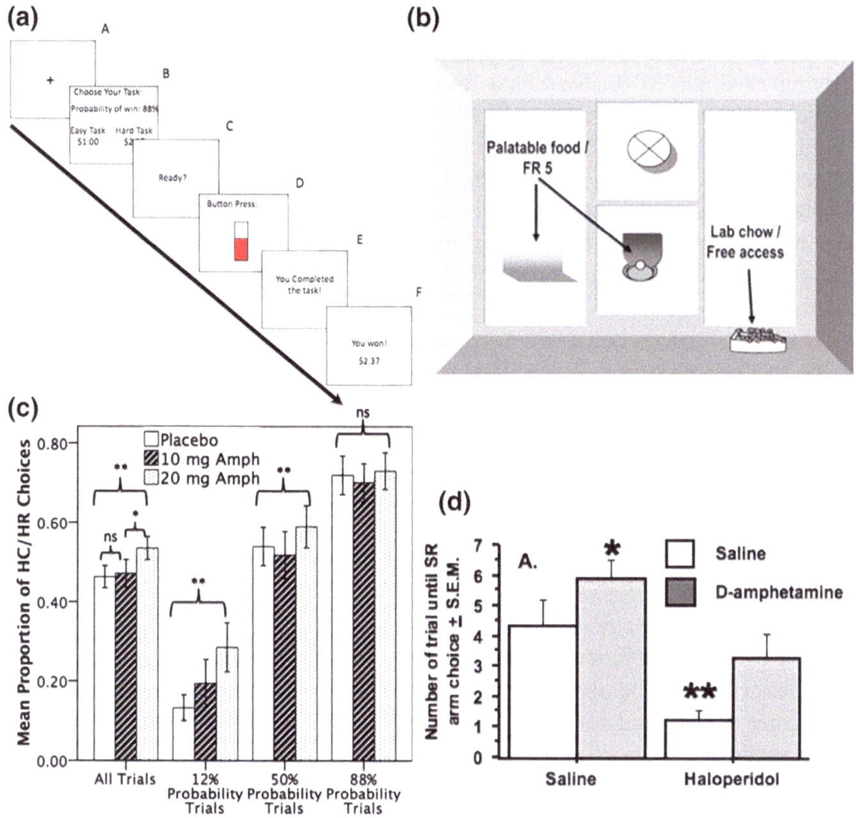

Fig. 1 Schematics of the human (**a**; EEfRT) and rat (**b**) effort-related choice tasks used to assess motivated behavior. The psychostimulant amphetamine (Amph) increased preference for the high-cost/high-reward (HC/HR) choice over the low-cost/low-reward option in humans (**c**) and rats (**d**). *FR5* fixed ratio 5 reinforcement schedule; *SR* small reward (figures were reproduced with permissions from Bardgett et al. 2009; Salamone et al. 2007; Treadway et al. 2009; Wardle et al. 2011)

achieve high break points. Decreased break points may also reflect intolerance for the progressively increasing time delay between reward presentations. Conversely, increased break points may stem from increased perseverative responding. Additionally, it is difficult to dissociate the hedonic from the motivational aspects of a reinforcer using a progressive ratio task alone. Addressing these caveats will help determine whether altered break points reflect changes in motivation or another factor.

In rodents, several manipulations decrease break points for a reward in a progressive ratio task. Withdrawal from chronic administration of several drugs of abuse, such as cocaine (Carroll and Lac 1987), amphetamine (Barr and Phillips 1999; Der-Avakian and Markou 2010), nicotine (LeSage et al. 2006), and morphine (Zhang et al. 2007) decreased responding for a sucrose reward on a progressive ratio schedule of reinforcement. Similarly, rhesus monkeys chronically treated with escalating methylphenidate doses showed decreased break points for a food reward (Rodriguez et al. 2010). Rats bred for congenital learned helplessness, a genetic animal model of behavioral despair, also showed decreased break points for sucrose (Vollmayr et al. 2004). However, not all manipulations typically used to produce depression-like behaviors in non-human animals alter motivated behavior using the progressive ratio task. For example, chronic mild stress (Barr and Phillips 1998) and neonatal maternal separation (Shalev and Kafkafi 2002), two procedures commonly used to model depression-like behavior, failed to alter break points for sucrose in rats.

The effort-related choice task was developed to address some of the limitations of the progressive ratio task described above (Fig. 1b) (Salamone et al. 1991). This task is similar to the progressive ratio task in that increasing effort is required to obtain a reward. However, an added cost/benefit component is implemented in which the choice of a lesser reward is concurrently offered and can be obtained with little or no effort. With effort requirements being equal, individuals tend to prefer rewards of greater versus lesser value. However, as the effort required to obtain the greater reward increases, preference eventually shifts toward the lesser reward requiring less effort (Salamone et al. 1997, 2007). Thus, the effort-related choice between two different rewards addresses several of the caveats limiting the progressive ratio task described above. Notably, the effort-related choice task allows for a dissociation to be observed between motivation (i.e., exerting effort to obtain the greater reward) and consummatory pleasure (i.e., opting for the freely available lesser reward).

In rats, sucrose pellets or food pellets with high carbohydrate content may be used as a highly palatable reward, whereas standard lab chow is used as the non-preferred reward. The effort required to obtain the highly palatable reward may involve a greater number of lever presses or climbing a larger barrier as compared to obtaining the non-preferred reward (Salamone et al. 1991, 1994). Regardless of the details of the procedure used, manipulating dopamine neurotransmission has been consistently shown to shift preference from high-cost/high-reward choices to low-cost/low-reward choices. For example, dopamine depletion in the nucleus accumbens or blocking the dopamine D1 or D2 receptor increases preference for freely available lab chow over lever pressing (Salamone et al. 1991) or climbing a barrier (Salamone et al. 1994; Yohn et al. 2015) for a more palatable reward. Similarly, dopamine D2 receptor over-expression in the striatum also increases preference for a low-cost/low-reward option in an effort-related choice task (Ward et al. 2012).

4.4.3 Convergence of Human and Non-human Animal Assessments of Motivation/Effort

Both human and animal versions of the progressive ratio task have been used to investigate the neurobiology of motivated behavior, particularly relating to drug and other addictions, although results have been mixed across species. For example, acute phenylalanine and tyrosine depletion in humans, which acts to decrease dopamine synthesis, decreased break points for alcohol (Barrett et al. 2008), tobacco (Venugopalan et al. 2011), and money (Cawley et al. 2013), suggesting that the increased motivation for these rewards is related to elevated brain dopamine levels. In rats, dopamine depletion produced by 6-OHDA injected directly into the nucleus accumbens decreased break points for a food reward (Aberman et al. 1998; Hamill et al. 1999), but increased break points for apomorphine (Roberts 1989), indicating that motivated behavior is mediated by dopamine D1-like and D2-like receptor signaling. Consistent with this view, administration of a dopamine D1-like or D2-like receptor antagonist in rhesus monkeys decreased break points for a food pellet (Von Huben et al. 2006).

Not all manipulations have produced congruent results across species. Pretreatment with aripiprazole, a partial agonist at dopamine D2 and serotonin (5-HT) 1A receptors and an antagonist at 5-HT 2A receptors, decreased break points for methamphetamine in recreational users (Stoops et al. 2013). Conversely, administration of a 5-HT 2A receptor antagonist failed to alter break points for cocaine, nicotine, or food in rats (Fletcher et al. 2002, 2012). Thus, while the human and animal versions of the progressive ratio task are procedurally similar, parallel cross-species testing using similar manipulations (e.g., global vs. targeted dopamine depletion) and rewards (e.g., drug vs. non-drug reinforcers) will help elucidate the level of congruence between both tasks.

The effort-related choice task in humans (i.e., EEfRT) has only recently been developed, and thus, data on congruence with the rodent version of the task is limited, but promising. The findings that effort-related choice is impaired in MDD and schizophrenia (Barch et al. 2014; Fervaha et al. 2013; Gold et al. 2013; Treadway et al. 2012a, 2015), psychiatric disorders characterized by disrupted reward-related decision making that is argued to reflect deficits in striatal dopamine transmission, support rodent studies indicating that effort-related choice is regulated by dopaminergic mechanisms (Salamone et al. 1991, 1994). Perhaps even more convincing, acute amphetamine administration, which increases striatal dopamine levels (Kuczenski et al. 1991), increased preference for the high-cost/high-reward choice in both humans (Fig. 1c) (Wardle et al. 2011) and rats (Fig. 1d) (Bardgett et al. 2009) in an effort-related choice task. Moreover, willingness to expend effort for larger rewards in humans was positively correlated with increased dopamine in the striatum and ventromedial prefrontal cortex (PFC) in PET imaging studies (Treadway et al. 2012b), consistent with the argument for dopamine involvement in effort-related choice.

4.5 Reward Learning

Reward learning is a process by which individuals experience, learn, and repeat goal-directed actions that maximize the probability of receiving future rewards.

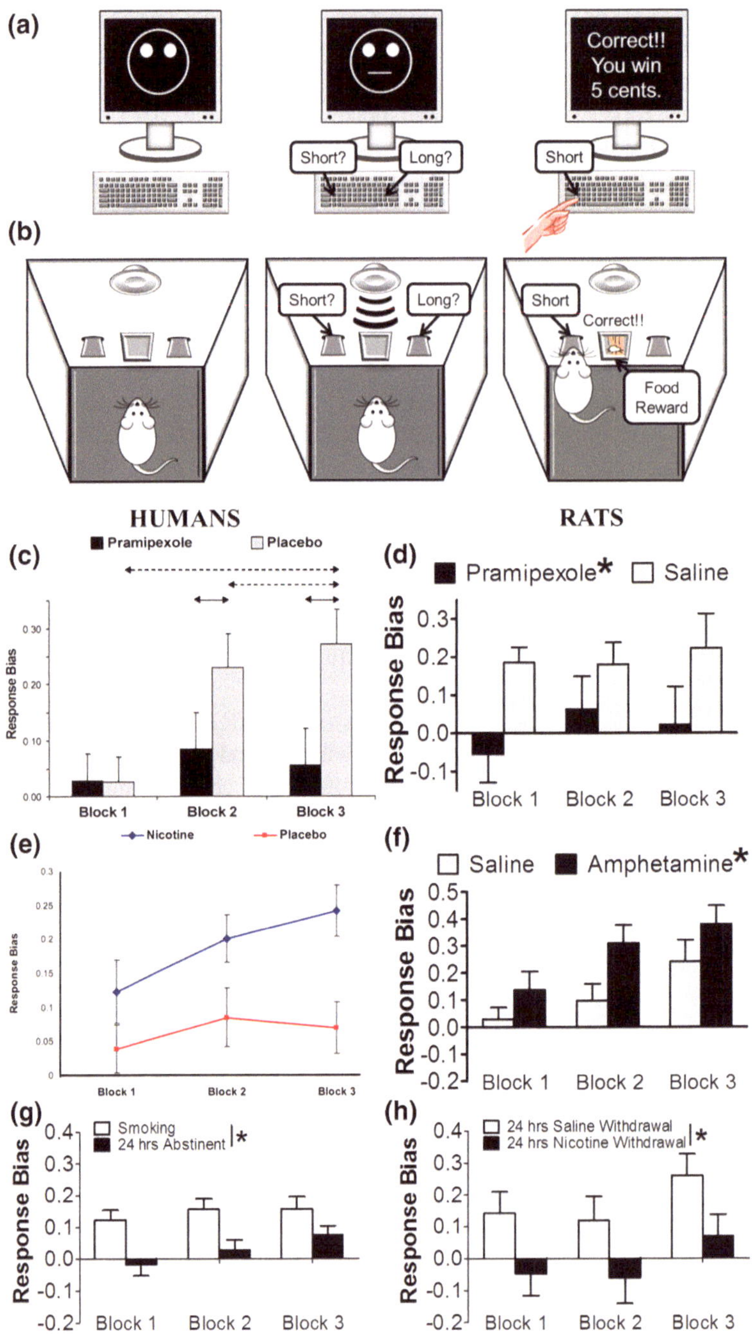
(a)
Correct!! You win 5 cents.
Short?
Long?
Short
(b)
Short?
Long?
Short
Correct!!
Food Reward
HUMANS
RATS
(c)
Pramipexole
Placebo
Response Bias
Block 1
Block 2
Block 3
(d)
Pramipexole*
Saline
Response Bias
Block 1
Block 2
Block 3
(e)
Nicotine
Placebo
Response Bias
Block 1
Block 2
Block 3
(f)
Saline
Amphetamine*
Response Bias
Block 1
Block 2
Block 3
(g)
Smoking
24 hrs Abstinent
Response Bias
Block 1
Block 2
Block 3
(h)
24 hrs Saline Withdrawal
24 hrs Nicotine Withdrawal
Response Bias
Block 1
Block 2
Block 3

◀ **Fig. 2** Schematics of human (**a**) and rat (**b**) versions of the Response Bias Probabilistic Reward task used to assess reward learning. A low dose of the dopamine D2/D3 receptor agonist pramipexole—thought to decrease DA signaling by means of autoreceptor activation—blunted response bias in humans (**c**) and rats (**d**). The psychostimulants nicotine and amphetamine increased response bias in humans (**e**) and rats (**f**), respectively. Withdrawal from chronic nicotine exposure blunted response bias in humans (**g**) and rats (**h**). Altogether, these data suggest a high level of concordance between the human and rat versions of this reward learning task (figures were reproduced with permissions from Barr et al. 2008; Der-Avakian et al. 2013; Pergadia et al. 2014; Pizzagalli et al. 2008a)

Similarly, learning can occur to avoid actions that do not result in a reward. Conversely, impaired reward learning results in decisions that are made without regard for reward feedback. Thus, reward learning may involve several other reward processes, including pleasure, motivation, and decision making.

Because of the cognitive aspects associated with reward learning, it is perhaps expected that areas of the prefrontal cortex underlie reward learning. Indeed, increased activity in the dorsal anterior cingulate and orbitofrontal cortices, areas involved in integrating and utilizing reinforcement history to guide behavior, is correlated with elevated reward learning (Frank and Claus 2006; Santesso et al. 2008). Additionally, dopaminergic mechanisms in the striatum are also involved in reward learning (Holroyd and Coles 2002; Jocham et al. 2011, 2014; O'Doherty et al. 2004; Santesso et al. 2009; Vrieze et al. 2013a), suggesting some overlap with mechanisms responsible for anticipatory pleasure and motivation.

4.5.1 Human Assessments of Reward Learning

The Response Bias Probabilistic Reward task (RBPRT) is a laboratory-based task developed initially in humans to objectively assess normal and disrupted reward responsiveness, defined as the propensity to modulate future behavioral choices based on prior reward experiences (Fig. 2a) (Pizzagalli et al. 2005; modified after Tripp and Alsop 1999). The RBPRT combines aspects of a signal-detection task and a probabilistic reward task. In the signal-detection component, participants must identify which of two ambiguous stimuli (e.g., a short or long mouth on a schematic face) is briefly (e.g., 10 ms) presented on a computer screen in order to receive monetary feedback. Unbeknownst to participants, a probabilistic reward component is implemented in the reinforcement schedule, whereby correct identification of one stimulus is reinforced three times more frequently (i.e., rich stimulus) than correct identification of the other stimulus (i.e., lean stimulus). Healthy participants (i.e., without a psychiatric diagnosis) develop a response bias for the rich stimulus, reflecting a shift from accurate responding when either stimulus is presented to increased responding on the key associated with the rich stimulus, regardless of which stimulus was presented. This pattern of change in behavior suggests that reward feedback from the differential reinforcement schedule was effective in modulating subsequent choices and that healthy participants will tend to bias their responding to try to maximize the rewards received.

Individuals with MDD and bipolar disorder did not develop a response bias for the rich stimulus in the RBPRT (Pizzagalli et al. 2008b, c). These individuals instead tended to respond accurately when either rich or lean stimulus was presented, suggesting that the differential reinforcement of the two stimuli was ineffective in promoting a bias for the rich stimulus. Euthymic individuals with remitted depression (Pechtel et al. 2013; Whitton et al. 2016), individuals without any history of psychiatric illness, but with high trait levels of anhedonia based on a self-report questionnaire (Pizzagalli et al. 2005), and healthy individuals exposed to stress, a precipitating factor for several psychiatric disorders (Bogdan and Pizzagalli 2006; Pizzagalli et al. 2007), also showed a blunted response bias compared to controls. Moreover, lower response biases predicted treatment outcome in MDD (Vrieze et al. 2013b). Conversely, response bias was not impaired in patients with schizophrenia (Ahnallen et al. 2012; Heerey et al. 2008).

Another reward learning task, the Probabilistic Stimulus Selection Task (PSST), was developed in humans to assess relative learning associated with positive versus negative reinforcement (Frank et al. 2004). In the task, participants are presented with one of three pairs of discrete symbols and instructed to select one of the two stimuli presented in each pair. Correct identification of one stimulus in the pair is rewarded at either an 80, 70, or 60 % rate (i.e., depending on the pair that is presented). Incorrect identification is rewarded 20, 30, and 40 % of the time, respectively. Participants train on this procedure until the relative reinforcement probabilities of each pair of stimuli are learned. Participants are then tested by being presented with the stimulus that was reinforced 80 % of the time paired with one of the other four stimuli (i.e., stimuli that were reinforced 70, 60, 40, and 30 % of the time). Participants are also presented with the stimulus that was reinforced 20 % of the time paired with the same four stimuli as above. Greater performance on the pairs that include the 80 % reinforced stimulus reflects better learning from positive feedback, whereas greater performance on the pairs that include the 20 % reinforced stimulus reflects better learning from negative feedback.

Healthy participants performed equally well when learning from positive and negative feedback using the PSST (Frank et al. 2004, 2007). However, when exposed to stress (e.g., threat of shock), healthy participants with high physiological and subjective responses to stress showed impairments in learning from positive, but not negative, feedback compared to non-stressed controls (Berghorst et al. 2013). Similarly, women with a history of childhood sexual assault were impaired in learning from positive, but not negative, feedback (Pechtel and Pizzagalli 2013). Patients with schizophrenia (Waltz et al. 2007) or Parkinson's disease (Frank et al. 2004) were also impaired in learning from positive, but not negative, feedback. Interestingly, impairment in Parkinson's disease was reversed with dopamine-enhancing medication (Frank et al. 2004). Collectively, these findings indicate that pathological or experimental conditions hypothesized to affect dopaminergic neurotransmission have negative effects on the ability to learn from positive feedback.

4.5.2 Non-human Animal Assessments of Reward Learning

The human RBPRT is an example of a clinical behavioral assessment of reward processing that does not rely on subjective self-report of behavior. Accordingly, a rat version of the RBPRT was developed based on the instructions and testing parameters used in the human version to assess reward learning in rats (Fig. 2b) (Der-Avakian et al. 2013). The rat version of the task is conducted in operant boxes equipped with two levers, a speaker and tone generator, and a food dispenser. Rats are trained to distinguish between two stimuli (e.g., long and short tones) by pressing a lever associated with either stimulus, which results in delivery of a food pellet for some, but not all, correct responses. As in the human task, correct identification of one tone is reinforced three times more frequently (rich) than correct identification of the other tone (lean). Data collected from the RBPRT (i.e., response bias, discriminability, accuracy, and reaction time) are also analyzed identically between humans and rats. Because most of the studies using the rat version of the RBPRT were conducted either in parallel with the human RBPRT or in order replicate previous findings from the human RBPRT, details concerning the results of these studies are described in the next section.

A similar reward bias task was recently developed using ambiguous odor cues (Wang et al. 2013) based on a similar task previously developed for use with monkeys (Rorie et al. 2010; Samejima et al. 2005). In this task, rats respond on either a left or right nose-poke hole in response to one of two odor cues. Each odor is either presented separately or in different mixture combinations (i.e., ambiguous odors). When presented with an ambiguous odor cue, rats tend to be biased toward the nose-poke hole that dispensed a relatively greater reward. This procedure allows for assessment of response vigor as well, whereby time to initiation of subsequent trials is decreased as the net value of rewards is increased. Response vigor (i.e., time to trial initiation) was shown to depend on the dorsomedial striatum (Wang et al. 2013).

Like the RBPRT, a rat version of the PSST has been recently developed (Trecker et al. 2012). Using touch screen monitors, rats are presented with three pairs of discrete stimuli and trained to select one of the two stimuli in each pair. Correct identification results in delivery of a food pellet on 90, 80, and 70 % of trials for each pair, whereas incorrect identification results in reward delivery on 10, 20, and 30 % of trials, respectively. During test trials, rats are presented with either: (1) the stimulus associated with 90 % reward paired with one of the other four novel stimuli; or (2) the stimulus associated with 10 % reward paired with one of the other four novel stimuli, in order to assess learning from positive and negative feedback, respectively. Preliminary findings indicate that rats were able to learn the probabilistic contingencies for different stimulus pairs and learned from both positive and negative feedback during test trials. Future studies may explore the roles of stress and dopaminergic transmission in the striatum and prefrontal cortex to determine whether positive feedback learning is impaired, as was observed in humans (see above).

4.5.3 Convergence of Human and Non-human Animal Assessments of Reward Learning

Perhaps because of their novelty, the human reward learning tasks described above have only been modified for use in rats, and parallel data from humans and rats are only available for the RBPRT. As in humans, healthy rats developed a response bias for the rich stimulus, reflected by increasing accuracy for the rich stimulus and decreasing accuracy for the lean stimulus throughout the test session (Der-Avakian et al. 2013; Pizzagalli et al. 2005). Modulation of reward learning was similar between rats and humans in response to several manipulations. First, an acute, low-dose administration of the dopamine D2/D3 receptor agonist pramipexole blunted response bias in healthy humans (Fig. 2c) (Pizzagalli et al. 2008a) and rats (Fig. 2d) (Der-Avakian et al. 2013). The low doses of pramipexole used in both studies, which putatively decrease striatal dopamine levels by means of presynaptic autoreceptor activation, suggest that reward learning can be modulated by altering dopamine neurotransmission in the striatum. Second, and in support of the claim above, acute administration of the psychostimulant nicotine, which acts to increase striatal dopamine levels, increased response bias in humans (Fig. 2e) (Barr et al. 2008). Similarly, administration of the psychostimulant amphetamine, which also increases striatal dopamine levels, increased response bias in rats (Fig. 2f) (Der-Avakian et al. 2013). Third, withdrawal from chronic nicotine exposure blunted response bias in humans (Fig. 2g) and rats (Fig. 2h) (Pergadia et al. 2014). In rats, re-exposure to acute nicotine after withdrawal dose-dependently increased response bias, suggesting a mechanism by which nicotine-induced enhancement of reward learning in abstinent smokers may contribute to relapse to smoking.

5 Conclusions and Future Considerations

Identifying and treating reward deficits in psychiatric and neurological disorders has become increasingly important given not only the pervasiveness of the deficits across several disorders, but also our increasing understanding of the precise reward processes that are involved, such as pleasure, anticipation, valuation, motivation, and reward learning. In order to understand the neurobiological mechanisms underlying these distinctive reward processes that will ultimately facilitate discovery of treatment targets, animal procedures are necessary that may be readily translated into analogous human tasks. Similarly, novel human tasks that do not already have a non-human analogue should be designed to assess behavior objectively using nonverbal communication. Ideally, moving forward, animal and human behavioral assessments that are developed in parallel will ensure that task parameters and psychometric properties are as analogous as possible.

Furthermore, many of the investigations probing the neurobiological mechanisms underlying different reward processes involved procedures that are not directly comparable across species. For example, human studies often relied on fMRI and electroencephalography (EEG), whereas non-human animal studies typically used brain lesion, intracranial microinjection, gene knock-out, and, more recently, optogenetic techniques. While inferences may be made about similarities in brain function tied to behavior between humans and animals using different techniques, ideally, similar approaches to measuring brain function in different species would help strengthen and validate observed similarities in behavior. For example, recent advances in human electrophysiological techniques and animal brain imaging techniques now allow for such parallel cross-species comparisons. Such an approach would not only facilitate discovery of novel neurobiological mechanisms subserving different reward processes, but, importantly, would help determine whether the reward behavior, or disruptions in reward behavior, being assessed are mediated by similar neurobiological mechanisms across different species.

A limitation of the procedures described above is that as reward processes become more narrowly defined, tasks designed to assess aspects of those reward constructs tend to become more complicated. The level of task complication may not be problematic for assessments in humans, where detailed verbal instructions may be given to study participants. However, verbal instructions must be translated to training protocols for animal tasks that may require several weeks or months to train for the most complicated tasks. Such high levels of sophistication are necessary if tasks are required to probe increasingly discrete aspects of the reward spectrum that are not confounded by other behavioral factors. Moreover, enhancing the specificity of the reward construct being assessed improves the likelihood of discovering more focused, discrete neurobiological mechanisms that underlie a given reward process.

Ultimately, the value of this approach is in being able to use animal procedures to make specific testable hypotheses regarding novel treatment strategies that have a high degree of successfully translating to the clinic. It is anticipated that this new approach in cross-species translational research will lead to the development of safe and effective medications for the treatment of reward deficits in psychiatric and neurological disorders that have thus far eluded researchers.

Acknowledgments AM was partially supported by RO1MH087989 and R56DA011946. DAP was partially supported by R01 MH068376 and R01 MH101521.

Disclosures

Over the past three years, Dr. Markou received contract research support from Astra-Zeneca and Forest Laboratories, and honorarium/consulting fee from AbbVie, Germany for activities unrelated to this chapter. Over the past three years, Dr. Pizzagalli has received honoraria/consulting fees Otsuka America Pharmaceutical and Pfizer for activities unrelated to this chapter. Drs. Der-Avakian and Barnes have no financial disclosures to report.

References

Aberman JE, Ward SJ, Salamone JD (1998) Effects of dopamine antagonists and accumbens dopamine depletions on time-constrained progressive-ratio performance. Pharmacol Biochem Behav 61(4):341–348

Adams CD, Dickinson A (1981) Instrumental responding following reinforcer devaluation. Q J Exp Psychol B 33(May):109–121

Ahnallen CG, Liverant GI, Gregor KL, Kamholz BW, Levitt JJ, Gulliver SB et al (2012) The relationship between reward-based learning and nicotine dependence in smokers with schizophrenia. Psychiatry Res 196(1):9–14

American Psychiatric Association (2013) Diagnostic and statistical manual of mental disorders: DSM-V. American Psychiatric Association, Washington, DC

Anderson MH, Hardcastle C, Munafo MR, Robinson ES (2012) Evaluation of a novel translational task for assessing emotional biases in different species. Cogn Affect Behav Neurosci 12 (2):373–381

Balleine B, Dickinson A (1992) Signalling and incentive processes in instrumental reinforcer devaluation. Q J Exp Psychol B 45(4):285–301

Balleine BW, Dickinson A (1998) Goal-directed instrumental action: contingency and incentive learning and their cortical substrates. Neuropharmacology 37(4–5):407–419

Balleine BW, O'Doherty JP (2010) Human and rodent homologies in action control: corticostriatal determinants of goal-directed and habitual action. Neuropsychopharmacology 35(1):48–69

Barbano MF, Cador M (2006) Differential regulation of the consummatory, motivational and anticipatory aspects of feeding behavior by dopaminergic and opioidergic drugs. Neuropsychopharmacology 31(7):1371–1381

Barbano MF, Cador M (2007) Opioids for hedonic experience and dopamine to get ready for it. Psychopharmacology 191(3):497–506

Barch DM, Dowd EC (2010) Goal representations and motivational drive in schizophrenia: the role of prefrontal-striatal interactions. Schizophr Bull 36(5):919–934

Barch DM, Treadway MT, Schoen N (2014) Effort, anhedonia, and function in schizophrenia: reduced effort allocation predicts amotivation and functional impairment. J Abnorm Psychol 123(2):387–397

Barch DM, Pagliaccio D, Luking K (2015) Mechanisms underlying motivational deficits in psychopathology: similarities and differences in depression and schizophrenia. Curr Top Behav Neurosci May 31. [Epub ahead of print]

Bardgett ME, Depenbrock M, Downs N, Points M, Green L (2009) Dopamine modulates effort-based decision making in rats. Behav Neurosci 123(2):242–251

Barnes SA, Der-Avakian A, Markou A (2014) Anhedonia, avolition, and anticipatory deficits: assessments in animals with relevance to the negative symptoms of schizophrenia. Eur Neuropsychopharmacol 24(5):744–758

Barr AM, Phillips AG (1998) Chronic mild stress has no effect on responding by rats for sucrose under a progressive ratio schedule. Physiol Behav 64(5):591–597

Barr AM, Phillips AG (1999) Withdrawal following repeated exposure to d-amphetamine decreases responding for a sucrose solution as measured by a progressive ratio schedule of reinforcement. Psychopharmacology 141(1):99–106

Barr AM, Phillips AG (2002) Increased successive negative contrast in rats withdrawn from an escalating-dose schedule of D-amphetamine. Pharmacol Biochem Behav 71(1–2):293–299

Barr AM, Fiorino DF, Phillips AG (1999) Effects of withdrawal from an escalating dose schedule of d-amphetamine on sexual behavior in the male rat. Pharmacol Biochem Behav 64(3):597–604

Barr RS, Pizzagalli DA, Culhane MA, Goff DC, Evins AE (2008) A single dose of nicotine enhances reward responsiveness in nonsmokers: implications for development of dependence. Biol Psychiatry 63(11):1061–1065

Barrett SP, Pihl RO, Benkelfat C, Brunelle C, Young SN, Leyton M (2008) The role of dopamine in alcohol self-administration in humans: individual differences. Eur Neuropsychopharmacol 18(6):439–447

Bentosela M, Jakovcevic A, Elgier AM, Mustaca AE, Papini MR (2009) Incentive contrast in domestic dogs (Canis familiaris). J Comp Psychol 123(2):125–130

Berghorst LH, Bogdan R, Frank MJ, Pizzagalli DA (2013) Acute stress selectively reduces reward sensitivity. Front Hum Neurosci 7:133

Berlin I, Givry-Steiner L, Lecrubier Y, Puech AJ (1998) Measures of anhedonia and hedonic responses to sucrose in depressive and schizophrenic patients in comparison with healthy subjects. Eur Psychiatry 13(6):303–309

Berridge KC, Robinson TE (2003) Parsing reward. Trends Neurosci 26(9):507–513

Bogdan R, Pizzagalli DA (2006) Acute stress reduces reward responsiveness: implications for depression. Biol Psychiatry 60(10):1147–1154

Brebion G, Bressan RA, Pilowsky LS, David AS (2009) Depression, avolition, and attention disorders in patients with schizophrenia: associations with verbal memory efficiency. J Neuropsychiatry Clin Neurosci 21(2):206–215

Brenes JC, Schwarting RK (2015). Individual differences in anticipatory activity to food rewards predict cue-induced appetitive 50-kHz calls in rats. Physiol Behav 149:107–118

Buck CL, Malavar JC, George O, Koob GF, Vendruscolo LF (2014a) Anticipatory 50 kHz ultrasonic vocalizations are associated with escalated alcohol intake in dependent rats. Behav Brain Res 271:171–176

Buck CL, Vendruscolo LF, Koob GF, George O (2014b) Dopamine D1 and mu-opioid receptor antagonism blocks anticipatory 50 kHz ultrasonic vocalizations induced by palatable food cues in Wistar rats. Psychopharmacology 231(5):929–937

Burke SN, Thome A, Plange K, Engle JR, Trouard TP, Gothard KM et al (2014) Orbitofrontal cortex volume in area 11/13 predicts reward devaluation, but not reversal learning performance, in young and aged monkeys. J Neurosci 34(30):9905–9916

Carroll ME, Lac ST (1987) Cocaine withdrawal produces behavioral disruptions in rats. Life Sci 40(22):2183–2190

Carver CS, White TL (1994) Behavioral-inhibition, behavioral activation, and affective responses to impending reward and punishment—the Bis Bas scales. J Pers Soc Psychol 67(2):319–333

Catanese F, Freidin E, Cuello MI, Distel RA (2011) Devaluation of low-quality food during early experience by sheep. Animal 5(6):938–942

Cawley EI, Park S, aan het Rot M, Sancton K, Benkelfat C, Young SN et al (2013) Dopamine and light: dissecting effects on mood and motivational states in women with subsyndromal seasonal affective disorder. J Psychiatry Neurosci 38(6):388–397

Chapman LJ, Chapman JP, Raulin ML (1976) Scales for Physical and Social Anhedonia. J Abnorm Psychol 85(4):374–382

Chevallier C, Grezes J, Molesworth C, Berthoz S, Happe F (2012) Brief report: selective social anhedonia in high functioning autism. J Autism Dev Disord 42(7):1504–1509

Corbit LH, Balleine BW (2003) The role of prelimbic cortex in instrumental conditioning. Behav Brain Res 146(1–2):145–157

Corbit LH, Balleine BW (2005) Double dissociation of basolateral and central amygdala lesions on the general and outcome-specific forms of pavlovian-instrumental transfer. J Neurosci 25 (4):962–970

Corbit LH, Muir JL, Balleine BW (2001) The role of the nucleus accumbens in instrumental conditioning: Evidence of a functional dissociation between accumbens core and shell. J Neurosci 21(9):3251–3260

Couvillon PA, Bitterman ME (1984) The overlearning-extinction effect and successive negative contrast in honeybees (Apis-Mellifera). J Comp Psychol 98(1):100–109

Crombag HS, Johnson AW, Zimmer AM, Zimmer A, Holland PC (2010) Deficits in sensory-specific devaluation task performance following genetic deletions of cannabinoid (CB1) receptor. Learn Mem 17(1):18–22

Damiano CR, Aloi J, Treadway M, Bodfish JW, Dichter GS (2012) Adults with autism spectrum disorders exhibit decreased sensitivity to reward parameters when making effort-based decisions. J Neurodev Disord 4:13

Damiano CR, Aloi J, Burrus C, Garbutt JC, Kampov-Polevoy AB, Dichter GS (2014) Intact hedonic responses to sweet tastes in autism spectrum disorder. Res Autism Spectr Disord 8 (3):230–236

Davis C, Woodside DB (2002) Sensitivity to the rewarding effects of food and exercise in the eating disorders. Compr Psychiatry 43(3):189–194

de Wit S, Corlett PR, Aitken MR, Dickinson A, Fletcher PC (2009) Differential engagement of the ventromedial prefrontal cortex by goal-directed and habitual behavior toward food pictures in humans. J Neurosci 29(36):11330–11338

de Wit S, Barker RA, Dickinson AD, Cools R (2011) Habitual versus goal-directed action control in parkinson disease. J Cognitive Neurosci 23(5):1218–1229

de Wit S, Watson P, Harsay HA, Cohen MX, van de Vijver I, Ridderinkhof KR (2012) Corticostriatal connectivity underlies individual differences in the balance between habitual and goal-directed action control. J Neurosci 32(35):12066–12075

Der-Avakian A, Markou A (2010) Withdrawal from chronic exposure to amphetamine, but not nicotine, leads to an immediate and enduring deficit in motivated behavior without affecting social interaction in rats. Behav Pharmacol 21(4):359–368

Der-Avakian A, Markou A (2012) The neurobiology of anhedonia and other reward-related deficits. Trends Neurosci 35(1):68–77

Der-Avakian A, D'Souza MS, Pizzagalli DA, Markou A (2013) Assessment of reward responsiveness in the response bias probabilistic reward task in rats: implications for cross-species translational research. Transl Psychiatry 3:e297

Dias-Ferreira E, Sousa JC, Melo I, Morgado P, Mesquita AR, Cerqueira JJ et al (2009) Chronic stress causes frontostriatal reorganization and affects decision-making. Science 325 (5940):621–625

Dichter GS, Smoski MJ, Kampov-Polevoy AB, Gallop R, Garbutt JC (2010) Unipolar depression does not moderate responses to the Sweet Taste Test. Depress Anxiety 27(9):859–863

Fawcett J, Clark DC, Scheftner WA, Gibbons RD (1983) Assessing anhedonia in psychiatric-patients—the pleasure scale. Arch Gen Psychiat 40(1):79–84

Fervaha G, Graff-Guerrero A, Zakzanis KK, Foussias G, Agid O, Remington G (2013) Incentive motivation deficits in schizophrenia reflect effort computation impairments during cost-benefit decision-making. J Psychiatr Res 47(11):1590–1596

Fletcher PJ, Grottick AJ, Higgins GA (2002) Differential effects of the 5-HT(2A) receptor antagonist M100907 and the 5-HT(2C) receptor antagonist SB242084 on cocaine-induced locomotor activity, cocaine self-administration and cocaine-induced reinstatement of responding. Neuropsychopharmacology 27(4):576–586

Fletcher PJ, Rizos Z, Noble K, Soko AD, Silenieks LB, Le AD et al (2012) Effects of the 5-HT2C receptor agonist Ro60-0175 and the 5-HT2A receptor antagonist M100907 on nicotine self-administration and reinstatement. Neuropharmacology 62(7):2288–2298

Forbes NF, Stewart CA, Matthews K, Reid IC (1996) Chronic mild stress and sucrose consumption: validity as a model of depression. Physiol Behav 60(6):1481–1484

Frank MJ, Claus ED (2006) Anatomy of a decision: striato-orbitofrontal interactions in reinforcement learning, decision making, and reversal. Psychol Rev 113(2):300–326

Frank MJ, Seeberger LC, O'Reilly RC (2004) By carrot or by stick: cognitive reinforcement learning in parkinsonism. Science 306(5703):1940–1943

Frank MJ, Samanta J, Moustafa AA, Sherman SJ (2007) Hold your horses: impulsivity, deep brain stimulation, and medication in parkinsonism. Science 318(5854):1309–1312

Gard DE, Gard MG, Kring AM, John OP (2006) Anticipatory and consummatory components of the experience of pleasure: a scale development study. J Res Pers 40(6):1086–1102

Gard DE, Kring AM, Gard MG, Horan WP, Green MF (2007) Anhedonia in schizophrenia: distinctions between anticipatory and consummatory pleasure. Schizophr Res 93(1–3):253–260

Geyer MA, Markou A (2002) The role of preclinical models in the development of psychotropic drugs. In: Davis KL, Charney D, Coyle JT, Nemeroff CB (eds) Neuropsychopharmacology: the fifth generation of progress. Lippincott Williams & Wilkins, pp 445–455
Glover AC, Roane HS, Kadey HJ, Grow LL (2008) Preference for reinforcers under progressive- and fixed-ratio schedules: a comparison of single and concurrent arrangements. J Appl Behav Anal 41(2):163–176
Gold JM, Waltz JA, Prentice KJ, Morris SE, Heerey EA (2008) Reward processing in schizophrenia: a deficit in the representation of value. Schizophr Bull 34(5):835–847
Gold JM, Strauss GP, Waltz JA, Robinson BM, Brown JK, Frank MJ (2013) Negative symptoms of schizophrenia are associated with abnormal effort-cost computations. Biol Psychiatry 74 (2):130–136
Gooding DC, Pflum MJ (2014a) The assessment of interpersonal pleasure: introduction of the Anticipatory and Consummatory Interpersonal Pleasure Scale (ACIPS) and preliminary findings. Psychiatry Res 215(1):237–243
Gooding DC, Pflum MJ (2014b) Further validation of the ACIPS as a measure of social hedonic response. Psychiatry Res 215(3):771–777
Hamill S, Trevitt JT, Nowend KL, Carlson BB, Salamone JD (1999) Nucleus accumbens dopamine depletions and time-constrained progressive ratio performance: effects of different ratio requirements. Pharmacol Biochem Behav 64(1):21–27
Harris RB, Zhou J, Youngblood BD, Smagin GN, Ryan DH (1997) Failure to change exploration or saccharin preference in rats exposed to chronic mild stress. Physiol Behav 63(1):91–100
Haslam J (1809) Observations on madness and melancholy; including practical remarks on those diseases, together with cases, and an account of the morbid appearances on dissection, 2nd edn. Callow, London, vii, 345 pp
Heerey EA, Gold JM (2007) Patients with schizophrenia demonstrate dissociation between affective experience and motivated behavior. J Abnorm Psychol 116(2):268–278
Heerey EA, Bell-Warren KR, Gold JM (2008) Decision-making impairments in the context of intact reward sensitivity in schizophrenia. Biol Psychiatry 64(1):62–69
Hilario MR, Clouse E, Yin HH, Costa RM (2007) Endocannabinoid signaling is critical for habit formation. Front Integr Neurosci 1:6
Hodos W (1961) Progressive ratio as a measure of reward strength. Science 134(3483):943–944
Hodos W, Kalman G (1963) Effects of increment size and reinforcer volume on progressive ratio performance. J Exp Anal Behav 6:387–392
Hogarth L, Attwood AS, Bate HA, Munafo MR (2012) Acute alcohol impairs human goal-directed action. Biol Psychol 90(2):154–160
Holroyd CB, Coles MG (2002) The neural basis of human error processing: reinforcement learning, dopamine, and the error-related negativity. Psychol Rev 109(4):679–709
Hsu CT, Patton DF, Mistlberger RE, Steele AD (2010) Palatable meal anticipation in mice. Plos One 5(9):e12903
Hyman SE (2012). Revolution stalled. Sci Transl Med 4(155):155cm111
Hyman SE, Fenton WS (2003) Medicine. What are the right targets for psychopharmacology? Science 299(5605):350–351
Insel T, Cuthbert B, Garvey M, Heinssen R, Pine DS, Quinn K et al (2010) Research domain criteria (RDoC): toward a new classification framework for research on mental disorders. Am J Psychiatry 167(7):748–751
Insel TR, Voon V, Nye JS, Brown VJ, Altevogt BM, Bullmore ET et al (2013) Innovative solutions to novel drug development in mental health. Neurosci Biobehav Rev 37(10 Pt 1):2438–2444
Isella V, Iurlaro S, Piolti R, Ferrarese C, Frattola L, Appollonio I et al (2003) Physical anhedonia in Parkinson's disease. J Neurol Neurosurg Psychiatry 74(9):1308–1311
Jilge B (1992) Restricted feeding—a nonphotic zeitgeber in the rabbit. Physiol Behav 51(1):157–166
Jocham G, Klein TA, Ullsperger M (2011) Dopamine-mediated reinforcement learning signals in the striatum and ventromedial prefrontal cortex underlie value-based choices. J Neurosci 31 (5):1606–1613

Jocham G, Klein TA, Ullsperger M (2014) Differential modulation of reinforcement learning by D2 dopamine and NMDA glutamate receptor antagonism. J Neurosci 34(39):13151–13162
Kamal A, Van der Harst JE, Kapteijn CM, Baars AJ, Spruijt BM, Ramakers GM (2010) Announced reward counteracts the effects of chronic social stress on anticipatory behavior and hippocampal synaptic plasticity in rats. Exp Brain Res 201(4):641–651
Kasch KL, Rottenberg J, Arnow BA, Gotlib IH (2002) Behavioral activation and inhibition systems and the severity and course of depression. J Abnorm Psychol 111(4):589–597
Killcross S, Coutureau E (2003) Coordination of actions and habits in the medial prefrontal cortex of rats. Cereb Cortex 13(4):400–408
Klein DF (1974) Endogenomorphic depression—conceptual and terminological revision. Arch Gen Psychiatry 31(4):447–454
Klossek UMH, Russell J, Dickinson A (2008) The control of instrumental action following outcome devaluation in young children aged between 1 and 4 years. J Exp Psychol-Gen 137 (1):39–51
Knutson B, Burgdorf J, Panksepp J (2002) Ultrasonic vocalizations as indices of affective states in rats. Psychol Bull 128(6):961–977
Kraepelin E (1921) Dementia praecox and paraphrenia. J Nerv Ment Dis 54(4):384
Kuczenski R, Segal DS, Aizenstein ML (1991) Amphetamine, cocaine, and fencamfamine: relationship between locomotor and stereotypy response profiles and caudate and accumbens dopamine dynamics. J Neurosci 11(9):2703–2712
Leboyer M, Bellivier F, Nosten-Bertrand M, Jouvent R, Pauls D, Mallet J (1998) Psychiatric genetics: search for phenotypes. Trends Neurosci 21(3):102–105
Leibenluft E, Charney DS, Pine DS (2003) Researching the pathophysiology of pediatric bipolar disorder. Biol Psychiatry 53(11):1009–1020
LeSage MG, Burroughs D, Pentel PR (2006) Effects of nicotine withdrawal on performance under a progressive-ratio schedule of sucrose pellet delivery in rats. Pharmacol Biochem Behav 83 (4):585–591
Li FZ (1999) The exercise motivation scale: Its multifaceted structure and construct validity. J Appl Sport Psychol 11(1):97–115
Llerena K, Park SG, McCarthy JM, Couture SM, Bennett ME, Blanchard JJ (2013) The motivation and pleasure scale-self-report (MAP-SR): reliability and validity of a self-report measure of negative symptoms. Compr Psychiat 54(5):568–574
Loas G, Duru C, Godefroy O, Krystkowiak P (2014) Hedonic deficits in Parkinson's disease: is consummatory anhedonia specific? Front Neurol 5:24
Ma ST, Maier EY, Ahrens AM, Schallert T, Duvauchelle CL (2010) Repeated intravenous cocaine experience: development and escalation of pre-drug anticipatory 50-kHz ultrasonic vocalizations in rats. Behav Brain Res 212(1):109–114
Markou A, Kosten TR, Koob GF (1998) Neurobiological similarities in depression and drug dependence: a self-medication hypothesis. Neuropsychopharmacology 18(3):135–174
Markou A, Chiamulera C, Geyer MA, Tricklebank M, Steckler T (2009) Removing obstacles in neuroscience drug discovery: the future path for animal models. Neuropsychopharmacology 34 (1):74–89
Markou A, Salamone JD, Bussey TJ, Mar AC, Brunner D, Gilmour G et al (2013) Measuring reinforcement learning and motivation constructs in experimental animals: relevance to the negative symptoms of schizophrenia. Neurosci Biobehav Rev 37(9 Pt B):2149–2165
Matthews K, Forbes N, Reid IC (1995) Sucrose consumption as an hedonic measure following chronic unpredictable mild stress. Physiol Behav 57(2):241–248
McFarland BR, Shankman SA, Tenke CE, Bruder GE, Klein DN (2006) Behavioral activation system deficits predict the six-month course of depression. J Affect Disorders 91(2–3):229–234
Meehl PE (1962) Schizotaxia, schizotypy, schizophrenia. Am Psychol 17(12):827–838
Mendelson SD, Pfaus JG (1989) Level searching: a new assay of sexual motivation in the male rat. Physiol Behav 45(2):337–341

Mendoza J, Angeles-Castellanos M, Escobar C (2005) Entrainment by a palatable meal induces food-anticipatory activity and c-Fos expression in reward-related areas of the brain. Neuroscience 133(1):293–303
Meyer B, Johnson SL, Carver CS (1999) Exploring behavioral activation and inhibition sensitivities among college students at risk for bipolar spectrum symptomatology. J Psychopathol Behav 21(4):275–292
Meyer-Lindenberg A, Weinberger DR (2006) Intermediate phenotypes and genetic mechanisms of psychiatric disorders. Nat Rev Neurosci 7(10):818–827
Mistlberger RE (1994) Circadian food-anticipatory activity: formal models and physiological mechanisms. Neurosci Biobehav Rev 18(2):171–195
Moore D, Siegfried D, Wilson R, Rankin MA (1989) The influence of time of day on the foraging behavior of the honeybee, Apis mellifera. J Biol Rhythms 4(3):305–325
Mote J, Minzenberg MJ, Carter CS, Kring AM (2014) Deficits in anticipatory but not consummatory pleasure in people with recent-onset schizophrenia spectrum disorders. Schizophr Res 159(1):76–79
Nawijn L, van Zuiden M, Frijling JL, Koch SB, Veltman DJ, Olff M (2015) Reward functioning in PTSD: a systematic review exploring the mechanisms underlying anhedonia. Neurosci Biobehav Rev 51:189–204
Nelson A, Killcross S (2006) Amphetamine exposure enhances habit formation. J Neurosci 26 (14):3805–3812
Nestler EJ, Hyman SE (2010) Animal models of neuropsychiatric disorders. Nat Neurosci 13 (10):1161–1169
O'Doherty J, Dayan P, Schultz J, Deichmann R, Friston K, Dolan RJ (2004) Dissociable roles of ventral and dorsal striatum in instrumental conditioning. Science 304(5669):452–454
Opiol H, Pavlovski I, Michalik M, Mistlberger RE (2015) Ultrasonic vocalizations in rats anticipating circadian feeding schedules. Behav Brain Res 284:42–50
Paul IA, English JA, Halaris A (2000) Sucrose and quinine intake by maternally-deprived and control rhesus monkeys. Behav Brain Res 112(1–2):127–134
Pechtel P, Pizzagalli DA (2013) Disrupted reinforcement learning and maladaptive behavior in women with a history of childhood sexual abuse: a high-density event-related potential study. JAMA Psychiatry 70(5):499–507
Pechtel P, Dutra SJ, Goetz EL, Pizzagalli DA (2013) Blunted reward responsiveness in remitted depression. J Psychiatr Res 47(12):1864–1869
Pelletier LG, Tuson KM, Fortier MS, Vallerand RJ, Briere NM, Blais MR (1995) Toward a new measure of intrinsic motivation, extrinsic motivation, and amotivation in sports—the Sport Motivation Scale (Sms). J Sport Exerc Psychol 17(1):35–53
Penrod B, Wallace MD, Dyer EJ (2008) Assessing potency of high- and low-preference reinforcers with respect to response rate and response patterns. J Appl Behav Anal 41(2):177–188
Pergadia ML, Der-Avakian A, D'Souza MS, Madden PA, Heath AC, Shiffman S et al (2014) Association between nicotine withdrawal and reward responsiveness in humans and rats. JAMA Psychiatry 71(11):1238–1245
Pfaus JG, Phillips AG (1991) Role of dopamine in anticipatory and consummatory aspects of sexual behavior in the male rat. Behav Neurosci 105(5):727–743
Phelps CE, Mitchell EN, Nutt DJ, Marston HM, Robinson ES (2015) Psychopharmacological characterisation of the successive negative contrast effect in rats. Psychopharmacology (Berl) 232:2697–2709
Pizzagalli DA, Jahn AL, O'Shea JP (2005) Toward an objective characterization of an anhedonic phenotype: a signal-detection approach. Biol Psychiatry 57(4):319–327
Pizzagalli DA, Bogdan R, Ratner KG, Jahn AL (2007) Increased perceived stress is associated with blunted hedonic capacity: potential implications for depression research. Behav Res Ther 45(11):2742–2753
Pizzagalli DA, Evins AE, Schetter EC, Frank MJ, Pajtas PE, Santesso DL et al (2008a) Single dose of a dopamine agonist impairs reinforcement learning in humans: behavioral evidence from a laboratory-based measure of reward responsiveness. Psychopharmacology 196(2):221–232

Pizzagalli DA, Goetz E, Ostacher M, Iosifescu DV, Perlis RH (2008b) Euthymic patients with bipolar disorder show decreased reward learning in a probabilistic reward task. Biol Psychiatry 64(2):162–168
Pizzagalli DA, Iosifescu D, Hallett LA, Ratner KG, Fava M (2008c) Reduced hedonic capacity in major depressive disorder: evidence from a probabilistic reward task. J Psychiatr Res 43(1):76–87
Reid I, Forbes N, Stewart C, Matthews K (1997) Chronic mild stress and depressive disorder: a useful new model? Psychopharmacology (Berl) 134(4):365–367; discussion 371–367
Ribot T (1896) La psychologie des sentiments. F. Alcan, Paris, xi, 443 pp
Roane HS, Lerman DC, Vorndran CM (2001) Assessing reinforcers under progressive schedule requirements. J Appl Behav Anal 34(2):145–166
Roberts DC (1989) Breaking points on a progressive ratio schedule reinforced by intravenous apomorphine increase daily following 6-hydroxydopamine lesions of the nucleus accumbens. Pharmacol Biochem Behav 32(1):43–47
Rodriguez JS, Morris SM, Hotchkiss CE, Doerge DR, Allen RR, Mattison DR et al (2010) The effects of chronic methylphenidate administration on operant test battery performance in juvenile rhesus monkeys. Neurotoxicol Teratol 32(2):142–151
Rorie AE, Gao J, McClelland JL, Newsome WT (2010) Integration of sensory and reward information during perceptual decision-making in lateral intraparietal cortex (LIP) of the macaque monkey. PLoS ONE 5(2):e9308
Salamone JD, Steinpreis RE, McCullough LD, Smith P, Grebel D, Mahan K (1991) Haloperidol and nucleus accumbens dopamine depletion suppress lever pressing for food but increase free food consumption in a novel food choice procedure. Psychopharmacology 104(4):515–521
Salamone JD, Cousins MS, Bucher S (1994) Anhedonia or anergia? Effects of haloperidol and nucleus accumbens dopamine depletion on instrumental response selection in a T-maze cost/benefit procedure. Behav Brain Res 65(2):221–229
Salamone JD, Cousins MS, Snyder BJ (1997) Behavioral functions of nucleus accumbens dopamine: empirical and conceptual problems with the anhedonia hypothesis. Neurosci Biobehav Rev 21(3):341–359
Salamone JD, Correa M, Farrar A, Mingote SM (2007) Effort-related functions of nucleus accumbens dopamine and associated forebrain circuits. Psychopharmacology 191(3):461–482
Salamone JD, Correa M, Farrar AM, Nunes EJ, Pardo M (2009) Dopamine, behavioral economics, and effort. Front Behav Neurosci 3:13
Samejima K, Ueda Y, Doya K, Kimura M (2005) Representation of action-specific reward values in the striatum. Science 310(5752):1337–1340
Santesso DL, Dillon DG, Birk JL, Holmes AJ, Goetz E, Bogdan R et al (2008) Individual differences in reinforcement learning: behavioral, electrophysiological, and neuroimaging correlates. Neuroimage 42(2):807–816
Santesso DL, Evins AE, Frank MJ, Schetter EC, Bogdan R, Pizzagalli DA (2009) Single dose of a dopamine agonist impairs reinforcement learning in humans: evidence from event-related potentials and computational modeling of striatal-cortical function. Hum Brain Mapp 30 (7):1963–1976
Schultz W (2002) Getting formal with dopamine and reward. Neuron 36(2):241–263
Schultz W, Dayan P, Montague PR (1997) A neural substrate of prediction and reward. Science 275(5306):1593–1599
Schwabe L, Wolf OT (2010) Socially evaluated cold pressor stress after instrumental learning favors habits over goal-directed action. Psychoneuroendocrinology 35(7):977–986
Shalev U, Kafkafi N (2002) Repeated maternal separation does not alter sucrose-reinforced and open-field behaviors. Pharmacol Biochem Behav 73(1):115–122
Sherdell L, Waugh CE, Gotlib IH (2012) Anticipatory pleasure predicts motivation for reward in major depression. J Abnorm Psychol 121(1):51–60
Simpson EH, Waltz JA, Kellendonk C, Balsam PD (2012) Schizophrenia in translation: dissecting motivation in schizophrenia and rodents. Schizophr Bull 38(6):1111–1117
Snaith RP, Hamilton M, Morley S, Humayan A, Hargreaves D, Trigwell P (1995) A scale for the assessment of hedonic tone—the Snaith-Hamilton Pleasure Scale. Br J Psychiatry 167:99–103

Soares JM, Sampaio A, Ferreira LM, Santos NC, Marques F, Palha JA et al (2012) Stress-induced changes in human decision-making are reversible. Transl Psychiatry 2:e131
Specht SM, Twining RC (1999) Human taste contrast and self-reported measures of anxiety. Percept Mot Skills 88(2):384–386
Starkstein SE, Jorge R, Mizrahi R, Robinson RG (2005) The construct of minor and major depression in Alzheimer's disease. Am J Psychiatry 162(11):2086–2093
Stoops WW (2008) Reinforcing effects of stimulants in humans: sensitivity of progressive-ratio schedules. Exp Clin Psychopharmacol 16(6):503–512
Stoops WW, Bennett JA, Lile JA, Sevak RJ, Rush CR (2013) Influence of aripiprazole pretreatment on the reinforcing effects of methamphetamine in humans. Prog Neuropsychopharmacol Biol Psychiatry 47:111–117
Sulzman FM, Fuller CA, Mooreede MC (1977) Feeding time synchronizes primate circadian-rhythms. Physiol Behav 18(5):775–779
Treadway MT, Buckholtz JW, Schwartzman AN, Lambert WE, Zald DH (2009) Worth the 'EEfRT'? The effort expenditure for rewards task as an objective measure of motivation and anhedonia. PLoS ONE 4(8):e6598
Treadway MT, Bossaller NA, Shelton RC, Zald DH (2012a) Effort-based decision-making in major depressive disorder: a translational model of motivational anhedonia. J Abnorm Psychol 121(3):553–558
Treadway MT, Buckholtz JW, Cowan RL, Woodward ND, Li R, Ansari MS et al (2012b) Dopaminergic mechanisms of individual differences in human effort-based decision-making. J Neurosci 32(18):6170–6176
Treadway MT, Peterman JS, Zald DH, Park S (2015) Impaired effort allocation in patients with schizophrenia. Schizophr Res 161(2–3):382–385
Trecker A, Robbins TW, Mar AC (eds) (2012) Development of a visual-guided probabilistic selection task for rats. In: Proceedings of measuring behavior 2012, 28–31 Aug 2012, Utrecht, The Netherlands
Tripp G, Alsop B (1999) Sensitivity to reward frequency in boys with attention deficit hyperactivity disorder. J Clin Child Psychol 28(3):366–375
Trosclair-Lasserre NM, Lerman DC, Call NA, Addison LR, Kodak T (2008) Reinforcement magnitude: an evaluation of preference and reinforcer efficacy. J Appl Behav Anal 41(2):203–220
Valentin VV, Dickinson A, O'Doherty JP (2007) Determining the neural substrates of goal-directed learning in the human brain. J Neurosci 27(15):4019–4026
Vallerand RJ, Pelletier LG, Blais MR, Briere NM, Senecal C, Vallieres EF (1992) The academic motivation scale—a measure of intrinsic, extrinsic, and amotivation in education. Educ Psychol Meas 52(4):1003–1017
van der Harst JE, Baars AM, Spruijt BM (2005) Announced rewards counteract the impairment of anticipatory behaviour in socially stressed rats. Behav Brain Res 161(2):183–189
Venugopalan VV, Casey KF, O'Hara C, O'Loughlin J, Benkelfat C, Fellows LK et al (2011) Acute phenylalanine/tyrosine depletion reduces motivation to smoke cigarettes across stages of addiction. Neuropsychopharmacology 36(12):2469–2476
Vollmayr B, Bachteler D, Vengeliene V, Gass P, Spanagel R, Henn F (2004) Rats with congenital learned helplessness respond less to sucrose but show no deficits in activity or learning. Behav Brain Res 150(1–2):217–221
Von Huben SN, Davis SA, Lay CC, Katner SN, Crean RD, Taffe MA (2006) Differential contributions of dopaminergic D1- and D2-like receptors to cognitive function in rhesus monkeys. Psychopharmacology 188(4):586–596
Vrieze E, Ceccarini J, Pizzagalli DA, Bormans G, Vandenbulcke M, Demyttenaere K et al (2013a) Measuring extrastriatal dopamine release during a reward learning task. Hum Brain Mapp 34 (3):575–586
Vrieze E, Pizzagalli DA, Demyttenaere K, Hompes T, Sienaert P, de Boer P et al (2013b) Reduced reward learning predicts outcome in major depressive disorder. Biol Psychiatry 73(7):639–645

Waltz JA, Frank MJ, Robinson BM, Gold JM (2007) Selective reinforcement learning deficits in schizophrenia support predictions from computational models of striatal-cortical dysfunction. Biol Psychiatry 62(7):756–764
Wang AY, Miura K, Uchida N (2013) The dorsomedial striatum encodes net expected return, critical for energizing performance vigor. Nat Neurosci 16(5):639–647
Ward RD, Simpson EH, Richards VL, Deo G, Taylor K, Glendinning JI et al (2012) Dissociation of hedonic reaction to reward and incentive motivation in an animal model of the negative symptoms of schizophrenia. Neuropsychopharmacology 37(7):1699–1707
Wardle MC, Treadway MT, Mayo LM, Zald DH, de Wit H (2011) Amping up effort: effects of d-amphetamine on human effort-based decision-making. J Neurosci 31(46):16597–16602
Watson D, Clark LA, Tellegen A (1988) Development and validation of brief measures of positive and negative affect—the Panas Scales. J Pers Soc Psychol 54(6):1063–1070
Weber DN, Spieler RE (1987) Effects of the light-dark cycle and scheduled feeding on behavioral and reproductive rhythms of the cyprinodont fish, Medaka, Oryzias-Latipes. Experientia 43 (6):621–624
Wenger D, Biebach H, Krebs JR (1991) Free-running circadian-rhythm of a learned feeding pattern in starlings. Naturwissenschaften 78(2):87–89
West EA, DesJardin JT, Gale K, Malkova L (2011) Transient inactivation of orbitofrontal cortex blocks reinforcer devaluation in macaques. J Neurosci 31(42):15128–15135
Whitton AE, Treadway MT, Pizzagalli DA (2015) Reward processing dysfunction in major depression, bipolar disorder and schizophrenia. Curr Opin Psychiatry 28(1):7–12
Whitton AE, Kakani P, Foti D, Van't Veer A, Haile A, Crowley DJ et al. (2016). Blunted neural responses to reward in remitted Major Depression: A high-density event-related potential study. BP:CNNI 1(1):87–95
Willey AR, Spear LP (2012) Development of anticipatory 50 kHz USV production to a social stimuli in adolescent and adult male Sprague-Dawley rats. Behav Brain Res 226(2):613–618
Willner P (2005) Chronic mild stress (CMS) revisited: consistency and behavioural-neurobiological concordance in the effects of CMS. Neuropsychobiology 52(2):90–110
Willner P, Towell A, Sampson D, Sophokleous S, Muscat R (1987) Reduction of sucrose preference by chronic unpredictable mild stress, and its restoration by a tricyclic antidepressant. Psychopharmacology 93(3):358–364
Willner P, Muscat R, Papp M (1992) Chronic mild stress-induced anhedonia: a realistic animal model of depression. Neurosci Biobehav Rev 16(4):525–534
Wolf DH, Satterthwaite TD, Kantrowitz JJ, Katchmar N, Vandekar L, Elliott MA et al (2014) Amotivation in schizophrenia: integrated assessment with behavioral, clinical, and imaging measures. Schizophr Bull 40(6):1328–1337
World Health Organization (1992) The ICD-10 classification of mental and behavioural disorders: clinical descriptions and diagnostic guidelines. World Health Organization, Geneva, xii, 362 pp
Yin HH, Knowlton BJ, Balleine BW (2005a) Blockade of NMDA receptors in the dorsomedial striatum prevents action-outcome learning in instrumental conditioning. Eur J Neurosci 22 (2):505–512
Yin HH, Ostlund SB, Knowlton BJ, Balleine BW (2005b) The role of the dorsomedial striatum in instrumental conditioning. Eur J Neurosci 22(2):513–523
Yohn SE, Santerre JL, Nunes EJ, Kozak R, Podurgiel SJ, Correa M et al (2015) The role of dopamine D1 receptor transmission in effort-related choice behavior: effects of D1 agonists. Pharmacol Biochem Behav 135:217–226
Zhang D, Zhou X, Wang X, Xiang X, Chen H, Hao W (2007) Morphine withdrawal decreases responding reinforced by sucrose self-administration in progressive ratio. Addict Biol 12 (2):152–157

Affective Biases in Humans and Animals

E.S.J. Robinson and J.P. Roiser

Abstract Depression is one of the most common but poorly understood psychiatric conditions. Although drug treatments and psychological therapies are effective in some patients, many do not achieve full remission and some patients receive no apparent benefit. Developing new improved treatments requires a better understanding of the aetiology of symptoms and evaluation of novel therapeutic targets in pre-clinical studies. Recent developments in our understanding of the basic cognitive processes that may contribute to the development of depression and its treatment offer new opportunities for both clinical and pre-clinical research. This chapter discusses the clinical evidence supporting a cognitive neuropsychological model of depression and antidepressant efficacy, and how this information may be usefully translated to pre-clinical investigation. Studies using neuropsychological tests in depressed patients and at risk populations have revealed basic negative emotional biases and disrupted reward and punishment processing, which may also impact on non-affective cognition. These affective biases are sensitive to antidepressant treatments with early onset effects observed, suggesting an important role in recovery. This clinical work into affective biases has also facilitated back-translation to animals and the development of assays to study affective biases in rodents. These animal studies suggest that, similar to humans, rodents in putative negative affective states exhibit negative affective biases on decision-making and memory tasks. Antidepressant treatments also induce positive biases in these rodent tasks, supporting the translational validity of this approach. Although still in the early stages of development and validation, affective biases in depression have the potential to offer new insights into the clinical condition, as well as facilitating the development of more translational approaches for pre-clinical studies.

E.S.J. Robinson (✉)
School of Physiology and Pharmacology, Medical Sciences Building,
University Walk, Bristol BS8 4PX, UK
e-mail: pmesjr@bris.ac.uk

J.P. Roiser
Institute of Cognitive Neuroscience, University College London,
17 Queen Square, London WC1N 3AR, UK

Curr Topics Behav Neurosci (2016) 28: 263–286
DOI 10.1007/7854_2015_5011

Keywords Emotion • Reward • Rodents • Animal model • Major depressive disorder • Antidepressants

Contents

1 Introduction 264
2 Disrupted "Hot" and "Cold" Processing in Depression 267
3 Pharmacological Effects on "Hot" Processing in Depression 269
4 A Cognitive Neuropsychological Model of Depression and Its Treatment 270
5 Limitations Associated with Standard Animal Models of Depression 271
6 Measuring Affective Biases in Pre-clinical Depression Research 273
7 Ambiguous Cue Interpretation/Judgement Bias Task 275
8 Validation of Interpretation/Judgment Task as Model of Affective Biases in Depression 276
9 Reward Processing Tests in Rodent Depression Models 277
10 The Rodent Affective Bias Test 278
11 Conclusions 279
References 280

1 Introduction

Depression is a debilitating condition that affects hundreds of millions of people worldwide. Its associated economic and social costs are enormous. For example, depression has been estimated to cost the EU economy £77 billion annually (Wittchen et al. 2011), much of which comprises reduced productivity and out-of-work benefits; and in the UK twice as many people die specifically as a result of suicide in depression than are killed on its roads each year (UK Office for National Statistics), representing a profound degree of personal tragedy for patients and their families.

At first glance, these statistics may seem at odds with the fact that depression is highly treatable, with robust evidence supporting both antidepressant medications and psychological therapies from large randomised controlled trials. However, effect sizes in these trials are generally small to medium relative to control conditions (Fournier et al. 2010), and when averaged across individuals represent just a few points on standard depression scales. In fact, such aggregate results mask great variability, as revealed by studies such as the STAR*D (Trivedi et al. 2006), which found that only around one-third of patients recovered fully on the first antidepressant they were prescribed, with another third subsequently recovering with either an alternate medication or psychotherapy. However, a third of patients remained unwell even following several courses of medication, representing months if not years of unsuccessful treatment.

Therefore, despite the existence of dozens of approved medications for depression, developing novel, better treatments remain a priority. In order to achieve this, pharmaceutical companies require valid animal models on which to test potential new treatments during early development. However, over the past decade this endeavour has been largely unsuccessful in identifying compounds that succeed in human trials (Agid et al. 2007), a major reason for the large-scale withdrawal of the pharmaceutical industry from research and development in psychiatry.

Whilst there were several factors that precipitated this troubling development, a commonly cited issue is the relatively weak ability of animal models to recapitulate human psychiatric diagnoses. The latter are defined solely descriptively, on the basis of symptom clusters, because the pathological mechanisms that drive these symptoms in humans remain largely unknown. Instead, as discussed below and in the previous chapter, pre-clinical scientists working in psychiatry drug development often employed variants of stress exposure models developed from the 1960s onwards (e.g. learned helplessness, proposed by Seligman and colleagues), because these models demonstrated sensitivity to (serendipitously discovered) early antidepressants (McArthur and Borsini 2006). However, an important corollary of this approach is that the drugs it yielded (often serotonin specific re-uptake inhibitors (SSRI) and serotonin and noradrenaline re-uptake inhibitor (SNRI)) largely converged around the same mechanism of action, specifically to increase monoamine transmission in the projection sites of these systems either through reuptake blockade or inhibiting metabolism. As discussed above, a substantial proportion of depressed patients do not respond to these classes of drugs, likely reflecting the heterogeneity in mechanisms driving depressive symptoms. New approaches to translation are therefore required in order to drive the development of novel treatments targeted at specific mechanisms, suited to specific subgroups of patients.

Classic animal models of depression derived from exposure to uncontrollable stress were considered to have strong face validity because (1) stress is known to precipitate depression in humans and (2) the behavioural features ("read-outs") induced in rodents were superficially similar to certain symptoms in humans. For example, the "behavioural despair" (giving up) evoked by the forced swim test (FST) (and other variants of learned helplessness) seemed to mimic depressive hopelessness and passivity; whilst reduced preference for sucrose bears a superficial similarity with anhedonia (Vollmayr et al. 2004). However, this approach could potentially be misleading, because different cognitive and neural processes could potentially give rise to the same apparent symptoms. Moreover, different individuals (rodent or human) may respond to stress in very different ways. Indeed, the high degree of individual variability elicited by uncontrollably stressful situations in human experiments, and the descriptions of the attendant thought processes that participants reported provided the rationale for Seligman and colleagues to move away from using ideas inspired by animal learned helplessness to explain depression (Abramson et al. 1978), towards high-level psychological descriptions of attributional style. Since such high-level constructs can only be accessed through introspection, and measured through verbal report or written questionnaires, impossible in animals, this marked a watershed parting of ways between "biological" and "psychological" explanations of depression, which to a large extent remains to this day.

In this paper, we intend to show how this gap can be re-bridged. Though the conceptual distance between classic cognitive and animal models of depression is great, modern research has demonstrated that there exist several types of disrupted basic cognitive processes in depression. Importantly, unlike high-level psychological constructs such as attributional style or dysfunctional negative schemata, these processes can be measured in animals, allowing back-translation in a much more straightforward manner. Contemporary theoretical accounts of depression, which are consistent with extensive neurocognitive and psychopharmacological data in depressed patients, propose that basic cognitive processes in depression play a causal role in the development of both high-level psychological constructs and symptoms (Harmer et al. 2009a; Roiser et al. 2012) and, at least in some patients, are critically related to effective treatment.

We initially provide an overview of disrupted basic cognitive processes in depression, including both "hot" (emotionally laden) and "cold" (emotionally independent) cognition, emphasising the existence of negative processing biases in both reward and emotional processing (Roiser and Sahakian 2013). We then review psychopharmacological data that suggest that altered monoamine neurotransmission may have a critical role to play in driving these biases and outline the cognitive neuropsychological model of depression, which incorporates both high-level psychological constructs and basic cognitive processes. Moving onto work in animals, we briefly review the limitations of current methods used to study depression-like

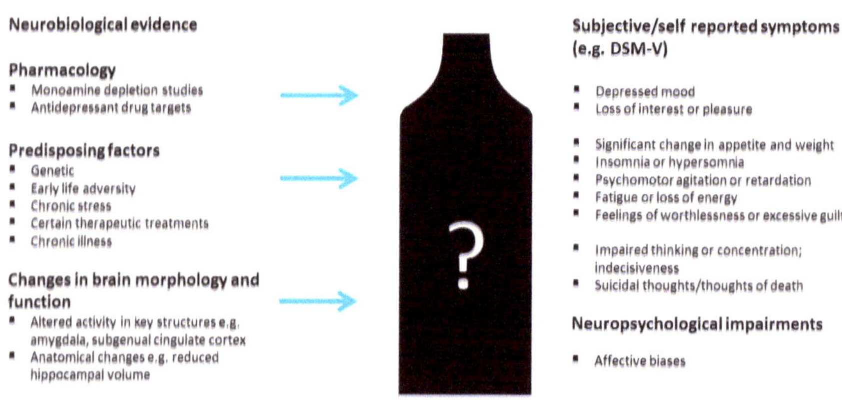

Fig. 1 Limitations in the ability to translate neurobiological data to clinical symptoms in depression have led to a bottleneck and lack of development of novel treatments. Self-report of symptoms is the standard approach used in diagnosis of depression and evaluation of clinical outcomes during clinical trials. These subjective symptoms cannot be directly translated to animal studies, which severely limits our understanding of how neurobiological changes linked to depression relate to symptoms, and how this knowledge can be utilised to develop new treatments. Methods used to quantify objective changes in neuropsychological processes such as affective biases in depression may provide the translational, objective measures needed to facilitate new understanding of the neurobiological processes underlying symptoms and the development of novel treatments

behaviour, before explaining how basic cognitive processes derived from the human depression literature can be assessed in animals. We illustrate this concept with several examples, in particular focusing on interpretational bias and reinforcement learning. Importantly, we explain how such basic cognitive processes in animals are altered through psychopharmacological manipulations in a manner consistent with that observed in humans, paving the way for a translational approach to drug development grounded in cognitive neuroscience, with high potential for developing personalised treatment (Fig. 1).

2 Disrupted "Hot" and "Cold" Processing in Depression

Demonstrations of negative processing biases in depression have a long history, going back to early cognitive theorists. Beck, amongst others, noted that depressed patients have a tendency to focus on current and past negative events (Beck 1967), which he ascribed to "negative schemata"; in other words, biased models of the environment instantiated through early experience, which colour the processing of external sensory inputs. This idea of a "top-down" bias—that depressed individuals focus on negative stimuli because these accord with their expectations—forms an important part of traditional cognitive approach to understanding and treating depression. However, as explained in more detail below, recent cognitive models of depression incorporate the additional possibility that these processing biases are "bottom-up" in nature; that is, the inputs themselves are negatively biased.

Whether caused by "top-down" or "bottom-up" processing, or some combination of the two, the existence of mood-congruent negative processing biases on "hot" (i.e. emotionally laden) cognitive tasks in depression is well established (reviewed in Roiser et al. 2012). By contrast, a consistent finding in never-depressed individuals is the presence of a small positive processing bias. Negatively biased processing in depressed groups, relative to controls, has been reported on tests of perception (e.g. face categorisation), memory (e.g. autobiographical narrative and free recall), attention (e.g. go/no-go or dot-probe tasks) and working memory (e.g. maintaining representations online in the face of distractors) (Erickson et al. 2005; Joormann and Gotlib 2006, 2008; Matt et al. 1992; Murphy et al. 1999). For example, one common approach is to present faces expressing positive or negative emotion (perhaps using morphing software to create gradations of intensity), and instructing participants to categorise them according to the six basic emotions: happy, sad, fearful, angry, disgusted, and surprised. Depressed individuals exhibit a negative bias, miscategorising happy faces as negative, especially at lower intensity levels when emotion is more ambiguous (reviewed in Gotlib and Joormann 2010). Although beyond the scope of this article, it is worth noting that individuals suffering from anxiety (which is highly comorbid with depression) also exhibit negative processing biases, particularly during the early stages of stimulus processing (Teachman et al. 2012).

Another aspect of "hot" processing that is disrupted in depression is reinforcement processing. This encompasses tasks on which participants may gain or lose money, points, or more basic reinforcers such as food, water, or pain. These kinds of measures are highly relevant to symptoms such as anhedonia and difficulty in decision-making (Eshel and Roiser 2010; Huys et al. 2015). Although this literature is more recent, and therefore less extensive than that on emotion processing, a consistent pattern of results has emerged. Whilst basic hedonic responses (i.e. ratings of reward pleasantness) appear to be intact in depression (Treadway and Zald 2011), a wealth of data supports the notion that depressed patients are hyposensitive to rewards (positive reinforcers) and hypersensitive to punishments (negative reinforcers) in terms of their influence on behaviour.

For example, several studies report that depressed individuals fail to adapt response patterns during a difficult, asymmetrically rewarded, perceptual decision task (Henriques and Davidson 2000; Henriques et al. 1994; Pizzagalli et al. 2008b) (such failure is also associated with poor outcome following treatment: Vrieze et al. 2013) and that they are unwilling to exert physical effort to obtain rewards (Treadway et al. 2012). Other studies reported reduced learning about stimuli associated with rewards in depression (Chase et al. 2010), especially in individuals with marked anhedonic symptoms. Finally, there have been several reports of impaired reward-related decision-making in depression, using tasks in which learning is not required. A consistent finding is that depressed individuals are relatively unwilling to place high bets on their decisions when they are very likely to win (Murphy et al. 2001; Roiser et al. 2009). Interestingly, reward processing abnormalities are also present in non-depressed individuals at high risk of developing symptoms, because either they have previously been depressed (Pizzagalli et al. 2008a) or they are closely related to a depressed person (Rawal et al. 2013), underscoring the likely causal nature of these reinforcement processing abnormalities in the development of symptoms.

Fewer studies have focused specifically on punishment processing in depression. Early studies on this topic explored the hypothesis that the reliable "cold" (i.e. non-emotion dependent) cognitive impairment in depressed groups relative to matched controls, equivalent to approximately half of one standard deviation on most neuropsychological tests (Rock et al. 2013; Snyder 2012), could be related to a "catastrophic response to perceived negative feedback" (Beats et al. 1996). These studies examined the pattern of responding across trials, finding that on tests during which participants received feedback, depressed individuals were far more likely to make an error if they had received an error on the previous trial (Elliott et al. 1996, 1997). Therefore, it may be that an ostensibly "cold" cognitive impairment is partly driven by exaggerated punishment processing, leading patients simply to give up when they make a mistake. An additional possibility may be that low motivation contributes to poor performance, though the extant data do not support this hypothesis unequivocally (see Austin et al. 2001 for a detailed discussion). Supporting the notion of exaggerated responses to punishment, probabilistic reversal learning tasks, on which misleading negative feedback is occasionally provided when the correct stimulus is selected, have revealed an exaggerated tendency to switch to the less frequently rewarded stimulus immediately following negative feedback in depressed individuals (Murphy et al. 2003; Taylor Tavares et al. 2008). However, disrupted reward and punishment processing cannot completely account for "cold" cognitive impairments in depression, for two reasons: (1) cognitive impairments remain even after symptoms have remitted (Rock et al. 2013), suggesting that they are not driven solely by current symptoms such as low motivation; (2) impairments are often observed on tests that do not feature explicit feedback.

3 Pharmacological Effects on "Hot" Processing in Depression

As discussed above, "hot" processing biases in depression have usually been ascribed to "top-down" influences, such as dysfunctional negative schemata. However, over the past decade a wealth of data has emerged from human experimental psychopharmacology studies suggesting that this explanation is likely to be incomplete. Specifically, this literature shows that manipulations that either boost or dampen transmission in the monoamine systems (dopamine, noradrenaline, and serotonin) can shift "hot" processing biases over timescales on the order of hours, in both depressed patients and healthy volunteers. Since dysfunctional negative schemata are proposed to be stable, inflexible representations of the environment, which take months if not years to change, and are not thought to be directly affected by pharmacological manipulations, traditional cognitive models of depression cannot easily account for these effects. Instead, a new cognitive neuropsychological model of depression has been proposed, in which "bottom-up" biases, driven by disrupted monoamine transmission, play a critical role in the development of schemata, symptoms, and their treatment (Harmer et al. 2009a; Roiser et al. 2012).

The first direct support for the hypothesis that monoamine transmission may play a role in depressive symptoms (other than the mood effects of antidepressant drugs themselves, which was the original basis for the monoamine hypothesis) was derived from studies using the acute tryptophan depletion (ATD) method (Ruhe et al. 2007). This dietary manipulation, which acutely restricts the supply of the precursor of serotonin to the brain, was used in a series of experiments that identified a temporary recurrence of some depressive symptoms in remitted patients during the low tryptophan period (lasting a few hours), which were resolved following resumption of a normal diet (Delgado et al. 1990; Smith et al. 1997). However, later work suggested that pronounced effects on mood were mainly observed in patients maintained on serotonergic antidepressant medication (Ruhe et al. 2007) leading to the criticism that ATD may simply have reversed a treatment effect. Interestingly, mood effects of ATD were rarely observed in healthy volunteers. Around the same time, complementary research using positron emission tomography (PET) to measure serotonin receptors reported substantial alterations in depressed patients, particularly decreased 5-HT1A receptors, which were elevated following a variety of antidepressant treatments (Savitz et al. 2009).

The above investigations did not measure basic cognitive processing, but instead focused on symptoms. However, from the early 2000s onwards several studies in healthy subjects reported that ATD, as well as similar methods that deplete dopamine or noradrenaline, could elicit negative processing biases on "hot" processing tests, including emotional and reinforcement processing (Cools et al. 2005; Firk and Markus 2008; Hasler et al. 2009a, b; McLean et al. 2004; Murphy et al. 2002; Robinson et al. 2011; Rogers et al. 2003; Roiser et al. 2006, 2008); importantly, mood was generally unaffected. Studies in remitted depressed individuals showed broadly similar results (Booij et al. 2005; Hayward et al. 2005; Munafo et al. 2006; Roiser et al. 2005), and in some cases the negative biases observed appeared remarkably similar to those observed in currently depressed patients, independent of any changes in mood. Complementing these findings, a series of studies taking the opposite approach, inducing a short-term boost in monoamine transmission with either antidepressant medication or tryptophan supplementation, found that emotional biases could also be shifted positively, in both healthy volunteers (Harmer et al. 2003, 2004; Murphy et al. 2006) and depressed individuals (Bhagwagar et al. 2004; Harmer et al. 2009b), again, usually in the absence of mood changes (reviewed in Harmer 2008). In some cases, early enhancement of positive emotional processing during antidepressant treatment preceded and predicted later symptomatic relief in depressed individuals (Tranter et al. 2009).

4 A Cognitive Neuropsychological Model of Depression and Its Treatment

The above studies have motivated a reconsideration and extension of traditional cognitive models of depression in order to incorporate pharmacological effects (Harmer et al. 2009a; Roiser et al. 2012). Whilst "top-down" biases, such as dysfunctional negative schemata, are still considered to play a critical role in the development and particularly the maintenance of depression, this new account additionally emphasises the contribution of biased "bottom-up" processing. In particular, it provides a cognitive framework for understanding how antidepressant drugs exert their effects, and a mechanistic explanation for the genesis of negative schemata.

In the cognitive neuropsychological model, "bottom-up" negative biases, which may be distally caused by either genetic or environmental influences that alter monoamine transmission (in particular psychosocial stress, which is thought to affect serotonin through its effects on cortisol: Dinan 1994), form a basis for the development of "top-down" biases. In other words, prolonged and consistent exposure to negatively biased inputs ("bottom-up" processing biases) causes the brain to develop negatively biased expectations ("top-down" processing biases). Since negative inputs accord with negative expectations, this state of combined "top-down" and "bottom-up" bias can become extremely stable. If it is experienced over a long period, dysfunctional negative schemata will become entrenched, eliciting high-level negative cognitions and low mood.

In the cognitive neuropsychological model, different treatment modalities (pharmacological and psychological) are proposed to target the different mechanisms driving symptoms ("bottom-up" and "top-down", respectively), providing a theoretical basis for prescribing different treatments to different patients. Antidepressant medications are proposed to target "bottom-up" biases and should be effective when "top-down" biases are weaker, since less fixed schemata are more likely to resolve spontaneously (i.e. without the assistance of a therapist) when negative inputs are removed. For example, the model would predict better response to medication in patients with shorter episodes and fewer dysfunctional attitudes, which is consistent with clinical data (Kohler et al. 2015; Riedel et al. 2011). By contrast, psychotherapy, for example CBT, is proposed to target "top-down" biases directly and should be effective when "bottom-up" biases are weaker, since the absence of negatively biased inputs should enable schemata to resolve more easily. Indeed, there are some preliminary data supporting this prediction (see Fig. 2 in Roiser et al. 2012). Finally, combining pharmacological and psychological treatment modalities should be more effective on average than either alone, for two reasons: (1) both "bottom-up" and "top-down" biases may operate simultaneously in some patients, meaning that both mechanisms need to be targeted for treatment to be effective; (2) in patients for whom either "top-down" or "bottom-up" biases are strong, combining treatment modalities provides the best chance that the relevant mechanism will be targeted. Again, this prediction is consistent with clinical data, at least in severely depressed individuals (Hollon et al. 2014).

In summary, the cognitive neuropsychological model of depression provides a useful theoretical framework for understanding how basic negative processing biases drive the development and maintenance of symptoms, as well as a cognitive account of how antidepressant drugs exert their effects. As outlined in the following sections, the discovery that basic affective biases can be directly modulated by pharmacological treatments in humans (both positively and negatively) has inspired pre-clinical researchers to develop novel models of emotional disturbance in depression. In a departure from previous approaches, these are focused not on symptoms themselves, but instead on the affective biases thought to underpin them.

5 Limitations Associated with Standard Animal Models of Depression

Recapitulating in animals the symptoms associated with psychiatric disorders is challenging and potentially an impossible task, particularly when the major species used for basic research are rodents. The human condition of depression is characterised by a heterogeneous range of symptoms, often assessed via self-report. In order to relate these symptoms to measurements in animals, a degree of anthropomorphism is inevitable. Methods used in animals to study depression-related behaviours have been strongly criticised; although perhaps it is not the methods which warrant criticism, but rather the way researchers have used the approaches and interpreted the resulting data.

Studies in animals provide researchers with the opportunity to test hypotheses about the cause and treatment of illness and are a key component in studies to understand the aetiology of depression and the development of new treatments. Patients' symptoms and associated psychopathology and neuropathology are often complex and further complicated by prolonged periods of illness. Unravelling the cause versus consequence of symptoms is therefore challenging. Animals offer a "blank canvas" in which specific hypotheses can be tested. However, these studies depend on the use of translatable endpoints and quantification of affective biases in animals may provide one methodological step needed to achieve this goal.

Animal models used in depression research are considered in detail in the preceding chapter. Therefore, this section focuses on the assays used to quantify depression-related behaviours in rodents (Table 1). When considering animal models of depression, the term "model" is often used to describe both the methods to induce a depression-like phenotype and those methods used to assay depression-like behaviour. The two key areas tested in depression-related studies are behavioural despair/hopelessness and anhedonia (Table 1). Neither of these behaviours directly translate to measures made in humans, but a degree of validity has been achieved through studies using known antidepressant treatments (predictive validity) and stress-related manipulations (face validity) (Geyer and Markou 1995; Cryan and Slattery 2007). Behavioural despair, quantified using either the FST or tail suspension test (TST), was initially developed as an assay to detect novel monoaminergic antidepressants (Porsolt et al. 1977) although it is now widely used in fundamental research and aetiological studies. Despite its high predictive validity for monoaminergic agents, the FST/TST is thought to have limited translational validity (see Nestler and Hyman 2010; Pollak et al. 2010; Berton et al. 2012; O'Leary and Cryan 2013 for further discussion). Its efficacy for the acute vs delayed effects of treatment, as well as its lack of sensitivity to non-monoaminergic manipulations has been highly criticised. Learned helplessness also measures a form of behavioural despair, where exposure to an inescapable stressor (usually footshock) induces a specific deficit in escape behaviour during subsequent presentations in an escapable environment (Overmier and Seligman 1967; Seligman and Beagley 1975; Maier 1984). The induction of learned helplessness not only produces a specific behavioural deficit which can be quantified, but also results in changes in neurobiology, suggesting it provides a phenotypic model (Pryce et al. 2011). Interestingly, not all animals treated in this paradigm develop learned helplessness, which has underpinned the development of the congenital learned helpless and non-learned helpless strains used as models of depression and resistance to depression (Henn and Vollmayr 2005; Pryce et al. 2011).

As discussed above, changes in reward processing are also commonly observed mood disorders (for a review see Eshel and Roiser 2010), and anhedonia has been widely used as a measure of depression-related behaviour in animals (Table 1). The approaches used in animals are usually based on consummatory tests such as the

Table 1 Assays used to quantify depression-related behaviour in rodents*

Animal behavioural assay	Proposed relationship to clinical condition	Key references
Forced swim test (FST)—rat and mouse; or tail suspension test (TST)—mouse	Hopelessness/despair	Porsolt et al. (1977) Detke et al. (1995) Steru et al. (1985); Cryan et al. (2005)
Learned helplessness	Hopelessness/despair	Maier (1984)
Sucrose preference test	Anhedonia	Willner et al. (1987)
Intracerebral self-stimulation (ICSS)	Anhedonia	Vogel et al. (1986); Zacharko and Anisman (1991)
Differential reinforcement of low-rate of responding 72 s (DRL-72)	Undefined but predictive of antidepressant efficacy	McGuire and Seiden (1980); Seiden et al. (1985)

*also see Markou and Pizzigalli, Depression models, this edition, for discussion about methods used to induce a depression-like phenotype

sucrose preference test (Willner et al. 1987). An alternative approach has been to look at reward threshold using intracerebral self-stimulation (ICSS) of reward centres in the brain (Vogel et al. 1986; Zacharko and Anisman 1991). Although these anhedonia tasks have better translational validity in principle, the measures used in patients are still largely based on self-report and the subjective experience of pleasure (Treadway and Zald 2011). Recent data from neuropsychological tests of reward processing may, however, provide a closer link between animal and human work, as discussed in more detail in the subsequent sections.

6 Measuring Affective Biases in Pre-clinical Depression Research

Back-translating human neuropsychological tests, used to measure affective biases in depression, requires the modification of the task to species-appropriate cues and behaviours (Paul et al. 2005). In almost all human studies, affective biases are investigated using emotional processing or interpretation tasks, which feature stimuli that are either language-dependent or facial expressions. Although animals are unable to perform tasks built around these cues, the principles that underlie such tasks can be developed for use in animals (Paul et al. 2005). Cues can be presented as either tones, lights, or spatial locations, and one of the most useful set-ups to achieve this is an operant chamber (as shown in Fig. 2). The presentation of the cues, the animal's responses, and resulting outcomes are all fully automated, enabling efficient and consistent methods to be used across laboratories. An alternative method for presenting animals with distinct cues is the bowl digging set-up (shown in Fig. 2). Here, the animals are trained to associate a particular cue, for example, the digging substrate or odour, with the presence of a hidden reward. More recently, touch screen equipment has been developed for rodents offering the potential to also develop tasks using visual cues (Bussey et al. 2012).

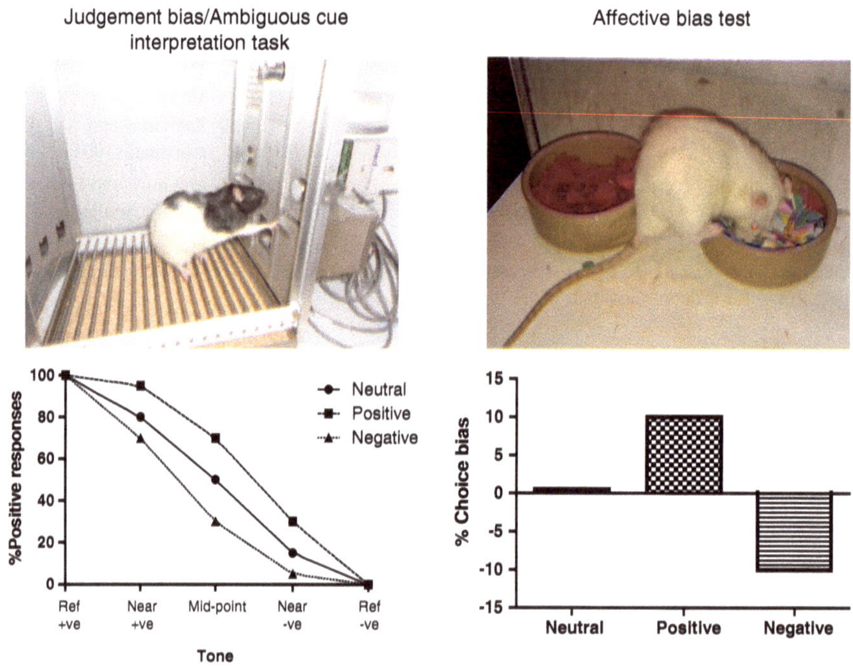

Fig. 2 Examples of two different approaches used to quantify affective biases in animals and hypothetical data illustrating the impact of different affective states. The judgement bias/ambiguous cue interpretation tasks (*left panels*) are the most widely used method to assess affective biases in animals. The tasks are hypothesised to quantify behavioural changes induced by different affective states, which reflect similar interpretational biases to those observed in humans with anxiety and/or depression (see Paul et al. 2005 for review and discussion). Animals are trained to associate two distinct cues with a positive and a negative outcome. Animals are then presented with intermediate, ambiguous cues (near positive, mid-point, and near negative) and their response selection is used to assess judgement bias. Positive biases are reflected in an upward shift in the graph with an increase in the number of responses in anticipation of positive events, whilst negative biases are reflected in a reduced number of responses during ambiguous cue presentation. The affective bias test (*right panel*) is designed to measure affective biases associated with the learning and recall of positive experiences (the association of a food reward with a specific digging medium). In this task, animals learn to associate one digging medium with finding a food reward under neutral conditions and a second digging medium during an affective state or drug manipulation. Using a choice test, the animal's subsequent preference for one of the substrates over the other is tested. If both experiences are encountered during a neutral affective state, no bias is observed; but both positive and negative affective state manipulations induce a choice bias during the preference test, reflected in an increase or decrease in the number of times the reward-paired substrate is chosen, respectively (*lower right panel*)

7 Ambiguous Cue Interpretation/Judgement Bias Task

The first empirical data to suggest that animals exhibit affective biases, similar to those observed in humans, were published by Harding and colleagues in 2004. Their experiment was designed to investigate whether animals in a putative negative affective state would exhibit pessimistic behaviours similar to those seen in human disorders such as depression. Building on the observations that patients with depression were more likely to anticipate negative outcomes when presented with neutral or ambiguous cues (Wright and Bower 1992), the group developed a rodent task which would enable similar ambiguous cue interpretations to be quantified in non-humans. In this pre-clinical "cognitive bias task", animals were initially trained to make an approach behaviour and press a lever to obtain a food reward, or to refrain from lever pressing to avoid a punishment. Each behaviour was trained using a specific tone frequency and animals were trained until they could distinguish the two cues. When these same animals were exposed to chronic mild stress and then presented with intermediate ambiguous tone frequencies, the animals in the putative negative affective state made fewer reward approach responses and were slower to respond. These data were interpreted as indicative of a "pessimistic" phenotype resulting from the negative affective state. A detailed discussion of the theoretical framework that underpins this task and how different behavioural response profiles may map onto human emotions is provided in Mendl et al. (2010).

The concepts from this original work have developed over the last decade, and a number of different variations to the original "cognitive bias task" have been reported (for review see Hales et al. 2014). The format of the task can involve either active choice, where the animal is required to make a response to either obtain reward or avoid punishment, or a go/no-go where only the reward response requires a response and avoidance of punishment is achieved by refraining from making a response. To investigate the underlying decision-making behaviour leading to either an optimistic or pessimistic choice, a go/go format is preferred over a go/no-go format, in which a reduction in approach behaviour could arise from either changes in reward-related motivation or enhanced anticipation of punishment. Hypothetical data illustrating either a positive or negative affective bias in an active choice task are shown in Fig. 2. Studies may use a mid-point-only cue or multiple intermediate cues to include a near positive and near negative. By using multiple intermediate ambiguous cues, it may also be possible to dissociate further between affective biases associated with reduced anticipation of reward versus increased anticipation of punishment, though further work is needed in this area. The majority of studies that have used this task to study depression-related neurobiology have used active choice formats with a footshock used as the punisher (Enkel et al. 2010; Papciak et al. 2013; Anderson et al. 2013). This does introduce a further potential confound, however, as animal are exposed to multiple footshocks over the course of training and testing, which may impact on affective state. The active choice tasks also have a long training period of ~3 months.

The original task utilised auditory cues where the animals were trained to associate distinct tone frequencies with either reward or avoidance of punishment (white noise) (Harding et al. 2004). Subsequent studies have also tested whether the same concept can be tested using spatial cues (Burman et al. 2008; Richter et al. 2012). In the spatial task, animals are trained to associate a specific location with obtaining a reward and a second location with either a lower value or aversive reward, e.g. quinine-flavoured pellet. The animals are trained until they show differential latencies between to two locations and are slower either to approach to low-value location or do not approach it at all. Judgement bias is then tested by placing goal posts in intermediate locations between the high-value reward and low-value locations. A more pessimistic judgement is reflected in a slower latency to approach the intermediate location. However, whilst this format is much easier to train, it may not engage the same neural processes as an active choice task which measures anticipation of reward and punishment avoidance (for further discussion, see Hales et al. 2014).

8 Validation of Interpretation/Judgment Task as Model of Affective Biases in Depression

The initial work undertaken with these tasks primarily focussed on animal welfare through validation of an objective measure of affective state. The number of studies where pharmacological manipulations have been used to test the validity of the judgement bias task methodology is limited. Overall, the studies where acute drug treatments have been used are not consistent with their predicted antidepressant or pro-depressant profile in man (Anderson et al. 2013; Rygula et al. 2014). Interestingly, treatment with a noradrenaline reuptake inhibitor, either with or without co-administration of corticosterone, has been reported to produce a negative bias by three different research groups (Enkel et al. 2010; Anderson et al. 2013; Rygula et al. 2014). The SSRIs appear to have mixed or no effect whilst one study has found that amphetamine, which is not an antidepressant, induces a positive bias (Anderson et al. 2013; Rygula et al. 2014). One possible explanation for the lack of positive bias effects seen with acute antidepressant treatments may relate to the apparent delayed onset of clinical efficacy. In the one study where fluoxetine was administered chronically, a tendency for a positive shift was observed, although this effect was not statistically robust (Anderson et al. 2013). Studies in animal models of depression include those in congenitally helpless rats, which were shown to exhibit a pessimistic phenotype (Enkel et al. 2010). The effect of chronic stress in normal animals has also been investigated, with one study reporting that chronic stress increases negative bias, an effect that was associated with baseline vulnerability (Rygula et al. 2013). A subsequent study from the same group showed a comparable finding in relation to chronic social defeat stress (Papciak et al. 2013).

Together, the data published for the judgement bias tasks suggest that the method offers a novel and translational approach to investigate affective biases associated with decision-making in rodents. Potential confounds relating to motivation and hedonic changes need to be taken into consideration in the design and interpretation of the experiments and further work is needed to gain a broader insight into the task's validity for depression research. The ability to use the task design in many different species, including honey bees (Bateson et al. 2011) is particularly appealing as it could facilitate studies in species such as drosophila where more complex genetic analyses are achievable.

9 Reward Processing Tests in Rodent Depression Models

Disrupted processing of reward information may be an additional important contributory factor to several symptoms of depression, including core features such as anhedonia and fatigue. In animals, anhedonia is a widely used endpoint for depression research (Table 1). Translation of this work to human studies is complicated by the fact that animal studies generally use primary rewards (food or electrical stimulation of reward pathways) whereas human studies use either hypothetical rewards (assessed via questionnaire) or secondary rewards (money). Additionally, studies in depressed patients suggest that reinforcement processing impairments may involve higher-order cognitive processes, such as learning and value-based choice, as opposed to simple consummatory responses (Eshel and Roiser 2010; Elliott et al. 2011; Roiser et al. 2012; Roiser and Sahakian 2013). Evidence to support this hypothesis includes data showing that patients with depression do not show altered hedonic responses to a primary reward in tests akin to the sucrose preference test (Dichter et al. 2012; McCabe et al. 2009; although also see Berlin et al. 1998).

Few animal studies have directly investigated reward learning and motivation in the context of depression models. These have generally quantified acquisition of a response–reward association such as a lever press task and/or motivation to respond for reward using a progressive ratio schedule, in which the effort required to obtain reward increases with each trial. In animals where a putative depression-like phenotype has been induced, reduced motivation for reward and/or impaired learning has been observed (Olausson et al. 2013; Leventopoulos et al. 2009; Gourley et al. 2008; Rüedi-Bettschen et al. 2005). In the congenitally helpless rat, reduced motivation for reward was detected using a progressive ratio task, with the same animals shown to have no deficits in learning in a non-food motivated task, the Morris Water Maze (Vollmayr et al. 2004). These changes may relate directly to the hedonic value of the reward and reflect similar neurobiological deficits as those observed using the sucrose preference test. It is, however, also possible that these changes involve a more complex process where affective biases influence learning and memory and, in turn, the subsequent recall of those associations, thereby directing and modifying subsequent behavioural responses.

An interesting translational task which may also provide a valuable approach to understanding the processing of both positive and negative information is the probabilistic reversal learning task (Dickstein et al. 2009; Hasler et al. 2009a, b; Eshel and Roiser 2010). Again, few studies have been carried out in animals, but the potential value in studying behavioural responses following both positive and negative feedback is appealing. An initial pharmacological characterisation of the task found effects associated with the serotonergic system including treatments with antidepressants (Bari et al. 2010). A role for serotonin was further supported by a subsequent study in mice (Ineichen et al. 2012). More recently, impairments in performance in animals exposed to isolation rearing have also been reported (Amitai et al. 2013).

10 The Rodent Affective Bias Test

The rodent affective bias test (ABT: Stuart et al. 2013, 2015) was developed to investigate the hypothesis that the cognitive processes associated with reward-related learning and memory may be modified by affective states. These affective biases then influence the animal's subsequent choice when the reward-associated cues are re-encountered. The ABT uses a discrimination learning phase where animals learn the association between a specific cue (a digging substrate) with a positive outcome (finding a food reward) (Fig. 2). The animals acquire these two independent experiences on different days with one learned during control conditions and the other during treatment. Affective bias is then quantified using a preference test where both the rewarded substrates are presented together and the animal's choices recorded. A bias score is then calculated from the number of choices made for the treatment-paired substrate versus choice for the control-paired substrate. An increase in choices for the treatment-paired substrate is interpreted as a positive bias, and a decrease as a negative bias (hypothetical data illustrated in Fig. 2). The value of the experience is kept constant, and all factors are counter-balanced so that any arising bias can be attributed to a relative shift in the perceived value of the memory of that experience. This task builds on clinical data which have shown that depression is strongly associated with both disrupted reward processing and affective biases associated with memory retrieval (Mathews and MacLeod 2005; Clark et al. 2009; Gotlib and Joormann 2010; Pringle et al. 2011; Roiser et al. 2012).

The ABT is limited to acute drug or affective state manipulations as it requires a within-subject study design, together with alternate presentations of each of the treatment–substrate–reward associations (Stuart et al. 2013). However, initial validation data revealed both antidepressant and pro-depressant drug treatments induce affective biases in this task that are consistent with similar treatments in healthy human volunteers performing emotional processing tasks (Stuart et al. 2013; Pringle et al. 2011). Antidepressant drugs from both the re-uptake inhibitor classes and receptor-blocking agents have been shown to induce positive biases, whilst drugs thought to have pro-depressants effects in man induce negative biases on this task (Stuart et al. 2013). Using a highly enriched social environment as a manipulation induced a positive bias towards experiences encountered during this enriched period (Stuart et al. 2013). Although still in the early stages of validation, the ABT appears to exhibit a high degree of translational validity in terms of pharmacological, physiological, and psychological manipulations of affective state (Stuart et al. 2013).

In a recent study, the rates of onsets of conventional antidepressants versus the NMDA antagonist ketamine were compared using the ABT. The results suggest that the rapid onset of action of ketamine may be related to its ability to modify previously learned biases, whereas the conventional antidepressant venlafaxine only modified new learning (Stuart et al. 2015). One potentially interesting idea arising from the observation that affective states appear to bias memories associated with rewarding experiences is the potential impact this may have on subsequent behaviour and motivation. The animals' choices are biased by their affective states *at the time the information was learned.* In the ABT, the animal makes a decision about which of the two experiences it encountered it prefers. In the context of a more naturalistic setting, these biases may influence motivation, i.e. negative biases leading to reduced motivation to re-engage in the associated behaviours.

11 Conclusions

Affective biases in depression provide an important opportunity for translational studies. The ability to quantify these biases using objective rather than subjective measures means a more direct comparison between human and animal studies can be made. In animals, judgement biases can be tested through the presentation of ambiguous information. These have been reported for a range of species and some degree of validation in the context of human depression has been demonstrated, for example through biases in processing mildly emotional faces. Negative judgement biases have been observed following chronic manipulations that are thought to induce negative affective states, but pharmacological studies using either acute or chronic administration have produced less clear results. The rodent ABT focuses on biases associated with reward learning and memory in rodents. This task does not have a direct human equivalent, but initial studies suggest a reasonable degree of face validity and good predictive and translational validity. The task is also sensitive to stress manipulations, suggesting a degree of construct validity. Taken together with findings of basic affective processing biases in depression that motivated the cognitive neuropsychological model, which appear to be directly modified by antidepressant treatment, these novel pre-clinical approaches raise great promise for the development of a truly translational novel paradigm for drug discovery.

Acknowledgments and Disclosures ESJR currently holds research funding from the MRC, BBSRC, Wellcome Trust, and Eli Lilly. Previous support which has contributed to the development of this work includes funding from RCUK and the British Pharmacological Society Integrative Pharmacology Fund. JPR is funded by the Wellcome Trust and previously received support from the Medical Research Council.

JPR is a consultant for Cambridge Cognition.

References

Abramson LY, Seligman ME, Teasdale JD (1978) Learned helplessness in humans: critique and reformulation. J Abnorm Psychol 87:49–74

Agid Y, Buzsaki G, Diamond DM, Frackowiak R, Giedd J, Girault JA, Grace A, Lambert JJ, Manji H, Mayberg H, Popoli M, Prochiantz A, Richter-Levin G, Somogyi P, Spedding M, Svenningsson P, Weinberger D (2007) How can drug discovery for psychiatric disorders be improved? Nat Rev Drug Discov 6:189–201

Amitai N, Young J, Higa K, Sharp R, Geyer M, Powell S (2013). Isolation rearing effects on probabilistic learning and cognitive flexibility in rats. Cogn Affect Behav Neurosci 1–19

Anderson MH, Munafò M, Robinson EJ (2013) Investigating the psychopharmacology of cognitive affective bias in rats using an affective tone discrimination task. Psychopharmacology 226:601–613

Austin MP, Mitchell P, Goodwin GM (2001) Cognitive deficits in depression: possible implications for functional neuropathology. Br J Psychiatry 178:200–206

Bari A, Theobald DE, Caprioli D, Mar AC, Aidoo-Micah A, Dalley JW et al (2010) Serotonin modulates sensitivity to reward and negative feedback in a probabilistic reversal learning task in rats. Neuropsychopharmacology 35:1290–1301

Bateson M, Desire S, Gartside SE, Wright GA (2011) Agitated honeybees exhibit pessimistic cognitive biases. Curr Biol 21:1070–1073

Beats BC, Sahakian BJ, Levy R (1996) Cognitive performance in tests sensitive to frontal lobe dysfunction in the elderly depressed. Psychol Med 26:591–603

Beck AT (1967) Depression: clinical, experimental and theoretical aspects. Harper & Row, New York

Berlin I, Givry-Steiner L, Lecrubier Y, Puech AJ (1998) Measures of anhedonia and hedonic responses to sucrose in depressive and schizophrenic patients in comparison with healthy subjects. Eur J Psychaitry 13:303–309

Berton O, Hahn CG, Thase ME (2012) Are we getting closer to valid translational models for major depression? Science 338:75–79

Bhagwagar Z, Cowen PJ, Goodwin GM, Harmer CJ (2004) Normalization of enhanced fear recognition by acute SSRI treatment in subjects with a previous history of depression. Am J Psychiatry 161:166–168

Booij L, Van der Does AJ, Haffmans PM, Riedel WJ, Fekkes D, Blom MJ (2005) The effects of high-dose and low-dose tryptophan depletion on mood and cognitive functions of remitted depressed patients. J Psychopharmacol 19:267–275

Burman OHP, Parker R, Paul ES, Mendl M (2008) A spatial judgement task to determine background emotional state in laboratory rats, Rattus norvegicus. Anim Behav 76:801–809

Bussey TJ, Holmes A, Lyon L, Mar AC, McAllister KA, Nithianantharajah J, Oomen CA, Saksida LM (2012) New translational assays for pre-clinical modelling of cognition in schizophrenia: the touchscreen testing method for mice and rats. Neuropharmacology 62 (3):1191–1203

Clark L, Chamberlain SR, Sahakian BJ (2009) Neurocognitive mechanisms in depression: implications for treatment. Annu Rev Neurosci 32:57–74

Chase HW, Frank MJ, Michael A, Bullmore ET, Sahakian BJ, Robbins TW (2010) Approach and avoidance learning in patients with major depression and healthy controls: relation to anhedonia. Psychol Med 40:433–440

Cools R, Blackwell A, Clark L, Menzies L, Cox S, Robbins TW (2005) Tryptophan depletion disrupts the motivational guidance of goal-directed behavior as a function of trait impulsivity. Neuropsychopharmacology 30:1362–1373

Cryan JF, Slattery DA (2007) Animal models of mood disorders: recent developments. Curr Opin Psychiatry 20:1–7

Cryan JF, Mombereau C, Vassout A (2005) The tail suspension test as a model for assessing antidepressant activity: review of pharmacological and genetic studies in mice. Neurosci Biobehav Rev 29:571–625

Delgado PL, Charney DS, Price LH, Aghajanian GK, Landis H, Heninger GR (1990) Serotonin function and the mechanism of antidepressant action. Reversal of antidepressant-induced remission by rapid depletion of plasma tryptophan. Arch Gen Psychiatry 47:411–418

Detke MJ, Rickels M, Lucki I (1995) Active behaviors in the rat forced swimming test differentially produced by serotonergic and noradrenergic antidepressants. Psychopharmacology 121:66–72

Dichter GS, Smoski MJ, Kampov-Polevoy AB, Gallop R, Garbutt JC (2012) Unipolar depression does not moderate responses to the Sweet Taste Test. Depress Anxiety 27:859–863

Dickstein DP, Finger EC, Brotman MA, Rich BA, Pine DS, Blair JR, Leibenluft E (2009) Impaired probabilistic reversal learning in youths with mood and anxiety disorders. Psychol Med 40(7):1089–1100

Dinan TG (1994) Glucocorticoids and the genesis of depressive illness. A psychobiological model. Br J Psychiatry 164:365–371

Elliott R, Sahakian BJ, Herrod JJ, Robbins TW, Paykel ES (1997) Abnormal response to negative feedback in unipolar depression: evidence for a diagnosis specific impairment. J Neurol Neurosurg Psychiatry 63:74–82

Elliott R, Sahakian BJ, McKay AP, Herrod JJ, Robbins TW, Paykel ES (1996) Neuropsychological impairments in unipolar depression: the influence of perceived failure on subsequent performance. Psychol Med 26:975–989

Elliott R, Zahn R, Deakin JF, Anderson IM (2011) Affective cognition and its disruption in mood disorders. Neuropsychopharmacology 36:153–182

Enkel T, Gholizadeh D, von Bohlen Und Halbach O, Sanchis-Segura C, Hurlemann R, Spanagel R et al. (2010). Ambiguous-cue interpretation is biased under stress- and depression-like states in rats. Neuropsychopharmacology 35:1008–1015

Erickson K, Drevets WC, Clark L, Cannon DM, Bain EE, Zarate CA Jr, Charney DS, Sahakian BJ (2005) Mood-congruent bias in affective go/no-go performance of unmedicated patients with major depressive disorder. Am J Psychiatry 162:2171–2173

Eshel N, Roiser JP (2010) Reward and punishment processing in depression. Biol Psychiatry 68 (2):118–124

Firk C, Markus CR (2008) Effects of acute tryptophan depletion on affective processing in first-degree relatives of depressive patients and controls after exposure to uncontrollable stress. Psychopharmacology 199:151–160

Fournier JC, DeRubeis RJ, Hollon SD, Dimidjian S, Amsterdam JD, Shelton RC, Fawcett J (2010) Antidepressant drug effects and depression severity: a patient-level meta-analysis. JAMA 303:47–53

Geyer M, Markou A (1995) Animal models of psychiatric disorders. In: Bloom F, Kupfer D (eds) Psychopharmacology: the fourth generation of progress. Raven Press, New York, pp 787–798

Gotlib IH, Joormann J (2010) Cognition and depression: current status and future directions. Annu Rev Clin Psychol 6:285–312

Gourley SL, Kiraly DD, Howell JL, Olausson P, Taylor JR (2008) Acute hippocampal brain-derived neurotrophic factor restores motivational and forced swim performance after corticosterone. Biol Psychiatry 64(10):884–890

Hales CA, Stuart SA, Anderson MH, Robinson ES (2014) Modelling cognitive affective biases in major depressive disorder using rodents. Br J Pharmacol 171(20):4524–4538

Harding EJ, Paul ES, Mendl M (2004) Animal behaviour: cognitive bias and affective state. Nature 427:312

Harmer CJ (2008) Serotonin and emotional processing: does it help explain antidepressant drug action? Neuropharmacology 55:1023–1028

Harmer CJ, Goodwin GM, Cowen PJ (2009a) Why do antidepressants take so long to work? A cognitive neuropsychological model of antidepressant drug action. Br J Psychiatry 195:102–108

Harmer CJ, Hill SA, Taylor MJ, Cowen PJ, Goodwin GM (2003) Toward a neuropsychological theory of antidepressant drug action: increase in positive emotional bias after potentiation of norepinephrine activity. Am J Psychiatry 160:990–992

Harmer CJ, O'Sullivan U, Favaron E, Massey-Chase R, Ayres R, Reinecke A, Goodwin GM, Cowen PJ (2009b) Effect of acute antidepressant administration on negative affective bias in depressed patients. Am J Psychiatry 166:1178–1184

Harmer CJ, Shelley NC, Cowen PJ, Goodwin GM (2004) Increased positive versus negative affective perception and memory in healthy volunteers following selective serotonin and norepinephrine reuptake inhibition. Am J Psychiatry 161:1256–1263

Hasler G, Luckenbaugh DA, Snow J, Meyers N, Waldeck T, Geraci M, Roiser J, Knutson B, Charney DS, Drevets WC (2009a) Reward processing after catecholamine depletion in unmedicated, remitted subjects with major depressive disorder. Biol Psychiatry 66:201–205

Hasler G, Mondillo K, Drevets WC, Blair JR (2009b) Impairments of probabilistic response reversal and passive avoidance following catecholamine depletion. Neuropsychopharmacology 34(13):2691–2698

Hayward G, Goodwin GM, Cowen PJ, Harmer CJ (2005) Low-dose tryptophan depletion in recovered depressed patients induces changes in cognitive processing without depressive symptoms. Biol Psychiatry 57:517–524

Henn FA, Vollmayr B (2005) Stress models of depression: forming genetically vulnerable strains. Neurosci Biobehav Rev 29(4–5):799–804

Henriques JB, Davidson RJ (2000) Decreased responsiveness to reward in depression. Cogn Emot 14:711–724

Henriques JB, Glowacki JM, Davidson RJ (1994) Reward fails to alter response bias in depression. J Abnorm Psychol 103:460–466

Hollon SD, DeRubeis RJ, Fawcett J, Amsterdam JD, Shelton RC, Zajecka J, Young PR, Gallop R (2014) Effect of cognitive therapy with antidepressant medications vs antidepressants alone on the rate of recovery in major depressive disorder: a randomized clinical trial. JAMA Psychiatry 71:1157–1164

Huys QJ, Daw ND, Dayan P (2015) Depression: a decision-theoretic analysis. Annu Rev Neurosci 38:1–23

Ineichen C, Sigrist H, Spinelli S, Lesch KP, Sautter E, Seifritz E, Pryce CR (2012) Establishing a probabilistic reversal learning test in mice: evidence for the processes mediating reward-stay and punishment-shift behaviour and for their modulation by serotonin. Neuropharmacology 63 (6):1012–1021

Joormann J, Gotlib IH (2006) Is this happiness I see? Biases in the identification of emotional facial expressions in depression and social phobia. J Abnorm Psychol 115:705–714

Joormann J, Gotlib IH (2008) Updating the contents of working memory in depression: interference from irrelevant negative material. J Abnorm Psychol 117:182–192

Kohler S, Unger T, Hoffmann S, Mackert A, Ross B, Fydrich T (2015) Dysfunctional cognitions of depressive inpatients and their relationship with treatment outcome. Compr Psychiatry 58:50–56

Leventopoulos M, Russig H, Feldon J, Pryce CR, Opacka-Juffry J (2009) Early deprivation leads to long-term reductions in motivation for reward and 5-HT1A binding and both effects are reversed by fluoxetine. Neuropharmacology 56(3):692–701

Maier SF (1984) Learned helplessness and animal models of depression. Prog Neuropsychopharmacol Biol Psychiatry 8(3):435–446

Mathews A, MacLeod C (2005) Cognitive vulnerability to emotional disorders. Annu Rev Clin Psychol 1:167–195

Matt GE, Vazquez C, Campbell WK (1992) Mood-congruent recall of affectively toned stimuli—a meta- analytic review. Clin Psychol Rev 12:227–255

McArthur R, Borsini F (2006) Animal models of depression in drug discovery: a historical perspective. Pharmacol Biochem Behav 84:436–452

McCabe C, Cowen PJ, Harmer CJ (2009) Neural representation of reward in recovered depressed patients. Psychopharmacology 205:667–677

McLean A, Rubinsztein JS, Robbins TW, Sahakian BJ (2004) The effects of tyrosine depletion in normal healthy volunteers: implications for unipolar depression. Psychopharmacology 171:286–297

McGuire PS, Seiden LS (1980) Differential effects of imipramine in rats as a function of DRL schedule value. Pharmacol Biochem Behav 13(5):691–694

Mendl M, Burman OHP, Paul ES (2010) An integrative and functional framework for the study of animal emotion and mood. Proc Roy Soc B Biol Sci 277:2895–2904

Munafo MR, Hayward G, Harmer C (2006) Selective processing of social threat cues following acute tryptophan depletion. J Psychopharmacol 20:33–39

Murphy FC, Michael A, Robbins TW, Sahakian BJ (2003) Neuropsychological impairment in patients with major depressive disorder: the effects of feedback on task performance. Psychol Med 33:455–467

Murphy FC, Rubinsztein JS, Michael A, Rogers RD, Robbins TW, Paykel ES, Sahakian BJ (2001) Decision-making cognition in mania and depression. Psychol Med 31:679–693

Murphy FC, Sahakian BJ, Rubinsztein JS, Michael A, Rogers RD, Robbins TW, Paykel ES (1999) Emotional bias and inhibitory control processes in mania and depression. Psychol Med 29:1307–1321

Murphy FC, Smith KA, Cowen PJ, Robbins TW, Sahakian BJ (2002) The effects of tryptophan depletion on cognitive and affective processing in healthy volunteers. Psychopharmacology 163:42–53

Murphy SE, Longhitano C, Ayres RE, Cowen PJ, Harmer CJ (2006) Tryptophan supplementation induces a positive bias in the processing of emotional material in healthy female volunteers. Psychopharmacology 187:121–130

Nestler EJ, Hyman SE (2010) Animal models of neuropsychiatric disorders. Nat Neurosci 13:1161–1169

O'Leary OF, Cryan JF (2013) Towards translational rodent models of depression. Cell Tissue Res 354:141–153

Olausson P, Kiraly DD, Gourley SL, Taylor JR (2013) Persistent effects of prior chronic exposure to corticosterone on reward-related learning and motivation in rodents. Psychopharmacology 225(3):569–577

Overmier JB, Seligman ME (1967) Effects of inescapable shock upon subsequent escape and avoidance responding. J Comp Physiol Psychol. 63:28–33

Papciak J, Popik P, Fuchs E, Rygula R (2013) Chronic psychosocial stress makes rats more 'pessimistic' in the ambiguous-cue interpretation paradigm. Behav Brain Res 256:305–310

Paul ES, Harding EJ, Mendl M (2005) Measuring emotional processes in animals: the utility of a cognitive approach. Neurosci Biobehav Rev 29:469–491

Pizzagalli DA, Goetz E, Ostacher M, Iosifescu DV, Perlis RH (2008a) Euthymic patients with bipolar disorder show decreased reward learning in a probabilistic reward task. Biol Psychiatry 64:162–168
Pizzagalli DA, Iosifescu D, Hallett LA, Ratner KG, Fava M (2008b) Reduced hedonic capacity in major depressive disorder: evidence from a probabilistic reward task. J Psychiatr Res 43:76–87
Pollak DD, Rey CE, Monje FJ (2010) Rodent models in depression research: classical strategies and new directions. Ann Med 42:252–264
Porsolt RD, Le Pichon M, Jalfre M (1977) Depression: a new animal model sensitive to antidepressant treatments. Nature 266(5604):730–732
Pringle A, Browning M, Cowen PJ, Harmer CJ (2011) A cognitive neuropsychological model of antidepressant drug action. Prog Neuropsychopharmacol Biol Psychiatry 35:1586–1592
Pryce CR, Azzinnari D, Spinelli S, Seifritz E, Tegethoff M, Meinlschmidt G (2011) Helplessness: a systematic translational review of theory and evidence for its relevance to understanding and treating depression. Pharmacol Ther 132(3):242–267
Rawal A, Collishaw S, Thapar A, Rice F (2013) 'The risks of playing it safe': a prospective longitudinal study of response to reward in the adolescent offspring of depressed parents. Psychol Med 43:27–38
Richter SH, Schick A, Hoyer C, Lankisch K, Gass P, Vollmayr B (2012) A glass full of optimism: enrichment effects on cognitive bias in a rat model of depression. Cogn Affect Behav Neurosci 12:527–542
Riedel M, Moller HJ, Obermeier M, Adli M, Bauer M, Kronmuller K, Brieger P, Laux G, Bender W, Heuser I, Zeiler J, Gaebel W, Schennach-Wolff R, Henkel V, Seemuller F (2011) Clinical predictors of response and remission in inpatients with depressive syndromes. J Affect Disord 133:137–149
Robinson OJ, Cools R, Sahakian BJ (2011) Tryptophan depletion disinhibits punishment but not reward prediction: implications for resilience. Psychopharmacology (Berl) 2:599–605
Rock PL, Roiser JP, Riedel WJ, Blackwell AD (2013) Cognitive impairment in depression: a systematic review and meta-analysis. Psychol Med 1–12
Rogers RD, Tunbridge EM, Bhagwagar Z, Drevets WC, Sahakian BJ, Carter CS (2003) Tryptophan depletion alters the decision-making of healthy volunteers through altered processing of reward cues. Neuropsychopharmacology 28:153–162
Roiser JP, Blackwell AD, Cools R, Clark L, Rubinsztein DC, Robbins TW, Sahakian BJ (2006) Serotonin transporter polymorphism mediates vulnerability to loss of incentive motivation following acute tryptophan depletion. Neuropsychopharmacology 31:2264–2272
Roiser JP, Cannon DM, Gandhi SK, Tavares JT, Erickson K, Wood S, Klaver JM, Clark L, Zarate CA Jr, Sahakian BJ, Drevets WC (2009) Hot and cold cognition in unmedicated depressed subjects with bipolar disorder. Bipolar Disord 11:178–189
Roiser JP, Elliott R, Sahakian BJ (2012) Cognitive mechanisms of treatment in depression. Neuropsychopharmacology 37:117–136
Roiser JP, Levy J, Fromm SJ, Wang H, Hasler G, Sahakian BJ, Drevets WC (2008) The effect of acute tryptophan depletion on the neural correlates of emotional processing in healthy volunteers. Neuropsychopharmacology 33:1992–2006
Roiser JP, McLean A, Ogilvie AD, Blackwell AD, Bamber DJ, Goodyer I, Jones PB, Sahakian BJ (2005) The subjective and cognitive effects of acute phenylalanine and tyrosine depletion in patients recovered from depression. Neuropsychopharmacology 30:775–785
Roiser JP, Sahakian BJ (2013) Hot and cold cognition in depression. CNS Spectr 18:139–149
Rüedi-Bettschen D, Pedersen EM, Feldon J, Pryce CR (2005) Early deprivation under specific conditions leads to reduced interest in reward in adulthood in Wistar rats. Behav Brain Res 156 (2):297–310
Ruhe HG, Mason NS, Schene AH (2007) Mood is indirectly related to serotonin, norepinephrine and dopamine levels in humans: a meta-analysis of monoamine depletion studies. Molecular Psychiatry 12:331–359
Rygula R, Papciak J, Popik P (2013) Trait pessimism predicts vulnerability to stress-induced anhedonia in rats. Neuropsychopharmacology 38:2188–2196

Rygula R, Papciak J, Popik P (2014) The effects of acute pharmacological stimulation of the 5-HT, NA and DA systems on the cognitive judgement bias of rats in the ambiguous-cue interpretation paradigm. Eur Neuropsychopharmacol 24(7):1103–1111
Savitz J, Lucki I, Drevets WC (2009) 5-HT(1A) receptor function in major depressive disorder. Prog Neurobiol 88:17–31
Seiden LS, Dahms JL, Shaughnessy RA (1985) Behavioral screen for antidepressants: the effects of drugs and electroconvulsive shock on performance under a differential-reinforcement-of-low-rate schedule. Psychopharmacology 86(1–2):55–60
Seligman ME, Beagley G (1975) Learned helplessness in the rat. J Comp Physiol Psychol 88 (2):534–541
Smith KA, Fairburn CG, Cowen PJ (1997) Relapse of depression after rapid depletion of tryptophan. Lancet 349:915–919
Snyder HR (2012) Major depressive disorder is associated with broad impairments on neuropsychological measures of executive function: a meta-analysis and review. Psychol Bull 139:81
Steru L, Chermat R, Thierry B, Simo P (1985) The tail suspension test: a new method for screening antidepressants in mice. Psychopharmacology 85:367–370
Stuart SA, Butler P, Munafò MR, Nutt DJ, Robinson ESJ (2013) A translational rodent assay of affective biases in depression and antidepressant therapy. Neuropsychopharmacology 38:1625–1635
Stuart SA, Butler P, Munafò MR, Nutt DJ, Robinson ESJ (2015) Distinct neuropsychological mechanisms may explain delayed-versus rapid-onset antidepressant efficacy. Neuropsychopharmacology. doi:10.1038/npp.2015.59
Taylor Tavares JV, Clark L, Furey ML, Williams GB, Sahakian BJ, Drevets WC (2008) Neural basis of abnormal response to negative feedback in unmedicated mood disorders. Neuroimage 42:1118–1126
Teachman BA, Joormann J, Steinman SA, Gotlib IH (2012) Automaticity in anxiety disorders and major depressive disorder. Clin Psychol Rev 32:575–603
Tranter R, Bell D, Gutting P, Harmer C, Healy D, Anderson IM (2009) The effect of serotonergic and noradrenergic antidepressants on face emotion processing in depressed patients. J Affect Disord 118:87–93
Treadway MT, Bossaller NA, Shelton RC, Zald DH (2012) Effort-based decision-making in major depressive disorder: a translational model of motivational anhedonia. J Abnorm Psychol 121:553–558
Treadway MT, Zald DH (2011) Reconsidering anhedonia in depression: lessons from translational neuroscience. Neurosci Biobehav Rev 35:537–555
Trivedi MH, Rush AJ, Wisniewski SR, Nierenberg AA, Warden D, Ritz L, Norquist G, Howland RH, Lebowitz B, McGrath PJ, Shores-Wilson K, Biggs MM, Balasubramani GK, Fava M (2006) Evaluation of outcomes with citalopram for depression using measurement-based care in STAR*D: implications for clinical practice. Am J Psychiatry 163:28–40
Vogel GW, Minter K, Woolwine B (1986) Effects of chronically administered antidepressant drugs on animal behavior. Physiol Behav 36:659–666
Vollmayr B, Bachteler D, Vengeliene V, Gass P, Spanagel R, Henn F (2004) Rats with congenital learned helplessness respond less to sucrose but show no deficits in activity or learning. Behav Brain Res 150(1–2):217–221
Vrieze E, Pizzagalli DA, Demyttenaere K, Hompes T, Sienaert P, de Boer P, Schmidt M, Claes S (2013) Reduced reward learning predicts outcome in major depressive disorder. Biol Psychiatry 73:639–645
Willner P, Towell A, Sampson D, Sophokleous S, Muscat R (1987) Reduction of sucrose preference by chronic unpredictable mild stress, and its restoration by a tricyclic antidepressant. Psychopharmacology 93:358–364
Wittchen HU, Jacobi F, Rehm J, Gustavsson A, Svensson M, Jonsson B, Olesen J, Allgulander C, Alonso J, Faravelli C, Fratiglioni L, Jennum P, Lieb R, Maercker A, van Os J, Preisig M,

Salvador-Carulla L, Simon R, Steinhausen HC (2011) The size and burden of mental disorders and other disorders of the brain in Europe 2010. Eur Neuropsychopharmacol 21:655–679
Wright WF, Bower GH (1992) Mood effects on subjective probability assessment. Organ Behav Hum Decis Process 52:276–291
Zacharko RM, Anisman H (1991) Stressor-induced anhedonia in the mesocorticolimbic system. Neurosci Biobehav Rev 15:391–405

Locomotor Profiling from Rodents to the Clinic and Back Again

Jared W. Young, Arpi Minassian and Mark A. Geyer

Abstract The quantification of unconditioned motoric activity is one of the oldest and most commonly utilized tools in behavioral studies. Although typically measured in reference to psychiatric disorders, e.g., amphetamine-induced hyperactivity used as a model of schizophrenia, bipolar disorder (BD), and Tourette's syndrome, the motoric behavior of psychiatric patients had not been quantified similarly to rodents until recently. The rodent behavioral pattern monitor (BPM) was reverse-translated for use in humans, providing the quantification of not only motoric activity but also the locomotor exploratory profile of various psychiatric populations. This measurement includes the quantification of specific exploration and locomotor patterns. As an example, patients with BD, schizophrenia, and those with history of methamphetamine dependence exhibited unique locomotor profiles. It was subsequently determined that reducing dopamine transporter function selectively recreated the locomotor profile of BD mania patients and not any other patient population. Hence, multivariate locomotor profiling offers a first-step approach toward understanding the neural mechanism(s) underlying abnormal behavior in patients with psychiatric disorders. Advances in wearable technology will undoubtedly enable similar multivariate assessments of exploratory and locomotor behavior in "real-world" contexts. Furthermore, trans-diagnostic studies of locomotor activity profiles will inform about essential brain-based functions that cut across diagnostic nosologies.

Keywords Bipolar disorder · Schizophrenia · Dopamine transporter · Dopamine · Translational

J.W. Young (✉) · A. Minassian · M.A. Geyer
Department of Psychiatry, University of California San Diego,
9500 Gilman Drive MC 0804, La Jolla, CA 92093–0804, USA
e-mail: jaredyoung@ucsd.edu

J.W. Young · M.A. Geyer
Research Service, VA San Diego Healthcare System, San Diego, CA, USA

Curr Topics Behav Neurosci (2016) 28: 287–303
DOI 10.1007/7854_2015_5015

Contents

1 Introduction 288
2 Current Findings Investigating Locomotor Profiles in Humans 290
3 Other Rodent Studies with Etiological Relevance to Human Studies 294
4 Novel Additional Approaches Using Wearable Technology 295
5 Summary of Approach and Future Directions 296
References 298

1 Introduction

Measuring unconditioned motor activity remains one of the oldest and most commonly utilized tools in rodent studies. It is in fact one of the simplest forms of behavior measured across all species. Despite numerous studies across species focusing on activity alone, motor activity does not constitute a unitary class of behavior. Motor activity is the sum of multiple aspects of behavior requiring a multivariate approach to its quantification (Paulus and Geyer 1996a). Since each aspect of behavior could be subserved by various neural circuits, differentially affecting single systems, knock-on effects on other measures could occur, resulting in vastly different patterns of behavior. Applying quantitative analyses of these behaviors enables the characterization of behavioral patterns from distinct manipulations. These patterns of behavior—hereafter named locomotor profiles—can therefore provide greater information on experimental effects compared to examining single measures alone.

The need for multivariate assessment tools to adequately characterize locomotor profiles has been recognized for some time (Eilam and Shefer 1997; Gould et al. 2009; Lat 1965). Some of these approaches were based on observer ratings, while others automated the entire measurement process (File and Wardill 1975; Geyer 1990; Geyer and Paulus 1996; Paulus and Geyer 1996b). The approach we have taken has followed the development of the behavioral pattern monitor (BPM; Fig. 1), a computerized photobeam activity chamber system designed to provide multivariate measures of exploratory motor activity. Beyond activity alone, the BPM is also equipped with sensors to measure rearing behavior in addition to eleven holes around the chamber to measure specific investigation. Along with these measures, the X–Y pattern of activity can be used to generate measures of path patterns using nonlinear dynamical system methods and fractal geometry to assess important aspects of the hierarchical and sequential organization of behavior at high levels of temporal and spatial resolution [detailed below; (Geyer et al. 1986)]. Thus, the BPM equipment produces >50 measures (Table 1) of exploration used to investigate the neural circuits underlying various locomotor profiles.

When investigating locomotor profiles derived from normal male rat behavior, a factor analysis revealed three primary factors [Table 1; Paulus and Geyer (1993)].

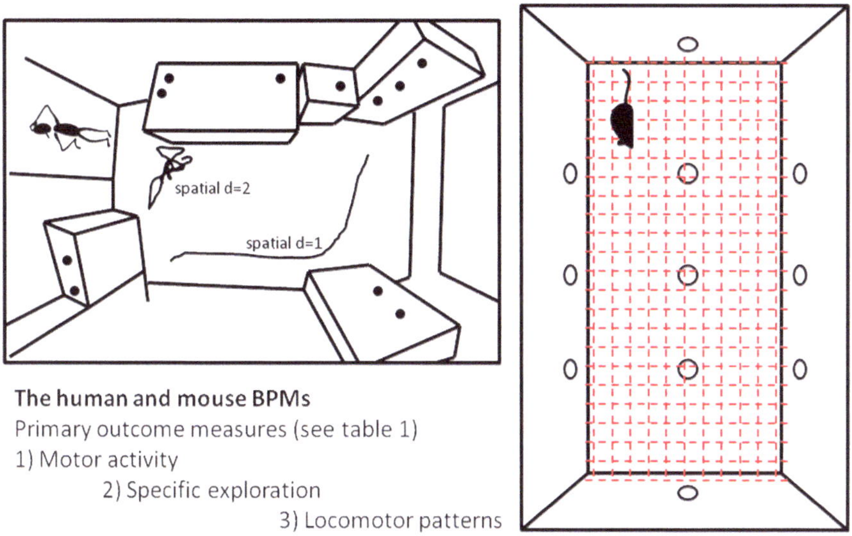

Fig. 1 The human and rodent behavioral pattern monitors (BPM). A schematic of the human BPM is above. The human BPM (*left*) was created by clearing a room, providing limited furnishing (no chair) and placing 11 objects (indicated by *filled circle*) throughout the room. Subjects are brought into the room and asked to remain there while equipment is set up elsewhere. The subject was left for 15 min during which time their locomotor and exploratory activity were monitored using a recessed video camera. The rodent BPM (*right*) was created using a 30.5 by 61 cm chamber, with 11 recessed holes to monitor specific exploration, with general activity monitored using *infrared beams*. Spatial *d* provides one quantitative measure of locomotor patterns represented above, with values closer to 2 describing meandering-like behavior, while values closer to one represent straight-line movement through space. Primary outcome measures include activity, specific exploration, and locomotor patterns (see Table 1)

These three factors included the amount of activity (e.g., distance traveled, transitions, and counts); specific exploration (e.g., rearing and holepoking); and locomotor path patterns [spatial scaling exponent *d*, dynamical entropy *h*, spatial CV, temporal CV, (Drai and Golani 2001; Eilam and Golani 1988; Geyer and Paulus 1996; Kafkafi et al. 2003)]. Together, these three factors accounted for 77 % of the variance in the measures. Importantly, when a subsequent factor analysis was conducted in 268 male C57BL/6 J mice, similar factors were reported, albeit that rearing and holepoking were identified as separate factors (Tanaka et al. 2012). Hence, activity is not univariate but reflects multiple behavioral processes that are expressed in interacting combinations (Berlyne 1966). The relevance of these multivariate locomotor profiles to human psychiatric disorders is detailed below.

Table 1 Measures and their associated factors in the rodent behavioral pattern monitor

Factor	Measure	Description
Motor activity measures	Counts	Cumulative number of distinct behaviors during testing
	Transitions	Number of movements from one pre-defined area to another
	Center entries	Number of entries into the area defined as the center
	Distance traveled	Total distance traveled while exploring the chamber measured in cm
Specific exploration	Hole pokes	Total number of investigatory holepokes across all 11 holes
	Repeated poking	Number of times a hole was repeatedly poked before poking into another hole
	Varied poking	Number of bouts of holepoking behavior
	Rear	Number of times the mouse reared against the side walls or into the air
	Center duration	Total time (s) spent in the area designated as the center of the chamber
Locomotor patterns	Spatial *d*	Spatial scaling exponent d measures the hierarchical and geometric organization of behavior
	Entropy *h*	Quantifies the sequential aspects of behavior, specifically measuring the degree to which behavior is observed along a continuum between complete order and disorder
	Spatial CV	Measure of the consistency by which the animal moves from one pre-defined region to another
	Temporal CV	Measure of the consistency of the dwell time in each of the pre-defined regions

2 Current Findings Investigating Locomotor Profiles in Humans

The multivariate quantitative measurement of exploratory behavior has provided abundant information on the underlying neural mechanisms of rodents utilizing pharmacological, genetic, and other means. While this information is useful, especially as it pertains to controls for complex behavioral studies, its relation to understanding mechanisms related to psychiatric disorders had been limited. For example, despite studies measuring hyperactivity of rodents in an experimental setting to model mania (Henry et al. 2010; Young and Geyer 2011; Young et al. 2011a) due to the common belief that patients with bipolar disorder (BD) mania are hyperactive, few studies prior to 2009 had quantified their activity to confirm this belief. Some initial activity studies have been conducted in humans using a wrist or leg actigraph to primarily measure circadian and activity rhythms of stroke, dementia, and brain tumor sufferers (Teicher 1995a; Teicher et al. 1986; Wolff et al. 1985). These portable actigraphs also indicate that children with attention deficit hyperactivity disorder (ADHD) exhibit higher levels of everyday activity than healthy children (Porrino et al. 1983). This work has been extended to measure activity levels, while ADHD children perform cognitive tasks such as the continuous performance test [CPT; Halperin et al. (1993), Teicher et al. (2004)]. Results are consistent with expectations (increased localized and general activity in patients with ADHD and BD); however, these data are as limited by their univariate assessment as are studies in rodents in which only the amount of locomotor activity is measured. A multivariate assessment of exploratory behavior in autistic children was conducted in 2001 by rating videotapes of subjects left in a room with colorful and interactive objects for eight minutes. Decreased activity and object interactions were observed in children with autism compared to healthy controls (Pierce and Courchesne 2001b). These ratings of decreased active exploration correlated with MRI-based measures of altered brain volumes in the children. Therefore, developing multivariate assessments of exploratory behavior in clinical populations could provide a window toward detecting behavioral deficits associated with dysfunctional neural mechanisms.

To determine whether quantitative multivariate assessment of locomotor profiles in psychiatric populations could provide insight into their altered neural mechanisms, we developed a human version of the BPM. This human BPM was designed to be analogous to the rodent BPMs described above wherein activity, specific exploration, and path patterns could be quantified in humans (Young et al. 2007). The human BPM is essentially a novel and unfamiliar room that the human participant had not previously been exposed to, consistent with the rodent BPM. As with holepokes in the rodent BPM and containers in the autism study (Pierce and Courchesne 2001a), 11 objects deemed safe, tactile, manipulable, and colorful were dispersed throughout the room on items of furniture. The subject is asked to remain in the room until the experimenter returns, with sessions being 15 min long to date. X–Y coordinate locations for the subject in the room were identified using a ceiling-mounted camera and automated tracking software. Using X–Y coordinates, the activity, specific exploration, and path patterns of human subjects have been quantified consistent with that of the rodent studies described above.

Development and implementation of the human BPM has enabled us to quantify and compare the locomotor profiles of a range of conditions with putative dopamine (and catecholamine) dysregulation. These psychiatric populations include acutely ill patients with BD mania and schizophrenia (Minassian et al. 2009; Perry et al. 2009), stabilized patients with BD (Henry et al. 2013a), individuals with a history of methamphetamine dependence (Henry et al. 2011), and healthy subjects administered a one-time dose of D-amphetamine. These groups demonstrate strikingly different locomotor profiles, again highlighting the importance of multivariate assessment of activity. From a diagnostic perspective, these signatures of exploration provide an obvious and meaningful difference between acutely ill populations that are difficult to distinguish because they present with psychotic and mood symptoms (Pini et al. 2004). Furthermore, the uniqueness of the exploratory signatures characteristic of BD, schizophrenia, and substance use conditions may shed light on distinct versus overlapping biological mechanisms.

Consistent with the differentiation of stimulants in rodents described above, patients with schizophrenia, BD, and methamphetamine dependence exhibit unique locomotor profiles. For example, patients in the manic phase of BD exhibited hyperactivity in the first 5 min in the hBPM, with rapid habituation, while patients with schizophrenia exhibited hyperactivity only in the last 5 min (Perry et al. 2009). In contrast, methamphetamine-dependent subjects did not exhibit hyperactivity at any timepoint (Henry et al. 2011). Interestingly, both patients with BD and schizophrenia exhibited increased entropy h and reduced spatial d, the latter being more prominent in patients with BD (Perry et al. 2009). These data indicate that both schizophrenia and BD patients exhibited more disorganized but straighter line movement through space when compared to controls (Perry et al. 2009), while methamphetamine-dependent subjects did not differ from control subjects in terms of spatial d (Henry et al. 2011). Methamphetamine-dependent subjects did exhibit significant differences from controls however, in terms of increased specific exploration (object interactions), consistent with euthymic and manic BD patients (Henry et al. 2010), while patients with schizophrenia did not differ from healthy comparison subjects (Perry et al. 2009, 2010) in terms of specific exploration. These specific locomotor profiles provide the opportunity to dissect the abnormal neural circuits that underlie these psychiatric population differences.

A cross-species paradigm such as the BPM enables one to assess the validity of animal model of psychiatric diseases. For example, the BPM locomotor profile specific to BD mania (hyperactivity, increased specific exploration, straighter movements) was also observed in our two mouse models of mania, the DAT knockdown (KD) mouse, and in mice given the DAT inhibitor GBR 12909 (Queiroz et al. 2015; Young et al. 2010a, b). Importantly, these effects are consistent irrespective of background strain (e.g., C57BL/6 or 129/s mice; (van Enkhuizen et al. 2012, 2014a; Young et al. 2010b)]. Thus, the human BD mania profile can be replicated in both a constitutive (DAT KD) and an acute pharmacological mouse model that share reduced function of the DAT. Patients in the manic phase of BD continue to show heightened exploration even as their manic symptoms begin to stabilize (Minassian et al. 2011), and individuals in a non-manic phase of BD also show this characteristic phenotype of heightened activity, increased specific exploration, and straighter movements, though in attenuated form (Henry et al. 2013a). BD patients taking mood stabilizers evidence less motor activity than those not on mood stabilizers, but specific exploration was relatively unaffected by mood stabilizer use (Henry et al. 2013a). Again, a similar phenotype was observed in DAT KD mice treated with valproate chow (15 % over 28 days); chow-treated mice had decreased transitions in the BPM but not significantly reduced holepokes (van Enkhuizen et al. 2012). Chronic (10 day) treatment with lithium in water (1 g/L) also attenuated the BD mania locomotor profile in terms of activity and specific exploration, although not spatial *d* (van Enkhuizen et al. 2015). Both treatments resulted in serum levels at therapeutic levels used to treat BD mania. Hence, this series of studies again underscored the relevance of the mechanism of the DAT in BD.

Cross-species studies can also quantify and differentiate among substance-induced changes in locomotor profiles. For example, individuals with a history of chronic methamphetamine dependence exhibited increased specific exploration as described above (Henry et al. 2011). Importantly, mice treated with a chronic methamphetamine regimen designed to mimic "binge" patterns in humans also exhibited elevated specific exploration (Henry et al. 2013a). Interestingly, locomotor profiles in healthy volunteers given a one-time dose of 20 mg D-amphetamine are strikingly different from that of methamphetamine-dependent subjects, suggesting unique biological mechanisms underlying chronic versus acute amphetamine exposure. A one-time dose of amphetamine in healthy volunteers increased locomotor activity without increasing specific exploration, which was also what we observed in mice administered a one-time dose of amphetamine (Minassian et al. 2016; Perry et al. 2009). Amphetamine-induced hyperactivity has been used extensively to model mania in mice, but these studies support earlier concerns that it does not recreate locomotor profiles of BD mania (Silverstone et al. 1998; Young et al. 2011a). The cross-species work summarized here further emphasizes that the paucity of amphetamine recreating mania-relevant behaviors likely results from its non-specificity to the DAT [relatively greater affinity for the norepinephrine transporter (NET) in humans and mice; (Henry et al. 2010)]. By detailing locomotor profiles across species, we have therefore provided evidence for greater selectivity of reduced DAT function driving locomotor profiles in BD mania as opposed to greater reduced functioning of NET.

Earlier, it was described that children with autism exhibited lower active exploration in a room which correlated with cerebellar volume levels (Pierce and Courchesne 2001a, b). Likewise, links between exploration in the human BPM have been investigated in relation to other behaviors. For example, the increased object interaction of methamphetamine-dependent subjects correlated with the number of perseverative errors they made on the Wisconsin Card Sorting Task, a frontal lobe-mediated cognitive task (Henry et al. 2011). Consistent with these observations, people with lesioned frontal lobes exhibit inappropriate "grasping" behavior of objects that are within reach (Lhermitte 1983). There may therefore be a link between this simplistic measure of exploratory behavior and frontal functioning. Such links could prove to be an extremely useful screen given that frontal dysfunctions are often linked to psychiatric disorders (Arnone et al. 2009; Goto et al. 2010; Koenigs and Grafman 2009). This link has also been observed in mice, whereby mice with higher specific exploration also exhibited a greater risk preference in a mouse version of the human Iowa Gambling Task (van Enkhuizen et al. 2014b; Young et al. 2011c), another behavior requiring intact frontal functioning (Bechara et al. 1994). Therefore, there is growing evidence for the utility of quantifying profiles of locomotor profiles and exploratory behavior and their relevance to other domains of functioning in psychiatric disorders.

3 Other Rodent Studies with Etiological Relevance to Human Studies

The locomotor profiles of other animals with putative etiological validity to various disease states have been tested. For example, another model of BD mania, created with a mutation in the *Clock* gene, exhibited similarities with the behavioral phenotype of patients in the BPM, but failed to completely recreate the locomotor profile (van Enkhuizen et al. 2013). In contrast to patients with BD, these *Clock*Δ19 mice exhibited more circumscribed, small-scale movements (low spatial *d*) instead of direct, more linear paths (van Enkhuizen et al. 2013). Hence, despite circadian rhythm abnormalities being linked to BD mania onset (although not in this particular gene) and with these mice exhibiting altered sleep patterns [see McClung (2013)], *Clock*Δ19 mice did not recreate the locomotor profile observed in BD mania patients (Minassian et al. 2011; Perry et al. 2009). Finally, although primarily prescribed as an anti-narcoleptic and investigated as a treatment for cognitive deficits in schizophrenia, the effects of modafinil have also been tested in the BPM. Modafinil exerts inhibitory effects on the DAT, more so than the NET. Importantly, modafinil recreated the locomotor profile of BD mania, an effect that was mediated in part by dopamine D_1 receptors, while the specific exploratory effects were also mediated by dopamine D_4 receptors (Young et al. 2011b). Hence, these data further support the premise that recreating the BD mania locomotor profile requires selective inhibition of the DAT.

In terms of schizophrenia, hallucinogens have often been used as acute manipulations to model this disease (Geyer and Vollenweider 2008). Hallucinogens such as psilocin (the active metabolite of psilocybin) also induced more circumscribed (lower spatial d) behavior in mice, an effect mediated by the 5-HT2C and 5-HT1A receptors, respectively (Halberstadt et al. 2011; Krebs-Thomson et al. 1998). This hallucinogen treatment reduced activity and exploration in mice, resulting in a locomotor profile distinct from that of schizophrenia and BD patients. Another mouse model of schizophrenia includes Chakragati mice (Dawe and Ratty 2007; Young et al. 2014). When tested in the BPM however, these mice did exhibit hyperactivity, but it was characterized by more circumscribed behavior [lower spatial *d*; Young et al. (2014)], likely as a result of imbalanced hemispheric dopamine release in these mice (Ratty et al. 1990). Interestingly, acute phencyclidine treatment has also been used to model schizophrenia (Neill et al. 2010), and it also results in hyperactivity yet more circumscribed (lower spatial *d*) behavior (Lehmann-Masten and Geyer 1991). Importantly, rearing rats in isolation have been used to model aspects of schizophrenia (Geyer et al. 1993), and indeed, these rats exhibited a reduced locomotor habituation and elevated specific exploration compared to those reared socially (Amitai et al. 2014). Naturally, further studies are required to delineate the neural mechanisms underlying the locomotor profiles of patients with schizophrenia.

4 Novel Additional Approaches Using Wearable Technology

As recently as a decade ago, the technologies available to quantify locomotor activity in humans were relatively limited. Much of the work in this area was conducted with wrist or leg actigraphy (Klein et al. 1992; Sims et al. 1999; Teicher 1995b), with some innovative applications [e.g., the McLean Motion Analysis Test (Faedda and Teicher 2005)]. Subsequently, ambulatory monitoring advanced to the point of being able to better capture whole-body movement via the use of a centrally mounted accelerometer and additional sensors that could capture important indices such as respiration and heart rate variability (Grossman 2004; Keenan and Wilhelm 2005; Wilhelm et al. 2003), but these devices were typically large and sometimes bulky garments. Such devices provided useful information but precisely quantifying locomotor activity was challenging, particularly with psychiatrically ill populations. Actigraphs have normally been used over multiple days (from 7 to 28) as opposed to single experimental sessions as typical in rodent work. Such assessment enabled the measurement of circadian rhythms and sleep patterns with a meta-analysis confirming sleep disturbances in BD patients (Geoffroy et al. 2015). Sleep disturbances have predicted relapse in BD, albeit with small sample sizes (Novak et al. 2014). People designated as at high risk of hypomania exhibited sleep disturbance and altered rhythmicity (Ankers and Jones 2009). Hence, these studies support the premise that altered sleep/rhythm may be an underlying manifestation of BD, not just a result of acute disease states. Despite the importance of these circadian rhythm findings, in terms of general activity little has been observed with actigraphs, with self-rating scales not strongly correlating with actigraphy-derived measures of motor activity (Walther et al. 2009). The limitations of self-report and observer-rated symptom rating scales support the necessity of quantifying activity and locomotor profiles using more objective and finite measures.

In recent years, the development of wearable devices has markedly propelled our ability to monitor physiological, behavioral, and even psychological activity in real time and with high sensitivity. These advancements have revolutionized medicine in a number of dimensions (Steinhubl et al. 2015). First, they fulfilled the original promise of older ambulatory monitoring devices to acquire assessments of multiple physiological and behavioral domains as people go about their daily lives and routines (Grossman 2004). This capability enables physicians to assess and even remotely monitor their patients' vital functions and symptomatology (see Turakhia et al. (2015) for a salient example). Ambulatory monitoring of locomotor activity has tremendous implications for the assessment and treatment of conditions such as Parkinson's disease (Cavanaugh et al. 2015), cerebral palsy (Claridge et al. 2014), and medical conditions where changes in duration or intensity of activity can be symptoms of worsening and/or improvement such as pulmonary disease (McNamara et al. 2014), obesity (DeLany et al. 2014), and myriad others.

The commercial availability of wearable sensors [e.g., FitBit®, Diaz et al. (2015)] and their relative ease of use by consumers have advanced the popularity of mobile health (mHealth) interventions targeting lifestyle changes such as increasing exercise and losing weight. mHealth applications have advanced hand in hand with smartphone capabilities, and more than one in four people in the USA currently use smartphones to track their physical activity (Topol 2015). This consumer trend has not excluded people with psychiatric disease; a recent study provided physical activity monitoring devices to patients with schizophrenia, BD, and depression, who reported high satisfaction with the utility and helpfulness of these devices for setting and achieving weight loss goals (Naslund et al. 2015). Combining their use with patient monitoring could be an important step forward in locomotor profile research relevant to psychiatric patients.

These wearable monitoring devices have facilitated research applications and the real-time study of locomotor activity and other relevant behavior domains in ways that a decade ago may have been difficult to envision. Rapid advances in engineering have led to miniaturization of devices that previously would have been unwieldy and invasive. A dramatic example of such technology includes epidermal electronics, e.g., very thin sensors that adhere to the skin and measure multiple dimensions of brain, heart, and skeletal activity (Kim et al. 2011). As these devices become more ubiquitous and presumably more affordable, the value of assessing locomotor activity to better understand the biology of neurocognitive disorders, as well as testing new treatment approaches, will grow by leaps and bounds.

5 Summary of Approach and Future Directions

The reverse translation of the rodent BPM to a human BPM has enabled multivariate locomotor profiles to be established for several psychiatric disorders including BD mania/euthymia, schizophrenia, and people with methamphetamine abuse (Table 2). Future studies with patient populations such as ADHD, autism, and juvenile-onset BD may reveal other unique locomotor profiles. With the generation of these unique profiles, work has begun mimicking them in rodents and determining whether the manipulations used to recreate them have predictive relevance to the disease state. The most studied example of this work is the reduced DAT function model recreating the BD mania profile, whereby BD mania treatments attenuate some, but not all of the profile. These data therefore indicate that reduced DAT functioning—or generally catecholaminergic clearance—may underlie the behavioral manifestations of BD mania (Table 2). Further support for this premise comes from evidence of more long-term rodent testing indicating that reduced DAT functioning also recreates other mania-relevant behaviors such as elevated motivation to gain rewards (Young and Geyer 2010) and increased risk preference in the Iowa Gambling Task (van Enkhuizen et al. 2014b). Locomotor profiling therefore offers a first-step approach toward understanding the neural mechanism(s) underlying abnormal behavior in patients with psychiatric disorders.

Table 2 Locomotor activity patterns in the behavioral pattern monitor in human conditions and corresponding mouse models

Condition	Human sample	Human results	Model mouse	Animal results
Bipolar disorder (BD)-manic	Hospitalized manic BD (Minassian et al. 2009, 2011; Perry et al. 2009)	↑motor activity, ↑object interactions, ↓spatial *d*	DAT KD (Perry et al. 2009) GBR 12909 16 mg/kg (Young et al. 2010a, b)	↑motor activity, ↑holepokes, ↓spatial *d*
BD-not manic	Outpatient BD in non-manic phase, on versus off mood stabilizers (Henry et al. 2013b)	Mood stabilizers ↓motor activity, =object interactions compared with no mood stabilizers	DAT KD treated with 15 g/kg valproate chow versus vehicle (van Enkhuizen et al. 2012)	Valproate ↓motor activity, =holepokes
Schizophrenia	Hospitalized SCZ (Perry et al. 2009)	=no activity habituation, =object interactions, ↑entropy *h*, ↓spatial *d*	None replicated to date	None replicated to date
Methamphetamine dependence	History of methamphetamine dependence, in remission (Henry et al. 2011)	=motor activity, ↑object interactions, =spatial *d*	14-day methamphetamine binge regimen (Henry et al. 2013a)	=motor activity, ↑holepokes, =spatial *d*
Acute amphetamine	Healthy subjects, one-time dose 20 mg D-amphetamine (Minassian et al. in preparation)	↑motor activity, =object interactions, =spatial *d*	1.4, 2.5, 4.5, or 7.9 mg/kg D-amphetamine (Minassian et al. in preparation)	↑motor activity, ↓holepokes, ↓spatial *d*

BD bipolar disorder, *DAT* dopamine transporter, *KD* knockdown, *SCZ* schizophrenia. Unless otherwise specified, *arrows* and = *symbols* represent significant increase, decrease, or equivalence relative to healthy comparison subjects or wildtype/vehicle-treated mice

Despite the use of locomotor profiles in specific populations, this approach could also be used to elaborate on the new National Institute of Mental Health (NIMH) Research Domain Criteria (RDoC) initiative. This RDoC initiative was designed to bypass diagnostic categories and focuses instead on classifying psychopathology based on dimensions of functioning in patients (Cuthbert and Insel 2010; Insel 2014; Morris and Cuthbert 2012). By basing categorization on domains of function, it is theorized that the circuitry underlying the behaviors will be more closely linked to the abnormalities (Segal and Geyer 1985) and that any drug treatments designed to target such circuitry would prove more efficacious. For example, upon closer inspection, overlap exists in motor behavior between subjects from one diagnostic category and another (e.g., few but several patients with schizophrenia exhibiting high object interactions consistent with mania patients). Therefore, it may be the case that reduced DAT activity recreates this locomotor profile irrespective of disease state. Examining DAT levels and the behavior of patients across diagnostic categories using these translationally relevant paradigms could be a useful future direction for neural mechanistic studies. In fact, one domain identified by RDoC is Arousal and Regulatory Systems, specifically identified as the construct Arousal. Actigraphy was identified a paradigm useful for measuring Arousal, consistent with a call by Bernard and Mittal (2015) that a motor dimension be included in RDoC. The multivariate approach identified within this chapter, however, suggests that a more fruitful approach would be to utilize a paradigm such as the BPM, providing more than a unitary measure of arousal. The use of a behavioral paradigm such as the BPM enables us to better investigate the biological mechanisms of essential cognitive and behavioral functions that transcend classical diagnostic nosologies, pushing psychiatry toward more personalized and efficacious treatments.

Acknowledgments We thank Drs. Perry, Henry, and van Enkhuizen for their support. This work was supported by NIH grants R01-MH071916, R01-MH104344, and the Veteran's Administration VISN 22 Mental Illness Research, Education, and Clinical Center.

References

Amitai N, Young JW, Higa K, Sharp RF, Geyer MA, Powell SB (2014) Isolation rearing effects on probabilistic learning and cognitive flexibility in rats. Cogn Affect Behav Neurosci 14(1):388–406

Ankers D, Jones SH (2009) Objective assessment of circadian activity and sleep patterns in individuals at behavioural risk of hypomania. J Clin Psychol 65(10):1071–1086

Arnone D, Cavanagh J, Gerber D, Lawrie SM, Ebmeier KP, McIntosh AM (2009) Magnetic resonance imaging studies in bipolar disorder and schizophrenia: meta-analysis. Br J Psychiatry 195(3):194–201

Bechara A, Damasio AR, Damasio H, Anderson SW (1994) Insensitivity to future consequences following damage to human prefrontal cortex. Cognition 50(1–3):7–15

Berlyne DE (1966) Curiosity and exploration. Science 153(3731):25–33

Bernard JA, Mittal VA (2015) Updating the research domain criteria: the utility of a motor dimension. Psychol Med 1–5

Cavanaugh JT, Ellis TD, Earhart GM, Ford MP, Foreman KB, Dibble LE (2015) Toward understanding ambulatory activity decline in parkinson disease. Phys Ther
Claridge EA, McPhee PG, Timmons BW, Ginis KA, MacDonald MJ, Gorter JW (2014) Quantification of physical activity and sedentary time in adults with cerebral palsy. Med Sci Sports Exerc
Cuthbert BN, Insel TR (2010) Toward new approaches to psychotic disorders: the NIMH Research Domain Criteria project. Schizophr Bull 36(6):1061–1062
Dawe GS, Ratty AK (2007) The chakragati mouse: a mouse model for rapid in vivo screening of antipsychotic drug candidates. Biotechnol J 2(11):1344–1352
DeLany JP, Dube JJ, Standley RA, Distefano G, Goodpaster BH, Stefanovic-Racic M, Coen PM, Toledo FG (2014) Racial differences in peripheral insulin sensitivity and mitochondrial capacity in the absence of obesity. J Clin Endocrinol Metab 99(11):4307–4314
Diaz KM, Krupka DJ, Chang MJ, Peacock J, Ma Y, Goldsmith J, Schwartz JE, Davidson KW (2015) Fitbit(R): an accurate and reliable device for wireless physical activity tracking. Int J Cardiol 185:138–140
Drai D, Golani I (2001) SEE: a tool for the visualization and analysis of rodent exploratory behavior. Neurosci Biobehav Rev 25(5):409–426
Eilam D, Golani I (1988) The ontogeny of exploratory behavior in the house rat (Rattus rattus): the mobility gradient. Dev Psychobiol 21(7):679–710
Eilam D, Shefer G (1997) The developmental order of bipedal locomotion in the jerboa (Jaculus orientalis): pivoting, creeping, quadrupedalism, and bipedalism. Dev Psychobiol 31(2):137–142
Faedda GL, Teicher MH (2005) Objective measures of activity and attention in the differential diagnosis of psychiatric disorders of childhood. Essent Psychopharmacol 6(5):239–249
File SE, Wardill AG (1975) The reliability of the hole-board apparatus. Psychopharmacologia 44 (1):47–51
Geoffroy PA, Scott J, Boudebesse C, Lajnef M, Henry C, Leboyer M, Bellivier F, Etain B (2015) Sleep in patients with remitted bipolar disorders: a meta-analysis of actigraphy studies. Acta Psychiatr Scand 131(2):89–99
Geyer MA (1990) Approaches to the characterization of drug effects on locomotor activity in rodents. In: Cowan A, Adler MW (eds) Testing and evaluation of drugs of abuse New York. Wiley-Liss Inc, USA
Geyer MA, Paulus M (1996) Multivariate analyses of locomotor and investigatory behavior in rodents. In: Ossenkopp K, Kavaliers M, Sanberg PR (eds) Mearusing movement and locomotion: from invertebrates to humans, Chapman & Hall, New York
Geyer MA, Vollenweider FX (2008) Serotonin research: contributions to understanding psychoses. Trends Pharmacol Sci 29(9):445–453
Geyer MA, Russo PV, Masten VL (1986) Multivariate assessment of locomotor behavior: pharmacological and behavioral analyses. Pharmacol Biochem Behav 25(1):277–288
Geyer MA, Wilkinson LS, Humby T, Robbins TW (1993) Isolation rearing of rats produces a deficit in prepulse inhibition of acoustic startle similar to that in schizophrenia. Biol Psychiatry 34(6):361–372
Goto Y, Yang CR, Otani S (2010) Functional and dysfunctional synaptic plasticity in prefrontal cortex: roles in psychiatric disorders. Biol Psychiatry 67(3):199–207
Gould TD, Dao DT, Kovacsics CE (2009) The open field test. In: Gould TD (ed) Mood and anxiety related phenotypes in mice: characterization using behavioral tests. Humana Press, New York, pp 1–20
Grossman P (2004) The LifeShirt: a multi-function ambulatory system monitoring health, disease, and medical intervention in the real world. Stud Health Technol Inform 108:133–141
Halberstadt AL, Koedood L, Powell SB, Geyer MA (2011) Differential contributions of serotonin receptors to the behavioral effects of indoleamine hallucinogens in mice. J Psychopharmacol 25(11):1548–1561

Halperin JM, Newcorn JH, Matier K, Sharma V, McKay KE, Schwartz S (1993) Discriminant validity of attention-deficit hyperactivity disorder. J Am Acad Child Adolesc Psychiatry 32(5):1038–1043

Henry BL, Minassian A, Young JW, Paulus MP, Geyer MA, Perry W (2010) Cross-species assessments of motor and exploratory behavior related to bipolar disorder. Neurosci Biobehav Rev 34(8):1296–1306

Henry BL, Minassian A, van Rhenen M, Young JW, Geyer MA, Perry W (2011) Effect of methamphetamine dependence on inhibitory deficits in a novel human open-field paradigm. Psychopharmacology 215(4):697–707

Henry BL, Geyer MA, Buell M, Perry W, Young JW, Minassian A (2013a) Behavioral effects of chronic methamphetamine treatment in HIV-1 gp120 transgenic mice. Behav Brain Res 236 (1):210–220

Henry BL, Minassian A, Patt VM, Hua J, Young JW, Geyer MA, Perry W (2013b) Inhibitory deficits in euthymic bipolar disorder patients assessed in the human behavioral pattern monitor. J Affect Disord 150(3):948–954

Insel TR (2014) The NIMH research domain criteria (RDoC) project: precision medicine for psychiatry. Am J Psychiatry 171(4):395–397

Kafkafi N, Lipkind D, Benjamini Y, Mayo CL, Elmer GI, Golani I (2003) SEE locomotor behavior test discriminates C57BL/6 J and DBA/2 J mouse inbred strains across laboratories and protocol conditions. Behav Neurosci 117(3):464–477

Keenan DB, Wilhelm FH (2005) Classification of locomotor activity by acceleration measurement: validation in Parkinson disease. Biomed Sci Instrum 41:329–334

Kim DH, Lu N, Ma R, Kim YS, Kim RH, Wang S, Wu J, Won SM, Tao H, Islam A, Yu KJ, Kim TI, Chowdhury R, Ying M, Xu L, Li M, Chung HJ, Keum H, McCormick M, Liu P, Zhang YW, Omenetto FG, Huang Y, Coleman T, Rogers JA (2011) Epidermal electronics. Science 333(6044):838–843

Klein E, Lavie P, Meiraz R, Sadeh A, Lenox RH (1992) Increased motor activity and recurrent manic episodes: predictors of rapid relapse in remitted bipolar disorder patients after lithium discontinuation. Biol Psychiatry 31(3):279–284

Koenigs M, Grafman J (2009) The functional neuroanatomy of depression: distinct roles for ventromedial and dorsolateral prefrontal cortex. Behav Brain Res 201(2):239–243

Krebs-Thomson K, Paulus MP, Geyer MA (1998) Effects of hallucinogens on locomotor and investigatory activity and patterns: influence of 5-HT2A and 5-HT2C receptors. Neuropsychopharmacology 18(5):339–351

Lat J (1965) The spontaneous exploratory reactions as a tool for psychopharmacological studies. A contribution towards a theory of contradictory results in psychopharmacology. In: Mikhelson MY, Long VG, Votava Z (eds) Pharmacology of conditioning, learning and retention. Pergamon Press, Oxford, pp. 47–66

Lehmann-Masten VD, Geyer MA (1991) Spatial and temporal patterning distinguishes the locomotor activating effects of dizocilpine and phencyclidine in rats. Neuropharmacology 30 (6):629–636

Lhermitte F (1983) 'Utilization behaviour' and its relation to lesions of the frontal lobes. Brain 106 (Pt 2):237–255

McClung CA (2013) How might circadian rhythms control mood? Let me count the ways. Biol Psychiatry 74(4):242–249

McNamara RJ, McKeough ZJ, McKenzie DK, Alison JA (2014) Physical comorbidities affect physical activity in chronic obstructive pulmonary disease: a prospective cohort study. Respirology 19(6):866–872

Minassian A, Henry BL, Geyer MA, Paulus MP, Young JW, Perry W (2009) The quantitative assessment of motor activity in mania and schizophrenia. J Affect Disord 120(1–3):200–206

Minassian A, Henry BL, Young JW, Masten V, Geyer MA, Perry W (2011) Repeated assessment of exploration and novelty seeking in the human behavioral pattern monitor in bipolar disorder patients and healthy individuals. PLoS ONE 6(8):e24185

Minassian A, Young JW, Cope ZA, Henry BL, Geyer MA, Perry W (2016) Amphetamine increases activity but not exploration in humans and mice. Psychopharmacology 233(2): 225–233

Morris SE, Cuthbert BN (2012) Research domain criteria: cognitive systems, neural circuits, and dimensions of behavior. Dialogues Clin Neurosci 14(1):29–37

Naslund JA, Aschbrenner KA, Barre LK, Bartels SJ (2015) Feasibility of popular m-health technologies for activity tracking among individuals with serious mental illness. Telemed J E Health 21(3):213–216

Neill JC, Barnes S, Cook S, Grayson B, Idris NF, McLean SL, Snigdha S, Rajagopal L, Harte MK (2010) Animal models of cognitive dysfunction and negative symptoms of schizophrenia: focus on NMDA receptor antagonism. Pharmacol Ther

Novak D, Albert F, Spaniel F (2014) Analysis of actigraph parameters for relapse prediction in bipolar disorder: a feasibility study. Conf Proc IEEE Eng Med Biol Soc 2014:4972–4975

Paulus MP, Geyer MA (1993) Three independent factors characterize spontaneous rat motor activity. Behav Brain Res 53(1–2):11–20

Paulus M, Geyer MA (1996a) Assessing the organization of motor behavior: new applroaches based on the behavior of complex physical systems. In: Ossenkopp KP, Kavaliers M, Sanberg PR (eds) Measuring movement and locomotion: From invertebrate animals to humans. Chapman & Hall, New York

Paulus M, Geyer MA (1996b) Assessing the organization of motor behavior: new approaches based on the behavior of complex physical systems. In: Ossenkopp K, Kavaliers M, Sanberg PR (eds) Measuring movement and locomotion: from invertebrates to humans. Chapman & Hall, New York

Perry W, Minassian A, Paulus MP, Young JW, Kincaid MJ, Ferguson EJ, Henry BL, Zhuang X, Masten VL, Sharp RF, Geyer MA (2009) A reverse-translational study of dysfunctional exploration in psychiatric disorders: from mice to men. Arch Gen Psychiatry 66(10):1072–1080

Perry W, Minassian A, Henry B, Kincaid M, Young JW, Geyer MA (2010) Quantifying over-activity in bipolar and schizophrenia patients in a human open field paradigm. Psychiatry Res 178(1):84–91

Pierce K, Courchesne E (2001a) Evidence for a cerebellar role in reduced exploration and stereotyped behavior in autism. Biol Psychiatry 49(8):655–664

Pierce K, Courchesne E (2001b) Evidence for a cerebellar role in reduced exploration and stereotyped behavior in autism. Biol Psychiatry 49:655–664

Pini S, de Queiroz V, Dell'Osso L, Abelli M, Mastrocinque C, Saettoni M, Catena M, Cassano GB (2004) Cross-sectional similarities and differences between schizophrenia, schizoaffective disorder and mania or mixed mania with mood-incongruent psychotic features. Eur Psychiatry 19(1):8–14

Porrino LJ, Rapoport JL, Behar D, Sceery W, Ismond DR, Bunney WE Jr (1983) A naturalistic assessment of the motor activity of hyperactive boys. I. Comparison with normal controls. Arch Gen Psychiatry 40(6):681–687

Queiroz AI, de Araujo MM, da Silva Araujo T, de Souza GC, Cavalcante LM, de Jesus Souza Machado M, de Lucena DF, Quevedo J, Macedo D (2015) GBR 12909 administration as an animal model of bipolar mania: time course of behavioral, brain oxidative alterations and effect of mood stabilizing drugs. Metab Brain Dis

Ratty AK, Fitzgerald LW, Titeler M, Glick SD, Mullins JJ, Gross KW (1990) Circling behavior exhibited by a transgenic insertional mutant. Brain Res Mol Brain Res 8(4):355–358

Segal DS, Geyer MA (1985) Animal models of psychopathology. In: Groves PM, Judd LL (eds) Psychobiological foundations of clinical psychiatry. J.B. Lippincott Co, Philadelphia

Silverstone PH, Pukhovsky A, Rotzinger S (1998) Lithium does not attenuate the effects of D-amphetamine in healthy volunteers. Psychiatry Res 79(3):219–226

Sims J, Smith F, Duffy A, Hilton S (1999) The vagaries of self-reports of physical activity: a problem revisited and addressed in a study of exercise promotion in the over 65s in general practice. Fam Pract 16(2):152–157

Steinhubl SR, Muse ED, Topol EJ (2015) The emerging field of mobile health. Sci Transl Med, 7 (283):283rv283

Tanaka S, Young JW, Halberstadt AL, Masten VL, Geyer MA (2012) Four factors underlying mouse behavior in an open field. Behav Brain Res 233(1):55–61

Teicher MH (1995a) Actigraphy and motion analysis: new tools for psychiatry. Harvard Rev Psychiatry 3(1):18–35

Teicher MH (1995b) Actigraphy and motion analysis: new tools for psychiatry. Harv Rev Psychiatry 3(1):18–35

Teicher MH, Lawrence JM, Barber NI, Finklestein SP, Lieberman H, Baldessarini RJ (1986) Altered locomotor activity in neuropsychiatric patients. Prog Neuropsychopharmacol Biol Psychiatry 10:755–761

Teicher MH, Lowen SB, Polcari A, Foley M, McGreenery CE (2004) Novel strategy for the analysis of CPT data provides new insight into the effects of methylphenidate on attentional states in children with ADHD. J Child Adolesc Psychopharmacol 14(2):219–232

Topol E (2015) The patient will see you now: the future of medicine is in your hands. Basic Books, New York (NY)

Turakhia MP, Ullal AJ, Hoang DD, Than CT, Miller JD, Friday KJ, Perez MV, Freeman JV, Wang PJ, Heidenreich PA (2015) Feasibility of extended ambulatory electrocardiogram monitoring to identify silent atrial fibrillation in high-risk patients: the screening study for undiagnosed atrial fibrillation (STUDY-AF). Clin Cardiol 38(5):285–292

van Enkhuizen J, Geyer MA, Kooistra K, Young JW (2012) Chronic valproate attenuates some, but not all, facets of mania-like behaviour in mice. Int J Neuropsychopharmacol: 1–11

van Enkhuizen J, Minassian A, Young JW (2013) Further evidence for ClockDelta19 mice as a model for bipolar disorder mania using cross-species tests of exploration and sensorimotor gating. Behav Brain Res 249:44–54

van Enkhuizen J, Geyer MA, Halberstadt AL, Zhuang X, Young JW (2014a) Dopamine depletion attenuates some behavioral abnormalities in a hyperdopaminergic mouse model of bipolar disorder. J Affect Disord 155:247–254

van Enkhuizen J, Henry BL, Minassian A, Perry W, Milienne-Petiot M, Higa KK, Geyer MA, Young JW (2014b) Reduced dopamine transporter functioning induces high-reward risk-preference consistent with bipolar disorder. Neuropsychopharmacology 39(13):3112–3122

van Enkhuizen J, Milienne-Petiot M, Geyer MA, Young JW (2015) Modeling bipolar disorder in mice by increasing acetylcholine or dopamine: chronic lithium treats most, but not all features. Psychopharmacology (Berl)

Walther S, Koschorke P, Horn H, Strik W (2009) Objectively measured motor activity in schizophrenia challenges the validity of expert ratings. Psychiatry Res 169(3):187–190

Wilhelm FH, Roth WT, Sackner MA (2003) The lifeShirt. An advanced system for ambulatory measurement of respiratory and cardiac function. Behav Modif 27(5):671–691

Wolff EA, Putnam FW, Post RM (1985) Motor activity and affective illness. Arch Gen Psychiatry 42:288–294

Young JW, Geyer MA (2010) Action of modafinil-increased motivation via the dopamine transporter inhibition and D1 receptors? Biol Psychiatry 67(8):784–787

Young JW, Geyer MA (2011) Using behavioral patterns across species in mood disorder research. In: Gould TD (ed) Mood and anxiety releated phenotypes in mice. Humana Press, New York, pp 21–41

Young JW, Minassian A, Paulus MP, Geyer MA, Perry W (2007) A reverse-translational approach to bipolar disorder: rodent and human studies in the behavioral pattern monitor. Neurosci Biobehav Rev 31(6):882–896

Young JW, Goey AK, Minassian A, Perry W, Paulus MP, Geyer MA (2010a) GBR 12909 administration as a mouse model of bipolar disorder mania: mimicking quantitative assessment of manic behavior. Psychopharmacology 208(3):443–454

Young JW, Goey AK, Minassian A, Perry W, Paulus MP, Geyer MA (2010b) The mania-like exploratory profile in genetic dopamine transporter mouse models is diminished in a familiar environment and reinstated by subthreshold psychostimulant administration. Pharmacol Biochem Behav 96(1):7–15

Young JW, Henry BL, Geyer MA (2011a) Predictive animal models of mania, hits, misses, and future directions. Br J Pharmacol

Young JW, Kooistra K, Geyer MA (2011b) Dopamine receptor mediation of the exploratory/hyperactivity effects of modafinil. Neuropsychopharmacology 36(7):1385–1396

Young JW, van Enkhuizen J, Winstanley CA, Geyer MA (2011c) Increased risk-taking behavior in dopamine transporter knockdown mice: further support for a mouse model of mania. J Psychopharmacol 25(7):934–943

Young JW, Ratty A, Dawe GS, Geyer MA (2014) Altered exploration and sensorimotor gating of the chakragati mouse model of schizophrenia. Behav Neurosci 128(4):460–467

Animal Models of Deficient Sensorimotor Gating in Schizophrenia: Are They Still Relevant?

Neal R. Swerdlow and Gregory A. Light

Abstract Animal models of impaired sensorimotor gating, as assessed by prepulse inhibition (PPI) of startle, have demonstrated clear validity at face, predictive, and construct levels for schizophrenia (SZ) therapeutics, neurophysiological endophenotypes, and potential causative insults for this group of disorders. However, with the growing recognition of the heterogeneity of the schizophrenias, and the less sanguine view of the clinical value of antipsychotic (AP) medications, our field must look beyond "validity," to assess the actual utility of these models. At a substantial cost in terms of research support and intellectual capital, what has come from these models, that we can say has actually helped schizophrenia patients? Such introspection is timely, as we are reassessing not only our view of the genetic and pathophysiological diversity of these disorders, but also the predominant strategies for SZ therapeutics; indeed, our field is gaining awareness that we must move away from a "find what's broke and fix it" approach, toward identifying spared neural and cognitive function in SZ patients, and matching these residual neural assets with learning-based therapies. Perhaps, construct-valid models that identify evidence of "spared function" in neural substrates might reveal opportunities for future therapeutics and allow us to study these substrates at a mechanistic level to maximize opportunities for neuroplasticity. Such an effort will require a retooling of our models, and more importantly, a re-evaluation of their utility. For animal models to remain relevant in the search for schizophrenia therapeutics, they will need to focus less on what is valid and focus more on what is useful.

Keywords Biomarker · Cognitive remediation · Mismatch negativity · Neurocognition · Prepulse inhibition · Schizophrenia

---

N.R. Swerdlow (✉) · G.A. Light
Department of Psychiatry, School of Medicine, University of California, San Diego, 9500 Gilman Dr. La Jolla, San Diego, CA 92093-0804, USA
e-mail: nswerdlow@ucsd.edu

Curr Topics Behav Neurosci (2016) 28: 305–325
DOI 10.1007/7854_2015_5012

Contents

1 Introduction 306
2 The Evolution of Prepulse Inhibition as a Validated Animal Model for Schizophrenia-Linked Neurobehavioral Deficits 306
3 Where Are We Now? 312
4 What's Next: A Paradigm Shift in the Use of Cross-Species PPI Models for Enhancing Schizophrenia Therapeutics? 313
4.1 Biomarkers to the rescue? 314
4.2 Drug-Enhanced PPI as a Biomarker for PACT? 315
5 Conclusion 318
References 318

1 Introduction

It is a fair assumption that for much of our history, and certainly since the emergence of our brain's capacity for introspection, humans have looked to infrahuman species for clues to understanding the complexities of our own thoughts, feelings, and behaviors. This "reverse anthropomorphism" reflects the compelling data from evolutionary biology that man's nervous system is fashioned on the neurobiological foundations of lower organisms (MacLean 1954; Karten 1991) as well as the perhaps less-compelling assumption that the infrahuman features retained in man's more advanced neural workings are informative about a brain that has acquired new and potentially emergent properties in abstract thought and complex emotions. It might be hard to pinpoint the first use of simple behaviors in infrahumans to understand human disorders. But studies of the late 1960s and early 1970s, from among others Michael Davis and his laboratory at Yale University (cf. Davis 1984), took one simple rodent behavior—the startle reflex—and developed a powerful laboratory-based assay for a simple cross-species behavior of relevance to human brain disorders. In this chapter, we review and critically evaluate the use of this simple behavior as part of a model with face, predictive, and construct validity for sensorimotor gating deficits in schizophrenia, and speculate on future applications of this model in the development of novel therapeutics for this disorder.

2 The Evolution of Prepulse Inhibition as a Validated Animal Model for Schizophrenia-Linked Neurobehavioral Deficits

The startle reflex is a constellation of responses to sudden, relatively intense stimuli. In humans, the blink reflex component of startle is measured using electromyography of orbicularis oculi; in laboratory animals, whole-body startle is quantified by assessing the downward force resulting from the contraction of the skeletal muscles. Prepulse inhibition (PPI) occurs when a weak prestimulus 30–500 ms prior to the startling stimulus inhibits the startle response; this inhibition is an operational measure of sensorimotor gating (Graham 1975). While the inhibitory effect of the prepulse on the startle reflex is exerted in the pons, studies have described the limbic forebrain circuitry and descending pontine projections that regulate the inhibitory "tone" within the pons and determine the degree to which the prepulse inhibits the subsequent motor response (cf. Swerdlow et al. 1992a, 2001a, 2008). PPI thus appears to reflect the activation of "hardwired," centrally mediated inhibitory processes that are regulated by forebrain neural circuitry.

PPI is a useful experimental measure for understanding brain mechanisms for a number of reasons (Davis 1984). It is tested in an automated apparatus, under tight stimulus control, and stimulus parameters can be easily modified by the experimenter to elicit optimal response characteristics for studying a number of different aspects of this measure. Because PPI is a form of startle plasticity, it is measured using a "fight-or-flight" behavior that is simple, robust, and exhibited across all mammalian species tested to date. Of relevance to the present discussion, PPI is easily studied across species and has been investigated in mice (Carter et al. 1999; Francis et al. 2003; Frankland et al. 2004), rats (Swerdlow et al. 2001a), guinea pigs (Vaillancourt and Boksa 2000), pigs (Lind et al. 2004), and infrahuman primates (Linn et al. 2003), using stimulus parameters and equipment for stimulus delivery and response acquisition that are similar or identical to what are used in humans. This cross-species similarity in the appearance of the behavior and its response to parametric manipulations is the basis for the *face validity* of animal models that use PPI. While there appear to be differences in the neurochemical regulation of PPI across species (cf. Swerdlow et al. 2008), the basic parametric properties of PPI exhibit striking similarities from rodents to humans (e.g., Swerdlow et al. 1994a, b). Furthermore, PPI is under significant genetic control in both rodents (Francis et al. 2003) and humans (Greenwood et al. 2007).

Despite its advantages as a laboratory measure of simple brain processes, PPI would likely be a scientific footnote were it not for the fact that it is reduced in humans afflicted with any one of several different brain disorders. Compared with matched controls, PPI is deficient in patients with schizophrenia (e.g., Braff et al. 1978; Swerdlow et al. 2006), Huntington's disease (Swerdlow et al. 1995; Valls-Sole et al. 2004), obsessive-compulsive disorder (OCD) (Swerdlow et al. 1993; Hoenig et al. 2005; Ahmari et al. 2012), nocturnal enuresis (Ornitz et al. 1992), Asperger's syndrome (McAlonan et al. 2002), 22q11 syndrome (Sobin et al. 2005), Kleinfelter syndrome (Van Rijn et al. 2011), fragile X syndrome (Frankland et al. 2004), blepharospasm (Gomez-Wong et al. 1998), and Tourette syndrome (Castellanos et al. 1996; Swerdlow et al. 2001b).

Development and applications of PPI in animal models: While it is clear that PPI deficits are not clinically specific, the real catalyst behind the intense investigation of PPI came from the initial reports of PPI deficits in schizophrenia patients (Braff et al. 1978). With this 1978 study and its subsequent replication in almost 40 reports in the literature (cf. Swerdlow et al. 2014), investigators have viewed the cross-species similarities in startle and PPI as an opportunity to leverage animal model studies to explicate the biology of this disorder. In the first connection of this initial report of PPI deficits in schizophrenia (Braff et al. 1978) with findings in experimental animals, evidence that startle inhibition by pulsating tactile tail pressure was eliminated after ablation of the nucleus accumbens (NAC; Sorenson and Swerdlow 1982) was viewed as potential evidence that accumbens dysfunction might contribute to the loss of startle inhibition by acoustic prepulses in schizophrenia; this suggestion has been substantiated by the number of subsequent reports, and 30+ years later, the NAC remains a central structure in current models for the regulation and dysregulation of PPI (e.g., Ma and Leung 2014).

A focus on the PPI-regulatory role of NAC dopaminergic systems (Swerdlow et al. 1986) and dopamine activity more broadly (Mansbach et al. 1988) was initially motivated by the prevailing hypothesis of a causative role of DA hyperfunction in the etiology of schizophrenia. The finding that PPI was disrupted in rodents by DA agonists (Swerdlow et al. 1986; Mansbach et al. 1988) was applied in a manner prescribed for animal models of that era, i.e., by assessing the ability of this pharmacological effect to predict the antipsychotic (AP) potential and potency of established and novel compounds (cf. Swerdlow et al. 1991, 1994b; Swerdlow and Geyer 1993). This approach differed from preexisting predictive models, such as apomorphine-induced canine emesis (Janssen and Niemegeers 1959), primarily because the behavior being measured (PPI) as a predictive index was analogous, if not homologous, across species. Thus, known AP compounds prevented the PPI-disruptive effects of DA agonists, and their potency in this assay correlated highly ($R = 0.99$) with their clinical AP potency (Swerdlow et al. 1994a, b). This compelling relationship is the basis for the *predictive validity* of this PPI model and led to the identification or validation of compounds with novel AP properties [e.g., ICI 204, 636 (quetiapine; Swerdlow et al. 1994a, b)].

The predictive model was expanded significantly by the observation that putative APs with novel chemical properties were distinguished by their ability to block the PPI-disruptive effects of NMDA antagonists (Johansson et al. 1994; Bakshi et al. 1994). Indeed, the prevailing wisdom of the early 1990s was that the ability to prevent the PPI-disruptive effects of NMDA antagonists such as phencyclidine and ketamine might predict the properties unique to "atypical" or second-generation APs (SGAPs) and thereby identify agents that would be both more clinically effective and better tolerated than first-generation APs. Over time, this approach ran into some experimental and clinical headwind. First, the ability to prevent NMDA antagonist-induced PPI deficits was not always specific to SGAPs [e.g., chlorpromazine blocks the PPI-disruptive effects of ketamine (Swerdlow et al. 1998)] or particularly sensitive to SGAPs (e.g., several studies reported either marginal or no ability of clozapine to prevent the PPI-disruptive effects of phencyclidine in rats). Second, and more importantly, clinical experience revealed that the benefits of SGAPs over older, first-generation APs were not robust, and in fact SGAPs carried a new and non-trivial list of adverse properties. Thus, while the predictive validity of these PPI models for antipsychotics were further extended in many informative ways as reviewed previously (e.g., Geyer et al. 2001; Swerdlow et al. 2008), they ultimately must be seen in the more humbling context of the clinical reality that APs of any generation are not well-tolerated and have limited ability to enhance the function and improve the quality of life in schizophrenia patients (Lieberman et al. 2005). This is not to say that APs lack clinical value: In fact, APs appear to have utility in blunting the severity of acute psychotic symptoms, and their use is associated with a lower risk of adverse consequences of schizophrenia—from hospitalization to suicide (Palmer et al. 1999; Meltzer et al. 2003; Sun et al. 2007). Nonetheless, 20 years of experimentation with PPI as a model predicting AP efficacy and potency has done little to advance us toward treatments that achieve either greater clinical improvement or fewer significant adverse effects than those that predated this model.

One obvious advantage of animal models of a human behavior is that they make it feasible to study neural substrates and extrapolate from these substrates to corresponding circuitry in humans. Indeed, extending from the initial findings of a nucleus accumbens locus of forebrain PPI regulation (Sorenson and Swerdlow 1982; Swerdlow et al. 1986; Kodsi and Swerdlow 1994), this approach was applied to understand the neural basis of PPI deficits in schizophrenia and revealed that the forebrain substrates regulating PPI overlap somewhat with those implicated in the pathophysiology of this disorder. Thus, disturbances in prefrontal cortex (PFC), basal forebrain dopamine (DA) function, and thalamic and mesial temporal lobe function figure prominently in current models of schizophrenia neuropathology; similarly, PPI is potently reduced by experimentally induced manipulations of the medial PFC, ventral striatum, pallidum, thalamus, and mesial temporal lobe (cf. Swerdlow et al. 1992a, b, 2001a, 2008; Rohleder et al. 2014). The apparent overlap in the neural substrates regulating PPI, with those implicated in the pathophysiology of schizophrenia, is part of the support for the *construct validity* of animal models for impaired PPI in schizophrenia and has been used in an iterative cross-species strategy. In this strategy, PPI changes after neural circuit manipulations in laboratory animals have been used to develop and then test hypotheses about specific circuit disturbances in patients (e.g., Kumari et al. 2003), and in some cases, circuit-based therapeutics are being modeled based on PPI deficits in rats (e.g., Posch et al. 2012; Angelov et al. 2014; Ma and Leung 2014). Often, when substrates have been demonstrated to regulate PPI in rodents, the fact that PPI is deficient in schizophrenia patients has been used as the basis for justifying a fine grain analysis of those substrates in rats, in terms of their anatomical, neurochemical, and molecular properties. In turn, information about the detailed characteristics of this circuitry derived from studies in rodents has been used to support, develop, or test hypotheses regarding the nature of neural circuit disturbances in schizophrenia (e.g., Hines et al. 2013; Miller et al. 2010).

The construct validity of PPI models in rodents for PPI deficits in schizophrenia is also strengthened by the fact that experimental manipulations in rodents that are thought to model some of the suspected pathogenic insults contributing to schizophrenia also produce adult rodents with deficient PPI. Of the more studied models of this kind—social isolation rearing and neonatal ventral hippocampal lesions—the former model was the subject of a recent review (Powell and Swerdlow 2015), and we will briefly review the latter model here.

In schizophrenia patients, the integrity of the hippocampal-PFC connection is reduced, and this deficiency predicts both neurocognitive and functional impairment (Hanlon et al. 2012). Lesions of the ventral hippocampus in neonatal rats (NVHLs) have been shown to recreate a number of deficits associated with schizophrenia (Lipska et al. 1993; Marquis et al. 2006; Angst et al. 2007; Marquis et al. 2008; cf. O'Donnell 2012), including reductions in PPI (Lipska et al. 1995; Le Pen and Moreau 2002; Le Pen et al. 2003; Daenen et al. 2003; Swerdlow et al. 2012a, b). To the degree that some forms of schizophrenia are characterized by aberrant ventral hippocampal development and connectivity, the NVHL model has been used to identify the expected "neuromaladaptive" consequences of such pathology and thereby help focus studies of pathophysiology and even therapeutics in this disorder. The model has been extended to demonstrate that a variety of early developmental insults to the mesial temporal lobe are accompanied by PPI deficits that emerge in adulthood, including immune/inflammatory activation of the VH (e.g., Zhu et al. 2014a, b; Ribeiro et al. 2013), neonatal pilocarpine-induced seizures (Labbate et al. 2014), and neonatal lesions of the basolateral amygdala (Vázquez-Roque et al. 2012). Other in utero or neonatal neurotoxic manipulations also produce PPI deficits in adult rats, including methylazoxymethanol (MAM) exposure (Le Pen et al. 2006), elevated neonatal allopregnanolone (Darbra et al. 2014), and neonatal administration of NMDA antagonists (Uehara et al. 2010). In some cases, the expression of PPI deficits induced by these early developmental manipulations can be blocked by acute treatments during adulthood, using antipsychotics (e.g., clozapine: Ribeiro et al. 2013), putative neuroprotective agents (e.g., minocycline: Zhu et al. 2014b), and glycinergic agents (Le Pen et al. 2003). Thus, it appears that PPI deficits are a common adult behavioral response to a wide range of perturbations in early rodent brain development, and particularly those that impact the mesial temporal lobe by various mechanisms. In total, this literature is consistent with the empirical evidence that PPI deficits are detected in many clinically and etiologically distinct brain disorders, as well as the prevailing wisdom that schizophrenia (and by extension its accompanying PPI deficits) reflects a heterogeneous neuropathology induced by any one or combination of a number of different possible early developmental insults.

Presumably, the failure to develop normal levels of PPI in these variations of the NVHL model could reflect many different underlying mechanisms. One potential mechanism implicated in recent studies is a developmental "hypercoupling" of forebrain regions (Chambers et al. 2010; Swerdlow et al. 2013a, b)—including PFC and nucleus accumbens (NAC)—due to the loss of their normal innervation by the ventral hippocampus (VH) after experimentally induced VH damage. Thus, the VH innervates both the PFC and the NAC, and conditions fostering greater PFC-NAC interconnectivity might be created by NVHLs via reduced competition at a synaptic level, or by the loss of a differentiating signal normally provided by VH innervation of either structure. NVHLs result in restructuring and electrophysiological changes within the PFC (Ryan et al. 2013), and hyper-correlated expression of schizophrenia-linked genes in the PFC and NAC (Swerdlow et al. 2013a, b). Others have reported aberrant limbic–cortical connectivity associated with both endogenous (Anticevic et al. 2013) and drug-induced psychosis (Driesen et al. 2013) in humans; similarly, excessive fronto-striatal metabolic correlation ["Brain Lock" (Schwartz 1997)] has been demonstrated in other disorders associated with the reduced PPI, such as OCD. Importantly, in OCD, therapeutic response to medication or psychotherapy is associated with a metabolic "uncoupling" of fronto-striatal regions (Schwartz et al. 1996; Schwartz 1998). Perhaps, the most speculative but exciting concept to emerge from the NVHL/"hypercoupling model" is the possibility that an "uncoupling" of fronto-striatal circuitry might provide an avenue for early therapeutic interventions in schizophrenia. That such an "uncoupling" can be produced in OCD via cognitive interventions (Schwartz et al. 1996) may suggest such a therapeutic option in schizophrenia, as discussed below.

One approach to capitalize on the validity of PPI models has been to explore the genetic underpinnings of impaired PPI in rodents, to generate or support hypotheses related to the genetic basis of impaired PPI in schizophrenia [and other disorders (e.g., Castellan Baldan et al. 2014; Charles et al. 2014; Renoux et al. 2014)]. Given the numerous brain regions and interconnections known to regulate PPI, it is not surprising that these studies have identified a long list of genes that, by their deletion, suppression, or differential expression, lead to a modification in PPI or its sensitivity to pharmacologic disruption (cf. Swerdlow et al. 2008). A number of creative strategies have been used to understand this complex genetic landscape and its overlap with brain circuitry, via assessing the PPI-altering effects of gene knockouts, humanized gene insertions (e.g., Risbrough et al. 2014), strain differences in regional gene expression (e.g., Shilling et al. 2008), drug-induced changes in regional expression of genes identified in postmortem schizophrenia brain tissue (e.g., Dietz et al. 2014), and pharmacogenetic manipulations of neural activity in targeted neuron populations via the use of Designer Receptors Exclusively Activated by Designer Drugs (DREADDs) (e.g., Nguyen et al. 2014), among other techniques. These strategies are not without potential pitfalls, including the importance of assessing hearing loss in mutant animals as a potential basis for reduced inhibitory effects of auditory prepulses. More generally, the long list of candidate genes for which modification reduces PPI in rodents suggests limitations to the utility of this approach in clarifying the genetic basis of reduced PPI in schizophrenia. The use of genetic manipulations to understand the role of regionally selective cell populations and proteins in the regulation of PPI, however, continues to be a promising and informative experimental strategy.

But, just as the sobering news about the limited clinical value of APs limit the utility of PPI as a predictive model, there is sobering news about the heterogeneity of the neural and genetic substrates of schizophrenia that may limit the utility of PPI as a construct model. As noted below, published reports now catalog over twenty different brain regions with identifiable abnormalities in different cohorts of schizophrenia patients (cf. Levitt et al. 2010). Furthermore, current estimates suggest that over 100 loci explain 7 % of the risk for the development of schizophrenia (e.g., Schizophrenia Working Group of the Psychiatric Genomics Consortium

2014; Stefansson et al. 2014), and it is likely that only after we identify gene × gene and gene × environment interactions among these many risk variants will we ever account for a significant amount of the variance in the expression of the schizophrenia phenotype. Thus, though there is substantial basic scientific value in understanding brain circuits and candidate gene effects on behavior, it is not clear that the construct validity of PPI will bring us substantially closer to an understanding of the complex and heterogeneous neural and genetic bases for schizophrenia.

Of course, the neural and genetic heterogeneity of schizophrenia reflects, at least in part, the fact that this diagnosis is defined by clinical criteria that do not map neatly onto any single biological substrate. Perhaps, it does not make sense to judge the ultimate utility of a biological model, like PPI, based on its ability to clarify the treatments or neural basis for such an imprecise, non-biologically defined clinical entity. One could even argue that sensorimotor gating is a meaningful domain of brain function and that by identifying the neural substrates of PPI and its deficiency in subgroups of patients, we will establish a basis for categorizing brain disorders that is ultimately more valid and useful than the clinical nosology by which schizophrenia has been characterized to date. Clarity on whether such a use of PPI is feasible, or sensible, will need to await the substantial continued evolution of this model.

3 Where Are We Now?

Three decades after the first use of PPI in cross-species models for impaired sensorimotor gating in schizophrenia, we have substantial evidence supporting three levels of validity for these models. With these models, we have gained a reasonable understanding at a regional and circuit level of the neural regulation of PPI in rodents, and we have several pieces of evidence supporting the translation of this circuit "blueprint" onto the human brain and its regulation of PPI. Circuit models are being magnified within several brain regions—particularly the PFC, NAC, and VH—to explicate the regulation of PPI by these regions at the cellular and molecular level. This "circuit biology of PPI" is perhaps the most productive and still promising application of this cross-species model. But one great hope for PPI models, based on their strong predictive validity, has not yet materialized, as evidenced by the substantial limitations in the clinical impact of APs on neurocognition, function, and quality of life in schizophrenia populations. Indeed, it is in some ways the greatest failing of this animal model—that PPI studies in rodents do such an excellent job identifying compounds that reproduce the disappointing clinical impact of existing AP agents. One could argue that this failing is not unique to PPI models, and to some degree, it reflects a greater failing of modern psychopharmacology in its approach to therapeutics for complex polygenic disorders of neurodevelopmental origin with dispersed and heterogeneous neuropathology, like the schizophrenias. Perhaps, the most dispassionate assessment is that in our extensive studies of PPI across species, we have developed models for which validity is clear, and yet utility is not.

4 What's Next: A Paradigm Shift in the Use of Cross-Species PPI Models for Enhancing Schizophrenia Therapeutics?

One unspoken assumption behind the anticipated utility of PPI as a model with predictive and construct validity is as follows: Because we can identify in rodents the neural circuitry regulating PPI and its deficiencies, we can determine ways to intervene within this circuitry to restore normal function, using PPI as a "readout." And, more importantly, we can then apply these restorative interventions, or derivatives thereof, to "fix what's broken" in the PPI-regulatory circuitry in schizophrenia patients and thereby impart therapeutic change. The failings in this "fix what's broken" assumption are apparent, once we review our current understanding of this disorder.

As it is currently conceptualized, the root cause of schizophrenia is an in utero and childhood developmental interruption and tangling of neural connections (Weinberger 1987; Murray et al. 1991; Lewis and Levitt 2002) that are orders of magnitude too complex to restore or replace. Failures of cell migration and axonal guidance begin early, and this compounds the unpredictability of forebrain disorganization, like a mechanical delay in the first of many tightly connecting trains. The absurdity of trying to "fix what's broken" is further appreciated by considering what happens when cells or fibers do not get to where they are supposed to be, at the time they are supposed to be there. When these passengers fail to arrive at their "final destinations," like the PFC, this triggers pre- and postsynaptic compensatory changes among many functionally distinct subregions and cell types, and convergent influences of neurotransmitters, peptides, and other neuromodulators, all within adjacent lamina. But it is not *just* the PFC: As noted above, the preponderance of findings in different schizophrenia cohorts support significant volumetric and/or morphometric abnormalities in over 20 brain regions (cf. Levitt et al. 2010). Calculate the permutations of synaptic interactions in the simplest cartoon schematic, the number of different risk genes, and the epigenetic events, and multiply by orders of magnitude, and one can easily appreciate the futility of expecting even the smartest drugs to "fix what's broken." The fundamental error in this "fix what's broken" approach to the development of pharmacotherapies for schizophrenia is that regardless of how valid the PPI animal model (or any other model, for that matter) may be, the drugs that it produces will not be able to reach backward two decades through a variable web of absent and misguided neural connections, and replace missing and improper ones with healthy ones. The sooner that we acknowledge that prefrontal and limbic-cortico-striato-pallido-thalamic dysfunction and dysmorphogenesis in schizophrenia are too widely distributed, complex, and variable to be "fixed" with medications and that the strategies for gene therapies would require interventions so early in brain development as to present insurmountable ethical and logistical barriers for the foreseeable decades, the sooner we will be able to consider alternative strategies for applying animal models to the development of more successful therapeutics for this disorder. We do not presume to have found such a strategy, but we hope to begin the discussion about one approach that may warrant some attention.

4.1 Biomarkers to the rescue?

Biomarkers are objective measures that can be informative about a variety of different clinical characteristics, such as an individual's normal biology, their pathology including the trajectory of illness, or their response to a therapeutic intervention. They offer the hope that despite great heterogeneity and multivariate interactions in the pathogenesis of brain disorders, meaningful clusters of individuals can be associated with an objective measure and then reliably stratified in terms of the cause, course, and/or treatment sensitivity of a given disorder (Perez et al. 2014).

An assumption driving the search for psychiatric biomarkers is that the biology of these biomarkers will be simpler, more easily understood, and less heterogeneous than the biology of clinical psychiatric syndromes. But if the pathogenic pathways leading to schizophrenia are highly heterogeneous, we might expect that the biomarkers for these pathways might also be highly heterogeneous. Importantly, biomarkers might also be used to identify neural resources that *remain intact and functional in schizophrenia*. These functional "assets" might then be used to compensate for those lost to the aberrant developmental processes in this disorder. Such a model is applied successfully to stroke rehabilitation, where interventions are designed not to regrow brain circuitry that is lost or damaged, but rather to engage the normal physiological and anatomical properties of healthy brain circuits (e.g., in neighboring regions or parallel circuits) to restore or subsume the function of damaged ones (cf. Taub et al. 2002). In many forms of psychotherapy, the therapist's task is to identify an individual's psychological strengths (ego, intellectual, social, or otherwise) and then to engage them to overcome damaging thoughts or behaviors that are otherwise sustained by areas of psychological weakness. At a neural level, both stroke rehabilitation and psychotherapy engage viable and healthy systems to compensate for, or re-establish, functions lost to illness. Similarly, biomarkers of "health" that reveal a patient's neural "assets" can then be leveraged in the service of therapy.

In keeping with this model of using biomarkers to identify residual intact neural "assets," it is reasonable to consider whether *intact PPI* can be used as a biomarker of schizophrenia patients who might be capable of marshaling adequate neural resources to meet the demands of and reap the benefits of a particular therapeutic intervention. Consistent with such a model, Kumari et al. (2012) demonstrated that baseline PPI levels positively predicted the therapeutic response to cognitive–behavioral therapy (CBT) ($r = 0.69$ between pretreatment PPI (120 ms) and pre- versus post-CBT change in PANSS score). Schizophrenia patients who exhibited the highest pre-therapy PPI levels were the ones who benefitted most from CBT, in terms of reductions in symptom severity. This finding supports the notion that higher PPI provides evidence of intact, functioning neural mechanisms, that positively predicts the therapeutic response to a cognitive intervention; it also harkens to the fronto-striatal "hypercoupling" state associated with PPI deficits in the NVHL model (above), since CBT has been demonstrated to metabolically "uncouple" fronto-striatal circuits in other clinical conditions (Schwartz et al. 1996).

Perhaps more importantly, this finding suggests that neural elements contributing to intact PPI in any given schizophrenia patients might enhance that individual's sensitivity to the therapeutic benefits of CBT. To the degree that intact sensorimotor gating reflects a generally "healthy brain," it is not surprising that patients with more intact brains would benefit more from learning-based therapies. *An unanswered question is whether a pharmacology for enhancing PPI in relatively intact nervous systems, applied to patients whose PPI is then enhanced by these agents, might be able to augment the therapeutic benefit of cognitive therapies in schizophrenia.* In other words, can a pharmacologically induced increase of sensorimotor gating serve as "readout" of a change in brain function that makes a patient more able to benefit from the therapeutic features of a cognitive therapy? This general paradigm called "PACT" (pharmacologic augmentation of cognitive therapies) has been utilized effectively in the treatment of anxiety disorders (e.g., Ressler et al. 2004) and is in the very early stages of development for application to schizophrenia patients, as described below.

4.2 Drug-Enhanced PPI as a Biomarker for PACT?

While many pharmacological agents are capable of disrupting PPI in intact rodents, relatively fewer are known to consistently enhance PPI. This may reflect the fact that, at baseline, mechanisms for sensorimotor gating function at their optimal levels; additionally, experimental stimulus parameters (in particular, prepulse intervals) are typically selected to maximize inhibitory effects of prepulses and thereby are most sensitive for detecting drug-induced reductions in inhibition. However, strains of both mice and rats have been identified with relatively low basal PPI levels, and investigators have also taken the strategy of identifying "low gating" rats within a particular strain, and in both cases, these strains and substrains have been shown to be more sensitive to PPI-enhancing effects of drugs or brain stimulation (Acheson et al. 2012; Angelov et al. 2014; Swerdlow et al. 2006). Roussos et al. (2008) reported parallel findings in humans, in which healthy subjects homozygous for the Val allele of the rs4680 COMT polymorphism exhibited low basal PPI levels and PPI-enhancing effects of the COMT inhibitor, tolcapone, while individuals homozygous for the MET allele of rs4680 exhibited high basal PPI and PPI-reducing effects of tolcapone. There are also rat strain differences in the sensitivity to PPI-enhancing versus disruptive effects of the same drugs, even among commonly used outbred rat strains (e.g., Swerdlow et al. 2004), that are independent of basal

PPI levels, and are associated with the differential expression of several genes, including COMT, within PPI-regulatory circuitry (Shilling et al. 2008).

Conceivably, by developing models sensitive to detecting the PPI-enhancing effects of drugs, we might identify candidates suitable for assessment in a PACT paradigm. A number of different drug classes have already been identified that enhance PPI, such as nicotinic agonists and certain SGAPs, but under specific experimental conditions, even psychostimulants can be shown to enhance PPI (cf. Swerdlow et al. 2008). Of course, these various drug effects might reflect sites of action anywhere from the PFC (Swerdlow et al. 2012a, b) to the pons (Pinnock et al. 2015) that might be more or less relevant to the ability of a drug to enhance the therapeutic impact of a cognitive therapy.

It is important to emphasize that, in the PACT model—unlike the traditional use of PPI as a predictive screen for AP efficacy—the ability of a drug to enhance PPI does not predict that giving that drug to an individual with schizophrenia will, by itself, have any therapeutic value. Indeed, our expectation would be that if a patient is treated with such a drug without the concomitant delivery of a cognitive therapy, this treatment will have little value. Cognitive therapies place demands on patients to develop compensatory strategies for learning and remembering information. In so doing, they specifically activate prefrontal regions subserving working memory and attention (Kumari et al. 2009; Haut et al. 2010). Patients will benefit most from cognitive therapies if they are able to meet the cognitive demands of these therapies, and drugs that facilitate this process—e.g., via the enhancement of sensorimotor gating, or activation of circuitries that lead to an enhancement of sensorimotor gating—should augment the benefits of cognitive therapies. Conversely, we would not predict that patients would benefit by taking these drugs and returning to an environment that lacks engagement with an active learning process.

We have begun to assess PPI-enhancing drug effects in rats as a predictor of utility in a PACT paradigm, using the low- to moderate-affinity NMDA-receptor antagonist, memantine. While NMDA antagonists are generally reported to disrupt PPI in rodents, PPI is actually increased in healthy subjects (HS) by NMDA antagonists such as ketamine (Duncan et al. 2001; Abel et al. 2003) and by the mixed NMDA antagonist/dopamine agonist, amantadine (Swerdlow et al. 2002). In intact rats, we detected PPI-enhancing effects of memantine, using relatively short (10–30 ms) prepulse intervals (Swerdlow et al. 2009). Based on this PPI enhancement, and reports of PPI-enhancing effects of ketamine and amantadine in healthy subjects (HS), we speculated that memantine would potentiate PPI in HS. Indeed, we reported that 20 mg memantine (po) enhanced PPI modestly across all HS (Swerdlow et al. 2009) and that this effect was most robust among HS with low basal PPI levels (Fig. 1a), and among HS scoring high on personality scales for novelty seeking, sensation seeking, and disinhibition. This set of findings provided us with a cross-species model in which PPI is enhanced by a drug within neurologically intact rodents and HS. Similar findings had been reported using the SGAPs, quetiapine (Swerdlow et al. 2006), and clozapine (Vollenweider et al. 2006).

Based on these findings in HS, we assessed the effects of memantine on PPI in schizophrenia patients (Chou et al. 2013a; Swerdlow et al. 2016). Our findings

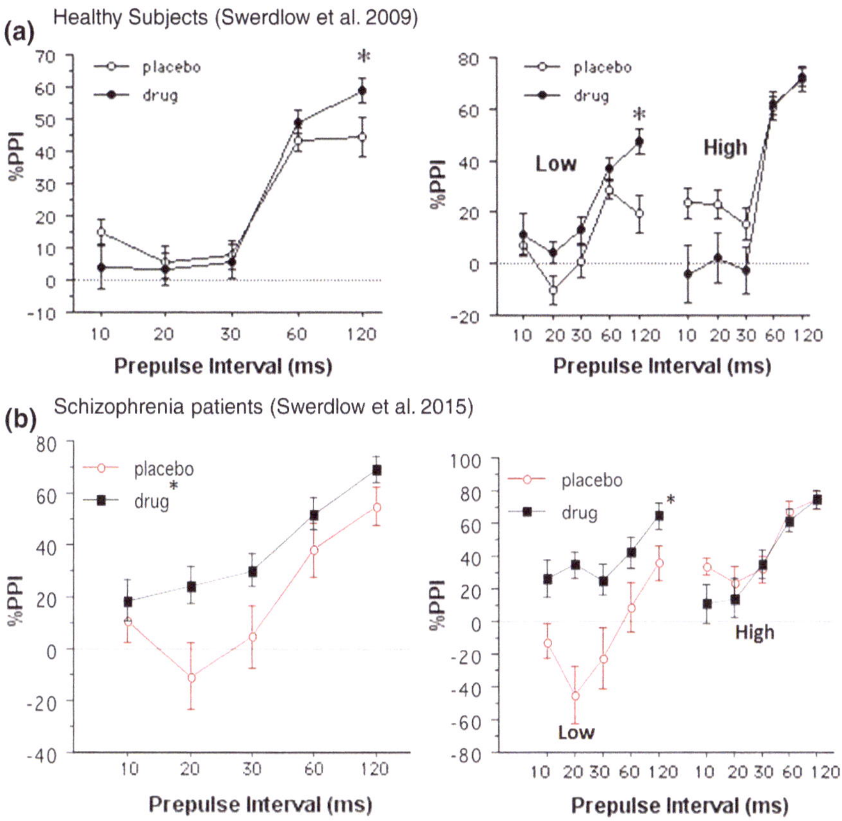

Fig. 1 **a** PPI in healthy subjects (HS) tested in a double-blind, placebo-controlled study of memantine (0 vs. 20 mg po), reported in Swerdlow et al. (2009). Data from all subjects are at *left*; at *right*, results are divided to show individuals with low versus high baseline PPI (grouped based on a median split). Memantine significantly enhanced PPI for 120-ms intervals in the inclusive group of HS (*asterisk*), but these effects were pronounced in "low gating" HS (*asterisk*) and absent in "high gating" HS. **b** Data from an identical paradigm in schizophrenia patients (Swerdlow et al. 2016). Again, memantine (20 mg po) significantly enhanced PPI in an inclusive group of schizophrenia patients (*left*; *asterisk*), and these effects were pronounced in "low gating" patients (*asterisk*) and absent in "high gating" patients. The next important question being assessed is whether PPI-enhancing effects of memantine predict properties beneficial to the therapeutic impact of a cognitive intervention in schizophrenia

suggest that schizophrenia patients are very sensitive to the PPI-enhancing effects of memantine (Fig. 1b), particularly among patients with low basal PPI levels; studies in progress are examining other potential predictors of memantine-enhanced PPI, as well as memantine-enhanced neurocognition in schizophrenia patients. These findings would suggest that the circuitry responsible for sensorimotor gating remains sufficiently intact and dynamic in schizophrenia patients to permit an increase in PPI in response to an acute drug challenge. Conceivably, this plasticity may represent a neural resource that could be engaged in a therapeutic capacity, which is a core tenet of the "PACT" strategy (Swerdlow 2011a, b). This is not to say that a single dose of memantine would be expected to have therapeutic effects in schizophrenia patients; however, the neural signal elicited by this drug challenge provides evidence that mechanisms can be accessed that lead to neurobehavioral evidence of enhanced sensorimotor gating. Memantine engaged the "target" circuitry regulating PPI, and the resulting signal provides a metric of specific available neural resources within any given individual. The ultimate test of this "PACT" predictive model will be to determine whether memantine-enhanced PPI predicts sensitivity to the ability of memantine to augment the therapeutic benefits of a cognitive intervention in these patients. We are pursuing a similar design with other PPI-enhancing drugs from different chemical classes (Chou et al. 2013b; Swerdlow et al. 2013a, b; Bhakta et al. 2014).

5 Conclusion

Observations of deficient PPI in schizophrenia patients, and in patients with a number of other brain disorders, stimulated the development and extension of cross-species models deficient in PPI. Variations of these models have achieved face, predictive, and construct validity for the loss of PPI in schizophrenia patients. Predictive validity has confirmed AP potential in a number of established drugs and novel compounds, but has not yielded any "breakthrough" therapies for schizophrenia. Construct validity has been used to understand the neurobiology of developmental insults and genes that lead to deficient PPI in rodents, but there is no clear pathway from this new information to a deeper understanding of the anatomically and genetically heterogenous underpinnings of the schizophrenias. More generally, the fact that pathogenesis of the schizophrenias appears to begin very early in the brain development and is associated with variable abnormalities in perhaps dozens of brain regions makes it unclear how—despite their 3 levels of validity—PPI models will prove useful in identifying the causes of, or effective treatments for, these disorders. We have described our preliminary experience with an alternative use of cross-species measures of PPI, to identify plasticity within PPI-regulatory neural mechanisms, that might be leveraged toward augmenting the therapeutic impact of cognitive therapies. It is clearly too early to suggest an abandonment of other efforts to develop and apply other animal models of PPI, but at some point, it becomes worthwhile to move beyond models that are valid, in search of ones that might have clinical utility for our patients.

Acknowledgements NRS is supported by NIMH awards MH59803, MH93453, MH42228, and MH094320. GL is supported by MH42228, MH065571, MH094151, MH093453, MH094320, UL1TR000100, MH081944, NARSAD, the Veterans Medical Research Foundation, and the VISN-22 Mental Illness, Research, Education, and Clinical Center. GL has served as a consultant for Astellas Inc, Forum, and Neuroverse for work unrelated to this chapter. The authors have no conflicts of interest.

References

Abel KM, Allin MP, Hemsley DR, Geyer MA (2003) Low doses of ketamine increases prepulse inhibition in healthy men. Neuropharmacology 44:729–737

Acheson DT, Stein MB, Paulus MP, Geyer MA, Risbrough VB (2012) The effect of pregabalin on sensorimotor gating in 'low' gating humans and mice. Neuropharmacology 63:480–485

Ahmari SE, Risbrough VB, Geyer MA, Simpson HB (2012) Impaired sensorimotor gating in unmedicated adults with obsessive-compulsive disorder. Neuropsychopharmacology 37:1216–1223

Angelov SD, Dietrich C, Krauss JK, Schwabe K (2014) Effect of deep brain stimulation in rats selectively bred for reduced prepulse inhibition. Brain Stimul 7(4):595–602

Angst MJ, Macedo CE, Guiberteau T, Sandner G (2007) Alteration of conditioned emotional response and conditioned taste aversion after neonatal ventral hippocampus lesions in rats. Brain Res 1143:183–192

Anticevic A, Brumbaugh MS, Winkler AM, Lombardo LE, Barrett J, Corlett PR, Kober H, Gruber J, Repovs G, Cole MW, Krystal JH, Pearlson GD, Glahn DC (2013) Global prefrontal and fronto-amygdala dysconnectivity in bipolar I disorder with psychosis history. Biol Psychiatry 73:565–573

Bakshi VP, Swerdlow NR, Geyer MA (1994) Clozapine antagonizes phencyclidine-induced deficits in sensorimotor gating of the startle response. J Pharmacol Exp Ther 271:787–794

Bhakta SG, Talledo JA, Lamb SN, Balvaneda B, Chou HH, Rana B, Young J, Light G, Swerdlow NR (2014) Effects of Tolcapone on neurocognitive and neurophysiological measures in healthy adults. Neuropsychopharmacology 39:S514

Braff D, Stone C, Callaway E, Geyer M, Glick I, Bali L (1978) Prestimulus effects on human startle reflex in normals and schizophrenics. Psychophysiology 15:339–343

Carter RJ, Lione LA, Humby T, Mangiarini L, Mahal A, Bates GP, Dunnett SB, Morton AJ (1999) Characterization of progressive motor deficits in mice transgenic for the human Huntington's disease mutation. J Neurosci 19:3248–3257

Castellan Baldan L, Williams KA, Gallezot JD, Pogorelov V, Rapanelli M, Crowley M, Anderson GM, Loring E, Gorczyca R, Billingslea E, Wasylink S, Panza KE, Ercan-Sencicek AG, Krusong K, Leventhal BL, Ohtsu H, Bloch MH, Hughes ZA, Krystal JH, Mayes L, de Araujo I, Ding YS, State MW, Pittenger C (2014) Histidine decarboxylase deficiency causes tourette syndrome: parallel findings in humans and mice. Neuron 81:77–90

Castellanos FX, Fine EJ, Kaysen DL, Marsh WL, Rapoport JL, Hallett M (1996) Sensorimotor gating in boys with Tourette's Syndrome and ADHD: preliminary results. Biol Psychiatry 39:33–41

Chambers RA, Sentir AM, Conroy SK, Truitt WA, Shekhar A (2010) Cortical-striatal integration of cocaine history and prefrontal dysfunction in animal modeling of dual diagnosis. Biol Psychiatry 67:788–792

Charles R, Sakurai T, Takahashi N, Elder GA, Gama Sosa MA, Young LJ, Buxbaum JD (2014) Introduction of the human AVPR1A gene substantially alters brain receptor expression patterns and enhances aspects of social behavior in transgenic mice. Dis Model Mech 7:1013–1022

Chou HH, Bhakta SG, Talledo JA, Lamb SN, Balvaneda B, Light GA, Twamley EW, Swerdlow NR (2013a) Memantine effects on MATRICS consensus cognitive performance battery in healthy adults and schizophrenia patients. Biol Psychiatry 73:273S

Chou HH, Talledo J, Lamb S, Thompson WK, Swerdlow NR (2013b) Amphetamine effects on MATRICS consensus cognitive battery performance in healthy adults. Psychopharmacology 227:165–176

Daenen EW, Wolterink G, Van Der Heyden JA, Kruse CG, Van Ree JM (2003) Neonatal lesions in the amygdala or ventral hippocampus disrupt prepulse inhibition of the acoustic startle response; implications for an animal model of neurodevelopmental disorders like schizophrenia. Eur Neuropsychopharmacol 13:187–197

Darbra S, Modol L, Llido A, Casas C, Vallee M, Pallares M (2014) Neonatal allopregnanolone levels alteration: effects on behavior and role of the hippocampus. Prog Neurobiol 113:95–105
Davis M (1984) The mammalian startle response. In Eaton RC (ed) Neural mechanisms of startle behavior. Springer Science, New York, pp 287–351
Dietz DM, Kennedy PJ, Sun H, Maze I, Gancarz AM, Vialou V, Koo JW, Mouzon E, Ghose S, Tamminga CA, Nestler EJ (2014) ΔFosB induction in prefrontal cortex by antipsychotic drugs is associated with negative behavioral outcomes. Neuropsychopharmacology 39:538–544
Driesen NR, McCarthy G, Bhagwagar Z, Bloch M, Calhoun V, D'Souza DC, Gueorguieva R, He G, Ramachandran R, Suckow RF, Anticevic A, Morgan PT, Krystal JH (2013) Relationship of resting brain hyperconnectivity and schizophrenia-like symptoms produced by the NMDA receptor antagonist ketamine in humans. Mol Psychiatry 18:1199–1204
Duncan EJ, Madonick SH, Parwani A, Angrist B, Rajan R, Chakravorty S et al (2001) Clinical and sensorimotor gating effects of ketamine in normals. Neuropsychopharmacology 25:72–83
Francis DD, Szegda K, Campbell G, Martin WD, Insel TR (2003) Epigenetic sources of behavioral differences in mice. Nat Neurosci 6:445–446
Frankland PW, Wang Y, Rosner B, Shimizu T, Balleine BW, Dykens EM, Ornitz EM, Silva AJ (2004) Sensorimotor gating abnormalities in young males with fragile X syndrome and Fmr1-knockout mice. Mol Psychiatry 9:417–425
Geyer MA, Krebs-Thomson K, Braff DL, Swerdlow NR (2001) Pharmacological studies of prepulse inhibition models of sensorimotor gating deficits in schizophrenia: a decade in review. Psychopharmacology 156:117–154
Gomez-Wong E, Marti MJ, Tolosa E, Valls-Solé J (1998) Sensory modulation of the blink reflex in patients with blepharospasm. Arch Neurol 55:1233–1237
Graham F (1975) The more or less startling effects of weak prestimuli. Psychophysiology 12:238–248
Greenwood TA, Braff DL, Light GA, Cadenhead KS, Calkins ME, Dobie DJ, Freedman R, Green MF, Gur RE, Gur RC, Mintz J, Nuechterlein KH, Olincy A, Radant AD, Seidman LJ, Siever LJ, Silverman JM, Stone WS, Swerdlow NR, Tsuang DW, Tsuang MT, Turetsky BI, Schork NJ (2007) The consortium on the genetics of schizophrenia (COGS): initial heritability analyses of endophenotypic measures for schizophrenia. Arch Gen Psychiatry 64:1242–1250
Hanlon FM, Houck JM, Klimaj SD, Caprihan A, Mayer AR, Weisend MP, Bustillo JR, Hamilton DA, Tesche CD (2012) Frontotemporal anatomical connectivity and working-relational memory performance predict everyday functioning in schizophrenia. Psychophysiology 49:1340–1352
Haut KM, Lim KO, MacDonald A (2010) Prefrontal cortical changes following cognitive training in patients with chronic schizophrenia: effects of practice, generalization and specificity. Neuropsychopharmacology 35:1850–1859
Hines RM, Hines DJ, Houston CM, Mukherjee J, Haydon PG, Tretter V, Smart TG, Moss SJ (2013) Disrupting the clustering of GABAA receptor α2 subunits in the frontal cortex leads to reduced γ-power and cognitive deficits. Proc Natl Acad Sci USA 110(41):16628–16633
Hoenig K, Hochrein A, Quednow BB, Maier W, Wagner M (2005) Impaired prepulse inhibition of acoustic startle in obsessive-compulsive disorder. Biol Psychiatry 57:1153–1158
Janssen PAJ, Niemegeers CJE (1959) Chemistry and pharmacology of compounds related to 4-(4-hydroxy-Cphenyl-piperidino)-butyrophenone. Part II. Inhibition of apomorphine vomiting in dogs. Arzneimittel-Forsch 9:765–767
Johansson C, Jackson DM, Svensson L (1994) The atypical antipsychotic, remoxipride, blocks phencyclidine-induced disruption of prepulse inhibition in the rat. Psychopharmacology 116:437–442
Karten HJ (1991) Homology and evolutionary origins of the 'Neocortex'. Brain Behav Evolution 38:264–272
Kodsi MH, Swerdlow NR (1994) Quinolinic acid lesions of the ventral striatum reduce sensorimotor gating of acoustic startle in rats. Brain Res 643:59–65

Kumari V, Gray JA, Geyer MA, Soni W, Mitterschiffthaler MT, Vythelingum GN, Simmons A, Williams SC, Sharma T (2003) Neural correlates of tactile prepulse inhibition: a functional MRI study in normal and schizophrenic subjects. Psychiatry Res 122(2):99–113

Kumari V, Peters ER, Fannon D, Antonova E, Premkumar P, Anilkumar AP, Williams SC, Kuipers E (2009) Dorsolateral prefrontal cortex activity predicts responsiveness to cognitive-behavioral therapy in schizophrenia. Biol Psychiatry 66:594–602

Kumari V, Premkumar P, Fannon D, Aasen I, Raghuvanshi S, Anilkumar AP, Antonova E, Peters ER, Kuipers E (2012) Sensorimotor gating and clinical outcome following cognitive behaviour therapy for psychosis. Schizophr Res 134:232–238

Labbate GP, da Silva AV, Barbosa-Silva RC (2014) Effect of severe neonatal seizures on prepulse inhibition and hippocampal volume of rats tested in early adulthood. Neurosci Lett 568:62–66

Le Pen G, Moreau JL (2002) Disruption of prepulse inhibition of startle reflex in a neurodevelopmental model of schizophrenia: reversal by clozapine, olanzapine and risperidone but not by haloperidol. Neuropsychopharm 27:1–11

Le Pen G, Kew J, Alberati D, Borroni E, Heitz MP, Moreau JL (2003) Prepulse inhibition deficits of the startle reflex in neonatal ventral hippocampal-lesioned rats: reversal by glycine and a glycine transporter inhibitor. Biol Psychiatry 54:1162–1170

Le Pen G, Gourevitch R, Hazane F, Hoareau C, Jay TM, Krebs MO (2006) Peri-pubertal maturation after developmental disturbance: a model for psychosis onset in the rat. Neuroscience 143:395–405

Levitt JJ, Bobrow L, Lucia D, Srinivasan P (2010) A selective review of volumetric and morphometric imaging in schizophrenia. In: Swerdlow NR (ed) Behavioral neurobiology of schizophrenia and its treatment. Current Topics in Behavioral Neuroscience, Springer, pp 243–282

Lewis DA, Levitt P (2002) Schizophrenia as a disorder of neurodevelopment. Annu Rev Neurosci 25:409–432

Lieberman JA, Stroup TS, McEvoy JP, Swartz MS, Rosenheck RA, Perkins DO, Keefe RS, Davis SM, Davis CE, Lebowitz BD, Severe J, Hsiao JK (2005) clinical antipsychotic trials of intervention effectiveness (CATIE) investigators. Effectiveness of antipsychotic drugs in patients with chronic schizophrenia. N Engl J Med 353:1209–1223

Lind NM, Arnfred SM, Hemmingsen RP, Hansen AK (2004) Prepulse inhibition of the acoustic startle reflex in pigs and its disruption by d-amphetamine. Behav Brain Res 155:217–222

Linn GS, Negi SS, Gerum SV, Javitt DC (2003) Reversal of phencyclidine-induced prepulse inhibition deficits by clozapine in monkeys. Psychopharmacology 169:234–239

Lipska BK, Jaskiw GE, Weinberger DR (1993) Postpubertal emergence of hyperresponsiveness to stress and to amphetamine after neonatal excitotoxic hippocampal damage: a potential animal model of schizophrenia. Neuropsychopharm 9:67–75

Lipska BK, Swerdlow NR, Geyer MA, Jaskiw GE, Braff DL, Weinberger DR (1995) Neonatal excitotoxic hippocampal damage in rats causes post-pubertal changes in prepulse inhibition of startle and its disruption by apomorphine. Psychopharmacology 122:35–43

Ma J, Leung LS (2014) Deep brain stimulation of the medial septum or nucleus accumbens alleviates psychosis-relevant behavior in ketamine-treated rats. Behav Brain Res 266:174–182

MacLean PD (1954) The limbic system and its hippocampal formation; studies in animals and their possible application to man. J Neurosurg 11(1):29–44

Mansbach RS, Geyer MA, Braff DL (1988) Dopaminergic stimulation disrupts sensorimotor gating in the rat. Psychopharmacology 94:507–514

Marquis JP, Goulet S, Doré FY (2006) Neonatal lesions of the ventral hippocampus in rats lead to prefrontal cognitive deficits at two maturational stages. Neuroscience 140:759–767

Marquis JP, Goulet S, Doré FY (2008). Neonatal ventral hippocampus lesions disrupt extra-dimensional shift and alter dendritic spine density in the medial prefrontal cortex of juvenile rats. Neurobiol Learn Mem 90:339–346

McAlonan GM, Daly E, Kumari V, Critchley HD, van Amelsvoort T, Suckling J, Simmons A, Sigmundsson T, Greenwood K, Russell A, Schmitz N, Happe F, Murphy DG (2002) Brain anatomy and sensorimotor gating in Asperger's syndrome. Brain 125:1594–1606

Meltzer HY, Alphs L, Green AI, Altamura AC, Anand R, Bertoldi A, Bourgeois M, Chouinard G, Islam MZ, Kane J, Krishnan R, Lindenmayer JP, Potkin S, International Suicide Prevention Trial Study Group (2003) Clozapine treatment for suicidality in schizophrenia: international suicide prevention trial (InterSePT). Arch Gen Psychiatry 60:82–91

Miller EJ, Saint Marie LR, Breier MR, Swerdlow NR (2010) Pathways from the ventral hippocampus and caudal amygdala to forebrain regions that regulate sensorimotor gating in the rat. Neuroscience 165(2):601–611

Murray RM, Jones P, O'Callaghan E (1991) Fetal brain development and later schizophrenia. Ciba Found Symp 156:155–163

Nguyen R, Morrissey MD, Mahadevan V, Cajanding JD, Woodin MA, Yeomans JS, Takehara-Nishiuchi K, Kim JC (2014) Parvalbumin and GAD65 interneuron inhibition in the ventral hippocampus induces distinct behavioral deficits relevant to schizophrenia. J Neurosci 34:14948–14960

O'Donnell P (2012) Cortical disinhibition in the neonatal ventral hippocampal lesion model of schizophrenia: New vistas on possible therapeutic approaches. Pharmacol Ther 133:19–25

Ornitz EM, Hanna GL, de Traversay J (1992) Prestimulation-induced startle modulation in attention deficit hyperactivity disorder and nocturnal enuresis. Psychophysiology 29:437–451

Palmer DD, Henter ID, Wyatt RJ (1999) Do antipsychotic medications decrease the risk of suicide in patients with schizophrenia? J Clin Psychiatry 60(Suppl 2):100–103

Perez VB, Swerdlow NR, Braff DL, Naatanen R, Light GA (2014) Using biomarkers to inform diagnosis, guide treatments and track response to interventions in psychotic illnesses. Biomark Med 8(1):9–14

Pinnock F, Bosch D, Brown T, Simons N, Yeomans JR, DeOliveira C, Schmid S (2015) Nicotine receptors mediating sensorimotor gating and its enhancement by systemic nicotine. Front Behav Neurosci 9:30. doi:10.3389/fnbeh.2015.00030

Posch DK, Schwabe K, Krauss JK, Lütjens G (2012) Deep brain stimulation of the entopeduncular nucleus in rats prevents apomorphine-induced deficient sensorimotor gating. Behav Brain Res 232:130–136

Powell SB, Swerdlow NR (2015) Social isolation rearing and sensorimotor gating in rat models of relevance to schizophrenia: what we know, and what we don't. In: Pletnikov M, Waddington J (eds) Modeling psychopathological dimensions of schizophrenia. Handbooks of Behavioral Neuroscience, vol 23. Elsevier, Amsterdam (in press)

Renoux AJ, Sala-Hamrick KJ, Carducci NM, Frazer M, Halsey KE, Sutton MA, Dolan DF, Murphy GG, Todd PK (2014) Impaired sensorimotor gating in Fmr1 knock out and Fragile X premutation model mice. Behav Brain Res 267:42–45

Ressler KJ, Rothbaum BO, Tannenbaum L, Anderson P, Graap K, Zimand E, Hodges L, Davis M (2004) Cognitive enhancers as adjuncts to psychotherapy: use of D-cycloserine in phobic individuals to facilitate extinction of fear. Arch Gen Psychiatry 61:1136–1144

Ribeiro BM, do Carmo MR, Freire RS, Rocha NF, Borella VC, de Menezes AT, Monte AS, Gomes PX, de Sousa FC, Vale ML, de Lucena DF, Gama CS, Macedo D (2013) Evidences for a progressive microglial activation and increase in iNOS expression in rats submitted to a neurodevelopmental model of schizophrenia: reversal by clozapine. Schizophr Res 151:12–19

Risbrough V, Ji B, Hauger R, Zhou X (2014) Generation and characterization of humanized mice carrying COMT158 Met/Val alleles. Neuropsychopharmacology 39:1823–1832

Rohleder C, Jung F, Mertgens H, Wiedermann D, Sué M, Neumaier B, Graf R, Leweke FM, Endepols H (2014) Neural correlates of sensorimotor gating: a metabolic positron emission tomography study in awake rats. Front Behav Neurosci 8:178

Roussos P, Giakoumaki SG, Rogdaki M, Pavlakis S, Frangou S, Bitsios P (2008) Prepulse inhibition of the startle reflex depends on the catechol O-methyltransferase Val158Met gene polymorphism. Psychol Med 38:1651–1658

Ryan RT, Bhardwaj SK, Tse YC, Srivastava LK, Wong TP (2013) Opposing alterations in excitation and inhibition of layer 5 medial prefrontal cortex pyramidal neurons following neonatal ventral hippocampal lesion. Cereb Cortex 23:1198–1227

Schizophrenia Working Group of the Psychiatric Genomics Consortium (2014) Biological insights from 108 schizophrenia-associated genetic loci. Nature 511:421–427
Schwartz JM (1997) Brain lock: free yourself from obsessive-compulsive behavior. Regan Books, New York, p 1997
Schwartz JM (1998) Neuroanatomical aspects of cognitive-behavioural therapy response in obsessive-compulsive disorder. An evolving perspective on brain and behaviour. Br J Psychiatry Suppl 35:38–44
Schwartz JM, Stoessel PW, Baxter LR Jr, Martin KM, Phelps ME (1996) Systematic changes in cerebral glucose metabolic rate after successful behavior modification treatment of obsessive-compulsive disorder. Arch Gen Psychiatry 53:109–113
Shilling PD, Saint Marie RL, Shoemaker JM, Swerdlow NR (2008) Strain differences in the gating-disruptive effects of apomorphine: relationship to gene expression in nucleus accumbens signaling pathways. Biol Psychiatry 63:748–758
Sobin C, Kiley-Brabeck K, Karayiorgou M (2005) Lower prepulse inhibition in children with the 22q11 deletion syndrome. Am J Psychiatry 162:1090–1099
Sorenson CA, Swerdlow NR (1982) The effect of tail pinch on the acoustic startle response in rats. Brain Res 247:105–113
Stefansson H, Meyer-Lindenberg A, Steinberg S, Magnusdottir B, Morgen K, Arnarsdottir S, Bjornsdottir G, Walters GB, Jonsdottir GA, Doyle OM, Tost H, Grimm O, Kristjansdottir S, Snorrason H, Davidsdottir SR, Gudmundsson LJ, Jonsson GF, Stefansdottir B, Helgadottir I, Haraldsson M, Jonsdottir B, Thygesen JH, Schwarz AJ, Didriksen M, Stensbøl TB, Brammer M, Kapur S, Halldorsson JG, Hreidarsson S, Saemundsen E, Sigurdsson E, Stefansson K (2014) CNVs conferring risk of autism or schizophrenia affect cognition in controls. Nature 505:361–366
Sun SX, Liu GG, Christensen DB, Fu AZ (2007) Review and analysis of hospitalization costs associated with antipsychotic nonadherence in the treatment of schizophrenia in the United States. Curr Med Res Opin 23:2305–2312
Swerdlow NR (2011a) Are we studying and treating schizophrenia correctly? Schizophr Res 130:1–10
Swerdlow NR (2011b) Beyond antipsychotics: Pharmacologically-augmented cognitive therapies (PACTs) for schizophrenia. Neuropsychopharmacology 37:310–311
Swerdlow NR, Geyer MA (1993) Clozapine and haloperidol in an animal model of sensorimotor gating deficits in schizophrenia. Pharmacol Biochem Behav 44:741–744
Swerdlow NR, Braff DL, Geyer MA, Koob GF (1986) Central dopamine hyperactivity in rats mimics abnormal acoustic startle response in schizophrenics. Biol Psychiatry 21:23–33
Swerdlow NR, Keith VA, Braff DL, Geyer MA (1991) Effects of spiperone, raclopride, SCH 23390 and clozapine on apomorphine inhibition of sensorimotor gating of the startle response in the rat. J Pharmacol Exp Ther 256:530–536
Swerdlow NR, Caine SB, Braff DL, Geyer MA (1992a) The neural substrates of sensorimotor gating of the startle reflex: a review of recent findings and their implications. J Psychopharmacol 6:176–190
Swerdlow NR, Caine SB, Geyer MA (1992b) Regionally selective effects of intracerebral dopamine infusion on sensorimotor gating of the startle reflex in rats. Psychopharmacology 108:189–195
Swerdlow NR, Benbow CH, Zisook S, Geyer MA, Braff DL (1993) A preliminary assessment of sensorimotor gating in patients with obsessive compulsive disorder. Biol Psychiatry 33:298–301
Swerdlow NR, Braff DL, Taaid N, Geyer MA (1994a) Assessing the validity of an animal model of deficient sensorimotor gating in schizophrenic patients. Arch Gen Psychiatry 51:139–154
Swerdlow NR, Zisook D, Taaid N (1994b) Seroquel (ICI 204, 636) restores prepulse inhibition of acoustic startle in apomorphine-treated rats: similarities to clozapine. Psychopharmacology 114:675–678

Swerdlow NR, Paulsen J, Braff DL, Butters N, Geyer MA, Swenson MR (1995) Impaired prepulse inhibition of acoustic and tactile startle response in patients with Huntington's disease. J Neurol Neurosurg Psychiatry 58:192–200

Swerdlow NR, Bakshi V, Waikar M, Taaid N, Geyer MA (1998) Seroquel, clozapine and chlorpromazine restore sensorimotor gating in ketamine-treated rats. Psychopharmacology 140:75–80

Swerdlow NR, Geyer MA, Braff DL (2001a) Neural circuit regulation of prepulse inhibition of startle in the rat: current knowledge and future challenges. Psychopharmacology 156:194–215

Swerdlow NR, Karban B, Ploum Y, Sharp R, Geyer MA, Eastvold A (2001b) Tactile prepuff inhibition of startle in children with Tourette's syndrome: in search of an "fMRI-friendly" startle paradigm. Biol Psychiatry 50:578–585

Swerdlow NR, Stephany N, Shoemaker JM, Ross L, Wasserman LC, Talledo J, Auerbach PP (2002) Effects of amantadine and bromocriptine on startle and sensorimotor gating: parametric studies and cross-species comparisons. Psychopharmacology 164:82–92

Swerdlow NR, Shoemaker JM, Platten A, Pitcher L, Goins J, Auerbach PP (2004) Heritable differences in the dopaminergic regulation of sensorimotor gating. I. Apomorphine effects on startle gating in albino and hooded outbred rat strains and their F1 and N2 progeny. Psychopharmacology 174:441–451

Swerdlow NR, Talledo J, Sutherland AN, Nagy D, Shoemaker JM (2006) Antipsychotic effects on prepulse inhibition in normal 'low gating' humans and rats. Neuropsychopharmacology 31:2011–2021

Swerdlow NR, Weber M, Qu Y, Light GA, Braff DL (2008) Realistic expectations of prepulse inhibition in translational models for schizophrenia research. Psychopharmacology 199:331–388

Swerdlow NR, van Bergeijk DP, Bergsma F, Weber E, Talledo J (2009) The effects of memantine on prepulse inhibition. Neuropsychopharmacology 34:1854–1864

Swerdlow NR, Light GA, Breier MR, Shoemaker JM, Saint Marie RL, Neary AC, Geyer MA, Stevens KE, Powell SB (2012a) Sensory and sensorimotor gating deficits after neonatal ventral hippocampal lesions in rats. Dev Neurosci 34:240–249

Swerdlow NR, Shilling PD, Breier M, Trim RS, Light GA, Marie RS (2012b) Fronto-temporal-mesolimbic gene expression and heritable differences in amphetamine-disrupted sensorimotor gating in rats. Psychopharmacology 224:349–362

Swerdlow NR, Bhakta SG, Talledo JA et al (2013a) Sensorimotor gating predicts sensitivity to pro-attentional effects of amphetamine in healthy adults. Society for Neuroscience, San Diego, 9–13 Nov 2013

Swerdlow NR, Powell SB, Breier MR, Hines SR, Light GA (2013b) Coupling of gene expression in medial prefrontal cortex and nucleus accumbens after neonatal ventral hippocampal lesions accompanies deficits in sensorimotor gating and auditory processing in rats. Neuropharmacology 75:38–46

Swerdlow NR, Light GA, Sprock J, Calkins ME, Green MF, Greenwood TA, Gur RE, Gur RC, Lazzeroni LC, Nuechterlein KH, Radant AD, Ray A, Seidman LJ, Siever LJ, Silverman JM, Stone WS, Sugar CA, Tsuang DW, Tsuang MT, Turetsky BI, Braff DL (2014) Deficient prepulse inhibition in schizophrenia detected by the multi-site COGS. Schizophr Res 152:503–512

Swerdlow NR, Bhakta S, Chou HH, Talledo JA, Balvaneda B, Light GA (2016) Memantine effects on sensorimotor gating and mismatch negativity in patients with chronic psychosis. Neuropsychopharmacology 41:419–430. doi:10.1038/npp.2015.162)

Taub E, Uswatte G, Elbert T (2002) New treatments in neurorehabilitation founded on basic research. Nat Rev Neurosci 3:228–236

Uehara T, Sumiyoshi T, Seo T, Matsuoka T, Itoh H, Suzuki M, Kurachi M (2010) Neonatal exposure to MK-801, an N-methyl-D-aspartate receptor antagonist, enhances methamphetamine-induced locomotion and disrupts sensorimotor gating in pre- and postpubertal rats. Brain Res 1352:223–230

Vaillancourt C, Boksa P (2000) Birth insult alters dopamine-mediated behavior in a precocial species, the guinea pig. Implications for schizophrenia. Neuropsychopharmacology 23:654–666

Valls-Sole J, Munoz JE, Valldeoriola F (2004) Abnormalities of prepulse inhibition do not depend on blink reflex excitability: a study in Parkinson's disease and Huntington's disease. Clin Neurophysiol 115:1527–1536
van Rijn S, Swaab H, Magnée M, van Engeland H, Kemner C (2011) Psychophysiological markers of vulnerability to psychopathology in men with an extra X chromosome (XXY). PLoS ONE 6:e20292
Vazquez-Roque RA, Solis O, Camacho-Abrego I, Rodriguez-Moreno A, Cruz Fde L, Zamudio S, Flores G (2012) Dendritic morphology of neurons in prefrontal cortex and ventral hippocampus of rats with neonatal amygdala lesion. Synapse 66:73–382
Vollenweider FX, Barro M, Csomor PA, Feldon J (2006) Clozapine enhances prepulse inhibition in healthy humans with low but not with high prepulse inhibition levels. Biol Psychiatry 60:597–603
Weinberger DR (1987) Implications of normal brain development for the pathogenesis of schizophrenia. Arch Gen Psychiatry 44:660–669
Zhu F, Zhang L, Ding YQ, Zhao J, Zheng Y (2014a) Neonatal intrahippocampal injection of lipopolysaccharide induces deficits in social behavior and prepulse inhibition and microglial activation in rats: Implication for a new schizophrenia animal model. Brain Behav Immun 38:166–174
Zhu F, Zheng Y, Ding YQ, Li Y, Zhang X, Wu R, Guo X, Zhao J (2014b) Minocycline and risperidone prevent microglia activation and rescue behavioral deficits inducted by neonatal intrahippocampal injection of lipopolysaccharide in rats. PLoS ONE 9:393966

Attention and the Cholinergic System: Relevance to Schizophrenia

Cindy Lustig and Martin Sarter

Abstract Traditional methods of drug discovery often rely on a unidirectional, "bottom-up" approach: A search for molecular compounds that target a particular neurobiological substrate (e.g., a receptor type), the refinement of those compounds, testing in animal models using high-throughput behavioral screening methods, and then human testing for safety and effectiveness. Many attempts have found the "effectiveness" criterion to be a major stumbling block, and we and others have suggested that success may be improved by an alternative approach that considers the neural circuits mediating the effects of genetic and molecular manipulations on behavior and cognition. We describe our efforts to understand the cholinergic system's role in attention using parallel approaches to test main hypotheses in both rodents and humans as well as generating converging evidence using methods and levels of analysis tailored to each species. The close back-and-forth between these methods has enhanced our understanding of the cholinergic system's role in attention both "bottom-up" and "top-down"—that is, the basic neuroscience identifies potential neuronal circuit-based mechanisms of clinical symptoms, and the patient and genetic populations serve as natural experiments to test and refine hypotheses about its contribution to specific processes. Together, these studies have identified (at least) two major and potentially independent contributions of the cholinergic system to attention: a neuromodulatory component that influences cognitive control in response to challenges from distractors that either make detection more difficult or draw attention away from the distractor, and a phasic or transient cholinergic signal that instigates a shift from ongoing behavior and the activation of cue-associated response. Right prefrontal cortex appears to play a particularly important role in the neuromodulatory component integrating motivational and cognitive influences for top-down control across populations, whereas the transient cholinergic signal involves orbitofrontal regions associated with shifts between internal and external attention. Understanding how these two modes of

C. Lustig (✉) · M. Sarter
Department of Psychology, University of Michigan, 530 Church Street, Ann Arbor, MI 48103, USA
e-mail: clustig@umich.edu

Curr Topics Behav Neurosci (2016) 28: 327–362
DOI 10.1007/7854_2015_5009

cholinergic function interact and are perturbed in schizophrenia will be an important prerequisite for developing effective treatments.

Keywords Schizophrenia · Acetylcholine · Attentional effort · Right prefrontal cortex · Cross-species

Contents

1 Attentional CRUNCH points in Schizophrenia: A Special Role for Right PFC? ... 330
2 Cholinergic-Attentional Control Deficits: Anatomical Foundations and Evidence from Rodent Studies ... 335
2.1 Cortical Cholinergic Projections: Anatomical Aspects Consistent with Top-down Functions ... 335
2.2 Multiple Time Scales of Cholinergic Signaling Mediate Distinct Attentional Processing Steps ... 337
3 Cholinergic Dysregulation in Schizophrenia: Possible Causes and Consequences ... 338
4 Parallel Rodent–Human Studies Implicating Right PFC ACh Dysfunction in Impaired Schizophrenia Responses to Attentional Challenge ... 339
5 Cholinergic and Dopaminergic Interactions with Right PFC: Integrating Attention and Motivation? ... 341
6 Cholinergic Transients: Spared, Impaired, or Overactive? ... 345
7 Implications for Treatment Development and Translational Research ... 347
8 Conclusions ... 349
References ... 352

Deficits in controlled attention are a primary cognitive symptom of schizophrenia. They are present before the first psychotic episode and in putatively healthy relatives, are not well-treated by standard dopaminergic medications, and persist even in remission states (Cornblatt and Keilp 1994; Nuechterlein and Dawson 1984; Wohlberg and Kornetsk 1973; see Lesh et al. 2011 for a recent review). These deficits are critical to address, as in both cross-sectional and longitudinal studies, impairments in controlled attention and closely-related executive functions are among the strongest predictors of real-world outcomes including social functioning and work skills (e.g., Bowie et al. 2008; Green 1996; Torgalesboen et al. 2014, 2015). The cholinergic system is an attractive target for potential pharmaceutical interventions because it plays a critical role in attention (see Hasselmo and Sarter 2011; Sarter et al. 2014; Sahakian et al. 1989, 1993) and is disrupted in schizophrenia. Here, we review some of the difficulties in developing such interventions, emerging approaches and results that may help overcome those difficulties, and promising avenues for future research.

Despite strong a priori reasons to believe that cholinergic treatments should benefit cognition in schizophrenia, attempts to develop them have been largely disappointing (see reviews by Foster et al. 2014; Money et al. 2010; Rowe et al. 2015; but see Meltzer 2015 for a more optimistic view). Several factors play into this. One is that in contrast to most antipsychotics, where drug development has relied heavily on "me too" variations of early discoveries, there is a lack of initial effective agents to serve as a starting point (Young et al. 2010). Another, common to almost all areas of psychiatric drug research, is the paucity of well-validated paradigms that allow preclinical research in animal models to translate to effectiveness in treating human neuropathology (Markou et al. 2009; Wong et al. 2010). A third is that most existing drugs are based on theoretical models that treat the cholinergic system primarily as a diffuse neuromodulator (e.g., Briand et al. 2007; Dani and Bertrand 2007; Picciotto et al. 2012). Burgeoning evidence points to both regionally and temporally specific cholinergic functions that interact with each other but are independent and dissociable (Hasselmo and Sarter 2011; Sarter et al. 2009b).

With those factors in mind, the limited success of cholinergic procognitive treatments is understandable. However, the more positive news is that the field has become acutely aware of these issues and is working to address them. In particular, the MATRICS (Measurement and Treatment Research to Improve Cognition in Schizophrenia) and CNTRICS (Cognitive Neuroscience Treatment Research to Improve Cognition in Schizophrenia) initiatives have specifically targeted the goal of producing standardized, reliable assessments that can be used in both preclinical and clinical research to assess specific aspects of neurocognitive function. In addition, there is an increasing recognition that there are multiple neurobiological pathways to the same behavioral outcome, and that understanding and targeting the correct pathways will be a critical link between manipulations of cellular mechanisms and successful behavioral change. Taking things a step further, if the ultimate target is improved cognitive function that will generalize to real-world behavior, then one also has to establish the degree to which the behavioral test measures that function. That is, it is not enough for a test to translate from animal model to patient, it also has to translate from the lab or clinic into everyday life.

In discussing these issues and the problems plaguing psychiatric drug discovery more generally, Sarter and Tricklebank (2012) suggested that most drug discovery efforts leap from genetic and molecular targets to changes in behavior in standardized, even reified preclinical behavioral paradigms (e.g., Morris water maze measures spatial learning/memory; Y-maze measures anxiety) without much consideration of the neural circuits that mediate such behaviors, or the cognitive operations that they reflect. They suggested that a more integrated consideration of the neural circuits level would likely lead to better success (Fig. 1).

Others have reviewed the genetic and cellular mechanisms of cholinergic function and their effects on neural circuits (e.g., Bentley et al. 2011; Bloem et al. 2014a, b; Hasselmo and Sarter 2011; Rowe et al. 2015; Riedel et al. 2015; Sarter 2015; Thiele 2013) as well as what is known about schizophrenia-related dysfunction at these levels of analysis (e.g., Gagne et al. 2015; Nikolaus et al. 2014; Seo et al. 2014). Here we begin by describing some of the deficits in attention performance observed in schizophrenia and their neuroimaging correlates in humans. Next, we describe the evidence for cholinergic modulation of neural circuits contributing to those neuroimaging and behavioral findings. Together, these guide hypotheses about the cognitive operations reflected by behavior, how they may be disrupted in schizophrenia and other conditions, and potential avenues for treatment.

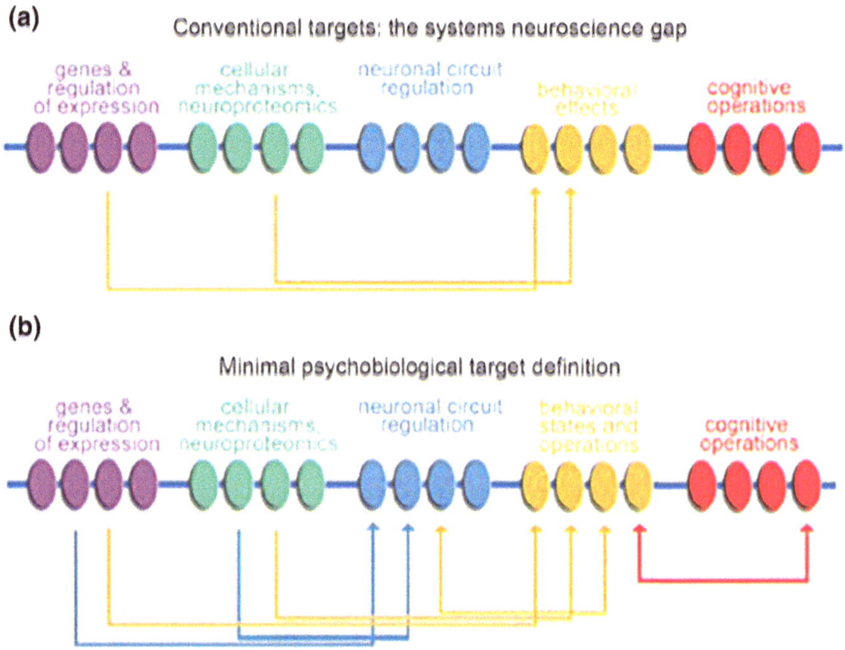

Fig. 1 Conventional and psychobiological target definition (from Sarter and Tricklebank 2012). **a** Current efforts at drug development have defined targets based on the findings that manipulation of genetic or cellular mechanisms influence behavior, but often fail to consider how neural circuits mediate those influences. As behavior is multiply determined, this is a likely contributor to the low success rate of pharmaceutical development. **b** A recommended approach is to establish whether an intervention affects the activity of neural circuits, and then in turn whether those circuits mediate the behavior in question. Further, consideration of the cognitive operations driving behavior is required to avoid misinterpretation of the behavior [e.g., does failure to respond to a signal represent a failure to see it (perceptual), a failure to activate the task set and responses associated with the signal (cognitive), or a lack of interest or motivation (motivational/emotional)] and facilitate generalization. The present review focuses on cholinergic regulation of frontoparietal circuits involved in attention and the consequences for behavior and cognition in schizophrenia (reproduced with permission)

1 Attentional CRUNCH points in Schizophrenia: A Special Role for Right PFC?

Attention is a very broad-based concept, and the use of this single term to describe what in all likelihood is a wide range of cognitive abilities and processes can lead to significant confusion. Recognizing this, CNTRICS recommended restricting its use in the context of schizophrenia to selective attention, and specifically to the "input selection" function. The related concept of "rule selection" was assigned to executive function. However, the distinction is somewhat arbitrary and made primarily to allow a more manageable grouping of different tasks within the CNTRICS framework (Luck et al. 2012). For example, one might consider selecting a target out of an array of distractors as an example of input selection, but of course rules are required to determine what constitutes the target. At the other extreme, choosing the correct sorting rule in the Wisconsin Card Sort would be an example of rule selection, but the process of subsequently keeping attention on the correct dimension (color, shape, or number) might be considered input selection.

Input selection is defined as restricting processing to a subset of inputs, and determining which inputs are sent to memory and/or response systems (Luck et al. 2012). For example, as you are reading this, you are receiving various other sensory inputs, such as ambient noise from the environment or the pressure from your chair on your back and legs. Well-functioning input selection processes ensure that the words on this page are selected for further processing, rather than those irrelevant inputs. The construct of input selection can be subdivided into those processes responsible for controlling attention versus those involved in implementing that control by increasing the strength of the relevant signal and/or decreasing the strength of irrelevant inputs. (See Box 1 for a description of human and animal paradigms selected by CNTRICS as relevant to input control.)

Patients are thought to have impaired top-down control, but relatively preserved implementation. For example, they are impaired if there is a conflict inherent in a cue [e.g., in anti-saccade tasks, where the task is to attend in the opposite direction of the cue; Fukushima et al. (1990)] or if the cue itself is nonspecific [e.g., identifies several potential target locations that must be simultaneously monitored; Hahn et al. (2012)]. In contrast, if the cue is simple (as in a prosaccade task), patients and controls typically show equivalent benefits to response time and accuracy—that is, they are equally able to implement the modulation of attention in response to that cue. Furthermore, even though sustaining attention over time is typically considered a controlled-attention or executive process, patients do not typically show exaggerated time-on-task declines unless the task that is to be maintained over time puts high demands on input selection (e.g., Demeter et al. 2013; Egeland et al. 2003 c.f., Hahn et al. 2012 for an example with high input-selection demands).

However, equivalent performance at the behavioral level does not necessarily imply equivalence at either the neural circuit or cognitive operations levels. Patients have dysfunctional sensory processing starting at very early stages, and both that dysfunction and attempts to compensate for it lead to widespread, interactive effects. Several studies have shown that patients with schizophrenia have reduced retinal nerve fiber thickness compared to controls, and it is speculated that dopamine-glutamate dysregulation may further affect the processing of remaining cells, e.g., by altering ganglionic receptive fields (see review by Gracitelli et al. 2015). Contrast sensitivity deficits are frequently associated with schizophrenia, although there is some controversy as to whether the magno- and parvo-cellular systems are affected equally or if not, which one is affected more. The answer to this question may differ depending on whether one is considering first-episode versus chronic patients, and on medication status (Shoshina et al. 2014).

Furthermore, adaptations that are beneficial in some circumstances can be detrimental in others. For example, Leonard et al. (2014) found that despite presumed impairments in magnocellular pathways, patients were more vulnerable to distractors designed to activate the magnocellular system than those designed to selectively activate the parvocellular system, whereas healthy controls showed the opposite pattern. The authors speculate that this may occur if later-stage top-down processing attempts to compensate for degraded magnocellular pathways by giving more weight to their inputs. Although this may aid the processing of relevant magnocellular inputs, it has the side effect of also increasing vulnerability to distractors with components (e.g., movement or flickering, contrast rather than hue differences from background) that stimulate this system.

Regardless of their source, such reductions in sensory processing add noise to the inputs that are to be selected among, creating an increased burden on input-selection processes. In addition, modulatory feedback from top-down attention systems affects processing even at these early stages (Laycock et al. 2007; Skottun and Skoyles 2007). In other words, patients may suffer from a "triple-hit": (1) their bottom-up sensory inputs are impaired starting at very early (e.g., retinal) stages, (2) impaired top-down control systems are inefficient at modulating these sensory inputs (e.g., Dima et al. 2010; Silverstein et al. 1996; see discussion by Robinson et al. 2011; Sarter et al. 2005), (3) both of these combine to result in noisier representations that further tax already-impaired top-down control, including input selection. (See Lustig and Jantz 2015 for a parallel argument applied to aging.)

The increased demands on an already-impaired top-down attentional system in schizophrenia may help explain what at first appear to be contradictory findings regarding schizophrenia-related abnormal activation patterns in the frontoparietal regions thought to support top-down control. That is, abnormalities in both structural and functional neuroimaging measures of these regions are typically found in both patients and their putatively healthy relatives, but some activation studies find that patients and their relatives show more activation in these regions than do controls, whereas others find the opposite (see meta-analyses by Gogarhi 2011; Minzenberg et al. 2009; Scognamiglio and Houenou 2014). Thus, there is general agreement that prefrontal processing is abnormal in schizophrenia, but apparent contradiction across studies as to the direction of that abnormality.

Manoach (2003) suggested that these contradictions might be resolved by considering load-performance interactions. This hypothesis suggests that at low loads, patients will show equivalent performance but increased frontoparietal activation (i.e., hyperactivation) compared to controls. As load increases, both patients and controls increase activation in response, but patients may reach a functional ceiling at lower loads than controls. Once that ceiling is reached, patients will show lower performance and activation (i.e., hypoactivation) compared to controls. After that point, activation may decline not just relative to controls but absolutely, as patients become demotivated or disorganized due to poor performance at the highest loads, or perhaps try alternative strategies. Controls are hypothesized to show a similar performance x activation function, but shifted to the right (Fig. 2).

Although it appears to have been missed in the Manoach (2003) review, an early PET study by Fletcher et al. (1998) provides what is to our knowledge the most direct evidence supporting this hypothesis. Healthy controls and patients with either

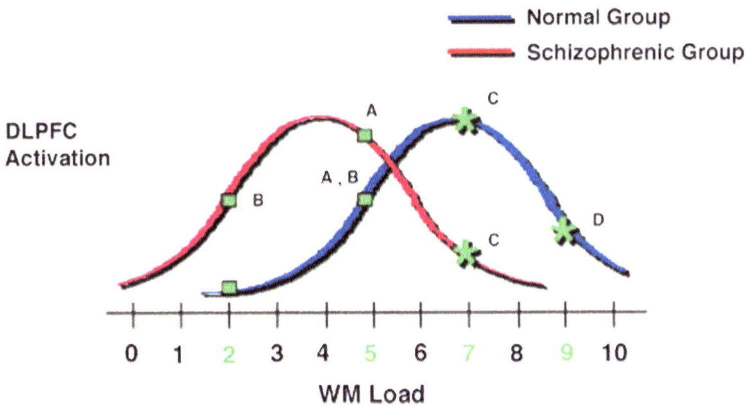

Fig. 2 CRUNCH in schizophrenia? Idealized relationship between working memory load and dorsolateral prefrontal (DLPFC) activation in schizophrenia (Figure from Manoach 2003; squares indicate empirical data points, asterisks indicate hypothesized results). As load increases, so does DLPFC activation, until load supersedes ability and both performance and activation decline. This pattern occurs for both patients and healthy controls, but the curve for patients is shifted left, so that at low objective loads (2 items) patients exhibit higher activation and similar performance as controls. At high loads (7 items) patients are expected to show decrease in activation and performance, whereas healthy controls continue to activate DLPFC to meet with the increased demand. This pattern has also been observed when comparing healthy older adults to young adults, where it is described under the CRUNCH (Compensation Related Utilization of Neural Circuits Hypothesis) framework (Reuter-Lorenz and Lustig 2005; Reuter-Lorenz and Cappell 2008)

high or low degrees of memory impairment heard and then attempted to retrieve word lists of 1–12 items. All groups showed an initial increase in prefrontal activation up to about 7 items. After this point, activation levels for high-impairment patients leveled off and began to drop, with activation at 12 items being roughly the same as at 4 items; for less-impaired patients this function was shifted right, with increasing activation up to about 9 items. Controls did not show a downturn, but this may have been because the task was not sufficiently demanding (and they may have switched to long-term rather than short-term memory retrieval), as their performance remained relatively stable following an initial drop at the introduction of supra-span lists.

Subsequent studies by other groups (e.g., Cairo et al. 2004; Manoach et al. 1999, 2000) likewise found that patients showed comparative hyperactivation at low working memory loads and hypoactivation at higher ones, but did not test even higher loads that might have allowed observation of the absolute downturn. Similar results were reported in a meta-analysis by Van Snellenberg et al. (2006). A later empirical study from this group failed to find shifted functions for patients compared to controls in a procedure previously shown to produce an inverted-U function in healthy adults, but interpretation is complicated by a failure to observe an effect of memory load in any brain regions in patients, suggesting they may have approached the task very differently (Van Snellenberg et al. 2013, 2015).

Notably, the shifted inverted-U pattern hypothesized for schizophrenia has been observed in several studies of healthy older adults, suggesting a common mechanism. The CRUNCH (Compensation Related Utilization of Neural Circuits Hypothesis) framework similarly suggests that older adults will show preserved performance but higher prefrontal activation at low loads, and that as load increases, individuals reach a "CRUNCHpoint" or functional ceiling, after which performance and activation decline (Reuter-Lorenz and Lustig 2005; Reuter-Lorenz and Cappell 2008). As in schizophrenia, most tests of this hypothesis have used working memory tasks, since load is easy to manipulate, but at least one study also found support for the hypothesis when manipulating executive-attention demands (Sebastian et al. 2013). Several of the studies in aging have found the hypothesized downturn in activation at high levels of demand, suggesting that these paradigms may be useful for rigorous tests of the hypothesis in schizophrenia.

Further supporting the hypothesis of a common pathway potentially supporting compensation in patients with schizophrenia and older adults, and possibly other populations including multiple sclerosis (e.g., Cader et al. 2006; Parry et al. 2003), right dorsolateral prefrontal cortex along middle and inferior frontal gyrus appears to be a common locus for the inverted U-curve with load in healthy adults, the CRUNCH pattern in older adults, and abnormality in schizophrenia (Fig. 3). As we will describe in the next section, parallel rodent–human studies suggest that the cholinergic system plays a critical role in modulating this region's response to demand, and may also provide insight into associated cognitive operations.

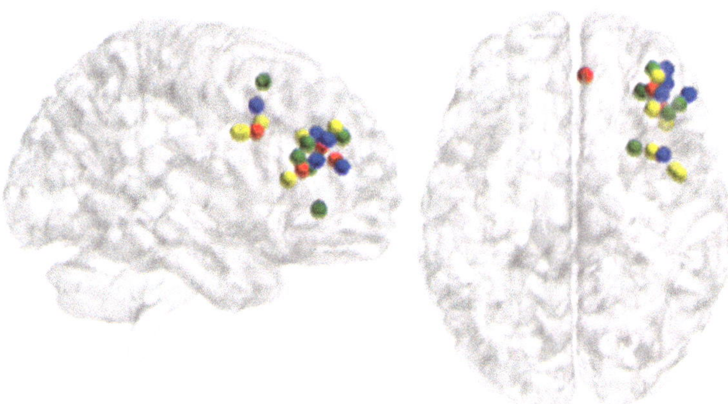

Fig. 3 Overlap between CRUNCH, schizophrenia, and dSAT. Right middle and frontal gyrus is a common site for activation peaks reported in studies of U-shaped demand-activation curves in young adults (*green*; Callicott et al. 1999; Van Snellenberg et al. 2015), shifted U-curves (the CRUNCH pattern) in older adults (*blue*; Cappell et al. 2010; Mattay et al. 2006; Schneider-Garces et al. 2010; Sebastian et al. 2013); schizophrenia-related abnormalities in working memory tasks (*red;* Fletcher et al. 1998 PET study cited in text; Minzenberg et al. 2009 meta-analysis), and distractor-related activation in the dSAT (*yellow*; Berry et al. 2014; Demeter et al. 2011)

2 Cholinergic-Attentional Control Deficits: Anatomical Foundations and Evidence from Rodent Studies

The neuronal circuitry-based foundations of the attentional impairments in schizophrenia remain largely unclear. However, the considerable evidence for the influence of basal forebrain cholinergic innervation on the frontoparietal networks that support attention and cognitive control suggests that abnormalities in this system likely play a major role. Below we briefly describe key aspects of the functional anatomy of the basal forebrain cholinergic system, including the main cortical effects of cholinergic neurotransmission and the influence of cortical feedback to the basal forebrain and, indirectly, via mesolimbic regions. The potential contributions of cholinergic abnormalities to attention problems in schizophrenia are discussed in light of our current understanding of abnormalities in the development and regulation of forebrain circuitry schizophrenia.

2.1 Cortical Cholinergic Projections: Anatomical Aspects Consistent with Top-down Functions

The basal forebrain cholinergic projections to the cortex historically have been considered a component of the brain's ascending systems. The attribution of broad, undefined functions, such as enhancing arousal, wakefulness, and gating the cortical processing of stimuli has been characteristic of such traditional hypotheses (e.g., Castro-Alamancos and Gulati 2014). Consistent with its largely ascending projections to telencephalic regions, such conventional conceptualizations of the cholinergic system are strictly bottom-up, meaning that cholinergic activation has been considered stimulus-driven or under the control of circadian mechanisms. In other words, cholinergically-induced cortical states have been viewed as secondary to the presence of a stimulus or endogenous clocks.

Our alternative view of the cholinergic system conceptualizes these ascending projections as a major arm of the brain's top-down machinery. In other words, we describe the recruitment of cholinergic neurons as occurring in the service of cognitive control functions such as task maintenance, particularly maintaining rules "on-line" for responding flexibly to changing stimulus (or cue) configurations, sustaining attention to defined cue sources and cue probabilities, integrating levels of motivation with performance, and with the last function possibly the calculation of the value of continued performance over alternative action.

Importantly, recent analyses of the anatomical organization of cholinergic projections to cortical regions increasingly reject the conventional notion of a "diffusely organized" or "reticular" projection system. Instead, they document evidence for a highly topographic, cluster-based organization of cholinergic neurons in the basal forebrain, with associated projection patterns that dissociate cortical subregions, columns, and even patches coinciding with other afferent cortical projection systems (Zaborszky et al. 2005, 2008, 2012, 2015a, b; Bloem et al. 2014a, b; Ji et al. 2015). Although there is still much debate about the extent of the cortical terminal field of individual cholinergic projections and clusters of basal forebrain cholinergic neurons (Mechawar et al. 2000; Umbriaco et al. 1994; Muñoz and Rudy 2014), recordings of acetylcholine (ACh) release in behaving animals increasingly demonstrate that cholinergic activity is at least in part highly localized and supports defined behavioral/cognitive operations (below), consistent with the contemporary description of basal forebrain cholinergic projections as a highly topographic, clusterized projection system (see also Xiang et al. 1998).

Understanding the cortical cholinergic input system as a main branch of the brain's top-down machinery begins with evidence indicating that, in primates and rodents, prefrontal regions are the only cortical regions that project directly to the cholinergic basal forebrain (Gaykema et al. 1991; Zaborszky et al. 1997). Second, mesolimbic regions, including the nucleus accumbens and the ventral tegmentum target cholinergic neurons in the basal forebrain (Smiley et al. 1999; Zaborszky and Cullinan 1996). Prefrontal regions directly influence these mesolimbic dopaminergic activity (Carr and Sesack 2000; Brady and O'Donnell 2004; Belujon and Grace 2008), and thus the activation of cholinergic neurons as a function of demands on attention (below) likewise constitutes a prefrontally-supervised function.

Third, prefrontal circuitry can influence and even control cholinergic activity via projections that, directly and indirectly, target cholinergic terminals. Specifically, the generation of phasic cholinergic signaling has been hypothesized to be based primarily on cortical circuitry controlling cholinergic terminals. That is, the cortex integrates cholinergic terminals into its circuitry based on heteroreceptors expressed at cholinergic terminals. Such terminal regulation adds a tremendous degree of functional specification of cholinergic function as local cortical circuitry can control the release of ACh and thus specify cholinergic function regardless of the (disputed) degree of terminal arborization space and the "diffuseness" of the cortical cholinergic projection system (Sarter et al. 2014).

In addition to these three anatomical sources of top-down control of cholinergic activity, cholinergic activity in prefrontal regions per se influences, top-down, cholinergic activity elsewhere in the cortex, but the reverse does not occur (Nelson et al. 2005). These prefrontal efferent effects require the stimulation of muscarinic acetylcholine receptors (mAChRs), consistent with the general view that this group of receptors mediates the flow of information within and between cortical circuits (Hasselmo et al. 1992; Disney and Aoki 2008). Cholinergic stimulation of mAChRs in the frontal cortex generates high-frequency oscillatory activity (Sarter et al. 2015), considered an indication of cross-regional coordination of cortical processes (Bauer et al. 2012). Thus, in prefrontal cortex, cholinergic activity may foster the orchestration of cognitive mechanisms that support, top-down, the sustaining of attention, particularly in the presence of distractors. Cholinergic activity in parietal and other cortical regions serves as a component mediating these top-down effects, as it is controlled via prefrontal projections to basal forebrain and mesolimbic regions and, via one or multiple synapses, to cholinergic terminals in sensory and sensory-association regions.

2.2 *Multiple Time Scales of Cholinergic Signaling Mediate Distinct Attentional Processing Steps*

Recent evidence indicates that in addition to the spatial specificity described above, cholinergic activity is also temporally specific. At least three timescales have been observed in rodent prefrontal cortex. The first appears to operate in the seconds - to-minutes range, and is represented by the increases in prefrontal acetylcholine (ACh) levels seen as the animal moves from baseline to task performance, with further increases in response to the distractor condition or other challenges (Himmelheber et al. 2000; Kozak et al. 2006, 2007; St. Peters et al. 2011). This neuromodulatory component of cholinergic activity interacts with mesolimbic systems: Stimulation of the nucleus accumbens shell reduces distractor-related performance impairments, reflecting enhanced or compensatory mesolimbic recruitment of cholinergic-attentional mechanisms. Consistent with this hypothesis, the benefits of accumbens stimulation are eliminated by removing either prefrontal or parietal cholinergic inputs (St. Peters et al. 2011).

Recent technological developments allowed the identification of a second "transient" response system, operating at the seconds timescale (Parikh et al. 2007, 2010). Thus far, cholinergic transients have been observed in response to signals that occur either with a long temporal delay between trials (Parikh et al. 2007) or when a signal trial is proceeded by a perceived nonsignal (i.e., correct rejection or miss) trial (Howe et al. 2010). Cholinergic transients are initiated by a signal-evoked thalamic glutamatergic response that is itself modulated by the longer timescale subsystem described above. Notably, this thalamic signal is required but not sufficient to initiate the cholinergic response, as indicated by the absence of such transients during consecutive trials requiring cue-oriented responses. The thalamic signal occurs for every detected signal, but the cholinergic transient is governed by the temporal and/or sequence constraints described above. This suggests that the cholinergic response is involved in cognitive rather than sensory processing, a supposition further supported by the finding that the cholinergic transients are more closely associated in time with cue-triggered response initiation than with the cue per se.

Parikh et al. (2007) also found evidence for cholinergic functions at a third, intermediate timescale. Gradual decreases in cholinergic activity over the 20 s before signal presentation were associated with correct signal detection (hits), whereas gradual increases were associated with failures to detect the signal (misses). We have not further explored this subsystem, although one distinguishing feature is that these more gradual increases and decreases are seen in both PFC and motor cortex, whereas the transients described above are only seen in PFC. Although further testing using different interstimulus intervals is needed to determine if the 20 s value is meaningful or coincidental (that is, would similar drifts occur at longer or shorter interstimulus intervals?), there are some intriguing parallels at this timescale in the human cognitive neuroscience literature. Fluctuations in response–time variability, though to reflect fluctuations in attention, occur in about 20 s cycles in the Ericksen flanker task, and this pattern is exaggerated in subjects with ADHD (Castellanos et al. 2005). In another attention task, O'Connell et al. (2009) observed that increases in EEG measures of alpha activity occurred about 20 s before a missed target, a pattern very reminiscent of the increases in cholinergic activity before a miss observed by Parikh et al. (2007). The idea that these phenomena may be related is reinforced by the fact that alpha oscillations are cholinergically modulated. For example, in a spatial attention paradigm, physostigmine reduced alpha activity in the hemisphere contralateral to stimulus presentation and improved performance (Bauer et al. 2012).

3 Cholinergic Dysregulation in Schizophrenia: Possible Causes and Consequences

Understanding of how neuronal circuitry is structurally abnormal and functionally dysregulated in schizophrenia remains quite limited, and further complicated by the likelihood of multiple subtypes of the disorder that may manifest different neuropathologies. For example, in a *post mortem* study, Scarr et al. (2009) found that about 25 % of patients had especially marked deficiencies in a PET marker of muscarinic binding sites in middle and inferior frontal gyrus sites, and might constitute a separate genetic and behavioral subgroup. The hypothesis of cholinergic dysfunction in schizophrenia is indirectly supported by several lines of evidence including genetics, neuroimaging, and high rates of nicotine use that suggest attempts at self-medication, but the nature of that dysfunction has been difficult to define. This difficulty accrues from the lack of suitable in vivo methods for monitoring cholinergic activity in patients, the limited insights afforded by post-mortem analysis of cholinergic enzyme levels that are not rate-limiting steps in the synthesis or metabolism of ACh, the challenges associated with the interpretation of changes in receptor levels measured in vivo studies, and the scarcity of pharmacological tools to assess defined and selective aspects of cholinergic function (for detailed discussion of evidence on cholinergic function in schizophrenia and related pharmacological issues see Sarter et al. 2012; Sarter et al. 2009a; Hasselmo and Sarter 2011).

Current theories of schizophrenia focus on the development of cortical microcircuitry, in particular the wiring of inhibitory interneurons and abnormal functions of amino acid receptors and cytoskeletal proteins. Establishing relationships between such mechanisms and hypotheses about circuit-based neuronal aberrations has remained a difficult objective (see also Higley and Picciotto 2014). However, there is wide agreement that the clear neurotransmitter-related hallmark of schizophrenia, the hyperdopaminergic functions during active disease states (e.g., Howes and Kapur 2009; Kapur and Mamo 2003), eventually may be understood as a consequence of the diverse, largely telencephalic, developmental, and cellular abnormalities that all yield schizophrenia. Similarly, abnormal interneuronal contacts, GABAergic, and glutamatergic receptor function all may contribute to the low levels of cholinergic neuromodulation and cholinergic transient dysregulation that are candidate hypotheses for explaining the attentional control issues in these patients. Specifically, we hypothesize that GABAergic functions are essential for suppressing cholinergic transients in noncued trials and we demonstrated that glutamatergic synapses of thalamic afferents are necessary, but not sufficient, for the generation of cholinergic transients (reviewed in Sarter et al. 2014; Sarter 2015). Thus, dysregulation in front-parietal GABA and glutamatergic functions may readily disrupt the generation of cholinergic transients and yield ill-timed transients, failures to generate transients, or transients with temporal dynamics that are sufficient altered to cause nonadaptive and even invalid cue detection operations.

Furthermore, abnormal mesolimbic dopaminergic activity has been conceptualized as a consequence of abnormal frontal cortical circuitry (e.g., Brady and O'Donnell 2004; Sesack and Grace 2010). Mesolimbic activity is necessary for cholinergic activation and associated performance (Neigh et al. 2004) and abnormal mesolimbic dopaminergic activity therefore is likely to alter cholinergic function and thus attentional control and cue detection. These mesolimbic–cholinergic interactions are key to understanding the integration of motivational with attentional functions (Small et al. 2005; Krebs et al. 2012; Mendelsohn et al. 2014; Hungya et al. 2015). Thus, abnormal dopaminergic functions in schizophrenia may greatly impact cholinergic neuromodulation and the generation of cholinergic transients. Consistent with this view, animals with sensitized mesolimbic dopaminergic functions—which may model an acute disease state—exhibit cholinergic systems that remain "frozen" at baseline and unable to support attentional performance (Kozak et al. 2007). We know much less about the reactivity of the dopaminergic system outside active disease states but it would be expected that it is dysregulated. Thus, the cholinergic abnormalities deduced from experiments in rodents may be present in schizophrenia and secondary to frontoparietal dysmorphogenesis, altered amino acid receptor function, and mesolimbic dysregulation. Dysregulated cholinergic neurotransmission in the cortex likely further escalates dysregulation in distributed cortical-mesolimbic-basal forebrain circuitry (Zaborszky et al. 1997; Zaborszky and Cullinan 1992), rendering the identification of a primary "causal culprit" a difficult objective. Finally, we cannot exclude the possibility of a primary abnormality in the regulation of cholinergic neurons in the disease, akin to the choline transporter (CHT) regulation abnormalities found in sign-tracking rats (see below).

4 Parallel Rodent–Human Studies Implicating Right PFC ACh Dysfunction in Impaired Schizophrenia Responses to Attentional Challenge

Of the three components described above, the relatively long timescale neuromodulatory component has been the most extensively studied, and is the most potentially relevant to the CRUNCH pattern observed in aging and hypothesized in schizophrenia. As noted earlier, microdialysis studies indicate that right PFC acetylcholine increases as the animal moves from baseline to task performance, and further in the face of a distractor or other attentional challenge (e.g., St. Peters et al. 2011). These increases in prefrontal ACh are more closely related to demands on attention than to performance levels, and thus often occur when performance is impaired by the distractor or other challenge (e.g., Kozak et al. 2006; Sarter et al. 2006; St Peters et al. 2011). On the other hand, cholinergic lesions reduce performance, especially in conditions of attentional challenge, indicating that these increases play an important if not sufficient role in supporting performance (e.g., Kucinski et al. 2013; McGaughy and Sarter 1999).

Prefrontal ACh increases associated with attentional performance and in response to the distractor appear to be largely right-lateralized (Apparsundaram et al. 2005; Martinez and Sarter 2004; Parikh et al. 2013). Human fMRI studies of the dSAT likewise indicate a special role for right PFC. Across several studies, baseline task performance without the distractor typically elicits bilateral PFC activation, but the response to the distractor is right lateralized (Berry et al. in prep., 2014; Demeter et al. 2011). Again paralleling the rodent studies, greater right PFC activation is associated with greater vulnerability to the distractor (Berry et al. in prep.; Demeter et al. 2011). As illustrated in Fig. 3, distractor-related increases and correlations with performance are prominent in right middle and frontal gyrus, near locations associated with the CRUNCH pattern in aging and disruption in schizophrenia.

Definitive evidence that the right PFC ACh increases observed in rodents contribute to the right PFC activation increases observed in fMRI is difficult to obtain because of the restrictions on what studies can be ethically performed in humans. Indirect support derives in part from humans with a genetic polymorphism (Ile89Val variant of the CHT gene SLC5A7, rs1013940) that, when expressed in cells, reduces the capacity of cholinergic synapses to sustain ACh release. These individuals fail to show the typical distractor-related increase in right PFC activation (Berry et al. 2014). Likewise, mice with a heterozygous deletion of the choline transporter gene show normal basal ACh release but greatly reduced prefrontal ACh responses to either direct basal forebrain stimulation or task demands on attention (Paolone et al. 2013b; Parikh et al. 2013). While this evidence is indirect, the parallels between right PFC activation and genetic limits on cholinergic capacity in humans and right PFC ACh increases and genetic limits on cholinergic capacity in mice make a cholinergic contribution to the fMRI findings the most parsimonious explanation.

Whether patients show an abnormal right PFC response to the dSAT, and what the cholinergic contribution to any such abnormality might be, has not yet been established. However, behavioral studies in patients and measurements of right PFC ACh in a rodent model of the disorder are so far consistent with these ideas. Although patients show some deficits even in signal detection in the basic, no-distractor SAT—possibly due to perceptual difficulties with the brief, low-contrast stimulus used as the "signal"—they show a specific, differential vulnerability to the distractor condition (Demeter et al. 2013). Their distractor vulnerability does not reflect a generalized performance impairment in the face of all forms of attentional demand, as they were able to sustain performance over time just as well as controls. This contrasts with the results from children (age 8–11 yrs), who showed less distractor vulnerability than patients but greater time-on-task declines. Together with previous findings separately implicating right middle and inferior frontal gyrus in responses to the distractor and abnormalities in schizophrenia, these data predict that patients would show right PFC abnormalities in the dSAT.

This prediction is also supported by findings from a rodent model of attention deficits in both the acute and chronic states of the disease. Rats with sensitized mesolimbic dopaminergic functions tested under amphetamine challenge—thought to model an acute disease state—exhibit cholinergic systems that remain "frozen" at baseline, unable to support attentional performance (Kozak et al. 2007). When these animals are tested without acute dopaminergic challenge, conditions thought to model the remission state as in the patients tested by Demeter et al. (2013), they are able to perform normally in the baseline, no-distractor SAT but with much greater increases in right PFC ACh than those observed in control animals. This exaggeration of the right PFC ACh response suggests that these animals required increased attentional effort and top-down control. The distractor condition was not tested in these animals, but is predicted to result in greater performance impairments than seen in controls—similar to the greater distractor-related performance impairments exhibited by the patients in Demeter et al. (2013)—and a drop in right PFC ACh levels, similar to the right PFC CRUNCH pattern described above. The corresponding prediction for schizophrenia patients would be exaggerated right middle/inferior frontal gyrus in the baseline SAT, and performance deficits and reduced right middle/inferior frontal gyrus activation in response to the distractor challenge.

5 Cholinergic and Dopaminergic Interactions with Right PFC: Integrating Attention and Motivation?

These parallel rodent–human studies build a strong if still circumstantial case that cholinergic dysregulation makes an important contribution to the right PFC abnormalities consistently observed in schizophrenia (e.g., Minzenberg et al. 2009). However, they leave somewhat ambiguous what that contribution might be in terms of cognitive operations. The findings of increased right PFC ACh in rodents and increased right PFC activation in humans, and that lesions or blockade of the cholinergic system impair performance, suggest an important role in increasing top-down control. On the other hand, that suggestion is seemingly contradicted by the negative relationship between performance and right PFC ACh or activation across conditions or individuals. Further complicating matters, although mice genetically modified to have reduced cholinergic function and humans with a genetic polymorphism thought to reduce cholinergic function fail to show dSAT-related increases in right PFC ACh or activation, they show relatively preserved performance (Berry et al. 2015; Paolone et al. 2013b; Parikh et al. 2013).

One proposed explanation for these complex findings is that right PFC ACh and/or activation should be thought of in terms of "attentional effort" (Sarter et al. 2006; see also Raizada and Poldrack 2007). That is, rather the specific attentional processes or mechanisms needed to respond to the requirements of a particular task (e.g., target selection, inhibition, or shifting attention), right PFC activity may reflect the motivated recruitment of those mechanisms.

Put in terms of cognitive operations, by this view right PFC would be described as translating error, conflict, or uncertainty signals from anterior cingulate that indicate a need for increased attentional control into the recruitment of motivation and attention to meet those demands. Conceptualizing the role of right PFC as a critical hub for integrating demand signals, motivation, and attentional control (Watanabe and Sakagami 2007) may explain why it is so often the locus of the "inverted U" activation-demand function in healthy young adults (e.g., Callicott et al. 1999; Van Snellenberg et al. 2015), the observed shift of that function in aging (e.g., Reuter-Lorenz and Cappell 2008), and the hypothesized shift and observed dysregulation in schizophrenia (Fletcher et al. 1998; Manoach 2003; Minzenberg et al. 2009). That is, as demand increases, there is increased recruitment of motivated attention until the "crunchpoint", after which it falls. It is not yet clear whether the drop in activation (and performance) at the end of the demand curve reflects a loss of motivation, the abandonment of current task-goal representations to try alternative strategies, including shifts from top-down to bottom-up attention, or some combination.

Abnormalities in right PFC in schizophrenia may thus be related to the disorder's "amotivational" aspects and negative symptoms (e.g., Wolkin et al. 1992). Many discussions of reward processing and abnormalities in schizophrenia focus on orbitofrontal cortex (see Young and Markou 2015 for a recent review of translational animal paradigms). Orbitfrontal cortex appears to play an important role in representing reward value (hedonics) and updating stimulus-reward associations. Anterior cingulate and dorsolateral prefrontal cortex may be more involved in using that information to guide behavior. Anterior cingulate has been implicated in demand signals (including both error or conflict and the amount of effort needed to overcome it), and dorsolateral prefrontal cortex is associated with the translation of reward-value and performance information to task-goal representations and top-down control (see discussion by Barch and Dowd 2010). Evidence from both healthy populations and patients points to an especially prominent role for right middle and inferior frontal gyrus in this translation, and its impairment in schizophrenia.

For example, whereas other regions show sensitivity to incentive valence (reward/loss) or arousal, right middle frontal gyrus is specifically responsive to unexpected changes in reward or loss that may signal a need to shift task sets (Akitsuki et al. 2003). Jimura et al. (2010) found that the tendency to deploy this region proactively or reactively was related to reward sensitivity as assessed by an independent personality test: In a working memory task where some task blocks presented a mix of rewarded and unrewarded trials, individuals with high reward sensitivity showed sustained right middle frontal gyrus activity throughout the rewarded blocks, suggesting sustained top-down control that benefitted even non-rewarded trials within the block. In addition, they showed strong transient activity at the early stages of rewarded trials, suggesting proactive control. In contrast, low reward sensitivity was associated with low sustained activity and a larger late transient, suggesting a reactive control strategy. Such findings indicate that right middle frontal gyrus is not involved in the evaluation of incentive per se, but rather in the mobilization of control in response to incentive. Likewise, right PFC abnormalities in schizophrenia are associated with evaluation of reward outcomes, especially unexpected outcomes that may require top-down control to re-evaluate and possibly change task set (e.g., Koch et al. 2009; Nielsen et al. 2012).

While dopaminergic contributions are heuristically linked to reward and motivation, we suggest that cholinergic contributions can be thought of in terms of activating and maintaining task set representations. Specifically, neuromodulatory activity in right PFC may help to stabilize task-goal representations and protect them from competing influences. Multisynaptic projections from PFC, including through basal forebrain, to posterior parietal and somatosensory cortex, then act to optimize input processing in accordance with those goals. Inputs that are relevant to task goals (e.g., the central target in the dSAT or color information in Stroop) will be enhanced, whereas those that are irrelevant (e.g., the changing background in dSAT or word information in Stroop) will be suppressed.

Conceptualizing cholinergic neuromodulatory function in terms of stabilizing internal task representations may at first seem inconsistent with widely accepted computational models proposing that "acetylcholine enhances the response to afferent sensory input while decreasing the internal processing based on previously formed cortical representations" (Hasselmo and McGaughy 2004, p. 207; Hasselmo et al. 1992). However, such inconsistencies are largely superficial. These models (as well as empirical data) also support ACh's role in self-sustained persistent firing to support continued representation in memory and attention—what is suppressed is the spread of activation or associational processing (see discussion by Deco and Thiele 2011; Newman et al. 2012; Hasselmo and Sarter 2011).

Low-cholinergic neuromodulation would thus be predicted to engender increased processing of irrelevant inputs, a greater tendency to make inappropriate associations and less-specific representations of context, and increased intraindividual performance variability related to fluctuation of the task set and its control over behavior—all prominent cognitive symptoms of schizophrenia. These predictions play out in rats exhibiting stable individual differences in sign-tracking (ST) versus goal tracking (GT). Sign-trackers are screened from outbred populations using a Pavlovian approach procedure, and are distinguished by their strong tendency to approach and manipulate the reward-predicting cue or "sign" (e.g., pressing a lever whose appearance predicts reward delivery, even if lever pressing is not required for reward). In contrast, GTs orient behavior toward the reward delivery system (e.g., the food cup where reward will be delivered). STs are thought to attribute incentive salience to the cue while GTs' behavior is more controlled by "cold" goal–directed cognition (Flagel et al. 2009; Meyer et al. 2012). ST (but not GT) is strongly dependent on nucleus accumbens core dopaminergic function, reflecting its role in incentive salience (Flagel et al. 2011; Saunders and Robinson 2012).

In addition to the dopaminergic contributions to incentive salience, the increased processing of the irrelevant cue and tendency to inappropriately associate the cue with incentive salience could also reflect low cholinergic function. To test this hypothesis, STs and GTs were tested on the SAT (Paolone et al. 2013a). Compared to GTs, STs had lower task-related increases in right PFC ACh, and their performance showed a high degree of fluctuation between performance and chance levels. Importantly, however, STs exhibited bouts of high levels of performance that matched those seen in GTs, and did not omit more trials than GTs. In other words, rather than a fundamental inability or low motivation to perform, their performance was unstable, indicating a fluctuation of control rather than its absence. STs' attentional control dysregulation also manifests in impairments in executing complex movements across dynamic surfaces (Kucinski and Sarter 2015). Furthermore, the reduced task-related increase in right PFC cholinergic activity is associated with attenuated choline transporter (CHT) capacity to support synaptic ACh synthesis and release (Kucinski et al. 2015).

To our knowledge, it is not yet established whether patients with schizophrenia have differential tendencies toward sign- or goal-tracking. However, behavioral tests to assess variations in reinforcement learning have been recommended by CNTRICS for preclinical studies (Markou et al. 2013). It has also been hypothesized that inappropriate learning of associations to irrelevant stimuli may contribute to delusional symptoms (e.g., Jensen et al. 2008; see recent review by Deserno et al. 2013; Gilmour et al. 2015), and dysregulated associational learning in schizophrenia is associated with abnormal right middle frontal gyrus activation (e.g., Koch et al. 2010). More transparently related to the attention-control deficits seen in STs, schizophrenia is associated with exaggerated intraindividual performance variability in many laboratory tasks (e.g., Cole et al. 2011; Kaiser et al. 2008; Roche et al. 2015; see reviews by MacDonald et al. 2006; Matthysse et al. 1999).

To summarize, substantial evidence from rodent models, healthy young adults, older adults, and patients with schizophrenia point to right PFC, especially right middle and inferior frontal gyrus, as an important site for the integration of motivation and top-down control. In particular, dopaminergic interactions with nucleus accumbens shell and cholinergic neuromodulatory influences are hypothesized to support cognitive-behavioral vigor for staying on task, and the stable representation of which task to stay on, respectively (e.g., Floresco 2015). Both of these aspects may be impaired in schizophrenia, leading to both low motivational tone and distractible, erratic performance even when behavior is activated.

6 Cholinergic Transients: Spared, Impaired, or Overactive?

Although sustained cholinergic neuromodulation of right PFC is important for maintaining goal-directed behavior in challenging conditions, it has become increasingly clear both that alternative compensatory pathways can support performance under at least some conditions, and that cholinergic innervation acts on more than one timescale. In particular, although mice heterozygous for a deletion in the choline high affinity transporter gene (CHT $\pm$) and humans with a genetic polymorphism thought to reduce the efficiency of the CHT (I89 V allele; rs1013940 of SLC5A7) fail to show task-related increases in right PFC Ach and dSAT-related increases in right PFC activation, respectively, they show relatively preserved performance (Berry et al. 2015; Paolone et al. 2013b). However, data from the mouse model indicates that performance remains cholinergically dependent, demonstrated by a compensatory increase in nicotinic acetylcholine receptors (nAChRs) and larger performance declines in response to nAChR blockade by mecamylamine.

This apparent paradox can be resolved by noting that reduced CHT function would be expected to primarily affect the sustained, neuromodulatory component of cholinergic function. That is, low CHT efficiency limits the rate at which choline can be transported into the cell for the production of ACh. It therefore does not affect basal, pre-task levels, only the degree to which elevated neurotransmission can be sustained over time (Paolone et al. 2013b; Parikh et al. Parikh et al. 2013). Although it has not been directly tested, it might also be expected that intermittent transient responses might be less impaired (as long as sufficient time passed between them for the system to "restock", even if more slowly), and that these more minor reductions might be compensated for by the increase in nAChRs.

As described above, cholinergic transients appear to support orienting toward salient signals and activating the response sets associated with them. Thus, if transients are relatively intact in low-CHT groups, they may take a "reactive", bottom-up approach, relying on signal salience to drive performance, rather than top-down, proactive cognitive control (c.f., Braver 2012). Supporting this hypothesis, although humans with the Ile89V polymorphism thought to reduce cholinergic function fail to show right PFC responses to the dSAT, they show differential activation of regions associated with bottom-up signal salience and emotional-motivational processing (Berry et al. 2015; see also Gorka et al. 2014). Furthermore, in a behavioral paradigm where the target had very low bottom-up salience (slight duration changes from a standard) and the distractor had high salience (videos playing alongside the main task computer), Ile89V participants had normal no-distractor performance and ability to sustain performance over time, but showed a specific vulnerability to the distractor (Berry et al. 2014). Together, these findings suggest that although transients can support the detection of signals and the activation of their associated task sets, top-down control from the neuromodulatory component is required in situations with multiple salient stimuli to prevent transients and false alarms to nontargets.

The same trial sequences that yield transient cholinergic responses in rodents lead to transient right PFC fMRI activations in humans. The inference of a cholinergic contribution to this transient fMRI BOLD activation is supported by "back translation" using tissue-oxygen measures thought to parallel BOLD in rodents performing the task and under direct cholinergic stimulation (Howe et al. 2013). Notably, the activation site for these transient responses is not in middle or inferior frontal gyrus, but instead in a lateral orbitofrontal region hypothesized to serve as a "gateway" between externally-directed perceptual processing and internally-directed reflective processing (e.g., Burgess et al. 2007; Chun and Johnson 2011). Thus, the timescale, location, and associated cognitive processes of transients indicate their independence from the right PFC neuromodulatory effects.

Although neuromodulatory and transient cholinergic responses are dissociable, the neuromodulatory component influences the occurrence and sharpness of transients by stimulating α4β2* nAChRs expressed by glutamatergic terminals. Reduced cholinergic neuromodulation in patients would thus be predicted to show two abnormalities in the transient response when attentional demand is increased. First, patients would show attenuated neuromodulatory enhancement of transients compared to controls. Second, reduced top-down control would result in inappropriate transients to nontarget distractors.

As of this writing, neither cholinergic transients nor the putatively parallel fMRI response have been assessed in schizophrenic patients. Indirect evidence comes from the literature on event-related potentials (ERPs) in schizophrenia. The rare-target design of the SAT is similar to the "oddball" paradigms used to elicit the P300 response in ERP research, and preliminary evidence (Berry et al. in prep.; Demeter et al. 2015) indicates that target detection and specifically switch-hits elicit a P300 response. In traditional oddball paradigms, the P300 response is strongly influenced by cholinergic manipulations, and has been suggested as a biomarker for neuropsychiatric disorders involving cholinergic disruptions (Javitt et al. 2008; see discussion by Weinberger and Harrison 2011). Compared to controls, patients and first-order relatives generally show a reduced amplitude of P300 to targets, and a relatively exaggerated P300 amplitude to distractors (e.g., Grillon et al. 1990; Kogoj et al. 2005). Interestingly, it has been suggested that the apparent schizophrenia-related reduction in P300 amplitude to targets may be an artifact of increased variability in latency (Donchin et al. 1970; Callaway et al. 1970; Roth et al. 2007; Ford et al. 1994; Roschke et al. 1996). As described above, increases in variability may reflect fluctuations in top-down control, further supporting the hypothesis that transient responses are affected by longer timescale neuromodulation.

7 Implications for Treatment Development and Translational Research

As reviewed here and elsewhere, there is considerable evidence to suggest *that* the cholinergic system is involved in the cognitive deficits of schizophrenia. Still lacking however, is a detailed knowledge of *how* this system is disrupted, and how it interacts with other (also likely disordered) neuromodulatory systems. This is especially the case since most of the evidence for specific receptor deficits comes from post-mortem studies, and there is increasing appreciation that there are important individual differences in the course of the disease, most likely with a genetic component (see discussions by Jablensky 2015; Ross et al. 2010). This may include subgroups with specific patterns of muscarinic deficits (e.g., Seo et al. 2014). In light of these factors, it seems premature to recommend any specific targets; readers interested in recent efforts are referred to Rowe et al. (2015) for nicotinic targets, Kruse et al. (2013) for muscarinic receptor ligands, and Money et al. (2010) for a general overview.

Despite these difficulties, there is considerable interest in the possibility of cholinergic treatments for cognitive deficits in schizophrenia because of the evidence for cholinergic dysfunction, limited efficacy of antipsychotics in treating cognitive deficits, and extensive evidence for cholinergic modulation of the neural circuitry supporting the cognitive-behavioral functions that are impaired by the disease. Below we describe what the major findings covered in this review suggest are promising directions and methods of research.

First, cholinergic influences are pervasive, with effects from early levels of sensory processing at the retina through high levels of executive control, but also regionally and temporally specific. Understanding specific cholinergic subsystems and how they interact—for example, how reduced top-down control and filtering lead to noisier sensory representations, and the burden that these noisier representations create for later control operations such as discriminating targets from distractors—will be essential for improving cognition in schizophrenia and other conditions, including normal aging. One hypothesis is that relatively long timescale muscarinic signaling supports PFC top-down control of sensory and response systems (Hasselmo et al. 1992; Disney and Aoki 2008), whereas within the basal forebrain-PFC circuit, cholinergic neuromodulation influences the cortical circuitry that generates cholinergic transients primarily via stimulating $\alpha4\beta2^*$ nicotinic acetylcholine receptors (nAChR) that are expressed at glutamatergic terminals of thalamic afferents (Lambe et al. 2003; Parikh et al. 2008, 2010).

This suggests that interventions targeting muscarinic receptors could improve task-set stability as well as the quality of sensory representations, whereas those impacting nicotinic receptors, especially $\alpha4\beta2^*$ nAChRs, could improve cue detection, especially the activation of the cue-appropriate response set. In rodents, a selective $\alpha4\beta2^*$ nAChR agonist has been shown to enhance transients in attentional challenge conditions (Howe et al. 2010). Notably, neither nicotine nor a $\alpha7$ nAChR agonist had these beneficial effects; in fact, blocking nicotine from stimulating $\alpha7$

nAChR receptors amplified its pro-attentional effects. To our knowledge, $\alpha4\beta2^*$ nAChR agonists have not been tested in schizophrenia, although they have had some success in adult attention deficit disorder (Apostol et al. 2012; Bain et al. 2013).

Second, the value of cross-species research using parallel tasks is illustrated in part by the human data demonstrating the regional specificity of right PFC responses to attentional demand (right middle and inferior frontal gyrus) versus transients (right orbitofrontal cortex). The human data reveal this distinction, which would not have been obvious from the animal studies. Furthermore, the human neuroimaging data began to link the cognitive processes associated with each region to observe structural, functional, and behavioral abnormalities in schizophrenia, as well as other conditions. The animal data indicate the likely neurotransmitter influences on activation and behavior, with genetic populations and pharmacologic manipulations helping to establish (or potentially disconfirm) the hypothesized cholinergic contributions.

This brings us to two more subtle points: Although parallel paradigms can be very powerful, it is important to remember that apparently parallel behavior between species or even between individuals does not necessarily entail the same neurocognitive processes. Rodents of course have much less-developed frontoparietal control systems than do humans, and even the simple signal detection task of the SAT likely places a greater proportional demand on rodents' top-down control. Even within human populations, the perceptual difficulties associated with schizophrenia may increase the attentional demands of the task relative to healthy controls. Furthermore, as discussed previously, low-cholinergic populations may be more likely to adopt a reactive, bottom-up approach as opposed to the proactive, top-down attention engaged by healthy controls. Thus, parametric manipulations of attentional challenge and other task variables, as well as converging evidence from other tasks testing-related constructs, are critical for establishing parallels and testing their limits.

In addition, parametric manipulations are important for revealing cholinergic effects. In many cases described above, drug and genetic manipulations have little or no effect on behavior, ACh release, or fMRI activation patterns in baseline conditions; the differences only become apparent with increases in demand. This pattern is consistent with ACh's putative role in responding to attentional demands, but means that assessments without such manipulations may miss important differences between different animal models, treatment conditions or between patients and controls. Thus, it is critical to test any potential treatments in the context of behavior, and if possible, with parametric manipulation of demand on the targeted cognitive process. Some have recommended that any potential pharmacologic treatment be administered in combination with cognitive training (other methods, such as transcranial stimulation, may have similar effects); to do otherwise has been described as giving protein powder without exercise and expecting muscle growth (Keefe et al. 2011).

Third, any potential interventions also need to be evaluated in the context of a disordered system, not just in healthy subjects, and in consideration of potential interactions with other drugs such as antipsychotics and nicotine. Furthermore, while cholinergic mechanisms may be established in rodent studies using local administration of different agents to the brain structures of interest, most treatments are administered systemically, and interactions throughout the brain and entire body must be taken into account when considering both efficacy and side effects. New technologies may ultimately help to address the last point, including potential nanoparticle delivery to targeted regions (see discussion by Money et al. 2010).

As one example of such new technologies, transcranial stimulation may provide an interesting method to stimulate or mimic cholinergic activity in specific regions, either in isolation or in combination with cholinergic drugs. For example, Reinhart et al. (2015) recently reported that direct-current stimulation of medial PFC, including anterior cingulate, normalized theta-band coherence between medial and dorsolateral prefrontal cortex in patients, and improved their performance on a measure of top-down control so that it was equivalent to healthy controls at baseline. Furthermore, the degree of change in theta correlated with performance improvements. These findings are remarkable in that the initially disorganized theta coherence found in patients is consistent with the idea that fluctuations in the coordination of anterior cingulate—dorsolateral prefrontal (including right middle and inferior frontal gyrus) contributes to fluctuations in top-down control and performance. Further, muscarinic ACh receptor-signaling exerts a strong influence over theta oscillations (Blatow et al. 2003; Lukatch and MacIver 1997). Although Reinhart et al. did not speculate on the neurotransmitter changes underlying their effects, this suggests that modulation of cholinergic activity plays a role. Their findings obviously occurred under highly artificial conditions, but long-term one might imagine that combining such stimulation with training and biofeedback to teach patients to regulate such activity more autonomously, perhaps in combination with pharmacologic agents to enhance the "raw materials" in terms of cholinergic availability and receptor sensitivity, could provide a powerful intervention with real-world effects.

8 Conclusions

Although it is widely agreed that cholinergic disruptions play an important role in cognitive impairments in schizophrenia, our understanding of exactly what those disruptions and their consequences might remain at a rudimentary stage. This is not surprising, as understanding of the basic function of the cholinergic system and its contribution to attention and other cognitive functions is also quite incomplete. However, there have been recent major advances in understanding the wired, regionally localized aspects of cholinergic neurotransmission as opposed to unspecified volume transmission effects, and in the recognition that it acts on multiple, likely interacting, timescales.

We have suggested here that cholinergic abnormalities in schizophrenia begin at the earliest sensory stages, and contribute to a vicious cycle in which reduced top-down control contributes to noisier perceptual processing, which in turn creates an increased burden for later controlled attention processes such as target selection. In rodent studies of PFC ACh release and human neuroimaging studies, this may lead to a "shifted U", where patients and animal models of the disorder exhibit higher levels of PFC activation and/or ACh release compared to controls in order to maintain performance at relatively lower levels of demand, but hit a functional ceiling at high levels of demand, so that activation/ACh levels are lower than healthy controls and may even show absolute declines. Right middle and inferior frontal gyrus appears to be an important locus for these effects and for the integration of dopaminergic (motivational) and cholinergic (cognitive) influences for a number of populations (older adults, ADHD, and even healthy controls), suggesting a fundamental neurocognitive component. Interventions that improve cholinergic function here, hypothesized to stabilize the task-set representations that guide sensorimotor and attentional processing, could thus have widespread cognitive effects, not just in schizophrenia.

There is less basis for speculation regarding potential schizophrenia-related disruptions in the more recently-identified cholinergic transient system. To the degree that P300 ERP responses to oddball stimuli may reflect such transients, they suggest blunted responses to targets, a failure to downregulate responses to distractors, and an overall increase in variability that may reflect aberrant interactions with the cholinergic system's longer timescale neuromodulatory top-down control functions. However, these hypothesized connections require more direct tests.

Throughout, the integration of data across species, methods, and levels of analysis helps to constrain hypotheses and interpretations within each level. The complexity of the cholinergic system and its disruption in schizophrenia makes it a challenging target for translational research. However, given the central importance of cognitive symptoms to disease prognosis and real-world function, it remains a critically important one. Recent advances in our understanding of this system's modes of function as well as new methods of analysis and potential intervention hold promise for more targeted and ultimately more successful intervention.

Box 1. CNTRICS control of attention tasks.

The goal of the CNTRICS initiative is to develop measurement approaches from different areas of neuroscience so that they can be refined, assessed for psychometric quality and sensitivity to schizophrenia-related impairments in the construct of interest, and ultimately implemented for treatment research. Different meetings focused on defining the constructs and choosing biomarkers, human tasks, or animal tasks.

One important construct identified at the first meeting was the control of attention (Luck and Gold 2008). All of the tasks used for this construct focused on the "input selection" function of attention, or the ability to restrict processing to relevant inputs. (see text) Human tasks selected at the next meeting (Nuechterlein et al. 2009) were the Guided Search task (Gold et al. 2007), and the human version of the distractor condition Sustained Attention Task (dSAT; Demeter et al. 2008; McGaughy and Sarter 1995). The Guided

Search task requires searching for a target (e.g., a red square with a gap at the top) among a set of distractors (red and blue squares with gaps on different sides). The dSAT is a simple detection task in which a centrally-presented signal (small low-contrast gray square on computer screen for humans, illumination of center panel light for rodents) does or does not occur with 50 % probability on each trial; at the end of the trial, participants are to report whether or not a signal occurred. Uncertainty is added by varying the duration of both the signal and the monitoring period. In the distractor (dSAT) condition, the background rapidly changes (flashing computer screen for humans, flashing houselight for rodents), increasing attentional challenge.

The dSAT was also selected as a potential task for imaging biomarkers related to the control of attention (Luck et al. 2012), along with the attentional singleton task (Theeuwes 1992) and attentional cueing paradigms (e.g., Giesbrecht et al. 2003). In the attentional singleton task, the target is an item with a unique shape, but on some trials a highly-salient distractor of a different color is also presented along with other, same-colored distractors. Attentional cueing paradigms present a cue indicating the likely location of an upcoming target, neural activation between the cue and the target is thought to represent maintenance of the goal to move attention to that location.

The fourth meeting focused on the selection of animal paradigms with promise for preclinical research (Lustig et al. 2013). The dSAT was again chosen as a potential measure of input control, along with the 5-choice serial reaction time task (see Robbins 2002 for a review) and the 5-choice continuous performance task (Young et al. 2009). The 5-choice serial reaction time task is based off of a sustained attention task previously developed for humans and is part of the CANTAB battery used with both healthy and clinical populations (Alexander et al. 2005; Sahakain et al. 1993; Cambridge Cognition, camcog.com). The task is to detect a brief visual target presented briefly in a 5-choice array. The 5-choice continuous performance task is very similar, except that it adds nontargets to which the subject must inhibit responding.

A common theme throughout these tasks (with the attentional cueing paradigms perhaps being somewhat of an outlier) is the requirement to detect and respond to a target signal in the face of attentional challenge. The paradigms vary in how that challenge is implemented—the global distractor of the dSAT, competing options in the 5-choice tasks, and similar but incorrect competitors in the guided search and attentional singleton tasks, the latter of which also adds a high degree of saliency to the distractor. An important question going forward is the degree to which the neural circuitry for responding to these different types of attentional demand is overlapping versus distinct, and how those components are affected by schizophrenia.

Acknowledgments The authors' research was supported by PHS grants MH086530, MH101697, MH093888, DA031656, DA032259, NS091856, and NS078435. We thank Jessica Nicosia and Rohith Pentaparthy for assistance in preparing Fig. 3. We also thank Barbara Sahakian and Trevor Robbins for their comments on an earlier manuscript draft.

References

Akitsuki Y, Sugiura M, Watanabe J, Yamashita K, Sassa Y, Awata S, Kawashima R (2003) Context-dependent cortical activation in response to financial reward and penalty: an event-related fMRI study. Neuroimage 19(4):1674–1685. doi:10.1016/s1053-8119(03)00250-7

Alexander MP, Stuss DT, Shallice T, Picton TW, Gillingham S (2005) Impaired concentration due to frontal lobe damage from two distinct lesion sites. Neurology 65(4):572–579. doi:10.1212/01.wnl.0000172912.07640.92

Apostol G, Abi-Saab W, Kratochvil CJ, Adler LA, Robieson WZ, Gault LM, Saltarelli MD (2012) Efficacy and safety of the novel alpha(4)beta(2) neuronal nicotinic receptor partial agonist ABT-089 in adults with attention-deficit/hyperactivity disorder: a randomized, double-blind, placebo-controlled crossover study. Psychopharmacology 219(3):715–725. doi:10.1007/s00213-011-2393-2

Apparsundaram S, Martinez V, Parikh V, Kozak R, Sarter M (2005) Increased capacity and density of choline transporters situated in synaptic membranes of the right medial prefrontal cortex of attentional task-performing rats. J Neurosci 25(15):3851–3856. doi:10.1523/jinneurosci.0505.-05.2005

Bain EE, Robieson W, Pritchett Y, Garimella T, Abi-Saab W, Apostol G, Saltarelli MD (2013) A randomized, double-blind, placebo-controlled phase 2 study of alpha 4 beta 2 agonist ABT-894 in adults with ADHD. Neuropsychopharmacology 38(3):405–413. doi:10.1038/npp.2012.194

Barch DM, Dowd EC (2010) Goal representations and motivational drive in schizophrenia: the role of prefrontal-striatal interactions. Schizophr Bull 36(5):919–934. doi:10.1093/schbul/sbq068

Bauer M, Kluge C, Bach D, Bradbury D, Heinze HJ, Dolan RJ, Driver J (2012) Cholinergic enhancement of visual attention and neural oscillations in the human brain. Curr Biol 22 (5):397–402. doi:10.1016/j.cub.2012.01.022

Belujon P, Grace AA (2008) Critical role of the prefrontal cortex in the regulation of hippocampus-accumbens information flow. J Neurosci 28:9797–9805

Bentley P, Driver J, Dolan RJ (2011) Cholinergic modulation of cognition: insights from human pharmacological functional neuroimaging. Prog Neurobiol 94(4):360–388. doi:10.1016/j.pneurobio.2011.06.002

Berry AS, Demeter E, Sabhapathy S, English BA, Blakely RD, Sarter M, Lustig C (2014) Disposed to distraction: genetic variation in the cholinergic system influences distractibility but not time-on-task effects. J Cogn Neurosci 26(9):1981–1991. doi:10.1162/jocn_a_00607

Berry AS, Blakely RD, Sarter M, Lustig C (2015) Cholinergic capacity mediates prefrontal engagement during challenges to attention: evidence from imaging genetics. Neuroimage 108:386–395. doi:10.1016/j.neuroimage.2014.12.036

Berry AS, Sarter M, Lustig C (in prep) Frontoparietal correlates of attentional effort during challenges to attention

Berry AS, Sarter M, Gehring WJ, Lustig C (in prep) Going in: frontoparietal responses associated with shifts from perceptual to reflective attention

Blatow M, Rozov A, Katona I, Hormuzdi SG, Meyer AH, Whittington MA, Monyer H (2003) A novel network of multipolar bursting interneurons generates theta frequency oscillations in neocortex. Neuron 38(5):805–817. doi:10.1016/s0896-6273(03)00300-3

Bloem B, Poorthuis RB, Mansvelder HD (2014a) Cholinergic modulation of the medial prefrontal cortex: the role of nicotinic receptors in attention and regulation of neuronal activity. Frontiers Neural Circ 8. doi:10.3389/fncir.2014.00017

Bloem B, Schoppink L, Rotaru DC, Faiz A, Hendriks P, Mansvelder HD, van de Berg WD, Wouterlood FG (2014b) Topographic mapping between basal forebrain cholinergic neurons and the medial prefrontal cortex in mice. J Neurosci 34:16234–16246

Bowie CR, Leung WW, Reichenberg A, McClure MM, Patterson TL, Heaton RK, Harvey PD (2008) Predicting schizophrenia patients' real-world behavior with specific neuropsychological and functional capacity measures. Biol Psychiatry 63(5):505–511. doi:10.1016/j.biopsych.2007.05.022

Brady AM, O'Donnell P (2004) Dopaminergic modulation of prefrontal cortical input to nucleus accumbens neurons in vivo. J Neurosci 24:1040–1049

Braver TS (2012) The variable nature of cognitive control: a dual mechanisms framework. Trends Cogn Sci 16(2):106–113. doi:10.1016/j.tics.2011.12.010

Briand LA, Gritton H, Howe WM, Young DA, Sarter M (2007) Modulators in concert for cognition: modulator interactions in the prefrontal cortex. Prog Neurobiol 83(2):69–91. doi:10.1016/j.pneurobio.2007.06.007

Burgess PW, Dumontheil I, Gilbert SJ (2007) The gateway hypothesis of rostral prefrontal cortex (area 10) function. Trends Cogn Sci 11(7):290–298. doi:10.1016/j.tics.2007.05.004

Cader S, Cifelli A, Abu-Omar Y, Palace J, Matthews PM (2006) Reduced brain functional reserve and altered functional connectivity in patients with multiple sclerosis. Brain 129:527–537. doi:10.1093/brain/awh670

Cairo TA, Woodward TS, Ngan ETC (2004) An event-related fMRI study of maintenance brain activity in schizophrenia during the performance of a variable load working memory task. Schizophr Res 67(1):262

Callaway E III, Jones RT, Donchin E (1970) Auditory evoked potential variability in schizophrenia. Electroencephalogr Clin Neurophysiol 29(5):421–428

Callicott JH, Mattay VS, Bertolino A, Finn K, Coppola R, Frank JA, Weinberger DR (1999) Physiological characteristics of capacity constraints in working memory as revealed by functional MRI. Cereb Cortex 9(1):20–26. doi:10.1093/cercor/9.1.20

Cappell KA, Gmeindl L, Reuter-Lorenz PA (2010) Age differences in prefontal recruitment during verbal working memory maintenance depend on memory load. Cortex 46(4):462–473. doi:10.1016/j.cortex.2009.11.009

Carr DB, Sesack SR (2000) Projections from the rat prefrontal cortex to the ventral tegmental area: target specificity in the synaptic associations with mesoaccumbens and mesocortical neurons. J Neurosci 20:3864–3873

Castellanos FX, Sonuga-Barke EJS, Scheres A, Di Martino A, Hyde C, Walters JR (2005) Varieties of attention-deficit/hyperactivity disorder-related intra-individual variability. Biol Psychiatry 57(11):1416–1423. doi:10.1016/j.biopsych.2004.12.005

Castro-Alamancos MA, Gulati T (2014) Neuromodulators produce distinct activated states in neocortex. J Neurosci 34:12353–12367

Chun MM, Johnson MK (2011) Memory: enduring traces of perceptual and reflective attention. Neuron 72(4):520–535. doi:10.1016/j.neuron.2011.10.026

Cole VT, Weinberger DR, Dickinson D (2011) Intra-individual variability across neuropsychological tasks in schizophrenia: a comparison of patients, their siblings, and healthy controls. Schizophr Res 129(1):91–93. doi:10.1016/j.schres.2011.03.007

Cornblatt BA, Keilp JG (1994) Impaired attention, genetics, and the pathophysiology of schizophrenia. Schizophr Bull 20(1):31–46

Dani JA, Bertrand D (2007) Nicotinic acetylcholine receptors and nicotinic cholinergic mechanisms of the central nervous system. Annu Rev Pharmacol Toxicol 47:699–729

Deco G, Thiele A (2011) Cholinergic control of cortical network interactions enables feedback-mediated attentional modulation. Eur J Neurosci 34(1):146–157. doi:10.1111/j.1460-9568.2011.07749.x

Demeter E, Sarter M, Lustig C (2008) Rats and humans paying attention: cross-species task development for translational research. Neuropsychology 22(6):787–799. doi:10.1037/a0013712
Demeter E, Hernandez-Garcia L, Sarter M, Lustig C (2011) Challenges to attention: a continuous arterial spin labeling (ASL) study of the effects of distraction on sustained attention. Neuroimage 54(2):1518–1529. doi:10.1016/j.neuroimage.2010.09.026
Demeter E, Guthrie SK, Taylor SF, Sarter M, Lustig C (2013) Increased distractor vulnerability but preserved vigilance in patients with schizophrenia: evidence from a translational sustained attention task. Schizophr Res 144(1–3):136–141. doi:10.1016/j.schres.2013.01.003
Demeter E, de Alburquerque D, Woldorff M (2015) The effects of distraction on the brain processes underlying signal detection. Poster presented at the Cognitive Neuroscience Society Meeting, New York
Deserno L, Boehme R, Heinz A, Schlagenhauf F (2013) Reinforcement learning and dopamine in schizophrenia: dimensions of symptoms or specific features of a disease group? Frontiers Psychiatry 4:172. doi:10.3389/fpsyt.2013.00172
Dima D, Dietrich DE, Dillo W, Emrich HM (2010) Impaired top-down processes in schizophrenia: a DCM study of ERPs. Neuroimage 52(3):824–832. doi:10.1016/j.neuroimage.2009.12.086
Disney AA, Aoki C (2008) Muscarinic acetylcholine receptors in macaque v1 are most frequently expressed by parvalbumin-immunoreactive neurons. J Comp Neurol 507(5):1748–1762. doi:10.1002/cne.21616
Donchin E, Callaway E, Jones RT (1970) Auditory evoked potential variability in schizophrenia.2. Application of discriminant analysis. Electroencephalogr Clin Neurophysiol 29(5):429. doi:10.1016/0013-4694(70)90060-x
Egeland J, Rund BR, Sundet K, Landro NI, Asbjornsen A, Lund A, Hugdahl K (2003) Attention profile in schizophrenia compared with depression: differential effects of processing speed, selective attention and vigilance. Acta Psychiatr Scand 108(4):276–284. doi:10.1034/j.1600-0447.2003.00146.x
Flagel SB, Akil H, Robinson TE (2009) Individual differences in the attribution of incentive salience to reward-related cues: implications for addiction. Neuropharmacology 56:139–148. doi:10.1016/j.neuropharm.2008.06.027
Flagel SB, Clark JJ, Robinson TE, Mayo L, Czuj A, Willuhn I, Akil H (2011) A selective role for dopamine in stimulus-reward learning. Nature 469(7328):53–63. doi:10.1038/nature09588
Fletcher PC, McKenna PJ, Frith CD, Grasby PM, Friston KJ, Dolan RJ (1998) Brain activations in schizophrenia during a graded memory task studied with functional neuroimaging. Arch Gen Psychiatry 55:1001–1008. doi:10.1001/archpsyc.55.11.1001
Floresco SB (2015) The nucleus accumbens: an interface between cognition, emotion, and action. Annu Rev Psychol 66(66):25–52. doi:10.1146/annurev-psych-010213-115159
Ford JM, White P, Lim KO, Pfefferbaum A (1994) Schizophrenics have fewer and smaller p300s—a single-trial analysis. Biol Psychiatry 35(2):96–103. doi:10.1016/0006-3223(94)91198-3
Foster DJ, Choi DL, Conn PJ, Rook JM (2014) Activation of M-1 and M-4 muscarinic receptors as potential treatments for Alzheimer's disease and schizophrenia. Neuropsychiatr Dis Treat 10:183–191. doi:10.2147/ndt.s55104
Fukushima J, Morita N, Fukushima K, Chiba T, Tanaka S, Yamashita I (1990) Voluntary control of saccadic eye-movements in patients with schizophrenic and affective-disorders. J Psychiatr Res 24(1):9. doi:10.1016/0022-3956(90)90021-h
Gagne A-M, Hebert M, Maziade M (2015) Revisiting visual dysfunctions in schizophrenia from the retina to the cortical cells: a manifestation of defective neurodevelopment. Prog Neuropsychopharmacol Biol Psychiatry 62:29–34. doi:10.1016/j.pnpbp.2015.04.007
Gaykema RP, van Weeghel R, Hersh LB, Luiten PG (1991) Prefrontal cortical projections to the cholinergic neurons in the basal forebrain. J Comp Neurol 303:563–583
Giesbrecht B, Woldorff MG, Song AW, Mangun GR (2003) Neural mechanisms of top-down control during spatial and feature attention. Neuroimage 19(3):496–512. doi:10.1016/s1053-8119(03)00162-9

Gilmour G, Gastambide F, Marston H, Walton ME (2015) Using intermediate cogntiive endpoints to facilitate translational research in schizophrenia. Curr Opin Behav Sci 4:128–135

Goghari VM (2011) Executive functioning-related brain abnormalities associated with the genetic liability for schizophrenia: an activation likelihood estimation meta-analysis. Psychol Med 41 (6):1239–1252. doi:10.1017/s0033291710001972

Gold JM, Fuller RL, Robinson BM, Braun EL, Luck SJ (2007) Impaired top-down control of visual search in schizophrenia. Schizophr Res 94(1–3):148–155. doi:10.1016/j.schres.2007.04.023

Gorka AX, Hanson JL, Radtke SR, Hariri AR (2014) Reduced hippocampal and medial prefrontal gray matter mediate the association between reported childhood maltreatment and trait anxiety in adulthood and predict sensitivity to future life stress. Biol Mood Anxiety Disord 4:12. doi:10.1186/2045-5380-4-12

Gracitelli CPB, Abe RY, Diniz-Filho A, Vaz-de-Lima FB, Paranhos A Jr, Medeiros FA (2015) Ophthalmology issues in schizophrenia. Curr Psychiatr Rep 17(5). doi:10.1007/s11920-015-0569-x

Green MF (1996) What are the functional consequences of neurocognitive deficits in schizophrenia? Am J Psychiatry 153(3):321–330

Grillon C, Courchesne E, Ameli R, Geyer MA, Braff DL (1990) Increased distractibility in schizophrenic-patients—electrophysiologic and behavioral evidence. Arch Gen Psychiatry 47 (2):171–179

Hahn B, Robinson BM, Kaiser ST, Matveeva TM, Harvey AN, Luck SJ, Gold JM (2012) Kraepelin and bleuler had it right: people with schizophrenia have deficits sustaining attention over time. J Abnorm Psychol 121(3):641–648. doi:10.1037/a0028492

Hangya B, Ranade SP, Lorenc M, Kepecs A (2015) Central cholinergic neurons are rapidly recruited by reinforcement feedback. Cell 162(5):1155–1168. doi:10.1016/j.cell.2015.07.057

Hasselmo ME, McGaughy J (2004) High acetylcholine levels set circuit dynamics for attention and encoding and low acetylcholine levels set dynamics for consolidation. In: Descarries L, Krnjevic K, Steriade M (eds) Acetylcholine in the cerebral cortex 145:207–231

Hasselmo ME, Sarter M (2011) Modes and models of forebrain cholinergic neuromodulation of cognition. Neuropsychopharmacology 36(1):52–73. doi:10.1038/npp.2010.104

Hasselmo ME, Anderson BP, Bower JM (1992) Cholinergic modulation of cortical associative memory function. J Neurophysiol 67(5):1230–1246

Higley MJ, Picciotto MR (2014) Neuromodulation by acetylcholine: examples from schizophrenia and depression. Curr Opin Neurobiol 29C:88–95

Himmelheber AM, Sarter M, Bruno JP (2000) Increases in cortical acetylcholine release during sustained attention performance in rats. Brain Res Cogn Brain Res 9(3):313–325

Howes OD, Kapur S (2009) The dopamine hypothesis of schizophrenia: version III–the final common pathway. Schizophr Bull 35:549–562

Howe WM, Ji J, Parikh V, Williams S, Mocaer E, Trocme-Thibierge C, Sarter M (2010) Enhancement of attentional performance by selective stimulation of alpha 4 beta 2*nAChRs: underlying cholinergic mechanisms. Neuropsychopharmacology 35(6):1391–1401. doi:10.1038/npp.2010.9

Howe WM, Berry AS, Francois J, Gilmour G, Carp JM, Tricklebank M, Sarter M (2013) Prefrontal cholinergic mechanisms instigating shifts from monitoring for cues to cue-guided performance: converging electrochemical and fMRI evidence from rats and humans. J Neurosci 33(20):8742–8752. doi:10.1523/jneurosci.5809-12.2013

Jablensky A (2015) Schizophrenia or schizophrenias? The challenge of genetic parsing of a complex disorder. Am J Psychiatr 172(2):105–107. doi:10.1176/appi.ajp.2014.14111452

Javitt DC, Spencer KM, Thaker GK, Winterer G, Hajos M (2008) Neurophysiological biomarkers for drug development in schizophrenia. Nat Rev Drug Discov 7(1):68–83. doi:10.1038/nrd2463

Jensen J, Willeit M, Zipursky RB, Savina I, Smith AJ, Menon M, Kapur S (2008) The formation of abnormal associations in schizophrenia: neural and behavioral evidence. Neuropsychopharmacology 33(3):473–479. doi:10.1038/sj.npp.1301437

Ji W, Gămănuţ R, Bista P, D'Souza RD, Wang Q, Burkhalter A (2015) Modularity in the organization of mouse primary visual cortex. Neuron 87:632–643
Jimura K, Locke HS, Braver TS (2010) Prefrontal cortex mediation of cognitive enhancement in rewarding motivational contexts. Proc Nat Acad Sci USA 107(19):8871–8876. doi:10.1073/pnas.1002007107
Kaiser S, Roth A, Rentrop M, Friederich H-C, Bender S, Weisbrod M (2008) Intra-individual reaction time variability in schizophrenia, depression and borderline personality disorder. Brain Cogn 66(1):73–82. doi:10.1016/j.bandc.2007.05.007
Kapur S, Mamo D (2003) Half a century of antipsychotics and still a central role for dopamine D2 receptors. Prog Neuropsychopharmacol Biol Psychiatry 27:1081–1090
Keefe RSE, Vinogradov S, Medalia A, Silverstein SM, Bell MD, Dickinson D, Stroup TS (2011) Report from the working group conference on multisite trial design for cognitive remediation in schizophrenia. Schizophr Bull 37(5):1057–1065. doi:10.1093/schbul/sbq010
Koch K, Wagner G, Schultz C, Schachtzabel C, Nenadic I, Axer M, Schloesser RGM (2009) Altered error-related activity in patients with schizophrenia. Neuropsychologia 47(13):2843–2849. doi:10.1016/j.neuropsychologia.2009.06.010
Koch K, Schachtzabel C, Wagner G, Schikora J, Schultz C, Reichenbach JR, Sauer H, Schlosser RGM (2010) Altered activation in association with reward-related trial-and-error activity in patients with schizophrenia. NeuroImage 50:223–232
Kogoj A, Pirtosek Z, Tomori M, Vodusek DB (2005) Event-related potentials elicited by distractors in an auditory oddball paradigm in schizophrenia. Psychiatry Res 137(1–2):49–59. doi:10.1016/j.psychres.2005.07.017
Kozak R, Bruno JP, Sarter M (2006) Augmented prefrontal acetylcholine release during challenged attentional performance. Cereb Cortex 16(1):9–17. doi:10.1093/cercor/bhi079
Kozak R, Martinez V, Young D, Brown H, Bruno JP, Sarter M (2007) Toward a neuro-cognitive animal model of the cognitive symptoms of schizophrenia: disruption of cortical cholinergic neurotransmission following repeated amphetamine exposure in attentional task-performing, but not non-performing, rats. Neuropsychopharmacology 32(10):2074–2086. doi:10.1038/sj.npp.1301352
Krebs M, Boehler N, Roberts C, Song W, Woldorff G (2012) The involvement of the dopaminergic midbrain and cortico-striatal-thalamic circuits in the integration of reward prospect and attentional task demands. Cereb Cortex 22:607–615
Kruse AC, Weiss DR, Rossi M, Hu J, Hu K, Eitel K, Shoichet BK (2013) Muscarinic receptors as model targets and antitargets for structure-based ligand discovery. Mol Pharmacol 84(4):528–540. doi:10.1124/mol.113.087551
Kucinski A, Sarter M (2015) Modeling parkinson's disease falls associated with brainstem cholinergic systems decline. Behav Neurosci 129(2):96–104. doi:10.1037/bne0000048
Kucinski A, Paolone G, Bradshaw M, Albin RL, Sarter M (2013) Modeling fall propensity in parkinson's disease: deficits in the attentional control of complex movements in rats with cortical-cholinergic and striatal-dopaminergic deafferentation. J Neurosci 33(42):16522–16539. doi:10.1523/jneurosci.2545-13.2013
Kucinski AJ, Koshy Cherian A, Valuskova P, Yegla B, Parikh V, Robinson T, Sarter M (2015) Prone to addiction as well as to falls: poor attention in sign-tracking rats extends to complex movement control and is associated with regression of choline transporter capacity. Society for Neuroscience Annual Meeting, Chicago, IL
Lambe EK, Picciotto MR, Aghajanian GK (2003) Nicotine induces glutamate release from thalamocortical terminals in prefrontal cortex. Neuropsychopharmacology 28(2):216–225. doi:10.1038/sj.npp.1300032
Laycock R, Crewther SG, Crewther DP (2007) A role for the 'magnocellular advantage' in visual impairments in neuro developmental and psychiatric disorders. Neurosci Biobehav Rev 31 (3):363–376. doi:10.1016/j.neubiorev.2006.10.003
Leonard CJ, Robinson BM, Hahn B, Gold JM, Luck SJ (2014) Enhanced distraction by magnocellular salience signals in schizophrenia. Neuropsychologia 56:359–366. doi:10.1016/j.neuropsychologia.2014.02.011

Lesh TA, Niendam TA, Minzenberg MJ, Carter CS (2011) Cognitive control deficits in schizophrenia: mechanisms and meaning. Neuropsychopharmacology 36(1):316–338. doi:10.1038/npp.2010.156
Luck SJ, Gold JM (2008) The construct of attention in schizophrenia. Biol Psychiatry 64(1):34–39. doi:10.1016/j.biopsych.2008.02.014
Luck SJ, Ford JM, Sarter M, Lustig C (2012) CNTRICS final biomarker selection: control of attention. Schizophr Bull 38(1):53–61. doi:10.1093/schbul/sbr065
Lukatch HS, MacIver MB (1997) Physiology, pharmacology, and topography of cholinergic neocortical oscillations in vitro. J Neurophysiol 77(5):2427–2445
Lustig C, Jantz T (2015) Questions of age differences in interference control: When and how, not if? Brain Res 1612:59–69
Lustig C, Kozak R, Sarter M, Young JW, Robbins TW (2013) CNTRICS final animal model task selection: control of attention. Neurosci Biobehav Rev 37(9):2099–2110. doi:10.1016/j.neubiorev.2012.05.009
MacDonald SWS, Nyberg L, Backman L (2006) Intra-individual variability in behavior: links to brain structure, neurotransmission and neuronal activity. Trends Neurosci 29(8):474–480. doi:10.1016/j.tins.2006.06.011
Manoach DS (2003) Prefrontal cortex dysfunction during working memory performance in schizophrenia: reconciling discrepant findings. Schizophr Res 60(2–3):285–298. doi:10.1016/s0920-9964(02)00294-3
Manoach DS, Press DZ, Thangaraj V, Searl MM, Goff DC, Halpern E, Warach S (1999) Schizophrenic subjects activate dorsolateral prefrontal cortex during a working memory task, as measured by fMRI. Biol Psychiatry 45(9):1128–1137. doi:10.1016/s0006-3223(98)00318-7
Manoach DS, Gollub RL, Benson ES, Searl MM, Goff DC, Halpern E, Rauch SL (2000) Schizophrenic subjects show aberrant fMRI activation of dorsolateral prefrontal cortex and basal ganglia during working memory performance. Biol Psychiatry 48(2):99–109. doi:10.1016/s0006-3223(00)00227-4
Markou A, Chiamulera C, Geyer MA, Tricklebank M, Steckler T (2009) Removing obstacles in neuroscience drug discovery: the future path for animal models. Neuropsychopharmacology 34(1):74–89. doi:10.1038/npp.2008.173
Markou A, Salamone JD, Bussey TJ, Mar AC, Brunnerd D, Gilmour G, Balsam P (2013) Measuring reinforcement learning and motivation constructs in experimental animals: relevance to the negative symptoms of schizophrenia. Neurosci Biobehav Rev 37(9):2149–2165. doi:10.1016/j.neubiorev.2013.08.007
Martinez V, Sarter M (2004) Lateralized attentional functions of cortical cholinergic inputs. Behav Neurosci 118(5):984–991. doi:10.1037/0735-7044.118.5.984
Mattay VS, Fera F, Tessitore A, Hariri AR, Berman KF, Das S, Weinberger DR (2006) Neurophysiological correlates of age-related changes in working memory capacity. Neurosci Lett 392(1–2):32–37. doi:10.1016/j.neulet.2005.09.025
Matthysse S, Levy DL, Wu YN, Rubin DB, Holzman P (1999) Intermittent degradation in performance in schizophrenia. Schizophr Res 40(2):131–146. doi:10.1016/s0920-9964(99)00038-9
McGaughy J, Sarter M (1995) Behavioral vigilance in rats—task validation and effects of age, amphetamine, and benzodiazepine receptor ligands. Psychopharmacology 117(3):340–357. doi:10.1007/bf02246109
McGaughy J, Sarter M (1999) Effects of ovariectomy, 192 IgG-saporin-induced cortical cholinergic deafferentatino, and administration of estradiol on sustained attention performance in rats. Behav Neurosci 113:1216–1232
Mechawar N, Cozzari C, Descarries L (2000) Cholinergic innervation in adult rat cerebral cortex: a quantitative immunocytochemical description. J Comp Neurol 428:305–318
Meltzer HY (2015) Pharmacotherapy of cognition in schizophrenia. Curr Opin Behav Sci 4:115–121
Mendelsohn A, Pine A, Schiller D (2014) Between thoughts and actions: motivationally salient cues invigorate mental action in the human brain. Neuron 207–217. DOI: 10.1016/j.neuron.2013.10.019

Meyer PJ, Lovic V, Saunders BT, Yager LM, Flagel SB, Morrow JD, Robinson TE (2012) Quantifying individual variation in the propensity to attribute incentive salience to reward cues. Plos One 7(6). doi:10.1371/journal.pone.0038987
Minzenberg MJ, Laird AR, Thelen S, Carter CS, Glahn DC (2009) Meta-analysis of 41 functional neuroimaging studies of executive function in schizophrenia. Arch Gen Psychiatry 66(8):811–822
Money TT, Scarr E, Udawela M, Gibbons AS, Jeon WJ, Seo MS, Dean B (2010) Treating schizophrenia: novel targets for the cholinergic system. Cns Neurol Disord Drug Targets 9 (2):241–256
Muñoz W, Rudy B (2014) Spatiotemporal specificity in cholinergic control of neocortical function. Curr Opin Neurobiol 26C:149–160
Neigh GN, Arnold HM, Rabenstein RL, Sarter M, Bruno JP (2004) Neuronal activity in the nucleus accumbens is necessary for performance-related increases in cortical acetylcholine release. Neuroscience 123:635–645
Nelson CL, Sarter M, Bruno JP (2005) Prefrontal cortical modulation of acetylcholine release in posterior parietal cortex. Neuroscience 132:347–359
Newman EL, Gupta K, Climer JR, Monaghan CK, Hasselmo ME (2012) Cholinergic modulation of cognitive processing: insights drawn from computational models. Frontiers Behav Neurosci 6. doi:10.3389/fnbeh.2012.00024
Nielsen MO, Rostrup E, W S, Bak N, Broberg BV, Lublin H, Glenthoj B (2012) Improvement of brain reward abnormalities by antipsychotic monotherapy in schizophrenia. Arch Gen Psychiatry 69(12):1195–1204. doi:10.1001/archgenpsychiatry.2012.847
Nikolaus S, Hautzel H, Mueller H-W (2014) Neurochemical dysfunction in treated and nontreated schizophrenia—a retrospective analysis of in vivo imaging studies. Rev Neurosci 25(1):25–96. doi:10.1515/revneuro-2013-0063
Nuechterlein KH, Dawson ME (1984) Information-processing and attentional functioning in the developmental course of schizophrenic disorders. Schizophr Bull 10(2):160–203
Nuechterlein KH, Luck SJ, Lustig C, Sarter M (2009) CNTRICS final task selection: control of attention. Schizophr Bull 35(1):182–196. doi:10.1093/schbul/sbn158
O'Connell RG, Dockree PM, Robertson IH, Bellgrove MA, Foxe JJ, Kelly SP (2009) Uncovering the neural signature of lapsing attention: electrophysiological signals predict errors up to 20s before they occur. J Neurosci 29(26):8604–8611. doi:10.1523/jneurosci.5967-08.2009
Paolone G, Angelakos CC, Meyer PJ, Robinson TE, Sarter M (2013a) Cholinergic control over attention in rats prone to attribute incentive salience to reward cues. J Neurosci 33:8321–8335
Paolone G, Mallory CS, Cherian AK, Miller TR, Blakely RD, Sarter M (2013b) Monitoring cholinergic activity during attentional performance in mice heterozygous for the choline transporter: a model of cholinergic capacity limits. Neuropharmacology 75:274–285
Parikh V, Kozak R, Martinez V, Sarter M (2007) Prefrontal acetylcholine release controls cue detection on multiple timescales. Neuron 56(1):141–154. doi:10.1016/j.neuron.2007.08.025
Parikh V, Man K, Decker MW, Sarter M (2008) Glutamatergic contributions to nicotinic acetylcholine receptor agonist-evoked cholinergic transients in the prefrontal cortex. J Neurosci 28(14):3769–3780. doi:10.1523/jneurosci.5251-07.2008
Parikh V, Ji J, Decker MW, Sarter M (2010) Prefrontal beta 2 subunit-containing and alpha 7 nicotinic acetylcholine receptors differentially control glutamatergic and cholinergic signaling. J Neurosci 30(9):3518–3530. doi:10.1523/jneurosci.5712-09.2010
Parikh V, Peters MS, Blakely RD, Sarter M (2013) The presynaptic choline transporter imposes limits on sustained cortical acetylcholine release and attention. J Neurosci 33(6):2326–2337. doi:10.1523/jneurosci.4993-12.2013
Parry AMM, Scott RB, Palace J, Smith S, Matthews PM (2003) Potentially adaptive functional changes in cognitive processing for patients with multiple sclerosis and their acute modulation by rivastigmine. Brain 126:2750–2760. doi:10.1093/brain/awg284
Peters MS, Demeter E, Lustig C, Bruno JP, Sarter M (2011) Enhanced control of attention by stimulating mesolimbic-corticopetal cholinergic circuitry. J Neurosci 31(26):9760–9771. doi:10.1523/jneurosci.1902-11.2011

Picciotto MR, Higley MJ, Mineur YS (2012) Acetylcholine as a neuromodulator: cholinergic signaling shapes nervous system function and behavior. Neuron 76(1):116–129. doi:10.1016/j.neuron.2012.08.036

Raizada RDS, Poldrack RA (2007) Challenge-driven attention: interacting frontal and brainstem systems. Frontiers Hum Neurosci 1:3. doi:10.3389/neuro.09.003.2007

Reinhart RMG, Zhu J, Park S, Woodman GF (2015) Synchronizing theta oscillations with direct-current stimulation strengthens adaptive control in the human brain. Proc Nat Acad Sci USA 112(30):9448–9453. doi:10.1073/pnas.1504196112

Reuter-Lorenz PA, Lustig C (2005) Brain aging: reorganizing discoveries about the aging mind. Curr Opin Neurobiol 15(2):245–251. doi:10.1016/j.conb.2005.03.016

Reuter-Lorenz PA, Cappell KA (2008) Neurocognitive aging and the compensation hypothesis. Curr Dir Psychol Sci 17(3):177–182. doi:10.1111/j.1467-8721.2008.00570.x

Riedel G, Platt B, van der Zee E (eds) (2015) Special issue of brain research: the cholinergic system and brain function. Brain Res 221

Robbins TW (2002) The 5-choice serial reaction time task: behavioural pharmacology and functional neurochemistry. Psychopharmacology 163(3–4):362–380. doi:10.1107/s00213-002-1154-7

Robinson L, Platt B, Riedel G (2011) Involvement of the cholinergic system in conditioning and perceptual memory. Behav Brain Res 221(2):443–465. doi:10.1016/j.bbr.2011.01.055

Roche MW, Silverstein SW, Lenzenweger MF (2015) Intermittent degradation and schizotypy. Schizophr Res Cogn 2:100–104

Roschke J, Wagner P, Mann K, Fell J, Grozinger M, Frank C (1996) Single trial analysis of event related potentials: a comparison between schizophrenics and depressives. Biol Psychiatry 40 (9):844–852. doi:10.1016/0006-3223(95)00652-4

Ross RG, Stevens KE, Proctor WR, Leonard S, Kisley MA, Hunter SK, Adams CE (2010) Research review: cholinergic mechanisms, early brain development, and risk for schizophrenia. J Child Psychol Psychiatry 51(5):535–549. doi:10.1111/j.1469-7610.2009.02187.x

Roth A, Roesch-Ely D, Bender S, Weisbrod M, Kaiser S (2007) Increased event-related potential latency and amplitude variability in schizophrenia detected through wavelet-based single trial analysis. Int J Psychophysiol 66(3):244–254. doi:10.1016/j.ijpsycho.2007.08.005

Rowe AR, Mercer L, Casetti V, Sendt K-V, Giaroli G, Shergill SS, Tracy DK (2015) Dementia praecox redux: a systematic review of the nicotinic receptor as a target for cognitive symptoms of schizophrenia. J Psychopharmacol 29(2):197–211. doi:10.1177/0269881114564096

Sahakian BJ, Jones G, Levy R, Gray J, Warburton D (1989) The effects of nicotine on attention, information processing, and short-term memory in patients with dementia of the Alzheimer type. Br J Psychiatr 154(6):797–800. doi:10.1192/bjp.154.6.797

Sahakian BJ, Owen AM, Morant NJ, Eagger SA, Boddington S, Crayton L, Levy R (1993) Further analysis of the cognitive effects of tetrahydroaminoacridine (tha) in alzheimers-disease—assessment of attentional and mnemonic function using cantab. Psychopharmacology 110 (4):395–401. doi:10.1007/bf02244644

Sarter M (2015) Behavioral-cognitive targets for cholinergic enhancement. Curr Opin Behav Sci 4:22–26

Sarter M, Tricklebank M (2012) Revitalizing psychiatric drug discovery. Nat Rev Drug Discov 11 (6):423–424. doi:10.1038/nrd3755

Sarter M, Nelson CL, Bruno JP (2005) Cortical cholinergic transmission and cortical information processing in schizophrenia. Schizophr Bull 31(1):117–138. doi:10.1093/schbul/sbi006

Sarter M, Gehring WJ, Kozak R (2006) More attention must be paid: the neurobiology of attentional effort. Brain Res Rev 51(2):145–160. doi:10.1016/j.brainresrev.2005.11.002

Sarter M, Martinez V, Kozak R (2009a) A neurocognitive animal model dissociating between acute illness and remission periods of schizophrenia. Psychopharmacology 202(1–3):237–258. doi:10.1007/s00213-008-1216-6

Sarter M, Parikh V, Howe WM (2009b) PERSPECTIVES phasic acetylcholine release and the volume transmission hypothesis: time to move on. Nat Rev Neurosci 10(5):383–386. doi:10.1038/nrn2635

Sarter M, Lustig C, Taylor SF (2012) Cholinergic contributions to the cognitive symptoms of schizophrenia and the viability of cholinergic treatments. Neuropharmacology 62(3):1544–1553. doi:10.1016/j.neuropharm.2010.12.001
Sarter M, Lustig C, Howe WM, Gritton H, Berry AS (2014) Deterministic functions of cortical acetylcholine. Eur J Neurosci 39(11):1912–1920. doi:10.1111/ejn.12515
Sarter M, Howe WM, Gritton H (2015) Cortical cholinergic transients for cue detection and attentional mode shifts. In: Wilson GS, Michael AC (eds) Compendium of in vivo monitoring in real-time molecular neuroscience. World Scientific, Singapore, pp 27–44
Saunders BT, Robinson TE (2012) The role of dopamine in the accumbens core in the expression of Pavlovian-conditioned responses. Eur J Neurosci 36(4):2521–2532. doi:10.1111/j.1460-9568.2012.08217.x
Scarr E, Cowie TF, Kanellakis S, Sundram S, Pantelis C, Dean B (2009) Decreased cortical muscarinic receptors define a subgroup of subjects with schizophrenia. Mol Psychiatr 14 (11):1017–1023. doi:10.1038/mp.2008.28
Schneider-Garces NJ, Gordon BA, Brumback-Peltz CR, Shin E, Lee Y, Sutton BP, Fabiani M (2010) Span, CRUNCH, and beyond: working memory capacity and the aging brain. J Cogn Neurosci 22(4):655–669. doi:10.1162/jocn.2009.21230
Scognamiglio C, Houenou J (2014) A meta-analysis of fMRI studies in healthy relatives of patients with schizophrenia. Aust N Z J Psychiatr 48(10):907–916. doi:10.1177/0004867414540753
Sebastian A, Baldermann C, Feige B, Katzev M, Scheller E, Hellwig B, Kloeppel S (2013) Differential effects of age on subcomponents of response inhibition. Neurobiol Aging 34 (9):2183–2193. doi:10.1016/j.neurobiolaging.2013.03.013
Seo MS, Scarr E, Dean B (2014) An investigation of the factors that regulate muscarinic receptor expression in schizophrenia. Schizophr Res 158(1–3):247–254. doi:10.1016/j.schres.2014.06.039
Sesack SR, Grace AA (2010) Cortico-basal ganglia reward network: microcircuitry. Neuropsychopharmacology 35:27–47
Shoshina II, Shelepin YE, Semenova NB (2014) Frequency-contrast sensitivity of visual stimulus perception in patients with schizophrenia treated with atypical and typical antipsychotics. Hum Physiol 40(1):35–39. doi:10.1134/s0362119714010150
Silverstein SM, Matteson S, Knight RA (1996) Reduced top-down influence in auditory perceptual organization in schizophrenia. J Abnorm Psychol 105(4):663–667
Skottun BC, Skoyles JR (2007) Contrast sensitivity and magnocellular funciton in schizophrenia. Vision Res 47:2923–2933
Small DM, Gitelman D, Simmons K, Bloise SM, Parrish T, Mesulam MM (2005) Monetary incentives enhance processing in brain regions mediating top-down control of attention. Cereb Cortex 15:1855–1865
Smiley JF, Subramanian M, Mesulam MM (1999) Monoaminergic-cholinergic interactions in the primate basal forebrain. Neuroscience 93:817–829
Theeuwes J (1992) Perceptual selectivity for color and form. Percept Psychophys 51(6):599–606. doi:10.3758/bf03211656
Thiele A (2013) Muscarinic signaling in the brain. Annu Rev Neurosci 36(36):271–294. doi:10.1146/annurev-neuro-062012-170433
Torgalsboen A-K, Mohn C, Rund BR (2014) 2 Neurocognitive predictors of remission of symptoms and social and role functioning in the early course of first-episode schizophrenia. Psychiatr Res 216(1):1–5. doi:10.1016/j.psychres.2014.01.031
Torgalsboen A-K, Mohn C, Czajkowski N, Rund BR (2015) Relationship between neurocognition and functional recovery in first-episode schizophrenia: results from the second year of the Oslo multi-follow-up study. Psychiatr Res 227(2–3):185–191. doi:10.1016/j.psychres.2015.03.037

Umbriaco D, Watkins KC, Descarries L, Cozzari C, Hartman BK (1994) Ultrastructural and morphometric features of the acetylcholine innervation in adult rat parietal cortex: an electron microscopic study in serial sections. J Comp Neurol 348:351–373

Van Snellenberg JX, Torres IJ, Thornton AE (2006) Functional neuroimaging of working memory in schizophrenia: task performance as a moderating variable. Neuropsychology 20(5):497–510. doi:10.1037/0894-4105.20.5.497.supp

Van Snellenberg JX, Girgis RR, Read C, Thompson JL, Weber J, Wager TD, Slifstein M, Lieberman JA, Abi-Dargham A, Smith EE (2013) Individuals with schizophrenia fail to show normative inverted-U activation in response to fine-grained working memory load manipulation. Schizophr Bull 39:S251–S252 (Abstract)

Van Snellenberg JX, Slifstein M, Read C, Weber J, Thompson JL, Wager TD, Smith EE (2015) Dynamic shifts in brain network activation during supracapacity working memory task performance. Hum Brain Mapp 36(4):1245–1264. doi:10.1002/hbm.22699

Watanabe M, Sakagami M (2007) Integration of cognitive and motivational context information in the primate prefrontal cortex. Cereb Cortex 17:I101–I109. doi:10.1093/cercor/bhm067

Weinberger DR, Harrison P (2011) Schizophrenia, 3rd edn. Wiley, New York

Wohlberg GW, Kornetsk C (1973) Sustained attention in remitted schizophrenics. Arch Gen Psychiatr 28(4):533–537

Wolkin A, Sanfilipo M, Wolf AP, Angrist B, Brodie JD, Rotrosen J (1992) Negative symptoms and hypofrontality in chronic schizophrenia. Arch Gen Psychiatr 49(12):959–965

Wong EHF, Yocca F, Smith MA, Lee C-M (2010) Challenges and opportunities for drug discovery in psychiatric disorders: the drug hunters' perspective. Int J Neuropsychopharmacol 13(9):1269–1284. doi:10.1017/s1461145710000866

Xiang Z, Huguenard JR, Prince DA (1998) Cholinergic switching within neocortical inhibitory networks. Science 281:985–988

Young JW, Markou A (2015) Translational rodent paradigms to investigate neuromechanisms underlying behaviors relevant to amotivation and altered reward processing in schizophrenia. Schizophr Bull 41(5):1024–1034. doi:10.1093/schbul/sbv093

Young JW, Light GA, Marston HM, Sharp R, Geyer MA (2009) The 5-choice continuous performance test: evidence for a translational test of vigilance for mice. Plos One 4(1). doi:10.1371/journal.pone.0004227

Young JW, Zhou X, Geyer MA (2010) Animal models of schizophrenia. In: Swerdlow NR (ed) Behavioral neurobiology of schizophrenia and its treatment, vol 4, pp 391–433

Zaborszky L, Cullinan WE (1992) Projections from the nucleus accumbens to cholinergic neurons of the ventral pallidum: a correlated light and electron microscopic double-immunolabeling study in rat. Brain Res 570:92–101

Zaborszky L, Cullinan WE (1996) Direct catecholaminergic-cholinergic interactions in the basal forebrain. I. Dopamine-β-hydroxylase-and tyrosine hydroxylase input to cholinergic neurons. J Comp Neurol 374:535–554

Zaborszky L, Gaykema RP, Swanson DJ, Cullinan WE (1997) Cortical input to the basal forebrain. Neuroscience 79:1051–1078

Zaborszky L, Buhl DL, Pobalashingham S, Bjaalie JG, Nadasdy Z (2005) Three-dimensional chemoarchitecture of the basal forebrain: spatially specific association of cholinergic and calcium binding protein-containing neurons. Neuroscience 136:697–713

Zaborszky L, Hoemke L, Mohlberg H, Schleicher A, Amunts K, Zilles K (2008) Stereotaxic probabilistic maps of the magnocellular cell groups in human basal forebrain. Neuroimage 42:1127–1141

Zaborszky L, van den Pol A, Gyengesi E (2012) The basal forebrain cholinergic projection system in mice. In: The mouse nervous system. Elsevier, Pennsylvania, pp 684–718

Zaborszky L, Csordas A, Mosca K, Kim J, Gielow MR, Vadasz C, Nadasdy Z (2015a) Neurons in the basal forebrain project to the cortex in a complex topographic organization that reflects corticocortical connectivity patterns: an experimental study based on retrograde tracing and 3D reconstruction. Cereb Cortex 25:118–137

Zaborszky L, Duque A, Gielow M, Gombkoto P, Nadasdy Z, Somogyi J (2015b) Organization of the basal forebrain cholinergic projection system: specific or diffuse? In: Paxinos G (ed) The Rat Nervous System. Academic Press, San Diego, pp 491–507

Attentional Set-Shifting Across Species

Verity J. Brown and David S. Tait

Abstract Attentional set-shifting, as a measure of executive flexibility, has been a staple of investigations into human cognition for over six decades. Mediated by the frontal cortex in mammals, the cognitive processes involved in forming, maintaining and shifting an attentional set are vulnerable to dysfunction arising from a number of human neurodegenerative diseases (such as Alzheimer's, Parkinson's and Huntington's diseases) and other neurological disorders (such as schizophrenia, depression, and attention deficit/hyperactivity disorder). Our understanding of these diseases and disorders, and the cognitive impairments induced by them, continues to advance, in tandem with an increasing number of tools at our disposal. In this chapter, we review and compare commonly used attentional set-shifting tasks (the Wisconsin Card Sorting Task and Intradimensional/Extradimensional tasks) and their applicability across species. In addition to humans, attentional set-shifting has been observed in a number of other animals, with a substantial body of literature describing performance in monkeys and rodents. We consider the task designs used to investigate attentional set-shifting in these species and the methods used to model human diseases and disorders, and ultimately the comparisons and differences between species-specific tasks, and between performance across species.

Keywords Cognitive flexibility · Attention set-shifting · Reversal learning · Wisconsin card sorting test · Prefrontal cortex · Neurodegenerative disease · Schizophrenia · Affective disorder · ADHD

V.J. Brown · D.S. Tait (✉)
School of Psychology and Neuroscience, University of St Andrews,
St Mary's Quad, South St, St Andrews, Fife KY16 9JP, Scotland, UK
e-mail: dst@st-and.ac.uk

V.J. Brown
e-mail: vjb@st-and.ac.uk

Curr Topics Behav Neurosci (2016) 28: 363–395
DOI 10.1007/7854_2015_5002

Contents

1 Introduction 364
2 Humans and Other Primates—WCST and ID/ED Tasks 367
3 Rodents—ID/ED Tasks 370
4 Neuroanatomy of Attentional Set 372
5 Neurodegenerative Diseases and Attentional Set-Shifting 375
5.1 Parkinson's Disease: Humans and Animal Models 376
5.2 Huntington's Disease: Humans and Animals Models 377
5.3 Alzheimer's Disease: Humans and Animal Models 378
6 Affective Disorders and Attentional Set-Shifting 379
7 Schizophrenia and Attentional Set-Shifting 380
8 Attention Deficit/Hyperactivity Disorder and Attentional Set-Shifting 383
9 Summary and Future Directions 383
10 Conclusion 384
References 385

1 Introduction

The search for treatments to restore cognitive performance in human neuropsychiatric disorders—such as schizophrenia, attention deficit/hyperactivity disorder and addiction—and neurodegenerative diseases—such as Alzheimer's and Parkinson's diseases—has consistently highlighted the importance of boosting fronto-executive function. Executive functions such as working memory, response inhibition and cognitive flexibility are mediated in part by the prefrontal cortex in humans and other animals and can be modulated by pharmacological manipulation (Kesner and Churchwell 2011; Chudasama 2011; Dalley et al. 2004; Robbins 2000; Miller and Cohen 2001). To establish effective treatments for these frontal-dependent cognitive dysfunctions prevalent in so many human disorders and diseases, it is imperative that researchers seek not only to determine and treat the 'cause', but also to fully understand the 'effect'. Discerning the precise cognitive processes affected by any condition must enhance the chances of success in developing the most suitable treatment. For example, 'cognitive flexibility' is impaired in all the previously mentioned disorders and diseases—and such impairments can appear superficially similar. Yet, each disorder and disease exhibits a differing pathology—and therefore assessments of arising cognitive dysfunctions must be able to tease apart the subtleties of these 'effects' to better understand the 'causes'. The specific aspect of 'cognitive flexibility' that is dysfunctional in any given condition therefore determines the best 'tools for the job' when choosing how to assess and treat that dysfunction.

Cognitive flexibility can be assessed in a variety of ways, including attentional set-shifting, rule or strategy-switching and reversal learning (Brown and Tait 2010).

Although planning, sorting and categorisation tasks have been in use since the 1920s (Vigotsky and Kasanin 1934), the Wisconsin Card Sorting Test (WCST), became the early established method of measuring attentional set-shifting in humans (Berg 1948; Grant and Berg 1948). 'Sets' have been ascribed numerous definitions in the past and have meant 'rather different things to different psychologists' (Gibson 1941); however, in the context of attentional set-shifting, an 'attentional set' is the result of a predisposition for attention to be preferentially directed towards a relevant aspect of a stimulus (or group of stimuli). The result of an 'attentional set' is that processing of that relevant information is enhanced, and processing of irrelevant information is inhibited. The WSCT has been used to explore cognitive dysfunction in patients with frontal cortex damage (Milner 1963; Demakis 2003; Barcelo and Knight 2002), age-related cognitive impairments (West 1996; Ridderinkhof et al. 2002), schizophrenia (Nieuwenstein et al. 2001), neurodegenerative diseases such as Parkinson's disease (Paolo et al. 1995) and Alzheimer's disease (Nagahama et al. 2005), and attention deficit/hyperactivity disorder (Romine et al. 2004). The WCST requires subjects to sort a deck of cards, each bearing a number of coloured shapes, into piles according to a rule: the colour of the shapes; the number of the shapes; or the shapes themselves. The rule can change, and the subjects must learn, from experimenter-provided feedback on their choices, the new rule. Attentional set-shifting can also be measured in humans and other primates using a series of two-choice complex stimulus discriminations, with one perceptual aspect (dimension, e.g. shape) predicting the stimulus correct/incorrect status, and the other dimension (e.g. line) being irrelevant to the stimulus correct/incorrect status. Two-choice discrimination measures of attentional set-shifting can be similar to the WCST and simply involve a rule change (Settlage et al. 1956); or tasks can consist of a 'total change' design (Slamecka 1968)—a series of discriminations, including stages with completely novel exemplars solved based on either the previously relevant stimulus dimension (intradimensional (ID) acquisition) or the previously irrelevant dimension (extradimensional (ED) shift acquisition) (Settlage et al. 1956; Eimas 1966; Shepp and Schrier 1969). The Cambridge Neuropsychological Automated Test Battery (CANTAB) is one of the most prevalent current two-choice discrimination attentional set-shifting tasks (Sahakian and Owen 1992; Roberts et al. 1988).

In 'total change' tasks, novel discriminations based on the same dimension as the preceding discrimination (IDs) are solved more rapidly, if an attentional set has been formed, than those that require a shift of attention (EDs) to another dimension. Tasks with ID/ED comparisons therefore contain internal validation: an increase in number of trials required to solve an ED shift relative to those required to solve an ID likely arises from the cost incurred of shifting attention from one dimension to the other. When analysing data from ID/ED tasks, the relationship between ID and ED within a group is often as important as the between-group ED comparison: only if there is evidence of a cost to shift attentional set within a group can it be concluded that a between-group ED difference reflects a change in attentional set-shifting performance.

The WCST and ID/ED tasks differ in the measures that can be obtained from them. In the WCST, no particular exemplar is 'correct'—rather, each card contains a novel complex stimulus that must be matched to one of a set of piles dependent on the current rule. Therefore, when measuring attentional set-shifting using the WCST, whilst there is a constant change in stimuli, previously present exemplars remain in use (e.g. 'green' with varying shapes and/or colours), meaning that subjects may need to overcome exemplar-specific context as well as the abstract context of set in order to resolve the problem presented to them. This can present a confound when attempting to parse attentional set-shifting dysfunction from cognitive processes involving stimulus-reward associations. In contrast, in ID/ED tasks, as both the ID and the ED stages consist of novel exemplars, only the prior, abstract context of set can be applied to solving those discriminations. Thus, there may be additional processes involved, such as those underlying partial reinforcement, when sorting rules change in the WCST and WCST-like tasks that are not present in ID/ED tasks. The multiple stages in ID/ED tasks, which can be moved, replaced or added to, therefore present an opportunity to parse out different cognitive processes: reversal learning stages can be added that can explicitly compare new learning to reversals of exemplar correct/incorrect status, as well as overtraining effects (Sutherland and Mackintosh 1971); additional ID stages can be added, as well as probe stages, to investigate set-formation (Chase et al. 2012; Bissonette and Powell 2012; Crofts et al. 2001).

Whilst humans and other primates are well-suited to discriminating complex visual stimuli, rodents do so less intuitively. The earliest investigations of rat set-shifting used two-choice discriminations to compare reversal learning to attentional shifting (Kelleher 1956) with no ID to act as a comparator or used a between-subjects design with one group performing an ID and one group performing an ED shift (Shepp and Eimas 1964). Later studies relied on rule-switching strategies, either reversals of strategies such as non-match-to-sample to match-to-sample (Joel et al. 1997), or had rats switch between spatial and response discriminations in a plus-maze (Ragozzino et al. 1999). Whilst rule-switching tasks fail to represent the complex stimuli present in primate attentional set-shifting tasks, the strategy-shifting tasks (e.g. Floresco et al. 2008) bear more similarity to the WCST. In Ragozzino et al. (1999), the stimuli remain identical between the 'spatial' (allocentric) and 'response' (egocentric) strategies, and due to the four stimuli limit on the plus-maze, during any strategy-shift, one of the previously correct stimuli remains correct and one becomes incorrect (likewise, one of the previously incorrect stimuli becomes correct, and one remains incorrect): thus, partial reinforcement may affect performance during a shift.

In 2000, Birrell and Brown presented a rodent version of the ID/ED task, using baited bowls that could be discriminated by one of the three dimensions—odour, digging medium or outer texture. With a large number of potential exemplars, the bowl-digging paradigm, like the CANTAB ID/ED task, allows for multiple stages with different sets of exemplars. The numerous studies that have adopted, and adapted, the Birrell and Brown methodology (see Tait et al. 2014) have demonstrated that attentional set-shifting can be measured easily in rats, with relatively short training and testing durations—in comparison with training rats to perform visual discriminations (Bussey et al. 1997)—and that rats are a suitable animal model for investigating attentional set-shifting dysfunction. The rodent ID/ED task has also been adapted for use with mice (Bissonette et al. 2008)—although, with several different adaptations (e.g. Young et al. 2010; Garner et al. 2006; Papaleo et al. 2008), there is currently no established consensus on the best methodology for achieving consistent evidence of attentional set.

2 Humans and Other Primates—WCST and ID/ED Tasks

As discussed in the introduction, the methods most commonly used to investigate set-shifting ability in primates are the WCST and the ID/ED tasks. The ID/ED task for primates, however, has been largely standardised by the computerised touch-screen version (Roberts et al. 1988) supplied with CANTAB by Cambridge Cognition Ltd. The exact protocol for the WCST followed by different research groups is not always identical: for example, it can be run with physical cards (Berg 1948) or using a computerised version (e.g. Mattes et al. 1991).

The physical WCST typically consists of a pack of 128 cards, bearing a number of coloured shapes, with four cue cards functioning as placeholders for sorting piles (Fig. 1). Human participants are asked to place the sorting cards onto one of the piles and are given feedback indicating whether they were correct or incorrect. After ten consecutive correct trials, the rule is changed and the participant must work out from the correct/incorrect feedback what the new sorting category is. Participants continue to sort until they have either completed six categories, or they have used all 128 cards, whichever comes first (Milner 1963)—although in Berg's original 1948 study, nine categories were completed using 60 cards, requiring only five consecutive correct trials to solve a category. Computerised versions of the WCST simulate the cards on a screen, but otherwise follow a very similar methodology as the physical WCST (Mattes et al. 1991; Konishi et al. 1999). The WCST obtains measures of numbers of categories sorted, trials completed and errors—defining perseverative errors as those that would have been correct on the preceding sorting category (Heaton 1993).

The computerised WCST has also been adapted for use with monkeys—rhesus macaques (Moore et al. 2005) and baboons (Bonte et al. 2011)—presenting the subjects with three complex stimuli, using only two dimensions (shape and colour). As with human WCST versions, subjects are required to get ten consecutive correct trials, although they only need to learn four categories to complete the test.

The CANTAB ID/ED task (Fig. 2), for both humans and other primates, consists of a number of two-choice discriminations (between either differing lines or shapes)

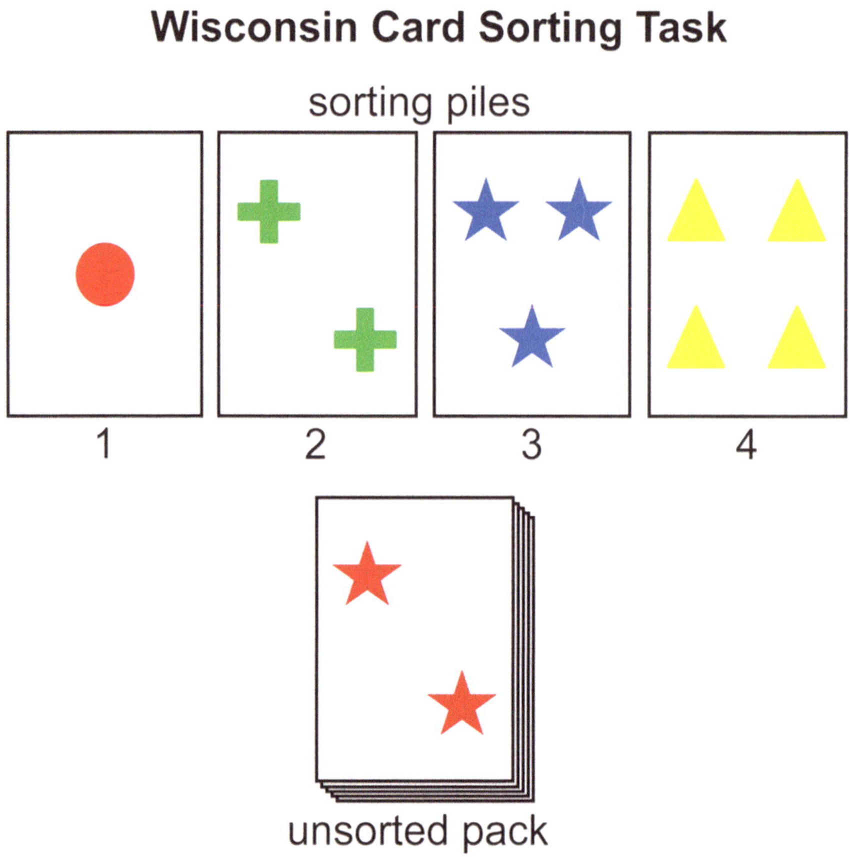

Fig. 1 Subjects sort a pack of 64 or 128 cards, each bearing a number of coloured shapes, into piles based on a perceptual rule—number, colour or shape—until they have reached a criterion, at which point the rule is changed. The subject must learn from feedback about their correct/incorrect choices what is the new perceptual rule

including simple discriminations (SD; where stimuli differ along only one dimension), compound discriminations (CD; where stimuli differ in both dimensions), reversals (where previous stimulus-reward contingencies are switched), ID acquisitions, ED shift acquisitions and probe stages (CDs where correct/incorrect stimuli remain the same, and the irrelevant dimension stimuli are changed). The order of these stages can be adjusted depending on what measures are of interest to specific research. Typically stages are ordered to shape the development of an attentional set, and then measure the cost of shifting from that set—an SD, followed by a CD, followed by an ID and followed by an ED shift. Reversal stages can be added after each of these stages, whilst probe stages are typically used after either CDs or IDs.

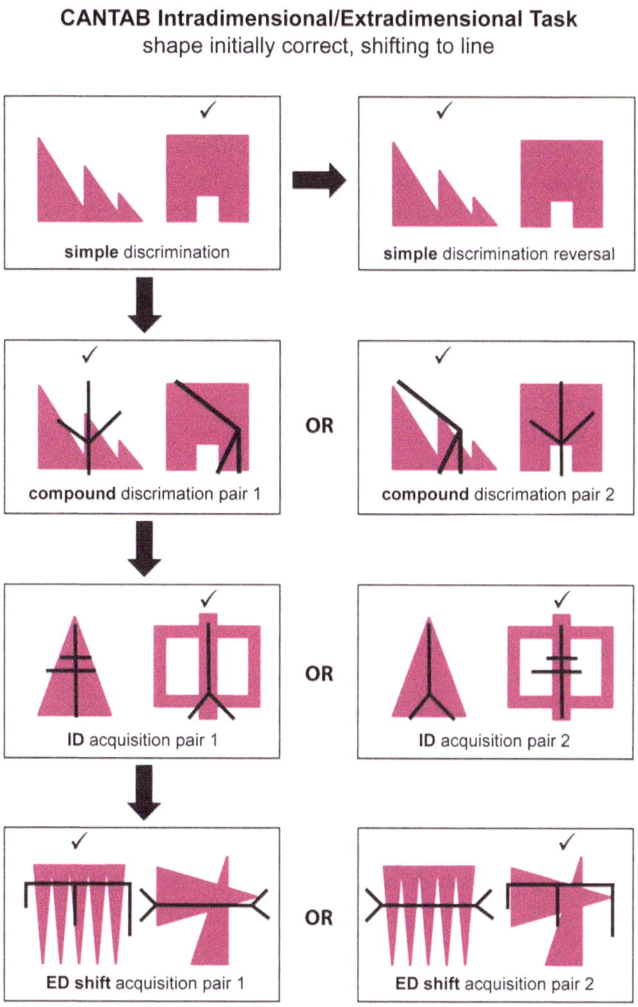

Fig. 2 Subjects learn two-choice visual discriminations using simple, then compound stimuli, where one of two perceptual dimensions—shape or line—is initially relevant, until they have reached a criterion. Subjects perform a number of stages, including reversals and at least one novel stimulus intradimensional (ID) acquisition, before they undertake an extradimensional (ED) shift, wherein novel stimuli are presented, but the previously relevant dimension is now irrelevant, and the previously irrelevant dimension is now relevant. More trials required to solve the ED shift than the ID acquisition indicates a cost of shifting attentional set from one dimension to the other

It has been argued that reversal stages enhance attentional set-formation by focusing attention on the relevant aspects of stimuli (Sutherland and Mackintosh 1971), although there is some evidence to the contrary. Marmosets become frustrated during reversal stages and this may disrupt attentional set-formation (Roberts, personal communication).

Probe stages can be used to confirm to what extent, if any, 'configural learning' (e.g. learning that AC and AD are correct, but BC and BD are incorrect) is being used to solve discriminations. Configural learning does not occur under normal cognitive conditions in the CANTAB ID/ED task (as evidenced by probe stages), although may explain differing 'shift-costs' (the difference between ID and ED shift performance) dependent on which dimension is being shifted from/to in rhesus macaques, when using colour and shape as dimensions (Baxter and Gaffan 2007), where it may not be possible to process each dimension independently (cf. Garner 1978).

Human subjects undertaking the computerised, touch-screen CANTAB ID/ED task must attain six consecutive correct trials (the probability of achieving this by chance is <0.05) to progress to the next stage (Downes et al. 1989; Owen et al. 1991). Criterion for monkeys varies considerably—e.g. 90 % correct within a single 60 trial/day block (Crofts et al. 2001), 90 % correct in each of two 60 trial/day block (Roberts et al. 1988), 90 % correct in a single 30 trial/day block (Clarke et al. 2007), or 12 correct out of 15 consecutive trials within a 60-min block (Weed et al. 2008).

3 Rodents—ID/ED Tasks

The most commonly used version of the rat ID/ED task (Fig. 3) has been described numerous times (Birrell and Brown 2000). Briefly, rats are trained to dig in sawdust-filled bowls containing a food reward (in apparatus with a waiting area and two sealable compartments that contain the bowls). Once rats are reliably digging, they perform a number of SDs (usually one for each perceptual dimension to be used during the test) to a criterion of six consecutively correct trials (as in the humans performing the CANTAB ID/ED task). The following day rats undertake a series of discriminations (usually seven: SD, CD, REV1, ID, REV2, ED, REV3), the order of which is designed, as with the primate ID/ED task, to facilitate the formation of attentional set and measure the cost of shifting from that attentional set.

There are numerous variants of the rodent ID/ED task, principally because the availability of material for use in the task varies considerably between research groups using it. Thus, whilst apparatus differences and exemplar differences are common, the majority of published research adheres to the basic methodology outlined in the original Birrell and Brown study (see Tait et al. 2014), although using only two dimensions—typically odour and digging medium. Variations beyond exemplar/apparatus constraints either involve the number of stages [some versions omit the REV between CD and ID (e.g. Rodefer and Nguyen 2008)], or the placement of the odours [most studies use herbs/spices mixed in with the digging media, although some use oils around the rims of the bowls (e.g. Goetghebeur and Dias 2009)].

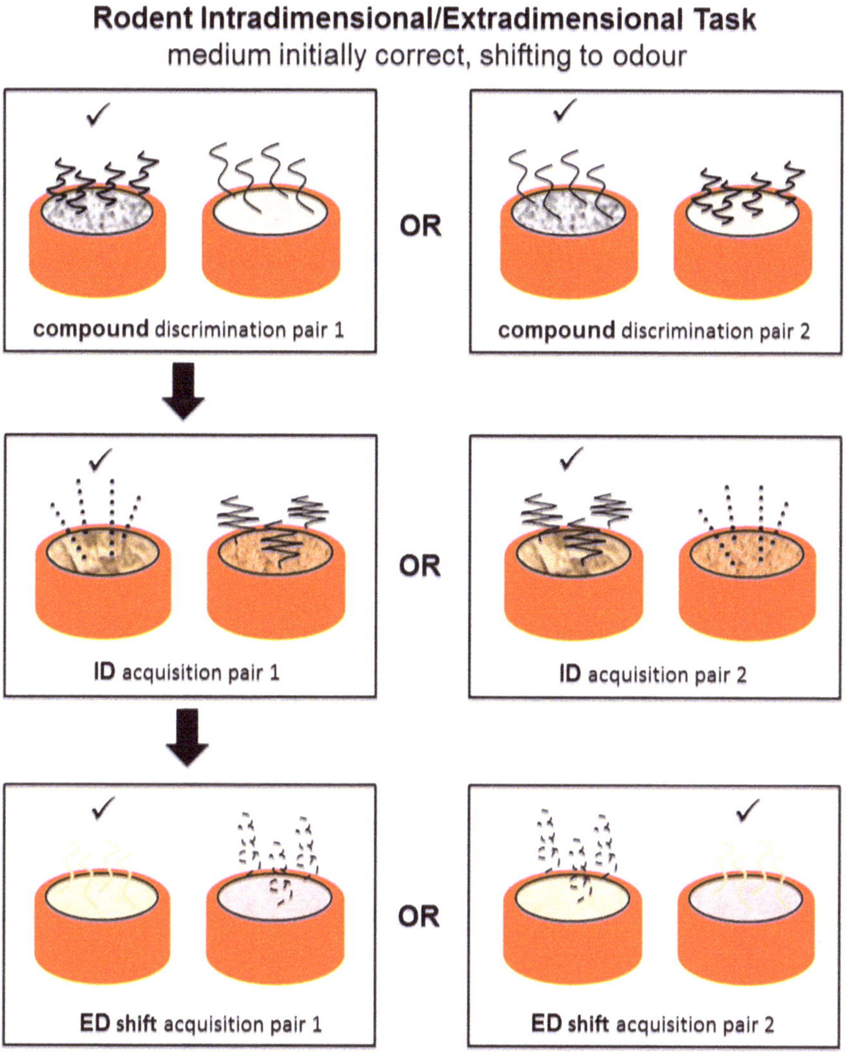

Fig. 3 By digging in bowls for reward, rodents learn two-choice discriminations, using simple, then compound stimuli, where one of two perceptual dimensions—typically odour or digging medium—is initially relevant. Rodents perform a number of stages, including reversals and at least one novel stimulus ID acquisition, before they undertake an ED shift, wherein novel stimuli are presented, but the previously relevant dimension is now irrelevant, and the previously irrelevant dimension is now relevant

Some studies also replace stages, either to encourage set-formation without the potential for reversal-induced frustration disrupting it [using ID1 and ID2 in place of REV1 and ID (e.g. Broberg et al. 2009)], or to specifically investigate the effects of those stages on set-formation/shifting (e.g. Chase et al. 2012).

4 Neuroanatomy of Attentional Set

Prior to the development of imaging techniques, investigations of the neuroanatomical substrates of attentional set in humans relied on data from subjects with brain injuries. Milner (1963) showed that patients with dorsolateral prefrontal cortex (DLPFC) lesions were impaired on set-shifting in the WCST—with no impairment arising from orbital prefrontal cortex (OFC) damage. Damage to the DLPFC and medial PFC (mPFC) is consistently reported as resulting in poor shifting performance (Stuss et al. 2000; Drewe 1974), even if effect size varies with a number of factors beyond just extent of brain injury (Demakis 2003).

Imaging studies have supported lesion studies in confirming the role of the PFC—specifically lateral and mPFC (e.g. Konishi et al. 2002; Monchi et al. 2001)—in attentional set-shifting, although there is a suggestion of left hemisphere dominance not always reported in the lesion literature (Demakis 2003; but see Drewe 1974). Brain regions beyond PFC are also recruited during attentional shifts—including (predominantly left hemisphere) posterior parietal cortex (PPC; Asari et al. 2005), which shares connectivity with left lateral PFC. A non-verbal adaptation of the WCST using shades and abstract shapes based on the Klingon alphabet (Specht et al. 2009) suggests that right hemisphere PFC activation was seen for goal-directed response selection, whilst left hemisphere PFC activity was found when inductive reasoning and feedback integration were required.

Further investigations of PFC region-specific function during WCST performance suggest that the DLPFC is involved in proactive interference—suppressing previously learned conditions—whereas the mPFC is involved in novel learning of conditions (Konishi et al. 2010). Yet, another study (Monchi et al. 2001), shows the activation of DLPFC after both positive and negative feedbacks, whereas the ventrolateral PFC (VLPFC) is only active during negative feedback (i.e. when a shift of set is required). The authors suggest that the VLPFC mediates the cognitive processes involved in attentional set-shifting—supported by a meta-analysis (Buchsbaum et al. 2005)—whilst the DLPFC is 'monitoring information in working memory'.

ID/ED tasks present further opportunity to explore the role of specific brain regions—not only in the process of shifting attentional set, but also in its formation/maintenance, and in reversal learning. Patients with frontal cortex damage are impaired during ED shift performance, but not during reversals or ID acquisition (Manes et al. 2002; Owen et al. 1991). The ED shift impairment in frontally injured patients is only apparent when perseveration to the previously relevant dimension is possible, and is absent when that dimension is switched out for a novel dimension in a 'learned irrelevance' condition ED shift (Owen et al. 1993). There is greater relative activation in the left caudate nucleus during reversal learning, and the right DLPFC and left polar PFC in ED shifting, than during ID acquisitions (Rogers et al. 2000). However, in a task designed to reduce the effect of having to search for the new rule, by repeatedly using the same dimensions, VLPFC was found to be active during ED shifting where DLPFC was not (Hampshire and Owen 2006). In the same task, medial OFC was activated during positive feedback, whereas lateral OFC activation was implicated in a response to negative feedback (rather than the feedback itself). The PPC was more active during response-mapping changes at reversal stages, but showed some activation at the beginning of shift stages—reducing as responses were repeated.

It is clear then that the involvement in PFC subregions in human attentional set-shifting is dependent on the task used. Use of stimuli that are difficult to verbalise in an ID/ED task is likely the best method for determining the role of distinct brain regions in attentional set-shifting without the confound of pre-existing stimulus-specific responses—whilst also permitting investigation of brain activity during novel onset learning and reversal learning.

The development of the CANTAB ID/ED task, equally applicable in humans and other primates, has allowed researchers to compare attentional set-shifting performance directly between species. Past research confirmed the involvement of PFC on shifting in rhesus macaques (Settlage et al. 1956), but the ID/ED task has been used to show a double dissociation between lateral PFC and OFC in marmosets: lateral PFC lesions impair attentional set-shifting, with no effect on reversal learning; OFC lesions impair reversal learning with no effect on attentional set-shifting (Dias et al. 1996). Further investigation of these deficits suggests that they are transitory in nature, with lesioned marmosets performing at control levels for both ED shift and reversal stages after the first (Dias et al. 1997).

Other targeted lesions in marmosets suggest that attentional set-shifting deficits are not dependent on either serotonin (5-HT) in PFC (Clarke et al. 2005), or cholinergic basal forebrain (BF) projections to PFC (Roberts et al. 1992)—although both lesions impair reversal learning. Dopaminergic depletion [via lesions with 6-hydroxydopamine (6-OHDA)] of PFC in marmosets may impair attentional set-formation, as subjects failed to show a difference between ID and ED shift stages (Roberts et al. 1994; Crofts et al. 2001)—although with no evidence of set-formation, it is impossible to conclude whether the lesions affected set-shifting itself. Drawing inference on this effect may be complicated by elevated dopamine (DA) in the caudate nucleus after 6-OHDA-induced PFC lesions (Roberts et al. 1994), and that 6-OHDA lesions in caudate nucleus result in reduced distraction by irrelevant information during probe stages (Crofts et al. 2001). It is clear then that PFC and caudate nucleus DA levels mediate cognitive processes involved in attentional set—but the interaction between the two areas may be as important as DA function in either region on its own.

Like the CANTAB task, the rodent ID/ED has provided great insight into attentional set-shifting in the rat, whilst also opening up research design space via species availability, and an easily adaptable paradigm in bowl digging. Rat mPFC mediates attentional set-shifting (Birrell and Brown 2000), whilst OFC mediates processes involved in reversal learning (McAlonan and Brown 2003). Unlike in primates, the mPFC and OFC lesion-induced deficits in rats are persistent (Chase et al. 2012; Tait et al. 2009), and rats typically do not improve in solving the various discriminations with successive tests (Tait et al. 2013; but see Wallace et al. 2014; Tait et al. 2009), meaning that, with appropriate counterbalancing, the rodent ID/ED task can be run repeatedly using the same subjects and stimuli.

The role of the OFC in attentional set-shifting tasks in rats is, however, not yet entirely clear. OFC lesions also impair set-formation, and in rats with evidence of an attentional set, ED shift performance (Chase et al. 2012)—and it has been argued that these impairments derive from the same underlying process that mediates the OFC lesion-induced reversal deficit. The fact that this set-formation/shifting impairment is not evident in OFC-lesioned monkeys may reflect a difference in the way the rodent task is run compared to the CANTAB task, rather than a difference in function. Specifically, in rodent tasks, during the first four trials of each stage, because the rodent can only experience one stimulus at a time, the animal is typically allowed to sample the correct bowl if its initial choice was incorrect. In the CANTAB ID/ED task, visual stimuli are presented simultaneously, allowing the subject to saccade between the two before making a response—with no within-trial opportunity to choose the correct stimulus after an initial incorrect choice. This results in greater perseveration in reversal learning stages in monkeys, because they typically repeat their errors a few times before choosing the previously incorrect stimulus—whereas, having self-corrected after the first error, a rodent will often dig in the correct bowl, particularly if it encounters it first, as early as the second trial of a reversal stage. This difference in how the stimuli are presented results in monkeys taking substantially longer to learn all stages of the task than rats can learn in the bowl-digging paradigm —compounded by the fact that foraging for food is an intuitive behaviour for rats, and learned more quickly than monkeys can learn visual discriminations. Given that the OFC lesion-induced deficit in set-formation is transient in rats, it might be sufficiently subtle that it is lost in the increased errors, and several-day testing period, typically required by monkeys in the ID/ED task. Indeed, although not significant, there is a small increase in errors during ID acquisitions in OFC-lesioned marmosets (Dias et al. 1996, 1997; and Roberts, personal communication).

As with humans, there is also evidence of PPC involvement in attentional set-shifting in rats (Fox et al. 2003). Rat anterior cingulate cortex may have a role in set-formation (Ng et al. 2007), with lesions abolishing the ID/ED difference, although task variants with multiple successive ID stages permit stronger conclusions about set-formation impairments (Chase et al. 2012). Likewise, dorsomedial striatal lesions (Lindgren et al. 2013) and subthalamic nucleus lesions (Xia, Dhawan, Tait and Brown, unpublished observations) impair set-formation in rats in a multiple ID task. The striatal lesions also produce a mild reversal learning impairment, although it is unclear, if, as with OFC lesion-induced impairments, this reversal learning impairment derives from dysfunction in the same cognitive processes that underlie the set-formation impairment.

As with marmosets, lesions damaging BF projections to PFC have no effect on attentional set-shifting (McGaughy et al. 2008), but nonspecific cell-body BF lesions do impair reversal learning (Tait and Brown 2008)—and also abolish ID/ED differences, not unlike OFC lesions. Noradrenergic lesions, whether directly into PFC, or via the dorsal ascending noradrenergic bundle, also impair attentional set-shifting (McGaughy et al. 2008; Tait et al. 2007)—suggesting that PFC catecholamines, DA and noradrenaline (NA), have an important role in attentional set-shifting in rats.

Whilst there have been considerably fewer publications on mouse attentional set-shifting, as with rats, lesions to mPFC and OFC impair set-shifting and reversal learning, respectively (Bissonette et al. 2008). Interestingly, despite using a multiple ID design, there was no evidence of an attentional set-formation impairment in OFC-lesioned mice. There are several difference between methodologies, however: the mice take four days to solve eight stages, whereas all stages in the rat task are completed in one day; the criterion for the mice to solve a discrimination is eight consecutive correct trials, whereas in the rat task, it is six consecutive trials; and unlike the typical rat task, this version of the mouse task does not counterbalance exemplar order. There has been little investigation into overtraining in the attentional set-shifting task—and it is the case that even six consecutive trials may result in overtraining under some circumstances—but it seems unlikely, given the variability within either six or eight consecutive correct trials at which the subject actually learns the discrimination, that eight consecutive correct trials has an impact on results. It is, however, probable that over the course of four days the mice have more time to consolidate information about attentional set—and thus, the mild and transient set-formation impairment seen in rats after OFC lesions (Chase et al. 2012) is not revealed in mice. This mouse task has revealed a set-formation impairment after the administration of the antiepileptic, topiramate (Bissonette and Powell 2012), but unlike the OFC lesion-induced deficits in rats, this impairment persisted across all ID stages. Thus, whilst there may appear to be discrepancies between monkey and rodent data, and between rodent species, principally in the roles of the OFC and striatum, it is more likely to arise from technique and/or design (including varied testing durations and criterion requirements) than it is a fundamental difference in function between species.

5 Neurodegenerative Diseases and Attentional Set-Shifting

Neurodegenerative diseases are typically associated with an ageing population—and normal ageing itself results in changes to cognitive function. It is not always easy to distinguish between the cognitive effects of early-stage neurodegenerative diseases and the effect of normal ageing—unless there is an associated motor dysfunction. Parkinson's, Huntington's and Alzheimer's diseases produce different patterns of neurodegeneration as they progress. Each disease results in frontal dysfunction however, and therefore, attentional set-shifting impairments, as well as deficits in other cognitive functions measured by the WCST and ID/ED tasks. The interactions between the distinct brain regions affected by neurodegeneration led to specific patterns of impairment that differ between the diseases.

Animal models of neurodegenerative diseases vary—from lesion techniques targeting the brain areas most affected by degeneration in the human diseases, to models arising from genetic alteration designed to mimic the progression of the disease. However, whilst animal models may simply be attempts to induce cognitive deficits, or take a more holistic approach to mimicking the disease itself, there are limitations to each method—with even the closest models unable to replicate the full pattern of cognitive and motor dysfunction observed in human patients. An advantage, then, of using attentional set-shifting tasks in animal models of human disease is that the cognitive processes measured can be identical between species—optimising the chance for translational interpretation of data.

5.1 Parkinson's Disease: Humans and Animal Models

Degeneration of dopaminergic pathways in Parkinson's disease (PD) results in both motor and cognitive impairments in patients (see Olanow and Tatton 1999 for review). A reduction in fronto-striatal dopaminergic function in PD is associated with impaired cognition—attentional set-shifting in the ID/ED task (Owen et al. 1993; Gauntlett-Gilbert et al. 1999; Downes et al. 1989; Owen et al. 1992) and the WCST (Paolo et al. 1995)—although inconsistently dependent on disease progression and other factors such as depression (Starkstein et al. 1989). Some data from PD patients undertaking the WCST show no evidence of either perseveration (i.e. repeatedly returning to the previously relevant dimension) or shifting impairment (Cooper et al. 1991). This inconsistency in data from PD patient WCST performance is not matched by data from the ID/ED task. Indeed, perhaps because of the WCST data inconsistency, efforts have been made using the ID/ED task to tease out the specific form of impairment present in PD patients. However, although there is a consistent ED shifting impairment in PD patients, the mechanism for this impairment has proven difficult to establish. Some data have suggested that medicated PD patients show increased 'learned irrelevance' (i.e. difficulty learning about a previously irrelevant dimension), and that it this mechanism that drives the ED shift impairment, whilst unmedicated patients are impaired at both perseverative and learned irrelevance ED shifts (Owen et al. 1993). Other data show that medicated PD patients show the same mechanisms of impairment as unmedicated patients (Gauntlett-Gilbert et al. 1999).

Modelling PD in animals has been limited to dopaminergic lesions and pharmacological manipulations, although the recent development of genetic models in the mouse presents a greater opportunity to investigate disease progression (Blesa et al. 2012). The lack of an effect on ID or ED shift performance after 6-OHDA lesions to caudate nucleus in the marmoset has already been noted—the only effect of the lesions being improved performance during a probe stage. And whilst this may seem at odds with data from PD patients, it is worth noting that disruption of new learning by 'intrusion' of previously learned sets in PD patients (Flowers and Robertson 1985) could be a mechanism by which distractibility to new irrelevant stimuli is reduced in probe stages. In contrast, another PD model, the neurotoxin 1-methyl-4-phenyl-1,2,3,6-tetrahydropyridine (MPTP), induces significant impairments in compound discrimination learning in rhesus macaques, although with no disproportionate effect on ED shifting (Decamp and Schneider 2004). It remains important, then, to consider the exact demands of any given task—and the possible effects of differences in methodology—and the weaknesses in any given model of human pathology.

There are few reports of attentional set-shifting in rat models of PD: most of the investigations of dopaminergic function have been related to schizophrenia-like dysfunction. Data from mice suggest that animal models of PD dementia are dependent on a combination of DA and acetylcholine (ACh) dysfunction—a failure to maintain an attentional set (and therefore no shift-cost) is only apparent after 6-OHDA lesions to mouse striatum in a mutant (heterozygous deletion of choline transporter) with reduced ACh. Neither the mutation nor the lesions on their own are sufficient to induce any change to performance on an ID/ED task with multiple ID stages (Zurkovsky et al. 2013)—although the authors report that their model induces mild loss in both DA ($\sim$33 %) and ACh (52 %). Interestingly, Zukovsky et al.'s data suggest that the mice in their combined DA/ACh loss group are capable of forming an attentional set over three consecutive IDs—although a reversal stage then disrupts this, and they fail to maintain that set over a subsequent three ID stages. However, the exposure to the exemplars during testing does not appear to have been counterbalanced (although direction of shift was), and thus the apparent formation of set, and then subsequent loss, could be an artefact of the specific stimuli and not an effect of the combined manipulation.

5.2 *Huntington's Disease: Humans and Animals Models*

Huntington's disease (HD) is typified by degeneration of subcortical brain areas, including the striatum and the substantia nigra in the basal ganglia, but also later on extending to cortex as well—resulting in both cognitive and motor impairments (see Imarisio et al. 2008 for review). HD patients show inconsistent impairment in the WCST (Peinemann et al. 2005; Brown et al. 2001)—with impairment seemingly dependent on whether there is atrophy affecting the whole striatum (Peinemann et al. 2005). Once an impairment is established however, performance appears to be consistent, at least over three years, as the disease progresses (Snowden et al. 2001).

Perhaps indicating greater sensitivity than the WCST to a specific impairment, ED shift deficits have been reported using the CANTAB ID/ED task in preclinical HD patients (Lawrence et al. 1998), in early-stage HD patients (Lawrence et al. 1996), and in those with further progression of HD (Lange et al. 1995). Further analysis of the mechanisms of the ED shift impairment shows that the performance deficit stems from perseveration to the previous dimension, and not from learned irrelevance of the new dimension (Lawrence et al. 1999). This, combined with very poor performance in an SD reversal stage (Lange et al. 1995) and perseverative-style responding in another visual reversal task (Lawrence et al.

1999), suggests a global perseveration impairment in HD patients. This evidence of perseveration in HD patients, similar to that observed after frontal brain damage (during ED shifting) and different to the patterns observed in PD patients (Owen et al. 1993; Gauntlett-Gilbert et al. 1999), exemplifies that, despite striatal degeneration in both HD and PD, the changes in fronto-striatal pathways are distinct between the two diseases. And further, as more information about the specifics of these differences arise, better understanding of the neuroanatomy underlying 'perseveration' and 'learned irrelevance' should become possible.

Data from animal models of HD are limited. A transgenic rhesus macaque model for HD has been developed (Yang et al. 2008; Chan et al. 2014), although to date, there are no data on attentional set-shifting abilities in this model. Neither a transgenic (Brooks et al. 2012b) nor a knock-in (Brooks et al. 2006) HD mouse model has conclusively shown effects in attentional set-shifting tasks. Transgenic HD model mice are impaired at strategy shifting in place/light maze task (Brooks et al. 2012b), whereas the knock-in HD model mice appear to have an ED shift deficit in an ID/ED task—although with no evidence of a shift-cost in the control animals, it is difficult to conclude that a between-group difference at the ED shift stage is actually caused by impaired attentional set-shifting (Brooks et al. 2006).

Investigation of attentional set-shifting in rat models of HD has been limited to quinolinic acid lesions of dorsomedial striatum (Lindgren et al. 2013), and results in a very mild reversal learning impairment and a set-formation impairment. Without evidence of set-formation, no effect of these lesions on set-shifting can be concluded.

The characterisation of attentional set-shifting and reversal learning in human HD patients is well-established, but there is now ample opportunity, with the macaque and mouse mutant HD models—and the CANTAB ID/ED task and the more reliable mouse ID/ED tasks (Bissonette et al. 2012)—for animal research to further elucidate patterns of cognition during disease progression.

5.3 *Alzheimer's Disease: Humans and Animal Models*

Degeneration of cholinergic pathways in Alzheimer's disease (AD) results in an increasing rate of decline in cognition during the lifetime of the patient (see Blennow et al. 2006 for review). Deficits in performance on the WCST have been reported in those suffering mild AD (Perry et al. 2000), with increased perseverative responding (Nagahama et al. 2003). Whilst imaging techniques looking at regional cerebral blood flow (rCBF) in AD patients have suggested reduced activity in rostrodorsal PFC is correlated with 'stuck-in-set' perseverative responding (continued responding to a previously correct dimension), whilst reduced activity in left parietal cortex is associated with 'recurrent' perseverative responding (returning to a previously correct dimension; Nagahama et al. 2005), it has also been suggested that more care must be taken when considering the relationship between the various measures obtained during the WCST—as this can significantly affect interpretation of data (Takeda et al. 2010). In the light of this, Terada et al. (2011) have demonstrated reduced rCBF in ventromedial PFC is associated with perseverative responding in AD patients.

ID/ED testing in AD patients has produced mixed results: a subgroup (~50 %) of mild/moderate AD patients failed to complete the ID stage of the task, and therefore, no measure of attentional set-shifting could be obtained—although the remainder of the patients were unimpaired on any stage of the task (Sahakian et al. 1990); a subgroup (~50 %) of mild/moderate AD patients were impaired at the ED shift stage of the task (Dorion et al. 2002). Despite these differences in performance, a consistent pattern has emerged from subgroups of high-functioning versus low-functioning AD patients—that also matches findings from WCST studies (e.g. Perry et al. 2000). This suggests distinctly different patterns of neurodegeneration in AD patients, with those suffering from more frontal pathology exhibiting worse performance in attentional set-shifting tasks (e.g. Terada et al. 2011).

Animal models of AD are limited by several factors, and even transgenic mouse models (Webster et al. 2014) fail to reflect the sporadic, rather than familial, aspect of most AD cases (LaFerla and Green 2012). There are no primate data on attentional set-shifting in AD models, and the few rat studies that could be comparable have only investigated normal ageing performance on the ID/ED task. Aged rats are impaired at reversal learning (Tait et al. 2013), with an ED shifting impairment (Barense et al. 2002) that may be manifest later in life than the reversal deficit (Tait et al. 2013). Unlike rats, aged mice have been reported unimpaired on any stage of the ID/ED task (Young et al. 2010). Transgenic mice (producing amyloid plaques in cortex and hippocampus in a similar pattern to AD patients) have been reported to show a general discrimination learning deficit by 14 months of age (Zhuo et al. 2007), with an earlier onset (between three to six months) reversal learning impairment (Zhuo et al. 2008). Discrepancies between results from AD mouse models and human AD patients—in the form of a failure to report an attentional set-shifting deficit in the mice—may result from differences in pathology between the sporadic AD and the transgenic model; or they may derive from the use of a mouse task design that is not as reliable in producing an attentional set as more current designs (e.g. Bissonette and Powell 2012).

6 Affective Disorders and Attentional Set-Shifting

There are numerous disorders of mental health, and it is not always a simple task to isolate them from each other. For example, depression can arise for a number of reasons, either as a symptom of a disorder, or as a by-product of another. Equally, affective disorders are associated with a wide range of symptoms, not all of which are necessarily present in each case. It is, therefore, not always easy to diagnose a disorder—or to fully dissociate one from another.

The lack of a full understanding of the pathology of affective disorders presents its own difficulties when investigating animal models—there is a clear cause and progression (although not always fully consistent between cases) in neurodegenerative diseases that is yet to be completely understood in affective disorders. Animal models often involve pharmacological manipulations, intended to simulate the changes reported in human patients, or environmental manipulations, and intended to induce a cognitive state in the animal similar to that in human patients. Both have their limitations, but the cross-species translational strength of tests of attentional set-shifting is as beneficial to research into affective disorders as it is neurodegenerative diseases.

Affective disorders represent a range of conditions with a variety of causes—single or recurrent major depressive disorders (MDD), bipolar disorder and substance-induced disorder (Davidson et al. 2002)—and can be comorbid with other disorders and diseases (Brown and Barlow 1992).

Patients with bipolar disorder are more impaired at the WCST than unipolar MDD patients, with errors arising from perseveration (Borkowska and Rybakowski 2001). Unipolar MDD patients are inconsistently impaired on the WCST however (Fossati et al. 2001; Moritz et al. 2002), and observation of an impairment may depend on the version of the WCST being used (Fossati et al. 2001), or the melancholic state of the patients (Austin et al. 1999). Deficits in unipolar MDD arise with reduced grey matter volume in hippocampus (Frodl et al. 2006) and reduced grey matter concentration in right medial and inferior frontal gyrus (Vasic et al. 2008).

Modelling depression in animals is limited, in that to determine a depressive state, a response to either a known antidepressant or to a stressor must be established (Yan et al. 2010). Some rat strains have been bred specifically to reflect some depression-like symptomology (Overstreet 2012), although models derived from stress, such as chronic unpredictable stress (CUS) and chronic intermittent cold stress (CIC), are also widely used. In the rodent ID/ED task, both CUS and CIC impair reversal learning (Bondi et al. 2008; Lapiz-Bluhm et al. 2009), but only CUS and restraint stress impair ED shifting (Bondi et al. 2008; Liston et al. 2006). All these stress-induced deficits are, however, ameliorated by antidepressants (Nikiforuk and Popik 2011; Danet et al. 2010; Bondi et al. 2008) in the form of selective serotonin reuptake inhibitors (SSRIs).

7 Schizophrenia and Attentional Set-Shifting

Schizophrenia is a disorder primarily associated with reduction in volume of PFC, hippocampus and temporal lobes, and a corresponding increase in ventricle size—as well as abnormal function in several neurotransmitter systems. Evidence of dopaminergic dysfunction initially led to the 'dopamine hypothesis' in the underlying aetiology of schizophrenia (Howes and Kapur 2009), although glutamate (e.g. Butler et al. 2005) and serotonin (e.g. Roth et al. 2004) are also implicated in mediating many of the symptoms.

There are numerous studies describing attentional set-shifting impairments in patients with schizophrenia—both in the WCST (e.g. Sullivan et al. 1993; Nieuwenstein et al. 2001; Haut et al. 1996) and in the CANTAB ID/ED task (e.g. Ceaser et al. 2008; Elliott et al. 1995; Pantelis et al. 1999). Schizophrenic patients exhibit perseverative errors in the WCST (Haut et al. 1996; Sullivan et al. 1993), and functional magnetic resonance imaging (fMRI) studies have shown reduced activity in right frontal areas in both medicated (Volz et al. 1997) and unmedicated (Riehemann et al. 2001) subjects. More recently, the DLPFC, in the context of working memory during WCST performance, was observed to exhibit reduced activity in the left hemisphere in schizophrenic patients, whilst the anterior cingulate cortex was active during set-shifting (Wilmsmeier et al. 2010). Right inferior frontal gyrus and bilateral caudate activation also increased in schizophrenic patients—which may appear to contradict data from Volz et al. and Riehemann et al.—however, it is worth noting that in Wilmsmeier et al.'s study, schizophrenia patients showed no difference in task performance to the controls, and the authors admit that patients were selected on their capability to perform the task in the fMRI.

Early data from schizophrenic patients undertaking the CANTAB ID/ED task show deficits in both ED shifting and reversal learning—both derived from perseveration (Elliott et al. 1995). Whilst a small number of patients failed the task at the ID stage and did not continue on, a significantly larger number of similarly chronically affected patients showed a failure to form set as well as to shift set (Pantelis et al. 1999). This general impairment in task performance seems tied to current IQ level (Ceaser et al. 2008), although ED shift performance is most strongly determined by current IQ (Jazbec et al. 2007). The impact of current IQ on ED shift performance exists independently of the schizophrenia itself, although the reversal learning deficit in schizophrenic patients is not dependent on IQ (Leeson et al. 2009). In contrast, first-episode schizophrenia patients do not suffer from a failure to complete the ID stage, and there is only a small (Hutton et al. 1998), sometimes insignificant (Hilti et al. 2010; Braw et al. 2008), difference between patients and controls in failure to complete the ED shift stage of the task. This implies a distinct progression in the way executive dysfunction changes from onset to chronic schizophrenia (Hutton et al. 1998).

In experimental animals, various methods have been used to model aspects of schizophrenia, or some of the deficits associated with the disorder. Administration of phencyclidine (PCP) has been considered one of the best models for schizophrenic frontal cortex dysfunction (Jentsch and Roth 1999): subchronic administration of PCP in both rats and monkeys reduces prefrontal DA transmission (Jentsch et al. 1997b, c), whereas acute PCP administration increases DA transmission in both rat and monkey PFC (Jentsch et al. 1997a, 1998).

In rats (see Tait et al. 2014 for review), both single dose (Egerton et al. 2005) and subchronic PCP regimes have been shown to impair attentional set-shifting in the ID/ED task (Rodefer et al. 2005, 2008; Egerton et al. 2008). Both the single dose, and one subchronic, PCP regimes produce a general discrimination learning impairment (Egerton et al. 2005), whilst another subchronic regime does not

(Rodefer et al. 2005). There is greater variability in the effects of other adult PCP regimes, with some reporting no effect (Deschenes et al. 2006), or only a general learning impairment (Fletcher et al. 2005).

Like subchronic PCP administration, amphetamine sensitisation also reduces prefrontal DA neurotransmission in rats (Hedou et al. 2001). An amphetamine sensitisation regime induces a general discrimination learning impairment, with the ED stage and some (Featherstone et al. 2008; Fletcher et al. 2005) or all (Fletcher et al. 2005) reversal stages being impaired relative to controls.

Neurodevelopment models for schizophrenia involve early life intervention or genetic manipulations. Methylazoxymethanol acetate (MAM) administration via injection into pregnant females on day 17 of gestation provides a different model for dopamine dysfunction in schizophrenia. MAM-treated rats are sensitive to amphetamine (Flagstad et al. 2004), indicating consistency with amphetamine sensitisation regimes as models of schizophrenia. Impaired reversal learning and ED shifting has been reported in MAM-treated rats (Featherstone et al. 2007). Neonatal PCP administration (days 7, 9 and 11) impairs ED shifting with no effect on general discrimination, or reversal, learning (Broberg et al. 2008) in rats tested as adults, whereas transient inaction of ventral hippocampus via infusion of tetrodotoxin in neonatal rats (day 7) induces both reversal learning and ED shifting deficits (Brooks et al. 2012a).

Genetic mouse models for schizophrenia exist (e.g. Stachowiak et al. 2013; Chen et al. 2006; Hikida et al. 2007), although data on attentional set-shifting are limited to a few studies investigating more targeted mutations that simulate some of the functional changes of schizophrenia. For example, a mutation increasing catechol-*O*-methyltransferase, leading to reduced PFC DA availability, induces an ED shift deficit (Papaleo et al. 2008), and reduced PFC GABAergic interneurons results in a failure to demonstrate set-formation (Bissonette et al. 2014). Earlier mouse studies investigating the effects of mutation on attentional set-shifting have suffered from no observable set-formation in control animals (e.g. Glickstein et al. 2005).

The ID/ED task, the rodent version of which having good test–retest reliability (Tait et al. 2009, 2013; Wallace et al. 2014), has been proposed as one of the core tests of executive function for the Cognitive Neuroscience Treatment Research to Improve Cognition in Schizophrenia (CNTRICS) test battery (Gilmour et al. 2013; Barch et al. 2009). The WCST has poor test–retest reliability and as such, is not included in the National Institute of Health's Measurement and Treatment Research to Improve Cognition in Schizophrenia (MATRICS) test battery (Nuechterlein et al. 2008; Barnett et al. 2010). In light of the desire to establish a standardised set of cognitive tests for schizophrenia research, and the availability of mouse genetic models, it seems necessary to achieve a consensus on the most effective means to measure attentional set-shifting in the mouse. To date, the most consistently successful method stems from Bissonette and colleagues (Bissonette et al. 2008, 2012, 2014; Bissonette and Powell 2012)—with mice requiring more than one ID stage to demonstrate set-formation reliably.

8 Attention Deficit/Hyperactivity Disorder and Attentional Set-Shifting

Attention deficit/hyperactivity disorder (ADHD) is a neurodevelopmental disorder, with an unclear aetiology, characterised by 'excessive motor activity, inattentiveness and impulsivity' (Lange et al. 2010). Children with ADHD are impaired on the WCST (Pineda et al. 1998), although, as in control subjects, performance improves with age (Seidman et al. 1997) and is not always apparent in adult ADHD patients (Rapport et al. 2001; Hervey et al. 2004)—although high-IQ controls exhibit fewer non-perseverative errors than high-IQ ADHD patients (Antshel et al. 2010). Recent data suggest that WCST impairments in adults with ADHD could arise from comorbid bipolar disorder rather than explicitly from the ADHD itself (Silva et al. 2014).

Data from the CANTAB ID/ED task on ADHD patients are varied—and effects may depend on sampling mechanisms (see Chamberlain et al. 2011 for review). Some studies on children indicate no effects of ADHD on reported measures (Goldberg et al. 2005; Corbett et al. 2009), whilst other suggest a general discrimination learning impairment with increased failure to complete stages at later reversal stages and the ED stage (Kempton et al. 1999; Mehta et al. 2004).

A rat model of ADHD, the spontaneously hypertensive rat (SHR), shows a general discrimination learning impairment and a failure to form set in one study (Cao et al. 2012). A second study also reports a failure to form set, with a transient reversal, but no general, learning impairment (Cheng and Li 2013). The data from the second study are limited, however, by the lack of a robust ID/ED difference in the control rats—so, it is difficult to conclude that the lack of set-formation in the SHRs is reliable. Given the availability of ADHD animal models (Sontag et al. 2010; Wickens et al. 2011), it seems there is opportunity for further investigating ADHD-related executive dysfunction in rats and mice using ID/ED tasks.

9 Summary and Future Directions

Attentional set-shifting using the WCST and ID/ED tasks is a valuable tool for exploring frontally mediated dysfunction in humans and other animals. The three major neurodegenerative diseases, AD, HD and PD, result in distinct patterns of PFC degradation and distinct patterns of executive dysfunction that can be assessed using attentional set-shifting. Equally, other neuropsychiatric disorders affect structure, neurotransmitter activity and PFC-dependent executive functions.

To date, animal models of human diseases and disorders have provided support for existing research, and opened up new avenues for progress in disease/disorder therapies. It remains necessary, however, in light of the non-standard methodologies of rodent ID/ED tasks, to draw towards a consensus on the best technique to measure rodent attentional set-shifting. In a globalised scientific community, the WCST and the CANTAB ID/ED tasks can be purchased and run according to an established set of instructions; but this is not the case for the rodent ID/ED tasks. Thus, whilst differences in performance in human studies can often be determined to arise from subject pool sampling—and indeed, we should take great care to consider such during interpretation of data—in rodents, we must consider not only strain, gender and age, but also the specific techniques used during testing (Tait et al. 2014). Data are remarkably consistent in pattern between the numerous research groups running the task—but odours, digging media, and textures differ—and indeed it is often impossible to fully replicate one group's apparatus because of simple lack of availability. Where possible, however, we should attempt to be consistent with previous published research, and provide clearer explanations for changes—and what those changes could mean to the data.

The answer to the problem of methodological inconsistency in rodent data is to standardise the task in a fashion that is accessible globally. And that must be the ultimate goal—as the WCST and the CANTAB ID/ED tasks can be fully automated, so a true ID/ED automated task for rodents should be our goal. There have been attempts to automate attentional set-shifting in rodents in the past, although each has its own flaws: no difference between ID and ED performance in a visual task (Brigman et al. 2005); a failure to use truly compounded stimuli in a multimodal task (Scheggia et al. 2014). It remains, then, a difficult, but not insurmountable task.

Whilst concern for interpreting rodent data typically derives from methodological differences, human data are often confounded by a verbal component to task solving. The CANTAB ID/ED task uses stimuli that are difficult to verbalise, but the WCST, unless specifically modified, does not: colour, number and shape are all easily translated to verbal descriptions. Some studies have made changes to the WCST protocol seeking to address this, although the vast majority do not. Parsing verbal from non-verbal (or minimally verbal) processing is important not only when considering data between humans and other animals, but also when considering data from human studies that have taken verbal components into account versus those that have not. Imaging studies investigating the involvement of precise PFC subregions in attentional set-shifting must be considered carefully in the context of the type of task used.

10 Conclusion

Attentional set-shifting is an important measure of executive function—and an important tool for exploring executive dysfunction—in both human diseases and animal models of human diseases. Although MATRICS does not include the WCST or ID/ED as a measure of executive function in its test battery for schizophrenia, CNTRICS does recommend ID/ED testing—and continued use of attentional set-shifting tasks in rodents in the search to develop new therapeutic tools for tackling human diseases and disorders, along with similar tests in humans to explore cognitive function/dysfunction, means that regardless of test–retest reliability, the current measures of attentional set-shifting in humans are vital components in the research process. We must continue our efforts to probe executive function using the best attentional set-shifting methodologies available to us—exposing parallels in function between species, and where necessary, elucidating the causes of differences.

References

Antshel KM, Faraone SV, Maglione K, Doyle AE, Fried R, Seidman LJ, Biederman J (2010) Executive functioning in high-IQ adults with ADHD. Psychol Med 40(11):1909–1918. doi:10.1017/S0033291709992273

Asari T, Konishi S, Jimura K, Miyashita Y (2005) Multiple components of lateral posterior parietal activation associated with cognitive set shifting. Neuroimage 26(3):694–702

Austin MP, Mitchell P, Wilhelm K, Parker G, Hickie I, Brodaty H, Chan J, Eyers K, Milic M, Hadzi-Pavlovic D (1999) Cognitive function in depression: a distinct pattern of frontal impairment in melancholia? Psychol Med 29(1):73–85

Barcelo F, Knight RT (2002) Both random and perseverative errors underlie WCST deficits in prefrontal patients. Neuropsychologia 40(3):349–356

Barch DM, Braver TS, Carter CS, Poldrack RA, Robbins TW (2009) CNTRICS final task selection: executive control. Schizophr Bull 35(1):115–135. doi:sbn154 [pii] 10.1093/schbul/sbn154

Barense MD, Fox MT, Baxter MG (2002) Aged rats are impaired on an attentional set-shifting task sensitive to medial frontal cortex damage in young rats. Learn Mem 9(4):191–201

Barnett JH, Robbins TW, Leeson VC, Sahakian BJ, Joyce EM, Blackwell AD (2010) Assessing cognitive function in clinical trials of schizophrenia. Neurosci Biobehav Rev 34(8):1161–1177. doi:10.1016/j.neubiorev.2010.01.012

Baxter MG, Gaffan D (2007) Asymmetry of attentional set in rhesus monkeys learning colour and shape discriminations. Q J Exp Psychol (Colchester) 60(1):1–8

Berg EA (1948) A simple objective technique for measuring flexibility in thinking. J Gen Psych 39:15–22

Birrell JM, Brown VJ (2000) Medial frontal cortex mediates perceptual attentional set shifting in the rat. J Neurosci 20(11):4320–4324

Bissonette GB, Powell EM (2012) Reversal learning and attentional set-shifting in mice. Neuropharmacology 62(3):1168–1174. doi:10.1016/j.neuropharm.2011.03.011

Bissonette GB, Martins GJ, Franz TM, Harper ES, Schoenbaum G, Powell EM (2008) Double dissociation of the effects of medial and orbital prefrontal cortical lesions on attentional and affective shifts in mice. J Neurosci 28(44):11124–11130

Bissonette GB, Lande MD, Martins GJ, Powell EM (2012) Versatility of the mouse reversal/set-shifting test: effects of topiramate and sex. Physiol Behav 107(5):781–786. doi:10.1016/j.physbeh.2012.05.018

Bissonette GB, Bae MH, Suresh T, Jaffe DE, Powell EM (2014) Prefrontal cognitive deficits in mice with altered cerebral cortical GABAergic interneurons. Behav Brain Res 259:143–151. doi:10.1016/j.bbr.2013.10.051

Blennow K, de Leon MJ, Zetterberg H (2006) Alzheimer's disease. Lancet 368(9533):387–403. doi:10.1016/S0140-6736(06)69113-7

Blesa J, Phani S, Jackson-Lewis V, Przedborski S (2012) Classic and new animal models of Parkinson's disease. J Biomed Biotechnol 2012:845618. doi:10.1155/2012/845618

Bondi CO, Rodriguez G, Gould GG, Frazer A, Morilak DA (2008) Chronic unpredictable stress induces a cognitive deficit and anxiety-like behavior in rats that is prevented by chronic antidepressant drug treatment. Neuropsychopharmacology 33(2):320–331

Bonte E, Flemming T, Fagot J (2011) Executive control of perceptual features and abstract relations by baboons (Papio papio). Behav Brain Res 222(1):176–182. doi:10.1016/j.bbr.2011.03.034

Borkowska A, Rybakowski JK (2001) Neuropsychological frontal lobe tests indicate that bipolar depressed patients are more impaired than unipolar. Bipolar Disord 3(2):88–94

Braw Y, Bloch Y, Mendelovich S, Ratzoni G, Gal G, Harari H, Tripto A, Levkovitz Y (2008) Cognition in young schizophrenia outpatients: Comparison of first-episode with multiepisode patients. Schizophr Bull 34(3):544–554. doi:10.1093/schbul/sbm115

Brigman JL, Bussey TJ, Saksida LM, Rothblat LA (2005) Discrimination of multidimensional visual stimuli by mice: intra- and extradimensional shifts. Behav Neurosci 119(3):839–842

Broberg BV, Dias R, Glenthoj BY, Olsen CK (2008) Evaluation of a neurodevelopmental model of schizophrenia–early postnatal PCP treatment in attentional set-shifting. Behav Brain Res 190(1):160–163

Broberg BV, Glenthoj BY, Dias R, Larsen DB, Olsen CK (2009) Reversal of cognitive deficits by an ampakine (CX516) and sertindole in two animal models of schizophrenia—sub-chronic and early postnatal PCP treatment in attentional set-shifting. Psychopharmacology 206(4):631–640

Brooks SP, Betteridge H, Trueman RC, Jones L, Dunnett SB (2006) Selective extra-dimensional set shifting deficit in a knock-in mouse model of Huntington's disease. Brain Res Bull 69 (4):452–457

Brooks JM, Pershing ML, Thomsen MS, Mikkelsen JD, Sarter M, Bruno JP (2012a) Transient inactivation of the neonatal ventral hippocampus impairs attentional set-shifting behavior: reversal with an alpha7 nicotinic agonist. Neuropsychopharmacology 37(11):2476–2486. doi:10.1038/npp.2012.106

Brooks SP, Janghra N, Higgs GV, Bayram-Weston Z, Heuer A, Jones L, Dunnett SB (2012b) Selective cognitive impairment in the YAC128 Huntington's disease mouse. Brain Res Bull 88 (2–3):121–129. doi:10.1016/j.brainresbull.2011.05.010

Brown TA, Barlow DH (1992) Comorbidity among anxiety disorders: implications for treatment and DSM-IV. J Consult Clin Psychol 60(6):835–844

Brown VJ, Tait DS (2010) Behavioral flexibility: attentional shifting, rule switching and response reversal

Brown RG, Redondo-Verge L, Chacon JR, Lucas ML, Channon S (2001) Dissociation between intentional and incidental sequence learning in Huntington's disease. Brain 124(Pt 11): 2188–2202

Buchsbaum BR, Greer S, Chang WL, Berman KF (2005) Meta-analysis of neuroimaging studies of the Wisconsin card-sorting task and component processes. Hum Brain Mapp 25(1):35–45. doi:10.1002/hbm.20128

Bussey TJ, Muir JL, Everitt BJ, Robbins TW (1997) Triple dissociation of anterior cingulate, posterior cingulate, and medial frontal cortices on visual discrimination tasks using a touchscreen testing procedure for the rat. Behav Neurosci 111(5):920–936

Butler PD, Zemon V, Schechter I, Saperstein AM, Hoptman MJ, Lim KO, Revheim N, Silipo G, Javitt DC (2005) Early-stage visual processing and cortical amplification deficits in schizophrenia. Arch Gen Psychiatry 62(5):495–504. doi:10.1001/archpsyc.62.5.495

Cao AH, Yu L, Wang YW, Wang JM, Yang LJ, Lei GF (2012) Effects of methylphenidate on attentional set-shifting in a genetic model of attention-deficit/hyperactivity disorder. Behav Brain Funct 8(1):10. doi:10.1186/1744-9081-8-10

Ceaser AE, Goldberg TE, Egan MF, McMahon RP, Weinberger DR, Gold JM (2008) Set-shifting ability and schizophrenia: a marker of clinical illness or an intermediate phenotype? Biol Psychiatry 64(9):782–788. doi:S0006-3223(08)00640-9 [pii] 10.1016/j.biopsych.2008.05.009

Chamberlain SR, Robbins TW, Winder-Rhodes S, Muller U, Sahakian BJ, Blackwell AD, Barnett JH (2011) Translational approaches to frontostriatal dysfunction in attention-deficit/ hyperactivity disorder using a computerized neuropsychological battery. Biol Psychiatry 69 (12):1192–1203. doi:10.1016/j.biopsych.2010.08.019

Chan AW, Xu Y, Jiang J, Rahim T, Zhao D, Kocerha J, Chi T, Moran S, Engelhardt H, Larkin K, Neumann A, Cheng H, Li C, Nelson K, Banta H, Zola SM, Villinger F, Yang J, Testa CM,

Mao H, Zhang X, Bachevalier J (2014) A two years longitudinal study of a transgenic Huntington disease monkey. BMC neuroscience 15:36. doi:10.1186/1471-2202-15-36
Chase EA, Tait DS, Brown VJ (2012) Lesions of the orbital prefrontal cortex impair the formation of attentional set in rats. Eur J Neurosci 36(3):2368–2375. doi:10.1111/j.1460-9568.2012.08141.x
Chen J, Lipska BK, Weinberger DR (2006) Genetic mouse models of schizophrenia: from hypothesis-based to susceptibility gene-based models. Biol Psychiatry 59(12):1180–1188. doi:10.1016/j.biopsych.2006.02.024
Cheng JT, Li JS (2013) Intra-orbitofrontal cortex injection of haloperidol removes the beneficial effect of methylphenidate on reversal learning of spontaneously hypertensive rats in an attentional set-shifting task. Behav Brain Res 239:148–154. doi:10.1016/j.bbr.2012.11.006
Chudasama Y (2011) Animal models of prefrontal-executive function. Behav Neurosci 125 (3):327–343. doi:2011-10778-002 [pii] 10.1037/a0023766
Clarke HF, Walker SC, Crofts HS, Dalley JW, Robbins TW, Roberts AC (2005) Prefrontal serotonin depletion affects reversal learning but not attentional set shifting. J Neurosci 25 (2):532–538
Clarke HF, Walker SC, Dalley JW, Robbins TW, Roberts AC (2007) Cognitive inflexibility after prefrontal serotonin depletion is behaviorally and neurochemically specific. Cereb Cortex 17 (1):18–27
Cooper JA, Sagar HJ, Jordan N, Harvey NS, Sullivan EV (1991) Cognitive impairment in early, untreated Parkinson's disease and its relationship to motor disability. Brain 114(Pt 5): 2095–2122
Corbett BA, Constantine LJ, Hendren R, Rocke D, Ozonoff S (2009) Examining executive functioning in children with autism spectrum disorder, attention deficit hyperactivity disorder and typical development. Psychiatry Res 166(2–3):210–222. doi:10.1016/j.psychres.2008.02.005
Crofts HS, Dalley JW, Collins P, Van Denderen JC, Everitt BJ, Robbins TW, Roberts AC (2001) Differential effects of 6-OHDA lesions of the frontal cortex and caudate nucleus on the ability to acquire an attentional set. Cereb Cortex 11(11):1015–1026
Dalley JW, Cardinal RN, Robbins TW (2004) Prefrontal executive and cognitive functions in rodents: neural and neurochemical substrates. Neurosci Biobehav Rev 28(7):771–784
Danet M, Lapiz-Bluhm S, Morilak DA (2010) A cognitive deficit induced in rats by chronic intermittent cold stress is reversed by chronic antidepressant treatment. Int J Neuropsychopharmacol 13(8):997–1009. doi:S1461145710000039 [pii] 10.1017/S1461145710000039
Davidson RJ, Pizzagalli D, Nitschke JB, Putnam K (2002) Depression: perspectives from affective neuroscience. Annu Rev Psychol 53:545–574. doi:10.1146/annurev.psych.53.100901.135148
Decamp E, Schneider JS (2004) Attention and executive function deficits in chronic low-dose MPTP-treated non-human primates. Eur J Neurosci 20(5):1371–1378. doi:10.1111/j.1460-9568.2004.03586.x
Demakis GJ (2003) A meta-analytic review of the sensitivity of the Wisconsin Card Sorting Test to frontal and lateralized frontal brain damage. Neuropsychology 17(2):255–264
Deschenes A, Goulet S, Dore FY (2006) Rule shift under long-term PCP challenge in rats. Behav Brain Res 167(1):134–140
Dias R, Robbins TW, Roberts AC (1996) Dissociation in prefrontal cortex of affective and attentional shifts. Nature 380(6569):69–72
Dias R, Robbins TW, Roberts AC (1997) Dissociable forms of inhibitory control within prefrontal cortex with an analog of the Wisconsin Card Sort Test: restriction to novel situations and independence from "on-line" processing. J Neurosci 17(23):9285–9297
Dorion AA, Sarazin M, Hasboun D, Hahn-Barma V, Dubois B, Zouaoui A, Marsault C, Duyme M (2002) Relationship between attentional performance and corpus callosum morphometry in patients with Alzheimer's disease. Neuropsychologia 40(7):946–956
Downes JJ, Roberts AC, Sahakian BJ, Evenden JL, Morris RG, Robbins TW (1989) Impaired extra-dimensional shift performance in medicated and unmedicated Parkinson's disease: evidence for a specific attentional dysfunction. Neuropsychologia 27(11–12):1329–1343

Drewe EA (1974) The effect of type and area of brain lesion on Wisconsin card sorting test performance. Cortex (a journal devoted to the study of the nervous system and behavior) 10 (2):159–170

Egerton A, Reid L, McKerchar CE, Morris BJ, Pratt JA (2005) Impairment in perceptual attentional set-shifting following PCP administration: a rodent model of set-shifting deficits in schizophrenia. Psychopharmacology 179(1):77–84

Egerton A, Reid L, McGregor S, Cochran SM, Morris BJ, Pratt JA (2008) Subchronic and chronic PCP treatment produces temporally distinct deficits in attentional set shifting and prepulse inhibition in rats. Psychopharmacology 198(1):37–49

Eimas PD (1966) Effects of overtraining and age on intradimensional and extradimensional shifts in children. J Exp Child Psychol 3(4):348–355

Elliott R, McKenna PJ, Robbins TW, Sahakian BJ (1995) Neuropsychological evidence for frontostriatal dysfunction in schizophrenia. Psychol Med 25(3):619–630

Featherstone RE, Rizos Z, Nobrega JN, Kapur S, Fletcher PJ (2007) Gestational methylazoxymethanol acetate treatment impairs select cognitive functions: parallels to schizophrenia. Neuropsychopharmacology 32(2):483–492

Featherstone RE, Rizos Z, Kapur S, Fletcher PJ (2008) A sensitizing regimen of amphetamine that disrupts attentional set-shifting does not disrupt working or long-term memory. Behav Brain Res 189(1):170–179

Flagstad P, Mork A, Glenthoj BY, van Beek J, Michael-Titus AT, Didriksen M (2004) Disruption of neurogenesis on gestational day 17 in the rat causes behavioral changes relevant to positive and negative schizophrenia symptoms and alters amphetamine-induced dopamine release in nucleus accumbens. Neuropsychopharmacology 29(11):2052–2064

Fletcher PJ, Tenn CC, Rizos Z, Lovic V, Kapur S (2005) Sensitization to amphetamine, but not PCP, impairs attentional set shifting: reversal by a D1 receptor agonist injected into the medial prefrontal cortex. Psychopharmacology 183(2):190–200

Floresco SB, Block AE, Tse MT (2008) Inactivation of the medial prefrontal cortex of the rat impairs strategy set-shifting, but not reversal learning, using a novel, automated procedure. Behav Brain Res 190(1):85–96

Flowers KA, Robertson C (1985) The effect of Parkinson's disease on the ability to maintain a mental set. J Neurol Neurosurg Psychiatry 48(6):517–529

Fossati P, Ergis AM, Allilaire JF (2001) Problem-solving abilities in unipolar depressed patients: comparison of performance on the modified version of the Wisconsin and the California sorting tests. Psychiatry Res 104(2):145–156

Fox MT, Barense MD, Baxter MG (2003) Perceptual attentional set-shifting is impaired in rats with neurotoxic lesions of posterior parietal cortex. J Neurosci 23(2):676–681

Frodl T, Schaub A, Banac S, Charypar M, Jager M, Kummler P, Bottlender R, Zetzsche T, Born C, Leinsinger G, Reiser M, Moller HJ, Meisenzahl EM (2006) Reduced hippocampal volume correlates with executive dysfunctioning in major depression. J Psychiatry Neurosci 31 (5):316–323

Garner WR (1978) Selective attention to attributes and to stimuli. J Exp Psychol Gen 107(3): 287–308

Garner JP, Thogerson CM, Wurbel H, Murray JD, Mench JA (2006) Animal neuropsychology: validation of the intra-dimensional extra-dimensional set shifting task for mice. Behav Brain Res 173(1):53–61

Gauntlett-Gilbert J, Roberts RC, Brown VJ (1999) Mechanisms underlying attentional set-shifting in Parkinson's disease. Neuropsychologia 37(5):605–616

Gibson JJ (1941) A critical review of the concept of set in contemporary experimental psychology. Psychol Bull 38(9):781–817

Gilmour G, Arguello A, Bari A, Brown VJ, Carter C, Floresco SB, Jentsch DJ, Tait DS, Young JW, Robbins TW (2013) Measuring the construct of executive control in schizophrenia: defining and validating translational animal paradigms for discovery research. Neurosci Biobehav Rev 37(9 Pt B):2125–2140. doi:10.1016/j.neubiorev.2012.04.006

Glickstein SB, Desteno DA, Hof PR, Schmauss C (2005) Mice lacking dopamine D2 and D3 receptors exhibit differential activation of prefrontal cortical neurons during tasks requiring attention. Cereb Cortex 15(7):1016–1024. doi:10.1093/cercor/bhh202

Goetghebeur P, Dias R (2009) Comparison of haloperidol, risperidone, sertindole, and modafinil to reverse an attentional set-shifting impairment following subchronic PCP administration in the rat-a back translational study. Psychopharmacology 202(1–3):287–293

Goldberg MC, Mostofsky SH, Cutting LE, Mahone EM, Astor BC, Denckla MB, Landa RJ (2005) Subtle executive impairment in children with autism and children with ADHD. J Autism Dev Disord 35(3):279–293

Grant DA, Berg EA (1948) A behavioral analysis of degree of reinforcement and ease of shifting to new responses in a Weigl-type card-sorting problem. J Exp Psychol 38(4):404–411

Hampshire A, Owen AM (2006) Fractionating attentional control using event-related fMRI. Cereb Cortex 16(12):1679–1689. doi:10.1093/cercor/bhj116

Haut MW, Cahill J, Cutlip WD, Stevenson JM, Makela EH, Bloomfield SM (1996) On the nature of Wisconsin Card Sorting Test performance in schizophrenia. Psychiatry Res 65(1):15–22

Heaton RK (1993) Wisconsin Card Sorting Test (WCST) (rev. and expanded. edn). Psychological Assessment Resources, Odessa

Hedou G, Homberg J, Feldon J, Heidbreder CA (2001) Expression of sensitization to amphetamine and dynamics of dopamine neurotransmission in different laminae of the rat medial prefrontal cortex. Neuropharmacology 40(3):366–382

Hervey AS, Epstein JN, Curry JF (2004) Neuropsychology of adults with attention-deficit/hyperactivity disorder: a meta-analytic review. Neuropsychology 18(3):485–503. doi:10.1037/0894-4105.18.3.485

Hikida T, Jaaro-Peled H, Seshadri S, Oishi K, Hookway C, Kong S, Wu D, Xue R, Andrade M, Tankou S, Mori S, Gallagher M, Ishizuka K, Pletnikov M, Kida S, Sawa A (2007) Dominant-negative DISC1 transgenic mice display schizophrenia-associated phenotypes detected by measures translatable to humans. Proc Natl Acad Sci USA 104(36):14501–14506. doi:10.1073/pnas.0704774104

Hilti CC, Delko T, Orosz AT, Thomann K, Ludewig S, Geyer MA, Vollenweider FX, Feldon J, Cattapan-Ludewig K (2010) Sustained attention and planning deficits but intact attentional set-shifting in neuroleptic-naive first-episode schizophrenia patients. Neuropsychobiology 61 (2):79–86. doi:000265133 [pii] 10.1159/000265133

Howes OD, Kapur S (2009) The dopamine hypothesis of schizophrenia: version III—the final common pathway. Schizophr Bull 35(3):549–562. doi:10.1093/schbul/sbp006

Hutton SB, Puri BK, Duncan LJ, Robbins TW, Barnes TR, Joyce EM (1998) Executive function in first-episode schizophrenia. Psychol Med 28(2):463–473

Imarisio S, Carmichael J, Korolchuk V, Chen CW, Saiki S, Rose C, Krishna G, Davies JE, Ttofi E, Underwood BR, Rubinsztein DC (2008) Huntington's disease: from pathology and genetics to potential therapies. Biochem J 412(2):191–209. doi:10.1042/BJ20071619

Jazbec S, Pantelis C, Robbins T, Weickert T, Weinberger DR, Goldberg TE (2007) Intra-dimensional/extra-dimensional set-shifting performance in schizophrenia: impact of distractors. Schizophr Res 89(1–3):339–349. doi:S0920-9964(06)00346-X [pii] 10.1016/j.schres.2006.08.014

Jentsch JD, Roth RH (1999) The neuropsychopharmacology of phencyclidine: from NMDA receptor hypofunction to the dopamine hypothesis of schizophrenia. Neuropsychopharmacology 20(3):201–225

Jentsch JD, Elsworth JD, Redmond DE Jr, Roth RH (1997a) Phencyclidine increases forebrain monoamine metabolism in rats and monkeys: modulation by the isomers of HA966. J Neurosci 17(5):1769–1775

Jentsch JD, Redmond DE Jr, Elsworth JD, Taylor JR, Youngren KD, Roth RH (1997b) Enduring cognitive deficits and cortical dopamine dysfunction in monkeys after long-term administration of phencyclidine. Science 277(5328):953–955

Jentsch JD, Tran A, Le D, Youngren KD, Roth RH (1997c) Subchronic phencyclidine administration reduces mesoprefrontal dopamine utilization and impairs prefrontal cortical-dependent cognition in the rat. Neuropsychopharmacology 17(2):92–99

Jentsch JD, Tran A, Taylor JR, Roth RH (1998) Prefrontal cortical involvement in phencyclidine-induced activation of the mesolimbic dopamine system: behavioral and neurochemical evidence. Psychopharmacology 138(1):89–95

Joel D, Weiner I, Feldon J (1997) Electrolytic lesions of the medial prefrontal cortex in rats disrupt performance on an analog of the Wisconsin Card Sorting Test, but do not disrupt latent inhibition: implications for animal models of schizophrenia. Behav Brain Res 85(2):187–201

Kelleher RT (1956) Discrimination learning as a function of reversal and nonreversal shifts. J Exp Psychol 51(6):379–384

Kempton S, Vance A, Maruff P, Luk E, Costin J, Pantelis C (1999) Executive function and attention deficit hyperactivity disorder: stimulant medication and better executive function performance in children. Psychol Med 29(3):527–538

Kesner RP, Churchwell JC (2011) An analysis of rat prefrontal cortex in mediating executive function. Neurobiol Learn Mem 96(3):417–431. doi:S1074-7427(11)00142-0 [pii] 10.1016/j.nlm.2011.07.002

Konishi S, Kawazu M, Uchida I, Kikyo H, Asakura I, Miyashita Y (1999) Contribution of working memory to transient activation in human inferior prefrontal cortex during performance of the Wisconsin Card Sorting Test. Cereb Cortex 9(7):745–753

Konishi S, Hayashi T, Uchida I, Kikyo H, Takahashi E, Miyashita Y (2002) Hemispheric asymmetry in human lateral prefrontal cortex during cognitive set shifting. Proc Natl Acad Sci USA 99(11):7803–7808. doi:10.1073/pnas.122644899

Konishi S, Hirose S, Jimura K, Chikazoe J, Watanabe T, Kimura HM, Miyashita Y (2010) Medial prefrontal activity during shifting under novel situations. Neurosci Lett 484(3):182–186. doi:10.1016/j.neulet.2010.08.047

LaFerla FM, Green KN (2012) Animal models of Alzheimer disease. Cold Spring Harbor Perspect Med 2(11). doi:10.1101/cshperspect.a006320

Lange KW, Sahakian BJ, Quinn NP, Marsden CD, Robbins TW (1995) Comparison of executive and visuospatial memory function in Huntington's disease and dementia of Alzheimer type matched for degree of dementia. J Neurol Neurosurg Psychiatry 58(5):598–606

Lange KW, Reichl S, Lange KM, Tucha L, Tucha O (2010) The history of attention deficit hyperactivity disorder. Attention Deficit Hyperactivity Disorders 2(4):241–255. doi:10.1007/s12402-010-0045-8

Lapiz-Bluhm MD, Soto-Pina AE, Hensler JG, Morilak DA (2009) Chronic intermittent cold stress and serotonin depletion induce deficits of reversal learning in an attentional set-shifting test in rats. Psychopharmacology 202(1–3):329–341. doi:10.1007/s00213-008-1224-6

Lawrence AD, Sahakian BJ, Hodges JR, Rosser AE, Lange KW, Robbins TW (1996) Executive and mnemonic functions in early Huntington's disease. Brain 119(Pt 5):1633–1645

Lawrence AD, Hodges JR, Rosser AE, Kershaw A, ffrench-Constant C, Rubinsztein DC, Robbins TW, Sahakian BJ (1998) Evidence for specific cognitive deficits in preclinical Huntington's disease. Brain 121(Pt 7):1329–1341

Lawrence AD, Sahakian BJ, Rogers RD, Hodge JR, Robbins TW (1999) Discrimination, reversal, and shift learning in Huntington's disease: mechanisms of impaired response selection. Neuropsychologia 37(12):1359–1374

Leeson VC, Robbins TW, Matheson E, Hutton SB, Ron MA, Barnes TR, Joyce EM (2009) Discrimination learning, reversal, and set-shifting in first-episode schizophrenia: stability over six years and specific associations with medication type and disorganization syndrome. Biol Psychiatry 66(6):586–593. doi:S0006-3223(09)00628-3 [pii] 10.1016/j.biopsych.2009.05.016

Lindgren HS, Wickens R, Tait DS, Brown VJ, Dunnett SB (2013) Lesions of the dorsomedial striatum impair formation of attentional set in rats. Neuropharmacology 71:148–153. doi:10.1016/j.neuropharm.2013.03.034

Liston C, Miller MM, Goldwater DS, Radley JJ, Rocher AB, Hof PR, Morrison JH, McEwen BS (2006) Stress-induced alterations in prefrontal cortical dendritic morphology predict selective impairments in perceptual attentional set-shifting. J Neurosci 26(30):7870–7874
Manes F, Sahakian B, Clark L, Rogers R, Antoun N, Aitken M, Robbins T (2002) Decision-making processes following damage to the prefrontal cortex. Brain 125(Pt 3): 624–639
Mattes R, Cohen R, Berg P, Canavan AG, Hopmann G (1991) Slow cortical potentials (SCPS) in schizophrenic patients during performance of the Wisconsin card-sorting test (WCST). Neuropsychologia 29(2):195–205
McAlonan K, Brown VJ (2003) Orbital prefrontal cortex mediates reversal learning and not attentional set shifting in the rat. Behav Brain Res 146(1–2):97–103
McGaughy J, Ross RS, Eichenbaum H (2008) Noradrenergic, but not cholinergic, deafferentation of prefrontal cortex impairs attentional set-shifting. Neuroscience 153(1):63–71
Mehta MA, Goodyer IM, Sahakian BJ (2004) Methylphenidate improves working memory and set-shifting in AD/HD: relationships to baseline memory capacity. J Child Psychol Psychiatry 45(2):293–305
Miller EK, Cohen JD (2001) An integrative theory of prefrontal cortex function. Annu Rev Neurosci 24:167–202. doi:10.1146/annurev.neuro.24.1.167 24/1/167 [pii]
Milner B (1963) Effect of different brain lesions on card sorting. Arch Neurol 9:90–100
Monchi O, Petrides M, Petre V, Worsley K, Dagher A (2001) Wisconsin Card Sorting revisited: distinct neural circuits participating in different stages of the task identified by event-related functional magnetic resonance imaging. J Neurosci 21(19):7733–7741
Moore TL, Killiany RJ, Herndon JG, Rosene DL, Moss MB (2005) A non-human primate test of abstraction and set shifting: an automated adaptation of the Wisconsin Card Sorting Test. J Neurosci Methods 146(2):165–173
Moritz S, Birkner C, Kloss M, Jahn H, Hand I, Haasen C, Krausz M (2002) Executive functioning in obsessive-compulsive disorder, unipolar depression, and schizophrenia. Arch Clin Neuropsychol 17(5):477–483
Nagahama Y, Okina T, Suzuki N, Matsuzaki S, Yamauchi H, Nabatame H, Matsuda M (2003) Factor structure of a modified version of the wisconsin card sorting test: an analysis of executive deficit in Alzheimer's disease and mild cognitive impairment. Dement Geriatr Cogn Disord 16(2):103–112. doi:10.1159/000070683
Nagahama Y, Okina T, Suzuki N, Nabatame H, Matsuda M (2005) The cerebral correlates of different types of perseveration in the Wisconsin Card Sorting Test. J Neurol Neurosurg Psychiatry 76(2):169–175
Ng CW, Noblejas MI, Rodefer JS, Smith CB, Poremba A (2007) Double dissociation of attentional resources: prefrontal versus cingulate cortices. J Neurosci 27(45):12123–12131
Nieuwenstein MR, Aleman A, de Haan EH (2001) Relationship between symptom dimensions and neurocognitive functioning in schizophrenia: a meta-analysis of WCST and CPT studies. Wisconsin Card Sorting Test. Continuous Performance Test. J Psychiatr Res 35(2):119–125
Nikiforuk A, Popik P (2011) Long-lasting cognitive deficit induced by stress is alleviated by acute administration of antidepressants. Psychoneuroendocrinology 36(1):28–39. doi:10.1016/j.psyneuen.2010.06.001
Nuechterlein KH, Green MF, Kern RS, Baade LE, Barch DM, Cohen JD, Essock S, Fenton WS, Frese FJ 3rd, Gold JM, Goldberg T, Heaton RK, Keefe RS, Kraemer H, Mesholam-Gately R, Seidman LJ, Stover E, Weinberger DR, Young AS, Zalcman S, Marder SR (2008) The MATRICS Consensus Cognitive Battery, part 1: test selection, reliability, and validity. Am J Psychiatry 165(2):203–213. doi:10.1176/appi.ajp.2007.07010042
Olanow CW, Tatton WG (1999) Etiology and pathogenesis of Parkinson's disease. Annu Rev Neurosci 22:123–144
Overstreet DH (2012) Modeling depression in animal models. Methods Mol Biol 829:125–144. doi:10.1007/978-1-61779-458-2_7

Owen AM, Roberts AC, Polkey CE, Sahakian BJ, Robbins TW (1991) Extra-dimensional versus intra-dimensional set shifting performance following frontal lobe excisions, temporal lobe excisions or amygdalo-hippocampectomy in man. Neuropsychologia 29(10):993–1006

Owen AM, James M, Leigh PN, Summers BA, Marsden CD, Quinn NP, Lange KW, Robbins TW (1992) Fronto-striatal cognitive deficits at different stages of Parkinson's disease. Brain 115(Pt 6):1727–1751

Owen AM, Roberts AC, Hodges JR, Summers BA, Polkey CE, Robbins TW (1993) Contrasting mechanisms of impaired attentional set-shifting in patients with frontal lobe damage or Parkinson's disease. Brain 116(Pt 5):1159–1175

Pantelis C, Barber FZ, Barnes TR, Nelson HE, Owen AM, Robbins TW (1999) Comparison of set-shifting ability in patients with chronic schizophrenia and frontal lobe damage. Schizophr Res 37(3):251–270. S092099649800156X [pii]

Paolo AM, Troster AI, Axelrod BN, Koller WC (1995) Construct validity of the WCST in normal elderly and persons with Parkinson's disease. Arch Clin Neuropsychol 10(5):463–473

Papaleo F, Crawley JN, Song J, Lipska BK, Pickel J, Weinberger DR, Chen J (2008) Genetic dissection of the role of catechol-O-methyltransferase in cognition and stress reactivity in mice. J Neurosci 28(35):8709–8723

Peinemann A, Schuller S, Pohl C, Jahn T, Weindl A, Kassubek J (2005) Executive dysfunction in early stages of Huntington's disease is associated with striatal and insular atrophy: a neuropsychological and voxel-based morphometric study. J Neurol Sci 239(1):11–19. doi:10.1016/j.jns.2005.07.007

Perry RJ, Watson P, Hodges JR (2000) The nature and staging of attention dysfunction in early (minimal and mild) Alzheimer's disease: relationship to episodic and semantic memory impairment. Neuropsychologia 38(3):252–271

Pineda D, Ardila A, Rosselli M, Cadavid C, Mancheno S, Mejia S (1998) Executive dysfunctions in children with attention deficit hyperactivity disorder. Int J Neurosci 96(3–4):177–196

Ragozzino ME, Detrick S, Kesner RP (1999) Involvement of the prelimbic-infralimbic areas of the rodent prefrontal cortex in behavioral flexibility for place and response learning. J Neurosci 19 (11):4585–4594

Rapport LJ, Van Voorhis A, Tzelepis A, Friedman SR (2001) Executive functioning in adult attention-deficit hyperactivity disorder. Clin Neuropsychol 15(4):479–491. doi:10.1076/clin.15.4.479.1878

Ridderinkhof KR, Span MM, van der Molen MW (2002) Perseverative behavior and adaptive control in older adults: performance monitoring, rule induction, and set shifting. Brain Cogn 49 (3):382–401

Riehemann S, Volz HP, Stutzer P, Smesny S, Gaser C, Sauer H (2001) Hypofrontality in neuroleptic-naive schizophrenic patients during the Wisconsin Card Sorting Test—a fMRI study. Eur Arch Psychiatry Clin Neurosci 251(2):66–71

Robbins TW (2000) Chemical neuromodulation of frontal-executive functions in humans and other animals. Exp Brain Res 133(1):130–138

Roberts AC, Robbins TW, Everitt BJ (1988) The effects of intradimensional and extradimensional shifts on visual discrimination learning in humans and non-human primates. Q J Exp Psychol B 40(4):321–341

Roberts AC, Robbins TW, Everitt BJ, Muir JL (1992) A specific form of cognitive rigidity following excitotoxic lesions of the basal forebrain in marmosets. Neuroscience 47(2):251–264

Roberts AC, De Salvia MA, Wilkinson LS, Collins P, Muir JL, Everitt BJ, Robbins TW (1994) 6-Hydroxydopamine lesions of the prefrontal cortex in monkeys enhance performance on an analog of the Wisconsin Card Sort Test: possible interactions with subcortical dopamine. J Neurosci 14(5 Pt 1):2531–2544

Rodefer JS, Nguyen TN (2008) Naltrexone reverses age-induced cognitive deficits in rats. Neurobiol Aging 29(2):309–313

Rodefer JS, Murphy ER, Baxter MG (2005) PDE10A inhibition reverses subchronic PCP-induced deficits in attentional set-shifting in rats. Eur J Neurosci 21(4):1070–1076

Rodefer JS, Nguyen TN, Karlsson JJ, Arnt J (2008) Reversal of subchronic PCP-induced deficits in attentional set shifting in rats by Sertindole and a 5-HT(6) receptor antagonist: comparison among antipsychotics. Neuropsychopharmacology 33(11):2657–2666

Rogers RD, Andrews TC, Grasby PM, Brooks DJ, Robbins TW (2000) Contrasting cortical and subcortical activations produced by attentional-set shifting and reversal learning in humans. J Cogn Neurosci 12(1):142–162

Romine CB, Lee D, Wolfe ME, Homack S, George C, Riccio CA (2004) Wisconsin Card Sorting Test with children: a meta-analytic study of sensitivity and specificity. Arch Clin Neuropsychol 19(8):1027–1041

Roth BL, Hanizavareh SM, Blum AE (2004) Serotonin receptors represent highly favorable molecular targets for cognitive enhancement in schizophrenia and other disorders. Psychopharmacology 174(1):17–24

Sahakian BJ, Owen AM (1992) Computerized assessment in neuropsychiatry using CANTAB: discussion paper. J R Soc Med 85(7):399–402

Sahakian BJ, Downes JJ, Eagger S, Evenden JL, Levy R, Philpot MP, Roberts AC, Robbins TW (1990) Sparing of attentional relative to mnemonic function in a subgroup of patients with dementia of the Alzheimer type. Neuropsychologia 28(11):1197–1213

Scheggia D, Bebensee A, Weinberger DR, Papaleo F (2014) The ultimate intra-/extra-dimensional attentional set-shifting task for mice. Biol Psychiatry 75(8):660–670. doi:10.1016/j.biopsych.2013.05.021

Seidman LJ, Biederman J, Faraone SV, Weber W, Ouellette C (1997) Toward defining a neuropsychology of attention deficit-hyperactivity disorder: performance of children and adolescents from a large clinically referred sample. J Consult Clin Psychol 65(1):150–160

Settlage P, Butler R, Odoi H (1956) Perseverative interface in monkeys following bilateral removal of the prefrontal areas. J Gen Psych 54:255–262

Shepp BE, Eimas PD (1964) Intradimensional and extradimensional shifts in the rat. J Comp Physiol Psychol 57:357–361

Shepp BE, Schrier AM (1969) Consecutive intradimensional and extradimensional shifts in monkeys. J Comp Physiol Psychol 67(2):199–203

Silva KL, Rovaris DL, Guimaraes-da-Silva PO, Victor MM, Salgado CA, Vitola ES, Contini V, Bertuzzi G, Picon FA, Karam RG, Belmonte-de-Abreu P, Rohde LA, Grevet EH, Bau CH (2014) Could comorbid bipolar disorder account for a significant share of executive function deficits in adults with attention-deficit hyperactivity disorder? Bipolar Disord 16(3):270–276. doi:10.1111/bdi.12158

Slamecka NJ (1968) A methodological analysis of shift paradigms in human discrimination learning. Psychol Bull 69(6):423–438

Snowden J, Craufurd D, Griffiths H, Thompson J, Neary D (2001) Longitudinal evaluation of cognitive disorder in Huntington's disease. J Int Neuropsychol Soc 7(1):33–44

Sontag TA, Tucha O, Walitza S, Lange KW (2010) Animal models of attention deficit/hyperactivity disorder (ADHD): a critical review. Attention Deficit Hyperactivity Disorders 2(1):1–20. doi:10.1007/s12402-010-0019-x

Specht K, Lie CH, Shah NJ, Fink GR (2009) Disentangling the prefrontal network for rule selection by means of a non-verbal variant of the Wisconsin Card Sorting Test. Hum Brain Mapp 30(5):1734–1743. doi:10.1002/hbm.20637

Stachowiak MK, Kucinski A, Curl R, Syposs C, Yang Y, Narla S, Terranova C, Prokop D, Klejbor I, Bencherif M, Birkaya B, Corso T, Parikh A, Tzanakakis ES, Wersinger S, Stachowiak EK (2013) Schizophrenia: a neurodevelopmental disorder—integrative genomic hypothesis and therapeutic implications from a transgenic mouse model. Schizophr Res 143(2–3):367–376. doi:10.1016/j.schres.2012.11.004

Starkstein SE, Bolduc PL, Preziosi TJ, Robinson RG (1989) Cognitive impairments in different stages of Parkinson's disease. J Neuropsychiatry Clin Neurosci 1(3):243–248

Stuss DT, Levine B, Alexander MP, Hong J, Palumbo C, Hamer L, Murphy KJ, Izukawa D (2000) Wisconsin Card Sorting Test performance in patients with focal frontal and posterior brain

damage: effects of lesion location and test structure on separable cognitive processes. Neuropsychologia 38(4):388–402
Sullivan EV, Mathalon DH, Zipursky RB, Kersteen-Tucker Z, Knight RT, Pfefferbaum A (1993) Factors of the Wisconsin Card Sorting Test as measures of frontal-lobe function in schizophrenia and in chronic alcoholism. Psychiatry Res 46(2):175–199
Sutherland NS, Mackintosh NJ (1971) Mechanisms of animal discrimination learning. Academic Press, New York
Tait DS, Brown VJ (2008) Lesions of the basal forebrain impair reversal learning but not shifting of attentional set in rats. Behav Brain Res 187(1):100–108
Tait DS, Brown VJ, Farovik A, Theobald DE, Dalley JW, Robbins TW (2007) Lesions of the dorsal noradrenergic bundle impair attentional set-shifting in the rat. Eur J Neurosci 25 (12):3719–3724
Tait DS, Marston HM, Shahid M, Brown VJ (2009) Asenapine restores cognitive flexibility in rats with medial prefrontal cortex lesions. Psychopharmacology 202(1–3):295–306
Tait DS, Chase EA, Brown VJ (2013) Tacrine improves reversal learning in older rats. Neuropharmacology 73C:284–289. doi:10.1016/j.neuropharm.2013.05.036
Tait DS, Chase EA, Brown VJ (2014) Attentional set-shifting in rodents: a review of behavioural methods and pharmacological results. Curr Pharm Des 20(31):5046–5059
Takeda N, Terada S, Sato S, Honda H, Yoshida H, Kishimoto Y, Kamata G, Oshima E, Ishihara T, Kuroda S (2010) Wisconsin card sorting test and brain perfusion imaging in early dementia. Dement Geriatr Cogn Disord 29(1):21–27. doi:10.1159/000261645
Terada S, Sato S, Honda H, Kishimoto Y, Takeda N, Oshima E, Yokota O, Uchitomi Y (2011) Perseverative errors on the Wisconsin Card Sorting Test and brain perfusion imaging in mild Alzheimer's disease. Int Psychogeriatr (IPA) 23(10):1552–1559. doi:10.1017/S104161021-1001463
Vasic N, Walter H, Hose A, Wolf RC (2008) Gray matter reduction associated with psychopathology and cognitive dysfunction in unipolar depression: a voxel-based morphometry study. J Affect Disord 109(1–2):107–116. doi:10.1016/j.jad.2007.11.011
Vigotsky LS, Kasanin JJ (1934) Thought in schizophrenia. Arch Neurol Psychiatry 31(5): 1063–1077. doi:10.1001/archneurpsyc.1934.02250050181009
Volz HP, Gaser C, Hager F, Rzanny R, Mentzel HJ, Kreitschmann-Andermahr I, Kaiser WA, Sauer H (1997) Brain activation during cognitive stimulation with the Wisconsin Card Sorting Test—a functional MRI study on healthy volunteers and schizophrenics. Psychiatry Res 75 (3):145–157
Wallace J, Marston HM, McQuade R, Gartside SE (2014) Evidence that the attentional set shifting test in rats can be applied in repeated testing paradigms. J Psychopharmacol 28(7):691–696. doi:10.1177/0269881114531663
Webster SJ, Bachstetter AD, Nelson PT, Schmitt FA, Van Eldik LJ (2014) Using mice to model Alzheimer's dementia: an overview of the clinical disease and the preclinical behavioral changes in 10 mouse models. Front Genet 5:88. doi:10.3389/fgene.2014.00088
Weed MR, Bryant R, Perry S (2008) Cognitive development in macaques: attentional set-shifting in juvenile and adult rhesus monkeys. Neuroscience 157(1):22–28. doi:S0306-4522(08) 01204-9 [pii] 10.1016/j.neuroscience.2008.08.047
West RL (1996) An application of prefrontal cortex function theory to cognitive aging. Psychol Bull 120(2):272–292
Wickens JR, Hyland BI, Tripp G (2011) Animal models to guide clinical drug development in ADHD: lost in translation? Br J Pharmacol 164(4):1107–1128. doi:10.1111/j.1476-5381.2011. 01412.x
Wilmsmeier A, Ohrmann P, Suslow T, Siegmund A, Koelkebeck K, Rothermundt M, Kugel H, Arolt V, Bauer J, Pedersen A (2010) Neural correlates of set-shifting: decomposing executive functions in schizophrenia. J Psychiatry Neurosci 35(5):321–329. doi:10.1503/jpn.090181
Yan HC, Cao X, Das M, Zhu XH, Gao TM (2010) Behavioral animal models of depression. Neurosci Bull 26(4):327–337. doi:10.1007/s12264-010-0323-7

Yang SH, Cheng PH, Banta H, Piotrowska-Nitsche K, Yang JJ, Cheng EC, Snyder B, Larkin K, Liu J, Orkin J, Fang ZH, Smith Y, Bachevalier J, Zola SM, Li SH, Li XJ, Chan AW (2008) Towards a transgenic model of Huntington's disease in a non-human primate. Nature 453 (7197):921–924. doi:10.1038/nature06975
Young JW, Powell SB, Geyer MA, Jeste DV, Risbrough VB (2010) The mouse attentional-set-shifting task: a method for assaying successful cognitive aging? Cogn Affect Behav Neurosci 10(2):243–251. doi:10/2/243 [pii] 10.3758/CABN.10.2.243
Zhuo JM, Prescott SL, Murray ME, Zhang HY, Baxter MG, Nicolle MM (2007) Early discrimination reversal learning impairment and preserved spatial learning in a longitudinal study of Tg2576 APPsw mice. Neurobiol Aging 28(8):1248–1257. doi:10.1016/j.neurobiolaging.2006.05.034
Zhuo JM, Prakasam A, Murray ME, Zhang HY, Baxter MG, Sambamurti K, Nicolle MM (2008) An increase in Abeta42 in the prefrontal cortex is associated with a reversal-learning impairment in Alzheimer's disease model Tg2576 APPsw mice. Curr Alzheimer Res 5(4): 385–391
Zurkovsky L, Bychkov E, Tsakem EL, Siedlecki C, Blakely RD, Gurevich EV (2013) Cognitive effects of dopamine depletion in the context of diminished acetylcholine signaling capacity in mice. Disease Models Mech 6(1):171–183. doi:10.1242/dmm.010363

Relating Translational Neuroimaging and Amperometric Endpoints: Utility for Neuropsychiatric Drug Discovery

Jennifer Li, Adam J. Schwarz and Gary Gilmour

Abstract Measures of neuronal activation are a natural and parsimonious translational biomarker to consider in the context of neuropsychiatric drug discovery studies. In this regard, functional neuroimaging using the BOLD fMRI technique is becoming more frequently employed to not only probe aberrant brain regions and circuits in disease, but also to assess the effects of novel pharmacological agents on these processes. In the ideal situation, these types of studies would first be conducted pre-clinically in rodents to confirm a measurable functional response on relevant brain circuits before seeking to replicate the findings in an analogous fMRI paradigm in humans. However, the need for animal immobilization during the scanning procedure precludes all but the simplest behavioural task-based paradigms in rodent BOLD fMRI. This chapter considers how in vivo oxygen amperometry may represent a viable and valid proxy for BOLD fMRI in freely moving rodents engaged in behavioural tasks. The amperometric technique and several examples of emerging evidence are described to show how the technique can deliver results that translate to pharmacological, event-related and functional connectivity variants of fMRI. In vivo oxygen amperometry holds great promise as a technique that may help to bridge the gap between basic drug discovery research in rodents and applied efficacy testing in humans.

Keywords Oxygen amperometry · Magnetic resonance imaging · BOLD signal · Reward-based learning · Pharmacological imaging · Functional connectivity

J. Li · G. Gilmour (✉)
In Vivo Pharmacology, Eli Lilly and Company, Erl Wood Manor,
Sunninghill Road, Windlesham, UK
e-mail: gilmour_gary@lilly.com

A.J. Schwarz
Translational Imaging, Eli Lilly and Company, Indianapolis, IN 46285, USA

Curr Topics Behav Neurosci (2016) 28: 397–421
DOI 10.1007/7854_2016_1

Contents

1 The Requirement for Translational Endpoints 398
1.1 Neuropsychiatric Drug Discovery in Perspective 398
1.2 Translation in Practice 399
1.3 The Ascent of Magnetic Resonance Imaging 399
1.4 Origins of the BOLD Signal 400
1.5 Temporal Evolution of the BOLD Signal 401
2 Measuring Functional Brain Changes in Rodents and Humans 403
2.1 The Spectrum of Established Methodologies 403
2.2 Limitations of FMRI BOLD Across Species 403
2.3 Addressing the Gap with O_2 Amperometry 404
3 Practical Applications of In Vivo O_2 Amperometry 406
3.1 The Translatable Axes of Neuroimaging and Amperometry 406
3.2 Pharmacological O_2 Amperometry 407
3.3 Event-Related O_2 Amperometry 408
3.4 Functional Connectivity and Resting-State O_2 Amperometry 410
4 Outstanding Issues and Future Directions 413
4.1 Physiological Equivalence of Signals 413
4.2 Working with the Right Types of Assay 413
4.3 Testing the Right Regions 414
5 Conclusion 415
References 416

1 The Requirement for Translational Endpoints

1.1 Neuropsychiatric Drug Discovery in Perspective

The identification and development of a new drug to treat unmet medical need in human disease is a notoriously long and complex process. The typical development time can range from 11 to 14 years at a cost recently estimated to be close to $2 billion per successful compound launched (Paul et al. 2010). For the majority of neuropsychiatric disorders, the hurdles are set even higher, principally due to a significant lack of knowledge of disease aetiology. While termination of molecules with undesirable profiles, and targets with unattractive properties, is an inevitable part of drug discovery, there is great scope to improve attrition rates. Recent critical analyses of failure rates in neuropsychiatric drug discovery identify "translation" of findings from rodent to human studies as a major issue (Morgan et al. 2012). In the move from positive preclinical biological effects into human testing, many drug development programs have simply lacked evidence in humans of either a measure of engagement of the intended target by the drug, or a measure of functional pharmacological activity as a consequence of target engagement. As a result, identifying such measures has become a vitally important objective of modern discovery efforts, often as important as identifying the potential drug itself. Having the ability to test hypotheses related to drug target engagement and functional activity as early as possible, and ideally before the time-consuming and expensive phase of efficacy testing begins, can substantially lower the risk of failure of a discovery project.

1.2 Translation in Practice

Translational science in the context of neuropsychiatric drug discovery therefore has two foci: measures relating to diseases of interest and measures relating to potential drugs/targets that might treat these diseases. From a disease perspective, translational research efforts typically focus on replication of what are considered important aspects of human pathology in animals, often referred to as disease "modelling". Such models are then used to test the effects of drugs, through which inferences are then made about their therapeutic utility. This is a field that is being approached with increasing caution and rigour (Markou et al. 2009), as most neuropsychiatric disorders are currently conceptualized at the diagnostic level as syndromes of often uniquely human symptoms (American Psychiatric Association 2014; WHO 1992). By definition, this makes it incredibly difficult in most cases to directly induce homologues of neuropsychiatric syndromes in animals, which has led researchers to concentrate on a deeper understanding of transdiagnostic, mechanistic constructs underlying the symptoms in question (Cuthbert 2015). From a drug/target perspective, translational research attempts to define biological measures ("biomarkers") in animals that index target engagement and/or predict therapeutic utility in defined human patient populations. In this regard, putative biomarkers can be considered from all domains of biology, from plasma and biofluid-based genetic, metabolic and biochemical markers, to behavioural endpoints and complex measures of integrated physiological systems. Given the reasons behind failure in neuropsychiatric drug discovery, the most critical issue in translational research therefore is determining whether a hypothesis of drug action has been satisfactorily tested if the primary endpoint of a clinical trial is negative. Attempts to establish biomarkers of neuronal activation have been a naturally parsimonious approach to solving this problem, which in practice has involved the investigation of electrophysiological, neurochemical and neuroimaging modalities. Translational validity thereby becomes a matter of demonstrating via neurophysiological measures that a drug has modulated brain activity in a consistent manner across species, and in the ideal situation that this activity directly relates to the alleviation of symptoms in question.

1.3 The Ascent of Magnetic Resonance Imaging

While the "gold standard" measurement—that of direct electrophysiological recording of neuronal cellular activity in the human brain—has been carried out in epilepsy and Parkinson's disease patients since the 1940s (see review by Engel et al. 2005), it has been limited to only those occasions of opportunity and consent that essential neurosurgical intervention provides. However, attempts to indirectly measure brain activity in humans are by no means new, and accounts of such work began to appear in the scientific literature from the late nineteenth century onwards, from Mosso's "human circulation balance" (Mosso 1884) and the first successful recordings of the human electroencephalogram (Berger 1933) through to the dawn of magnetic resonance technologies (Damadian et al. 1977; Lauterbur 1973; Mansfield and Maudsley 1976). Magnetic resonance imaging (MRI) is a technique that uses pulsed changes in magnetic field applied to the brain (or other tissues) in the presence of a strong static magnetic field to manipulate the magnetic alignment and resonant frequency of atomic nuclei, in particular the hydrogen atoms in water molecules, resulting in radiofrequency signals that can be deconvolved to resolve spatially distinct signal sources. Importantly, MRI methods are extremely flexible, allowing the influence of different tissue microenvironments on the MRI parameters to be reflected in different image intensities; for example, grey matter, white matter and cerebral spinal fluid can be discriminated in structural images of the brain (Gadian 1996). As MRI scanner technology and analytic methods have evolved, MRI-based paradigms have come to revolutionize neuroscience. Application of MRI in humans is now one of the key methodologies by which scientists non-invasively probe brain structure in awake humans.

1.4 Origins of the BOLD Signal

A hugely significant advance in nuclear magnetic resonance technology arrived with the ability to measure the blood oxygen-level-dependent (BOLD) signal in the brain (Bandettini et al. 1992; Kwong et al. 1992; Ogawa et al. 1990). This method takes advantage of the fact that the magnetic properties of haemoglobin in blood depend upon its oxygenation status. Oxygenated haemoglobin (HbO_2) is diamagnetic, whereas deoxygenated haemoglobin (Hbr) is paramagnetic, differences that can be readily detected using T_2^*-weighted MRI acquisition sequences. As oxygen is an essential substrate for the brain that is continually consumed by neurons and glia in an activity-dependent manner, the local consumption of O_2 drives a haemodynamic response to deliver the required oxygen, resulting in changes in cerebral blood flow (CBF), cerebral blood volume (CBV), as well as the haemoglobin oxygenation balance. The result is a change in BOLD signal that indirectly reflects regional levels of activity (Huettel et al. 2004), although the exact neuronal origins of the BOLD signal are still under debate. The BOLD signal is sensitive to several cerebrovascular factors, including CBF and CBV in addition to the cerebral metabolic rate of oxygen ($CMRO_2$) consumption. While $CMRO_2$ is the most direct physiological correlate of neuronal activity (Shulman et al. 2002; Smith et al. 2002; Hyder et al. 2002), there is convincing evidence that neuronal spiking activity per se is not a principal driver of the haemodynamic response in an activated brain region. Several laboratories have investigated the relationships between factors such as action potential production, synaptic activity, tissue oxygen, CBF and the BOLD signal itself. In this regard, one early and important finding was that overall regional spiking activity does not necessarily modulate CBF levels in a predictable manner, as the activity of local inhibitory networks can significantly dissociate these factors (Caesar et al. 2003; Offenhauser et al. 2005; Mathiesen et al. 1998). Rather than spiking activity, there appears to be a much stronger correlation between synaptic activity, local field potential (LFP) amplitude and CBF (Mathiesen et al. 2000). The importance of these findings for the BOLD signal itself has been most elegantly demonstrated in studies by the Logothetis laboratory. An assessment of visual cortex function in anaesthetized monkeys presented with a visual stimulus of a rotating checkerboard pattern showed that the BOLD contrast signal was most closely correlated with LFP (Logothetis et al. 2001; Goense and Logothetis 2008), in particular with those potentials in the gamma band of the frequency spectrum. In most situations, LFP corresponds to the underlying multi-unit activity of a region, such that both measures tend to correlate with the BOLD signal. However, this is not always the case and when LFP and multi-unit activity show divergence, the BOLD signal then appears to be driven by the field potential. BOLD signal activation therefore seems to be a reflection more of the input of neuronal activity to a region, rather than the more commonly regarded output function of spiking activity. A final level of complexity is offered by recent findings that the precise nature and magnitude of the BOLD signal may potentially vary across the brain as a function of regional differences in neurovascular architecture and neurovascular coupling processes (e.g. Ances et al. 2008).

1.5 Temporal Evolution of the BOLD Signal

In cognitive neuroscience, functional MRI (fMRI) refers to the experimental application of stimuli (e.g. sensory cues, motor tasks and cognitive demands) to measure resultant changes in BOLD signal across the brain, from which neuronal activity and thereby regional involvements in processing can be inferred. Current BOLD imaging approaches in humans enable the whole brain to typically be sampled at $\sim$3–5 mm resolution every $\sim$3 s. Following functional stimulation, an fMRI BOLD response displays a characteristic time course, as illustrated in Fig. 1. There is often (but not always) a fast decrease in the BOLD signal occurring within 1–2 s after stimulation onset, followed by an increase in the BOLD signal peaking around 4–7 s, which then falls back to baseline, sometimes accompanied by a "post-stimulus undershoot". The marker of neuronal activation used in nearly all fMRI studies is actually the increase in BOLD response rather than the initial dip

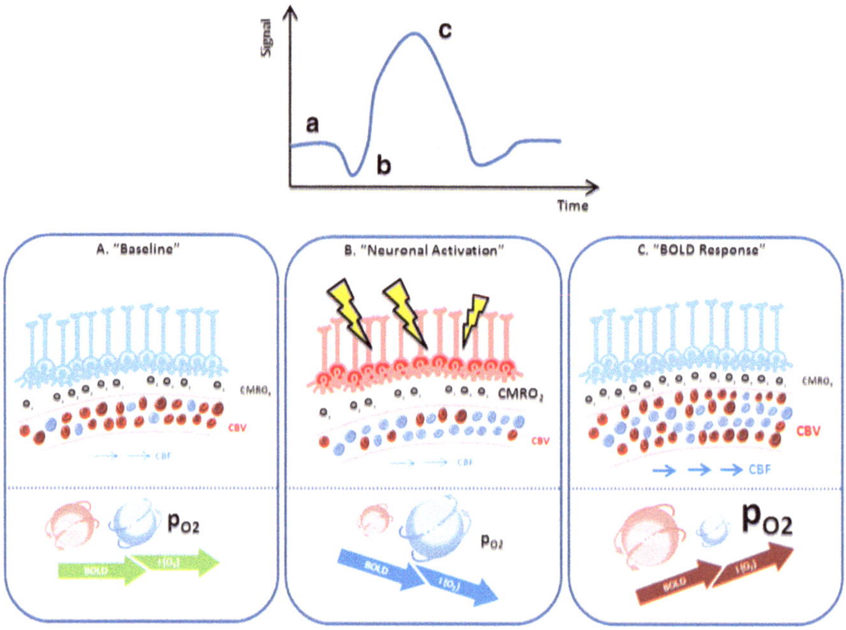

Fig. 1 A theoretical representation of the BOLD fMRI and oxygen amperometric responses. Both BOLD and O_2 amperometric signals change in intensity as a function of patterns of neuronal activation over time. At baseline levels of neuronal activity (**a**), there is a steady haemodynamic state and balanced ratio of oxygenated haemoglobin (HbO_2) to deoxygenated haemoblogin (Hbr) in local blood vessels, resulting in a stable BOLD signal. The partial pressure of oxygen in the parenchyma is stable, and thus, the oxygen amperometric signal is also stable. Upon neuronal activation (**b**), oxygen consumption ($CMRO_2$) by neurons increases, leading to a period where increases in Hbr:HbO_2 ratio in blood and decreased partial pressure of oxygen in parenchyma is detectable as an initial decrease in the BOLD signal, and reduction in O_2 amperometric current. At a certain point the drop in oxygen tension will trigger a haemodynamic response (**c**). This response results in increased CBV and CBF, replenishing and overcompensating HbO_2 and increasing the partial pressure of oxygen in the parenchyma above that of baseline. The increased HbO_2:Hbr ratio is detected as an increase in the BOLD signal, which is typically termed the "BOLD response". An increase in O_2 amperometric current is also measured, coincident with the BOLD response. Abbreviations: *BOLD* blood oxygen-level-dependent signal, *CBF* cerebral blood flow; *CBV* cerebral blood volume; $CMRO_2$ cerebral metabolic rate of oxygen consumption; $I(O_2)$ oxygen amperometric current; pO_2 partial pressure of oxygen

in signal, as despite this feature lagging behind the neuronal activation event, it is much more reliably and consistently evoked across studies. The post-stimulus undershoot is thought to be a form of volume artefact that can originate in specific circumstances when there are differences in the timescales over which CBV and CBF parameters recover to baseline conditions (Buxton et al. 1998; Mandeville et al. 1999). It tends not to be considered a reliable marker of neuronal activation.

2 Measuring Functional Brain Changes in Rodents and Humans

2.1 The Spectrum of Established Methodologies

The extensive use of fMRI methodologies in cognitive neuroscience and neuropsychiatric drug discovery settings has increased the need for equivalent, translational approaches in freely moving animals, such that homologous experimentally induced changes in brain activity can be measured. A great multitude of methods can be applied to index functional brain changes in rodents, all of which have slightly different attributes and therefore application. These include methodologies from electrophysiological (patch, unit activity, LFP, EEG and MEG), metabolic (2-DG and FDG), neurochemical (microdialysis and PET imaging), optical (2-photon and NIRS) and electrochemical (enzyme-based biosensors and voltammetry) disciplines. Depending on the experimental need, e.g. for spatial or temporal resolution of measurement, for assessment of discrete regional versus whole-brain activity, and potential limitations placed on the desirability of "one-shot" terminal measures, attributes of different methodologies will be more suited to some styles of study over others. All have potential to reveal insights into brain function. Importantly, however, none of the aforementioned techniques offer an adequate homologue of BOLD fMRI for awake animals.

2.2 Limitations of FMRI BOLD Across Species

Of the non-invasive functional imaging techniques, BOLD fMRI has moderately good spatial (3–6 mm) and temporal (1–4 s) resolution in humans. Initially, block designs were adopted where images of the resting/control and task/stimulus states were acquired and a subtraction analysis conducted to determine areas of task-specific activation. These activated regions were then superimposed onto structural MRI anatomical images to provide activation maps. However, by the end of the 1990s, the implementation of event-related fMRI was becoming more widespread. fMRI methods have also been employed in rodents—most commonly in the rat—since the early 1990s. Many early rat fMRI BOLD studies employed sensory stimuli, such as electrical stimulation of the forepaw or mechanical stimulation of the vibrissae, and elicited well-defined cue-locked BOLD signal changes in the appropriate somatosensory pathways (Mandeville et al. 1998). In this regard, the sensory cortical columns of the neocortex have been studied extensively in animals at a spatial resolution of up to 1 mm using BOLD fMRI-based methods (Duong et al. 2001; Zhao et al. 2005). From the late 1990s, pharmacologically induced changes in brain function have been increasingly studied in rodents using acute drug administration as a stimulus, an approach known as pharmacological MRI (phMRI) (Leslie and James 2000). More recently, fMRI measures of resting-state functional connectivity have been successfully back-translated to rodents (Lu et al. 2012; Pawela et al. 2008; Schwarz et al. 2013a, b). However, the majority of rodent fMRI studies to date have been performed under maintenance anaesthesia to both minimize motion in the scanner and maintain physiological stability. However, this comes at the cost of only being able to assess a limited range of functional responses. Several groups have now implemented conscious rat fMRI, where animals are habituated to the restraint and noise of the MRI scanner, and their stress levels (indexed by circulating corticosterone levels) are normalized before performing the actual experiment. However, whether anaesthetized or not, the need for complete immobilization in the scanner has precluded the combination of functional imaging with nearly all behavioural or cognitive tasks, in contrast to human fMRI. This greatly limits translational opportunities that would be valuable in a neuropsychiatric drug discovery context and makes it extremely difficult to combine functional imaging with experimental paradigms that require the rodent to move and interact with instruments in its environment. Thus, there remains a methodological gap in the arsenal of available rodent brain activity measures, namely that of a BOLD-like imaging technique that can be employed in freely moving, behaving rodents.

2.3 Addressing the Gap with O_2 Amperometry

Recent developments in a number of laboratories suggest that constant potential amperometry (CPA) can be used effectively in vivo to measure tissue oxygen levels in the brains of freely moving rodents, resulting in the ability to record brain oxygenation changes that are related to behavioural performance, akin to human task-based fMRI studies. This technique employs the use of electrodes implanted into pre-specified brain regions to measure local concentrations of oxygen. Oxygen is an inherently reactive, electron-accepting species critically driving the reactions of mitochondrial oxidative phosphorylation, a biochemical cycle at the heart of aerobic metabolism and the energy requirements of a cell. These electroactive properties of oxygen can also be exploited to measure its concentration in a biological matrix using a constant potential amperometric methodology. By holding an electrode at a negative constant potential of −650 mV (the peak of the reduction wave for oxygen), its electrochemical reduction occurs at the electrode surface by the 2-step reaction shown below. During this reduction process, electrons are transferred to oxygen, thereby producing a negative current proportional to the concentration of oxygen present at the electrode tip (Hitchman 1978).

$$\mathbf{O_2 + 2H^+ + 2e^- \leftrightarrow H_2O_2}$$

$$\mathbf{H_2O_2 + 2H^+ + 2e^- \leftrightarrow 2H_2O}$$

The basic unit of the electrochemical circuit for CPA consists of 3 electrodes: the working, reference and auxiliary electrodes. The working electrode is placed in the brain region of interest and is the electrode where the oxygen amperometric current is measured under the applied potential. The reference electrode is placed into the same biological matrix as the working electrode and acts as a relative measure for the current passing through the working electrode. No current flows through the reference electrode. The circuit is completed by an auxiliary electrode, which is typically wrapped around a screw placed into the skull. The three electrodes are connected via a potentiostat that ensures that the desired potential (−650 mV) is constantly applied between the working electrode and the reference electrode and that the auxiliary electrode is kept at a sufficient potential to drive the redox reaction of oxygen and balance the circuit (O'Neill et al. 1998). In practice, multiple working electrodes can be utilized in the same circuit, such that several different brain regions can be sampled within the same animal. While traditional CPA protocols typically use noble metal electrodes, several groups have reported more effective use of carbon-based electrodes when sampling from a biological matrix (Bolger and Lowry 2005; Lowry et al. 1996; Venton et al. 2003; Zimmerman et al. 1992). Carbon paste electrodes (CPEs) are now more commonly used as working electrodes for measurement of oxygen levels by CPA in the brain due to their in vivo stability and low propensity for surface poisoning (Kane and O'Neill 1998). By protecting the metal at the electrode tip from brain tissue with a layer of carbon paste, CPEs allow free diffusion of oxygen yet resist potential damage from brain proteins and lipids. The utility of CPEs for measurement of brain oxygen levels by CPA has been previously characterized (Bolger et al. 2011). At a potential of −650 mV, CPEs display high sensitivity to oxygen (−1.49 ± 0.01 nA/μM) with a low detection limit (ca. 0.1 μM). In vitro, they respond in a linear manner to oxygen concentration ($R^2 > 0.99$), and there is minimal interference from other potentially electroactive substances that are found in extracellular fluid. CPEs can be used to measure dissolved tissue oxygen in the brain with a subsecond time resolution (Lowry et al. 1996) and are stable in vivo for many months such that longitudinal behavioural studies can be performed (Bolger et al. 2011). The ideal working electrode size has to be greater than the diameter of an average blood capillary (~100 μm) to prevent direct blood sampling of dissolved oxygen (Bolger and Lowry 2005), yet not be too big that implantation causes a confounding level of damage. The signal produced by a working electrode is proportional to its active surface area and will diminish with tip diameter (Dale et al. 2005). Experience suggests that a working electrode diameter of 200 μm maintains a signal-to-noise ratio that enables effective and sensitive detection of O_2 and is very likely to be an averaged tissue oxygen response elicited by changes in a neuronal population. The field of sensitivity of Clark-type polarographic oxygen microelectrodes is known to be a sphere of double the tip diameter (Offenhauser et al. 2005), which although untested is also likely to be true of CPEs. This would suggest that BOLD fMRI and O_2 amperometric techniques offer fairly similar spatial resolutions for brain imaging studies. Simultaneous BOLD MRI and O_2 amperometry recordings have been conducted in anaesthetized rats, and the signals measured by each technique following basic experimental manipulation of brain oxygen levels were demonstrated to be very similar in response characteristics (Sibson et al. 2009).

3 Practical Applications of In Vivo O_2 Amperometry

3.1 The Translatable Axes of Neuroimaging and Amperometry

The wide range of applications of functional MRI methodologies has made it a critical tool for understanding brain regions and networks involved in normal behaviour, neuropsychiatric disease states and the therapeutic effects of drugs. **Pharmacological MRI (phMRI)** has been widely used to assess the direct effect of different drug classes on brain activation in both healthy and diseased states. In this approach, a systemic pharmacological challenge is used to elicit a central functional response (i.e. BOLD signal change) that can be informative of pharmacological mechanism, either alone or via interaction with other compounds. A large number of preclinical phMRI studies have now been performed (Jenkins 2012). At its inception, the technique was also applied in humans (Breiter et al. 1997; Stein et al. 1998); however, when used with a BOLD readout, the method requires fast pharmacokinetics and thus is typically limited to compounds that can be injected intravenously in human subjects. Several groups have nevertheless applied BOLD phMRI to the study of suitable compounds (Becerra et al. 2006; Deakin et al. 2008; Kufahl et al. 2005; Leppa et al. 2006; McKie et al. 2011; Upadhyay et al. 2011; Vollm et al. 2004), providing a strongly translatable methodology (Becerra et al. 2006). Task or **event-based fMRI** studies have also been critical in discovering the involvement of different brain regions in the performance of various cognitive and behavioural tasks. These have enabled the neural circuits underpinning a huge variety of brain functional states to be elucidated, and this approach has revolutionized our understanding of brain function. More recently, techniques of **resting-state fMRI (rsfMRI)** have revealed large-scale networks of intrinsic connectivity in the human brain. These have been back-translated from humans to rodents, and initial results suggest some degree of conservation of resting-state networks and patterns of functional connectivity across species (Lu et al. 2012). fMRI methods are now increasingly being applied in the mouse as well, opening the way to probing alterations in brain function associated with the many transgenic models available in this species (Ferrari et al. 2012; Liska et al. 2015; Sforazzini et al. 2014a). This section will consider how in vivo oxygen amperometry can be used as a proxy for the BOLD fMRI signal in freely moving animals and will describe how the three main areas of neuroimaging research discussed above (Fig. 2) can be recapitulated by this technique.

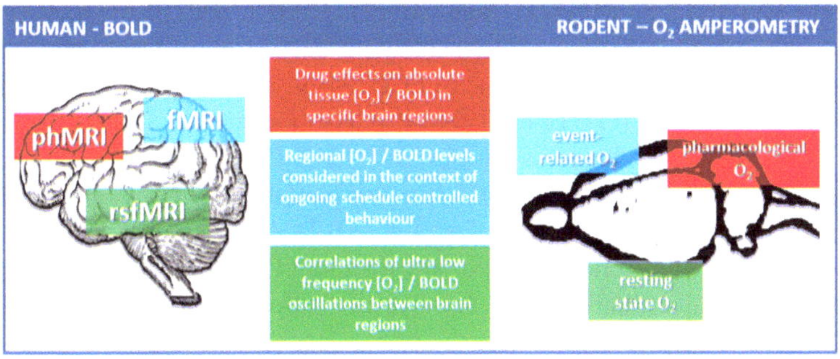

Fig. 2 Three main translational applications of BOLD imaging and oxygen amperometry techniques. Pharmacological MRI (phMRI)/O_2 is used to measure drug effects on specific regional activity, while standard task-based fMRI is used to record regional responses to the performance of schedule-controlled behaviours, similar to event-related O_2 amperometry. Resting-state fMRI (rsfMRI) and resting-state O_2 amperometry measure the correlation of low-frequency oscillations in the signals between brain regions within and between different resting-state networks

3.2 *Pharmacological O_2 Amperometry*

To date, the predominant use of functional neuroimaging for drug discovery has been the phMRI paradigm, which has been used to assess the effects of pharmacological agents (or their combination) on regional brain activity. During a BOLD phMRI study, the subject is placed into an MRI scanner and an fMRI time series (with—in most cases—no accompanying cognitive or sensory stimulus) initiated. The test compound is administered, typically via an intravenous route, partway through the time series. The effect of the test drug on regional brain activity is modelled with respect to the pre-injection baseline, and typically compared to a vehicle/placebo condition, or to study conditions where the subject has been pre-treated with a second (modulatory) compound. In an analogous fashion, the oxygen amperometry technique can also be used to assess the direct effect of test compounds on regional brain activation, and since it is carried out in freely moving rodents, the effect of the drug can be considered without the potentially confounding influence of anaesthesia or sedation in rodents.

As a proof of principle, several amperometry studies have been conducted investigating the effect of the NMDA receptor antagonist ketamine on regional O_2 signals, demonstrating robust drug-induced activation of several cortical and subcortical regions (Li et al. 2014; Francois et al. 2015). Very similar effects of ketamine and other NMDA receptor antagonists have been consistently seen in BOLD phMRI, relative cerebral blood volume (rCBV) and 2DG studies in anaesthetized animals (Littlewood et al. 2006; Chin et al. 2011; Gozzi et al. 2008; Duncan et al. 1998a, b, 2000) and healthy volunteers (Deakin et al. 2008; De Simoni et al. 2013; Langsjo et al. 2003, 2004), suggesting a strong degree of homology across techniques and species. Ketamine interaction studies have also been successfully completed in both BOLD phMRI and pharmacological O_2 amperometry studies. Several pharmacological classes have been shown to significantly attenuate regional ketamine or PCP responses in rodents, for example the atypical antipsychotics (O_2: Li et al. 2014; rCBV: Gozzi et al. 2008), mGluR 2/3 receptor agonists (O_2: Francois et al. 2015; BOLD: Chin et al. 2011; rCBV: Schwarz et al. 2013b) and the M1/M4 receptor partial agonist xanomeline (O_2: Li, unpublished observation; BOLD: Baker et al. 2012). Also, a recent phMRI study in healthy human volunteers described attenuation of the BOLD signal increase induced by ketamine challenge by the atypical antipsychotic risperidone and the anticonvulsant lamotrigine (Doyle et al. 2013a). Hopefully, future work in this field will continue to build a more comprehensive dataset of drug effects between techniques and species to gain more complete understanding of the full potential for equivalence.

An important point to consider is the level of interpretation that should be placed on drug effects on phMRI and pharmacological O_2 amperometric signals. Caution should be applied if such datasets are used in isolation as an indicator of potential efficacy in a disease state, and the NMDA receptor antagonist work cited here provides a perfect example of this. For example, both ketamine and PCP can be administered to humans and demonstrated to have psychotomimetic properties (Krystal et al. 1994; Malhotra et al. 1996; Newcomer et al. 1999). From this, it is common to surmise that the in vivo sequelae of administration of these compounds in animals must also be "psychosis-like" and further that attenuation of these effects might be an indication of antipsychotic efficacy. However, in the context of phMRI and pharmacological O_2 amperometry, antipsychotic modulation of NMDA receptor antagonist-induced signals need not imply an effect on "psychotic" symptoms (e.g. Lahti et al. 1995). A more conservative and appropriate usage of pharmacological MRI/O_2 amperometry signals is to consider them as pharmacodynamic biomarkers that can confirm the mechanism of action of a compound (e.g. reducing NMDA receptor antagonist elicited brain activity associated with a hyperglutamatergic state), or provide an indirect index of target engagement in the brain. If integrated into early-phase clinical development, such studies can inform dose selection for subsequent clinical proof of concept trials. Without corroborating evidence, such signals should not necessarily be considered to predict potential efficacy of the test agent against any specific behavioural domain of function.

3.3 Event-Related O_2 Amperometry

During a task-based fMRI study, subjects undergo scans while performing behavioural tests. The analysis of regional BOLD changes is then considered in the context of the various different events, cues or stimuli that are part of the task. These paradigms have great potential to elucidate how different regions in the brain engage and interact during different types of cognitive and behavioural processes. Such information can help provide a deeper, mechanistic understanding of disease states by determining whether specific neuropsychiatric symptoms are associated with alterations in brain activity associated with specific cognitive functions. Most importantly, the technique has great potential to offer a degree of construct validity concerning drug effects, by demonstrating that a test drug has the potential to modulate and hopefully normalize regional brain activity that has been disturbed in disease. While this is an enormously popular paradigm in human research, it is almost completely precluded as a rodent MRI paradigm by the need for significant immobilization of the animals during scans. Therefore, if O_2 amperometric signals can be shown to be an effective proxy for BOLD signal changes measured in event-related fMRI studies, this would represent a significant strategic advance for translational research. So far, evidence to date reveals strong empirical parallels between observations using rodent O_2 amperometry and findings from rat and human fMRI studies.

Several proof-of-principle studies have been conducted to demonstrate the potential equivalence of rodent event-related O_2 amperometric and human task-based fMRI signals. Where, for instance, a BOLD signal increase can be observed in the nucleus accumbens of humans during performance of a monetary incentive delay task (Knutson et al. 2000, 2001; Knutson and Cooper 2005), localized O_2 increases in the same region can also be observed in response to reward anticipation conditions in rats performing an similar task for food reward (Francois et al. 2012). The activation levels of the rat accumbens can be modulated by varying the magnitude and/or motivational incentive value of the reward, effects consistent with human imaging studies. This suggests that in the context of reward anticipation paradigms, the O_2 amperometric response is modulated by similar experimental manipulations in a manner equivalent to that observed for the human BOLD signal. Amperometric O_2 signals can also display regionally specific responses during performance of behavioural tasks, where for instance different patterns of activation of the nucleus accumbens and infralimbic cortex can be observed in rats engaged in reward-based learning tasks (Francois et al. 2014), as shown in Fig. 3. In a similar manner, O_2 amperometry has been used to demonstrate dissociated activity of dorsal and ventral hippocampus during spatial processing and anxiety behaviours (McHugh et al. 2011) and also between the amygdala and hippocampus as a response to different events during fear conditioning (McHugh et al. 2013). Amperometric O_2 signals measured from the amygdala during fear conditioning are consistent with temporal difference models of aversive learning (McHugh et al. 2014). Altogether, such findings help to bridge the gap between the types of mechanistic, neural hypotheses that are more typically tested in rodent studies with those of human studies. The translational power of the O_2 amperometric technique was recently highlighted by a study that measured PFC O_2 amperometric signals in rodents performing a sustained attention task, comparing these effects to both rat PFC choline and human BOLD signals (Howe et al. 2013). Sequence-specific transient O_2 responses were observed in the rat O_2 amperometric signal, paralleling the rodent cholinergic and human fMRI results. These three different lines of evidence converged to indicate that prefrontal cortical activity is engaged during attention-demanding tasks in shifts from monitoring to

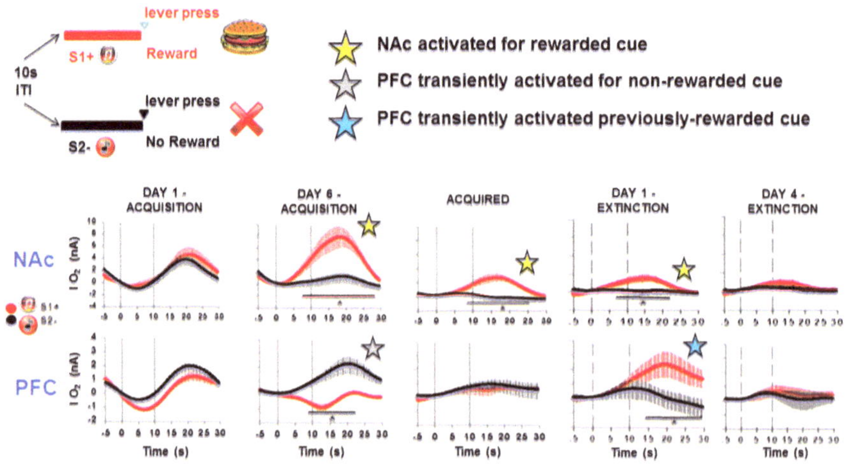

Fig. 3 Regionally specific O_2 responses during the performance of reward-based learning tasks. Different patterns of activation of the nucleus accumbens (NAc) and the infralimbic prefrontal cortex (PFC) were observed in rats performing acquisition and extinction of a cued lever-pressing task requiring discrimination between a rewarded and non-rewarded cue (*top left*). Activation of NAc was specifically observed following rewarded cue onset during the entire acquisition phase and also during the first days of extinction. In contrast, PFC activated only during the earliest periods of acquisition and extinction, more specifically to the non-rewarded cue (modified from Francois et al. 2014)

cue-associated processing and that human BOLD imaging findings can potentially be related back to rodent choline measurements via in vivo O_2 amperometry. Finally, as well as regionally specific dissociations of O_2 signals occurring during tasks, it has also been demonstrated that pharmacological and event-related O_2 signals can also dissociate (Francois et al. 2015). In a study where anterior cingulate cortex O_2 amperometric signal changes were measured during performance of a simple reaction time task, it was observed that the mGluR2/3 receptor agonist LY379268 could attenuate a task-independent increase in O_2 signal caused by ketamine, but could not attenuate simultaneously measured, task-dependent, event-related signals. A functional dissociation of drug effects on task-independent and event-related signals might prove to be very important for considerations of predictive efficacy of novel drug targets.

3.4 *Functional Connectivity and Resting-State O_2 Amperometry*

A focus of recent BOLD imaging paradigms has been that of intrinsic brain activity (Fox and Raichle 2007; Biswal et al. 1995) where functional connectivity is inferred between brain regions on the basis of patterns of correlation as subjects are scanned during a passive "resting" state. In such resting-state fMRI (rsfMRI) studies, temporal correlations in (low-frequency) BOLD signal fluctuations are readily observed and have led to the identification of several resting-state networks in the brain (Fox et al. 2005). Importantly, there is some evidence that resting-state networks and related measures of functional connectivity are impacted by a number of neuropsychiatric diseases (Fox and Greicius 2010) and may provide a novel parameter sensitive to pharmacological agents. Distinct task-independent intrinsic connectivity networks have also been demonstrated in rodents using BOLD fMRI (Lu et al. 2012; Becerra et al. 2011; Sforazzini et al. 2014b). Such work is now leading to cross-species studies of pharmacological modulation of regional functional connectivity (Zhu et al. 2013; Gass et al. 2014), most notably at present in the context of translational biomarkers of psychosis and antipsychotic efficacy (Smucny et al. 2014). The degree to which functional connectivity in the rodent can be modulated by specific behavioural states and drugs has been recently extended to the *in vivo* oxygen amperometry technique, allowing cross-species translation of functional connectivity measures beyond those fMRI BOLD measurements possible in anaesthetized animals. In this regard, O_2 amperometric functional connectivity measures have been shown to be differently affected by typical and atypical antipsychotics, where for instance haloperidol and clozapine have dissociable effects on ketamine-induced increases in O_2 signal magnitude and coherence between the medial prefrontal cortex and ventral striatum in the rat (Li et al. 2014). These findings are comparable to previous human studies (De Simoni et al. 2013; Driesen et al. 2013; Doyle et al. 2013b; Joules et al. 2015) and point to distinct mechanistic differences between different classes of antipsychotic drugs.

Resting-state networks are by no means static structures. A variety of distributed networks where preferential functional connectivity is exhibited between brain regions (such as the default mode network) spontaneously increase and decrease their activity and their within- and between-network connectivity dependent on ongoing events. However, dynamic aspects of resting-state networks that depend on ongoing behaviours are very difficult to assess in rodent BOLD fMRI studies, and the notion of a rest condition as a sufficient or appropriate control (Buckner and Vincent 2007), or perhaps even what constitutes a rest condition in a rodent is an important experimental question still to be answered. As a first step, O_2 amperometry has been used to demonstrate dynamic differences in functional connectivity between pairings of brain regions in freely moving rats as they engage in periods of instrumental responding for food reward (Li et al. 2015). In this study, the foci were a node pair in the rodent homologue of the default mode network (DMN) (Lu et al. 2012; Schwarz et al. 2013b; Sforazzini et al. 2014b), and a node pair in a distinct lateral cortical network (LCN) (Schwarz et al. 2013a). As illustrated in Fig. 4, baseline levels of functional connectivity were greater for those node pairs within the DMN and LCN compared to pairings measured between the networks, a finding consistent with the rsfMRI results. The ability to couple measurements of functional connectivity with behaviour using O_2 amperometry revealed that the DMN, but not the LCN, was sensitive to behavioural state, where decreases in functional connectivity occurred as animals engaged in instrumental action (compared to periods

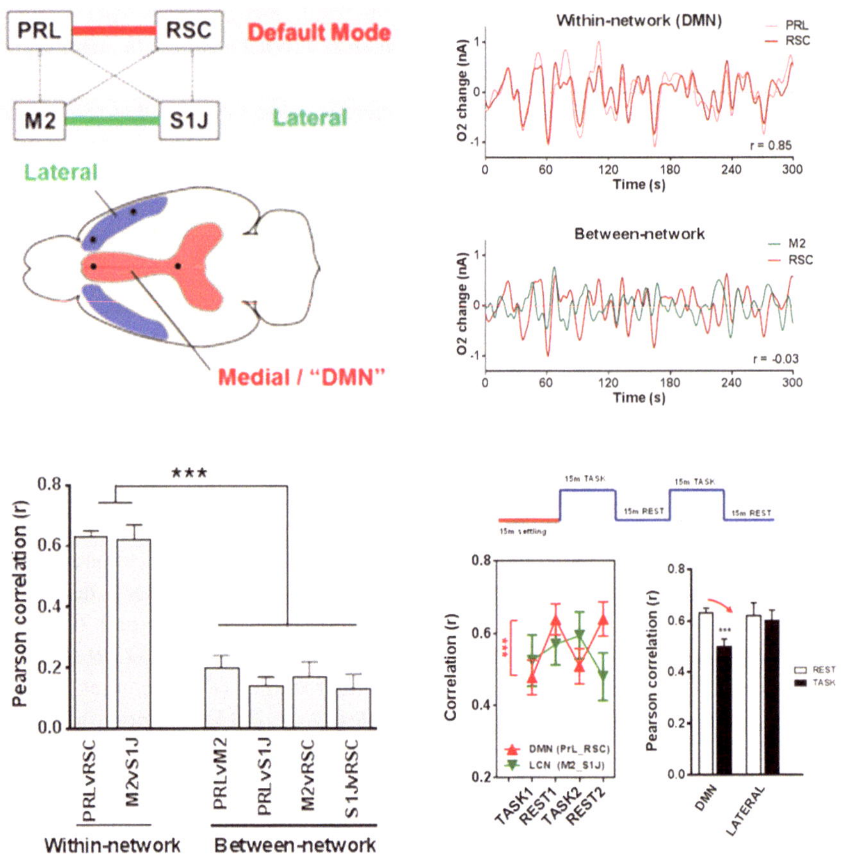

Fig. 4 Task-induced modulation of intrinsic O_2 functional connectivity networks in the behaving rat. O_2 probes were used to target two nodes of the default mode network and lateral cortical network (*top left*), allowing measurement of 2 within- and 4 between-network connectivity pairs. From observation of raw regional O_2 signals filtered between 0.01 and 0.1 Hz from one representative animal over 5 min at rest (*top right*), highly synchronous fluctuations can be observed from a "within-network" pairing (*top*), but not a "between-network" pairing. Within- and between-network differences can be neatly summarized through the use of Pearson's correlation values (*bottom left*). Task-induced modulation of functional connectivity is readily observed from the DMN pairing, but not the LCN pairing, using a 15-min task/rest block design (*bottom right*). Broadband correlations showed significant task versus rest differences selectively for DMN connectivity (modified from Li et al. 2015). Abbreviations: *DMN* default mode network; *LCN* lateral cortical network; *M2* secondary motor cortex; *PRL* prelimbinc cortex; *RSC* retrosplenial cortex; *S1J* primary somatosensory cortex, jaw representation

of spontaneous, unscheduled behaviour). There is great potential to further expand this work by utilizing comparisons of more complex task states and/or by probing node pairs within different networks. Such work should offer significant opportunities to translate between rodent functional connectivity and human rsfMRI studies.

4 Outstanding Issues and Future Directions

4.1 Physiological Equivalence of Signals

The application and validation of O_2 amperometry technology is still nascent and some important issues concerning its utilization for translational studies in neuropsychiatric drug discovery are important to bear in mind. Caution must be applied at present when considering the exact physiological equivalence of BOLD fMRI and O_2 amperometry and how this may be influenced by pharmacology. Both signals indirectly index regional neuronal activity via the haemodynamic processes of neurovascular coupling and therefore are susceptible to factors that might alter CBF or neurovascular coupling. These effects may be completely unrelated to neural activity. While it is recognized that drugs may interact and potentially confound a BOLD response at points in the neurovascular response distal to neural activity (Hyder et al. 2002; Iannetti and Wise 2007), an O_2 amperometry signal has a slightly different physiological basis which may not always be as equivalently responsive to these effects as BOLD. In addition, it may be important to consider how different disease states may also potentially disturb neurovascular coupling mechanisms in different brain regions or have divergent effects on the different physiologies underlying BOLD and O_2 amperometric signals.

4.2 Working with the Right Types of Assay

The requirement for translational validity of behavioural assays between rodents and humans has long been a recognized need for neuropsychiatric drug discovery research. Predictive validity of animal research often critically depends on construct validity, that is the ability to unequivocally assess the same psychological construct across species. Such efforts are particularly well-described in the field of schizophrenia, where exercises such as the "Cognitive Neuroscience Treatment Research to Improve Cognition in Schizophrenia (CNTRICS)" initiative identified viable translational assays for both humans and rodents capable of interrogating different cognitive domains impacted in schizophrenia (Carter and Barch 2007). Likewise, many aspects of commercially available human cognitive test batteries, such as CANTAB (Sahakian and Owen 1992), have also been back-translated for use in rodents (Horner et al. 2013; Mar et al. 2013; Oomen et al. 2013). Most, if not all, of the human tasks can be made "scanner-friendly", such that they can be performed by subjects using simple one-handed button or joystick controls to minimize movement confounds. The rodent homologues of these tasks, however, typically require the animal to perform actions in operant or touchscreen boxes, or in mazes or arenas. As multi-channel O_2 amperometry studies currently require tethered conditions, this may limit the extent to which some of the larger or more complex testing environments can currently be employed in this context. Also, while attempts to identify homologous cognitive tasks between species have rightfully focused on the equivalence of the underlying psychological constructs, neuroimaging and amperometry studies also require additional consideration of statistical and powering limitations. Both raw BOLD and O_2 amperometric signals are inherently noisy and it can be difficult to infer any event-related change from a single or low number of events presented during a test session. Accordingly, it is a necessary requirement that events of interest can be presented enough times to allow sufficient averaging to increase signal:noise ratio, yet still preserve the psychological characteristics of the task. For instance, for incentive delay and simple response latency tasks, both of which have been successfully used in event-related O_2 amperometry studies, cues of interest are typically presented 30–40 times within a single test session. Consideration of the need for additional control conditions in amperometry-friendly tasks is also important, for instance to rule out movement artefacts or confounds generated by appetitively driven behaviour. Tests of spontaneous behaviour (e.g. assays based on innate tendencies to explore environments or objects) are not especially suited to this type of study due to the lack of objective "event" with which to lock analysis of signals against.

Functional connectivity or resting-state O_2 amperometry studies raise other types of issues to consider with regard to the appropriateness of assays and study designs. By definition, determination of functional connectivity involves measurement of correlation of ultra-low frequencies (0.01–0.1 Hz) of O_2 signal fluctuation. In order to achieve this, animals ideally need to engage in sustained blocks of behavioural activity for 10–15 min at a time, where these blocks are contrasted to a control condition. Clearly, not all behavioural assays will be easy to administer in this context. So far in such block design studies, the control condition employed has been periods of unscheduled, spontaneous behaviour. Future work should consider the increased precision of interpretation that may be offered by the contrast of two task conditions that differ only (within practical reason) by the psychological parameter of interest and analysis methodologies that support dynamic changes in connectivity and shorter task-block frequencies. However, many cognitive tasks administered to rodents involve presentation of several trials in a short session with fairly regular inter-trial intervals, where it is not uncommon for a completion of a single trial cycle to fall in a 30–60 s range. It needs to be carefully considered therefore whether the architecture of a behavioural task may have the potential to cause an artefact in a frequency range that might be mistaken for a functional connectivity effect.

4.3 Testing the Right Regions

One of the main limitations of O_2 amperometry is that, unlike fMRI, it is not a whole-brain technique and so it is not possible to monitor activation or deactivation in all brain regions simultaneously. Presently, it has been feasible to implant a maximum of 4 working electrodes and therefore record tissue oxygen simultaneously from this number of different brain regions. While this number may increase slightly with future technical improvements, O_2 amperometry will in practice always remain a "region of interest" technique, where some degree of a priori knowledge is ideally required to decide where to place electrodes during any particular study. Also, an important difference from BOLD fMRI studies is that confounding signals from white matter or extraparenchymal tissue are not typically available, limiting the ability to correct for physiological, non-neural, contributions to the O_2 signals. For event-related studies, it may be important to use one or more of the working electrodes to record from control regions to allow more confident conclusions of regional specificity of effect. Similarly, in functional connectivity studies, it will not be possible at present to sample from all nodes of any identified resting-state network within the same animal, so judicious choice of appropriate node pairs within and between networks will have to be matched to the experimental hypothesis under question.

5 Conclusion

Concerns have been raised regarding the perceived failure rate of recent drug discovery efforts for neuropsychiatric disease. In particular, the translation of findings from rodent to human studies has been criticized as lacking cross-species measures that demonstrate test drugs of interest engage and stimulate their targets equivalently. fMRI methodologies seem well suited to potentially provide measures of neuronal activation in human studies that might assist in the determination of translational validity of drug effects. Back-translation of imaging findings from humans to rodents has therefore been of intense current interest. While several techniques are being explored in this context, in vivo O_2 amperometry represents a very promising methodology which may offer a surrogate of the BOLD fMRI signal in rats. The O_2 amperometry signal is effectively real-time and of comparable spatial resolution to the BOLD fMRI signal, yet can be utilized in freely moving, behaving rats. The technique opens up a new realm of preclinical translational imaging biomarker opportunities that would not have been possible otherwise in standard rat fMRI studies. Both BOLD fMRI and O_2 amperometric signals are indirect measures of neuronal activation that depend on neurovascular coupling responses. BOLD is a magnetic contrast measure of the ratio of haemoglobin: deoxyhaemoglobin in the local microvasculature, while the amperometric signal is an electrochemical current generated by the specific reduction of local oxygen in the brain parenchyma. Despite these slightly different origins, evidence generated to date suggests that O_2 amperometry can serve as a valid surrogate of the BOLD signal across a number of different testing paradigms: from pharmacological and event-related imaging, to fMRI measures of functional connectivity and resting-state networks. Ultimately, the best measure of success of the O_2 amperometry technique in the context of neuropsychiatric drug discovery would be the successful test of a hypothesis of drug target engagement or efficacy in a rodent disease model that anticipated the effect observed in a human trial. By acting as an experimental “missing link” between the basic science of rodent studies and the clinical outcome, the technique has potential to increase the probability of success of early-stage clinical trials and hopefully relegate “failed” trials to drug discovery history.

References

American Psychiatric Association (2014) Diagnostic and statistical manual of mental disorders: DSM-5, 5th edn. American Psychiatric Association

Ances BM, Leontiev O, Perthen JE, Liang C, Lansing AE, Buxton RB (2008) Regional differences in the coupling of cerebral blood flow and oxygen metabolism changes in response to activation: implications for BOLD-fMRI. NeuroImage 39(4):1510–1521

Baker S, Chin CL, Basso AM, Fox GB, Marek GJ, Day M (2012) Xanomeline modulation of the blood oxygenation level-dependent signal in awake rats: development of pharmacological magnetic resonance imaging as a translatable pharmacodynamic biomarker for central activity and dose selection. J Pharmacol Exp Ther 341(1):263–273

Bandettini PA, Wong EC, Hinks RS, Tikofsky RS, Hyde JS (1992) Time course EPI of human brain function during task activation. Magn Reson Med 25(2):390–397

Becerra L, Harter K, Gonzalez RG, Borsook D (2006) Functional magnetic resonance imaging measures of the effects of morphine on central nervous system circuitry in opioid-naive healthy volunteers. Anesth Analg 103(1):208–216 (table of contents)

Becerra L, Pendse G, Chang PC, Bishop J, Borsook D (2011) Robust reproducible resting state networks in the awake rodent brain. PLoS ONE 6(10):e25701

Berger H (1933) Über das Elektrenkephalogramm des Menschen. Eur Arch Psychiatry Clin Neurosci 98(1):231–254

Biswal B, Yetkin FZ, Haughton VM, Hyde JS (1995) Functional connectivity in the motor cortex of resting human brain using echo-planar MRI. Magn Reson Med 34(4):537–541

Bolger FB, Lowry JP (2005) Brain tissue oxygen: *in vivo* monitoring with carbon paste electrodes. Sensors 5:473–487

Bolger FB, McHugh SB, Bennett R, Li J, Ishiwari K, Francois J, Conway MW, Gilmour G, Bannerman DM, Fillenz M, Tricklebank M, Lowry JP (2011) Characterisation of carbon paste electrodes for real-time amperometric monitoring of brain tissue oxygen. J Neurosci Methods 195(2):135–142

Breiter HC, Gollub RL, Weisskoff RM, Kennedy DN, Makris N, Berke JD, Goodman JM, Kantor HL, Gastfriend DR, Riorden JP, Mathew RT, Rosen BR, Hyman SE (1997) Acute effects of cocaine on human brain activity and emotion. Neuron 19(3):591–611

Buckner RL, Vincent JL (2007) Unrest at rest: default activity and spontaneous network correlations. NeuroImage 37(4):1091–1096; discussion 1097–1099

Buxton RB, Wong EC, Frank LR (1998) Dynamics of blood flow and oxygenation changes during brain activation: the balloon model. Magn Reson Med 39(6):855–864

Caesar K, Thomsen K, Lauritzen M (2003) Dissociation of spikes, synaptic activity, and activity-dependent increments in rat cerebellar blood flow by tonic synaptic inhibition. Proc Natl Acad Sci USA 100(26):16000–16005

Carter CS, Barch DM (2007) Cognitive neuroscience-based approaches to measuring and improving treatment effects on cognition in schizophrenia: the CNTRICS initiative. Schizophr Bull 33(5):1131–1137

Chin CL, Upadhyay J, Marek GJ, Baker SJ, Zhang M, Mezler M, Fox GB, Day M (2011) Awake rat pharmacological magnetic resonance imaging as a translational pharmacodynamic biomarker: metabotropic glutamate 2/3 agonist modulation of ketamine-induced blood oxygenation level dependence signals. J Pharmacol Exp Ther 336(3):709–715

Cuthbert BN (2015) Research Domain Criteria: toward future psychiatric nosologies. Dialogues Clin Neurosci 17(1):89–97

Dale N, Hatz S, Tian F, Llaudet E (2005) Listening to the brain: microelectrode biosensors for neurochemicals. Trends Biotechnol 23(8):420–428

Damadian R, Goldsmith M, Minkoff L (1977) NMR in cancer: XVI. FONAR image of the live human body. Physiol Chem Phys 9(1):97–100, 108

De Simoni S, Schwarz AJ, O'Daly OG, Marquand AF, Brittain C, Gonzales C, Stephenson S, Williams SC, Mehta MA (2013) Test-retest reliability of the BOLD pharmacological MRI response to ketamine in healthy volunteers. NeuroImage 64:75–90

Deakin JF, Lees J, McKie S, Hallak JE, Williams SR, Dursun SM (2008) Glutamate and the neural basis of the subjective effects of ketamine: a pharmaco-magnetic resonance imaging study. Arch Gen Psychiatry 65(2):154–164

Doyle OM, De Simoni S, Schwarz AJ, Brittain C, O'Daly OG, Williams SC, Mehta MA (2013a) Quantifying the attenuation of the ketamine pharmacological magnetic resonance imaging response in humans: a validation using antipsychotic and glutamatergic agents. J Pharmacol Exp Ther 345(1):151–160

Doyle OM, De Simoni S, Schwarz AJ, Brittain C, O'Daly OG, Williams SC, Mehta MA (2013b) Quantifying the attenuation of the ketamine pharmacological magnetic resonance imaging response in humans: a validation using antipsychotic and glutamatergic agents. J Pharmacol Exp Ther 345(1):151–160

Driesen NR, McCarthy G, Bhagwagar Z, Bloch M, Calhoun V, D'Souza DC, Gueorguieva R, He G, Ramachandran R, Suckow RF, Anticevic A, Morgan PT, Krystal JH (2013) Relationship of resting brain hyperconnectivity and schizophrenia-like symptoms produced by the NMDA receptor antagonist ketamine in humans. Mol Psychiatry 18(11):1199–1204

Duncan GE, Leipzig JN, Mailman RB, Lieberman JA (1998a) Differential effects of clozapine and haloperidol on ketamine-induced brain metabolic activation. Brain Res 812(1–2):65–75

Duncan GE, Moy SS, Knapp DJ, Mueller RA, Breese GR (1998b) Metabolic mapping of the rat brain after subanesthetic doses of ketamine: potential relevance to schizophrenia. Brain Res 787(2):181–190

Duncan GE, Miyamoto S, Leipzig JN, Lieberman JA (2000) Comparison of the effects of clozapine, risperidone, and olanzapine on ketamine-induced alterations in regional brain metabolism. J Pharmacol Exp Ther 293(1):8–14

Duong TQ, Kim DS, Ugurbil K, Kim SG (2001) Localized cerebral blood flow response at submillimeter columnar resolution. Proc Natl Acad Sci USA 98(19):10904–10909

Engel AK, Moll CK, Fried I, Ojemann GA (2005) Invasive recordings from the human brain: clinical insights and beyond. Nat Rev Neurosci 6(1):35–47

Ferrari L, Turrini G, Crestan V, Bertani S, Cristofori P, Bifone A, Gozzi A (2012) A robust experimental protocol for pharmacological fMRI in rats and mice. J Neurosci Methods 204 (1):9–18

Fox MD, Greicius M (2010) Clinical applications of resting state functional connectivity. Frontiers Syst Neurosci 4:19

Fox MD, Raichle ME (2007) Spontaneous fluctuations in brain activity observed with functional magnetic resonance imaging. Nat Rev Neurosci 8(9):700–711

Fox MD, Snyder AZ, Vincent JL, Corbetta M, Van Essen DC, Raichle ME (2005) The human brain is intrinsically organized into dynamic, anticorrelated functional networks. Proc Natl Acad Sci USA 102(27):9673–9678

Francois J, Conway MW, Lowry JP, Tricklebank MD, Gilmour G (2012) Changes in reward-related signals in the rat nucleus accumbens measured by in vivo oxygen amperometry are consistent with fMRI BOLD responses in man. NeuroImage 60(4):2169–2181

Francois J, Huxter J, Conway MW, Lowry JP, Tricklebank MD, Gilmour G (2014) Differential contributions of infralimbic prefrontal cortex and nucleus accumbens during reward-based

learning and extinction. J Neurosci (the official journal of the Society for Neuroscience) 34 (2):596–607
Francois J, Gastambide F, Conway MW, Tricklebank M, Gilmour G (2015) Dissociation of mGlu2/3 agonist effects on ketamine-induced regional and event-related oxygen signals. Psychopharmacology 232:4219–4229
Gadian D (1996) NMR and its applications to living systems, 2nd edn. Oxford University Press, UK
Gass N, Schwarz AJ, Sartorius A, Schenker E, Risterucci C, Spedding M, Zheng L, Meyer-Lindenberg A, Weber-Fahr W (2014) Sub-anesthetic ketamine modulates intrinsic BOLD connectivity within the hippocampal-prefrontal circuit in the rat. Neuropsychopharmacology (official publication of the American College of Neuropsychopharmacology) 39(4):895–906
Goense JB, Logothetis NK (2008) Neurophysiology of the BOLD fMRI signal in awake monkeys. Curr Biol (CB) 18(9):631–640
Gozzi A, Large CH, Schwarz A, Bertani S, Crestan V, Bifone A (2008) Differential effects of antipsychotic and glutamatergic agents on the phMRI response to phencyclidine. Neuropsychopharmacology (official publication of the American College of Neuropsychopharmacology) 33(7):1690–1703
Hitchman ML (1978) Measurement of dissolved oxygen. Krieger Publishing Company
Horner AE, Heath CJ, Hvoslef-Eide M, Kent BA, Kim CH, Nilsson SR, Alsio J, Oomen CA, Holmes A, Saksida LM, Bussey TJ (2013) The touchscreen operant platform for testing learning and memory in rats and mice. Nat Protoc 8(10):1961–1984
Howe WM, Berry AS, Francois J, Gilmour G, Carp JM, Tricklebank M, Lustig C, Sarter M (2013) Prefrontal cholinergic mechanisms instigating shifts from monitoring for cues to cue-guided performance: converging electrochemical and fMRI evidence from rats and humans. J Neurosci (the official journal of the Society for Neuroscience) 33(20):8742–8752
Huettel SA, McKeown MJ, Song AW, Hart S, Spencer DD, Allison T,.McCarthy G (2004) Linking hemodynamic and electrophysiological measures of brain activity: evidence from functional MRI and intracranial field potentials. Cerebral Cortex (New York, 1991) 14(2):165–173
Hyder F, Rothman DL, Shulman RG (2002) Total neuroenergetics support localized brain activity: implications for the interpretation of fMRI. Proc Natl Acad Sci USA 99(16):10771–10776
Iannetti GD, Wise RG (2007) BOLD functional MRI in disease and pharmacological studies: room for improvement? Magn Reson Imaging 25(6):978–988
Jenkins BG (2012) Pharmacologic magnetic resonance imaging (phMRI): imaging drug action in the brain. NeuroImage 62(2):1072–1085
Joules R, Doyle OM, Schwarz AJ, O'Daly OG, Brammer M, Williams SC, Mehta MA (2015) Ketamine induces a robust whole-brain connectivity pattern that can be differentially modulated by drugs of different mechanism and clinical profile. Psychopharmacology 232:4205–4218
Kane DA, O'Neill RD (1998) Major differences in the behaviour of carbon paste and carbon fibre electrodes in a protein-lipid matrix: implications for voltammetry in vivo. Analyst 123 (12):2899–2903
Knutson B, Cooper JC (2005) Functional magnetic resonance imaging of reward prediction. Curr Opin Neurol 18(4):411–417
Knutson B, Westdorp A, Kaiser E, Hommer D (2000) FMRI visualization of brain activity during a monetary incentive delay task. NeuroImage 12(1):20–27
Knutson B, Adams CM, Fong GW, Hommer D (2001) Anticipation of increasing monetary reward selectively recruits nucleus accumbens. J Neurosci (the official journal of the Society for Neuroscience) 21(16):RC159
Krystal JH, Karper LP, Seibyl JP, Freeman GK, Delaney R, Bremner JD, Heninger GR, Bowers MB Jr, Charney DS (1994) Subanesthetic effects of the noncompetitive NMDA antagonist, ketamine, in humans. Psychotomimetic, perceptual, cognitive, and neuroendocrine responses. Arch Gen Psychiatry 51(3):199–214

Kufahl PR, Li Z, Risinger RC, Rainey CJ, Wu G, Bloom AS, Li SJ (2005) Neural responses to acute cocaine administration in the human brain detected by fMRI. NeuroImage 28(4):904–914
Kwong KK, Belliveau JW, Chesler DA, Goldberg IE, Weisskoff RM, Poncelet BP, Kennedy DN, Hoppel BE, Cohen MS, Turner R (1992) Dynamic magnetic resonance imaging of human brain activity during primary sensory stimulation. Proc Natl Acad Sci USA 89(12):5675–5679
Lahti AC, Koffel B, LaPorte D, Tamminga CA (1995) Subanesthetic doses of ketamine stimulate psychosis in schizophrenia. Neuropsychopharmacology (official publication of the American College of Neuropsychopharmacology) 13(1):9–19
Langsjo JW, Kaisti KK, Aalto S, Hinkka S, Aantaa R, Oikonen V, Sipila H, Kurki T, Silvanto M, Scheinin H (2003) Effects of subanesthetic doses of ketamine on regional cerebral blood flow, oxygen consumption, and blood volume in humans. Anesthesiology 99(3):614–623
Langsjo JW, Salmi E, Kaisti KK, Aalto S, Hinkka S, Aantaa R, Oikonen V, Viljanen T, Kurki T, Silvanto M, Scheinin H (2004) Effects of subanesthetic ketamine on regional cerebral glucose metabolism in humans. Anesthesiology 100(5):1065–1071
Lauterbur PC (1973) Image Formation by Induced Local Interactions: examples of employing nuclear magnetic resonance. Nature 242:190–191
Leppa M, Korvenoja A, Carlson S, Timonen P, Martinkauppi S, Ahonen J, Rosenberg PH, Aronen HJ, Kalso E (2006) Acute opioid effects on human brain as revealed by functional magnetic resonance imaging. NeuroImage 31(2):661–669
Leslie RA, James MF (2000) Pharmacological magnetic resonance imaging: a new application for functional MRI. Trends Pharmacol Sci 21(8):314–318
Li J, Ishiwari K, Conway MW, Francois J, Huxter J, Lowry JP, Schwarz AJ, Tricklebank M, Gilmour G (2014) Dissociable effects of antipsychotics on ketamine-induced changes in regional oxygenation and inter-regional coherence of low frequency oxygen fluctuations in the rat. Neuropsychopharmacology (official publication of the American College of Neuropsychopharmacology) 39(7):1635–1644
Li J, Martin S, Tricklebank MD, Schwarz AJ, Gilmour G (2015) Task-induced modulation of intrinsic functional connectivity networks in the behaving rat. J Neurosci (the official journal of the Society for Neuroscience) 35(2):658–665
Liska A, Galbusera A, Schwarz AJ, Gozzi A (2015) Functional connectivity hubs of the mouse brain. NeuroImage 115:281–291
Littlewood CL, Jones N, O'Neill MJ, Mitchell SN, Tricklebank M, Williams SC (2006) Mapping the central effects of ketamine in the rat using pharmacological MRI. Psychopharmacology 186 (1):64–81
Logothetis NK, Pauls J, Augath M, Trinath T, Oeltermann A (2001) Neurophysiological investigation of the basis of the fMRI signal. Nature 412(6843):150–157
Lowry JP, Boutelle MG, O'Neill RD, Fillenz M (1996) Characterization of carbon paste electrodes in vitro for simultaneous amperometric measurement of changes in oxygen and ascorbic acid concentrations in vivo. Analyst 121(6):761–766
Lu H, Zou Q, Gu H, Raichle ME, Stein EA, Yang Y (2012) Rat brains also have a default mode network. Proc Natl Acad Sci USA 109(10):3979–3984
Malhotra AK, Pinals DA, Weingartner H, Sirocco K, Missar CD, Pickar D, Breier A (1996) NMDA receptor function and human cognition: the effects of ketamine in healthy volunteers. Neuropsychopharmacology (official publication of the American College of Neuropsychopharmacology) 14(5):301–307
Mandeville JB, Marota JJ, Kosofsky BE, Keltner JR, Weissleder R, Rosen BR, Weisskoff RM (1998) Dynamic functional imaging of relative cerebral blood volume during rat forepaw stimulation. Magn Reson Med 39(4):615–624
Mandeville JB, Marota JJ, Ayata C, Moskowitz MA, Weisskoff RM, Rosen BR (1999) MRI measurement of the temporal evolution of relative CMRO(2) during rat forepaw stimulation. Magn Reson Med 42(5):944–951
Mansfield P, Maudsley AA (1976) Line scan proton spin imaging in biological structures by NMR. Phys Med Biol 21(5):847–852

Mar AC, Horner AE, Nilsson SR, Alsio J, Kent BA, Kim CH, Holmes A, Saksida LM, Bussey TJ (2013) The touchscreen operant platform for assessing executive function in rats and mice. Nat Protoc 8(10):1985–2005

Markou A, Chiamulera C, Geyer MA, Tricklebank M, Steckler T (2009) Removing obstacles in neuroscience drug discovery: the future path for animal models. Neuropsychopharmacology (official publication of the American College of Neuropsychopharmacology) 34(1):74–89

Mathiesen C, Caesar K, Akgoren N, Lauritzen M (1998) Modification of activity-dependent increases of cerebral blood flow by excitatory synaptic activity and spikes in rat cerebellar cortex. J Physiol 512 (Pt 2):555–566

Mathiesen C, Caesar K, Lauritzen M (2000) Temporal coupling between neuronal activity and blood flow in rat cerebellar cortex as indicated by field potential analysis. J Physiol 523(Pt 1):235–246

McHugh SB, Fillenz M, Lowry JP, Rawlins JN, Bannerman DM (2011) Brain tissue oxygen amperometry in behaving rats demonstrates functional dissociation of dorsal and ventral hippocampus during spatial processing and anxiety. Eur J Neurosci 33(2):322–337

McHugh SB, Marques-Smith A, Li J, Rawlins JN, Lowry J, Conway M, Gilmour G, Tricklebank M, Bannerman DM (2013) Hemodynamic responses in amygdala and hippocampus distinguish between aversive and neutral cues during Pavlovian fear conditioning in behaving rats. Eur J Neurosci 37(3):498–507

McHugh SB, Barkus C, Huber A, Capitao L, Lima J, Lowry JP, Bannerman DM (2014) Aversive prediction error signals in the amygdala. J Neurosci (the official journal of the Society for Neuroscience) 34(27):9024–9033

McKie S, Richardson P, Elliott R, Vollm BA, Dolan MC, Williams SR, Anderson IM, Deakin JF (2011) Mirtazapine antagonises the subjective, hormonal and neuronal effects of m-chlorophenylpiperazine (mCPP) infusion: a pharmacological-challenge fMRI (phMRI) study. NeuroImage 58(2):497–507

Morgan P, Van Der Graaf PH, Arrowsmith J, Feltner DE, Drummond KS, Wegner CD, Street SD (2012) Can the flow of medicines be improved? Fundamental pharmacokinetic and pharmacological principles toward improving Phase II survival. Drug Discovery Today 17 (9–10):419–424

Mosso A (1884) Applicazione della bilancia allo studio della circolazione sanguigna dell'uomo. Atti R Accad Lincei Mem Cl Sci Fis Mat Nat XIX:531–543

Newcomer JW, Farber NB, Jevtovic-Todorovic V, Selke G, Melson AK, Hershey T, Craft S, Olney JW (1999) Ketamine-induced NMDA receptor hypofunction as a model of memory impairment and psychosis. Neuropsychopharmacology (official publication of the American College of Neuropsychopharmacology) 20(2):106–118

Offenhauser N, Thomsen K, Caesar K, Lauritzen M (2005) Activity-induced tissue oxygenation changes in rat cerebellar cortex: interplay of postsynaptic activation and blood flow. J Physiol 565(Pt 1):279–294

Ogawa S, Lee TM, Kay AR, Tank DW (1990) Brain magnetic resonance imaging with contrast dependent on blood oxygenation. Proc Natl Acad Sci USA 87(24):9868–9872

O'Neill RD, Lowry JP, Mas M (1998) Monitoring brain chemistry in vivo: voltammetric techniques, sensors, and behavioral applications. Crit Rev Neurobiol 12(1–2):69–127

Oomen CA, Hvoslef-Eide M, Heath CJ, Mar AC, Horner AE, Bussey TJ, Saksida LM (2013) The touchscreen operant platform for testing working memory and pattern separation in rats and mice. Nat Protoc 8(10):2006–2021

Paul SM, Mytelka DS, Dunwiddie CT, Persinger CC, Munos BH, Lindborg SR, Schacht AL (2010) How to improve R&D productivity: the pharmaceutical industry's grand challenge. Nat Rev Drug Discovery 9(3):203–214

Pawela CP, Biswal BB, Cho YR, Kao DS, Li R, Jones SR, Schulte ML, Matloub HS, Hudetz AG, Hyde JS (2008) Resting-state functional connectivity of the rat brain. Magn Reson Med 59 (5):1021–1029

Sahakian BJ, Owen AM (1992) Computerized assessment in neuropsychiatry using CANTAB: discussion paper. J R Soc Med 85(7):399–402

Schwarz AJ, Gass N, Sartorius A, Risterucci C, Spedding M, Schenker E, Meyer-Lindenberg A, Weber-Fahr W (2013a) Anti-correlated cortical networks of intrinsic connectivity in the rat brain. Brain connectivity 3(5):503–511

Schwarz AJ, Gass N, Sartorius A, Zheng L, Spedding M, Schenker E, Risterucci C, Meyer-Lindenberg A, Weber-Fahr W (2013b) The low-frequency blood oxygenation level-dependent functional connectivity signature of the hippocampal-prefrontal network in the rat brain. Neuroscience 228:243–258

Sforazzini F, Bertero A, Dodero L, David G, Galbusera A, Scattoni ML, Pasqualetti M, Gozzi A (2014a) Altered functional connectivity networks in acallosal and socially impaired BTBR mice. Brain Struct Function

Sforazzini F, Schwarz AJ, Galbusera A, Bifone A, Gozzi A (2014b) Distributed BOLD and CBV-weighted resting-state networks in the mouse brain. NeuroImage 87:403–415

Shulman RG, Hyder F, Rothman DL (2002) Biophysical basis of brain activity: implications for neuroimaging. Q Rev Biophys 35(3):287–325

Sibson NR, McHugh SB, Lowry JP (2009) Combined fMRI and metabolic voltammetry in vivo: understanding the neurochemical basis of functional imaging signals. Br Neurosci Assoc Abstr 20:48

Smith AJ, Blumenfeld H, Behar KL, Rothman DL, Shulman RG, Hyder F (2002) Cerebral energetics and spiking frequency: the neurophysiological basis of fMRI. Proc Natl Acad Sci USA 99(16):10765–10770

Smucny J, Wylie KP, Tregellas JR (2014) Functional magnetic resonance imaging of intrinsic brain networks for translational drug discovery. Trends Pharmacol Sci 35(8):397–403

Stein EA, Pankiewicz J, Harsch HH, Cho JK, Fuller SA, Hoffmann RG, Hawkins M, Rao SM, Bandettini PA, Bloom AS (1998) Nicotine-induced limbic cortical activation in the human brain: a functional MRI study. Am J Psychiatry 155(8):1009–1015

Upadhyay J, Anderson J, Schwarz AJ, Coimbra A, Baumgartner R, Pendse G, George E, Nutile L, Wallin D, Bishop J, Neni S, Maier G, Iyengar S, Evelhoch JL, Bleakman D, Hargreaves R, Becerra L, Borsook D (2011) Imaging drugs with and without clinical analgesic efficacy. Neuropsychopharmacology (official publication of the American College of Neuropsychopharmacology) 36(13):2659–2673

Venton BJ, Zhang H, Garris PA, Phillips PE, Sulzer D, Wightman RM (2003) Real-time decoding of dopamine concentration changes in the caudate-putamen during tonic and phasic firing. J Neurochem 87(5):1284–1295

Vollm BA, de Araujo IE, Cowen PJ, Rolls ET, Kringelbach ML, Smith KA, Jezzard P, Heal RJ, Matthews PM (2004) Methamphetamine activates reward circuitry in drug naive human subjects. Neuropsychopharmacology (official publication of the American College of Neuropsychopharmacology) 29(9):1715–1722

WHO (1992) World Health Organisation: international statistical classification of diseases and related health problems, 10th Revision (ICD-10)

Zhao F, Wang P, Hendrich K, Kim SG (2005) Spatial specificity of cerebral blood volume-weighted fMRI responses at columnar resolution. NeuroImage 27(2):416–424

Zhu Y, Gao B, Hua J, Liu W, Deng Y, Zhang L, Jiang B, Zang Y (2013) Effects of methylphenidate on resting-state brain activity in normal adults: an fMRI study. Neurosci Bull 29(1):16–27

Zimmerman JB, Kennedy RT, Wightman RM (1992) Evoked neuronal activity accompanied by transmitter release increases oxygen concentration in rat striatum in vivo but not in vitro. J Cereb Blood Flow Metab (official journal of the International Society of Cerebral Blood Flow and Metabolism) 12(4):629–637

Cognitive Translation Using the Rodent Touchscreen Testing Approach

M. Hvoslef-Eide, S.R.O. Nilsson, L.M. Saksida and T.J. Bussey

Abstract The development of novel therapeutic avenues for the treatment of cognitive deficits in psychiatric and neurodegenerative disease is of high importance, yet progress in this field has been slow. One reason for this lack of success may lie in discrepancies between how cognitive functions are assessed in experimental animals and humans. In an attempt to bridge this translational gap, the rodent touchscreen testing platform is suggested as a translational tool. Specific examples of successful cross-species translation are discussed focusing on paired associate learning (PAL), the 5-choice serial reaction time task (5-CSRTT), the rodent continuous performance task (rCPT) and reversal learning. With ongoing research assessing the neurocognitive validity of tasks, the touchscreen approach is likely to become increasingly prevalent in translational cognitive research.

Keyword Translation · Cognition · Touchscreen · Rat · Mouse · Human

Contents

1 Introduction 424
1.1 The Touchscreen Testing Platform as a Translational Tool 424
1.2 Improved Identification of Cognitive Phenotypes Using a Touchscreen Test Battery 426
2 The Paired Associate Learning Task (PAL) 427
3 The 5-Choice Serial Reaction Time Task (5-CSRTT) 429
3.1 Alzheimer's Disease and Acetylcholine: Pharmacological and Genetic Translations 429
3.2 Serotonin Depletion Results in Parallel Deficits Across Species 430
3.3 Some Limitations of the 5-Choice Serial Reaction Time Task 431
4 The Rodent Continuous Performance Task (rCPT) 431
4.1 CPT and rCPT: Convergence by Functional Anatomy and Pharmacology 432
4.2 Animal Models on the rCPT: Parallels with Human CPT Data 432

M. Hvoslef-Eide (✉) · S.R.O. Nilsson · L.M. Saksida · T.J. Bussey
Department of Psychology, University of Cambridge, Cambridge CB2 3EB, UK
e-mail: mh473@cam.ac.uk

Curr Topics Behav Neurosci (2016) 28: 423–447
DOI 10.1007/7854_2015_5007

4.3 Vigilance Decrement in the rCPT ... 433
4.4 Relationship Between Task Parameters and Performance Consistent Across Species ... 433
5 Reversal Learning ... 434
5.1 Orbitofrontal and Dorsal Striatal Perturbations Cause Impairments of Touchscreen Reversal Learning in Rodents, Non-Human Primates, and Humans ... 435
5.2 Serotonergic Manipulations Affect Both Rodent and Human Touchscreen Reversal Learning ... 435
5.3 Cross-Species Impairments in Touchscreen Reversal Learning Following a Disc Large 2 Mutation ... 436
6 Limitations and Challenges ... 436
7 Conclusion ... 437
References ... 437

1 Introduction

Cognitive impairments are prevalent in most neuropsychiatric and neurodegenerative disorders, and these impairments can be predictive of the efficacy of therapeutic interventions (O'Carroll 2000; Barnett et al. 2005; Green 2006; Simon et al. 2007). Although the importance of cognitive dysfunction as a target for decreasing and preventing psychiatric morbidity has been repeatedly emphasised (e.g. Geyer and Tamminga 2004), few pharmacotherapeutic options have been developed (Weiss et al. 2002; Green 1996, 2006; Holthausen et al. 2007; Keefe et al. 2007). Preclinical research, including the use of behavioural paradigms in experimental animals, continues to indicate the relevance and putative efficacy of an ever-growing number of targets for cognitive enhancement in neuropsychiatric and neurodegenerative disorders. However, these targets are typically 'lost in translation'; targets with preclinical promise predominantly fail to yield success in clinical trials (Insel 2012). Furthermore, little is known about the mechanisms underpinning the few existing pharmacological treatments and their relation to clinical outcome, and attempts to link distinct cognitive dysfunctions with genetic loci have so far been largely unsuccessful (Insel and Wang 2010; Insel et al. 2013). The absence of data connecting mechanistic action to therapeutic efficacy remains a severe shortcoming, as such knowledge is crucial for developing more refined agents and establishing novel therapeutic avenues.

1.1 The Touchscreen Testing Platform as a Translational Tool

One source of this lack of success may be the mismatch between how cognitive functions are evaluated in experimental animals and humans. Increasingly in human studies, different cognitive domains are assessed with a battery of visual touchscreen-based tasks using similar stimuli and responses, thus facilitating between-task comparison and reducing confounds (Sahakian and Owen 1992; Levaux et al. 2007; Barnett et al. 2010). Conversely, in the rodent, comprehensive cognitive batteries are seldom employed, and different cognitive domains are assessed by tasks that typically vary widely in the nature of stimuli, responses, reinforcers, and testing environment. Moreover, these tasks often bear no resemblance to the tests used with humans. These factors reduce the likelihood of rodent and human studies assessing comparable cognitive functions, thus compromising the efficacy of translation.

In the light of these considerations, we and others have introduced and developed the touchscreen cognitive testing method, which has the potential to achieve more accurate, efficient, and reproducible phenotyping of rodents, and help bridge the 'translational divide' between animal and human studies of cognition. This method—as of now used by more than 130 laboratories—uses an automated operant chamber with a computer monitor for the presentation of visual stimuli and an infrared touchscreen assembly to record the animal's responses (Fig. 1). Liquid reinforcers or food pellets can be used as rewards. A battery of rodent touchscreen behavioural tasks that closely parallel human touchscreen tests has been developed (e.g. Hvoslef-Eide et al. 2015). How these touchscreen tasks can be used to assess a wide variety of cognitive domains is discussed elsewhere (Bussey et al. 2012;

Fig. 1 The rodent touchscreen operant chamber allows for the flexible presentation of visual stimuli at any location on the screen (image courtesy of Campden Instruments). Screen masks are used to frame the response locations relevant to a particular task. Rats or mice respond directly to the stimuli by breaking the infrared beams overlaying the touchscreen with their nose. Appetitive reinforcers are delivered in a reward magazine to the rear of the chamber

Horner et al. 2013; Mar et al. 2013; Oomen et al. 2013; Hvoslef-Eide et al. 2015). The tasks have high translational face validity. Although face validity does not guarantee translational neurocognitive validity (the same cognitive constructs and circuits mediating the tasks across species), minimising the methodological differences enables back- and forward-translational opportunities and improves the likelihood of neurocognitive validity of touchscreen tasks (Romberg et al. 2011; Nithianantharajah et al. 2012a, b; Talpos and Steckler 2013).

1.2 Improved Identification of Cognitive Phenotypes Using a Touchscreen Test Battery

The use of multiple tests is an important approach for the unconfounded identification of cognitive phenotypes. Performance across rodent behavioural tests depends on a variety of overlapping as well as more distinct cognitive, motoric, and motivational functions. Traditionally, rodent phenotyping has been accomplished by employing multiple tests in multiple different environments. This can include mazes and open fields to assess memory and spatial learning, bowl-digging protocols to assess discrimination learning and cognitive flexibility, and operant boxes to probe contingency learning, attentional function, and impulsive behaviours. This traditional approach differs from that used in computerised batteries of human cognitive evaluation in which all peripheral variables (environment, context, reward, response requirement, and sensory domain) are controlled, as far as possible, across tasks. Moreover, as opposed to the computer-controlled tests used with human participants, prevalent rodent paradigms are often experimenter-controlled and therefore time-consuming and labour-intensive (limiting throughput) and sensitive to unintended experimenter interference (reducing the likelihood of reproducibility). Thus, the aim of identifying translatable, reliable rodent cognitive phenotypes should benefit from automated high-throughput platforms where multiple cognitive domains can be assessed in a unified environment, similar to the approaches used in humans. The touchscreen platform provides this unified environment, and a comprehensive and expanding battery of tests is available (Bussey et al. 2012; Horner et al. 2013; Mar et al. 2013; Oomen et al. 2013; Hvoslef-Eide et al. 2015). Thus, the interpretation of impaired performance on, say, a spatial working memory task as a memory deficit can remain tentative until other interpretations, for example attentional function, can be ruled out. The inclusion of a touchscreen attentional task in the battery provides the required control, and interpretation is clear because the two tasks use identical stimuli, responses, and rewards; only the cognitive demand varies. Similarly, impairments on a paired associate learning (PAL) task could be due to deficits in visual processing rather than memory if performance on other visual tasks is compromised. In this way, a less confounded interpretation of the findings from all of the tasks in the battery can be made. This is important in the context of translation, as confidence in the cognitive phenotype is critical.

Here, we outline some examples of successful translation between rodent and human touchscreen studies. We have chosen examples from four individual touchscreen tests: paired associate learning (PAL), the 5-choice serial reaction time task (5-CSRTT), the more recently developed rodent continuous performance task (rCPT), and reversal learning.

2 The Paired Associate Learning Task (PAL)

The CANTAB paired associate learning (PAL) task measures the ability to remember object-location associations (see chapter by Barnett et al., this volume). The task is sensitive to performance decrements in patients with mild cognitive impairment (MCI) and has predictive power in identifying the MCI patients that will go on to develop Alzheimer's disease (AD; Swainson et al. 2001; Blackwell et al. 2005; De Jager et al. 2005; Égerházi et al. 2007), which has lead to the use of CANTAB-PAL as a screening tool for AD in general practice clinics. Patients with established schizophrenia are also impaired on this task (Wood et al. 2002; Chouinard et al. 2007; Donohoe et al. 2008; Aubin et al. 2009) as are patients in the prodromal phase or at-risk individuals (Bartók et al. 2005; Barnett et al. 2005). Furthermore, performance on the CANTAB-PAL task correlates with normal daily functioning and the severity of symptoms (Prouteau 2004; Barnett et al. 2005; Prouteau et al. 2005; Ritsner and Blumenkrantz 2007; Aubin et al. 2009). Based on this clinically valuable human paradigm, a rodent touchscreen PAL task was developed (Talpos et al. 2009; Bartko et al. 2010) in which the formation of three specific object-location associations are required for high levels of performance. More specifically, three visual patterns (A, B, and C) are associated with touchscreen locations; 1, 2, and 3, respectively. On a given trial, the rodent must choose between a stimulus presented in its correct location (e.g. A1) and a stimulus presented in one of its two incorrect locations (e.g. B3).

Rodent PAL is fundamentally similar to CANTAB-PAL in that it taps into object-location associations. However, it does differ in that in rodent PAL, three object-location associations remain relevant throughout training, whereas in CANTAB-PAL novel object-location associations are used on each trial, requiring the encoding and maintenance of individual associations across a delay. Despite this difference, the rodent PAL task has generated patterns of results very much in line with observations from human research using CANTAB-PAL. In an important demonstration of neurocognitive validity, task performance in both rodents and humans depends on analogous neuroanatomical regions. The hippocampus is differentially activated during CANTAB-PAL in MCI patients and healthy controls (de Rover et al. 2011), while damage to the temporal or frontal lobe, or the amygdala and hippocampus, impairs performance equally (Owen et al. 1995), in line with the idea that concurrent prefrontal and hippocampal activation in humans supports encoding and retrieval of paired associates (Eichenbaum 1999, 2000; Milner et al. 1997; Simons and Speirs 2003). Similarly, the hippocampus (Talpos et al. 2009; Kim et al. 2015; Delotterie et al. 2015) and the prefrontal cortex (McAllister et al. 2013) are critical for performance in rodents. In addition, both performance and acquisition of novel paired associates are compromised following damage to the rat retrosplenial cortex (Oomen et al. 2012), an area analogous to the posterior cingulate cortex in humans, which is activated during CANTAB-PAL performance (de Rover et al. 2011). Furthermore, patients with AD can show facilitation of CANTAB-PAL performance after the administration of a cholinesterase inhibitor (Greig et al. 2005), and C57Bl/6 mice show enhanced performance on PAL following systemic donepezil administration (Bartko et al. 2010).

In a further comparison of the rodent and CANTAB-PAL, a recent cross-species study assessed both mice and human carriers of schizophrenia-relevant mutations to the discs large homolog (*Dlg*) family of synaptic scaffold proteins (Xu et al. 2008) using a battery of touchscreen assays, showing a clear impairment in the acquisition of PAL (Nithianantharajah et al. 2012a, b). Interestingly, four human participants with copy number variant (CNV) mutations in the coding region of the *Dlg2* gene were similarly impaired on CANTAB-PAL (Nithianantharajah et al. 2012a, b). However, as described above, there are differences between rodent and CANTAB-PAL. To address these differences, the human CNV carriers were assessed on the rodent PAL task. Provided with no explicit instructions–just like the mice—the human mutation carriers displayed an analogous impairment in the acquisition of rodent PAL (Fig. 2; Nithianantharajah et al. 2015). Reducing the

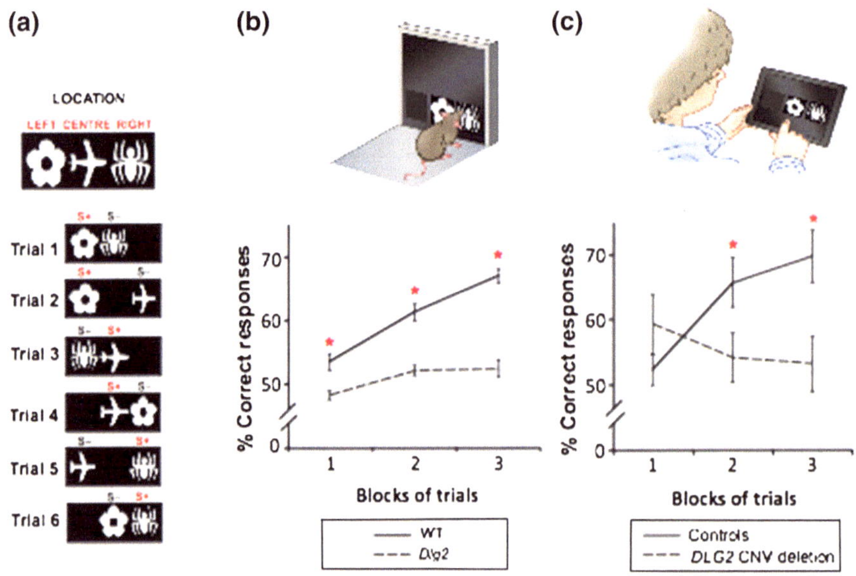

Fig. 2 The *Dlg2* deletion results in impaired acquisition of three object-location paired associates on the touchscreen rodent PAL task. **a** The PAL task utilises three objects and three locations, whereby each object is correct in a particular location only (as seen in the top panel). On each trial, an object is presented in its correct location alongside another object presented in one of its two incorrect locations, giving rise to six possible trial types. **b** $Dlg2^{-/-}$ mice were impaired compared to wild-type (WT) mice on the acquisition of the PAL task. **c** Using the identical task as in the mouse, four human participants with copy number variant (CNV) deletions in the *Dlg2* (three of whom have been diagnosed with schizophrenia and one of whom is an unaffected relative) were impaired on the acquisition of the rodent PAL task (from Nithianantharajah et al. 2015)

methodological differences between tasks when evaluating the behavioural effect of a genetic mutation in mice and humans–as opposed to the use of different versions of a touchscreen task for different species—increases the likelihood that the same underlying cognitive constructs are being measured, thereby demonstrating the potential power of this approach for translation. The forward translation of rodent touchscreen tasks to human studies may prove an increasingly useful approach in the assessment of cognitive constructs beyond PAL (see also 5-CSRTT below).

3 The 5-Choice Serial Reaction Time Task (5-CSRTT)

Attentional dysfunction represents an endophenotypic marker for a wide range of psychopathologies and has therefore become of principal interest for efforts towards the alleviation of symptoms and therefore translational approaches (Robbins 2002; Carter and Barch 2007; Demeter et al. 2008; Young et al. 2013). The rodent 5-CSRTT has been a predominant method for modelling attentional and inhibitory functions in rodents, and its translational utility is discussed at length elsewhere (Robbins 2002; Young et al. 2009b; Lustig et al. 2013). In this task, the animal is required to simultaneously monitor a number of potential stimulus inputs across a delay before responding within an appropriate time frame to a single, unpredictable stimulus location signalled by the brief onset of a light source. The task allows the identification of a behavioural profile based on a range of indices—including accuracy, omissions, premature responses, perseverative responses, and latency measures (Robbins 2002).

The task was developed in a 9-hole operant chamber in which animals respond in a nosepoke hole following the onset of an LED. Variations of the 9-hole 5-CSRTT have also successfully been translated to the touchscreen platforms for humans (Sahakian et al. 1993; Sahakian and Coull 1993; Voon et al. 2014; Worbe et al. 2014) and rodents (Bartko et al. 2011; Romberg et al. 2011; Mar et al. 2013). A clear advantage of using the touchscreen as opposed to the 9-hole version of the 5-CSRTT is the potential to include the task in a battery in which direct comparisons can be made across tasks targeting a range of cognitive constructs. Furthermore, the recent use of the touchscreen 4-CSRTT in human studies (Voon et al. 2014; Worbe et al. 2014) facilitates comparison between future work and the extensive rodent literature generated using the 5-CSRTT.

3.1 Alzheimer's Disease and Acetylcholine: Pharmacological and Genetic Translations

Among other dysfunctions, the psychopathophysiology of Alzheimer's disease (AD) is underpinned by cortical accumulations of neurofibrillary tangles (Crystal et al. 1988) and extracellular B-amyloid plaques (Roth et al. 1966), as well as reduced efficacy of cholinergic signalling (Davies and Maloney 1976; Whitehouse et al. 1982). AD is also characterised by attentional impairments (Alexander 1973; Baudic et al. 2006; Belleville et al. 2007; Silveri et al. 2007) that are observable in computer-controlled tasks (Oken et al. 1994; Baddeley et al. 1999; Levinoff et al. 2005; Bentley et al. 2008; McGuinness et al. 2010) including the human 5-CSRTT (Sahakian et al. 1993; Sahakian and Coull 1993). These attentional impairments may also be produced by cholinergic dysfunction (Lawrence and Sahakian 1995; Klinkenberg et al. 2011) and may be ameliorated by pharmacological anti-cholinesterase inhibition (Sahakian et al. 1993; Sahakian and Coull 1993; Foldi et al. 2005; Gauthier et al. 2007). Similarly, transgenic mouse models with AD-like pathology also show anticholinesterase-sensitive attentional impairments in the rodent touchscreen 5-CSRTT (Romberg et al. 2011, 2013).

Sahakian et al. (1993) showed that the anticholinesterase tetrahydroaminoacridine improved accuracy and decreased reaction times in AD patients on the touchscreen 5-CSRTT. Similarly, the TgCRND8 mouse model with Swedish and Indiana mutations of the human APP gene showed impaired accuracy in the 5-CSRTT at shorter stimulus durations (Romberg et al. 2013), and the 3xTgAD mouse model (Tau-P301L, APP-Swe, and PS1-M146V) showed vigilance-related accuracy impairments and increased perseveration when tested under challenging conditions (Romberg et al. 2011). In a clear example of clinical predictive validity, the attentional impairment in the 3xTgAD model was also remediated by the acetylcholinesterase inhibitor donepezil (Romberg et al. 2011). Notably, Bartko et al. (2011) also showed that global knockout (KO) of the muscarinic (M1) acetylcholine receptor increased perseverative and premature responding and reduced percent omissions in the mouse touchscreen 5-CSRTT, further demonstrating the importance of cholinergic systems for performance on the touchscreen 5-CSRTT.

3.2 *Serotonin Depletion Results in Parallel Deficits Across Species*

Reduced serotonin transmission has been extensively linked to increased impulsive behaviour (Evenden 1999; Dalley et al. 2011). The rodent 5-CSRTT is a powerful tool for the study of motor impulsivity, as the subject is required to inhibit responding until the stimulus is present. Failure in response inhibition, as observed through prematurely made responses, is considered a marker of lack of impulse control. Serotonin depletion in rats as a result of the administration of 5,7-dihydroxytryptamine (5,7-DHT) increases impulsive responding in the 5-CSRTT (Harrison et al. 1997; Winstanley et al. 2004a), and serotonin depletion in human volunteers increases impulsive responding in the 4-CSRTT mentioned earlier (Worbe et al. 2014). This is another example of how the use of rodent touchscreen tasks in human populations can result in parallel findings across species.

3.3 *Some Limitations of the 5-Choice Serial Reaction Time Task*

The 5-CSRTT has established construct and predictive validity (Robbins 2002; Lustig et al. 2013) and has been the gold standard of attentional and impulsive assessment in the rodent for more than three decades (Carli et al. 1983). Residual concerns have nevertheless remained, specifically regarding the absence of non-target trials and the consequent resistance to signal detection analyses typically used to evaluate human attentional processes. Furthermore, the rodent 5-CSRTT assesses divided visuospatial attention, whereas human touchscreen assays of attentional functioning predominantly measure focused visual attention and employ complex visual stimuli. In response to some of these concerns, a rodent version of the continuous performance test (CPT), one of the most widely used tests of human attentional function (Rosvold et al. 1956; Perry and Hodges 1999; Stopford et al. 2012; Cornblatt et al. 1989; Cornblatt and Malhotra 2001) was developed.

4 The Rodent Continuous Performance Task (rCPT)

Although several rodent attentional paradigms have been translated to the human experimental and clinical settings (Sahakian et al. 1993; Demeter et al. 2008; Young et al. 2013), their translational utility is somewhat restricted as a large portion of human data on attentional functioning continues to be generated by visual or touchscreen variants of the CPT (e.g. Kofler et al. 2013). In these tasks, a single target or a non-target stimulus is presented across trials; the participant is required to respond when the target stimulus is presented and must withhold from responding when a non-target stimulus is presented. Performance on CPTs is evaluated by signal detection analyses based on composite measures derived from *hit rate* (the ratio of correct responses of the total number of target presentations) and *false alarm rate* (the ratio of incorrect responses of the total number of non-target presentations). These composite measures include discrimination sensitivity indices (such as *d'* or sensitivity index *SI*) and response criterion values (such as *c* or *b;* Frey and Colliver 1973; Green and Swets 1989; Stanislaw and Todorov 1999; Macmillan and Creelman 2004). Furthermore, performance on CPTs depends on complex visual discriminations as opposed to the spatial or visuospatial brightness discriminations employed in rodent attentional tasks (Carli et al. 1983; McGaughy and Sarter 1995; Young et al. 2009a) and their translated human versions (Sahakian et al. 1993; Demeter et al. 2008; Young et al. 2013). Like CPTs, the rCPT requires subjects to detect and respond or inhibit responding to a target stimulus and non-target stimuli, respectively, presented sequentially in a central location on a touchscreen. Attentional functioning and behavioural inhibition is evaluated using standard parametric manipulations also used in human CPTs: increasing the cognitive load through manipulation of task parameters such as stimulus duration, target ratio, inter-stimulus interval, stimulus contrasts, or the addition of flanking distractors. The ability to measure attentional function in rodents using a task nearly identical to the task used in clinical research for half a century provides promising opportunities for bridging the gap between rodent and human work.

4.1 CPT and rCPT: Convergence by Functional Anatomy and Pharmacology

Human CPTs appears to be contingent upon frontal cortical areas, including the activity along the medial wall of the prefrontal cortex (mPFC). Lesion (Salmaso and Denes 1982; Glosser and Goodglass 1990), imaging (Buchsbaum et al. 1990; Keilp et al. 1997; Fallgatter and Strik 1997; Carter et al. 1998; Adler et al. 2001; Toichi et al. 2004), and electrophysiological (Valentino et al. 1993; Fallgatter and Strik 1999; Müller and Anokhin 2012) studies show causal and correlative relationships between CPT performance and prefrontal cortical areas, including the dorsolateral prefrontal cortex. In broad agreement with these data, we have found that mPFC lesions in rats and mice impair rCPT performance. In the mouse, lesions centred on the prelimbic cortex impair performance when animals are challenged with increased attentional load through decreased stimulus durations, lower target probabilities, and longer inter-stimulus intervals (Hvoslef-Eide et al. in prep). These impairments are observed as higher false alarm rates and a more liberal response criterion, consistent with a role for the mPFC in inhibitory control (Chudasama and Robbins 2006; Pattij and Vanderschuren 2008; Dalley et al. 2008, 2011). Excitotoxic mPFC lesions also impair CPT performance in the rat (Mar et al. in prep), suggesting that performance on touchscreen CPTs depends on the integrity of the prefrontal cortex across mice, rats, and humans. Moreover, as in the 5-CSRTT (Sahakian et al. 1993; Romberg et al. 2011), cholinergic signalling modulators such as nicotine (Levin et al. 1998, 2001; White and Levin 1999) and donepezil (Friedman et al. 2002) can enhance vigilance in human CPTs. Similarly, donepezil can dose-dependently improve attentional function in DBA mice under some task parameters in the rCPT (Kim et al. 2015).

4.2 Animal Models on the rCPT: Parallels with Human CPT Data

Attention-related abnormalities in CPTs represent endophenotypes of many neuropsychiatric and neurodegenerative disorders (Alexander 1973; Cornblatt et al. 1989; Ursu et al. 2003; Corbett and Constantine 2006; Cubillo et al. 2012) perhaps most notably in schizophrenia (Wohlberg and Kornetsky 1973; Cornblatt and Keilp 1994; Keefe et al. 2007; Filbey et al. 2008; Kahn et al. 2012; Nuechterlein et al. 2015). In the rCPT, analogous impairments are observed in schizophrenia-relevant mouse models, thereby demonstrating the tasks' construct validity. For example, conditional knockout of the NR1 subunit in corticolimbic GABAergic neurons (Belforte et al. 2009) induces an acquisition deficit when the attentional load is manipulated through shorter stimulus durations in the CPT (Hvoslef-Eide et al. 2013). Moreover, a chromosomal microdeletion at locus 22q11.2 is associated with a high risk of developing schizophrenia (Schneider et al. 2014) and extensive attentional deficits (Sobin et al. 2004) including CPT impairments (Shashi et al. 2010, 2012; Hooper et al. 2013; Harrell et al. 2013; Schoch et al. 2014). Hit rate-related measures in the CPT can also predict the onset of prodromal psychotic symptoms in individuals with 22q11.2 deletion syndrome (Antshel et al. 2010). Critically, the Df(h22q11)/+ mouse model of the 22q11.2 microdeletion syndrome shows a touchscreen CPT deficit that parallels the deficits of 22q11.2 deletion syndrome patients. These impairments can be expressed on measures of hit rate, *d'*, and *c* challenged with decreased stimulus presentation duration increased inter-stimulus intervals, and extended session length (Nilsson et al. in preparation). Thus, the observation of hit rate impairments in the Df(h22q11)/+ mutant parallels the dysfunction of 22q11.2 deletion syndrome patients as measured by CPTs, indicating translational validity of the task for assessing attentional functioning.

4.3 Vigilance Decrement in the rCPT

Human CPTs measure vigilance as observed by performance decrements across session length (Rosvold et al. 1956; Nuechterlein 1983; Mass et al. 2000). Although not often seen in the 5-CSRTT (but see Romberg et al. 2011), within-session performance decrements have been reported for alternative rodent–human translational paradigms such as the 5-choice continuous performance task (Young et al. 2009a) and the sustained attention task (Peters et al. 2011), and are also observed in mice when using the rCPT (Kim et al. 2015). In mice with a C57BL/6 background, hit rates and false alarms typically decrease with session length, which produce an elevated response criterion towards the end of the session (Kim et al. 2015; Nilsson et al. in prep). In mice with a DB2/2J background, hit rates have been found to decrease with session length, while false alarm rate remains constant (Kim et al. 2015). Thus, similar to humans, mice show decreased hit rates as a function of session length, which suggests that the rCPT has translational utility as a measure of vigilance.

4.4 Relationship Between Task Parameters and Performance Consistent Across Species

Manipulations of several task parameters have similar effects on CPT performance in humans and mice. For example, manipulations of target probability (Berwid et al.

2005) and inter-stimulus intervals (Conners et al. 2003; Hervey et al. 2006; Epstein et al. 2007) in human CPTs suggest that these parameters positively correlate with false alarms rates and hit rates and hence have more pronounced effects on measures of response criterion (e.g. *c*) relative to discrimination sensitivity (e.g. *d'*; Macmillan and Creelman 2004). Conversely, stimulus degradation manipulations can produce larger perceptual sensitivity decrements relative to measures of response criterion (Nuechterlein 1983; Chee et al. 1989; Mass et al. 2000). Importantly, similar parametric manipulations of inter-stimulus intervals, stimulus contrasts, stimulus duration, and target probabilities in the rCPT have comparable effects on the performance of mice and rats (Kim et al. 2015; Nilsson et al. in prep; Mar et al. in prep), which indicate that the parametric variables have comparable relationships with performance measures across species. To summarise, performance of both rCPT and human CPT is sensitive to mPFC lesions and prefrontal GABAergic dysfunction, dependent on the 22q11.2 chromosomal locus (or its orthologous region in the mouse), and is influenced by manipulations of attentional and perceptual load.

5 Reversal Learning

Touchscreen reversal learning tests are included within the intradimensional/extradimensional (ID/ED) protocol in the CANTAB battery, which has been used to generate a wealth of data in neurodegenerative and neuropsychiatric patients (Downes et al. 1989; Hughes et al. 1994; Pantelis et al. 1999; Kempton et al. 1999; Shamay-Tsoory et al. 2007: Ceaser et al. 2008; Leeson et al. 2009). In reversal learning, subjects first learn a two-choice simultaneous discrimination and then, the stimulus contingencies are reversed (the correct stimulus becomes the incorrect stimulus, and vice versa). Whereas reversal learning in human and non-human primates most often is assessed using touchscreen methods with visual stimuli (Dias et al. 1996; Bussey et al. 2001; Clarke et al. 2004; Leeson et al. 2009), prevalent rodent analogues typically employ non-visual olfactory, somatosensory (Schoenbaum and Chiba 1999; Birrell and Brown 2000; McAlonan and Brown 2003), or spatial cues (Becker et al. 1981; Boulougouris et al. 2007). Although non-touchscreen visual reversal learning tasks have been used with rodents, these tests are typically non-automated (Gardner and Coate 1965; Stevens and Fechter 1968; Sasaki 1969; Mullins and Winefield 1979; Mason et al. 1980) or dependent on brightness discriminations (Abdul-Monim et al. 2003; Floresco et al. 2008) with limited resemblance to the automated tasks with complex visual stimuli used with human and non-human primate subjects. To address these concerns, we have developed a suite of rodent touchscreen reversal learning tasks and evaluated their neurocognitive validity on anatomical, pharmacological, and genetic levels. Here, we provide a few examples.

5.1 Orbitofrontal and Dorsal Striatal Perturbations Cause Impairments of Touchscreen Reversal Learning in Rodents, Non-Human Primates, and Humans

The orbitofrontal cortex (OFC) has been implicated in touchscreen reversal learning in the mouse (Graybeal et al. 2011), rat (Chudasama and Robbins 2003; Izquierdo et al. 2013; Alsiö et al. 2015), and non-human primate (Dias et al. 1996), mirroring the findings of human imaging studies (Cools et al. 2002; Hampshire and Owen 2006; Clatworthy et al. 2009; Robinson et al. 2010; Hornak et al. 2004). Chudasama and Robbins (2003) showed that quinolinic OFC lesions impair touchscreen reversal learning in the rat by increasing the number of early errors. Similarly, Alsiö and colleagues showed that temporary OFC inactivation impairs touchscreen visual reversal learning in the rat by selectively increasing the number of early errors and early response omissions (Alsiö et al. 2015). OFC lesions in the rat also impair touchscreen reversal learning, and this deficit was interpreted as more prominent when the animal had to adjust to negative feedback (Izquierdo et al. 2013). In terms of the dorsal striatum, both mouse and human studies have found the region to be critical for reversal learning (Graybeal et al. 2011; Robinson et al. 2010).

It is important to note that OFC dependency in the rodent has also been observed using more traditional tasks (Birrell and Brown 2000; Schoenbaum et al. 2002; McAlonan and Brown 2003; Kim and Ragozzino 2005; Boulougouris et al. 2007; Ghods-Sharifi et al. 2008; Bissonette et al. 2008; Castañé et al. 2010). This illustrates an important point: we are not arguing that it is impossible to achieve effective translation using non-touchscreen tests in rodents. However, we do believe that using similar, validated, and more reproducible tests across species can make successful translation much more likely.

5.2 Serotonergic Manipulations Affect Both Rodent and Human Touchscreen Reversal Learning

Demonstrations of cross-species parallels within reversal learning are not restricted to the involvement of specific brain regions, but extend to a variety of signalling systems (Frank and O'Reilly 2006; Robbins and Roberts 2007), perhaps most notably serotonin (5-hydroxytryptamine, or 5-HT; Daw et al. 2002; Roberts 2011). Acute tryptophan depletions in healthy subjects can impair touchscreen reversal learning (Park et al. 1994; Rogers et al. 1999; but see Evers et al. 2005; Talbot et al. 2005; Finger et al. 2007; Cools et al. 2008), and behavioural genetic studies suggest that 5-HT signalling efficacy is bidirectionally related to reversal learning performance in humans (den Ouden et al. 2013). Comparable effects have been observed in non-touchscreen assays in the rodent with systemic pharmacological manipulations to 5-HT levels affecting bowl-digging (Lapiz-Bluhm et al. 2009), go/no-go

(Masaki et al. 2006), and spatial probabilistic reversal learning (Bari et al. 2010) in the rat.

As in human studies, rodent touchscreen reversal learning has been shown to be sensitive to pharmacological and genetic manipulations of 5-HT levels. In the mouse, hetero- and homozygous deletions of the 5-HT transporter cause dose-dependent elevations in cortical and striatal 5-HT levels (Mathews et al. 2004; Daws 2006) and induce parallel dosage-dependent improvements in touchscreen reversal learning observed as decreases in trials and errors to criterion (Brigman et al. 2010). Moreover, pharmacologically induced elevation of cortical 5-HT through subchronic fluoxetine treatment (Cryan et al. 2004) decreases trials and incorrect responses to criterion during the early phase of learning when responding is biased towards the previously correct stimulus (Brigman et al. 2010). Taken together, there are clear demonstrations of translation between the rodent, monkey, and human with regard to both specific regions and neurotransmitter systems central for reversal learning.

5.3 *Cross-Species Impairments in Touchscreen Reversal Learning Following a Disc Large 2 Mutation*

As part of the touchscreen test battery assessment of the *Dlg2* gene knockout mouse previously mentioned, visual reversal learning was assessed (Nithianantharajah et al. 2012a, b). In a clear example of cross-species translation, both human carriers of the *Dlg2* mutation and $Dlg2^{-/-}$ mice showed comparable deficits in touchscreen visual reversal learning, observed as a decrease in accuracy during early sessions, suggesting impairments in behavioural inhibition/cognitive flexibility (Nithianantharajah et al. 2012a, b).

6 Limitations and Challenges

These examples demonstrate how touchscreen behavioural paradigms can be readily translatable between human participants and rodents. Like any method, these protocols nevertheless contain limitations—many of which are the focus of refinement in ongoing experiments. For example, some rodent touchscreen tasks may require extensive training (e.g. PAL, 5-CSRTT). However, many non-touchscreen tasks of complex cognition also require lengthy training, and furthermore, touchscreen automation allows the parallel testing of numerous animals at once, mitigating the time cost of extensive training. A major challenge for touchscreen translation is to experimentally demonstrate neurocognitive validity for every rodent touchscreen test and its human counterpart—but again, this challenge is not specific to touchscreens, and validation of the touchscreen tests is ongoing.

Finally, we note that although touchscreens have many advantages with respect to translation, as explained above this in no sense rules out other complementary methods, for example those using so-called ethological approaches (Gerlai and Clayton 1999).

7 Conclusion

In this chapter, we have presented some examples of successful translation using touchscreen paradigms. With ongoing research assessing the neurocognitive validity of these tasks, the touchscreen approach is likely to become increasingly prevalent in translational cognitive research.

References

Abdul-Monim Z, Reynolds GPG, Neill JC (2003) The atypical antipsychotic ziprasidone, but not haloperidol, improves phencyclidine-induced cognitive deficits in a reversal learning task in the rat. J Psychopharmacol 17:57–65

Adler CM, Sax KW, Holland SK et al (2001) Changes in neuronal activation with increasing attention demand in healthy volunteers: an fMRI study. Synapse 42:266–272. doi:10.1002/syn.1112

Alexander DA (1973) Attention dysfunction in senile dementia. Psychol Rep 32:229–230. doi:10.2466/pr0.1973.32.1.229

Alsiö J, Nilsson S, Gastambide F et al (2015) The role of 5-HT2C receptors in touchscreen visual reversal learning in the rat: a cross-site study. Psychopharmacology 1–15. doi:10.1007/s00213-015-3963-5

Antshel KM, Shprintzen R, Fremont W, et al (2010) new research cognitive and psychiatric predictors to psychosis in velocardiofacial syndrome: a 3-year follow-up study. 49:333–344. doi:10.1016/j.jaac.2010.01.010

Aubin G, Stip E, Gélinas I et al (2009) Daily activities, cognition and community functioning in persons with schizophrenia. Schizophr Res 107:313–318. doi:10.1016/j.schres.2008.08.002

Baddeley A, Cocchini G, Sala Della S et al (1999) Working memory and vigilance: evidence from normal aging and Alzheimer's disease. Brain Cogn 41:87–108. doi:10.1006/brcg.1999.1097

Bari A, Theobald DEH, Caprioli D et al (2010) Serotonin modulates sensitivity to reward and negative feedback in a probabilistic reversal learning task in rats. Neuropsychopharmacol 35:1290–1301. doi:10.1038/npp.2009.233

Barnett JH, Sahakian BJ, Werners U et al (2005) Visuospatial learning and executive function are independently impaired in first-episode psychosis. Psychol Med 35:1031–1041

Barnett JH, Robbins TW, Leeson VC et al (2010) Assessing cognitive function in clinical trials of schizophrenia. Neurosci Biobehav Rev 34:1161–1177. doi:10.1016/j.neubiorev.2010.01.012

Bartko SJ, Vendrell I, Saksida LM, Bussey TJ (2010) A computer-automated touchscreen paired-associates learning (PAL) task for mice: impairments following administration of scopolamine or dicyclomine and improvements following donepezil. Psychopharmacology 214:537–548. doi:10.1007/s00213-010-2050-1

Bartko SJ, Romberg C, White B et al (2011) Intact attentional processing but abnormal responding in M1 muscarinic receptor-deficient mice using an automated touchscreen method. Neuropsychopharm 61(8):1366–1378

Bartók E, Berecz R, Glaub T, Degrell I (2005) Cognitive functions in prepsychotic patients. Prog Neuropsychopharmacol Biol Psychiatry 29:621–625. doi:10.1016/j.pnpbp.2005.01.008
Baudic S, Barba GD, Thibaudet MC et al (2006) Executive function deficits in early Alzheimer's disease and their relations with episodic memory. Arch Clin Neuropsychol 21:15–21. doi:10.1016/j.acn.2005.07.002
Becker JT, Olton DS, Anderson CA, Breitinger ERP (1981) Cognitive mapping in rats: the role of the hippocampal and frontal systems in retention and reversal. Behav Brain Res 3:1–22
Belforte JE, Zsiros V, Sklar ER et al (2009) Postnatal NMDA receptor ablation in corticolimbic interneurons confers schizophrenia-like phenotypes. Nat Neurosci 13:76–83. doi:10.1038/nn.2447
Belleville S, Chertkow H, Gauthier S (2007) Working memory and control of attention in persons with Alzheimer's disease and mild cognitive impairment. Neuropsychology 21:458–469. doi:10.1037/0894-4105.21.4.458
Bentley P, Driver J, Dolan RJ (2008) Cholinesterase inhibition modulates visual and attentional brain responses in Alzheimer's disease and health. Brain 131:409–424. doi:10.1093/brain/awm299
Berwid OG, Curko Kera EA, Marks DJ et al (2005) Sustained attention and response inhibition in young children at risk for Attention Deficit/Hyperactivity Disorder. J Child Psychol Psychiatry 46:1219–1229. doi:10.1111/j.1469-7610.2005.00417.x
Birrell JM, Brown VJ (2000) Medial frontal cortex mediates perceptual attentional set shifting in the rat. J Neurosci 20:4320–4324
Bissonette GB, Martins GJ, Franz TM et al (2008) Double dissociation of the effects of medial and orbital prefrontal cortical lesions on attentional and affective shifts in mice. J Neurosci Res 28:11124–11130. doi:10.1523/JNEUROSCI.2820-08.2008
Blackwell AD, De Jager C, Budge M et al (2005) Preclinical detection and differential diagnosis of Alzheimer's disease using a visuospatial paired associates learning test (CANTAB PAL): Implications for screening and clinical trial sample enrichment. Alzheimer's & Dement 1:S80–S81. doi:10.1016/j.jalz.2005.06.286
Boulougouris V, Dalley JW, Robbins TW (2007) Effects of orbitofrontal, infralimbic and prelimbic cortical lesions on serial spatial reversal learning in the rat. Behav Brain Res 179:219–228. doi:10.1016/j.bbr.2007.02.005
Brigman JL, Mathur P, Harvey-White J et al (2010) Pharmacological or genetic inactivation of the serotonin transporter improves reversal learning in mice. Cereb Cortex 20:1955–1963. doi:10.1093/cercor/bhp266
Buchsbaum MS, Nuechterlein KH, Haier RJ et al (1990) Glucose metabolic rate in normals and schizophrenics during the continuous performance test assessed by positron emission tomography. Brit J Psychiat 156:216–227. doi:10.1192/bjp.156.2.216
Bussey TJ, Wise SP, Murray EA (2001) The role of ventral and orbital prefrontal cortex in conditional visuomotor learning and strategy use in rhesus monkeys (Macaca mulatta). Behav Neurosci 115:971–982
Bussey TJ, Holmes A, Lyon L et al (2012) New translational assays for preclinical modelling of cognition in schizophrenia: the touchscreen testing method for mice and rats. Neuropharmacology 62:1191–1203. doi:10.1016/j.neuropharm.2011.04.011
Carli M, Robbins TW, Evenden JL, Everitt BJ (1983) Effects of lesions to ascending noradrenergic neurones on performance of a 5-choice serial reaction task in rats; implications for theories of dorsal noradrenergic bundle function based on selective attention and arousal. Behav Brain Res 9:361–380
Carter CS, Barch DM (2007) Cognitive neuroscience-based approaches to measuring and improving treatment effects on cognition in schizophrenia: the CNTRICS initiative. Schizophr Bull 33:1131–1137. doi:10.1093/schbul/sbm081
Carter CS, Braver TS, Barch DM et al (1998) Anterior cingulate cortex, error detection, and the online monitoring of performance. Science 280:747–749

Castañé A, Theobald DEH, Robbins TW (2010) Selective lesions of the dorsomedial striatum impair serial spatial reversal learning in rats. Behav Brain Res 210:74–83. doi:10.1016/j.bbr.2010.02.017

Ceaser AE, Goldberg TE, Egan MF et al (2008) Set-shifting ability and schizophrenia: a marker of clinical illness or an immediate phenotype? Biol Psychiatry 64:782–788

Chee P, Logan G, Schachar R et al (1989) Effects of event rate and display time on sustained attention in hyperactive, normal, and control children. J Abnorm Child Psychol 17:371–391

Chouinard S, Stip E, Poulin J et al (2007) Rivastigmine treatment as an add-on to antipsychotics in patients with schizophrenia and cognitive deficits. Curr Med Res Opin 23:575–583. doi:10.1185/030079906X167372

Chudasama Y, Robbins TW (2003) Dissociable contributions of the orbitofrontal and infralimbic cortex to pavlovian autoshaping and discrimination reversal learning: further evidence for the functional heterogeneity of the rodent frontal cortex. J Neurosci 23:8771–8780

Chudasama Y, Robbins TW (2006) Functions of frontostriatal systems in cognition: comparative neuropsychopharmacological studies in rats, monkeys and humans. Biol Psychol 73:19–38. doi:10.1016/j.biopsycho.2006.01.005

Clarke HF, Dalley JW, Crofts HS et al (2004) Cognitive inflexibility after prefrontal serotonin depletion. Science 304:878–880. doi:10.1126/science.1094987

Clatworthy PL, Lewis SJG, Brichard L et al (2009) Dopamine release in dissociable striatal subregions predicts the different effects of oral methylphenidate on reversal learning and spatial working memory. J Neurosci 29:4690–4696. doi:10.1523/JNEUROSCI.3266-08.2009

Conners CK, Epstein JN, Angold A, Klaric J (2003) Continuous performance test performance in a normative epidemiological sample. J Abnorm Child Psychol 31:555–562

Cools R, Clark L, Owen AM, Robbins TW (2002) Defining the neural mechanisms of probabilistic reversal learning using event-related functional magnetic resonance imaging. J Neurosci 22:4563–4567

Cools R, Robinson OJ, Sahakian B (2008) Acute tryptophan depletion in healthy volunteers enhances punishment prediction but does not affect reward prediction. Neuropsychopharmacol 33:2291–2299. doi:10.1038/sj.npp.1301598

Corbett BA, Constantine LJ (2006) Autism and attention deficit hyperactivity disorder: assessing attention and response control with the integrated visual and auditory continuous performance test. NCNY 12:335–348. doi:10.1080/09297040500350938

Cornblatt BA, Keilp JG (1994) Impaired attention, genetics, and the pathophysiology of schizophrenia. Schizophr Bull 20:31–46

Cornblatt BA, Lenzenweger MF, Erlenmeyer-Kimling L (1989) The continuous performance test, identical pairs version: II. Contrasting attentional profiles in schizophrenic and depressed patients. Psychiatry Res 29(1):65–85

Cornblatt BA, Malhotra AK (2001) Impaired attention as an endophenotype for molecular genetic studies of schizophrenia. Am J Med Genet 105:11–15

Cryan JF, O'Leary OF, Jin S-H et al (2004) Norepinephrine-Deficient Mice Lack Responses to Antidepressant Drugs, Including Selective Serotonin Reuptake Inhibitors. Proc Natl Acad Sci USA 101:8186–8191. doi:10.2307/3372472?ref=no-x-route:ab20f928fa3830dcbfa7405ac3ccea45

Crystal H, Dickson D, Fuld P et al (1988) Clinico-pathologic studies in dementia: nondemented subjects with pathologically confirmed Alzheimer's disease. Neurology 38:1682–1687. doi:10.1212/WNL.38.11.1682

Cubillo A, Halari R, Smith A et al (2012) A review of fronto-striatal and fronto-cortical brain abnormalities in children and adults with attention deficit hyperactivity disorder (ADHD) and new evidence for dysfunction in adults with ADHD during motivation and attention. Cortex 48:194–215. doi:10.1016/j.cortex.2011.04.007

Dalley JW, Mar AC, Economidou D, Robbins TW (2008) Neurobehavioral mechanisms of impulsivity: Fronto-striatal systems and functional neurochemistry. Pharmacol Biochem Behav 90:250–260. doi:10.1016/j.pbb.2007.12.021

Dalley JW, Everitt BJ, Robbins TW (2011) Impulsivity, compulsivity, and top-down cognitive control. Neuron 69:680–694. doi:10.1016/j.neuron.2011.01.020

Davies P, Maloney AJ (1976) Selective loss of central cholinergic neurons in Alzheimer's disease. Lancet 2:1403

Daw ND, Kakade S, Dayan P (2002) Opponent interactions between serotonin and dopamine. Neural Netw 15:603–616. doi:10.1016/S0893-6080(02)00052-7

Daws LC (2006) Ethanol inhibits clearance of brain serotonin by a serotonin transporter-independent mechanism. J Neurosci 26:6431–6438. doi:10.1523/JNEUROSCI.4050-05.2006

De Jager C, Blackwell AD, Budge MM, Sahakian BJ (2005) Predicting cognitive decline in healthy older adults. Am J Geriatr Psychiatry 13:735–740. doi:10.1176/appi.ajgp.13.8.735

de Rover M, Pironti VA, McCabe JA, Acosta-Cabronero J, Arana FS, Morein-Zamir S, Hodges JR, Robbins TW, Fletcher PC, Nestor PJ, Sahakian BJ (2011) Hippocampal dysfunction in patients with mild cognitive impairment: a functional neuroimaging study of a visuospatial paired associates learning task. Neuropsychologia 49(7):2060–2070

Delotterie DF, Mathis C, Cassel JC, Rosenbrock H, Dorner-Ciossek C, Marti A (2015) Touchscreen tasks in mice to demonstrate differences between hippocampal and striatal functions. Neurobiol Learn Mem 120:16–27

Demeter E, Sarter M, Lustig C (2008) Rats and humans paying attention: cross-species task development for translational research. Neuropsychology 22:787–799. doi:10.1037/a0013712

den Ouden HEM, Daw ND, Fernandez G et al (2013) Dissociable effects of dopamine and serotonin on reversal learning. Neuron 80:1090–1100. doi:10.1016/j.neuron.2013.08.030

Dias R, Robbins TW, Roberts AC (1996) Dissociation in prefrontal cortex of affective and attentional shifts. Nature 380:69–72. doi:10.1038/380069a0

Donohoe G, Spoletini I, McGlade N et al (2008) Are relational style and neuropsychological performance predictors of social attributions in chronic schizophrenia? Psychiatry Res 161:19–27. doi:10.1016/j.psychres.2007.10.001

Downes JJ, Roberts AC, Sahakian BJ et al (1989) Impaired extra-dimensional shift performance in medicated and unmedicated Parkinson's disease: evidence for a specific attentional dysfunction. Neuropsychologica 27:1329–1343

Égerházi A, Berecz R, Bartók E, Degrell I (2007) Automated neuropsychological test battery (CANTAB) in mild cognitive impairment and in Alzheimer's disease. Prog Neuropsychopharmacol Biol Psychiatry 31:746–751. doi:10.1016/j.pnpbp.2007.01.011

Eichenbaum H (1999) The hippocampus and mechanisms of declarative memory. Behav Brain Res 103:123–133

Eichenbaum H (2000) A cortical-hippocampal system for declarative memory. Nat Rev Neursoci 1:41–50

Epstein JN, Casey BJ, Tonev ST et al (2007) ADHD- and medication-related brain activation effects in concordantly affected parent-child dyads with ADHD. J Child Psychol Psychiatry 48:899–913. doi:10.1111/j.1469-7610.2007.01761.x

Evenden JL (1999) Varieties of impulsivity. Psychopharmacology 146:348–361

Evers EAT, Cools R, Clark L et al (2005) Serotonergic modulation of prefrontal cortex during negative feedback in probabilistic reversal learning. Neuropsychopharmacol 30:1138–1147. doi:10.1038/sj.npp.1300663

Fallgatter AJ, Strik WK (1997) Right frontal activation during the continuous performance test assessed with near-infrared spectroscopy in healthy subjects. Neurosci Lett 223:89–92

Fallgatter AJ, Strik WK (1999) The NoGo-anteriorization as a neurophysiological standard-index for cognitive response control. Int J Psychophysiol 32:233–238

Filbey FM, Toulopoulou T, Morris RG et al (2008) Selective attention deficits reflect increased genetic vulnerability to schizophrenia. Schizophrn Res 101:169–175. doi:10.1016/j.schres.2008.01.019

Finger EC, Marsh AA, Buzas B et al (2007) The impact of tryptophan depletion and 5-HTTLPR genotype on passive avoidance and response reversal instrumental learning tasks. Neuropsychopharmacol 32:206–215. doi:10.1038/sj.npp.1301182

Floresco SB, Block AE, Tse MTL (2008) Inactivation of the medial prefrontal cortex of the rat impairs strategy set-shifting, but not reversal learning, using a novel, automated procedure. Behav Brain Res 190:85–96

Foldi NS, White REC, Schaefer LA (2005) Detecting effects of donepezil on visual selective attention using signal detection parameters in Alzheimer's disease. Int J Geriat Psychiatry 20:485–488. doi:10.1002/gps.1319

Frank MJ, O'Reilly RC (2006) A mechanistic account of striatal dopamine function in human cognition: Psychopharmacological studies with cabergoline and haloperidol. Behav Neurosci 120:497–517. doi:10.1037/0735-7044.120.3.497

Frey PW, Colliver JA (1973) Sensitivity and responsivity measures for discrimination learning. Learn Motiv 4:327–342. doi:10.1016/0023-9690(73)90023-4

Friedman JI, Adler DN, Howanitz E, Harvey PD, Brenner G, Temporini H, White L, Parrella M (2002) A double blind placebo controlled trial of donepezil adjunctive treatment to risperidone for the cognitive impairment of schizophrenia. Biol Psychiatry 51(5):349–357

Gardner RA, Coate WB (1965) Reward versus nonreward in a simultaneous discrimination. J Exp Psychol 69:579–582. doi:10.1037/h0022057

Gauthier S, Juby A, Rehel B, Schecter R (2007) EXACT: rivastigmine improves the high prevalence of attention deficits and mood and behaviour symptoms in Alzheimer's disease*. Int J Clin Pract 61:886–895. doi:10.1111/j.1742-1241.2007.01387.x

Gerlai R, Clayton NS (1999) Analysing hippocampal function in transgenic mice: an ethological persepctive. Trends Neurosci 22:47–51

Geyer MA, Tamminga CA (2004) Measurement and treatment research to improve cognition in schizophrenia: neuropharmacological aspects. Psychopharmacology 174:1–2. doi:10.1007/s00213-004-1846-2

Ghods-Sharifi S, Haluk DM, Floresco SB (2008) Differential effects of inactivation of the orbitofrontal cortex on strategy set-shifting and reversal learning. Neurobiol Learn Mem 89:567–573. doi:10.1016/j.nlm.2007.10.007

Glosser G, Goodglass H (1990) Disorders in executive control functions among aphasic and other brain-damaged patients. J Clin Exp Neuropsychol 12:485–501. doi:10.1080/01688639008400995

Graybeal C, Feyder M, Schulman E et al (2011) Paradoxical reversal learning enhancement by stress or prefrontal cortical damage: rescue with BDNF. Nat Neurosci 14:1507–1509. doi:10.1038/nn.2954

Green MF (1996) What are the functional consequences of neurocognitive deficits in schizophrenia? Am J Psychiatry 153:321–330

Green MF (2006) Cognitive impairment and functional outcome in schizophrenia and bipolar disorder. J Clin Psychiatry 67(supl 9):3–8

Green DM, Swets JA (1989) Signal detection theory and psychophysics. Peninsula Pub

Greig NH, Sambamurti K, Yu QS et al (2005) An overview of phenserine tartrate, a novel acetylcholinesterase inhibitor for the treatment of Alzheimer's disease. Curr Alzheimer Res 2:281–290

Hampshire A, Owen AM (2006) Fractionating attentional control using event-related fMRI. Cereb Cortex 16:1679–1689. doi:10.1093/cercor/bhj116

Harrell W, Eack S, Hooper SR et al (2013) Research in developmental disabilities feasibility and preliminary efficacy data from a computerized cognitive intervention in children with chromosome 22q11.2 deletion syndrome. Res Dev Disabil 34:2606–2613. doi:10.1016/j.ridd.2013.05.009

Harrison AA, Everitt BJ, Robbins TW (1997) Central 5-HT depletion enhances impulsive responding without affecting the accuracy of attentional performance: interactions with dopaminergic mechanisms. Psychopharmacology 133:329–342

Hervey AS, Epstein JN, Curry JF et al (2006) Reaction time distribution analysis of neuropsychological performance in an ADHD sample. NCNY 12:125–140. doi:10.1080/09297040500499081

Holthausen EA, Wiersma D, Cahn W et al (2007) Predictive value of cognition for different domains of outcome in recent-onset schizophrenia. Psychiatry Res 149:71–80. doi:10.1016/j.psychres.2005.07.037

Hooper SR, Curtiss K, Schoch K et al (2013) Research in developmental disabilities a longitudinal examination of the psychoeducational, neurocognitive, and psychiatric functioning in children with 22q11.2 deletion syndrome. Res Dev Disabil 34:1758–1769. doi:10.1016/j.ridd.2012.12.003

Hornak J, O'Doherty J, Bramham J et al (2004) Reward-related reversal learning after surgical excisions in orbito-frontal or dorsolateral prefrontal cortex in humans. J Cogn Neurosci 16:463–478. doi:10.1162/089892904322926791

Horner AE, Heath CJ, Hvoslef-Eide M et al (2013) The touchscreen operant platform for testing learning and memory in rats and mice. Nat Protoc 8:1961–1984. doi:10.1038/nprot.2013.122

Hughes C, Russell J, Robbins TW (1994) Evidence for executive dysfunction in autism. Neuropsychologia 32:477–492

Hvoslef-Eide M, Lyon L, Mar AC et al. (2013) Attentional deficits following postnatal NR1 knock-out in corticolimbic interneurons. Program No. 255.29. 2013 Neuroscience Meeting Planner. Society for Neuroscience, San Diego, (Online)

Hvoslef-Eide M, Mar AC, Nilsson S et al (2015) The NEWMEDS rodent touchscreen test battery for cognition relevant to schizophrenia. Psychopharmacology 232:3852–3872

Insel TR (2012) Next-generation treatments for mental disorders. Sci Transl Med 4:155ps19–155ps19. doi:10.1126/scitranslmed.3004873

Insel TR, Wang PS (2010) Rethinking mental illness. JAMA 303:1970–1971. doi:10.1001/jama.2010.555

Insel TR, Voon V, Nye JS et al (2013) Neuroscience and biobehavioral reviews. Neurosci Biobehav Rev 37:2438–2444. doi:10.1016/j.neubiorev.2013.03.022

Izquierdo A, Darling C, Manos N, Pozos H, Kim C, Ostrander S, Cazares V, Stepp H, Rudebeck PH (2013) Basolateral amygdala lesions facilitate reward choices after negative feedback in rats. J Neurosci 33(9):4105–4109

Kahn PV, Walker TM, Williams TS et al (2012) Standardizing the use of the continuous performance test in schizophrenia research: a validation study. Schizophr Res 1–6. doi:10.1016/j.schres.2012.09.009

Keefe RSE, Bilder RM, Davis SM et al (2007) Neurocognitive effects of antipsychotic medications in patients with chronic schizophrenia in the CATIE Trial. Arch Gen Psychiatry 64:633–647. doi:10.1001/archpsyc.64.6.633

Keilp JG, Herrera J, Stritzke P, Cornblatt BA (1997) The continuous performance test, identical pairs version (CPT-IP): III: brain functioning during performance of numbers and shapes subtasks. Psychiatry Res 74:35–45. doi:10.1016/S0925-4927(96)02881-8

Kempton S, Vance A, Maruff P et al (1999) Executive function and attention deficit hyperactivity disorder: stimulant medication and better executive function performance in children. Psychol Med 29:527–538

Kim J, Ragozzino ME (2005) The involvement of the orbitofrontal cortex in learning under changing task contingencies. Neurobiol Learn Mem 83:125–133. doi:10.1016/j.nlm.2004.10.003

Kim CH, Hvoslef-Eide M, Nilsson SR, Johnson MR, Herbert BR, Robbins TW, Saksida LM, Bussey TJ, Mar AC (2015) The continuous performance test (rCPT) for mice: a novel operant touchscreen test of attentional function. Psychopharmacology 232(21–22):3947–3966

Klinkenberg I, Sambeth A, Blokland A (2011) Acetylcholine and attention. Behav Brain Res 221:430–442. doi:10.1016/j.bbr.2010.11.033

Kofler MJ, Rapport MD, Sarver DE et al (2013) Reaction time variability in ADHD: a meta-analytic review of 319 studies. Clin Psychol Rev 33(6):795–811

Lapiz-Bluhm MDS, Soto-Piña AE, Hensler JG, Morilak DA (2009) Chronic intermittent cold stress and serotonin depletion induce deficits of reversal learning in an attentional set-shifting test in rats. Psychopharmacology 202:329–341. doi:10.1007/s00213-008-1224-6

Lawrence AD, Sahakian BJ (1995) Alzheimer disease, attention, and the cholinergic system. Alzheimer Dis Assoc Disord 9(Suppl 2):43–49

Leeson VC, Robbins TW, Matheson E et al (2009) Discrimination learning, reversal, and set-shifting in first-episode schizophrenia: stability over six years and specific associations with medication type and disorganization syndrome. Biol Psychiatry 66:586–593. doi:10.1016/j.biopsych.2009.05.016

Levaux M-N, Potvin S, Sepehry AA et al (2007) Computerized assessment of cognition in schizophrenia: promises and pitfalls of CANTAB. Eur Psychiatry 22:104–115. doi:10.1016/j.eurpsy.2006.11.004

Levin ED, Conners KC, Silva D et al (1998) Transdermal nicotine effect on attention. Psychopharmacology 140:135–141

Levin ED, Conners KC, Silva D et al (2001) Effects of chronic nicotine and methylphenidate in adults with attention deficit/hyperactivity disorder. Exp Clin Psychopharm 9(1):83–90

Levinoff EJ, Saumier D, Chertkow H (2005) Focused attention deficits in patients with Alzheimer's disease and mild cognitive impairment. Brain Cogn 57:127–130. doi:10.1016/j.bandc.2004.08.058

Lustig C, Kozak R, Sarter M et al (2013) CNTRICS final animal model task selection: control of attention. Neurosci Biobehav Rev 37:2099–2110. doi:10.1016/j.neubiorev.2012.05.009

Macmillan NA, Creelman CD (2004) Detection theory. Psychology Press

Mar AC, Horner AE, Nilsson S et al (2013) The touchscreen operant platform for assessing executive function in rats and mice. Nat Protoc 8:1985–2005. doi:10.1038/nprot.2013.123

Masaki D, Yokoyama C, Kinoshita S et al (2006) Relationship between limbic and cortical 5-HT neurotransmission and acquisition and reversal learning in a go/no-go task in rats. Psychopharmacology 189:249–258. doi:10.1007/s00213-006-0559-0

Mason JR, Stevens D, Wixon DR, Owens MP (1980) Assessment of the relative importance of S+ and S- in rats using differential training on intercurrent discriminations. Learn Motiv 11:49–60. doi:10.1016/0023-9690(80)90020-X

Mass R, Wolf K, Wagner M, Haasen C (2000) Differential sustained attention/vigilance changes over time in schizophrenics and controls during a degraded stimulus continuous performance test. Eur Arch Psychiatry Clin Neurosci 250:24–30. doi:10.1007/PL00007535

Mathews TA, Fedele DE, Coppelli FM et al (2004) Gene dose-dependent alterations in extraneuronal serotonin but not dopamine in mice with reduced serotonin transporter expression. J Neurosci Methods 140:169–181. doi:10.1016/j.jneumeth.2004.05.017

McAllister KAL, Saksida LM, Bussey TJ (2013) Dissociation between memory retention across a delay and pattern separation following medial prefrontal cortex lesions in the touchscreen TUNL task. Neurobiol Learn Mem 101:120–126. doi:10.1016/j.nlm.2013.01.010

McAlonan K, Brown VJ (2003) Orbital prefrontal cortex mediates reversal learning and not attentional set shifting in the rat. Behav Brain Res 146:97–103. doi:10.1016/j.bbr.2003.09.019

McGaughy J, Sarter M (1995) Behavioral vigilance in rats: task validation and effects of age, amphetamine, and benzodiazepine receptor ligands. Psychopharmacology 117:340–357. doi:10.1007/BF02246109

McGuinness B, Barrett SL, Craig D et al (2010) Attention deficits in Alzheimer's disease and vascular dementia. J Neurol Neurosurg Psychiatry 81:157–159. doi:10.1136/jnnp.2008.164483

Milner B, Jonhsrude I, Crane J (1997) Right medial temporal-lobe contribution to object-location memory. Philos Trans R Soc Lond B Biol Sci 352:1469–1474

Müller V, Anokhin AP (2012) Neural synchrony during response production and inhibition. PLoS ONE 7:e38931–11. doi:10.1371/journal.pone.0038931

Mullins GP, Winefield AH (1979) The relative importance of responses to S+ and S- in simultaneous discrimination learning. Q J Exp Psychol 31:329–338. doi:10.1080/14640747908400731

Nithianantharajah J, Komiyama NH, McKechanie A et al (2012a) Synaptic scaffold evolution generated components of vertebrate cognitive complexity. Nat Neurosci 16:16–24. doi:10.1038/nn.3276

Nithianantharajah J, McKechanie AG, Stewart TJ, Johnstone M et al (2012b) Bridging the translational divide: identical cognitive touchscreen testing in mice and humans carrying mutations in a disease-relevant homologous gene. Sci Rep 5:14613

Nithianantharajah J, McKechanie AG, Stewart TJ, Johnstone M, Blackwood DH, St Clair D, Grant SG, Bussey TJ, Saksida LM (2015) Bridging the translational divide: identical cognitive touchscreen testing in mice and humans carrying mutations in a disease-relevant homologous gene. Scientific reports 5

Nuechterlein KH (1983) Signal detection in vigilance tasks and behavioral attributes among offspring of schizophrenic mothers and among hyperactive children. J Abnorm Psychol 92:4–28

Nuechterlein KH, Green MF, Calkins ME et al (2015) Attention/vigilance in schizophrenia: performance results from a large multi-site study of the consortium on the genetics of schizophrenia (COGS). Schizophr Res 163:38–46. doi:10.1016/j.schres.2015.01.017

O'Carroll R (2000) Cognitive impairment in schizophrenia. Adv Psychiatry Treat 6:161–168. doi:10.1192/apt.6.3.161

Oken BS, Kishiyama SS, Kaye JA, Howieson DB (1994) Attention deficit in Alzheimer's disease is not simulated by an anticholinergic/antihistaminergic drug and is distinct from deficits in healthy aging. Neurology 44:657. doi:10.1212/WNL.44.4.657

Oomen CA, Bussey TJ, Saksida LM (2012) The retrosplenial cortex is involved in object-place paired-associates learning in the rat touchscreen operant chamber. Program No. 600.06. 2012 Neuroscience Meeting Planner. Society for Neuroscience, New Orleans (Online)

Oomen CA, Hvoslef-Eide M, Heath CJ et al (2013) The touchscreen operant platform for testing working memory and pattern separation in rats and mice. Nat Protoc 8:2006–2021. doi:10.1038/nprot.2013.124

Owen AM, Sahakian BJ, Semple J et al (1995) Visuo-spatial short-term recognition memory and learning after temporal lobe excisions, frontal lobe excisions or amydgalo-hippocampctomy in man. Neuropsychologia 33:1–24

Pantelis C, Barber FZ, Barnes TR et al (1999) Comparison of set-shifting ability in patients with chronic schizophrenia and frontal lobe damage. Schizophr Res 37:251–270

Park SB, Coull JT, McShane RH et al (1994) Tryptophan depletion in normal volunteers produces selective impairments in learning and memory. Neuropharmacology 33:575–588. doi:10.1016/0028-3908(94)90089-2

Pattij T, Vanderschuren L (2008) The neuropharmacology of impulsive behaviour. Trends Pharmacol Sci 29:192–199. doi:10.1016/j.tips.2008.01.002

Perry RJ, Hodges JR (1999) Attention and executive deficits in Alzheimer's disease: a critical review. Brain 122:383–404

Peters MS, Cherian AK, Bradshaw M, Sarter M (2011) Sustained attention in mice: expanding the translational utility of the SAT by incorporating the Michigan Controlled Access Response Port (MICARP). Behav Brain Res 225:574–583. doi:10.1016/j.bbr.2011.08.025

Prouteau A (2004) Self-assessed cognitive dysfunction and objective performance in outpatients with schizophrenia participating in a rehabilitation program. Schizophr Res 69:85–91. doi:10.1016/j.schres.2003.08.011

Prouteau A, Verdoux H, Briand C et al (2005) Cognitive predictors of psychosocial functioning outcome in schizophrenia: a follow-up study of subjects participating in a rehabilitation program. Schizophr Res 77:343–353. doi:10.1016/j.schres.2005.03.001

Ritsner MS, Blumenkrantz H (2007) Predicting domain-specific insight of schizophrenia patients from symptomatology, multiple neurocognitive functions, and personality related traits. Psychiatry Res 149:59–69. doi:10.1016/j.psychres.2006.01.002

Robbins TW (2002) The 5-choice serial reaction time task: behavioural pharmacology and functional neurochemistry. Psychopharmacology 163:362–380. doi:10.1007/s00213-002-1154-7

Robbins TW, Roberts AC (2007) Differential regulation of fronto-executive function by the monoamines and acetylcholine. Cereb Cortex 17(Suppl 1):i151–i160. doi:10.1093/cercor/bhm066

Roberts AC (2011) The importance of serotonin for orbitofrontal function. Biol Psychiatry 69:1185–1191. doi:10.1016/j.biopsych.2010.12.037

Robinson OJ, Frank MJ, Sahakian BJ, Cools R (2010) Dissociable responses to punishment in distinct striatal regions during reversal learning. NeuroImage 51:1459–1467. doi:10.1016/j.neuroimage.2010.03.036

Rogers RD, Blackshaw AJ, Middleton HC et al (1999) Tryptophan depletion impairs stimulus-reward learning while methylphenidate disrupts attentional control in healthy young adults: implications for the monoaminergic basis of impulsive behaviour. Psychopharmacology 146:482–491

Romberg C, Mattson MP, Mughal MR et al (2011) Impaired attention in the 3xTgAD mouse model of Alzheimer's disease: rescue by donepezil (aricept). J Neurosci 31:3500–3507. doi:10.1523/JNEUROSCI.5242-10.2011

Romberg C, Horner AE, Bussey TJ, Saksida LM (2013) A touch screen-automated cognitive test battery reveals impaired attention, memory abnormalities, and increased response inhibition in the TgCRND8 mouse model of Alzheimer's disease. NBA 34:731–744. doi:10.1016/j.neurobiolaging.2012.08.006

Rosvold HE, Mirsky AF, Sarason I et al (1956) A continuous performance test of brain damage. J Consult Psychol 20:343–350. doi:10.1037/h0043220

Roth M, Tomlinson BE, Blessed G (1966) Correlation between scores for dementia and counts of "senile plaques" in cerebral grey matter of elderly subjects. Nature 209:109–110. doi:10.1038/209109a0

Sahakian BJ, Coull JT (1993) Tetrahydroaminoacridine (THA) in Alzheimer's disease: An assessment of attentional and mnemonic function using CANTAB. Acta Neurol Scand 88:29–35. doi:10.1111/j.1600-0404.1993.tb04251.x

Sahakian BJ, Owen AM (1992) Computerized assessment in neuropsychiatry using CANTAB: discussion paper. J R Soc Med 85:399–402

Sahakian BJ, Owen AM, Morant NJ et al (1993) Further analysis of the cognitive effects of tetrahydroaminoacridine (THA) in Alzheimer's disease: assessment of attentional and mnemonic function using CANTAB. Psychopharmacology 110:395–401. doi:10.1007/BF02244644

Salmaso D, Denes G (1982) Role of the frontal lobes on an attention task: a signal detection analysis. Percept Mot Skill 54:1147–1150. doi:10.2466/pms.1982.54.3c.1147

Sasaki M (1969) Influence of prereversal experience on reversal learning in white rats. Ann Anim Psychol 19:17–28

Schneider M, Debbane M, Bassett AS et al (2014) Psychiatric disorders from childhood to adulthood in 22q11.2 deletion syndrome: results from the international consortium on brain and behavior in 22q11.2 deletion syndrome. Am J Psychiatry 171:627–639. doi:10.1176/appi.ajp.2013.13070864

Schoch K, Harrell W, Hooper SR et al (2014) Applicability of the nonverbal learning disability paradigm for children With 22q11.2 deletion syndrome. J Learn Disabil 47:153–166. doi:10.1177/0022219412443556

Schoenbaum G, Chiba AA (1999) Neural encoding in orbitofrontal cortex and basolateral amygdala during olfactory discrimination learning

Schoenbaum G, Nugent SL, Saddoris MP, Setlow B (2002) Orbitofrontal lesions in rats impair reversal but not acquisition of go, no-go odor discriminations. NeuroReport 13:885–890

Shamay-Tsoory SG, Shur S, Harari H et al (2007) Neurocognitive bias of impaired empathy in schizophrenia. Neuropsychology 21:431–438

Shashi V, Kwapil TR, Kaczorowski J et al (2010) Evidence of gray matter reduction and dysfunction in chromosome 22q11.2 deletion syndrome. Psychiatry Res-Neuroim 181:1–8. doi:10.1016/j.pscychresns.2009.07.003

Shashi V, Veerapandiyan A, Keshavan MS et al (2012) Altered development of the dorsolateral prefrontal cortex in chromosome 22q11.2 deletion syndrome: an in vivo proton spectroscopy study. Biol Psychiatry 72:684–691. doi:10.1016/j.biopsych.2012.04.023

Silveri MC, Reali G, Jenner C, Puopolo M (2007) Attention and memory in the preclinical stage of dementia. J Geriatr Psychiatry Neurol 20:67–75. doi:10.1177/0891988706297469

Simon AE, Cattapan-Ludewig K, Zmilacher S et al (2007) Cognitive functioning in the schizophrenia prodrome. Schizophr Bull 33:761–771. doi:10.1093/schbul/sbm018

Simons JS, Speirs HJ (2003) Preforntal and medial temporal lobe interactions in long-term memory. Nat Rev Neurosci 4:637–648

Sobin C, Kiley-Brabeck K, Daniels S et al (2004) Networks of attention in children with the 22q11 deletion syndrome. Dev Neuropsychol 26:611–626. doi:10.1207/s15326942dn2602_5

Stanislaw H, Todorov N (1999) Calculation of signal detection theory measures. Behav Res Methods Instrum Comput 31:137–149

Stevens DA, Fechter LD (1968) Relative strengths of approach and avoidance tendencies in discrimination learning of rats trained under two types of reinforcement. J Exp Psychol 76:489–491. doi:10.1037/h0025553

Stopford CL, Thomson JC, Neary D et al (2012) Working memory, attention, and executive function in Alzheimer's disease and frontotemporal dementia. Cortex 48:429–446

Swainson R, Hodges JR, Galton CJ et al (2001) Early detection and differential diagnosis of Alzheimer's disease and depression with neuropsychological tasks. Dement Geriatr Cogn Disord 12:265–280

Talbot PS, Watson DR, Barrett SL, Cooper SJ (2005) Rapid tryptophan depletion improves decision-making cognition in healthy humans without affecting reversal learning or set shifting. Neuropsychopharmacology 31:1519–1525. doi:10.1038/sj.npp.1300980

Talpos J, Steckler T (2013) Touching on translation. Cell Tissue Res. doi:10.1007/s00441-013-1694-7

Talpos JC, Winters BD, Dias R et al (2009) A novel touchscreen-automated paired-associate learning (PAL) task sensitive to pharmacological manipulation of the hippocampus: a translational rodent model of cognitive impairments in neurodegenerative disease. Psychopharmacology 205:157–168. doi:10.1007/s00213-009-1526-3

Toichi M, Findling RL, Kubota Y et al (2004) Hemodynamic differences in the activation of the prefrontal cortex: attention vs. higher cognitive processing. Neuropsychologia 42:698–706. doi:10.1016/j.neuropsychologia.2003.08.012

Ursu S, Stenger VA, Shear MK et al (2003) Overactive action monitoring in obsessive-compulsive disorder: evidence from functional magnetic resonance imaging. Psychol Sci 14:347–353

Valentino DA, Arruda JE, Gold SM (1993) Comparison of QEEG and response accuracy in good vs poorer performers during a vigilance task. Int J Psychophysiol 15:123–133. doi:10.1016/0167-8760(93)90070-6

Voon V, Irvine MA, Derbyshire K et al (2014) Measuring "waiting" impulsivity in substance addictions and binge eating disorder in a novel analogue of rodent serial reaction time task. Biol Psychiatry 75:148–155. doi:10.1016/j.biopsych.2013.05.013

Weiss EM, Bilder RM, Fleischhacker WW (2002) The effects of second-generation antipsychotics on cognitive functioning and psychosocial outcome in schizophrenia. Psychopharmacology 162:11–17. doi:10.1007/s00213-002-1053-y

White HK, Levin ED (1999) Four-week nicotine skin patch treatment effects on cognitive performance in Alzheimer's disease. Psychopharmacology 143:158–165

Whitehouse PJ, Price DL, Struble RG et al (1982) Alzheimer's disease and senile dementia: loss of neurons in the basal forebrain. Science 215:1237–1239. doi:10.2307/1688438?ref=no-x-route:b0a5c28d36e9e70b883b7c5ea2426388

Winstanley CA, Dalley JW, Theobald DE, Robbins TW (2004) Fractionating impulsivity: contrasting effects of central 5-HT depletion on different measures of impulsive behavior. Neuropsychopharmacology 29:1331–1343

Wohlberg GW, Kornetsky C (1973) Sustained attention in remitted schizophrenics. Arch Gen Psychiatry 28:533–537

Wood SJ, Proffitt T, Mahony K et al (2002) Visuospatial memory and learning in first-episode schizophreniform psychosis and established schizophrenia: a functional correlate of hippocampal pathology? Psychol Med 32:429–438

Worbe Y, Savulich G, Voon V et al (2014) Serotonin depletion induces "waiting impulsivity" on the human four-choice serial reaction time task: cross-species translational significance. Neuropsychopharmacology 39:1519–1526. doi:10.1038/npp.2013.351

Xu B, Roos JL, Levy S et al (2008) Strong association of de novo copy number mutations with sporadic schizophrenia. Nat Genet 40:880–885. doi:10.1038/ng.162

Young JW, Light GA, Marston HM et al (2009a) The 5-choice continuous performance test: evidence for a translational test of vigilance for mice. PLoS ONE 4:e4227–13. doi:10.1371/journal.pone.0004227

Young JW, Powell SB, Risbrough V et al (2009b) Using the MATRICS to guide development of a preclinical cognitive test battery for research in schizophrenia. Pharmacol Ther 122:150–202. doi:10.1016/j.pharmthera.2009.02.004

Young JW, Geyer MA, Rissling AJ, Sharp RF (2013) Translational psychiatry—reverse translation of the rodent 5C-CPT reveals that the impaired attention of people with schizophrenia is similar to scopolamine-induced deficits in mice

The Paired Associates Learning (PAL) Test: 30 Years of CANTAB Translational Neuroscience from Laboratory to Bedside in Dementia Research

Jennifer H. Barnett, Andrew D. Blackwell, Barbara J. Sahakian and Trevor W. Robbins

Abstract The origins and rationale of the Cambridge Neuropsychological Test Automated Battery (CANTAB) as a cross-species translational instrument suitable for use in human neuropsychopharmacological studies are reviewed. We focus on its use for the early assessment and detection of Alzheimer's disease, in particular the Paired Associates Learning (PAL) test. We consider its psychometric properties, neural validation, and utility, including studies on large samples of healthy volunteers, patients with mild cognitive impairment (MCI), and Alzheimer's disease. We demonstrate how it can be applied in cross-species studies using experimental animals to bridge the cross-species translational 'gap'. We also show how the CANTAB PAL has bridged a second translational 'gap' through its application to the early detection of memory problems in primary care clinics, using iPad technology.

Keywords Memory · Associative learning · Alzheimer's disease · Mild cognitive impairment · Translational · Drug development

J.H. Barnett (✉) · A.D. Blackwell · B.J. Sahakian
Department of Psychiatry, University of Cambridge, Cambridge, UK
e-mail: jenny.barnett@camcog.com

J.H. Barnett · A.D. Blackwell
Cambridge Cognition, Cambridge, UK

B.J. Sahakian · T.W. Robbins
Medical Research Council/Wellcome Trust Behavioural and Clinical Neuroscience Institute, Cambridge, UK

T.W. Robbins
Department of Psychology, University of Cambridge, Cambridge, UK

Curr Topics Behav Neurosci (2016) 28: 449–474
DOI 10.1007/7854_2015_5001

Contents

1 The Origins of CANTAB 450
1.1 The Translational Approach 451
2 Development of PAL, a Non-Verbal Test of Cued Recall, and an Exemplar of CANTAB 453
2.1 Neural Validity of the CANTAB PAL 455
3 PAL as a Flexible Tool in Alzheimer's Drug Development 457
3.1 PAL in Proof of Concept Phases of Drug Development: Sensitivity to the Effects of Drugs 458
3.2 PAL in Prodromal and Early-Stage Alzheimer's Trials: Sensitivity to Burden of Neuropathology 460
4 Clinical Utility: PAL in Clinical Research and Specialist Settings 462
4.1 Differentiating Normal and Abnormal Cognition in Older Adults: Early Detection of Memory Problems by CANTAB PAL 462
4.2 Predicting Progression from MCI to Dementia: Relationship of CANTAB PAL to Functionality in Daily Life 464
4.3 Differentiating MCI and Depression 465
4.4 Relative Specificity of PAL for Other Disorders, Including Non-Alzheimer's Dementias 466
5 To Bedside: Adaptation of CANTAB PAL for Mainstream Healthcare and Beyond 466
5.1 The Development of CANTAB Mobile 467
5.2 The Future of Cognitive Health 468
References 469

1 The Origins of CANTAB

CANTAB was developed with the aid of a major award from the Wellcome Trust in the 1980s. The basic idea was to utilise what were, at the time, new technologies for psychological research, such as the computerised control of experiments and touch-sensitive screens. The latter had already proven useful in studies of animal cognition where monkeys could interact directly with stimuli presented on the computer screen simply by touching them, a procedure which drastically reduces the difficulty of training on what are otherwise quite complex tasks. Taking the same logic to the neuropsychological assessment of patients was at the time highly innovative, though today it is considerably easier due to the availability of cheap, reliable touchscreen devices such as the iPad.

The other main principle in the construction of CANTAB was to employ tests inspired from the extensive animal neuropsychological literature (such as 'delayed (non-)matching-to-sample') in a form suitable for testing humans. Moreover, what had been sensitive but complex human tests, such as the Wisconsin Card Sort Test, were decomposed to versions that could be employed in testing both animals (e.g. marmoset monkeys) and patients. Eventually, variants of this paradigm emerged for testing rodents also, but ingeniously using different modalities, smell and touch, in a 'digging through sand' paradigm—resulting in its wholesale use for screening drug effects and more recently phenotyping genetic murine mutants (see Brown and Tait, this volume).

Finally, some useful and ingenious tests from classical human neuropsychology, such as the Corsi Blocks test of short-term spatial span and the Tower of London test of planning, were made more powerful by computerisation, through the addition of sensitive timed performance measures and the potential to randomise or prespecify the presentation of complex stimuli and problem sets.

The CANTAB tests were originally designed for use in elderly patients with neurodegenerative diseases such as Parkinson's and Alzheimer's diseases. Since then, their use has broadened considerably to include neuropsychiatric disorders such as depression, schizophrenia, OCD, and addiction, other neurological disorders such as multiple sclerosis and traumatic brain injury, developmental disorders such as ADHD and autism, and for testing the effects of drugs in healthy volunteers as well as patient populations. As such, the CANTAB bibliography (www.cambridgecognition.com/bibliography) now lists more than 1500 articles on the use of CANTAB, including papers detailing the impact of clinical disorders ranging from anxiety to cancer, early life trauma, cardiovascular, and viral disorders. A wide range of interventions have also been studied, ranging from antenatal care to neurostimulation, nutritional and surgical treatments, as well as many classes of pharmacological compounds.

1.1 The Translational Approach

The CANTAB exemplifies translation in two ways. First, the tests originated in part from translating neuropsychological tests in rodents and monkeys to humans, and the use of a touch-sensitive screen and computer-generated visual stimuli has also allowed back-translation to monkeys and rodents through the invention of 'monkey CANTAB' by TW Robbins and AC Roberts (see Weed et al. 1999) and the more recent application of computerised touchscreen paradigms to measure attention, learning, and memory in rats and mice (Hvoslef-Eide et al. 2015 and this volume). Second, the tests developed initially for laboratory-based research have ultimately been widely adopted for use in clinical trials and clinical practice. In this chapter, we review both of these translational themes.

The development of CANTAB was therefore intended to be particularly conducive to translation across the spectrum from basic neuroscience to human clinical research, and from clinical research to clinical practice (Cooksey 2006). An especially instructive example in terms of Alzheimer's disease has been the testing and assessment of the (now largely outmoded) cholinergic hypothesis. There was, at the time, considerable emphasis on the possibility that dementia could be treated effectively with cholinergic compounds because of evidence that cortical acetylcholine depletion occurred fairly early in the disease as a consequence of degeneration of the cholinergic nucleus basalis of Meynert (nbM). Consequently, much attention was focused on producing models of this specific pathology by destroying the homologue of the structure, the nucleus basalis magnocellularis in rodents and monkeys, to produce reliable cognitive deficits that could then be remediated by cholinergic agents such as the anticholinesterases. In fact, achieving this model was much more problematic than first appeared, as it proved difficult to lesion selectively the cholinergic neurons of the nbM without also damaging non-cholinergic cells and also the adjacent globus pallidus (Dunnett et al. 1991). Ultimately, it became clear that selective lesions of the nbM did not produce reliable impairments in long-term learning and memory, but mainly affected attentional and perhaps working memory functions. This position was reached partly through the use of more selective cholinergic lesioning agents such as AMPA and, subsequently, the cholinergic immunotoxin 192-IgG-saporin, but also by the use of sensitive behavioural measures, including tests of sustained attention (Robbins 2002; see Lustig and Sarter, this volume). The 5-choice serial reaction time test (5CSRTT) for rodents was developed originally from a human test of sustained attention used to test effects of stressors and drugs on human performance (Robbins 2002). In this test, rats or mice are trained to detect brief flashes of light presented unpredictably in space, in order to earn food rewards. Cholinergic depletion of the nbM produces substantial impairments in the accuracy of doing this in rats, especially in lengthy test sessions with brief stimuli presented at a fast rate, which may induce vigilance deficits through fatigue (Dalley et al. 2004). Such impairments are remediated by treatment with cholinergic drugs (e.g. Muir et al. 1995). More recently, in the touchscreen version of this task for mice (see Bussey et al., this volume), Romberg et al. (2011) found that the impaired performance of triple transgenic Alzheimer's model mice challenged with excessively brief stimulus durations in this task was rescued with low doses of the cholinesterase inhibitor donepezil.

These data are very consistent with evidence from an early trial of the anticholinesterase tacrine in patients with Alzheimer's disease which showed clinical benefit, but with improvement being mainly limited to the human CANTAB version of the 5-choice task, and not in tests of short-term memory such as delayed matching-to-sample (Sahakian et al. 1993). The human outcome was therefore readily predicted and paralleled by studies using similar behavioural tests in animals. The interpretation of the human data was that treatment with cholinergic agents would be of limited, although sometimes significant, therapeutic benefit in patients diagnosed with Alzheimer's disease for whom there was evidence of cognitive deficit correlated with massive neural degeneration of the brain (Howard et al. 2012). This damage would limit the neuromodulatory boost provided from cholinergic agents to the performance of simple tasks such as the 5CSRTT not requiring extensive networking functions. In fact, it appears that these data do parallel the current clinical experience with cholinergic agents in Alzheimer's disease. It may be that these drugs are even more effective in the case of Lewy body or Parkinsonian dementias where there is greater evidence of cortical cholinergic loss (Emre et al. 2007).

While tacrine was approved by the US Food and Drug Administration for use in 1993, it was eventually withdrawn due to safety concerns. Other cholinesterase inhibitors were approved, followed in 2003 by the NMDA receptor antagonist memantine, but the effects of all these compounds are symptomatic, improving cognition in a manner that is usually time-limited and which does not affect the underlying progress of disease. Since 2003, no additional treatments have been approved, and the clinical experience of many symptomatic Alzheimer's development programmes has paralleled that of tacrine and thrown into relief the importance of carrying out clinical trials on patients early in the course of the disease before severe neurodegeneration occurs. The holy grail of AD drug development remains the search for a compound with a disease-modifying effect, which would arrest the spread of neuropathology with a neuroprotective strategy that either prevents or slows the progress of disease or, ideally, reverses existing neural damage. One effect of this shift in emphasis has been to increase the attention being paid to screening the initial cognitive deficits in Alzheimer's patients, particularly among those considered in a prodromal state of disease (i.e. with mild cognitive impairment; MCI). This strategy of identifying early cases has required the invention of tests sensitive to this early stage, which includes, most obviously, impairments in episodic memory.

2 Development of PAL, a Non-Verbal Test of Cued Recall, and an Exemplar of CANTAB

The development of the Paired Associates Learning (PAL) test as an instrument for detecting dementia is a particularly good example of the CANTAB approach to translation. Over the course of 25 years, its use in humans has extended from basic research and classic neuropsychology through the testing of drugs for Alzheimer's and other CNS disorders, and now into mainstream healthcare, where it is used in the triage of patients presenting with memory complaints in primary care. Simultaneously, versions of the PAL test have been developed for other species, including rodent (Talpos et al. 2009) and primate (Taffe et al. 2004) versions which have been used successfully in a range of experimental settings, contributing particularly to translational psychopharmacological studies.

Episodic memory is the ability to learn, store, and retrieve information about past experiences. In the animal literature, it is often described as remembering 'what, when and where', although human episodic memory rarely relies on all three. The CANTAB PAL consists of multiple trials in which the participant learns the location of one or more visual patterns on the screen (Fig. 1). This 'object-location memory' (Parkinson et al. 1988) is particularly dependent on integrity of the entorhinal and transentorhinal cortex and hippocampal areas (Smith and Milner 1981; Parkinson et al. 1988; McDonald and White 1993; Owen et al. 1996; Miyashita et al. 1998; Maguire et al. 1998; De Rover et al. 2011). At the easiest stages of PAL, there is a single pattern to remember the location of, and at the most difficult there are eight patterns. Participants are allowed multiple attempts at each trial, with automatic re-presentation of all stimuli after each attempt. The test is

Fig. 1 Encoding and retrieval stages of the CANTAB Paired Associates Learning test. During encoding, participants watch as the location of various patterns is revealed and must remember which pattern is in which *white box*. During retrieval, patterns are presented in the centre of the screen and the participant must press the box in which they previously saw the pattern

adaptive so that if a trial is not completed despite multiple attempts, the test automatically terminates and the error score calculated includes an adjustment for stages that were not attempted. This graded nature allows both cognitively impaired and cognitively intact populations to be assessed on the same test. There are multiple equivalent forms of test stimuli, to reduce the risk of learning effects across successive assessments, and the patterns are designed to be difficult to verbalise, thus discouraging the use of rehearsal strategies.

Many reliable and well-validated human episodic memory tests exist, which differ in important theoretical and practical aspects (Lezak et al. 2004). The two most common forms are verbal recall tests such as story- or wordlist-learning, and non-verbal tests such as the CANTAB PAL (Sahakian et al. 1988). Verbal tests are very widely used in clinical practice, but their reliance on language may be problematic in some situations, including in patients with difficulties in speaking, reading/hearing, or for whom the test language is not their native language. In contrast, visual memory tests such as the CANTAB PAL offer the advantage of being largely cultural and language neutral, allowing easy use across different countries and the opportunity for parallel use in non-human primates and other species.

As well as presenting test stimuli in different ways, memory tests differ in the way that participants' responses are collected. Traditional neuropsychological tests usually require either verbal or written responses; like the rest of the CANTAB, the PAL test captures responses through presses on the touchscreen. This was originally in order to facilitate use in patients with motoric problems and non-human species but its ease of use, and the increasing availability and reliability of touchscreen devices, has facilitated its adoption. Over the past 25 years, we estimate it has been used with ease by hundreds of thousands of participants ranging from very young children to the elderly (Robbins et al. 1994; Rhodes et al. 2006), including patients with motor difficulties in neurodegenerative disorders, such as Parkinson's (Owen et al. 1993) and Huntington's diseases (Lange et al. 1995), and in AD and other dementias (Sahakian et al. 1994; Fowler et al. 1995; Barson et al. 2000; Deakin et al. 2004).

2.1 Neural Validity of the CANTAB PAL

The initial inspiration for the PAL test came from studies by Brenda Milner on object–location memory (Smith and Milner 1981). In her charmingly everyday, but nevertheless cogent and original test of memory, patients were confronted by a tray of toy-like objects which required a verbal evaluation of their likely cost (to ensure adequate attention to and study of the test items). Then, towards the end of the test session, when testing on other tasks had intervened, subjects were given the tray with the objects and asked (without warning) to place where the toy-like objects had been situated on the tray. This test differs from the PAL task in involving an essentially incidental initial learning stage, but both test the ability to learn object–location associations. Patients with right-sided hippocampal damage were especially impaired on this task (Smith and Milner 1981).

The design of CANTAB PAL was also influenced by a study of effects of large hippocampal lesions in rhesus monkeys on object–location association learning (Parkinson et al. 1988). In this test, monkeys were required to study and recall where objects had been located. On study trials, monkeys would receive rewards in certain locations underneath characteristic objects. After a short delay, they were then required to discriminate at which location they had previously seen this object when confronted with identical objects in different locations. Again, large ablative lesions aimed at the hippocampal formation produced substantial impairments in memory performance (Parkinson et al. 1988).

Translation of the human PAL task for rhesus monkeys was achieved by requiring rhesus monkeys to learn the location of complex visual stimuli on the touch-sensitive screen (Weed et al. 1999; Taffe et al. 2002). Early studies addressed the utility of the task for modelling deficits in Alzheimer's disease (Taffe et al. 2004), including deleterious effects of scopolamine, an NMDA receptor antagonist, and the nicotinic receptor antagonist mecamylamine (Taffe et al. 1999, 2002; Katner et al. 2004). Some of these experiments were relevant to parallel human studies, e.g. with scopolamine (Robbins et al. 1997a, b; Taffe et al. 1999) and nicotine (Katner et al. 2004). More recently, the monkey PAL task has mainly been utilised to assess deleterious effects of drugs of abuse such as cannabis (Taffe 2012) and alcohol (Wright and Taffe 2014). However, its relevance to remediating cognitive deficits has been shown from surprising cognitive enhancing effects on visuospatial memory performance by mephedrone and d-methamphetamine (Wright et al. 2012), using cannabidiol to reverse the detrimental effects of cannabis (Wright et al. 2013). However, there have been, as yet, no studies of which neural systems are necessary for PAL performance in macaques.

Likely analogues of CANTAB PAL in novel touchscreen tests have also been developed for rats and mice that similarly require learning the location of a visual object on the touchscreen (Talpos et al. 2009; Bartko et al. 2011; Horner et al. 2013, see chapter by Hvoslef-Eide et al. this volume). The neural basis of performance on these tasks has been investigated with the aid of conventional lesioning methods in both mice and rats and has implicated the medial prefrontal cortex as well as the hippocampus. Thus, performance on PAL requires activation of both NMDA and AMPA glutamate receptor subtypes in the dorsal hippocampus to maintain performance (Talpos et al. 2009), whereas the integrity of the medial prefrontal cortex appears necessary for rats to acquire the task (McAllister et al. 2015). In mice, genes controlling both NMDA and AMPA receptor function also impair acquisition of PAL when knocked out in mice (Coba et al. 2012). There appears to be a similar requirement for intact dorsal hippocampal function during performance though not acquisition of the task (Kim et al. 2015). The antimuscarinic drug scopolamine impairs PAL performance in mice, whereas donepezil enhances it (Bartko et al. 2011). In an exciting study, both mouse and human carriers of the Dlg2 mutation implicated in schizophrenia were shown to have impaired PAL performance, again illustrating the translational potential of the paradigm (Nithianantharajah et al. 2013).

Demonstrating some commonalities across species in the neural basis of PAL performance is important because they suggest the use of common processing strategies in experimental animals and humans. The central requirements of PAL are to associate ventral visual stream information concerning visual objects with spatial information (i.e. 'what' and 'where'). These streams coincide in the entorhinal cortex and adjoining hippocampal formation, where some of the earliest pathology of Alzheimer's disease occurs. Nevertheless, there may well be some differences between the animal and human versions because both monkey and rodent PAL take a considerable period to learn and require continuous feedback in the form of food reinforcement, whereas the human PAL task is a learning task where there is no feedback. Thus, other processes, such as reinforcement learning, likely contribute to the acquisition of animal PAL, and this may entail the recruitment of other neural systems (such as the striatum).

The equivalence of the neural circuitry involved in humans has been confirmed by lesion and neuroimaging studies. In healthy older adults, functional neuroimaging studies confirm that the CANTAB PAL task activates the bilateral hippocampus in a load-dependent manner such that activation increases as the number of patterns to be encoded increases (De Rover et al. 2011). Activation is also seen in other brain areas associated with hippocampal–prefrontal connectivity, such as the precuneus and the medial prefrontal cortex during successful performance on the task, making it a good overall test of hippocampal connectivity (De Rover et al. 2011). As would be expected from this profile, lesion studies confirm that successful performance of the PAL test is dependent on functional integrity of the temporal lobe, particularly the entorhinal cortex, but is also affected by resections of the frontal lobe (Owen et al. 1995). In patients with schizophrenia, the extent of impairment seen on the CANTAB PAL test correlates with the observed loss in hippocampal volume (Kéri et al. 2012).

3 PAL as a Flexible Tool in Alzheimer's Drug Development

In older adults, deficits in episodic memory are often a symptom of AD, the most common cause of dementia. These memory impairments arise early in the clinical presentation of AD because they rely heavily on areas affected during the early stages of Alzheimer's neuropathology, i.e. the spread of neurofibrillary tangles and neuropil threads initially in the transentorhinal cortex and then in the entorhinal cortex and hippocampus (Braak and Braak 1991). A major motivator in the design of the CANTAB PAL was to design a test enabling the earliest possible detection of AD-related cognitive impairment (Fig. 2), both for use in clinic and, importantly, to assist the development of drugs to treat AD.

The computerised and non-verbal nature of the CANTAB PAL gives it advantages over other forms of memory assessment for use in the drug development process. It provides a directly comparable assay with which to look for positive or negative effects of a new compound on cognition in preclinical models and later clinical development, giving the best possible chance of predicting human cognitive signals from preclinical data. In the later stages of clinical development, where multinational trials are now the norm, PAL and other computerised tests offer operational benefits including a reduction in potential sources of error through standardised and automated administration, data capture, and scoring and the ability to remotely monitor cognitive data captured at sites all round the world. The non-verbal and culture-free nature of the CANTAB PAL also reduces any potential variance introduced by language and cultural translations of stimuli in word-based tests.

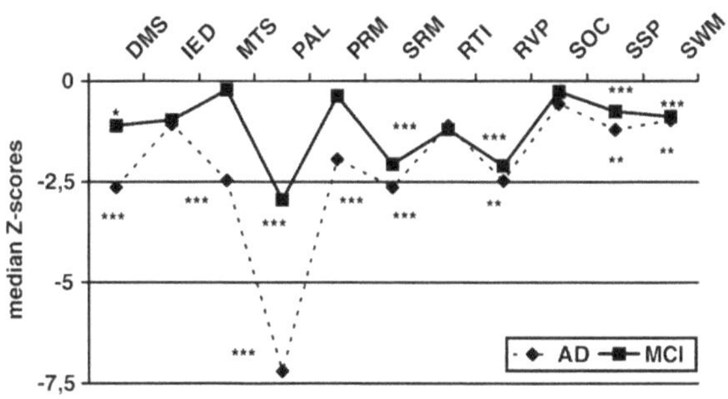

Fig. 2 The profile of cognitive impairments across a range of CANTAB tests among patients with mild cognitive impairment (MCI) and mild–moderate Alzheimer's disease (AD). Substantial and relatively selective impairments are seen in MCI on the Paired Associates Learning (PAL) test, with further worsening evident in patients with AD. Reproduced from Egerházi et al. (2007), with permission. CANTAB tests: *DMS* delayed match to sample; *IED* intra-/extra dimensional shifting; *MTS* match to sample visual search; *PAL* Paired Associates Learning; *PRM* pattern recognition memory; *SRM* spatial recognition memory; *RTI* reaction time; *RVP* rapid visual information processing; *SOC* Stockings of Cambridge; *SSP* spatial span; *SWM* spatial working memory (see www.cantab.com for details of each test)

3.1 *PAL in Proof of Concept Phases of Drug Development: Sensitivity to the Effects of Drugs*

Alongside other CANTAB tests, PAL is commonly used in the early stages of clinical drug development programmes, including phase 1 studies. These typically involve a small number of healthy volunteers randomised to receive either a single or a small number of doses of placebo or active drug in a crossover or parallel design. They are, by definition, primarily concerned with safety and not efficacy, as such they often involve protocols that are suboptimal with respect to testing cognitive efficacy. Cognitive results from phase 1 studies are often unpublished or mentioned only briefly in a summary paper describing all of early phase development. Thus, it is difficult to judge either from published literature, or indeed, from in-depth knowledge of any one drug development programme, the extent to which phase 1 studies of cognition produce reliable and replicable results. An alternative means of assessing their utility is to consider comparably powered studies published with already approved drugs in healthy volunteers. Figure 3 summarises PAL effect sizes gleaned from mostly small-scale ($n < 40$) single-dose studies in healthy volunteers of compounds with a wide variety of mechanism of actions.

Clearly, the test is sensitive to a wide range of compounds with both positive and negative effects on cognition. Individual studies of this size can be difficult to interpret, because statistical power is low and multiple indices of cognitive function

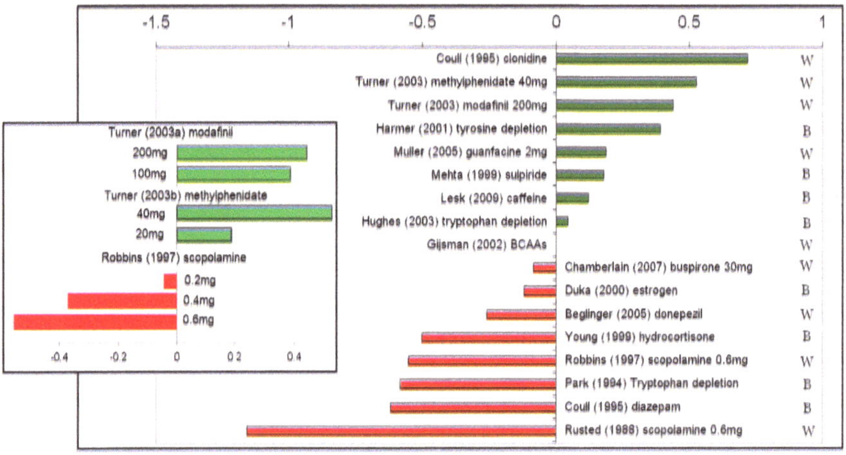

Fig. 3 Magnitude of drug effects as measured by the CANTAB Paired Associates Learning (PAL) in small healthy volunteer studies (Rusted and Warburton 1988; Park et al. 1994; Coull et al. 1995; Robbins et al. 1997a, b; Young et al. 1999; Mehta et al. 1999; Duka et al. 2000; Harmer et al. 2001; Gijsman et al. 2002; Turner et al. 2003a, b; Hughes et al. 2003; Beglinger et al. 2005; Müller et al. 2005; Chamberlain et al. 2007; Lesk et al. 2009). *W* within-subjects (crossover) designs; *B* between-subjects (parallel) designs. *Bars* reflect effect size (Cohen's d); within- and between-subjects effect sizes are comparable where the within-subject correlation is 0.5 (Morris and DeShon 2002). *Green* reflects improvement, *red* impairment, in PAL performance

may have been assessed. However, there are a number of signals that can increase confidence that the observed effect is replicable and not due to chance. These include the observation that the same effects are seen on multiple, theoretically related measures (i.e. reaction time measures taken from more than one test) or that there are obvious dose–response effects, as seen in the inset box of Fig. 3.

The existence of such a body of published work allows comparisons of effects of novel compounds with established treatments, providing both real-world benchmarking and a means to investigate analogous perturbations across different neurochemical systems. For example, in one early study, Robbins et al. (1997a, b) directly compared the effects of the muscarinic antagonist scopolamine and the benzodiazepine diazepam. Intriguingly, although scopolamine produced dose-dependent deficits in PAL performance, they were even more pronounced for delayed matching-to-sample (DMS; i.e. visual recognition memory) and were shown not simply to reflect perceptual deficits. Of even greater significance, diazepam produced greater effect on PAL than DMS, suggesting that GABA-ergic neurotransmission (possibly in the hippocampus) is of particular importance to PAL performance. This study demonstrates the point that cholinergic systems may have preferential neuromodulatory influences on the neural networks underlying representational memory, consistent with their attentional role, whereas GABA-ergic effects may be part of the circuitry that directly mediate such representations.

The majority of the compounds currently in the development for AD have actions which are unlikely to produce positive effects on cognition in the healthy individuals taking part in phase 1 trials. Phase 2 studies thus become crucial in offering the first opportunity to assess a compound's likely effect on memory in the target population. Phase 2 study designs are often more amenable to cognitive assessment than phase 1 and have considerably greater statistical power. Determining which cognitive domains to test in a phase 2 trial is not always straightforward, though episodic memory is an obvious front runner for MCI and AD trials. Cognitive tests that tap specific neural or neurochemical systems are likely to be more sensitive to compounds that act on relatively selective receptor systems or brain targets; conversely, global measures of cognitive function may be more sensitive to the effect of drugs with multiple actions. Drug–placebo differences would therefore be expected to be maximally detected by a cognitive test that is both optimised to this stage of disease (i.e. would show maximum placebo decline) and is matched to the treatment in terms of the selectivity of cognitive functions measured. For example, Greig et al. (2005) reported on an early-stage clinical trial of the anticholinesterase phenserine in patients with Alzheimer's disease which reported significant benefit of the drug on PAL performance, which was not found using the broader clinical measure of cognitive performance, the ADAS-Cog. A more recent example of the successful targeting of cognitive tests to treatment in a relatively short and small-scale phase 2a trial was reported by Pfizer, in which a 5HT-6 receptor antagonist showed a dose–response relationship with two measures from the PAL and also on pattern recognition memory, a second CANTAB memory test, over 4 weeks in patients with mild to moderate Alzheimer's disease (Brisard et al. 2010).

3.2 PAL in Prodromal and Early-Stage Alzheimer's Trials: Sensitivity to Burden of Neuropathology

In recent years, new approvals for AD drugs have been notably absent and the field has suffered several disappointing failures from promising symptomatic and disease-modifying compounds. A common explanation for these failures is that the development programmes were aimed too far along the disease progress, in patients who had dementia and therefore had already suffered considerable neurological damage. Many researchers now believe that in order to arrest or reverse the disease progress, compounds will need to be administered much earlier, years prior to the emergence of clinical symptoms of AD. As such, much clinical trial activity has now switched to patients with predementia stages of the disease, particularly the stage of 'prodromal AD' or 'MCI due to AD' where biomarkers of AD are present, along with a relatively selective memory impairment, but where cognitive function remains relatively intact and independence is maintained.

Following some early indications in studies by Fowler et al. (1995, 1997, 2002), Swainson et al. (2001) showed that CANTAB PAL is highly sensitive to the memory impairments that are characteristic of this phase; it successfully differentiated patients with 'questionable dementia' into two subgroups more selectively than a wide range of other neuropsychological tests or the ADAS-cog. The test is thus potentially useful in prodromal AD trials both as a means of identifying patients suitable for recruitment into the trial and as a sensitive measurement of the efficacy of a treatment. CANTAB PAL appears to be a particularly useful screening or inclusion criteria for prodromal AD not only as a sensitive memory measure, but also because scores on the test show significant correlations with the amount of amyloid pathology present in this population (see Fig. 4). Recent data from one of the largest prodromal AD trials thus far show that among people who all meet clinical criteria for MCI, those who have CSF amyloid levels suggestive of Alzhiemer's pathology score around half a standard deviation lower on the CANTAB PAL (Scheltens et al. 2014).

As the field moves towards pivotal trials for prodromal AD, and potentially even presymptomatic phases of AD, there is little precedent available as to the optimal endpoints at either a scientific or regulatory level. As noted in recent reports (Vellas et al. 2013), in the absence of reliable surrogate biomarkers, a reduction in the expected age-related decline on sensitive measures of memory such as the CANTAB PAL is as yet the only robust means of judging efficacy in presymptomatic AD populations.

One challenge in using cognitive tests, as opposed to fluid or imaging biomarkers in clinical trials, is the potential for practice effects in patients tested repeatedly. It is therefore crucial that the selected cognitive test has psychometric properties that make it suitable for repeated administration. These are typically achieved through the existence of multiple matched forms of the test which use different stimuli sets in order to minimise carry-over from previous assessments. One advantage of computerised tests such as the PAL is that essentially unlimited

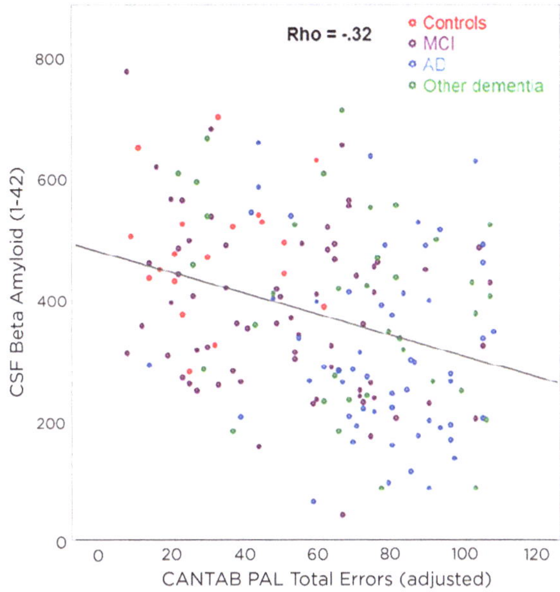

Fig. 4 CANTAB Paired Associates Learning total error scores and levels of cerebrospinal fluid beta amyloid 1-42 in individuals with Alzheimer's and other dementias, mild cognitive impairment, and healthy elderly controls from the EDAR study (www.edarstudy.eu). Figure adapted from (Barnett et al. 2011)

numbers of equivalent versions can be produced by enforcing random selection of stimuli from a very large bank of potential patterns for use at each assessment. This probably helps both to reduce practice effects and increase test–retest reliability. For example, one-month retesting using the CANTAB PAL in healthy elderly volunteers was reported to produce relatively minor practice effects of around 0.2 standard deviations in magnitude and good test–retest reliabilities of 0.7–0.9 for the two major outcome measures (Lowe and Rabbitt 1998).

Detection of cognitive change in clinical trials requires the test to be appropriately positioned relative to the spectrum of disorder for which it is being used. In the same way that inches and miles cannot be measured on the same ruler, it is unlikely that the tools optimally sensitive to detecting cognitive improvement among those with minimal impairment will be also appropriate for measuring cognitive deterioration among patients with rapid cognitive decline. The optimal sensitivity of the ADAS-cog, the most common endpoint for mild-moderate AD trials, is an MMSE score of fourteen (Hobart 2010), i.e. moderate dementia. In contrast, PAL is better suited to detecting more subtle impairments and changes of impairment in the presymptomatic, MCI, and mild AD stages.

For studies of potentially disease-modifying compounds, a significant decline must be measurable on the chosen cognitive endpoint during the time span of the trial. This is absolutely crucial in order to be able to detect mitigation in this rate of decline due to disease-modifying effects of the treatment. The CANTAB PAL may therefore be particularly useful in clinical trials of individuals in prodromal or presymptomatic stages of AD where 'natural' rates of decline on many other cognitive tests will be minimal. In the ADNI data, Ito and colleagues (Ito et al.

2010) report that the annual rate of decline in ADAS-cog points among 75-year-olds is less than one point in cognitively normal individuals (those with MMSE > 27), 2 points per year at MMSE = 25, 5 points at MMSE = 22, 7 points at MMSE = 19, and 9 points in patients with MMSE of 16 or less. Deterioration of performance on the PAL test is approximately double this rate, at around 1.2 points per year among cognitively intact older adults (Blackwell et al. 2010), 5 points per year among unselected individuals with subjective memory complaints (Chamberlain et al. 2011), 20 points per year for individuals with subjective memory complaints who subsequently convert to dementia (Chamberlain et al. 2011), and 24 points per year among patients with AD (Fowler et al. 2002).

4 Clinical Utility: PAL in Clinical Research and Specialist Settings

Since its conception, the CANTAB PAL has been used in specialist clinics to demonstrate and investigate the earliest stages of cognitive impairment in AD and related disorders. While no cognitive tests offer perfect differential diagnoses, the test has proved highly sensitive in even the very early stages of AD and often specific enough to accurately differentiate patients with AD from other groups, such as depression (Swainson et al. 2001).

Patients presenting to specialist clinical settings such as memory clinics typically comprise a heterogeneous mix of people with mild cognitive impairment, dementia of various forms, and cognitive impairments due to other reversible or irreversible causes such as hormonal imbalance, infection, or trauma. In some settings, people with concerns about their memory but no objective deficit, the so-called worried well, may also present in considerable numbers. The utility of PAL in such clinics has included differentiating healthy individuals from those with MCI, AD or depression, and providing some predictive validity among individuals presenting with memory impairments. Here, we selectively review data describing the use of PAL in contributing to these aims.

4.1 Differentiating Normal and Abnormal Cognition in Older Adults: Early Detection of Memory Problems by CANTAB PAL

It is not always obvious to clinicians or the public to what extent memory decline should be expected as a normal part of healthy ageing. For episodic memory, the characteristic pattern of life course performance is that the ability reaches a peak early in life and remains stable through young adulthood, before beginning to worsen around the fifth decade—albeit with considerable interindividual variation

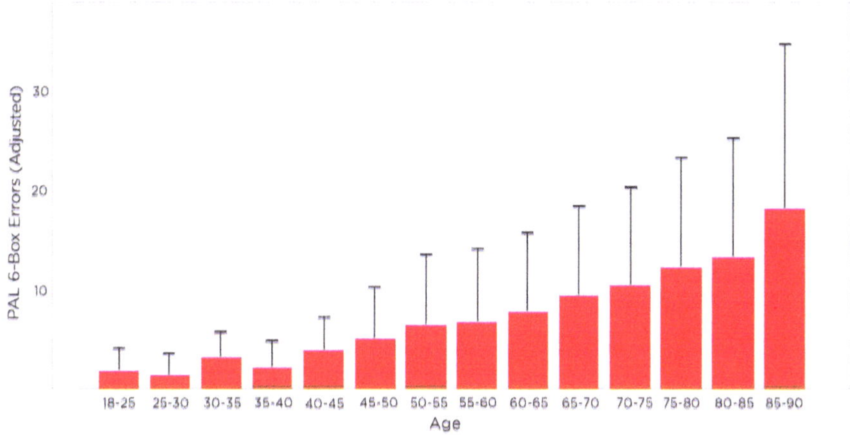

Fig. 5 Mean (SD) number of errors made on the CANTAB Paired Associates Learning test by adults aged 18–90 years old with no current psychological or neurological symptoms (Cambridge Cognition, internal data)

in the extent and rate of decline. Figure 5 shows the mean and standard deviation of performance (total error scores, adjusted for stages successfully completed) on the CANTAB PAL among more than 5000 non-demented individuals aged 18–90 years. Age-related decline can be seen from perhaps the early 40 s, and while other demographic factors such as gender and education have additional effects, these are largely static and minor when compared with the effect of ageing.

CANTAB PAL shows excellent sensitivity and a high specificity in differentiating mild–moderate Alzheimer's dementia from age-matched controls: a cut-off of greater than 20 errors at the six-pattern stage shows 100 % sensitivity and 92 % specificity when aggregated across three case–control studies (Cambridge Cognition, internal data). By definition, a patient with MCI has impairment in one or more domains of cognition and will consequently have a high probability of being in a neurodegenerative disease process that will ultimately result in dementia. The extent to which any cognitive test or other biomarker can demonstrate that it truly detects MCI is currently limited by the absence of a definitive diagnostic test for MCI. Nonetheless, PAL shows reasonable sensitivity and specificity, for example the OPTIMA study reported a sensitivity of 0.83 and a specificity of 0.82 for the CANTAB PAL in differentiating clinically defined MCI from age-matched healthy controls (Chandler et al. 2008). These data are quite impressive for a non-invasive, inexpensive measure, which takes only a few minutes to administer.

4.2 *Predicting Progression from MCI to Dementia: Relationship of CANTAB PAL to Functionality in Daily Life*

For older adults experiencing memory problems, the most immediate concern is often whether they are developing dementia, so a test's ability to predict the likelihood of progression from MCI to dementia is particularly clinically important. Mitchell et al. (2009) reported longitudinal data from patients with MCI suggesting that the two tests most able to discriminate at baseline whether or not a patient will convert from MCI to AD were the CANTAB PAL and the Addenbrooke's Cognitive Examination (ACE; Mathuranath et al. 2000), a bedside assessment of five cognitive domains (attention, language, fluency, memory, and visuospatial). Used together, these test scores accounted for more than 99 % of variance in outcome, with sensitivity to conversion of 94 %.

Early studies (Fowler et al. 1995, 1997, 2002) using PAL in patients with subjective memory complaints found that test scores fell into two clusters: one that predominantly overlapped with healthy elderly individuals and the other with AD patients. Two-year follow-up of these individuals found that all of the individuals in the cluster characterised by poor and declining performance went on to have a diagnosis of NINCDS-ADRDA probable AD, while those with initially better and longitudinally stable performance remained unimpaired (Fowler et al. 2002). A similar study found that among patients with questionable dementia, those who were relatively unimpaired at baseline on both the PAL and the Graded Naming Test (McKenna and Warrington 1980), remained dementia-free at 32 months, while all of those identified as impaired on these two tests at baseline had diagnoses of 'probable AD' by 32 months (Blackwell et al. 2004).

It is important to calibrate what these changes mean in real-world terms to a patient. Scores on the CANTAB PAL correlate well with self-reported everyday memory failures, as measured by the Cambridge Behavioural Inventory ($r = 0.69$; see Fig. 6), and with the independent completion of activities of daily living ($r = 0.58$) suggesting that changes on the PAL test do reflect a real deterioration in memory-related daily experience.

Test scores from a single assessment are inherently limited in their ability to infer how much an individual's current memory performance is likely to reflect a pathological cognitive decline. Memory tests should therefore be suitable for repeated administration, so that subtle decline or improvement can be directly measured and interpreted over time. Psychometric properties that make a test suitable for repeated assessment include good test–retest reliability reflecting consistency of measurement over time and minimal practice effects. The PAL uses multiple matched stimuli sets and random location of stimuli in order to minimise carry-over from previous assessments. Lowe and Rabbitt (1998) reported test–retest reliability of the PAL test of 0.68 for first trial memory scores and 0.86 for average trials to success in healthy adults aged between 60 and 82 years. Since cognitive performance is typically stable over short time periods in AD, good psychometric

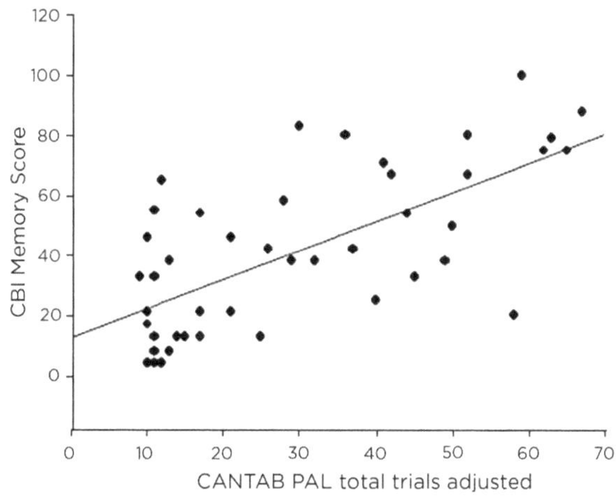

Fig. 6 Correlation between number of errors made on the CANTAB PAL test and self-reported memory function (Cambridge Behavioural Inventory score) in patients with a diagnosis of Alzheimer's disease (unpublished data, A Blackwell; $r = 0.69$; $p < 0.001$)

properties may actually be easier to achieve in this patient group than in healthy volunteers. For example, Fowler and colleagues report one-month test–retest reliabilities of 0.64 for healthy controls, 0.71 for MCI, and 0.88 for AD patients for CANTAB PAL total error scores (Fowler et al. 1995).

4.3 Differentiating MCI and Depression

Detecting cases of AD-related MCI is complicated by the high prevalence of undiagnosed depression among elderly populations (Kay et al. 1985; Livingston et al. 1990), which can produce cognitive impairments that are often confused with early symptoms of dementia or MCI. In the Cambridge Memory Clinic, for example, the memory complaints of 5 % of referrals were ultimately attributed to depression (Alladi et al. 2006). If the cognitive symptoms of dementia are inaccurately attributed to the depression or vice versa, a potentially treatable problem may go undetected; thus, cognitive tests which are differentially sensitive to depression and dementia can be particularly valuable in elderly patients. Swainson et al. (2001) compared cognitive test scores between individuals with dementia, depression, and healthy elderly controls. They found no differences between depressed and healthy individuals on a number of cognitive measures including the CANTAB PAL, suggesting it would be useful in differentiating the early signs of dementia from depression. In contrast, patients with depression scored significantly worse than healthy controls on the ADAS-cog, on logical memory, and on measures of sustained attention and executive function, suggesting that poor results on these tests may be due either to depression or to the early signs of dementia.

4.4 Relative Specificity of PAL for Other Disorders, Including Non-Alzheimer's Dementias

Of course that is not to say that PAL is exclusively impaired in AD. Just as the hippocampal and temporal lobe memory systems can be subject to many forms of insult, so too is PAL performance susceptible to other disorders, such as traumatic brain injury, non-Alzheimer's dementias, or some forms of psychosis (Sahgal et al. 1991; Owen et al. 1995; Barnett et al. 2005). However, Lee et al. (2003) found relatively minor deficits in PAL performance in fronto-temporal dementia, consistent with the greater episodic memory deficits in Alzheimer's disease. The mild impairments were significant for the semantic dementia though not the frontal variant subgroups. For dementia of the Lewy Body type, large impairments in PAL have been reported (Galloway et al. 1992); however, these patients often additionally have demonstrable fluctuating attentional deficits confirmed by other CANTAB tests (Sahgal et al. 1992), which may contribute to the apparent memory loss. Severe cases of Parkinson's disease may exhibit relatively mild impairments in PAL performance (Owen et al. 1993), and such deficits are part of the profile of the 'mild cognitive impairment' seen in this disease (Yarnall et al. 2014).

Impaired performance on this memory test cannot therefore be diagnostic of Alzheimer's disease, although it may be suggestive, especially if an elderly patient has exhibited progressive deterioration in performance, and other possible causes have been excluded.

5 To Bedside: Adaptation of CANTAB PAL for Mainstream Healthcare and Beyond

From the beginning, one intention in developing a test particularly sensitive to the early signs of AD was to ultimately encourage timely diagnosis and treatment of AD in mainstream healthcare. Here, we describe recent efforts to cross the 'T2 translational gap' (Cooksey 2006), by formulating CANTAB PAL into a tool that can be used by non-specialists to aid the early detection of possible AD in general practice. We also describe possible future uses of the test in encouraging better public understanding and awareness of cognitive health.

5.1 The Development of CANTAB Mobile

The CANTAB PAL's good sensitivity and specificity in detecting AD make it superior to most tests available to non-specialist clinicians such as general practitioners, who are probably the healthcare professionals most likely to interact with at-risk older adults. The cognitive tests most familiar to GPs are pencil and paper or interview-based tests which show a good sensitivity to frank dementia but poor sensitivity to earlier stages of disease, including MCI. A systematic review of cognitive screening measures commonly used in primary care (including the Mini Mental State Exam, MMSE, and the General Practitioner Assessment of Cognition, GP-COG) found that sensitivity to detect mild to severe dementia ranged from 69 to 100 % and specificity from 75 to 98 % in GP, community, or population samples (Brodaty et al. 2006).

The properties that make a test optimally useful in academic research or clinical trials are somewhat different from those required for widespread testing in mainstream healthcare. In order to address some of the practical barriers limiting the usefulness of very sensitive cognitive tests in mainstream healthcare, the CANTAB PAL software has recently been adapted for non-specialist use in the form of CANTAB Mobile, an iPad-based assessment tool. The aim in doing so was to create a tool that is easy to use by a healthcare professional with no specialist training and that takes as short a time as possible to administer. In the UK, where general practitioners act as gatekeepers to dementia diagnostic services, the major use of cognitive tests in primary care is to provide a means of deciding whether an individual is likely to be suffering from cognitive decline, and hence, whether they would benefit from further assessment (i.e. referral to memory services). At the time of writing, CANTAB Mobile has been used in this way at around 400 sites across the UK, largely primary care practices. Analysis of the first 10,000 tests completed found that in only 25 % of assessments did the test detect clinically significant memory problems. This is important because it suggests that the real-world utility of such tests is not only in identifying those patients who do have a significant memory impairment, but also in reassuring the very large number of concerned older adults who do not.

Several technical changes were made in order to better suit the CANTAB PAL test for the primary care environment. The CANTAB Mobile app is adaptive so that if a patient is struggling at a particular level of difficulty, the test automatically terminates, limiting the duration of testing to around 5 min. The non-verbal nature of the test makes it particularly useful in non-literate or ethnic minority groups, and voiceover instructions in multiple languages makes it possible to accurately assess patients from ethnic minorities in their native language without the need for a translator. NICE guidelines for cognitive assessment in dementia state that relevant demographic factors such as age, gender, and educational level should be taken into account when deciding whether a person's score is outside their personal expected range (National Collaborating Centre for Mental Health 2007). In practice, this is extremely difficult for non-specialists to achieve, so CANTAB Mobile results are compared with a large normative database and then automatically adjusted for age, gender, and level of education. The scores are output in a report which indicates whether memory performance was as expected for someone of these demographics, or outside of the normal range and thus a potential clinical concern. These adaptations have made the PAL test of considerable use to GPs as an aid to clinical triage and decision-making.

5.2 *The Future of Cognitive Health*

As touchscreen technologies become evermore ubiquitous in businesses, shops, and homes, it is interesting to consider how touchscreen tests such as PAL can best contribute to improving the cognitive health of the population.

Tackling the large and increasing public health burden that Alzheimer's and other dementias represent probably requires two parallel efforts. In one, a better understanding of basic brain biology, and better use of translational tools such as CANTAB to predict drug effects from animal studies into phase 1 and from early-stage into late-stage clinical development, should hopefully result in increased success in the development of new therapies. In particular, CANTAB's use in phase 2 studies allows accurate early decisions to be made about efficacy, based on more reliable data about the effect of new compounds in the clinical population. In the second track, making CANTAB and other cognitive tools more widely available and simple to use can contribute to tackling the overall impact of Alzheimer's and other cognitive disorders by making it easier to conduct large-scale cognitive monitoring, for example in longitudinal prevention studies, or for determination by the government of healthcare provision.

A technological analogy may be found here in the development of the sphygmomanometer (blood pressure monitor). Over many decades, blood pressure monitoring moved from a specialist procedure to one that became standard in outpatient and then primary care facilities, usable by healthcare providers in home visits, and is now available directly to consumers for self-assessment. This progress was driven by a number of factors, including the development of the technology itself, such that devices became cheaper, more reliable, and more mobile, but also increasing awareness of the benefits of monitoring blood pressure—including the benefits of measurements taken at multiple time points, over the course of the day, and in contexts other than the clinic. Alongside developments in the availability of effective interventions to treat hypertension, the distal effect of the availability of accurate and widespread monitoring of blood pressure has been a massive reduction in incidence of coronary heart disease and stroke.

We believe that cognitive health is now primed for a similar increase in public awareness, consumer, and home-based intervention, and, hopefully, consequent impacts on public health (Sahakian 2014). One appealing vision for the future of CANTAB is therefore that it will be one of the technological innovations that allow an equivalent of blood pressure monitoring for the mind—simplifying cognitive assessment for non-specialist clinicians, increasing public awareness of the links between behaviours such as diet, exercise, smoking, and brain health, and ultimately contributing to the pathway by which people at high risk of dementia-related diseases are identified, monitored, diagnosed, and treated.

Like blood pressure monitors, there may well be a case for making sensitive well-validated cognitive tests such as the PAL available in a wider range of healthcare environments such as pharmacies, nursing homes, and other places where people at high risk of undetected AD may be accessible, and ultimately to consumers themselves. Such tests would need to be presented in a way that appropriately takes into account aspects of the at-home testing environment, including the capabilities and limitations of users' own computers or mobile devices, they should use test feedback as a means to encourage appropriate utilisation of healthcare systems and increase healthy behaviour, and they should empower patients to take responsibility for and control of their own brain health, without increasing worry or unnecessary interventions.

As technology moves on further, we also see a role for well-established cognitive tests such as the PAL in benchmarking new methods of cognitive assessment and monitoring which will inevitably develop. This might include means of inferring cognitive performance and risk of cognitive disorders from the day-to-day use of devices such as computers, smartphones, and, eventually, ubiquitous monitoring of patterns of behaviour throughout a wired home environment.

Declaration of Interest and Acknowledgements JHB is an employee of and ADB a director of Cambridge Cognition, the company that develops and markets the CANTAB cognitive testing system. JB, AB, and BJS own shares and/or share options in Cambridge Cognition. BJS consults for Cambridge Cognition, Peak, Otsuka, Lundbeck, and Servier, holds a grant from Janssen/J&J, and owns shares in CeNeS. TWR consults for Cambridge Cognition, Eli Lilly, Shire, Lundbeck, Teva, Otsuka, and Chempartners. He holds grants from Lilly, Lundbeck, and GSK and receives royalties from Cambridge Cognition for CANTAB.

References

Alladi S, Arnold R, Mitchell J et al (2006) Mild cognitive impairment: applicability of research criteria in a memory clinic and characterization of cognitive profile. Psychol Med 36:507–515. doi:10.1017/S0033291705006744

Barnett JH, Damian M, Verhey FRJ et al (2011) Cognitive and CSF biomarkers in older adults, MCI, and dementia. Alzheimer's Dement 7:S129–S130. doi:10.1016/j.jalz.2011.05.340

Barnett JH, Sahakian BJ, Werners U et al (2005) Visuospatial learning and executive function are independently impaired in first-episode psychosis. Psychol Med 35:1031–1041

Barson FP, Kinsella GJ, Ong B, Mathers SE (2000) A neuropsychological investigation of dementia in motor neurone disease (MND). J Neurol Sci 180:107–113

Bartko SJ, Vendrell I, Saksida LM, Bussey TJ (2011) A computer-automated touchscreen Paired Associates Learning (PAL) task for mice: impairments following administration of scopolamine or dicyclomine and improvements following donepezil. Psychopharmacology 214:537–548. doi:10.1007/s00213-010-2050-1

Beglinger LJ, Tangphao-Daniels O, Kareken DA et al (2005) Neuropsychological test performance in healthy elderly volunteers Before and after Donepezil administration. J Clin Psychopharmacol 25:159–165. doi:10.1097/01.jcp.0000155822.51962.b4
Blackwell AD, Barnett JH, Hayat S et al (2010) The effect of age, sex and education on visuospatial Paired Associates Learning ability: preliminary data from a British population study. Alzheimers Dement 6:S485
Blackwell AD, Sahakian BJ, Vesey R et al (2004) Detecting dementia: novel neuropsychological markers of preclinical Alzheimer's disease. Dement Geriatr Cogn Disord 17:42–48
Braak H, Braak E (1991) Neuropathological stageing of Alzheimer-related changes. Acta Neuropathol 82:239–259
Brisard C, Safirstein B, Booth K et al (2010) Safety, tolerability, and preliminary efficacy of SAM-531, a 5HT-6 antagonist, in subjects with mild-to-moderate Alzheimer's disease: results from a phase 2a study. Alzheimer's Dement 6:S311. doi:10.1016/j.jalz.2010.05.1013
Brodaty H, Low L-F, Gibson L, Burns K (2006) What is the best dementia screening instrument for general practitioners to use? Am J Geriatr Psychiatry 14:391–400. doi:10.1097/01.JGP.0000216181.20416.b2
Chamberlain SR, Blackwell AD, Nathan PJ et al (2011) Differential cognitive deterioration in dementia: a two year longitudinal study. J Alzheimers Dis 24:125–136. doi:10.3233/JAD-2010-100450
Chamberlain SR, Muller U, Deakin JB et al (2007) Lack of deleterious effects of buspirone on cognition in healthy male volunteers. J Psychopharmacol 21:210–215. doi:10.1177/0269881106068066
Chandler J, Marsico M, Harper-Mozley L et al (2008) Cognitive assessment: Discrimination of impairment and detection of decline in Alzheimer's disease and mild cognitive impairment. Alzheimer's Dement 4:T551–T552. doi:10.1016/j.jalz.2008.05.1676
Coba MP, Komiyama NH, Nithianantharajah J et al (2012) TNiK is required for postsynaptic and nuclear signaling pathways and cognitive function. J Neurosci 32:13987–13999. doi:10.1523/JNEUROSCI.2433-12.2012
Cooksey D (2006) A review of UK health research funding. Norwich, UK
Coull JT, Middleton HC, Robbins TW, Sahakian BJ (1995) Contrasting effects of clonidine and diazepam on tests of working memory and planning. Psychopharmacology 120:311–321
Dalley JW, Theobald DE, Bouger P et al (2004) Cortical cholinergic function and deficits in visual attentional performance in rats following 192 IgG-saporin-induced lesions of the medial prefrontal cortex. Cereb Cortex 14:922–932. doi:10.1093/cercor/bhh052
De Rover M, Pironti VA, McCabe JA, et al (2011) Hippocampal dysfunction in patients with mild cognitive impairment: a functional neuroimaging study of a visuospatial Paired Associates Learning task. Neuropsychologia 49:2060–70. doi:10.1016/j.neuropsychologia.2011.03.037
Deakin JB, Rahman S, Nestor PJ et al (2004) Paroxetine does not improve symptoms and impairs cognition in frontotemporal dementia: a double-blind randomized controlled trial. Psychopharmacol 172:400–408
Duka T, Tasker R, McGowan JF (2000) The effects of 3-week estrogen hormone replacement on cognition in elderly healthy females. Psychopharmacology 149:129–139
Dunnett SB, Everitt BJ, Robbins TW (1991) The basal forebrain-cortical cholinergic system: interpreting the functional consequences of excitotoxic lesions. Trends Neurosci 14:494–501
Egerházi A, Berecz R, Bartók E, Degrell I (2007) Automated Neuropsychological Test Battery (CANTAB) in mild cognitive impairment and in Alzheimer's disease. Prog Neuropsychopharmacol Biol Psychiatry 31:746–751. doi:10.1016/j.pnpbp.2007.01.011
Emre M, Cummings JL, Lane RM (2007) Rivastigmine in dementia associated with Parkinson's disease and Alzheimer's disease: similarities and differences. J Alzheimers Dis 11:509–519
Fowler KS, Saling MM, Conway EL et al (1995) Computerized delayed matching to sample and paired associate performance in the early detection of dementia. Appl Neuropsychol 2:72–78. doi:10.1207/s15324826an0202_4
Fowler KS, Saling MM, Conway EL et al (1997) Computerized neuropsychological tests in the early detection of dementia: prospective findings. J Int Neuropsychol Soc 3:139–146

Fowler KS, Saling MM, Conway EL et al (2002) Paired associate performance in the early detection of DAT. J Int Neuropsychol Soc 8:58–71

Galloway P, Sahgal A, McKeith IG et al (1992) Visual pattern recognition memory and learning deficits in senile dementias of the Alzheimer and Lewy Body types. Dementias 3:101–107

Gijsman HJ, Scarnà A, Harmer CJ et al (2002) A dose-finding study on the effects of branch chain amino acids on surrogate markers of brain dopamine function. Psychopharmacology 160: 192–197. doi:10.1007/s00213-001-0970-5

Greig NH, Sambamurti K, Yu Q et al (2005) An overview of phenserine tartrate, a novel acetylcholinesterase inhibitor for the treatment of Alzheimer's disease. Curr Alzheimer Res 2:281–290

Harmer CJ, McTavish SFB, Clark L et al (2001) Tyrosine depletion attenuates dopamine function in healthy volunteers. Psychopharmacology 154:105–111. doi:10.1007/s002130000613

Hobart J (2010) Developing the ADAS-Cog to enhance its measurement performance for clinical trials of MCI. In: Clinical trials on Alzheimer's disease. Toulouse, France

Horner AE, Heath CJ, Hvoslef-Eide M et al (2013) The touchscreen operant platform for testing learning and memory in rats and mice. Nat Protoc 8:1961–1984. doi:10.1038/nprot.2013.122

Howard R, McShane R, Lindesay J et al (2012) Donepezil and memantine for moderate-to-severe Alzheimer's disease. N Engl J Med 366:893–903. doi:10.1056/NEJMoa1106668

Hughes JH, Gallagher P, Stewart ME et al (2003) The effects of acute tryptophan depletion on neuropsychological function. J Psychopharmacol 17:300–309. doi:10.1177/0269881103017301 2

Hvoslef-Eide M, Mar A, Nilsson S, et al (2015) The NEWMEDS rodent touchscreen test battery for cognition relevant to schizophrenia. Psychopharmacology 232:3853-3872

Ito K, Corrigan B, Zhao Q et al (2010) Disease progression model for cognitive deterioration from Alzheimer's disease neuroimaging initiative database. Alzheimers Dement 7:151–160. doi:10.1016/j.jalz.2010.03.018

Katner SN, Davis SA, Kirsten AJ, Taffe MA (2004) Effects of nicotine and mecamylamine on cognition in rhesus monkeys. Psychopharmacology 175:225–240. doi:10.1007/s00213-004-1804-z

Kay DW, Henderson AS, Scott R et al (1985) Dementia and depression among the elderly living in the Hobart community: the effect of the diagnostic criteria on the prevalence rates. Psychol Med 15:771–788

Kéri S, Szamosi A, Benedek G, Kelemen O (2012) How does the hippocampal formation mediate memory for stimuli processed by the magnocellular and parvocellular visual pathways? Evidence from the comparison of schizophrenia and amnestic mild cognitive impairment (aMCI). Neuropsychologia 50:3193–3199. doi:10.1016/j.neuropsychologia.2012.10.010

Kim CH, Heath CJ, Kent BA et al (2015) The role of the dorsal hippocampus in two versions of the touchscreen automated Paired Associates Learning (PAL) task for mice. Psychopharmacology. doi:10.1007/s00213-015-3949-3

Lange KW, Sahakian BJ, Quinn NP et al (1995) Comparison of executive and visuospatial memory function in Huntington's disease and dementia of Alzheimer type matched for degree of dementia. J Neurol Neurosurg Psychiatry 58:598–606

Lee ACH, Rahman S, Hodges JR et al (2003) Associative and recognition memory for novel objects in dementia: implications for diagnosis. Eur J Neurosci 18:1660–1670

Lesk VE, Honey TEM, de Jager CA (2009) The effect of recent consumption of caffeine-containing foodstuffs on neuropsychological tests in the elderly. Dement Geriatr Cogn Disord 27:322–328. doi:10.1159/000207445

Lezak MD, Howieson DB, Loring DW (2004) Neuropsychological Assessment, 4th edn. Oxford University Press, Oxford

Livingston G, Hawkins A, Graham N et al (1990) The gospel oak study: prevalence rates of dementia, depression and activity limitation among elderly residents in inner London. Psychol Med 20:137–146

Lowe C, Rabbitt P (1998) Test/re-test reliability of the CANTAB and ISPOCD neuropsychological batteries: theoretical and practical issues. Cambridge neuropsychological test automated battery. International study of post-operative cognitive dysfunction. Neuropsychologia 36:915–923

Maguire EA, Frith CD, Burgess N et al (1998) Knowing where things are: parahippocampal involvement in encoding object locations in virtual large-scale space. J Cogn Neurosci 10:61–76. doi:10.1162/089892998563789

Mathuranath PS, Nestor PJ, Berrios GE et al (2000) A brief cognitive test battery to differentiate Alzheimer's disease and frontotemporal dementia. Neurology 55:1613–1620

McAllister K, Mar A, Theobald DE, Saksida LM, Bussey TJ (2015) Comparing the effects of subchronic phencyclidine and medial prefrontal cortex lesions on cognitive tests relevant to schizophrenia. Psycopharmacology 232:3883–3898

McDonald RJ, White NM (1993) A triple dissociation of memory systems: hippocampus, amygdala, and dorsal striatum. Behav Neurosci 107:3–22

McKenna P, Warrington EK (1980) Testing for nominal dysphasia. J Neurol Neurosurg Psychiatry 43:781–788

Mehta MA, Sahakian BJ, McKenna PJ, Robbins TW (1999) Systemic sulpiride in young adult volunteers simulates the profile of cognitive deficits in Parkinson's disease. Psychopharmacology 146:162–174. doi:10.1007/s002130051102

Mitchell J, Arnold R, Dawson K et al (2009) Outcome in subgroups of mild cognitive impairment (MCI) is highly predictable using a simple algorithm. J Neurol 256:1500–1509. doi:10.1007/s00415-009-5152-0

Miyashita Y, Kameyama M, Hasegawa I, Fukushima T (1998) Consolidation of visual associative long-term memory in the temporal cortex of primates. Neurobiol Learn Mem 70:197–211. doi:10.1006/nlme.1998.3848

Morris SB, DeShon RP (2002) Combining effect size estimates in meta-analysis with repeated measures and independent-groups designs. Psychol Methods 7:105–125

Muir JL, Everitt BJ, Robbins TW (1995) Reversal of visual attentional dysfunction following lesions of the cholinergic basal forebrain by physostigmine and nicotine but not by the 5-HT3 receptor antagonist, ondansetron. Psychopharmacology 118:82–92

Müller U, Clark L, Lam ML et al (2005) Lack of effects of guanfacine on executive and memory functions in healthy male volunteers. Psychopharmacology 182:205–213. doi:10.1007/s00213-005-0078-4

National Collaborating Centre for Mental Health (2007) Dementia: a NICE-SCIE Guideline on supporting people with dementia and their carers in health and social care. The British Psychological Society

Nithianantharajah J, Komiyama NH, McKechanie A et al (2013) Synaptic scaffold evolution generated components of vertebrate cognitive complexity. Nat Neurosci 16:16–24. doi:10.1038/nn.3276

Owen AM, Beksinska M, James M et al (1993) Visuospatial memory deficits at different stages of Parkinson's disease. Neuropsychologia 31:627–644

Owen AM, Milner B, Petrides M, Evans AC (1996) A specific role for the right parahippocampal gyrus in the retrieval of object-location: a positron emission tomography study. J Cogn Neurosci 8:588–602. doi:10.1162/jocn.1996.8.6.588

Owen AM, Sahakian BJ, Semple J et al (1995) Visuo-spatial short-term recognition memory and learning after temporal lobe excisions, frontal lobe excisions or amygdalo-hippocampectomy in man. Neuropsychologia 33:1–24

Park S, Coull JT, Mcshane R et al (1994) Tryptophan depletion in normal volunteers produces selective impairments in learning and memory. Neuropharmacology 33:575–588. doi:10.1016/0028-3908(94)90089-2

Parkinson J, Murray E, Mishkin M (1988) A selective mnemonic role for the hippocampus in monkeys: memory for the location of objects. J Neurosci 8:4159–4167

Rhodes SM, Coghill DR, Matthews K (2006) Acute neuropsychological effects of methylphenidate in stimulant drug-naive boys with ADHD II–broader executive and non-executive domains. J Child Psychol Psychiatry 47:1184–1194

Robbins TW (2002) The 5-choice serial reaction time task: behavioural pharmacology and functional neurochemistry. Psychopharmacol 163:362–380

Robbins TW, James M, Owen AM et al (1994) Cambridge neuropsychological test automated battery (CANTAB): a factor analytic study of a large sample of normal elderly volunteers. Dementia 5:266–281

Robbins TW, Semple J, Kumar R et al (1997a) Effects of scopolamine on delayed-matching-to-sample and paired associates tests of visual memory and learning in human subjects: comparison with diazepam and implications for dementia. Psychopharmacol 134:95–106

Robbins TW, Semple J, Kumar R et al (1997b) Effects of scopolamine on delayed-matching-to-sample and paired associates tests of visual memory and learning in human subjects: comparison with diazepam and implications for dementia. Psychopharmacology 134:95–106

Romberg C, Mattson MP, Mughal MR et al (2011) Impaired attention in the 3xTgAD mouse model of Alzheimer's disease: rescue by donepezil (Aricept). J Neurosci 31:3500–3507. doi:10.1523/JNEUROSCI.5242-10.2011

Rusted JM, Warburton DM (1988) The effects of scopolamine on working memory in healthy young volunteers. Psychopharmacology 96:145–152

Sahakian BJ (2014) What do experts think we should do to achieve brain health? Neurosci Biobehav Rev 43:240–258. doi:10.1016/j.neubiorev.2014.04.002

Sahakian BJ, Coull JJ, Hodges JR (1994) Selective enhancement of executive function by idazoxan in a patient with dementia of the frontal lobe type. J Neurol Neurosurg Psychiatry 57:120–121

Sahakian BJ, Morris RG, Evenden JL, et al (1988) A comparative study of visuospatial memory and learning in Alzheimer-type dementia and Parkinson's disease. Brain 111 (Pt 3:695–718)

Sahakian BJ, Owen AM, Morant NJ et al (1993) Further analysis of the cognitive effects of tetrahydroaminoacridine (THA) in Alzheimer's disease: assessment of attentional and mnemonic function using CANTAB. Psychopharmacol 110:395–401

Sahgal A, Galloway P, McKeith IG et al (1992) A comparative study of attentional deficits in senile dementias of the Alzheimer and lewy body types. Dementia 3:350–354

Sahgal A, Sahakian BJ, Robbins TW et al (1991) Detection of visual memory and learning deficits in Alzheimer's disease using the Cambridge neuropsychological test automated battery. Dementia 2:150–158

Scheltens P, Boada M, Dorflinger E et al (2014) Baseline patient characteristics from the Phase 3 SCarlet RoAD trial, a study of gantenerumab in patients with prodromal AD. JPAD 1:231–232

Smith ML, Milner B (1981) The role of the right hippocampus in the recall of spatial location. Neuropsychologia 19:781–793

Swainson R, Hodges JR, Galton CJ et al (2001) Early detection and differential diagnosis of Alzheimer's disease and depression with neuropsychological tasks. Dement Geriatr Cogn Disord 12:265–280

Taffe MA (2012) Δ^9 Tetrahydrocannabinol impairs visuo-spatial associative learning and spatial working memory in rhesus macaques. J Psychopharmacol 26:1299–1306. doi:10.1177/0269881112443743

Taffe MA, Weed MR, Gold LH (1999) Scopolamine alters rhesus monkey performance on a novel neuropsychological test battery. Brain Res Cogn Brain Res 8:203–212

Taffe MA, Weed MR, Gutierrez T et al (2004) Modeling a task that is sensitive to dementia of the Alzheimer's type: individual differences in acquisition of a visuo-spatial Paired Associate Learning task in rhesus monkeys. Behav Brain Res 149:123–133

Taffe MA, Weed MR, Gutierrez T et al (2002) Differential muscarinic and NMDA contributions to visuo-spatial Paired Associate Learning in rhesus monkeys. Psychopharmacol 160:253–262

Talpos JC, Winters BD, Dias R et al (2009) A novel touchscreen-automated Paired Associate Learning (PAL) task sensitive to pharmacological manipulation of the hippocampus: a translational rodent model of cognitive impairments in neurodegenerative disease. Psychopharmacology 205:157–168. doi:10.1007/s00213-009-1526-3

Turner DC, Robbins TW, Clark L et al (2003a) Cognitive enhancing effects of modafinil in healthy volunteers. Psychopharmacology 165:260–269

Turner DC, Robbins TW, Clark L et al (2003b) Relative lack of cognitive effects of methylphenidate in elderly male volunteers. Psychopharmacology 168:455–464. doi:10.1007/s00213-003-1457-3

Vellas B, Carrillo MC, Sampaio C et al (2013) Designing drug trials for Alzheimer's disease: what we have learned from the release of the phase III antibody trials: a report from the EU/US/CTAD Task Force. Alzheimers Dement 9:438–444. doi:10.1016/j.jalz.2013.03.007

Weed MR, Taffe MA, Polis I et al (1999) Performance norms for a rhesus monkey neuropsychological testing battery: acquisition and long-term performance. Brain Res Cogn Brain Res 8:185–201

Wright MJ, Taffe MA (2014) Chronic periadolescent alcohol consumption produces persistent cognitive deficits in rhesus macaques. Neuropharmacology 86:78–87. doi:10.1016/j.neuropharm.2014.07.003

Wright MJ, Vandewater SA, Angrish D et al (2012) Mephedrone (4-methylmethcathinone) and d-methamphetamine improve visuospatial associative memory, but not spatial working memory, in rhesus macaques. Br J Pharmacol 167:1342–1352. doi:10.1111/j.1476-5381.2012.02091.x

Wright MJ, Vandewater SA, Taffe MA (2013) Cannabidiol attenuates deficits of visuospatial associative memory induced by Δ(9) tetrahydrocannabinol. Br J Pharmacol 170:1365–1373. doi:10.1111/bph.12199

Yarnall AJ, Breen DP, Duncan GW et al (2014) Characterizing mild cognitive impairment in incident Parkinson disease: the ICICLE-PD study. Neurology 82:308–316. doi:10.1212/WNL.0000000000000066

Young AH, Sahakian BJ, Robbins TW, Cowen PJ (1999) The effects of chronic administration of hydrocortisone on cognitive function in normal male volunteers. Psychopharmacology 145:260–266

Experimental Medicine in Psychiatry New Approaches in Schizophrenia, Depression and Cognition

Gerard R. Dawson

Abstract The use of experimental medicine studies to bridge the gap between Phase 1 and 2 drug trials and so to enhance translation of basic neuroscience studies using experimental animals to the clinic is proposed. Illustrative examples are provided for affective disorders and schizophrenia in relation also to cognitive dysfunction.

Keywords Experimental medicine · Schizotypy · Depression · Cognition · Clinical trials

Contents

1 Introduction 475
2 Experimental Medicine and Schizophrenia 477
3 Schizotypy and Its Prevalence in the General Population 480
3.1 Effects of Schizotypy and Treatment on the Performance of Cognitive Tasks 483
3.2 Effects of Schizotypy and Treatment on Signal Detection, Biconditional Learning and Aberrant Salience 484
3.3 Effects of Schizotypy and Treatment on Eye Movements 485
4 The Neuropharmacology of Depression and Cognition 488
5 Summary and Conclusions 493
References 493

1 Introduction

It is often stated that in the last 50 years, the rate of production of new treatments for psychiatric diseases by pharmaceutical companies has been relatively constant and despite large increases in investment is now actually on the decline. In the last

G.R. Dawson (✉)
P1vital LTD, Manor House, Howbery Park, Wallingford, Oxfordshire OX10 8BA, UK
e-mail: gdawson@p1vital.com

Curr Topics Behav Neurosci (2016) 28: 475–497
DOI 10.1007/7854_2015_5016

three decades, industry and academia have invested strongly in biomedical research aimed at discovering new pathways and mechanisms that might lead to new treatments for anxiety, cognitive disorders and depression. They have often worked together under new initiatives, such as the Innovative Medicines Initiative (IMI), much of it directly aimed at elucidating the mechanisms underlying disease and brain function. Indeed, our knowledge of basic brain systems and functions increases by the day with sometimes remarkable new findings. New experimental techniques have been developed, and a wealth of new drug targets aimed at improving the treatment of psychiatric and other brain disorders has been suggested and investigated at enormous expense to the pharma industry. However, the clinical translation of these research findings to the benefit of psychiatric patients has been disappointing. These results and other late-stage failures has led to pressure from investors and shareholders that has resulted in a number of pharmaceutical companies withdrawing from the active development of new psychiatric treatments. Late-stage, Phase III failures such as Roche's Bitopertin (RO-4917838) a glycine reuptake inhibitor, developed as an adjunct to antipsychotics for the treatment of persistent negative symptoms, is a case in point. Results from a Phase II proof-of-concept study were promising showing a significant reduction in negative symptoms and a trend towards improved functioning (Umbricht et al. 2014). On the strength of these results, six Phase III trial studies were initiated, but development was terminated when early results from two of the studies failed to reach their primary endpoints. These results, other late-stage failures and the complexity of developing drugs for diseases that have no clear pathology, have led pharmaceutical companies to reprioritise or withdraw from drug development in psychiatry. This is particularly problematic as there remains a huge unmet need for new treatments for psychiatric disorders as many existing treatments are only partially effective and many have significant side effects that led to non-compliance among patients.

The past 20 years has shown that it is clearly very difficult to develop new drugs for the treatment of psychiatric disorders. New drug classes such as selective inhibitors of glycine reuptake and phosphodiesterase inhibitors of brain enzyme have been identified. They have shown promising results in animal assays and cleared safety hurdles in animals and humans. However, while potent molecules can be generated, the gap between their effects in animal assays and their efficacy in patients remains. The failure in patients often raises the question of whether in diseases such as major depressive disorder (MDD) and schizophrenia that have specific but also overlapping functional constructs, the compound was assessed for efficacy in the right patient population. Moreover, patients with schizophrenia are usually treated with multiple drugs which may mask or nullify the effects of a new compound under investigation. Similarly, patients with MDD may have a history of treatment resistance or partial response or may be misdiagnose as unipolar when they are in fact bipolar with low levels of mania (Angst et al. 2011). Consequently, there are now significantly more human experimental and translational medicine studies that are conducted between Phase 1 and Phase 2 clinical trials that focus on detecting the efficacy of new compounds before large patient trials. These studies often employ tasks that activate specific brain circuits that may be modulated by compounds of interest or surrogate patient populations that are drug free but have symptoms or a subset of symptoms that are present in patient populations and are hypothesised to respond to the compound of interest (Dawson and Goodwin 2005). The hope is that experimental medicine studies will provide more detail and precision on the effects of new treatments on brain circuits and systems rather than symptoms per se. This is seen as essential to bridging the gap between animal and human studies and to increase the probability of success in this particularly difficult area of drug development (Gould and Manji 2004). A second aim of experimental medicine is to use existing drugs to probe neural systems with existing or new drugs to provide insight to the aetiology of a disease and potential methods of treatment. The focus of this chapter is the latest experimental medicine approaches in psychiatry particularly those that relate to depression, schizophrenia and cognition and the new light they have shone on the neuropharmacology of these disorders. Much of the focus is on new methods and surrogate's populations for determining the role of brain systems in these disorders and the neuropharmacology underpinning them.

Experimental medicine provides a path forward at the interface between Phase 1 and 2 trials that may bridge the gap between animal and human studies and provide an early indication of efficacy for a particular psychiatric disorder or a symptom that is prevalent across psychiatric disorders. The classic solutions are to: (a) measure the biomarkers in healthy volunteer or "surrogate patient" models of illness rather than in clinical groups and (b) measure treatment effects on biomarkers for the underlying disease process rather than treatment effects on symptoms. Consequently, early trials in surrogate patient populations, such as health volunteers with high schizotypy (see below for a description), incorporate objective measures that can bridge to symptom ratings and provide insight into the neuropharmacology unpinning these symptoms. In addition, new biomarkers that reveal the interplay between neuropharmacology and the brain systems that subserve motivation and reward may identify new therapeutic targets.

2 Experimental Medicine and Schizophrenia

Schizophrenia is a common and typically lifelong disorder, and despite drug treatment, most patients continue to experience symptoms with impaired quality of life. Positive symptoms (delusions, hallucinations and thought disorder) respond to antipsychotic drugs, but the benefit is often incomplete. The second generation drugs may not have the advantages initially claimed (Lieberman 2007), and only clozapine has some limited efficacy in treating the negative symptoms of anhedonia and flattened affect. It is also recognised increasingly that impaired cognitive function is a third and separate domain of symptoms (Kremen et al. 2000) with attention, memory and executive function being the most affected (Heinrichs 1998; Saykin et al. 1991). These impairments probably contribute to poor social functioning and quality of life although the precise connection is unclear. Current antipsychotics are also generally ineffective against cognitive impairment in schizophrenia. However, recent research suggests that brain training can improve functioning in patients (Sahakian et al. 2015). The National Institute for Mental Health in the USA has recognised this unmet need in their initiative "Measurement and Treatment Research to Improve Cognition in Schizophrenia" (MATRICS) to stimulate development of new drug treatments for neurocognitive impairment associated with the disorder (Green et al. 2004).

The aetiology of schizophrenia remains unknown and makes the focus of appropriate targets for drug development especially difficult. Although several neurotransmitter systems have been implicated in the pathogenesis of psychosis, only a few drugs targeting non-dopaminergic transmitters have any efficacy (Miyamoto et al. 2005). Given the probable complex components in schizophrenia, attributing the disease to a disturbance in a single neurochemical system is likely to prove too simplistic (Roth et al. 2004). Thus, the development of novel and more effective drugs for schizophrenia is hampered by the absence of identified pathology on which to target drugs. Paradoxically, this has led to a plethora of potential drug targets and compounds being developed as antipsychotic drugs. However, the identification of reliable and rapid methods for detecting the potential drug efficacy remains challenging for several reasons. The heterogeneous nature of the illness, its onset, duration and fluctuating symptomatology, comorbidities with mood disorders and substance use and the fact that most patients are multiply medicated makes the design and execution of a clinical trial very difficult.

Much of the current research in schizophrenia is guided by the neuroleptics which were first discovered in the early 1950s (Stip 2002). Numerous studies since then have confirmed that reducing the acute psychotic symptoms of schizophrenia critically depends on antagonising dopamine receptors (Klein and Davis 1969; Matthysse 1973) and it is still widely accepted that dopamine antagonism is key in the treatment and pathophysiology of schizophrenia (Kapur and Mamo 2003; Laruelle and Abi-Dargham 1999; Seeman 2013). More recently, positron emission tomography (PET) measuring dopamine displacement of D2 radioligand binding (Abi-Dargham et al. 2000; Laruelle et al. 1996) has confirmed that higher than expected dopamine release occurs in patients with schizophrenia. It has also been shown that increased uptake of $[^{18}F]$fluorodopa in both acute and prodromal patients suggests that presynaptic dopamine neurones play an important role in the symptoms of schizophrenia (Howes et al. 2009; McGowan et al. 2004). Although only subcortical striatal dopamine function has been quantifiable using these techniques, some indirect evidence suggests that frontal cortical dopamine release may be reduced in patient with schizophrenia. Nonetheless, in developing both animal and human translational models of schizophrenia, the administration of dopamine-releasing agents such as amphetamine and methylphenidate has been used widely to model its symptoms. Connell (1958) was one of the first to observe an acute schizophrenia-like syndrome in heavy users of stimulant drugs. Subsequently, these drugs and others have been used to induce positive symptoms in healthy volunteers. Although acute, high doses are required to induce paranoid ideation and hallucinations in healthy subjects, they do not induce the cognition impairments and negative symptoms of schizophrenia (Steeds et al. 2015).

Dopamine also has an important role in reward. For example, the presentation of unexpected rewards (reward prediction error) leads to the learning of new associations in animals (Schultz 2015). Subsequently, Pessiglione et al. (2006) showed during instrumental learning, the magnitude of reward prediction error expressed in the striatum was modulated by the administration of drugs enhancing (3,4-dihydroxy-L-phenylalanine; L-DOPA) or reducing (haloperidol) dopaminergic function. They showed that subjects treated with L-DOPA were more likely to choose the most rewarding action relative to subjects treated with haloperidol. Since then, a number of theories regarding probabilistic learning and reward and the tasks to probe these theories have emerged. It has been suggested for example that in psychosis unregulated dopamine release results in an inappropriate prediction error signal during the processing of irrelevant stimuli. Kapur (2003) suggested that this can result in the attribution of inappropriate salience to irrelevant external stimuli and internal representations such as thoughts or memories. Thus, delusions and hallucinations can form from the continual confusion emerging from the repeated and unusual experience of inappropriately salient stimuli. For example, to test the potential of Kapur's hypothesis, Roiser et al. (2009) devised a salience attribution test (SAT), in which coloured images of household objects or animals were presented just before a subject has the possibility to earn money, as a reward, over a large number of trials. The likelihood that money is available on a trial is signalled by one of four conditioned stimuli (CSs). CSs vary on two different visual dimensions: colour (e.g. blue or red) and shape (e.g. animal or household object irrelevant). Therefore, there are four different types of CS: blue animals; red animals; blue household objects and red household objects. One of these dimensions (e.g. colour) is task-relevant so that one level of the dimension is reinforced on 28/32 (87.5 %) of the trials, while only 4/32 (12.5 %) trials of the other are reinforced. For example, if "colour" is the reinforced dimensions, 14/16 blue animals and 14/16 blue household objects will be reinforced, compared to only 2/16 red animals and 2/16 red household objects. The other dimensions, in this example "shape", are task-irrelevant, so that 16/32 (50 %) of both levels are reinforced. The subjects are not informed of these contingencies, and this behaviour is driven by learning during the task. Roiser et al. (2009) showed that in patients with schizophrenia, there was elevated aberrant salience in patients with delusions, but not in those without. This is consistent with the aberrant salience hypothesis (Kapur 2003). However, functional magnetic imaging (fMRI) studies in patients with schizophrenia report reduced ventral striatal hemodynamic responses during reward cue conditions (Juckel et al. 2006; Schlagenhauf et al. 2008); this effect also correlated with negative symptoms and does not support the aberrant salience hypothesis. One explanation for this is that reward-related cues in an ambiguous setting may not differentially activate the ventral striatum sufficiently to motivate behaviour (Roiser et al. 2009; Ziauddeen and Murray 2010). These findings underline the importance of the distinction between negative symptoms, such as anhedonia, that are secondary to the psychotic state, and enduring negative symptoms in the absence of psychosis (primary negative symptoms) that is not responsive to antipsychotic drugs suggesting that they have a separate pathogenesis.

They also underline how difficult it is to test these hypotheses in patients that are treated with multiple drugs.

We therefore investigate the utility of using healthy volunteers with high, medium and low schizotypy as subjects who are medication free and mirror some of the symptoms of schizophrenia, albeit with much reduced severity. With this surrogate population, it may be possible to evaluate the effects of novel treatments for schizophrenia, thus providing a rapid indication of a drug's potential efficacy at an early phase of drug development. However, the first question to be addressed is as follows: Are there sufficient numbers of healthy volunteers with high schizotypy in the population to support clinical studies?

3 Schizotypy and Its Prevalence in the General Population

Schizotypy is a multidimensional construct that is closely linked to schizophrenia on phenomenological, genetic and neural levels. There is growing awareness that schizotypy is a continuously distributed trait that can add valuable insights into the aetiology, treatment, and prevention of schizophrenia (Raine 1991). The term schizotypy, first coined by Rado (1953), is a set of personality traits seen in the general population that are similar to those seen in patients with schizophrenia. Individuals with high levels of schizotypy have some or all of the behavioural traits, cognitive deficits and emotional disturbances seen in patients with schizophrenia. Schizotypy can assessed in the general population using psychometric self-report questionnaires including the Schizotypal Personality Questionnaire (SPQ) (Raine 1991), the Rust Inventory of Schizotypal Cognitions (RISC) (Rust 1988), the Chapman scales (Chapman and Kwapil 1995) and the Oxford–Liverpool Inventory of Feelings and Experiences (O-LIFE) (Mason and Claridge 2006). In general, schizotypy has three dimensions, positive, negative and disorganised, and resembles the factor structure of the symptoms reported in schizophrenia (Liddle 1987).

Schizotypy scales such as the SPQ capture schizophrenia-related characteristics in the general population including cognitive-perceptual and interpersonal disturbances as well as disorganised behaviour and speech and represent clinically relevant features of schizotypal personality disorder (SPD; Raine 1991). Healthy volunteers with high SPQ scores typically >43 (high schizotypes) have cognitive, structural and functional brain deficits similar to those seen in patient with schizophrenia (Raine 1991). Neuroleptic medication has been reported to reduce psychotic-like symptoms and symptom severity in SPD (Koenigsberg et al. 2002), suggesting that antipsychotic drug response in high schizotypes is similar to schizophrenia. High schizotypes from these non-clinical populations also resemble patients with schizophrenia in having increased rates of non-localising (soft) neurological signs (Barkus et al. 2006a) and increased psychotic experiences after cannabis use (Barkus et al. 2006b), cognitive biases (Stirling et al. 2007) and self-reported auditory hallucinations with a similar pattern of fMRI brain activation to those reported in auditory hallucinations (Barkus et al. 2007).

In order to determine whether the performance of high schizotypes is impaired in biomarker assays that are sensitive to the impairments found in patients with schizophrenia, we conduct a large-scale trial over three study sites (University of Manchester UK, University of Oxford UK and the Institute of Psychiatry, London UK). The predictions for this study were that performance on biomarker measures will be impaired in high schizotypy relative to subject with average levels of schizotypy individuals. In addition, some or all of biomarker performance may be improved by antipsychotic and cognition modulating drugs (risperidone, amisulpride and nicotine, respectively) in participants with high schizotypy scores, whereas little or no change will be detected in those with average scores. For this study, we chose to compare participants with average or mean rather than low schizotypy scores on the grounds that those with mean scores were more representative of a normal population. However, subsequent analysis suggests that participants with low schizotypy scores would be a more appropriate control group (see discussion on this issue later in the chapter). We used the SPQ and the brief version of the SPQ-(B) to determine schizotypy scores. On this questionnaire, high schizotypes are defined as those with scores ≥1 standard deviation above the mean (that is a score of >43.27). The average schizotypes are defined as having a score in the range of 0.5 standard deviations either side of the mean (that is scores of 21–36 inclusive) and for these participants served as a control group. Participants had to

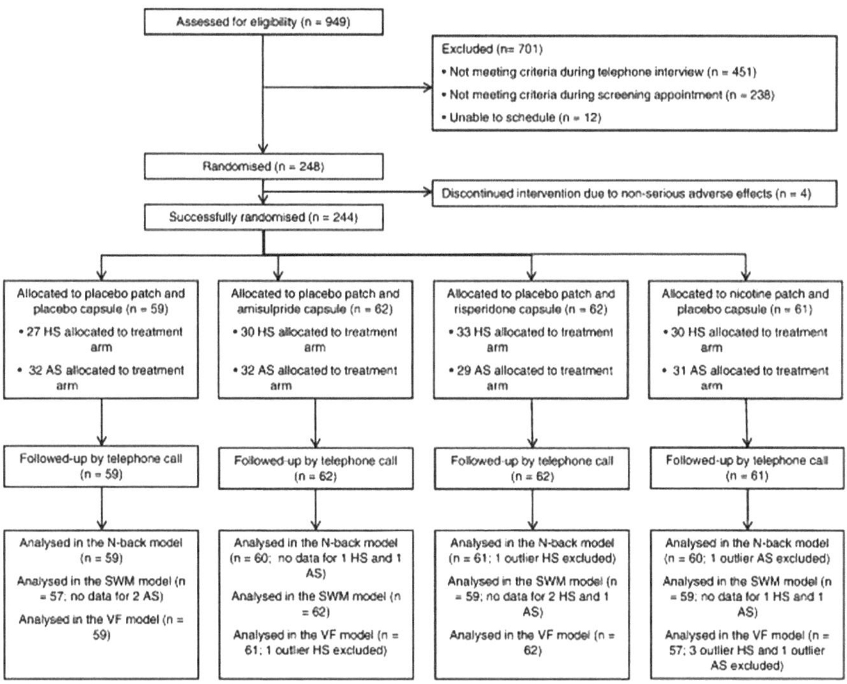

Fig. 1 A schematic of the recruitment of approximately 240 subjects with average and high schizotypy

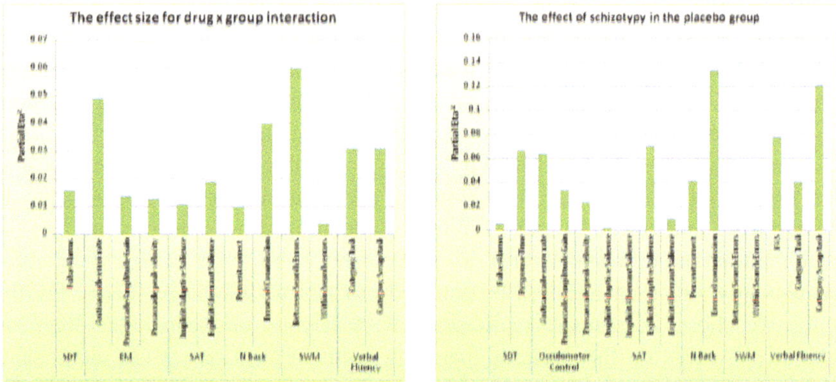

Fig. 2 Summary of the effect sizes of schizotypy and drug main effects and the schizotypy × drug interaction. (*SDT* signal detection test; *EM* eye movements; *SAT* salience attribution test; *SWM* spatial working memory)

meet both the SPQ-B and SPQ entry criteria for the study. During screening, if any potential participants had clinically significant psychiatric disorders, they were excluded from the study and were referred to the appropriate services.

Figure 1 shows a schematic of the recruitment for this study (adapted from Koychev et al. 2012). Schizotypal personality has four main components: cognitive disorganisation, anhedonia asocial behaviour and aberrant perceptions and beliefs (unusual experiences that can give rise to belief in the occult and the supernatural (Raine 1991). Thus, we used innovative recruitment techniques, such as social media-based advertising targeted at subjects with interests in the supernatural and the occult, in an effort to identify high schizotypes. As a result >22,000, SPQ and SPQ-(B) questionnaires were completed from which 244 (122 with high and 122 with average SPQ scores) were randomised to receive a single dose of risperidone, amisulpride, nicotine or placebo in a double-blind, between-subject design (approximately equal numbers in each group received these treatment, see Fig. 1). The subjects were then assessed on seven biomarker tasks (see Fig. 2 for a summary of tasks and results.

As cognitive deficits such as impairments in executive function, attention and memory are core features of schizophrenia and predict functional outcome and treatment adherence (Burton 2005; Green 2006; Green et al. 2000). We employed three standard cognitive tasks of working memory (N-back), spatial working memory (SWM) and verbal fluency (VF) tasks. We also included the salience attribution task described above and the signal detect task described by Barkus et al. and a biconditional learning task (Haddon et al. 2008). In addition, various oculomotor tasks were included as they have provided particularly useful biomarkers in schizophrenia and in schizophrenia spectrum populations (Reilly et al. 2008;

Levy et al. 2010). In these tasks, schizophrenia patients have impairments such as increased error rates, increased latencies, reduced spatial accuracy and lower pursuit (for review see Hutton and Ettinger 2006). Similar effects are also seen in participants with high schizotypy scores (Ettinger et al. 2005a; Gooding 1999; Holahan and O'Driscoll 2005; O'Driscoll et al. 1998). Finally, in a subset of participants, we evaluated the performance of a subset of average and high-scoring participants in the Arena task, a human virtual reality version of the Morris water maze (Parslow et al. 2004, Antonova 2009, 2011).

3.1 *Effects of Schizotypy and Treatment on the Performance of Cognitive Tasks*

Two of the tasks were sensitive to the schizotypy phenotype, the N-back task and VF. Consistent with a subtle working memory deficit, the HS group had more errors of commission (3-back) but performed equally as well of the AS group on per cent correct. The working memory deficit is a consistent feature of the performance of both schizotypy subjects and in patient with schizophrenia in neurocognitive tasks (Lee and Park 2005; Raine 2006). In general, the performance of HS subjects is intermediate between that of healthy controls and schizophrenia patients (Trestman et al. 1995). In the category switch condition of the VF task, the HS produced a significantly lower number of words, but not the "F", "A" and "S" or category naming ones. This suggests that the differences between HS and AS subjects are not due to VF impairment per se. Rather, they appear unable to inhibit irrelevant cues suggesting a deficit in central executive control. Thus, these two biomarkers, N-back and VS, are sensitive to the effect of schizotypy on executive function in its working memory and inhibitory domains. The deficits we found in working memory on the N-back task, but not the SWM task, suggest that there are key procedural differences between the tasks. One possibility is that early information processing is a key component in the origin and development of the cognitive deficits observed in patients with schizophrenia.

In terms of ameliorating the effect of schizotypy by drug treatment, amisulpride had the largest effect. In the HS group, it reduced the number of errors of commission on the N-back task and interestingly worsened the performance in the AS group. Although not quite reaching statistical significance in the SWM task, the same effect was evident on the number of between-search errors. Similarly, on the VF task, AS participants treated with amisulpride produced significantly fewer correct words, whereas performance tended to improve in the HS group. These data tend to support the proposal that the relationship between dopamine and executive function is best described by a U-shaped curve (Barch 2004). Thus, the prefrontal cortex of HS may be in a mild hypodopaminergic state which was improved with amisulpride, while the AS group may have had optimal dopamine function which was impaired by this treatment.

Risperidone treatment impaired spatial working memory in HS and VF performance of the AS group. The effects of risperidone may have been caused by its sedative effects (Miller 2004), an observation that was supported by an increase in reaction times in the N-back task. The effects of nicotine on cognitive processes were somewhat mixed. Overall, it tended to improve the performance of the HS group and to impair the AS subjects. In the N-back working memory task, it had reduced errors in the HS group but did not influence AS performance. In VF task, however, nicotine did not improve performance in the HS group but impaired it in the AS control group. This suggests that the two tasks may engage different cognitive processes. The beneficial effect of nicotine on the N-back task may have been to improve attention, an effect that has been wildly replicated (Newhouse et al. 2004), but may have reduced the inhibition required in the category swap condition of the VF task.

In summary, although there are no previously published studies that examine the effect of amisulpride on cognition in high schizotypes, the results of this study are consistent with previous studies showing improved functioning with amisulpride treatment in prodromal psychotic states (Ruhrmann et al. 2007). Similarly, high schizotypes have been reported to demonstrate preferential improvements in latent inhibition after haloperidol (Williams et al. 1997). Relatives of patients with schizophrenia displaying schizotypal features have also been reported to have improved daily functioning after treatment with risperidone (Tsuang et al. 1999; Rybakowski et al. 2007). The errors of commission in the N-back task proved to be more sensitive to schizotypy and drug effects than accuracy (errors of omission). We conclude that errors of commission are a more sensitive measure than the ability to respond to a cue. It could be speculated that increased errors of commission might reflect increased impulsivity in high schizotypes although this would tend to increase accuracy rather than decrease it as was observed. Therefore, increased errors of commission may reflect a more specific cognitive processing deficit involving decision-making.

Although these results are encouraging in that drug and phenotypic effects were observed in this study, further validation will be required when drugs known to reliably improve cognition in schizophrenia have been developed. However, the role that these experimental medicine methods will play in the development of such drugs remains to be seen.

3.2 *Effects of Schizotypy and Treatment on Signal Detection, Biconditional Learning and Aberrant Salience*

The signal detection task is directed at the putative impairment of perception and a "jumping to conclusions" cognitive style implicated in the pathogenesis of hallucinations and delusions. Unfortunately, no effects of schizotypy, drug or their interaction were observed in the present study. Previously, effects of schizotypy were reported in a group defined not only by high SPQ scores but also by high scores on the LSHS measure of hallucinatory tendency (Barkus et al. 2007). The lack of effect of schizotypy and therefore of drug by schizotypy interactions suggests that a more marked difference in the level of schizotypy between groups may be required to observe such effects, such as subjects with low rather than average schizotypy scores. In addition, there were significant effects of sex and site on signal detection measures and this may have weakened the power of the study to detect meaningful schizotypy and drug effects.

The salience attribution task is based on the aberrant salience hypothesis of schizophrenia. These postulates that dysregulated dopamine release result in aberrant learning of the significance or salience of environmental or internal cues and a failure to discriminate which cues predict rewards (adaptive salience). Patients with schizophrenia and individuals with high schizotypy have greater aberrant salience on the salience attribution task (Roiser et al. 2009), and D_2 dopamine receptor antagonists are predicted to normalise performance by reducing aberrant dopamine release and improving adaptive salience. Conversely, in individuals with average schizotypy, D_2 dopamine receptor antagonists are predicted to interfere with adaptive salience without affecting aberrant salience. However, we found no effect of schizotypy on aberrant salience. Risperidone did reduce the explicit measure of adaptive salience as would be expected from its dopamine receptor antagonist action. More surprisingly, amisulpride did not reduce implicit or explicit measures of adaptive salience despite a similar mechanism of action to risperidone and its effects on the N-back task. Different pre- and post-synaptic receptor actions or regional selectivity of these two drugs may explain these differences. For example, amisulpride can induce disinhibitory effects at low doses (50–300 mg) (Schoemaker et al. 1997). The improvement we saw in explicit adaptive salience following nicotine administration is consistent with preclinical studies suggesting that nicotine stimulates dopamine release in the ventral striatum, thus enhancing reward signalling. A perhaps stronger test of the dopamine theory of aberrant salience would be to select subjects for high aberrant salience and determine whether this response is normalised by dopamine antagonists.

The biconditional learning task suffered from a technical failure in our study which reduced power and may also have reduced the ability to replicate an earlier observation of an effect of schizotypy. However, this problem only affected the control condition, and an effect of schizotypy should have been observed in the intact biconditional task. A more likely explanation is that in the previous study reporting an effect of schizotypy (Haddon et al. 2011), the schizotypy measure was defined by high O-LIFE and not SPQ scores. Although the high schizotypy group had greater O-LIFE scores then the average schizotypy group in the present study, the group differences were less marked than in the previous study.

3.3 *Effects of Schizotypy and Treatment on Eye Movements*

Impairments in oculomotor control are among the most widely replicated neurocognitive deficits in schizophrenia. Schizophrenia patients display robust yet selective impairments in smooth pursuit eye movements (when following a slowly moving visual target) and antisaccades (looking away from a prepotent visual target) but not in basic oculomotor control conditions, such as visual fixation or simple reflexive saccades. Consequently, eye movements are widely studied as tools in the assessment of cognitive and brain function in this patient group (Ettinger and Kumari 2005). Their well-known neural correlates (Leigh and Zee 1999), involving fronto-striato-parietal neural circuitry, as well as the ease of administration and the availability of parametrically variable levels of task difficulties and appropriate control conditions make these tasks particularly attractive for pharmacological challenge studies. Smooth pursuit deficits have been linked to enduring primary negative symptoms in schizophrenia, called the deficit syndrome, thought to involve frontal dysfunction (Ross et al. 2000). It has also been shown that antisaccade performance is similarly related to negative symptoms in first-episode (Ettinger et al. 2004) and chronic (Ettinger et al. 2006) schizophrenia. In our study, eye movement performance proved to be a sensitive measure of the effects of schizotypy and treatment effects. Firstly, risperidone caused a slowing of antisaccades, and prosaccades peak saccade velocity. It also reduced prosaccade spatial accuracy and impaired the subjects' ability to match eye velocity to target velocity during smooth pursuit eye movement. The effects of risperidone and amisulpride were also modulated by schizotypy in that they tended to disrupt saccade inhibition in AS subjects and improve it in HS subjects. That finding that the risperidone had differential effects on antisaccade performance in AS and HS subjects may be due to underlying difference in dopamine control or signalling in the two groups and echoes the effects observed in the N-back task.

Finally, we evaluated the performance of AS and HS subjects in a human analogue of the Morris water maze (MWM; Morris 1981), the latter a standard preclinical test used to test memory in rats. This analogue (Parslow et al. 2004), termed the ARENA, demonstrates bilateral hippocampal activation during allocentric, but not egocentric, spatial memory encoding in healthy male participants (Parslow et al. 2004). Recently, we replicated the hippocampal activation during encoding in a group of healthy young participants and showed that the effect is attenuated in healthy older adults (Antonova et al. 2009). Thus, the ARENA paradigm produces replicable hippocampal activation associated with allocentric memory, making it useful for the study of normal effects of ageing as well as for testing the effect of drugs on human memory. In our most recent ARENA maze experiment, we evaluated the effects of acute treatment with 0.4 mg scopolamine in healthy young volunteers and found that it also impaired allocentric memory and hippocampal activation. Interestingly, the magnitude of the scopolamine-induced impairment in allocentric memory was not as large as that observed in elderly subjects. Thus, we were interested how AS and HS would perform in this task given the role that the hippocampus may play in the pathogenies of schizophrenia. As in previous experiments, we evaluated the performance of the subjects while they performed the task in a 3T scanner. Although there were no performance deficits in the AS or HS subjects, there were significant differences in hippocampal activation between the groups. Surprisingly, the hippocampus was more strongly activated in

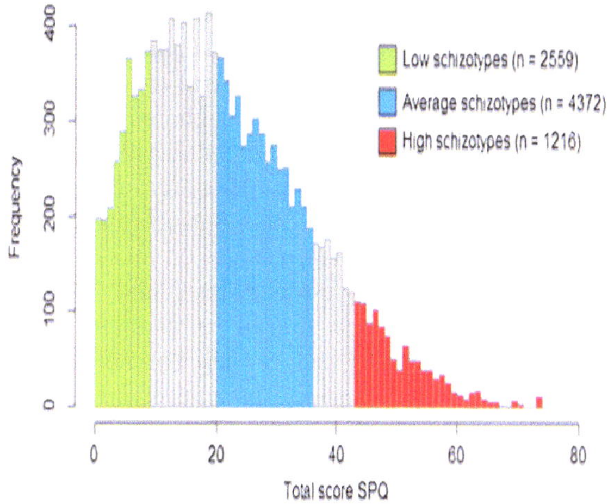

Fig. 3 Distribution of schizotypy scores from the Schizotypy personality questionnaire in a sample of approximately 13,000 subjects from the general population

the HS group compared to the AS when encoding the location of a visible object and retrieving the location of the object when it was no longer visible. These results suggest that although the HS could match the performance of AS subjects, they had to recruit significantly more resources to do so. As we described below, increases in brain processing as measure using functional magnetic resonance imaging during cognitive tasks may be a particularly useful biomarker in determining the effects of drugs on cognitive processing.

Finally, until recently, it had been assumed that levels of schizotypy were normally distributed in the population and indeed we selected average scoring schizotypes to as our comparison group in the studies described above. However, when the study had been completed, the distribution of scores was tabulated for those completing the SPQ and SPQ-B questionnaire and showed that the distribution was skewed to the left and that low scores <20 are the in fact the norm (see Fig. 3). Moreover, in a subsequent study in which the performance of a group of subjects with low schizotypy scores (LS) was assessed in the N-back task, they were found to have superior performance than both the AS and HS subjects. Taken together, this suggests that the appropriate comparison groups for subsequent studies would be HS and LS subjects to maximise the performance differences between groups on tasks and to enhance the ability to detect drug effects.

In summary, seven biomarker tasks were evaluated for their ability to detect (i) effects of schizotypy, (ii) effects of reference antipsychotics and nicotine and (iii) greater drug effects in the high schizotypy group. Over 22,000 people completed an online schizotypy questionnaires, of which 240 were entered into the study suggesting that selecting participants via the Internet for high schizotypy is fast and reliable. In addition, multicentre recruitment accelerates the progress of biomarker studies in healthy participants without a significant cost of increased variance. However, it is likely that greater biomarker sensitivity to drug effects relevant to schizophrenia could be achieved by recruiting from the lowest and highest extremes on schizotypy questionnaires such as upper and lower quartiles and requiring high scores on specific aspects of the schizophrenia spectrum. Risperidone impaired performance on a number of tasks, and this was most evident in average schizotypy groups and on latency measures. However, drugs with sedative actions may mask cognitive enhancing effects. Subjects with high schizotypy scores show reliable schizophrenia-like abnormalities on four cognitive biomarker tasks: salience attribution, antisaccade error, N-back and VF. The cholinergic agonist nicotine enhanced performance on three of the tasks sensitive to schizotypy: salience attribution, antisaccade error and N-back but not on VF. Moreover, nicotine specifically enhanced performance in subjects with high schizotypy scores on the antisaccade and N-back tasks. Amisulpride tended to improve performance only in high schizotypes and to impair it in average schizotypes, and this was statistically significant on the N-back task although apparent in the antisaccade task in the form of a group x drug interaction. The biconditional learning and signal detection tasks were not affected by schizotypy or by drugs with the exception that risperidone slowed biconditional learning. Thus, the most effective tasks for detecting differential drug effects in average vs high schizotypes are the antisaccade error and N-back tasks. Salience attribution, anti-saccade error and N-back and VF tasks are sensitive to schizotypy and show the most promise as biomarkers for detecting drugs that could improve cognitive function in schizophrenia. Biconditional learning and signal detection are not sensitive to schizotypy or to cognitive enhancement by nicotine. Future studies, use of a low schizotypy group as a comparator, may increase the possibility of detecting differential drug by schizotypy effects when drug responses in a high schizotypy group may not be significant. Such differential effects may also increase confidence that drugs are acting on processes relevant to schizophrenia. Overall, the data were consistent with the inverted "U"-shaped effect of dopamine in that in some of the biomarker tasks, both antipsychotics enhance performance in the subjects with high schizotypy scores and impaired in those with average scores. Although these effects are subtle, this experimental medicine approach offers many benefits over patient clinical trials.

4 The Neuropharmacology of Depression and Cognition

Major depressive disorder (MDD) is a severe and common psychiatric disorder, with a lifetime prevalence of about 15 %, but can be as high as 25 % in women. Although MDD primarily involves mood disturbances, patients also usually present with alterations in cognitive function. Between 66 and 94 % of patients, MDD has cognitive impairments (Conradi et al. 2011; Rock et al. 2013). These include problems with attention, memory and executive functioning, and they persist into remission (Rock et al. 2013). Beck (1967) originally suggested that cognitive deficits are a consequence of the depression syndrome and a core symptom. However, it is now widely recognised that MDD is associated with clinically significant deficits in many aspects of cognitive processing including attention, concentration and learning. There have been numerous suggestions regarding the cause of these cognitive deficits; for example, Channon et al. (1993) have proposed that abnormalities in executive function, in particular working memory, are responsible. An alternative suggestion is that the core cognitive symptoms are expressed as negative biases in thinking with lower thresholds for perceiving negative emotions and higher thresholds for perceiving positive emotions leading to a bias to attend and remember negative items rather than positive ones (Harmer et al. 2009; Roiser et al. 2012; Roiser and Sahakian 2013). Finally, it has also been suggested that in the depressive state, memory and executive function deficits exist independent of depressed mood (Liddle 1987). Cognitive dysfunction in MDD may be a symptom of depressive illness and persist, as residual symptoms, despite otherwise effective antidepressant therapy.

Cognitive dysfunction in subjects recovered from depression has been widely reported in several studies including systematic review and meta-analysis by Hasselbalch et al. (2011) and more recently by Rock et al. (2013). Whether this cognitive dysfunction is a state or a trait phenomenon or an intermediate marker for recurrent unipolar depression, rather than "scars" caused by past episodes, is currently a significant matter of debate. Gorwood et al. (2008) suggested that depression has "toxic effects" on brain function, particularly the hippocampus. They found that in patients with MDD, an impairment of delayed recall was related to the cumulative length of depressive disorder. In a meta-analysis of studies that used the CANTAB battery to assess cognitive deficits, Rock et al. (2014) showed that while depressed patients had deficits across a number of cognitive domains including executive function, remitted depressed patients had moderate and significant deficits within the domains of executive function and attention. However, in the domain of memory, remitted depressed patients showed only a tendency towards small/moderate deficits in memory suggesting a recovery of function in the memory domain. Taken together, these data suggest that (i) depression impairs cognitive function across a number of domains, (ii) repeated episodes of depression increase the severity of the cognitive dysfunction across domains, (iii) between episodes of depression, there is some recovery of memory function, but executive function remains impaired. Interestingly, there are a number of reliable reports that hippocampal volume is decreased during long untreated episodes of depression. However, during treatment with antidepressants, hippocampal volume recovers or is protected (Sheline et al. 2003).

The neural basis and neuropharmacology of cognition, in particular working memory, have been well characterised in imaging studies. There are strong grounds for suggesting that the data from fMRI studies may be more informative than behaviour (performance tasks) as an indicator of underlying neurocognitive dysfunction and may shed light on the nature of the neural dysfunction in cognitive processing induced by depression and its after-effects. Kerestes et al. (2011) used blood-oxygen-level dependent (BOLD) fMRI to measure neural activity during a working memory N-back task with faces with emotional expressions as distracter stimuli. A group of subjects ($n = 19$) with at least two past major depressive episodes (remitted from depression; Hamilton Depression Rating scale, HAM-D17 score ≤ 7) and medication free were compared with a group of healthy controls with similar age and intelligence quotient (IQ) scores. The group of remitted subjects exhibited significantly greater activity relative to control group ($n = 20$) in the left dorsolateral prefrontal cortex in response to negative emotional distracters during high working memory load. This suggests that remitted subjects may continue to exhibit attentional biases towards negative emotional information, reflected by greater recruitment of prefrontal regions implicated in attentional control in the context of negative emotional information (Barrantes-Vidal et al. 2013). Thus, BOLD signal changes might be related to cognitive changes, such as adaptations of strategy formation or cognitive effort, that are not manifest in behavioural measures.

Andersson et al. (2010) also investigated the BOLD response following acute citalopram treatment on face emotion processing in subjects remitted from depression compared to controls. Compared with viewing neutral faces, citalopram enhanced left anterior cingulate response to happy faces, right posterior insula and right lateral orbitofrontal responses to sad faces and reduced amygdala responses bilaterally to fearful faces. In controls, relative to subjects remitted from depression, citalopram increased bilateral hippocampal responses to happy faces and increased right anterior insula response to sad faces. These results are consistent with previous findings showing 5-HT modulation of affective processing (Nelson et al. 2013), and the involvement of the hippocampus suggests a degree of memory or cognitive process in what is an apparently relatively passive task.

In addition to cognitive tasks activating brain areas of interest, alterations in resting-state connectivity assessed by fMRI have also been observed in depression across multiple networks including parts of the cognitive control network (anterior cingulate, prefrontal cortex) which are involved in decision-making, attention and resolving conflicts. Data from McCabe et al. (2011) suggest that antidepressant medications can decrease resting-state functional connectivity in healthy subjects with no history of mood disorders in areas known to mediate reward and emotional processing in the brain (orbitofrontal cortex, striatum, amygdala). These results support the proposition that antidepressant medications might work by normalising the elevated resting-state functional connectivity seen in depressed patients. Taken together, these data suggest that the interplay between depression and cognition is complex, but that impaired cognition may remain as a result of depression. However, although antidepressants may ameliorate cognitive symptoms and contribute to an increase in hippocampal volume during recovery, they do not appear to treat cognitive symptoms during bouts of depression or indeed in remission. Whether the cognitive deficits observed are solely related to reduce hippocampal function or to deficits in processing in a wider network remains an open question.

As described above, there is a widely accepted hypothesis that bouts of severe depression reduce hippocampal volume. In general, the more intense the history of depression, the greater the decrease in hippocampal volume (Shah et al. 1998, Sheline et al. 2003). Hippocampal size may also be related to other measures of illness intensity, such as the number of past hospitalisations and recurrence of the disorder (Rapp et al. 2005). Meta-analyses show that hippocampal volume reduction and the total number of depressive episodes may be particularly correlated with right hippocampal volume (Videbech and Ravnkilde 2004; Campbell et al. 2004). Gorwood et al. (2008) have convincingly shown that the number of previous depressive episodes is related to the degree of the cognitive deficits, i.e. the greater the number of previous episodes, the greater the cognitive deficit. Thus, there is a potential confound or potential synergistic effect of age and cognition. The older the patient, the greater the number of depressive episodes they are likely to have had and the greater their degree of age-related cognitive decline.

To further investigate the relationship between the number of previous depressive episodes and severity of cognitive deficits, McDermott and Ebmeier (2009) conducted a meta-analysis or relevant studies using the correlation (Pearson's r) between depression severity scores and neuropsychological test performance. Individual meta-analyses were conducted for composite measures of cognitive functional domains (episodic memory, executive function, processing speed, semantic memory and visuo-spatial memory). In total, sixty-nine studies were identified which met the inclusion criteria. This number was reduced to a total of fourteen studies which reported a correlation value between depression severity and individual neuropsychological test scores. Their results showed that severity of depression is related to cognitive performance in episodic memory, executive function and processing speed. Specifically, increased severity of depression was significantly associated with reduced cognitive performance across these domains. They also showed that the pattern of cognitive impairment and its relationship to depression severity differed across domains. They suggested that cognitive processing in the prefrontal cortex and associated systems may also be affected in depression. Taken together, these data suggest that although the relationship between depression and cognition is a complex one, the magnitude of the cognitive deficit does appears to increase with each episode of depression. The deficits in both hippocampal and prefrontal cortex do appear to be correlated with the previous number of episodes of depression. Moreover, the deficits in cognitive processing appear to persist between episodes of depression and increase in magnitude with successive bouts of depression. Such deficits may create vulnerability or is a risk factor for recurring depression and may become independent of the depressive state. Consequently, we devised an experimental medicine approach to determining the nature of the cognitive deficits in remission and whether they would respond to a newly discovered treatment for depression that has shown to improve cognitive processing in subjects during treatment. However, it is not clear whether the improvements in cognitive processing observed in previous studies are due to improvements in mood or an improvement in cognitive processing per se. Thus, we developed an experimental medicine approach to determine whether the effects of treatment on cognition were independent of its effects on mood.

Brentillix is a recently approved treatment for depression. Its functional effects are multimodal modulating noradrenergic, serotonergic, cholinergic, dopaminergic histaminergic, GABAergic and glutamatergic neurotransmission. It has a mixed pharmacology and is a antagonist at 5-HT3, 5-HT7 and 5-HT1D, 5-HT1B receptors, a partial receptor agonist at 5-HT1A and an inhibitor of the serotonin (5-HT) transporter (SERT) (Sanchez et al. 2015). In animals, it has a superior procognitive function when compared to other antidepressants (Jensen et al. 2014). In patients with MDD, Brentillix is effective at a dose of 20 mg/d and lower (Alvarez et al. 2012; Boulenger et al. 2014; Henigsberg et al. 2012). The procognitive effects of Brentillix were first seen in elderly patients (Katona et al. 2012) and were confirmed in subsequent studies in patients with MDD showing that Brentillix significantly improved cognitive function (Mahableshwarkar et al. 2015; McIntyre et al. 2014). A detailed path analysis suggested that its effects on cognition were independent of its antidepressant effects. However, confirming this dissociation in patients with MMD is difficult, and we therefore devised an experimental medicine approach to determine whether Brentillix had beneficial effects on cognitive process that could not be accounted for by its effect on mood.

As suggest above, patients in remission from MDD may have deficits in cognition that last beyond the period of depression and that the magnitude of this deficit is related to the number of previous episodes that they had. We therefore recruited subjects that had a minimum of two confirmed episodes of depression and self-reported cognitive deficits and compared them to age-matched healthy controls in a double-blind placebo-controlled trial of neural and cognitive function. If we could show that subjects in remission form depression had cognitive deficits or deficits in neural circuits related to cognition, it would strongly suggest that the effects of Brentillix on cognition were independent of its effects on mood. We used several tests of cognitive function and found deficits in cognitive performance that were normalised by Brentillix. Behaviourally, we found improved performance during the trail making test (TMT) and digit symbol substitution test (DSST, Brown et al. 2016 submitted). Interestingly, we found to improve performance on a subjective measure of cognitive difficulties assessed in the perceived difficulty questionnaire (PDQ) suggesting that the healthy volunteers and remitted subjects receiving Brentillix were aware of a noticeable improvement in cognitive abilities compared to those receiving placebo. Intriguingly, we also found that Brentillix modulates the BOLD signal within neural structures previously identified as hyperactive in depressed patients with cognitive deficits. We found a normalisation of the pattern of BOLD response in areas of the prefrontal cortex and hippocampus during the performance of the N-back task. Specifically, Brentillix reduced neural activity in the right DLPFC and left hippocampus and across a network of temporal–parietal areas. These effects of Brentillix are opposite in direction to the increases in BOLD signal described in MDD (Fitzgerald et al. 2008; Harvey et al. 2010; Matsuo et al. 2007; Rose et al. 2006; Walter et al. 2007). Thus, these data suggest that Brentillix increased efficiency during the effortful working memory performance. The results suggest that during the N-back task, remitted subjects had to devote more resources to maintain performance on the task compared to health controls. This increased effort was reflected in significantly greater BOLD signals in remitted subjects during the N-back task which was reduced to health control levels by treatment with Brentillix. Taken together, these data support the hypothesis that Brentillix improves executive function in MDD that is not confounded by its effects on mood.

5 Summary and Conclusions

Although it is widely recognised that the introduction of experimental and translational medicine models can play a major role in the development of psychiatric drugs, the validation of these methods and strong evidence that they add value needs confirmation and strengthening. The strategy of using experimental medicine studies in bridging the gap between animal and human studies has also been funded through initiatives such as NEWMEDs (http://www.newmeds-europe.com/). As the studies described above show studies conducted across centres can be conducted quickly and with a high degree of replication. They can also facilitate the evaluation of early and late-stage drugs to elucidate their mechanism of action and the neural systems they modulate. When taken together, they provide not only a more rapid path for drug development but also shed light on new ways to treat psychiatric and brain disorders.

References

Abi-Dargham A, Rodenhiser J, Printz D, Zea-Ponce Y, Gil R, Kegeles LS, Weiss R, Cooper TB, Mann JJ, Van Heertum RL, Gorman JM, Laruelle M (2000) Increased baseline occupancy of D2 receptors by dopamine in schizophrenia. Proc Natl Acad Sci USA 97:8104–8109

Alvarez E, Perez V, Dragheim M, Loft H, Artigas F (2012) A double-blind, randomized, placebo-controlled, active reference study of Lu AA21004 in patients with major depressive disorder. Int J Neuropsychopharmacol/Official Scientific Journal of the Collegium Internationale Neuropsychopharmacologicum (CINP), 15(5):589–600. 10.1017/S1461145711001027

Andersson JLR, Jenkinson M, Smith S (2010) Non-linear registration, aka spatial normalisation. FMRIB technical report TR07JA2

Angst J, Azorin JM, Bowden CL, Perugi G, Vieta E, Gamma A, Young AH.BRIDGE Study Group (2011) Prevalence and characteristics of undiagnosed bipolar disorders in patients with a major depressive episode: the BRIDGE study. Arch Gen Psychiatry 68:791–798

Antonova E, Parslow D, Brammer M, Dawson GR, Jackson SH, Morris RG (2009) Age-related neural activity during allocentric spatial memory. Memory 2:125–143

Antonova E, Parslow D, Brammer M, Simmons A, Williams S, Dawson GR, Morris R (2011) Scopolamine disrupts hippocampal activity during allocentric spatial memory in humans: an fMRI study using a virtual reality analogue of the Morris Water Maze. J Psychopharmacol 9:1256–1265

Barch DM (2004) Pharmacological manipulation of human working memory. Psychopharmacology (Berl) 174:126–135

Barkus E, Stirling J, Hopkins R, Lewis S (2006a) The presence of neurological soft signs along the psychosis proneness continuum. Schizophr Bull 32:573–577

Barkus E, Stirling J, Hopkins R, McKie S, Lewis S (2007) Cognitive and neural processes in non-clinical auditory hallucinations. Brit. J Psychiatry 51:s76–s81

Barkus EJ, Stirling J, Hopkins RS, Lewis S (2006b) Cannabis-induced psychosis-like experiences are associated with high schizotypy. Psychopathology 39:175–178

Barrantes-Vidal N, Gross GM, Sheinbaum T, Mitjavila M, Ballespí S, Kwapil TR (2013) Positive and negative schizotypy are associated with prodromal and schizophrenia-spectrum symptoms. Schizophr Res 145:50–55

Boulenger J-P, Loft H, Olsen CK (2014) Efficacy and safety of vortioxetine (Lu AA21004), 15 and 20 mg/day: a randomized, double-blind, placebo-controlled, duloxetine-referenced study in the acute treatment of adult patients with major depressive disorder. Int Clin Psychopharmacol 29 (3):138–149

Burton SC (2005) Strategies for improving adherence to second-generation antipsychotics in patients with schizophrenia by increasing ease of use. J Psychiatric Pract 11(6):369–378.

Campbell S, Marriott M, Nahmias C, MacQueen GM (2004) Lower hippocampal volume in patients suffering from depression: a meta-analysis. Am J Psychiatry 161:598–607

Chapman LJ, Kwapil TR (1995) Scales for the measurement of schizotypy. In: Raine A, Lencz T, Mednick SA (eds) Schizotypal personality. Cambridge University Press, Cambridge, pp 79–106

Channon S, Baker JE, Robertson MM (1993) Working memory in clinical depression—an experimental study. Psychol Med 23:87–91

Connell PH (1958) Amphetamine psychosis. Chapman and Hall, London

Conradi HJ, Ormel J, de Jonge P (2011) Presence of individual (residual) symptoms during depressive episodes and periods of remission: a 3-year prospective study. Psychol Med 41:1165–1167

Dawson GR, Goodwin GM (2005) Experimental medicine in psychiatry. J Psychopharmacol 19:565–566

Ettinger U, Kumari V, Crawford TJ, Corr PJ, Das M, Zachariah E, Hughes C, Sumich AL, Rabe-Hesketh S, Sharma T (2004) Smooth pursuit and antisaccade eye movements in siblings discordant for schizophrenia. J Psychiatry Res 38:177–184.

Ettinger U, Picchioni M, Hall MH, Schulze K, Toulopoulou T, Landau S, Crawford TJ, Murray RM (2006) Antisaccade performance in monozygotic twins discordant for schizophrenia: the Maudsley twin study. Am J Psychiatry 163:543–545

Ettinger U, Kumari V, Crawford TJ, Flak V, Sharma T, Davis RE, Corr PJ (2005) Saccadic eye movements, schizotypy, and the role of neuroticism. Biol Psychol 68:61–78

Fitzgerald PB, Srithiran A, Benitez J, Daskalakis ZZ, Oxley TJ, Kulkarni J, Egan GF (2008) An fMRI study of prefrontal brain activation during multiple tasks in patients with major depressive disorder. Human Brain Mapping, 29(4):490–501. doi:10.1002/hbm.20414

Gooding DC (1999) Antisaccade task performance in questionnaire-identified schizotypes. Schizophr Res 35:157–166

Gorwood P, Corruble E, Falissard B, Goodwin GM (2008) Toxic effects of depression on brain function: impairment of delayed recall and the cumulative length of depressive disorder in a large sample of depressed outpatients. Am J Psychiatry 165:731–739

Gould TD, Manji HK (2004) The molecular medicine revolution and psychiatry: bridging the gap between basic neuroscience research and clinical psychiatry. J Clin Psychiatry 598-604

Green MF (2006) Cognitive impairment and functional outcome in schizophrenia and bipolar disorder. J Clin Psychiatry 67:e12

Green MF, Kern RS, Braff DL, Mintz J (2000) Neurocognitive deficits and functional outcome in schizophrenia: are we measuring the "right stuff"? Schizophr Bull 26:119–136

Green MF, Nuechterlein KH, Gold JM, Barch DM, Cohen J, Essock S, Fenton WS, Frese F, Goldberg TE, Heaton RK, Keefe RS, Kern RS, Kraemer H, Stover E, Weinberger DR, Zalcman S, Marder SR (2004) Approaching a consensus cognitive battery for clinical trials in schizophrenia: the NIMH-MATRICS conference to select cognitive domains and test criteria. Biol Psychiatry 56:301–307

Haddon JE, George DN, Grayson L, McGowan C, Honey RC, Killcross S (2011) Impaired conditional task performance in a high schizotypy population: relation to cognitive deficits. Q J Exp Psychol 64:1–9

Haddon JE, George DN, Killcross S (2008) Contextual control of biconditional task performance: evidence for cue and response competition in rats. Q J Exp Psychol 61:1307–1320

Harmer CJ, O'Sullivan U, Favaron E, Massey-Chase R, Ayres R, Reinecke et al (2009) Effect of acute antidepressant administration on negative affective bias in depressed patients. Am J Psychiatry 166:1178–1184

Harvey PD, Green MF, Nuechterlein KH (2010) Latest developments in the matrics process. Psychiatry (Edgmont) 7:49–52
Hasselbalch BJ, Knorr U, Kessing LV (2011) Cognitive impairment in the remitted state of unipolar depressive disorder: a systematic review. J Affect Disord 134:20–23
Heinrichs RW, Zakzanis KK (1998) Neurocognitive deficit in schizophrenia: a quantitative review of the evidence. Neuropsychology 12:426–445
Henigsberg N, Mahableshwarkar AR, Jacobsen P, Chen Y, Thase ME (2012) A randomized, double-blind, placebo-controlled 8-week trial of the efficacy and tolerability of multiple doses of Lu AA21004 in adults with major depressive disorder. J Clin Psychiatry 73(7):953–959. doi:10.4088/JCP.11m07470
Holahan AL, O'Driscoll GA (2005) Antisaccade and smooth pursuit performance in positive- and negative-symptom schizotypy. Schizophr Res 76:43–54
Howes OD, Montgomery AJ, Asselin MC, Murray RM, Valli I, Tabraham P, Bramon-Bosch E, Valmaggia L, Johns L, Broome M, McGuire PK, Grasby PM (2009) Elevated striatal dopamine function linked to prodromal signs of schizophrenia. Arch Gen Psychiatry 66:13–20
Hutton SB, Ettinger U (2006) The antisaccade task as a research tool in psychopathology: a critical review. Psychophysiology 43:302–313
Jensen JB, du Jardin KG, Song D, Budac D, Smagin G, Sanchez C, Pehrson AL (2014) Vortioxetine, but not escitalopram or duloxetine, reverses memory impairment induced by central 5-HT depletion in rats: evidence for direct 5-HT receptor modulation. Eur Neuropsychopharmacol 24:148–159. doi:10.1016/j.euroneuro.2013.10.011
Juckel G, Schlagenhauf F, Koslowski M, Wustenberg T, Villringer A, Knutson B, Wrase J, Heinz A (2006) Dysfunction of ventral striatal reward prediction in schizophrenia. Neuroimage 29:409–416
Kapur S, Mamo D (2003) Half a century of antipsychotics and still a central role for dopamine D2 receptors. Prog Neuropsychopharmacol Biol Psychiatry 27:1081–1090
Kapur S (2003) Psychosis as a state of aberrant salience: a framework linking biology, phenomenology, and pharmacology in schizophrenia. Am J Psychiatry 160:13–23
Katona C, Hansen T, Olsen CK (2012) A randomized, double-blind, placebo-controlled, duloxetine-referenced, fixed-dose study comparing the efficacy and safety of Lu AA21004 in elderly patients with major depressive disorder. Int Clin Psychopharmacol 27(4):215–223. doi:10.1097/YIC.0b013e3283542457
Kerestes R, Ladouceur CD, Meda S, Nathan PJ, Blumberg HP, Maloney K, Ruf B, Saricicek A, Pearlson GD, Bhagwagar Z, Phillips ML (2011) Abnormal prefrontal activity subserving attentional control of emotion in remitted depressed patients during a working memory task with emotional distracters. Psychological Medicine, Cambridge University Press
Klein D, Davis J (1969) Diagnosis and drug treatment of psychiatric disorders. Williams & Wilkins, Baltimore 4:24
Koenigsberg HW, Harvey PD, Mitropoulou V, Schmeidler J, New AS, Goodman M, Silverman JM, Serby M, Schopick F, Siever LJ (2002) Characterizing affective instability in borderline personality disorder. Am J Psychiatry 159:784–788
Koychev I, McMullen K, Lees J, Dadhiwala R, Grayson L, Perry C, Schmechtig A, Walters J, Craig KJ, Dawson GR, Dourish CT, Ettinger U, Wilkinson L, Williams S, Deakin JF, Barkus E (2012) A validation of cognitive biomarkers for the early identification of cognitive enhancing agents in schizotypy: a three-center double-blind placebo-controlled study. Eur Neuropsychopharmacol 22:469–481
Kremen WS, Seidman LJ, Faraone SV, Toomey R, Tsuang MT (2000) The paradox of normal neuropsychological function in schizophrenia. J Abnorm Psychol 109:743–752
Laruelle M, Abi-Dargham A, van Dyck CH, Gil R, D'Souza CD, Erdos J, McCance E, Rosenblatt W, Fingado C, Zoghbi SS, Baldwin RM, Seibyl JP, Krystal JH, Charney DS, Innis RB (1996) Single photon emission computerized tomography imaging of amphetamine-induced dopamine release in drug-free schizophrenic subjects. Proc Natl Acad Sci USA 93:9235–9240

Laruelle M, Abi-Dargham A (1999) Dopamine as the wind of the psychotic fire: new evidence from brain imaging studies. J Psychopharmacol 13:358–371
Lee J, Park S (2005) Working memory impairments in schizophrenia: a meta-analysis. J Abnorm Psychol 114:599–611
Leigh J and Zee DS (1999) The neurology of eye movements. Philadelphia, FA Davis Company
Levy DL, Sereno AB, Gooding DC, O'Driscoll GA (2010) Eye tracking dysfunction in schizophrenia: characterization and pathophysiology behavioural neurobiology of schizophrenia and its treatment. Springer, Berlin
Liddle PF (1987) The symptoms of chronic schizophrenia. A re-examination of the positive-negative dichotomy. Br J Psychiatry 151:145–151
Lieberman JA (2007) Effectiveness of antipsychotic drugs in patients with chronic schizophrenia: efficacy, safety and cost outcomes of CATIE and other trials. J Clin Psychiatry 68:e04
Mahableshwarkar AR, Zajecka J, Jacobson W, Chen Y, Keefe RS (2015) A Randomized, placebo-controlled, active-reference, double-blind, flexible-dose study of the efficacy of vortioxetine on cognitive function in major depressive disorder. Neuropsychopharmacology: Official Publication of the American College of Neuropsychopharmacology. doi:10.1038/npp.2015.52
Mason O, Claridge G (2006) The Oxford-Liverpool Inventory of Feelings and Experiences (O-LIFE): further description and extended norms. Schizophr Res 82:203–211
Matthysse S (1973) Antipsychotic drug actions: a clue to the neuropathology of schizophrenia? Fed Proc 32:200–205
McCabe C, Mishor Z (2011) Antidepressant medications reduce subcortical–cortical resting-state functional connectivity in healthy volunteers. NeuroImage 57:1317–1323
McDermott LM, Ebmeier KP (2009) A meta-analysis of depression severity and cognitive function. J Affect Disord 119:1–8
McGowan S, Lawrence AD, Sales T, Quested D, Grasby P (2004) Presynaptic dopaminergic dysfunction in schizophrenia: a positron emission tomographic [18F] fluorodopa study. Arch Gen Psychiatry 61:134–142
McIntyre RS, Lophaven S, Olsen CK (2014) A randomized, double-blind, placebo-controlled study of vortioxetine on cognitive function in depressed adults. Int J Neuropsychopharmacol/Official Scientific Journal of the Collegium Internationale Neuropsychopharmacologicum (CINP), 17 (10):1557–1567. doi:10.1017/S1461145714000546
Miller DD (2004) Atypical antipsychotics: sleep, sedation, and efficacy. Prim Care Companion J Clin Psychiatry 6(Suppl 2):3–7
Miyamoto S, Duncan GE, Marx CE, Lieberman JA (2005) Treatments for schizophrenia: a critical review of pharmacology and mechanisms of action of antipsychotic drugs. Mol Psychiatry 10:79–104
Morris RGM (1981) Spatial localisation does not depend on the presence of local cues. Learn and Motiv 12:239–260
Nelson MT, Seal ML, Pantelis C, Phillips LJ (2013) Evidence of a dimensional relationship between schizotypy and schizophrenia: a systematic review. Neurosci Biobehav Rev 37: 317–327
O'Driscoll GA, Lenzenweger MF, Holzman PS (1998) Antisaccades and smooth pursuit eye tracking and schizotypy. Arch Gen Psychiatry 55:837–843
Parslow DM, Rose D, Brooks B, Fleminger S, Gray JA, Giampietro V, Brammer MJ, Williams S, Gasston D, Andrew C, Vythelingum GN, Loannou G, Simmons A, Morris RG (2004) Allocentric spatial memory activation of the hippocampal formation measured with fMRI. Neuropsychology 18:450–461
Pessiglione M, Seymour B, Flandin G, Dolan RJ, Frith CD (2006) Dopamine-dependent prediction errors underpin reward-seeking behaviour in humans. Nature 442:1042–1045
Rado S (1953) Dynamics and classification of disordered behaviour. Am J Psychiatry 110: 406–416
Raine A (1991) The SPQ: a scale for the assessment of schizotypal personality based on DSM-III-R criteria. Schizophr Bull 17:555–564

Raine A (2006) Schizotypal personality: neurodevelopmental and psychosocial trajectories. Annu Rev Clin Psychol 2:291–326

Reilly JL, Lencer R, Bishop JR, Keedy S, Sweeney JA (2008) Pharmacological treatment effects on eye movement control. Brain Cogn 68:415–435

Rock PL, Roiser JP, Riedel WJ, Blackwell AD (2013) Cognitive impairment in depression: a systematic review and meta-analysis. Psychol Med 44:2029–2034

Roiser JP, Elliott R, Sahakian BJ (2012) Cognitive mechanisms of treatment in depression. Neuropsychopharmacology 37:117–136

Roiser JP, Sahakian BJ (2013) Hot and cold cognition in depression. CNS Spectr 18:139–149

Roiser JP, Stephan KE, den Ouden HE, Barnes TR, Friston KJ, Joyce EM (2009) Do patients with schizophrenia exhibit aberrant salience? Psychol Med 39:199–209

Ross RG, Harris JG, Olincy A (2000) Radant A.Eye movement task measures inhibition and spatial working memory in adults with schizophrenia, ADHD, and a normal comparison group. Psychiatry Res 95:35–42

Roth BL, Sheffler DJ, Kroeze WK (2004) Magic shotguns versus magic bullets: selectively non-selective drugs for mood disorders and schizophrenia. Nat Rev Drug Discov 3:353–359

Ruhrmann S, Bechdolf A, Kühn KU, Wagner M, Schultze-Lutter F, Janssen B, Maurer K, Häfner H, Gaebel W, Möller HJ, Maier W, Klosterkötter J (2007) LIPS study group. Acute effects of treatment for prodromal symptoms for people putatively in a late initial prodromal state of psychosis. Br J Psychiatry Suppl 51:s88–95

Rust J (1988) The rust inventory of schizotypal cognitions (RISC). Schizophr Bull 14:317–322

Rybakowski JK, Drozdz W, Borkowska A (2007) Low dose risperidone in the treatment of schizophrenia-like symptoms in high-risk subjects. Human Psychopharmacology 22:407–412

Sahakian BJ, Bruhl AB, Cook J, Killikelly C, Savulich G, Piercy T, Hafizi S, Perez J, Fernandez-Egea E, Suckling J, Jones PB (2015) The impact of neuroscience on society: cognitive enhancement in neuropsychiatric disorders and in healthy people. Philos Trans R Soc B 370

Sanchez C, Asin KE, Artigas F (2015) Vortioxetine, a novel antidepressant with multimodal activity: review of preclinical and clinical data. Pharmacol Ther 145:43–57

Saykin AJ, Gur RC, Gur RE, Mozley PD, Mozley LH, Resnick SM, Kester DB, Stafiniak P (1991) Neuropsychological function in schizophrenia. Selective impairment in memory and learning. Arch Gen Psychiatry 48:618–624

Schlagenhauf F, Juckel G, Koslowski M, Kahnt T, Knutson B, Dembler T, Kienast T, Gallinat J, Wrase J, Heinz A (2008) Reward system activation in schizophrenic patients switched from typical neuroleptics to olanzapine. Psychopharmacology 196:673–684

Schoemaker H, Claustre Y, Fage D, Rouquier L, Chergui K, Curet O, Oblin A, Gonon F, Carter C, Benavides J, Scatton B (1997) Neurochemical characteristics of amisulpride, an atypical dopamine D2/D3 receptor antagonist with both presynaptic and limbic selectivity. J Pharmacol Exp Ther 280:83–97

Schultz W (2015) Neuronal reward and decision signals: from theories to data. Physiol Rev 95:853–951

Seeman P (2013) Schizophrenia and dopamine receptors. Eur Neuropsychopharmacol 23: 999–1009

Shah PJ, Ebmeier KP, Glabus MF, Goodwin GM (1998) Cortical grey matter reductions associated with treatment-resistant chronic unipolar depression: controlled magnetic resonance imaging study. Br J Psychiatry 172:527–532

Sheline YI, Gado MH, Kraemer HC (2003) Untreated depression and hippocampal volume loss. Am J Psychiatry 160:1516–1518

Steeds H, Carhart-Harris RL, Stone JM (2015) Drug models of schizophrenia. Ther Adv Psychopharmacol 5:43–58

Stip E (2002) Happy birthday neuroleptics! 50 years later: la folie du doute. Eur Psychiatry 17:115–119

Stirling J, Barkus E, Lewis S (2007) Hallucination proneness, schizotypy and meta-cognition. Behav Res Ther 45:1401–1408

Trestman RL, Keefe RS, Mitropoulou V, Harvey PD, deVegvar ML, Lees-Roitman S, et al. (1995) Cognitive function and biological correlates of cognitive performance in schizotypal personality disorder. Psychiatry Res 59:127–136

Tsuang MT, Stone WS, Seidman LJ, Faraone SV, Zimmet S, Wojcik J, Kelleher JP, Green AI (1999) Treatment of nonpsychotic relatives of patients with schizophrenia: four case studies Biological Psychiatry 45:1412–1418

Umbricht D, Alberati D, Martin-Facklam M, Borroni E, Youssef EA, Ostland M, Santarelli L (2014) Effect of bitopertin, a glycine reuptake inhibitor, on negative symptoms of schizophrenia: a randomized, double-blind, proof-of-concept study. JAMA Psychiatry 71:637–646

Videbech P, Ravnkilde B (2004) Hippocampal volume and depression: a meta-analysis of MRI studies. Am J Psychiatry 161:1957–1966

Williams JH, Wellman NA, Geaney DP, Feldon J, Cowen PJ, Rawlins JN (1997) Haloperidol enhances latent inhibition in visual tasks in healthy people. Psychopharmacology (Berlin) 133 (3):262–268

Ziauddeen H, Murray GK (2010) The relevance of reward pathways for schizophrenia. Curr Opin Psychiatry 23:91–96

MIX
Papier aus verantwortungsvollen Quellen
Paper from responsible sources
FSC® C105338

If you have any concerns about our products,
you can contact us on
ProductSafety@springernature.com

In case Publisher is established outside the EU,
the EU authorized representative is:
Springer Nature Customer Service Center GmbH
Europaplatz 3, 69115 Heidelberg, Germany

Printed by Libri Plureos GmbH
in Hamburg, Germany